Lecture Notes in Artificial Intelligence 9468

Subseries of Lecture Notes in Computer Science

More information about this series at http://www.springer.com/series/1244

Rajendra Prasath · Anil Kumar Vuppala
T. Kathirvalavakumar (Eds.)

Mining Intelligence and Knowledge Exploration

Third International Conference, MIKE 2015
Hyderabad, India, December 9–11, 2015
Proceedings

Editors
Rajendra Prasath
Norwegian University of Science
and Technology
Trondheim
Norway

T. Kathirvalavakumar
V.H.N.S.N. College (Autonomous)
Virudhunagar
India

Anil Kumar Vuppala
International Institute of Information
Technology
Hyderabad
India

ISSN 0302-9743 ISSN 1611-3349 (electronic)
Lecture Notes in Artificial Intelligence
ISBN 978-3-319-26831-6 ISBN 978-3-319-26832-3 (eBook)
DOI 10.1007/978-3-319-26832-3

Library of Congress Control Number: 2015954624

LNCS Sublibrary: SL7 – Artificial Intelligence

Springer Cham Heidelberg New York Dordrecht London

Printed on acid-free paper

Springer International Publishing AG Switzerland is part of Springer Science+Business Media
(www.springer.com)

Preface

This volume contains the papers presented at MIKE 2015: The Third International Conference on Mining Intelligence and Knowledge Exploration held during December 09–11, 2015, at the International Institute of Information Technology, Hyderabad, India (http://www.mike.org.in/2015/). There were 185 submissions from 20 countries and each qualified submission was reviewed by a minimum of two Program Committee members using the criteria of relevance, originality, technical quality, and presentation. The committee accepted 48 full papers for oral presentation and eight papers for short presentation at the conference. We also had four presentations in the doctoral consortium. The overall acceptance rate is 32.43 %.

The International Conference on Mining Intelligence and Knowledge Exploration (MIKE) is an initiative focusing on research and applications on various topics of human intelligence mining and knowledge discovery. Human intelligence has evolved steadily over several generations, and today human expertise is excelling in multiple domains and in knowledge-acquiring artifacts. The primary goal was to focus on the frontiers of human intelligence mining towards building a body of knowledge in this key domain. The focus was also to present state-of-the-art scientific results, to disseminate modern technologies, and to promote collaborative research in mining intelligence and knowledge exploration. At MIKE 2015, specific focus was placed on the human cognition and speech processing theme.

The accepted papers were chosen on the basis of their research excellence, providing a body of literature for researchers involved in exploring, developing, and validating learning algorithms and knowledge-discovery techniques. Accepted papers were grouped into various subtopics including information retrieval, machine learning, pattern recognition, knowledge discovery, classification, clustering, image processing, network security, speech processing, natural language processing, language, cognition and computation, fuzzy sets, and business intelligence. Researchers presented their work and had an excellent opportunity to interact with eminent professors and scholars in their area of research. All participants benefitted from discussions that facilitated the emergence of new ideas and approaches. The authors of short papers presented their work during a special session and obtained feedback from thought leaders in the discipline.

A large number of distinguished professors, well-known scholars, industry leaders, and young researchers participated in making MIKE 2015 a great success. We express our sincere thanks to the International Institute of Information Technology, Hyderabad, for allowing us to host MIKE 2015. We were pleased to have Prof. Ramon Lopaz de Mantaras (Artificial Intelligence Research Institute, Spain), Prof. Mandar Mitra (Indian Statistical Institute, Kolkata, India), Prof. Pinar Ozturk and Prof. Bjorn Gamback (Norwegian University of Science and Technology, Norway), Prof. Sudeshna Sarkar and Prof. Niloy Ganguly (Indian Institute of Technology, Kharagpur, India), Prof. Philip O'Reilly (University College Cork, Ireland), Prof. Nirmalie Wiratunga (Robert Gordan University, Scotland, UK), Prof. Paolo Rosso (Universitat Politècnica de

València, Spain), Prof. Chaman L.Sabharwal (Missouri University of Science and Technology, USA), and Dr. Rajarshi Pal (IRDBT, Hyderabad) serving as advisory members for MIKE 2015.

Several eminent scholars — including Prof. B. Yegnanarayana, International Institute of Information Technology, Hyderabad, India; Prof. Agnar Aamodt, Norwegian University of Science and Technology (NTNU), Trondheim, Norway; Prof. Tapio Saramäki, Tampere University of Technology, Tampere, Finland; Prof. Anupam Basu, Indian Institute of Technology, Kharagpur, India; and Prof. N. Subba Reddy, Gyeongsang National University, Jinju, Korea — delivered invited talks on learning and knowledge exploration tasks in various interdisciplinary areas of artificial intelligence and machine learning.

This year, we organized the SAIL 2015 workshop that specifically focused on "Sentiment Analysis of Indian Language Tweets." In all, 22 teams participated in the shared task from 16 countries and submitted their official run. We selected 7 submissions from the official runs and the acceptance rate of this shared task was 31.82 %. Prof. Vasudeva Verma delivered an invited talk at SAIL 2015. Dr. Amitava Das, Indian Institute of Information Technology, Sri City, Andra Pradesh, India, Dr. Dipankar Das, Jadavpur University, India, Dr. Manish Shrivastava, IIIT Hyderabad, India, and Dr. Rajendra Prasath, Norwegian University of Science and Technology, Norway, served as the co-organizers of the shared task on "Sentiment Analysis of Indian Languages (SAIL) Tweets." Leading practitioners from top-tier technology organizations participated in this very successful workshop.

Dr. Manoj Chinnakotla from Microsoft (Bing), Dr. Amitava Das of IIIT, Sricity, Andra Pradesh, India, Dr. Rajendra Prasath (NTNU, Norway), and Dr. Maunendra Sankar Desarkar (IIT, Hyderabad) jointly organized a workshop on the recent advances in IR and the leading-edge R&D being undertaken in their organizations at present.

We thank the Technical Program Committee members and all reviewers/ subreviewers for their timely and thorough participation in the reviewing process.

We express our sincere gratitude to Prof. P.J. Narayanan, Director (IIIT Hyderabad), for allowing us to organize MIKE 2015 on the campus. We also thank Prof. Vasudeva Varma (Dean Research and Development) and Prof. Jayanthi Sivaswamy (Dean Academics) of IIIT Hyderabad for their valuable inputs. We thank Prof. B. Yegnanarayana, Prof. Peri Bhaskrarao and Prof. Dipti Misra Sharma of LTRC, IIIT Hyderabad, for their valuable support to MIKE 2015. We especially thank Dr. Suryakanth V. Gangashetty for helping MIKE 2015 as the local organizing chair. We appreciate the time and efforts put in by the members of the local organizing team at IIIT Hyderabad, especially the research scholars of the Speech and Vision Lab, volunteers, administrative staff, account section staff, outreach and hostel management staff of IIIT Hyderabad, who dedicated their time and efforts to MIKE 2015. We thank Prof. Rammurthy. G., of IIIT Hyderbad for helping us to invite Prof. Tapio Saramäki from Finland as an invited speaker. We especially thank Ms. K.K. Madhavi and Dr. Vijay of IIIT Hyderabad for helping us to get sponsorship. We thank Mr. Guruprasad for designing registration portal. We thank Mr. Sunil Kumar Vuppala, Infosys for his valuable suggestions and help. We extend our sincere thanks to Mr. Muthuvijayaraja of VHNSN College, Virudhunagar for his great support.

We are very grateful to all our sponsors, especially Microsoft and many other local supporters, for their generous support of MIKE 2015. We would also like to thank the Telangana government for their valuable support.

Finally, we acknowledge the help of EasyChair in the submission, review, and proceedings creation processes. We are very pleased to express our sincere thanks to Springer, especially Alfred Hofmann, Anna Kramer, Erika Siebert-Cole, and the editorial staff, for their support in publishing the proceedings of MIKE 2015.

December 2015

Rajendra Prasath
Anil Kumar Vuppala
T. Kathirvalavakumar

MIKE 2015: International Institute of Information Technology, Hyderabad, India

Program Committee

Agnar Aamodt	Norwegian University of Science and Technology, Trondheim, India
Ibrahim Adeyanju	Ladoke Akintola University of Technology, Nigeria
Rajendra Akerkar Akerkar	Vestlandsforsking, Sogndal, Norway
Gethsiyal Augasta	Sarah Tucker College, Tirunelveli, India
Zeyar Aung	Masdar Institute of Science and Technology, United Arab Emirates
Rakesh Balabantaray	International Institute of Information Technology, Bhubaneswar, India
Lavanya Balaraja	University of Madras, Chennai, India
Biswanath Barik	Norwegian University of Science and Technology, Trondheim, India
Anupam Basu	Indian Institute of Technology, Kharagpur, India
Prof. Kamal K. Bharadwaj	Jawaharlal Nehru University, New Delhi, India
Pinaki Bhaskar	IIT - CNR, Italy
Vasudha Bhatnagar	University of Delhi, New Delhi, India
Plaban Kumar Bhowmik	Indian Institute of Technology, Kharagpur, India
Isis Bonet	EIA, Antioquia School of Engineering, Colombia
Erik Cambria	Nanyang Technological University, Singapore
Tanmoy Chakraborty	Indian Institute of Technology, Kharagpur, India
Joydeep Chandra	Indian Institute of Technology, Patna, India
Sanjay Chatterji	Samsung R&D Institute India, Bangalore, India
Manoj Chinnakotla	Microsoft R&D Pvt. Ltd., India
Kamal Kumar Choudhary	Indian Institute of Technology, Ropar, India
Gladis Christopher	Presidency College, Chennai, India
Amélie Cordier	Université Claude Bernard Lyon 1 (LIRIS), France
Amitava Das	Indian Institute of Information Technology (IIIT), Sricity, Andhra Pradesh, India
Dipankar Das	Jadavpur University, Kolkata, India
Tirthankar Dasgupta	IBM Research, New Delhi, India
Maunendra Sankar Desarkar	Indian Institute of Technology, Hyderabad, India
Aidan Duane	Waterford Institute of Technology (WIT), Ireland
Shashidhar G. Koolagudi	National Institute of Technology Karnataka, Surathkal, India
Björn Gambäck	Norwegian University of Science and Technology, Trondheim, India

Vineet Gandhi	IIIT Hyderabad, India
Debasis Ganguly	Dublin City University, Dublin, Ireland
Niloy Ganguly	Indian Institute of Technology, Kharagpur, India
Saptarshi Ghosh	Max Planck Institute for Software Systems (MPI-SWS), Germany
Rob Gleasure	University College Cork, Ireland
Sumit Goswami	Defense Research and Development Organization, New Delhi, India
Devanur Guru	University of Mysore, Mysore, India
Wu Huayu	Institute for Infocomm Research (I2R), Singapore
Gloria Ines Alvarez Vargas	Pontificia Universidad Javeriana Cali, Colombia
Abdeldjalil Khelassi	Tlemcen University, Algeria
Sangwoo Kim	Yonsei University College of Medicine, Korea
P.V.V. Kishore	KL University, Guntur, Andhra Pradesh, India
Shruti Kohli	Birla Institute of Technology, Mesra, India
M. Kothandapani	Anna University College of Technology, Arni, India
Vamsi Krishna	Indian Space Research Organization, India
T. Kumaran	Government Arts College (Men), Krishnagiri, India
Uttama Lahiri	Indian Institute of Technology, Gandhinagar, Gujarat, India
Chaman Lal Sabharwal	Missouri University of Science and Technology, Rolla, USA
Jaiprakash Lallchandani	International Institute of Information Technology, Bangalore, India
Ramon Lopez De Mantaras	Artificial Intelligence Research Institute (IIIA - CSIC), Barcelona, Spain
Yutaka Maeda	Kansai University, Japan
Rajib Ranjan Maiti	IRDBT, Hyderabad, India
Prasenjit Majumder	DAIICT, Gandhinagar, India
Radhika Mamidi	IIT Hyderabad, India
Roshan Martis	St. Joseph Engineering College, Mangaluru, India
Mandar Mitra	Indian Statistical Institute, Kolkata, India
Pabitra Mitra	Indian Institute of Technology, Kharagpur, India
Vinay Kumar Mittal	IIIT Sricity, Andra Pradesh, India
Muthu Rama Krishnan Mookiah	Ngee Ann Polytechnic, Singapore
A. Murugan	Dr. Ambedkar Govt Arts College, Chennai, India
Jian-Yun Nie	University of Montreal, Canada
Philip O'Reilly	University College Cork, Cork, Ireland
Sylvester Olubolu Orimaye	Monash University, Sunway Campus, Malaysia
Inah Omoronyia	Glasgow University, Glasgow, Scotland
Pinar Ozturk	Norwegian University of Science and Technology, Trondheim, India
Rajkumar P.V.	University of Texas, San Antonio, USA
Jiaul Paik	University of Maryland, College Park, USA

Nirmalae Wiratunga	Robert Gordon University, Aberdeen, UK
Wei Lee Woon	Masdar Institute, United Arab Emirates
Xiaolong Wu	California State University, Long Beach, USA

Additional Reviewers

Alluri, Knrk Raju
Asoka Chakravarthi, Bharathi Raja
Augasta, Gethsiyal
Augasta, Gladis
Bandyopadhyay, Ayan
Bhat, Riyaz Ahmad
Chakma, Kunal
Christopher, Gladis
Dalmia, Ayushi
Giménez, Maite
Hernández Farias, Delia Irazú
Karfa, Dr. Chandan
Krishna, Amrith
Krishna, Hari
Lohar, Pintu
Manjunath, Shantharamu
Mondal, Anupam
Muralidaran, Vigneshwaran
P, Vijayavel
Paidi, Gangamohan
Pal, Dipasree
Patra, Braja Gopal
Prasath, Rajendra
Rudrapal, Dwijen
Sharma, Avinash
Singh, Mayank
Suhil, Mahamad
T, Kumaran
Thirumuru, Ramakrishna
Vegesna, Vishnu Vidyadhara Raju

Contents

Spreading Activation Way of Knowledge Integration

Shubhranshu Shekhar(✉), Sutanu Chakraborti, and Deepak Khemani

Department of Computer Science and Engineering,
Indian Institute of Technology Madras, Chennai 600036, India
{shekhars,sutanuc,khemani}@cse.iitm.ac.in

Abstract. Search and recommender systems benefit from effective integration of two different kinds of knowledge. The first is introspective knowledge, typically available in feature-theoretic representations of objects. The second is external knowledge, which could be obtained from how users rate (or annotate) items, or collaborate over a social network. This paper presents a spreading activation model that is aimed at a principled integration of these two sources of knowledge. In order to empirically evaluate our approach, we restrict the scope to text classification tasks, where we use the category knowledge of the labeled set of examples as an external knowledge source. Our experiments show a significantly improved classification effectiveness on hard datasets, where feature value representations, on their own, are inadequate in discriminating between classes.

1 Introduction

Machine Learning (ML) algorithms induce models from feature representations of the underlying complex real-world data such as images, videos or natural language. There is, however, a large class of problems where appropriate feature-value representations are hard to arrive at. At the current state of the art, it is hard, for example, to classify a movie as romantic or otherwise based on its content. "Romance" is an emergent notion, that arises out of nonlinear interactions of the pixels in each frame, the frames with each other, and the movie with the mental model of the observer. Collaborative recommender systems cleverly bypass this hard problem by exploiting the knowledge of how users rate items, and hypothesizing that movies liked by similar users can be reckoned to be similar. Thus, external knowledge in the form of user ratings (or annotations) can effectively complement feature based representations. With external knowledge sources like social networks increasingly becoming pivotal in large classes of search and recommender problems, we envisage that object relations can be exploited in several interesting ways that would complement the traditional feature-theoretic models. This also has the favorable effect of significantly reducing human effort involved in engineering an appropriate set of features tailored to a task.

R. Prasath et al. (Eds.): MIKE 2015, LNAI 9468, pp. 1–11, 2015.
DOI: 10.1007/978-3-319-26832-3_1

In this paper, we propose a spreading activation based architecture for principled integration of traditional feature based representation and external knowledge. The architecture is primarily a network of information units. There are two main components of this architecture – a component to capture introspective knowledge (feature representation) and a component to capture extrinsic knowledge. Spreading activation models have been effectively used in systems such as cognitive modeling [1], information retrieval [9] and case-based systems [15]. On the other hand, in [20] authors make an attempt to bring together distinct knowledge sources using spreading activation. All these methods use constraints such as distance constraints to terminate the spreading.

In [3], the authors show that without such heuristic constraints, spreading activation is an inadequate technique for querying information networks. Accumulation based spreading [3] overcomes the drawbacks of pure spreading activation model that we employ in our proposed architecture. In Sect. 3.1, we present an analysis of accumulation based spreading activation for the proposed architecture. We demonstrate the effectiveness of our approach for a document classification task on real world datasets. In this context we treat the category knowledge of documents as an external knowledge source, and show that our approach holds promise in improving classification effectiveness in hard domains where there is significant overlap in feature based representations of two classes.

2 Background

2.1 Case Retrieval Network

Lenz and Burkhard proposed Case Retrieval Network [15] as a representation formalism for Case Based Reasoning [13]. CRN (Fig. 1) facilitates spreading activation over a feature value representation and similarity knowledge among them. A CRN consists of information entities(IE) which are most basic unit of knowledge, such as a particular attribute-value pair and a set of nodes called case nodes. IEs constitute a case and IE nodes are related to each other by weighted symmetric *similarity arcs* (σ) denoting the semantic similarity between two IEs. IEs are linked to case nodes by weighted *relevance arcs* (ρ) which capture the strength of relevance of the IE to a document. CRNs are successfully applied for real world CBR systems [2,10,16]. However, the CRNs fail to capture relationship among cases (objects) independent of feature value similarity. As shown in Fig. 1, we observe that cases C_i are not connected. In this work, we relax the case independence assumption in CRNs, and allow cases to influence each other by incorporating background knowledge about them.

2.2 Spreading Activation

Spreading activation [8,19] facilitates querying networks of information. The Linear Spreading Activation model with distance constraint is used to process CRNs. The mechanics of the model is described below.

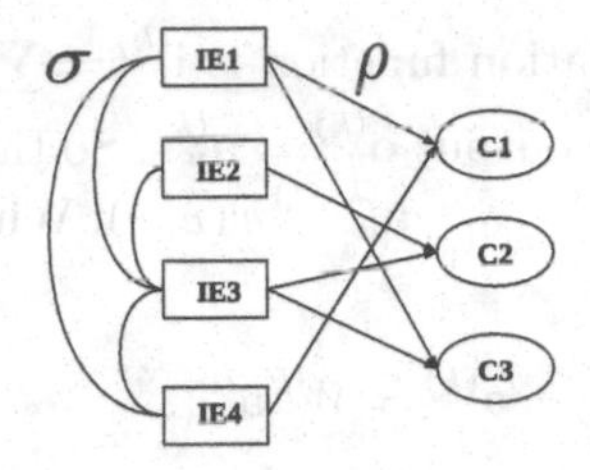

Fig. 1. A simple CRN. C_i represents case (document), and IE_j represents a word

Given a graph $G = (V, E; w)$ with edge weights $w : E \rightarrow \mathbb{R}$. We assume that G is undirected as shown in Fig. 2, but our results easily generalize to directed graphs. Following the conventions used in [3], $|V| = n$ and weight matrix $W \in \mathbb{R}^{n \times n}$ defined by $(W)_{u,v} = w(u,v)$; $w(u,v) = 0$ if $(u,v) \notin E$. $N(v) = \{u : (u,v) \in E\}$ denotes the neighbors of $v \in V$. The activation at time step $k > 0$ is denoted as $\mathbf{a}^{(k)} \in \mathbb{R}^V$. $\mathbf{a}_v^{(k)}$ denotes the activation for a node $v \in V$. The basic

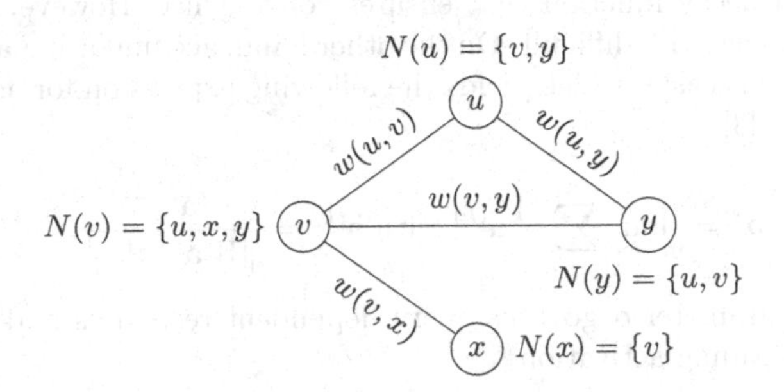

Fig. 2. A graph $G = \{V, E; w\}$ with $V = \{u, v, x, y\}$

framework is split into three parts – input, activation and output. Each part consists of a function and a state, where the function defines the transition into a state. Input function combines the outgoing activations of the adjacent nodes to an incoming activation. Output function computes the outgoing activation from a node. Activation function determines whether a node is activated or not.

Given a query, initial activation $\mathbf{a}^{(0)}$ is determined. These activated nodes spread the activation to the adjacent nodes in the network and activation propagates until a stopping criterion is met. The nodes are ranked according to final activation after the spread terminates. The query is answered based on the ranked activations of the nodes.

2.3 Linear Model

The general linear model constitutes a linear input function, and an identity activation and output function. Therefore the input for a node v at time step k

using summation as accumulation function is $\mathbf{i}_v^{(k)} = \sum_{u \in N(v)} \mathbf{o}_u^{(k-1)} w(u,v)$, the activation $\mathbf{a}_v^{(k)} = \mathbf{i}_v^{(k)}$ and the output $\mathbf{o}_v^{(k)} = \mathbf{a}_v^{(k)}$. So the simplified activation for a node v at time k is $\mathbf{a}_v^{(k)} = \sum_{u \in N(v)} \mathbf{a}_u^{(k-1)} w(u,v)$. With the weighted adjacency matrix $W \in \mathbb{R}^{nxn}$ this yields

$$\mathbf{a}^{(k)} = W^k \mathbf{a}^{(0)},$$

that corresponds to power iteration with matrix W. The power iteration converges [11] to the principal eigen vector of W, if W represents a connected graph. Berthold et al. [3] proposed the accumulation based model to avoid the limitation of the linear model. The iterations are modified to consider more than previous state that allows initial states to influence the final result. The accumulated activation is given as follows

$$\mathbf{a}^* = \sum_{k=0}^{\infty} \lambda(k) \cdot \mathbf{a}^{(k)} = \left(\sum_{k=0}^{\infty} \lambda(k) W^k \right) \mathbf{a}^{(0)},$$

where $\lambda(k)$ is a decay function that ensures convergence. However, choosing a suitable decay function is difficult. On the other hand, accumulating a normalized activation vector is easier which yields the following expression for accumulation based spreading [3].

$$\mathbf{a}^* = \lim_{m \to \infty} \sum_{k=0}^{m} \alpha^k \mathbf{a}^{(k)} \text{ with } \mathbf{a}^{(k)} = \frac{W \mathbf{a}^{(k-1)}}{\left\| W \mathbf{a}^{(k-1)l} \right\|}.$$

The choice of parameter α governs query dependent responses and the convergence of the spreading activation.

3 Our Framework

We introduce a structural and a behavioral change in the CRN model.

- Structural change – Introduce links among objects based on the background knowledge.
- Behavioral change – Use accumulation based spreading for processing the network. Additionally, we allow forward-backward spreading from features to objects, and from objects to features as explained in the next section.

3.1 Proposed Model

We construct a two layered network to represent objects and their attributes similar to networks proposed in [15,20]. The network is a graph $G = (V, E)$ where the nodes $V = D \cup T$ is the union of disjoint sets D and T. D represents the objects in the network and T represents attributes of the objects. Edges $E = E_E \cup E_D$ are sets of two disjoint edge sets that represent two kinds of relations in the network. Edges $E_E = \{\{d,t\} : d \in D, t \in T, w(d,t) > 0\}$ with weights

$w : D \times T \to \mathbb{R}$ describe the relations between the objects and their attributes. Edges $E_D = \{\{d, d\} : d \in D, m(d, d) > 0\}$ with weights $m : D \times D \to \mathbb{R}$ describe the relations between a pair of objects, independent of the attributes, based on external knowledge about the objects such as category knowledge or collaborative knowledge.

The spreading activation can be triggered for a query by defining a separate spreading function for each class of nodes (object and attribute nodes) in the network. The iterative process consists of three steps: (i) attributes activate objects; (ii) these activated objects further update the activation based on object relations; (iii) the objects activate the attributes back again. Spreading activation takes place in forward and backward direction between the two layers until the convergence. The activation for attributes in T at a time step k is given as

$$\mathbf{a}_t^{(k)} = \frac{\sum_{d \in N(t)} \mathbf{a}_d^{(k-1)} w(d,t)}{\left(\sqrt{\sum_{d \in N(t)} w(d,t)^2} \cdot \left\|\mathbf{a}_d^{(k-1)}\right\|\right)}$$

The activations for objects using the attribute activation from same iteration is a two step process – activation from attributes and activation from objects. Step 1 is:

$$\mathbf{a'}_d^{(k)} = \frac{\sum_{t \in N(d) \cap T} \mathbf{a}_t^{(k)} w(d,t)}{\left(\sqrt{\sum_{t \in N(d) \cap T} w(d,t)^2} \cdot \left\|\mathbf{a}_t^{(k)}\right\|\right)}$$

and step 2 is:

$$\mathbf{a}_d^{(k)} = \frac{\sum_{d_i \in N(d) \cap D} \mathbf{a'}_d^{(k)} m(d,d_i)}{\left(\sqrt{\sum_{d_i \in N(d) \cap D} m(d,d_i)^2} \cdot \left\|\mathbf{a'}_d^{(k)}\right\|\right)}.$$

Given a query q, it is treated as a virtual object and broken down to its constituent attributes and the initial activation $\mathbf{a}_d^{(0)}$ is established. Spreading activation is instantiated with this initial vector. When further iterations follow, the object relations activate certain new objects which in turn activate the attribute nodes. Thus the knowledge contained in object relation edges E_D spreads across and complements the traditional attribute based spreading.

3.2 Classification Task

We illustrate the effectiveness of the proposed framework on the task of document classification. An example network consisting documents and terms, and their relations is shown in Fig. 3. Unlike CRN (Fig. 1), this network has links that connect documents.

In our task, we integrate the class labels of the documents into the framework to obtain class-enriched activations with respect to a query. Graph $G = (V, E)$ is constructed as explained in the last section, the documents are the objects in

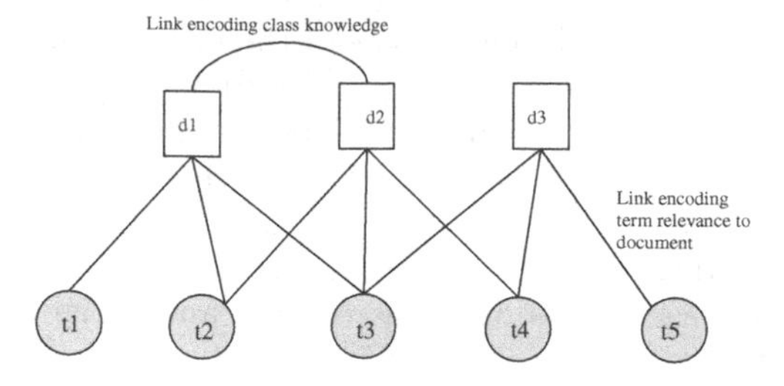

Fig. 3. A simple network where documents belonging to the same class are connected using E_D edges

the graph and the terms present in the document represent the attributes. Two documents are related if they share same class label. Therefore, the edges E_D are formed using the class knowledge. For $d_i, d_j \in D$, if d_i and d_j belong to the same class, $m(d_i, d_j) = 1$, else 0. Edges E_E are formed using the relevance of keywords to the document. We use simplistic notion of relevance encoded as $w(d,t) = 1$, if term $t \in T$ is present in the document $d \in D$, otherwise 0. Though we have used class knowledge, other sources of background knowledge can be used to model associations between documents. For example, if same set of people are reading two different pieces of news on an online news website, then an association can be formed between the pair of news articles. We, now, present the spreading activation formalism for the classification task.

Formal Representation

Let $W \in \mathbb{R}^{D\times T}$ and $M \in \mathbb{R}^{D\times D}$ be the weight matrices of the graph G such that $(W)_{d,t} = w(d,t)$ for $d \in D$, the document set, and $t \in T$, the terms. $(M)_{d_i,d_j} = m(d_i, d_j)$ for $d_i, d_j \in D$ encoding the class knowledge. Let $W_D \in \mathbb{R}^{D\times T}$ be the L_2 row normalization of W, $W_T \in \mathbb{R}^{T\times D}$ be the L_2 row normalization of transpose of W, and $M_D \in \mathbb{R}^{D\times D}$ be the L_2 normalization of M. In addition, let $\mathbf{a'}_D^{(k)}$, $\mathbf{a}_D^{(k)}$ and $\mathbf{a}_T^{(k)}$ be the document activation from terms, updated document activation after spreading though E_D, and term activations after k time steps respectively. $\mathbf{a}_D^{(k)}$ denotes the activations of two steps – activation from terms and activations from documents. The document and term activations are given as:

$$\mathbf{a}_D^{(k)} = \frac{M_D \cdot W_D \cdot \mathbf{a}_T^{(k)}}{\left\|\mathbf{a'}_D^{(k)}\right\| \left\|\mathbf{a}_T^{(k)}\right\|}, \qquad \text{and} \qquad \mathbf{a}_T^{(k)} = \frac{W_T \cdot \mathbf{a}_D^{(k-1)}}{\left\|\mathbf{a}_D^{(k-1)}\right\|}.$$

By combining the above two expressions, we obtain the power iteration representing the final document activation following three steps of spreading in an iteration at time step k, which is given as:

$$\mathbf{a}_D^{(k)} = \frac{M_D W_D (W_T M_D W_D)^k \mathbf{a}_T^{(0)}}{\left\|(W_T M_D W_D)^k \mathbf{a}_T^{(0)}\right\|} \tag{1}$$

The Perron-Frobenius Lemma [11] states that power iteration converges if G is connected and not bipartite. With the introduction of edges E_D, G is not bipartite. In addition, graph underlying $(W_T M_D W_D)$ is connected, therefore the power iteration given in Eq. 1 converges. We employ the accumulation based spreading using $\mathbf{a}_D^{(k)}$ and observe the behavior of convergence with respect to parameter α. The empirical results are discussed in the following section.

4 Empirical Evaluation

We show the effectiveness of our proposed architecture for the task of document classification. We demonstrate the importance of adding class-knowledge in the form of document-document edges E_D in the proposed network through spreading activation. We compare our spreading models against CRN [15] which does not incorporate the class knowledge and does not use backward spreading of activation from objects to attributes. We also compare the spreading models with kNN, a popular machine learning classifier, and SVM, one of the best classifiers known to perform well for text classification tasks [12] that uses class labels to learn the model.

We report the complexity analysis of datasets using $GAME_{class}$ [7] measure that uses introspective knowledge to estimate complexity, and show that for a more complex dataset, class knowledge improves performance significantly. $GAME_{class}$ indicates whether the nearest neighbors belong to the same class when documents are sorted in the order given by similarity computed in feature space. The models, with background knowledge, used for experiment are summarized below:

A1 We use simple two step spreading. Document relations with respect to class knowledge is encoded in form of document to document E_D edges. Here, we only have forward spreading propagation. Spreading terminates when documents with distance constraints update their activations. These activations are then used to rank relevant documents.

A2 The proposed model. We use accumulation based spreading activation on our proposed network with class knowledge incorporated in E_D edges. Spreading activation takes place in forward and backward directions between the terms and document layers until the convergence as described in Sect. 3.1.

In our experiments, we have used $k = 3$ for all the models, and have used binary weights for all the edges in the network as described in Sect. 3.1.

Dataset Description. We used the 20 Newsgroups and WebKB datasets for our experiments. The 20 Newsgroups dataset is a collection of approximately 20,000 newsgroup documents. We have procured four datasets, namely *RelPol, Hardware, Science*, and *Recreation*, from the original corpus. *RelPol* and *Hardware* are binary classification datasets, while *Recreation* and *Science* are 4-class datasets. The dataset is divided into training (60 %) and test (40 %).

Table 1. Classification accuracy of different models. A1 and A2 model class-knowledge. A1 is the model with distance constraint, and A2 is out proposed model.

Dataset	Text Classification (Accuracy)				
	kNN	CRN	A1	SVM	A2
20NewsGroup					
RelPol	91.4 %	91.28 %	89.0 %	93.2 %	92.75 %
Hardware	72.24 %	70.02 %	78.85 %	78.2 %	80.91 %
Recreation	77.92 %	77.92 %	88.62 %	88.8 %	88.75 %
Science	72.71 %	72.71 %	79.62 %	79.6 %	81.03 %
WebKB					
WebKB	93.26 %	93.62 %	88.14 %	92.4 %	93.45 %

The documents in the WebKB are 8145 webpages collected by the World Wide Knowledge Base project of the CMU text learning group, and were downloaded from university computer science departments. It is a multi-class dataset where the task is to label webpages as *student, faculty, course* or *project*. We randomly split dataset into train (80 %) and test (20 %) sets.

Evaluation

Table 1 shows the experimental results. For example, our proposed model A2 outperforms on *hardware, science* and *WebKB* datasets. In Table 2, we report $GAME_{class}$ score, where a lower score indicates hard dataset with respect to classification. It is interesting to note that for datasets with low $GAME_{class}$ scores emphasizing inadequate feature representation, our model shows conspicuous improvements. *RelPol* and *WebKB* are relatively easier to classify as per $GAME_{class}$ score, where our proposed method fares marginally. This indicates that for a hard dataset, integration of class knowledge improves performance significantly.

In Fig. 4, we also show how α influences the performance of our proposed method. We experiment with values in the range $0 - 1$ and note that a smaller value of α close to 0.1 performs consistently well for all the datasets indicating more query specific results [3].

Table 2. $GAME_{class}$ score for datasets

DataSet	RelPol	Hardware	Recreation	Science	WebKB
$GAME_{class}$	2.035	1.003	1.162	1.049	2.041

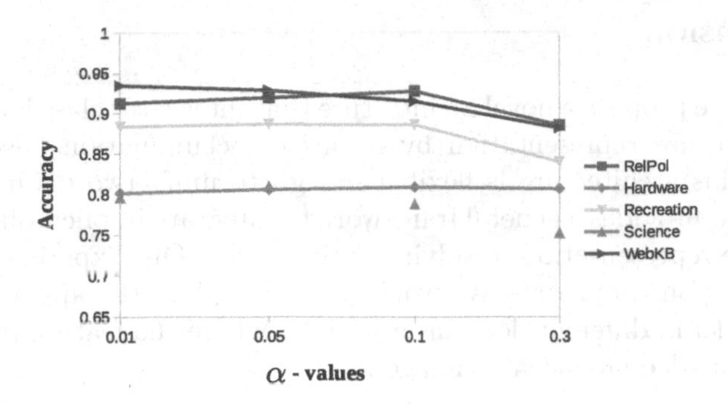

Fig. 4. Effect of α on accuracy of our model

5 Discussion and Related Work

There are several domains which have different knowledge sources available. For example, consider huge collection of public images on Flickr. Nearly half of those images are not tagged. Our proposed framework can be directly applied to find related tags for untagged images. The available tags can be treated as background knowledge. In this work we have demonstrated the effectiveness of background knowledge complementing feature representation.

Methods combining different sources of information are studied with growing interest. In particular, learning from linked data has received lot of attention. In web classification, combining link and content information is well studied [5, 18]. A common way in which the multiple information sources are combined is to treat each of them disjoint and independent feature sets. On the feature sets different classifiers are trained, and are combined [14] for prediction task. On the other hand, web classification problem is posed as relational learning problem [6, 17], where a classifier assigns class probability to each node, and each node then reevaluates its class probabilities based on its neighbors.

Co-training [4] is also effective in combining different knowledge sources. Co-training makes use of both labeled and unlabeled data to achieve better performance. Underlying idea is to train classifiers on different sets of features. The trained classifiers are used to predict labels of unlabeled instances. The prediction of each classifier is used to train the other.

However, in contrast to above approaches the presented work does not extract flat features from the links describing object relationships. With the help of spreading activation, we process the graph consisting of edges depicting relevance of a feature to an object, and relatedness of objects based on external knowledge source. Back propagation helps enrich features, but explicitly we do not extract any features. The framework is more intuitive and simple to follow, and novel in the way that the method unifies the two different knowledge sources in one framework.

6 Conclusion

In this paper we propose a novel architecture that integrates class knowledge with traditional feature representation by means of accumulation based spreading activation. This architecture is flexible enough to unify two distinct sources of knowledge and provides a general framework to integrate implicit object relations to aid feature representation in solving a given task. Our experiments show the effectiveness of our approach. We would like to explore the applicability of the proposed model in different domains especially where the feature representation is inherently inadequate such as images.

References

1. Anderson, J.R.: A spreading activation theory of memory. Journal of Verbal Learning and Verbal Behavior **22**(3), 261–295 (1983)
2. Balaraman, V., Chakraborti, S.: Satisfying varying retrieval requirements in case based intelligent directory assistance. In: FLAIRS Conference, pp. 160–165 (2004)
3. Berthold, M.R., Brandes, U., Kötter, T., Mader, M., Nagel, U., Thiel, K.: Pure spreading activation is pointless. In: Proceedings of the 18th ACM Conference on Information and Knowledge Management, pp. 1915–1918. ACM (2009)
4. Blum, A., Mitchell, T.: Combining labeled and unlabeled data with co-training. In: Proceedings of the Eleventh Annual Conference on Computational Learning Theory, pp. 92–100. ACM (1998)
5. Calado, P., Cristo, M., Moura, E., Ziviani, N., Ribeiro-Neto, B., Gonçalves, M.A.: Combining link-based and content-based methods for web document classification. In: Proceedings of the Twelfth International Conference on Information and Knowledge Management, pp. 394–401. ACM (2003)
6. Chakrabarti, S.: Mining the Web: Discovering Knowledge from Hypertext Data. Elsevier, San Francisco (2002)
7. Chakraborti, S., Cerviño Beresi, U., Wiratunga, N., Massie, S., Lothian, R., Khemani, D.: Visualizing and evaluating complexity of textual case bases. In: Althoff, K.-D., Bergmann, R., Minor, M., Hanft, A. (eds.) ECCBR 2008. LNCS (LNAI), vol. 5239, pp. 104–119. Springer, Heidelberg (2008)
8. Collins, A.M., Loftus, E.F.: A spreading-activation theory of semantic processing. Psychol. Rev. **82**(6), 407 (1975)
9. Crestani, F.: Application of spreading activation techniques in information retrieval. Artif. Intell. Rev. **11**(6), 453–482 (1997)
10. Fdez-Riverola, F., Iglesias, E.L., Díaz, F., Méndez, J.R., Corchado, J.M.: Applying lazy learning algorithms to tackle concept drift in spam filtering. Expert Syst. Appl. **33**(1), 36–48 (2007)
11. Golub, G.H., Van Loan, C.F.: Matrix Computations, 3rd edn. JHU Press, Baltimore (2012)
12. Joachims, T.: Text categorization with support vector machines: learning with many relevant features. In: Nédellec, C., Rouveirol, C. (eds.) ECML 1998. LNCS, vol. 1398. Springer, Heidelberg (1998)
13. Kolodner, J.: Case-Based Reasoning. Morgan Kaufmann, San Mateo (2014)
14. Kuncheva, L.I.: Combining Pattern Classifiers: Methods and Algorithms. John Wiley & Sons, Chichester (2004)

15. Lenz, M., Burkhard, H.D.: Case retrieval nets: basic ideas and extensions. In: Görz, G., Hölldobler, S. (eds.) KI 1996. LNCS, vol. 1137. Springer, Heidelberg (1996)
16. Lenz, M., Burkhard, H.D.: CBR for document retrieval: the fallq project. In: Leake, D.B., Plaza, E. (eds.) Case-Based Reasoning Research and Development. Lecture Notes in Computer Science, vol. 1266, pp. 84–93. Springer, Berlin Heidelberg (1997)
17. Lu, Q., Getoor, L.: Link-based classification. In: ICML, vol. 3, pp. 496–503 (2003)
18. Qi, X., Davison, B.D.: Knowing a web page by the company it keeps. In: Proceedings of the 15th ACM International Conference on Information and Knowledge Management, pp. 228–237. ACM (2006)
19. Quillian, M.: Semantic memory. Semantic information processing (1968)
20. Shekhar, S., Chakraborti, S., Khemani, D.: Linking cases up: an extension to the case retrieval network. In: Lamontagne, L., Plaza, E. (eds.) ICCBR 2014. LNCS, vol. 8765, pp. 450–464. Springer, Heidelberg (2014)

Class Specific Feature Selection Using Simulated Annealing

V. Susheela Devi(✉)

Department of Computer Science and Automation,
Indian Institute of Science, Bangalore 560 012, India
susheela@csa.iisc.ernet.in

Abstract. This paper proposes a method of identifying features which are important for each class. This entails selecting the features specifically for each class. This is carried out by using the simulated annealing technique. The algorithm is run separately for each class resulting in the feature subset for that class. A test pattern is classified by running a classifier for each class and combining the result. The 1NN classifier is the classification algorithm used. Results have been reported on eight benchmark datasets from the UCI repository. The selected features, besides giving good classification accuracy, gives an idea of the important features for each class.

1 Introduction

Whenever the training data is large or the dimensionality is large, feature selection helps to take care of the space and time complexity especially when classifiers such as the nearest neighbour are used. Besides, by reducing the number of features, the 'curse of dimensionality' problem is taken care of. Another benefit of feature selection would be facilitating data visualization and data understanding [6].

Feature selection is usually carried out using the wrapper or filter methods [3]. The filter method carries out feature selection based on the properties of the dataset itself. The wrapper method generates feature subsets by a method of search which are evaluated using a classification algorithm and the best feature subset found. Searching for the best feature subset cannot be done by exhaustive enumeration because of the time and space complexity involved and a number of methods are available to reduce this search space [10]. These include the branch and bound technique, the sequential forward and backward search and the min-max approach.

The search can also be carried out using soft computing techniques such as neural networks or evolutionary algorithms. Genetic algorithms [7,8], simulated annealing [4], tabu search [13] and particle swarm optimization [9] have all been used for feature selection.

Class-specific feature selection, finds a different set of features for each class. The features which help to identify instances of one class maynot be the ones used to identify the instances belonging to another class. For example, if the

R. Prasath et al. (Eds.): MIKE 2015, LNAI 9468, pp. 12–21, 2015.
DOI: 10.1007/978-3-319-26832-3_2

class labels are different animals, the class of elephants can be best identified by the presence of the trunk and a giraffe by the length of its neck. So, for every class, a specific feature subset maybe enough to identify instances of that class. It is therefore meaningful to carry out class specific feature selection. This means that the features which are important to classify a pattern belonging to one class are different from the features which are important to classify patterns belonging to another class.

The F-score or Fisher score measures the discrimination between features [12]. Given a training pattern $x_i, i = 1, ..., d$, the F-score of a feature f is calculated as follows :

$$F(f) = \frac{\sum_{k=1}^{c}(\bar{x}_f^k - \bar{x}_f)^2}{\sum_{k=1}^{c}\frac{1}{n_k - 1}\sum_{i=1}^{n_k}(x_{i,f}^k - \bar{x}_j^k)^2}$$

$\bar{x}$: Average if f^{th} feature in dataset
$\bar{x}_f^k$: Average of all patterns of fth feature belonging to class k
$x_{i,j}^k$: i^{th} element of fth feature of kth class
c : Number of classes
n_k : Number of elements of kth class

The numerator indicates the inter class variance, and the denominator indicates the sum of variances separately within each class. For a feature f if only those features which give large values in the numerator and small values in the denominator are used, F_j value can be increased. This is the intention of carrying out class-specific feature selection.

A few papers have worked on class-specific feature selection. In [1], a separate feature set is found for each class. A separate classifier is then built for each class based on its own feature set. To find the feature subset, the features are ranked according to a separability index such that higher the separability index, the higher the rank of the feature. Higher ranked features are combined together to form the feature subset. A hypersphere classification algorithm is used.

In [5], class-specific feature selection is used in multiclass support vector machines. Here k binary classifiers are constructed, each classifier being trained with the examples of one class with a positive label and all the other samples with a negative label. A new observation P is assigned to the class j which produces the largest distance between P and the functional margin of the classifier.

In [2], the sequential backward selection is used to select features. The evaluation of the feature subsets is done using a Naive-Bayes Classifier and a validation set.

In this paper, the simulated annealing technique has been used to find the best feature subset separately for each class. In addition, most of the papers have used small datasets with small number of classes. In this paper, some large datasets with large number of features and classes have been used. The rest of the paper is organized as follows : Sect. 1 explains the simulated annealing technique as used in the proposed method. Section 2 details the methodology. The implementation details are given in Sect. 3. Section 4 describes the results and Sect. 5 gives the Conclusion followed by the references.

2 Methodology

In the method proposed in this paper, the attempt is to find a feature subset for each class which is the best feature subset to separate this class from all the other classes. A test pattern is classified by using k different classifiers, one for each class using only the features chosen for that class. The notations used are given in Table 1.

Table 1. Notations used

Symbol	Description
T	Temperature
X_c	Current solution
J_c	Fitness evaluation of X_c
X_n	Neighbour of current solution
J_c	Fitness evaluation of X_n
α	Cooling rate
ϵ	A very small value say 0.01
p	Probability of accepting a worse solution
Δt	Difference between J_n and J_c

2.1 Feature Subset Selection for Each Class

To find the best feature subset, a simulated annealing(SA) algorithm has been used. The simulated annealing procedure is run k times if k is the number of classes. In each run i, the feature subset for the ith class is found. When carrying out the evaluation of the current string in the SA, all the patterns of the ith class are taken as belonging to one class and all the other patterns belong to the other class. This process results in a set of features which classify patterns as belonging to class i or not class i.

If there are d features, the current solution of the simulated annealing has d elements where each element is either 1 or 0. If the element j is 1, it means that the feature j is present in the feature subset and if the element j is 0, it means that the feature j is not present in the feature subset. Using the current feature subset, the classification of a verification set is carried out. When the feature subset for the ith class is being found, the classification is correct if one of the following conditions are satisfied:

1. If the actual class of the pattern is i and the classified class is also i.
2. If the actual class of the pattern is not i and the predicted class is also not i.

This means that if the class is i and the predicted class is not i or if the class is not i and the predicted class is i, it is a misclassification. The current solution is evaluated by finding the classification accuracy of the verification set using the feature subset selected in the current solution. The SA is used to find the best feature subset using the classification accuracy as the evaluation criterion.
The algorithm is as given below:
Feature Subset Selection Algorithm

1. X_c is the current solution. T=100.$it1$=5.
2. Set the value of the elements of X_c to 0 or 1 at random.
3. Evaluate the fitness J_c of *Validation* using *Train* using only the feature subset which has a value 1 in X_c. Also it=0;
4. $it = it + 1$
5. Find a neighbour of X_c.
6. Let the neighbour of X_c be X_n. Evaluate X_n to get the evaluation J_n.
7. If $J_n < J_c$, p = $\exp\frac{(J_c - J_n)}{T}$
8. if $((J_n \geq J_c)$ or $(random(0,1) < p))$ set $X_c = X_n$ and $J_c = J_n$.
9. if $((it \bmod it1) == 0)$ $T = T * \alpha$
10. If $(T > \epsilon)$ Go to 4.
11. The string which gave the best fitness upto this is chosen as the best solution and this string gives the feature subset chosen.

The simulated annealing algorithm used to find the right feature subset for each class is shown above. X_c is a vector which has d values if the dimensionality is d. The elements of X_c are 0 or 1 which is set at random when the algorithm is started. X_c is evaluated by classifying the patterns in *Validation* by using the patterns in *Train* but only using the features which correspond to a 1 in X_c. A neighbour of X_c is then generated. This is done by choosing a small number k of elements in X_c and changing their value to 0 if they were earlier 1 and changing their value to 1 if they were earlier 0. Here k is a user-specified value and a good value for k would be 5 % to 10 % of d. This is the new candidate solution X_n. X_n is then evaluated. If X_n is better than X_c, then X_n is made the current solution X_c. If X_n is worse than X_c, X_n can still be accepted as the current solution with a probability p where $p = \exp^{\frac{J_c - J_n}{T}}$. This process is repeated for a number of iterations and the solution with the best evaluation is chosen as the final solution. At the beginning, the value of T is large and p is large enough so that X_n maybe accepted as the current solution even if it is a worse solution. This procedure is done to avoid falling into a local minimum. But every few generations T is reduced by the cooling rate α which is around 0.95. If the value of α is closer to 1 then it takes longer for T to reduce and the algorithm will have more number of iterations. T is kept at one value for $it1$ iterations after which it is reduced using the cooling rate. As p comes down the probability of accepting a worse solution keeps coming down. This procedure is used to prevent the algorithm from converging to a local minimum. The algorithm is stopped when T reaches a very small value say 0.01.

2.2 Classification of a Test Pattern

When a test pattern is to be classified, k classifiers are used. The classifier for class 1 uses only the feature subset selected for Class 1 and classifies the pattern as either belonging to Class 1 or not. The same procedure is applied for each of the k classes. These results are combined to find the class label of the test pattern. If the algorithm assigns two(or more) classes to the test pattern, or if the pattern is not assigned to any class, then the closest neighbour of the pattern is found using all the features and the class label of the closest neighbour is assigned to the test pattern.

3 Implementation

The algorithm has been written using GCC in linux. It has been implemented for a number of datasets from the machine learning repository [11]. Table 2 shows the datasets on which the program has been implemented. For each of the datasets the feature subset has been found for every class. It can be seen that the first three datasets have a large number of classes. Dataset 1 and 2 have 10 classes whereas dataset 3 has 26 classes corresponding to the letters of the English alphabet. The number of instances indicated in Table 2 includes the training data, the verification data and the test data. The verification data is classified using the training data to evaluate the current solution of the simulated annealing procedure. The final feature subsets chosen for each class are used to classify the test patterns and the classification accuracy has been reported.

Table 2. Description of datasets used

Sl.No	Dataset	No.of classes	No. of instances	No. of features
1	Optical digit recog	10	5620	64
2	OCR	10	10003	192
3	Letter Recognition	26	20000	16
4	Breast Cancer	2	683	9
5	Iris	3	150	4
6	Wine	3	178	13
7	Glass	6	214	9
8	Seeds	3	210	7

Table 3 shows the division of the dataset in each case into the training set, verification set and test set.

Table 3. No. of patterns in the datasets

Sl.No	Dataset	No.train	No.verific	No.test
1	Optical digit recog	2523	1300	1797
2	OCR	4450	2220	3333
3	Letter recog	8000	5000	7000
4	Breast Cancer	298	153	232
5	Iris	60	30	60
6	Wine	80	40	58
7	Glass	96	46	72
8	Seeds	90	48	72

4 Results

For each dataset, the simulated annealing procedure was used to obtain the feature subset for each class. T was initially fixed at 100. The cooling rate α was taken to be 0.95. The value of ϵ was taken to be 0.01. For every class c, the feature subset which best classified patterns of class c and classified other patterns as not belonging to class c were found.

Table 4 shows the number of features chosen among the 64 features for each of the 10 classes for the Opt.Dig.Rec data. Table 5 shows the number of features selected among the 192 features for the OCR data for each of the 10 classes. Table 6 shows the number of features selected among the 16 features for the letter recognition for each of the 26 classes. The total number of features and the number of features selected for each class for the Breast Cancer, Iris, Wine, Glass and Seeds datasets are shown in Table 7. In the Breast Cancer dataset, there are totally 9 features and two classes. In Class 1, 5 features are chosen and in Class 2, 6 features are chosen.

These reduced class-specific features were used to classify a test set in each of the datasets. The results are presented in Table 8 and Fig. 1.

It can be seen that in most of the cases, the class based feature selection is helpful in improving the classification. In data sets where the number of classes

Table 4. No. of features chosen in each class for Opt.Dig.Rec

Tot.features	Cl1	Cl2	Cl3	Cl4	Cl5	Cl6	Cl7	Cl8	Cl9	Cl10
64	32	42	41	37	35	37	36	43	34	37

Table 5. No. of features chosen in each class for OCR data

Tot.features	Cl1	Cl2	Cl3	Cl4	Cl5	Cl6	Cl7	Cl8	Cl9	Cl10
192	103	104	107	95	94	98	98	93	106	92

Table 6. No. of features chosen in each class for letter recog. data

Tot.features	Cl 1	Cl 2	Cl 3	Cl 4	Cl 5	Cl 6	Cl 7	Cl 8	Cl 9	Cl 10
16	11	10	10	10	9	14	11	11	11	11
	Cl 11	Cl 12	Cl 13	Cl 14	Cl 15	Cl 16	Cl 17	Cl 18	Cl 19	Cl 20
	10	11	10	10	12	12	14	9	11	11
	Cl 21	Cl 22	Cl 23	Cl 24	Cl 25	Cl 26				
	11	15	9	13	12	10				

Table 7. No. of features chosen in each class for some datasets

Dataset	Tot.features	Cl 1	Cl 2	Cl 3	Cl 4	Cl 5	Cl 6
Breast Cancer	9	5	6	-	-	-	-
Iris	4	2	3	3	-	-	-
Wine	13	9	7	11	-	-	-
Glass	9	5	6	2	7	6	4
Seeds	7	3	5	5	-	-	-

Table 8. Comparison of class based feature selection(CBFS) with using all features

	CBFS Acc %	All features Acc %
Opt.Dig.Rec	97.22	97.50
OCR	92.56	91.33
Letter Recog	95.04	93.32
Breast Cancer	96.55	98.71
Iris	93.33	96.67
Wine	93.10	89.66.
Glass	47.22	44.44
Seeds	80.56	86.11

are high and the features are large, there is more improvement. It is only in the case of Iris and Seeds that the classification accuracy does not improve in the case of CBFS as compared to using all features. This is likely to be due to the small number of classes or attributes.

The class based feature selection has also been compared with feature selection which is not class based. A simulated annealing procedure similar to the one used for class based feature selection was used. Only one feature subset is found. The evaluation of the current solution of the simulated annealing is carried out by finding the classification accuracy of a validation set by carrying out 1NN on the training set using only the feature subset selected. The feature subset giving the best classification accuracy is used as the final feature subset selected.

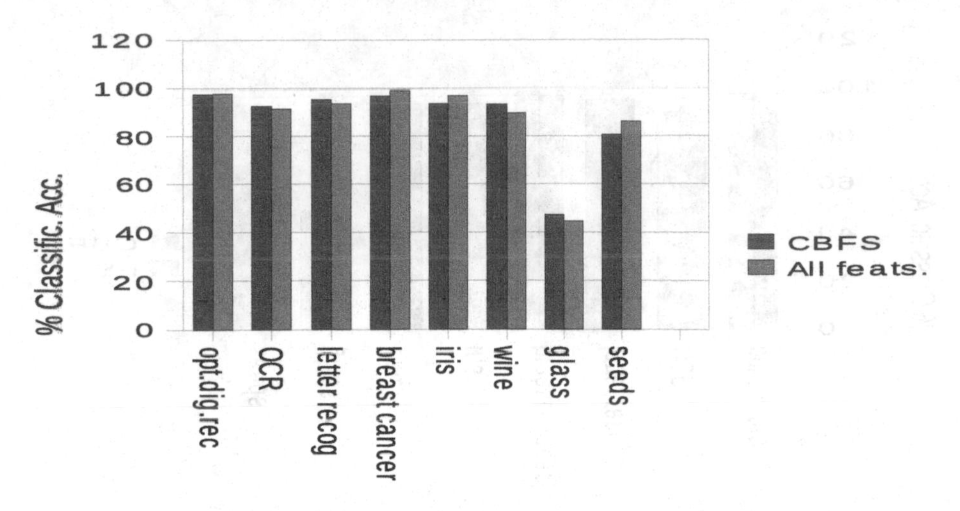

Fig. 1. Comparison of using all features Vs. class-based feature selection

Table 9 and Fig. 2 gives the classification accuracy obtained on the test dataset using feature selection and CBFS. It can be seen that for the datasets like Opt.Dig.Rec, OCR and Letter Recognition where the number of classes are 10 or more and the number of attributes is high, CBFS gives better results. Opt.Dig.Rec which has 10 classes and 64 features, shows an improvement using CBFS. In the case of OCR there are 192 features and 10 classes and here too, the improvement is evident. In Letter Recognition, there are 26 classes and 16 features and CBFS gives better results than FS. It is only in the case of datasets with only a few classes and few attributesi like Breast Cancer, Wine, Glass, and Seeds that FS does better than CBFS.

Table 9. Comparison of feature selection(FS) and class based feature selection(CBFS)

	FS Acc %	CBFS Acc %
Opt.Dig.Rec	96.99	97.22
OCR	89.62	92.56
Letter Recog	94.84	95.04
Breast Cancer	98.28	96.55
Iris	93.33	93.33
Wine	94.83	93.10
Glass	48.61	47.22
Seeds	81.94	80.56

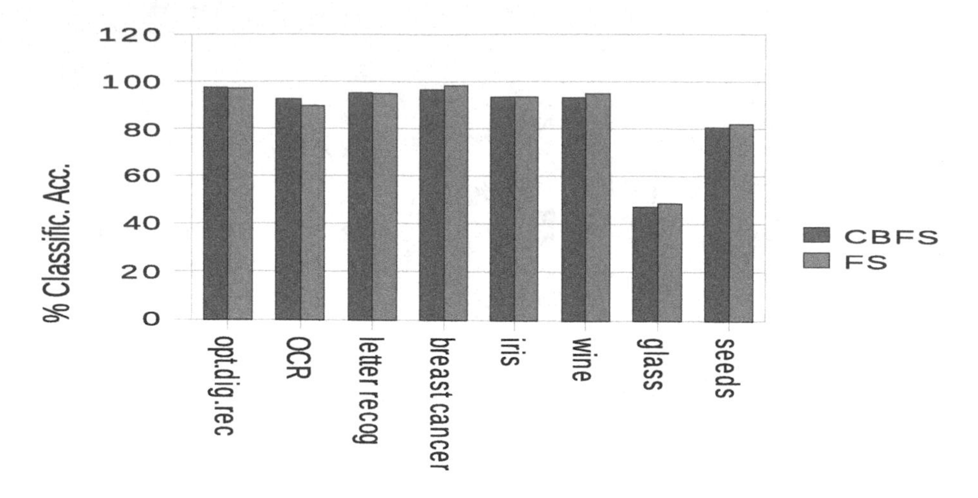

Fig. 2. Comparison of using feature selection Vs. class-based feature selection

5 Conclusion

Class-based feature selection is based on the principle that different classes have different features which are important for that class. A simulated annealing procedure is used to find the feature subset for each class so that patterns of this class are classified correctly. To classify a test pattern, a combination of k classifiers are used one for each class. While more bookkeeping is required as it is necessary to store the feature subset for each class, the feature selection is done only once and stored.

The use of class-based feature selection is found to give a good classification accuracy. This is specially true when the number of classes is large. In such a case only a few features maybe important for a particular class. In addition, the features selected for each class are an indication of which features are important for a class. It gives a description of a class. Class-based feature selection needs to be tried out with more data sets in the future, especially datasets where the number of features and the number of classes are large. It is also possible to use other classifiers besides the nearest neighbour bases approaches to carry out class specific feature selection. This is to be investigated.

References

1. Mackin, P.D., Roy, A., Mukhopadhyay, S.: Methods for pattern selection, class-specific feature selection and classification for automated learning. Neural Netw. (2013). doi:10.1016/j.neunet.2012.12.007
2. Gilbert, J.E., Soares, C., Williams, P., Dozier, G.: A class-specific ensemble feature selection approach for classification problems. In: ACMSE 2010 (2010)
3. Dash, M., Liu, H.: Feature selection for classification. Intell. Data Anal. **1**, 131–156 (1997)

4. Debuse, J.C.W., Rayward-Smith, V.J.: Feature subset selection within a simulated annealing data mining algorithm. J. Intell. Inf. Syst. **9**, 57–81 (1997)
5. Francois, D., de Lannoy, G., Verleysen, M.: Class-specific feature selection for one-against-all multiclass svms. In: ESANN 2011 Proceedings, European Symposium on Artificial Neural Networks, Computational Intelligence and Machine Learning, pp. 263–268 (2011)
6. Guyon, I., Eliseeff, A.: An introduction to variable and feature selection. J. Mach. Learn. Res. **3**, 1157–1182 (2003)
7. Oh, J.-S.L.I.-S., Moon, B.-R.: Hybrid genetic algorithms for feature selection. IEEE Trans. PAMI **26**(11), 1424–1437 (2004)
8. Lanzi, P.L.: Fast feature selection with genetic algorithms: a filter approach. In: IEEE International Conference on Evolutionary Computation, pp. 537–540 (1997)
9. Lie, Y., Wang, G., Chen, H., Dong, H., Zhu, X., Wang, S.: An improved particle swarm optimization for feature selection. J. Bionic Eng. **8**, 191–200 (2011)
10. Murty, M.N., Devi, V.S.: Pattern Recognition : An Algorithmic Approach. Undergraduate Topics in Computer Science. Springer, London (2011)
11. UCI Repository of Machine Learning Databases (1998). http://www.ics.uci.edu/mlearn/MLRepository.html
12. Chen, Y.W., Lin, C.J.: Combining svms with various feature selection strategies. Strat. **324**(1), 1–10 (2006)
13. Zhang, H., Sun, G.: Feature selection using tabu search method. Pattern Recog. **35**, 701–711 (2002)

A Redundancy Study for Feature Selection in Biological Data

Emna Mouelhi[1(✉)], Waad Bouaguel[2], and Ghazi Bel Mufti[3]

[1] ISG, University of Tunis, Tunis, Tunisia
Mouelhi.emmna@yahoo.fr
[2] LARODEC, ISG, University of Tunis, Tunis, Tunisia
bouaguelwaad@mailpost.tn
[3] LARIME, ESSEC, University of Tunis, Tunis, Tunisia
belmufti@yahoo.com

Abstract. The curse of dimensionality is one of the well known issues in Biological data bases. A possible solution to avoid this issue is to use feature selection approach. Filter feature selection are well know feature selection methods that selects the most significant features and discards the rest according to their significance level. In general The set of eliminated features may hide some useful information that may be valuable in further studies. Hence, this paper present a new approach for filter feature selection that uses redundant features to create new instances and avoid the curse of dimensionality.

Keywords: Curse of dimensionality · Relief · Feature selection · Filter

1 Introduction

Biological data are obtained from several scientific experiments, this makes the amount of data significantly enormous. Generally, biological databases are described within a large number (with hundreds or thousands of dimensions) and a reduced number of instances. When the dimensionality increases, time-consuming complexity increases while the performance of classification decreases. Thus, the problem can be viewed as a learning one called curse of dimensionality [1]. which might be solved by reducing the number of features. This phenomenon is known as dimensionality reduction [2]. This latter is splitted into feature extraction and feature selection.

The first one tends to decrease the dimension space by combining the given input data into reduced size data sets, whereas the second one selects the relevant features while preserving the principals mining of the original data set. Whereas the second one, feature selection is the one most frequently used in pre-analysis process and before accomplishing classification task since it keeps the same importance of each feature as it was features input.

[3] summarized feature selection process that includes three principles method: filter, wrapper, and embedded. The first one filters the pertinent features by calculating the weights of features which reduce the size of the data

R. Prasath et al. (Eds.): MIKE 2015, LNAI 9468, pp. 22–28, 2015.
DOI: 10.1007/978-3-319-26832-3_3

and consequently reduce its complexity. The second method selects the best variables based on classification algorithm to obtain optimal subsets. As for the latest method, embedded is based on wrapper method and its fundamental role is to reduce the complexity of classification.

Feature selection is a promising way to avoid the curse of dimensionality and increase classification performance however the reduced size of learning sample may remain a cause of possible degradation in classification performance. Hence, we have to find a way to reduce the dimensionality of the features and in the same time increase the number of observations in order to find a balance between the number of features and the size of the learning sample. A possible solution is to perform a feature selection and then use the eliminated features in order to create new instances. Thus, we propose a new feature selection method that eliminate unwanted features in order to increase the classification performance and avoid the curse of dimensionality. This paper is organized as follows. "New Method for Instance Feature Selection" present a two-stages: feature selection approach combining feature selection and instances generation. "Experimental investigations" describes the used datasets and the performance metrics and summarize the obtained results and conclusions are drawn in "Conclusion".

2 New Method for Instance Feature Selection

We propose in this section a new method based on filter framework. The new method decrease the space of features, perform the biological data processing and make the application of many classification or clustering methods an easy task to achieve. filter methods are known to select the most pertinent features and discard the remaining ones.

Our new method use filter in order to rank features by rate of pertinence. We start by dividing the obtained ranked list into two groups: the first one represent the most relevant features the second one represent the irrelevant features. Then the second set is eliminated and the first one is retained. In general filter methods does not consider the redundancy between the relevant features. If the feature weights are superior to a particular threshold, these features will b e selected even though many of them are highly correlated to each other. Therefore, the first set of features may contain redundant features that can replace others in a feature subset. They basically bring similar information as other features and they usually eliminated after a deep study. In fact, the discarded features my hide some information. Hence, reducing the dimension space of data using filter method can lead to remove some useful information. Consequently, another problem of information lose may. For this reason, we have proposed this approach to resolve the first problem which is the curse of dimension and the second problem related to the information lose.

Once the set of pertinent features is identified, we compute the similarity between all features. If we have one feature similar to another feature, we transform feature having less rate of pertinence into a new instance. Otherwise, if we have more than one feature similar to another feature, we keep the feature having the highest rate of pertinence and we calculate the average of each example of the rest of features and we transform this result into new instance.

For more details, this approach entails two principal steps: feature ranking and feature filtering. Figure 1 below summarizes this new method.

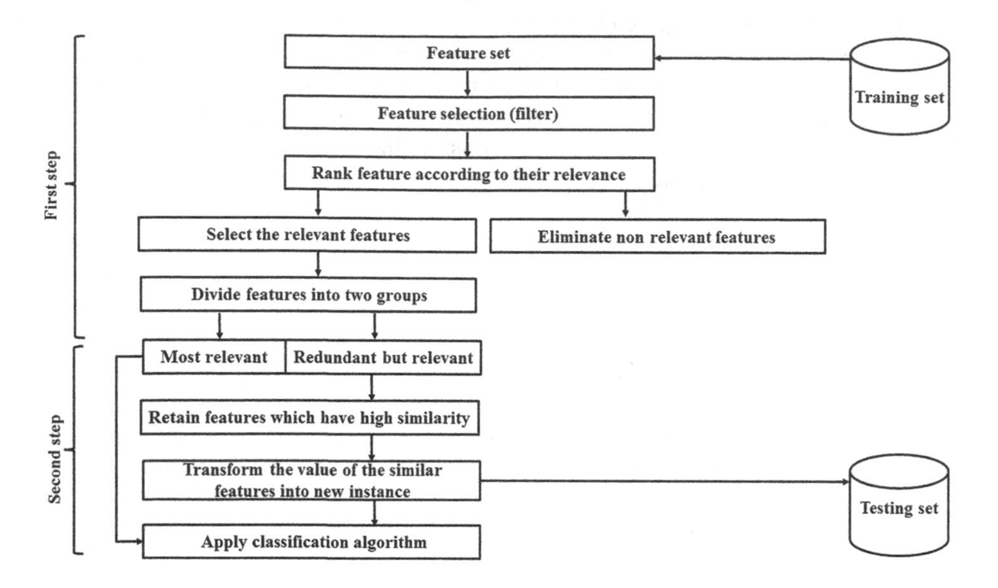

Fig. 1. Flowchart filter framework.

2.1 Step I: Feature Ranking

The first step aims to select the relevant features and to rank them according to their pertinent level for the target feature by applying filter method. We choose relief as a filter method to select the best features. Relief returns a sorted list representing the most relevant features. Nonetheless, this list of may include two features that provide similar information as date of birth and age. Typically, feature redundancy is defined in terms of feature correlation, where two features are redundant to each other if they are correlated. Therefore, this task can be based on the correlation function.

2.2 Step II: Feature Filtering

The second step consists of eliminating the redundant features, but such features may contain valuable information which could possibly lead to improve the model performance. Hence, our proposed method can solve this problem by transforming redundant and less important features into new instances in order to equilibrate the number of instances to the number of features. Let take an illustrative example: Assume we have ten features, ranked in terms of effectiveness. This means that the feature selection algorithm selected feature 1, then feature 2, and so on. Assume that we have selected up to five features and discard the rest. The discarded features provide some useful information and could be useful to further consideration. Hence, the new feature selection method use these eliminated features in order to increase the classification performance and avoid the curse of dimensionality. The underlining idea of the novel approach is to transform the value of the similar features into new instances for the retained features. Note that our goal is to reduce the feature space by performing feature selection and increasing the learning space by creating new instances using the redundant features.

Then, assume we have ten features, ranked in terms of effectiveness. This means that Relief algorithm selected feature x^1, then feature x^2, and so on. Assume that we have selected up to $K = 4$ features, so we use features $x^1; x^2; x^3; x^4$ and discard the rest. Assume that correlation results give: firstly x^1 redundant and similar to x^2 and x^3 redundant and similar to x^4. Features x^2 and x^4 could be eliminated but they may provide some useful information. Thus, we propose to create new instance based on the value of the redundant features. The result of our approach shows a duplication in the number of instances and a reduction in the number of features, but there is another case where for each feature there are more than one feature similar, then an aggregation method can solve this problem. We take the same example where we select up to $K = 4$ features, and $x^1; x^2; x^3; x^4$ are used while the rest is discarded. Additionally, we suppose that correlation results are set to x^3 and x^4 redundant and similar to x^2. Features x^3 and x^4 could be eliminated but they may provide some useful information. Thus, we propose to create new instance based on the value of the redundant features. This example is illustrated in Fig. 2.

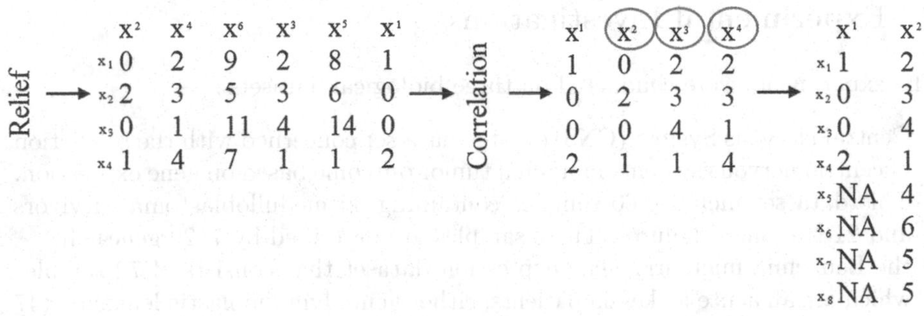

Relief →

	X^2	X^4	X^6	X^3	X^5	X^1
x_1	0	2	9	2	8	1
x_2	2	3	5	3	6	0
x_3	0	1	11	4	14	0
x_4	1	4	7	1	1	2

→ Correlation →

X^1	X^2	X^3	X^4
1	0	2	2
0	2	3	3
0	0	4	1
2	1	1	4

→

	X^1	X^2
x_1	1	2
x_2	0	3
x_3	0	4
x_4	2	1
x_5	NA	4
x_6	NA	6
x_7	NA	5
x_8	NA	5

Fig. 2. Illustrative example of transforming features into new instances.

In order to transform features into new instances, we proceed as follows: for each feature, we check wether it has redundant features. If the number of these latter is greater than 1, an aggregation step is performed employing the typical average function. Then, the redundant features are transformed into instances. Otherwise, the single redundant feature is directly transformed into a new instance.

The following flowchart presented in Fig. 3 models the different steps.

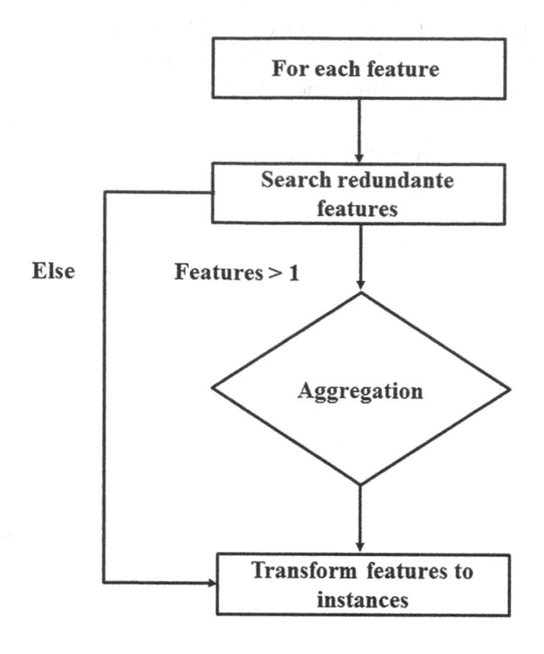

Fig. 3. Flowchart of the transformation process.

3 Experimental Investigations

The experiments were conducted on three biological datasets

- Central Nervous System (CNS), a large data set concerned with the prediction of central nervous system embryonal tumor outcome based on gene expression. This data set includes 60 samples containing 39 medulloblastoma survivors and 21 treatment failures. These samples are described by 7129 genes [4].
- the Leukemia microarry gene expression dataset that consists of 72 samples which are all acute leukemia patients, either acute lymphoblastic leukemia (47 ALL) or acute myelogenous leukemia (25 AML). The total number of genes to be tested is 7129 [5].
- Lung Cancer convers genes involved in a number of molecular and genetic events of the lung cancer including the chromosomal location, mutations and expression. There are a total unique entries in the database.

The performance of our proposed method is evaluated using the standard information retrieval performance measures: precision, recall and F-measure on support vector machine (SVM) and decision tree (DT) classifiers.

Table 1. Performances for CNS dataset.

	Precision	Recall	F-measure
DT			
New approach	0.636	0.619	0.624
Relief	0.510	0.485	0.489
With all features	0.386	0.408	0.393
SVM			
New approach	0.625	0.6	0.604
Relief	0.538	0.523	0.521
With all features	0.422	0.444	0.430

Table 2. Performances for leukemia dataset.

	Precision	Recall	F-measure
DT			
New approach	0.694	0.691	0.690
Relief	0.465	0.403	0.434
With all features	0.317	0.295	0.306
SVM			
New approach	0.618	0.614	0.667
Relief	0.550	0.533	0.541
With all features	0.493	0.418	0.455

Table 3. Performances for lung-cancer dataset.

	Precision	Recall	F-measure
DT			
New approach	0.508	0.534	0.603
Relief	0.348	0.380	0.364
With all features	0.273	0.390	0.332
SVM			
New approach	0.536	0.605	0.587
Relief	0.274	0.290	0.282
With all features	0.256	0.250	0.253

From Tables 1, 2 and 3 we notice the superiority of the result's obtained by our new approach. Tables 1, 2 and 3 summarize the obtained performances when the algorithms are trained over all the features, over relief result and on the set obtained by the new approach. From these results we notice that feature selection using relief or the new approach improve significatively the performance of the classification algorithms.

Comparing relief results with the new approach based on the simple SVM and DT classification rule, it is evident that the later performs significantly much better. This performance is due to the fact that redundant features are eliminated and transformed to new instance. Hence, we have more data to train our algorithm.

4 Conclusion

Feature selection is an important task in Biological data. We have proposed in this paper the use of filter method in our new approach in order to improve the result of this latter. In fact, this new method is divided into two steps: Feature Ranking and Feature Filtering. In each step we reduce the feature space and increase the instance space. Results on three biological databases show that combining feature selection with similarity study and instances study has a great impact on classification results.

References

1. Bellman, R.: Processus Adaptive Control: A Guided Tour. Princeton University Press, Princeton (1961)
2. Guerif, S.: Rduction de dimension en apprentissage numrique non supervise. Ph.D. thesis, Universit Paris 13 (2006)
3. Salvador, G., Julin, L., Francisco, H.: Data preprocessing in Data Mining. In: Kacprzyk, J., (ed.) Polish Academy of Sciences. Springer, Poland, Warsaw (2015)
4. Pomeroy, S.L., Tamayo, P., Gaasenbeek, M., Sturla, L.M., Angelo, M., McLaughlin, M.E., Kim, J.Y.H., Goumnerova, L.C., Black, P.M., Lau, C., Allen, J.C., Zagzag, D., Olson, J.M., Curran, T., Wetmore, C., Biegel, J.A., Poggio, T., Mukherjee, S., Rifkin, R., Califano, A., Stolovitzky, G., Louis, D.N., Mesirov, J.P., Lander, E.S., Golub, T.R.: Prediction of central nervous system embryonal tumour outcome based on gene expression. Nat. **415**(6870), 436–442 (2002)
5. Golub, T.R., Slonim, D.K., Tamayo, P., Huard, C., Gaasenbeek, M., Mesirov, J.P., Coller, H., Loh, M.L., Downing, J.R., Caligiuri, M.A., Bloomfield, C.D.: Molecular classification of cancer: class discovery and class prediction by gene expression monitoring. Sci. **286**, 531–537 (1999)

New Feature Detection Mechanism for Extended Kalman Filter Based Monocular SLAM with 1-Point RANSAC

Agniva Sengupta([✉]) and Shafeeq Elanattil

Kritikal Solutions Pvt. Ltd., Bangalore, India
{agniva.sengupta,shafeeq.elanattil}@kritikalsolutions.com,
{i.agniva,eshafeeqe}@gmail.com
http://www.kritikalsolutions.com/

Abstract. We present a different approach of feature point detection for improving the accuracy of SLAM using single, monocular camera. Traditionally, Harris Corner detection, SURF or FAST corner detectors are used for finding feature points of interest in the image. We replace this with another approach, which involves building non-linear scale space representation of images using Perona and Malik Diffusion equation and computing the scale normalized Hessian at multiple scale levels (KAZE feature). The feature points so detected are used to estimate the state and pose of a mono camera using extended Kalman filter. By using accelerated KAZE features and a more rigorous feature rejection routine combined with 1-point RANSAC for outlier rejection, short baseline matching of features are significantly improved, even with lesser number of feature points, especially in the presence of motion blur. We present a comparative study of our proposal with FAST and show improved localization accuracy in terms of absolute trajectory error.

Keywords: EKF · MonoSLAM · AKAZE · Localization

1 Introduction

Harris corner detection, SURF or FAST corner detector [8] are the usual feature descriptor of choice while detecting sensible landmarks for localization and mapping. Despite being fast and effective in most situations, they often exhibit poor repeatability in presence of motion blur. While mapping out areas with few corners or flat texture, the system often detects too few landmarks, resulting in poor localization accuracy of the system. We noticed many cases where sudden movement of the camera resulted in a series of motion-blurred frames. In those cases, Harris Corner or FAST does not detect any feature points (beyond an acceptable threshold) and the camera localization becomes significantly erroneous after such maneuvers.

There has been extensive research in MonoSLAM over the last two decades (or more). However, despite the stellar performance of approach like EKF based

R. Prasath et al. (Eds.): MIKE 2015, LNAI 9468, pp. 29–36, 2015.
DOI: 10.1007/978-3-319-26832-3_4

MonoSLAM, PTAM [10], DTAM [11] etc., the monocular camera based SLAM paradigm is yet to reach at par with stereo/RGB-D based SLAM frameworks in terms of accuracy.

Moreover, in case of extended Kalman filter based monoslam [6], the feature matching in subsequent frames is done using a normalized cross-correlation of image patches, instead of using a descriptor-to-descriptor comparison across image. This is done to ensure real-time operation of the algorithm. Hence, it is very important to ensure proper initialization of feature points, so that they can be identified easily in the subsequent frames.

1.1 Objective

While localizing and mapping a monocular camera using extended Kalman filter, two very specific areas for improvement (over and above the existing state-of-the-art) were identified. The first issue was observed with scenarios where camera exhibits a sudden motion, abruptly changing its pose over a short period of time. The movement induces blurry frames, short baseline matching goes wrong for a few frames and the camera localization suffers considerable loss of accuracy after every such situations. We analyze these conditions and propose an alternative solution for handling this situation better.

A secondary objective is to keep the feature vector size constant while maintaining same accuracy levels. This is done by aggressively pruning the number of feature points being tracked by the filter.

The main contribution of our work is the integration of accelerated KAZE features with EKF based mono SLAM. We show the possibility of obtaining better localization accuracy using AKAZE. We also use 1-point RANSAC for outlier rejection [5] and the combined output has been described in the results.

1.2 Related Work

All the filtering based monocular SLAM algorithms work in two recognizable steps: extract features from the image plane and track the features to update the state vector, which typically updates both the camera/robot state as well as the world map. Feature extraction is a key component of this algorithm and considerable research has been done to study the effect of various feature detection techniques on the outcome of the SLAM architecture. [9] compares the effect of SURF, SIFT, BRIEF and BRISK on visual SLAM. [13] proposed ORB as an efficient alternative to SURF and SIFT.

In the following sections, we describe our proposal and compare it with some of the existing techniques.

2 Method

We first briefly describe the usual steps associated with the conventional monoslam algorithm based on EKF. Then we present the feature detection mechanism that we incorporated into the process.

The state representation of the pose of the camera is a 13 dimensional vector [1]:

$$x_v = \begin{bmatrix} r^W \\ q^{WC} \\ v^W \\ \omega^C \end{bmatrix} \tag{1}$$

Which can be explained as a 3D position vector r^W, unit quaternion q^{WC}, velocity vector v^W, and angular velocity vector ω^C relative to a world frame W and a frame C fixed with the camera. Acted upon by an uniform angular and translational velocity, the state transition is formulated by:

$$g_v(\mu_{t-1}) = \begin{bmatrix} r_{t-1}^{WC} + v_{t-1}^{W}\Delta t \\ q_{t-1}^{WC} \times quat(\omega_{t-1}^{C}\Delta t) \\ v_{t-1}^{W} \\ \omega_{t-1}^{C} \end{bmatrix} \tag{2}$$

where μ_{t-1} is the previous mean and μ_t is the current mean. The motion model thus generated is non-linear in nature, since the linear and angular velocity driving the camera is random and cannot be properly predicted.

Given the non-linearity of the state transition, the extended Kalman filter formulation is used for simultaneous state estimation and prediction:

$$\bar{\mu_t} = g(a, \mu_{t-1}) \tag{3}$$

$$\bar{\Sigma_t} = G_t \Sigma_{t-1} G_t^T + R^t \tag{4}$$

$$K_t = \bar{\Sigma_t} H_t^T (H_t \bar{\Sigma_t} H_t^T + Q_t)^{-1} \tag{5}$$

$$\mu_t = \bar{\mu_t} + K_t(z_t - h(\bar{\mu_t})) \tag{6}$$

$$\Sigma_t = ((\mathbb{1}) - K_t H_t)\bar{\Sigma_t} \tag{7}$$

However, (1) does not represent the entire feature vector. The state space representation used here includes the state of the camera, as well as the entire set of feature points being tracked by the system.

Detecting feature points of interest is a key element of this algorithm. Traditionally, Harris corner detector or Features from Accelerated Segment Test (FAST) are used for detecting key points in an image.

We propose to introduce KAZE features [3] for detecting the landmarks in the image. The scale space is discretized in logarithmic increments and maintained in a series of O octaves and S sub-levels. These indices are mapped to their corresponding scale σ by:

$$\sigma_i(o,s) = \sigma_0 2^{\frac{o+s}{S}} \tag{8}$$

The scale space is converted to time units with the mapping:

$$t_i = 1/2\sigma_i^2 \tag{9}$$

Starting from the classic non-linear diffusion formulation:

$$\frac{\partial L}{\partial t} = div(c(x,y,t). \bigtriangledown L) \tag{10}$$

where the conductivity c is dependent on the gradient magnitude:

$$c(x,y,t) = g(| \bigtriangledown L_\sigma(x,y,t)|) \tag{11}$$

and the function g, as expressed by Perona and Malik [12], can have two different formulation:

$$g_1 = e^{(-\frac{|\bigtriangledown L_\sigma|^2}{k^2})}, g_2 = \frac{1}{(1+\frac{|\bigtriangledown L_\sigma|^2}{k^2})} \tag{12}$$

where k is the contrast factor that controls the level of diffusion.

There is no analytical solution for the PDEs involved in Eq. 10, so they are approximated using a semi-implicit scheme. Starting from Eq. 9 and the contrast parameter, the non linear scale space is defined as:

$$L^{i+1} = (I - (t_{i+1} - t_i). \sum_{l=1}^{m} A_l(L^i))^{-1} L^i \tag{13}$$

Over multiple scale levels, the response of scale normalized determinant of Hessian is used for detecting feature points of interest:

$$L_{hessian} = \sigma^2 (L_{xx} L_{yy} - L_{xy}^2) \tag{14}$$

where L_{xx} L_{yy} are the second order horizontal and vertical derivatives respectively. On a set of filtered image Li, a rectangular window of $\sigma^i \times \sigma^i$ is searched for the extrema. Sub-pixel accuracy is not searched for. We also skip the formation of feature descriptor, since the patch matching in subsequent frames will be done by cross correlation of image segments.

To speed up the operation, we use the Fast Explicit Diffusion [2] scheme by performing M cycles of n explicit diffusion steps with non-uniform steps τ_j that is formed by:

$$\tau_j = \frac{\tau_{max}}{\cos(\pi \frac{2j+1}{4n+2})} \tag{15}$$

where τ_max is the maximum step that does not violate the stability of the explicit scheme.

The discretization of the diffusion equation can be expressed as:

$$\frac{L^{i+1} - L^i}{\tau} = A(L^i)L^i \tag{16}$$

And given an apriori estimate of $L^{i+1,0} = L^i$, a FED cycle can be expressed as:

$$L^{i+1,j+1} = (I + \tau_j A(L^i))L^{i+1,j} \tag{17}$$

Using this step, we get a feature point (u,v), which needs to be converted to inverse depth parametrization [4]. Basically, the inverse depth parameters is a six dimensional vector, represented by (18):

$$y_i = (x_{c,i}\ y_{c,i}\ z_{c,i}\ \theta_i\ \phi_i\ \rho_i)^T \tag{18}$$

where $x_{c,i}, y_{c,i}, z_{c,i}$ represents the position of the camera w.r.t the world when the feature was first observed, θ_i, ϕ_i represents the azimuth and elevation of the feature point, when observed and ρ_i is the depth estimate of the feature point (which is usually initialized at 0.1).

The feature points, represented in inverse depth, are appended to the camera pose vector to form the state vector of the system. This vector is iteratively predicted and measured by EKF Eqs. (3) through (7).

The rest of the EKF measurement and update is done by standard formulation, with two step partial update for low and high innovation inliers in a RANSAC hypothesis [5].

Moreover, we do not allow any feature's inverse depth parameters to persist in the feature vector beyond 3 cycles of EKF, thereby reducing the rate of increase of feature vector size.

3 Results

We use Absolute Trajectory Error (ATE) as a means to validate our approach. ATE compares the trajectory of a robot/camera, as reconstructed by an algorithm using real sensor data as its input, to the actual trajectory (ground truth).

We benchmark our approach in the RGB-D SLAM dataset of TUM [14,15]. Only RGB data is used for the experiments, while the groundtruth trajectory provided in the dataset is used for validation. The EKF is implemented on MATLAB (which is based on the open source code provided by [10]) while the computation of AKAZE features is done in C++. We observed approximately 20–25 % decrease in root mean squared error of absolute trajectory over short sequences, using the technique we proposed in the previous section (Table 1).

The dataset we have used for the demonstration example is *freiburg1_room* from [14]. The image sequence has been captured using a Microsoft Kinect. The experiment where we use FAST feature descriptor along with existing filter based

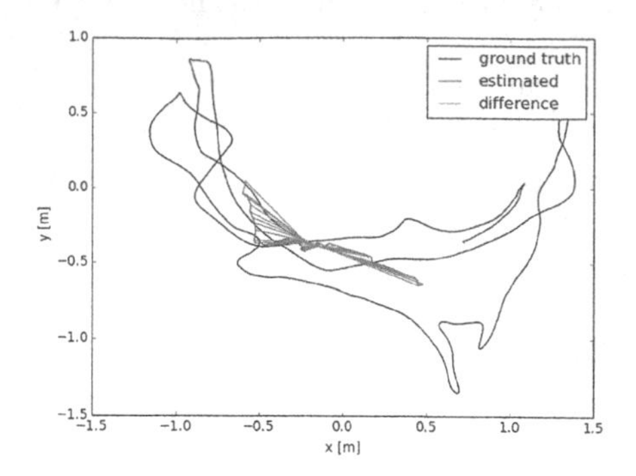

Fig. 1. Localization using FAST

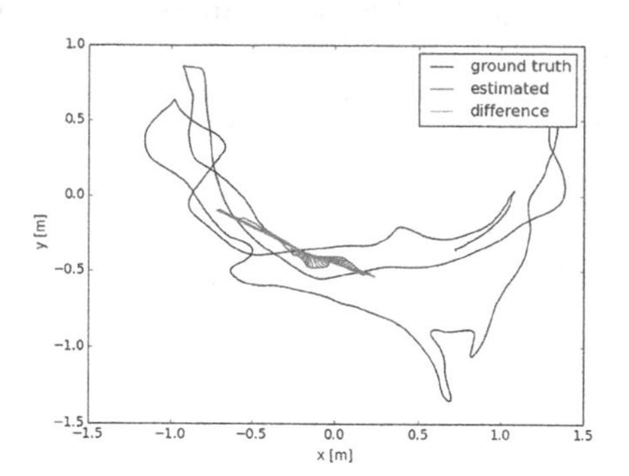

Fig. 2. Localization using AKAZE

Table 1. Comparison of results between EXP A (FAST) and EXP B (AKAZE)

	EXP A	EXP B
Root Mean Square Error	0.320698 m	0.243540 m
Mean	0.278879 m	0.206980 m
Median	0.232776 m	0.155320 m
Standard Deviation	0.158348 m	0.128339 m
Min. Error	0.092825 m	0.086545 m
Max. Error	0.619539 m	0.561998 m

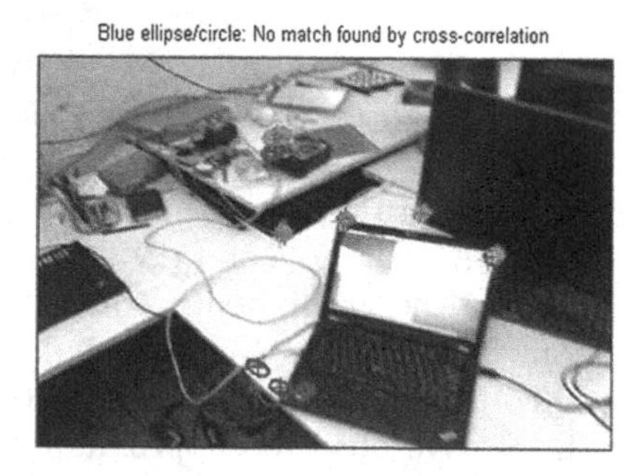

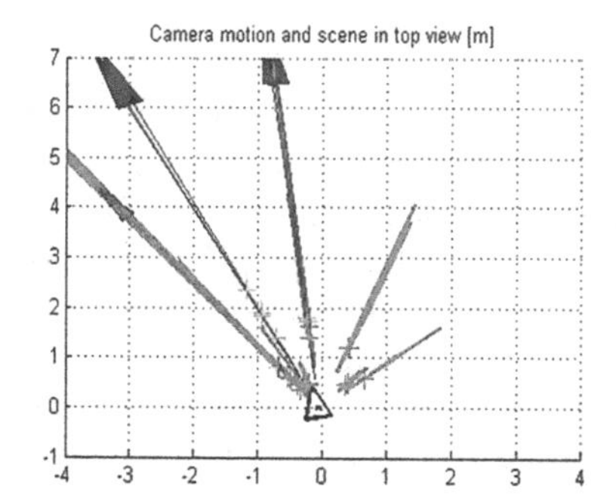

Fig. 3. The circles represent the feature points detected by AKAZE. The red ellipses are the matched points, the pink ones are those rejected by 1-point RANSAC (Color figure online)

Table 2. Comparison of results between EXP C (ORB) and EXP D (AKAZE)

	EXP C	EXP D
Root Mean Square Error	1.213417 m	1.150064 m
Mean	1.073303 m	1.023851 m
Median	1.160499 m	1.109093 m
Standard Deviation	0.566041 m	0.523809 m
Min. Error	0.171942 m	0.103387 m
Max. Error	2.306686 m	2.084978 m

monoslam algorithm has been denoted **EXP A** (Fig. 1). Our proposed approach has been denoted **EXP B** (Figs. 2 and 3).

For the sake of completeness, we also compared the proposed approach with feature detection using ORB. The results obtained are tabulated below. **EXP C** denotes the results obtained while using ORB as the feature detector. **EXP D** denotes the proposed framework using AKAZE. This was done on the data set *freiburg1_360*, which proved to be more error prone (in terms of short-baseline localization accuracy) due to the presence of heavy motion blur. Even in this experiment, AKAZE performed better than ORB. However, the advantage was slightly less pronounced (Table 2).

The time performance of AKAZE is better than SURF or SIFT, but not as efficient as FAST [7]. The extended Kalman filter based mono SLAM section of the proposed algorithm is mostly similar to [5].

4 Conclusion

Using accelerated KAZE features for feature point detection in monoslam is not documented in any of the literature we surveyed so far. It results in better localization accuracy in dataset involving motion – blurred frames. This has been validated in RGB-D dataset by comparison against ground truth values. MonoSLAM is a field of study which has immense scope for improvement in terms of accuracy and reliability. It is necessary to benchmark the performance of MonoSLAM using various feature detectors. Although both the original MonoSLAM algorithm and AKAZE runs in real time, this research work does not cover the time performance of the two combined. This needs to be analyzed further.

References

1. Albrecht, S.: An analysis of visual mono-slam. Diss. Master's Thesis. Universität Osnabrück, 2009 (2009)
2. Alcantarilla, P.F., Solutions, T.: Fast explicit diffusion for accelerated features in nonlinear scale spaces. IEEE Trans. Patt. Anal. Mach. Intell. **34**(7), 1281–1298 (2011)

3. Alcantarilla, P.F., Bartoli, A., Davison, A.J.: KAZE features. In: Fitzgibbon, A., Lazebnik, S., Perona, P., Sato, Y., Schmid, C. (eds.) ECCV 2012, Part VI. LNCS, vol. 7577, pp. 214–227. Springer, Heidelberg (2012)
4. Civera, J., Davison, A.J., Montiel, J.M.: Inverse depth parametrization for monocular slam. IEEE Trans. Rob. **24**(5), 932–945 (2008)
5. Civera, J., Grasa, O.G., Davison, A.J., Montiel, J.: 1-point ransac for extended kalman filtering: Application to real-time structure from motion and visual odometry. J. Field Robot. **27**(5), 609–631 (2010)
6. Davison, A.J., Murray, D.W.: Simultaneous localization and map-building using active vision. IEEE Trans. Pattern Anal. Mach. Intell. **24**(7), 865–880 (2002)
7. Feng, L., Wu, Z., Long, X.: Fast image diffusion for feature detection and description. Int. J. Comput. Theory Eng. **8**(1), 58–62 (2016)
8. Gauglitz, S., Höllerer, T., Turk, M.: Evaluation of interest point detectors and feature descriptors for visual tracking. Int. J. Comput. Vision **94**(3), 335–360 (2011)
9. Hartmann, J.M., Klussendorff, J.H., Maehle, E.: A comparison of feature descriptors for visual slam. In: 2013 European Conference on Mobile Robots (ECMR), pp. 56–61. IEEE (2013)
10. Klein, G., Murray, D.: Parallel tracking and mapping for small ar workspaces. In: 6th IEEE and ACM International Symposium on Mixed and Augmented Reality, ISMAR 2007, pp. 225–234. IEEE (2007)
11. Newcombe, R.A., Lovegrove, S.J., Davison, A.J.: DTAM: Dense tracking and mapping in real-time. In: 2011 IEEE International Conference on Computer Vision (ICCV), pp. 2320–2327. IEEE (2011)
12. Perona, P., Malik, J.: Scale-space and edge detection using anisotropic diffusion. IEEE Trans. Pattern Anal. Mach. Intell. **12**(7), 629–639 (1990)
13. Rublee, E., Rabaud, V., Konolige, K., Bradski, G.: ORB: an efficient alternative to SIFT or SURF. In: 2011 IEEE International Conference on Computer Vision (ICCV), pp. 2564–2571. IEEE (2011)
14. Sturm, J., Engelhard, N., Endres, F., Burgard, W., Cremers, D.: http://vision.in.tum.de/data/datasets/rgbd-dataset/download
15. Sturm, J., Engelhard, N., Endres, F., Burgard, W., Cremers, D.: A benchmark for the evaluation of RGB-D slam systems. In: 2012 IEEE/RSJ International Conference on Intelligent Robots and Systems (IROS), pp. 573–580. IEEE (2012)

Sequential Instance Based Feature Subset Selection for High Dimensional Data

Afef Ben Brahim[1(✉)] and Mohamed Limam[1,2]

[1] ISG, University of Tunis, Tunis, Tunisia
afef.benbrahim@yahoo.fr
[2] Dhofar University, Salalah, Sultanate of Oman

Abstract. Feature subset selection is a key problem in the data-mining classification task that helps to obtain more compact and understandable models without degrading their performance. This paper deals with the problem of supervised wrapper based feature subset selection in data sets with a very large number of attributes and a low sample size. In this case, standard wrapper algorithms cannot be applied because of their complexity. In this work we propose a new hybrid -filter wrapper-approach based on instance learning with the main goal of accelerating the feature subset selection process by reducing the number of wrapper evaluations. In our hybrid feature selection method, named Hybrid Instance Based Sequential Backward Search (HIB-SBS), instance learning is used to weight features and generate candidate feature subsets, then SBS and K-nearest neighbours (KNN) compose an evaluation system of wrappers. Our method is experimentally tested and compared with state-of-the-art algorithms over four high-dimensional low sample size datasets. The results show an impressive reduction in the execution time compared to the wrapper approach and that our proposal outperforms other methods in terms of accuracy and cardinality of the selected subset.

Keywords: Feature selection · Sequential · Hybrid · Wrapper · High dimensional data

1 Introduction

High technologies routinely produce large data sets characterized by high numbers of features [5,6]. Owing to the obstacles inherent in dealing with extremely large numbers of interacting variables, small samples create the need for feature selection, while at the same time making feature selection algorithms less reliable. Feature selection is required when the number of features is large with respect to the sample size because the use of a large number of features can result in overfitting the data: the designed classifier performs well on the sample data but not on the feature-label distribution from which the data have been drawn. Feature selection techniques differ from each other in the way they incorporate the search of the optimal subset of relevant features in the model

R. Prasath et al. (Eds.): MIKE 2015, LNAI 9468, pp. 37–46, 2015.
DOI: 10.1007/978-3-319-26832-3_5

selection. In the context of classification and based on whether the feature selection process involves employing a learning algorithm to guide the search, feature selection methods divide into wrappers and filters [1,8]. Filters select subsets of features as a pre-processing step, independently of the chosen predictor. Wrapper and embedded methods, on the other hand, generally use a specific learning algorithm to evaluate a specific subset of features. One major issue with wrapper methods is their high computational complexity due to the need to train a large number of classifiers. Many heuristic algorithms e.g., forward and backward selection [1] have been proposed to alleviate this issue. However, due to their heuristic nature, none of them can provide any guarantee of optimality. With tens of thousands of features, which is the case in gene expression microarray data analysis, a hybrid approach can be adopted wherein it follows filter model in the search step selecting small number of candidate subsets of features and then a wrapper method is applied to the reduced subsets to achieve the best possible performance with a particular learning algorithm which lead to less complex model. Accordingly, the hybrid model is more efficient than filter and less expensive than wrapper.

In this paper, we propose a new hybrid approach based on instance learning with the main goal of speeding up the feature subset selection process by reducing the number of wrapper evaluations while maintaining good performance in terms of accuracy and size of the obtained subset. In this approach, an instance feature weighting technique is used to generate candidate feature subsets. The candidate feature subsets are then integrated in a search procedure of the optimal feature subset, where SBS [1] and KNN [4] compose an evaluation system of wrappers. The best feature subset search technique is done with ten-fold cross validation. Our proposed approach takes advantage of the low sample size of the data as the number of candidate subsets is the number of instances and thus only a few blocks of variables are analyzed and so the number of wrapper evaluations decreases significantly. The approach shows a good performance not only in terms of classification accuracy, but also with respect to the number of features selected.

2 Feature Selection Methods

Feature selection methods use an evaluation function to search for most relevant features. Filter methods use a general relevance measure of the features to the prediction evaluation function, while wrapper methods rely on the performance of a classification algorithm.

Filter techniques [1] assess the relevance of features by looking only at the intrinsic properties of the data. In most cases a feature relevance score is calculated, and low scoring features are removed. Afterwards, this subset of features is presented as input to the classification algorithm. A sub-category of filter methods that are refer to as rankers, are methods that employ some criterion to score each feature and provide a ranking. The Relief algorithm is one of these rankers [2]. It assigns a relevance weight to each feature, which is meant to denote the relevance of the feature to the target concept. Relief samples instances randomly from the training set and updates the relevance values based on the difference between the selected instance and the two nearest instances of the same and opposite class. Another commonly used filter for feature selection is the statistical t-test. It is used in the form that defines the score of a feature as the ratio of the difference between its mean values for each of the two classes and the standard deviation. The weight of each feature is thus given by its computed absolute score. Filter models select features that are independent of the classifier and avoid the cross-validation step in a typical wrapper model, therefore they are computationally efficient. They are useful in most real-world problems and easily scaled to very high-dimensional data sets, at least for an initial stage of filtering out useless features. They are independent of the classification algorithm as the search in the feature subset space is separated from the search in the hypothesis space, but this presents in the same time a disadvantage of filter methods because they ignore the interaction with the classifier and that they ignore feature dependencies.

Wrapper approaches [8] use a performance measure of a learning algorithm along with a statistical re-sampling technique such as cross-validation to select the best feature subset. Wrappers search for features, usually using greedy search strategy, better suited to the learning algorithm aiming to improve learning performance and thus the generality of the selected features is limited and different learning algorithms could lead to different feature selection results. But it also tends to be more computationally expensive than the filter model because of repeatedly building learning models on each candidate subset of features, especially if building the learning algorithm has a high computational cost [12]. However, they usually provide the best performing feature set for that particular type of model and have the ability to take into account feature dependencies as they consider groups of features jointly. Generally, wrapper methods can be divided into two groups, deterministic and randomized methods. Deterministic wrapper methods search through the space of available feature either forwards or backwards. In the forward selection process, single attributes are added to an initially empty set of attributes. In the forward stage-wise selection technique only one variable can be added to the set at one step. Sequential backward elimination works in the opposite direction of forward selection. Starting from the full set, we sequentially remove the feature that results in the smallest decrease in the value of the objective function. Compared to deterministic wrapper methods, randomized wrapper algorithms search the next feature subset partly at random. Single features or several features at once can be added, removed, or replaced from the previous feature set. Famous representatives of randomized wrapper methods are genetic algorithms [1], stochastic search methods which are inspired by evolutionary biology and use techniques encouraged from evolutionary processes such as mutation, crossover, selection and inheritance to select a feature subset. Advantages of wrapper approaches include the interaction between feature subset search and model selection, and the ability to take into account feature dependencies. A common drawback of these techniques is that they are very computationally intensive.

3 Hybrid Instance Based Sequential Backward Search Method (HIB-SBS)

Hybrid feature selection takes advantages of combining feature space filtering and wrapper methods. It overcomes problems related to the two models cited before. In a first step, hybrid methods use a filter to generate few relevant candidate feature subsets. On a second step, a wrapper approach analyses the generated candidates to find the best features. [14] developed a stochastic algorithm based on the GRASP meta-heuristic. It is a multi-start constructive method which constructs a solution in its first stage and then runs an improving stage over that solution. [13] proposed a hybrid genetic algorithm with two stages of optimization where the mutual information between the predictive labels and the true classes serves as the fitness function for the genetic algorithm and where an improved estimation of the conditional mutual information acts in a filter manner.

In this paper, we propose a hybrid approach where a filter model is applied in the search step selecting small number of candidate subsets of features and then a wrapper method is applied to the reduced subsets to achieve the best possible performance with a particular learning algorithm which lead to less complex model.

3.1 Step1: Candidate Feature Subsets Selection

We have a data set X containing m training instances x_i and d features a_j, with $d > m$. In a first step of the feature selection process, the feature space is reduced to m candidate subsets. Each instance of the training data generates a candidate feature subset based on an instance feature weighting technique [7].

We search for the two nearest neighbors of each sample x_i for each feature a_j, one from the same class (called nearest hit or NH), and the other from the different class (called nearest miss or NM). The margin of x_{ni} is then computed as

$$W(x_{ij}) = d(x_{ij}, NM(x_{ij})) - d(x_{ij}, NH(x_{ij})) \tag{1}$$

using the Manhatan distance function to define a sample's margin and nearest neighbors while other distance functions may also be used. This weight definition is used in the Relief algorithm (using Euclidean distance) for the feature-selection purpose [2]. This weight is then projected on each feature a_j and we get the matrix $\mathbf{W}$ of feature weights $w_{j,i}$.

Then, features in the space of each instance are ranked based on their calculated weights. Note that a feature a_j may have different ranks depending on the instance considered. After this instance based feature ranking step, a candidate subset of cardinality n is chosen from the best ranked features of each instance. This pre-processing step leads to m candidate feature subsets $\{CFS_1, CFS_2, ...CFS_m\}$ of cardinality n.

3.2 Step 2: Sequential Backward Search

In this second step, the CFSs are integrated in a search process of the optimal feature subset, where the subset search technique and a kNN classifier consist an evaluation system of wrappers. This wrapper approach is based on sequential search, i.e. feature selection decisions of training instances are combined based on their effect on classification performance using an iterative process.

This step begins by considering all the features composing the m CFS in a single feature subset called FS_{Union}. A kNN classifier is trained on the projection of FS_{Union} on the training data and the classification error β_{init} is calculated on the test data set using 10-fold CV. Then, a kNN classifier is applied on the training data using $FS_{Union} \setminus \{CFS_i\}$ and its classification error rate β_i is calculated. Thus, the kNN classifier is applied m times and m classification error rates $ERR = \{\beta_1, \beta_2, ...\beta_m\}$ are obtained. The algorithm then finds in ERR the error rates which are smaller than β_{init}. The resulting error rates subset is $ERR\prime \subset ERR$. If $ERR\prime$ is empty, FS_{Union} is selected as the final feature subset as it gives the minimum error rate. If it is not the case, candidate subsets which their exclusion has resulted in error rates of $ERR\prime$ are rejected and FS_{Union} is updated to contain features of remaining CFS. That means that the K worst CFS are eliminated from the wrapper search procedure. Thus, the number of CFS decreases to $(m - K)$. The whole error β_{init} is also updated to be equal to β_i, the minimum classification error in $ERR\prime$. Hence, the optimal feature subset to be selected should reduce β_i the minimum error rate achieved in the last iteration. This SBS process is iterated until there is no decrease in the classification error rate, i.e. until $ERR\prime$ is empty. The resulting feature subset FS_{Union} is returned as the optimal feature subset. Note that 10-fold CV is used for the feature subset search. We call this version of our proposed approach Hybrid Instance Based SBS (HIB-SBS) and its algorithm is described in Algorithm 1 and illustrated in Fig. 1.

4 Experimental Study

In this section we report the experimental setup and results of our hybrid feature selection method proposed in Sect. 3 in comparison with four algorithms, Relief and t-test algorithms which are filters, Randomized which is a wrapper and another hybrid-sequential algorithm. The three first algorithms are described in Sect. 2. The hybrid-sequential algorithm used for our comparisons in addition to these algorithms uses a forward sequential feature selection with KNN algorithm in a wrapper fashion. It finds important features from a reduced set of features obtained using filter results of a t-test as a pre-processing step. We refer to this algorithm in our experiments as Hyb-Seq. The KNN classifier is used with all algorithms to evaluate classification performance. The considered algorithms are applied to several microarray data sets described in Sect. 4.1. Classification performance, final subset cardinality and execution time are used as metrics to evaluate our approach.

Algorithm 1. HIB-SBS

Input:
$[\mathbf{X}, CFS_{All}]$
Set $FS_{Union} = \forall a_j \in CFS_{All}$
β_{init} = Apply kNN (X, FS_{Union})
repeat
  β_i = Apply kNN $(X, (FS_{Union} \setminus \{CFS_i\}))$
  Obtain $ERR = \{\beta_1, \beta_2, ...\beta_m\}$
  Obtain $ERR\prime = \forall \beta_i < \beta_{init}$
  if $ERR\prime = \emptyset$ return selected features FS_{Union}
  else
  Find" Bad CFSs": the K CFSs resulting in $\beta_i \in ERR\prime$
  Update $CFS_{All} = (CFS_{All} \setminus$ Bad CFSs)
  Update $FS_{Union} = \forall a_j \in CFS_{All}$
  Update $m = m - K$
  $\beta_{min} = min(ERR\prime)$
  Update $\beta_{init} = \beta_{min}$
until $ERR\prime = \emptyset$
Output: FS_{Union}

4.1 Datasets

The experiments were conducted on four high dimensional low sample size data sets. The classification task in DLBCL, standing for diffuse large Bcells, is the prediction of the tissue types [9], where genes are used to discriminate DLBCL tissues from Follicular Lymphomas. It contains 77 samples described by 7029 genes. We consider also another Lymphoma data set which task is to discriminate between two types of Lymphoma based on gene expression measured by microarray technology [3]. It contains 45 samples described by 4026 genes. Another data set is the Breast cancer data set used by [10]. The Central Nervous System (CNS) data set is described in [11]. It is a large data set concerned with the prediction of central nervous system embryonal tumor outcome based on gene expression. Table 1 summarizes the characteristics of the four datasets, where the classification task is binary, namely the number of samples, the initial dimension of the input space and the number of classes.

Table 1. Datasets characteristics

Dataset	DLBCL	Lymphoma	Breast	CNS
No. of samples	77	45	97	60
No. of features	7029	4026	24482	7129

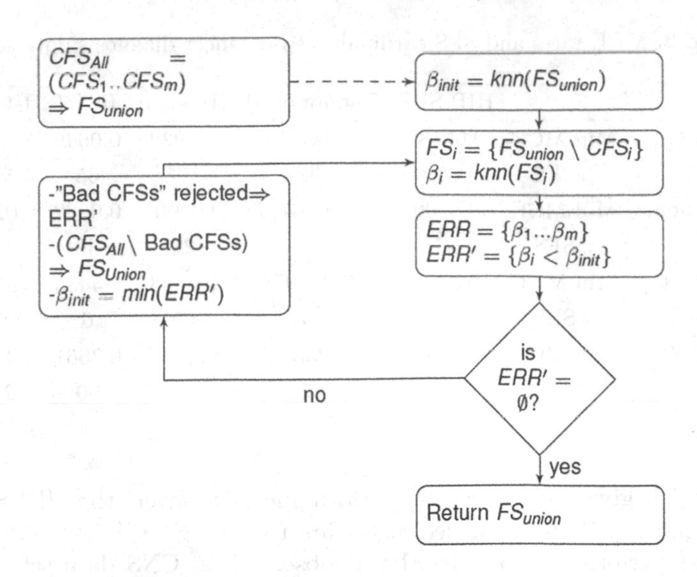

Fig. 1. Hybrid instance based sequential backward subset search

4.2 Evaluation Metrics

We use 10-fold stratified cross-validation to predict the classification performance of KNN in the SBS procedure on four data sets. The miss-classification error (MCE) rate of a classifier is defined as the proportion of misclassified instances over all classified instances. This metric is important and always used to evaluate feature selection algorithms for classification tasks. We also use as other metric the execution time (in seconds) to compare our proposed HIB-SBS to the other feature selection methods and evaluate also the final feature subset cardinality obtained.

5 Results

Ten initial candidate feature subset cardinalities ranging from 1 to 10 features were experimented for our proposed HIB-SBS approach. We tested performance of compared methods with final subset cardinalities ranging from 2 to 100 features. We recorded the minimum MCE and the corresponding final selected feature subset (SFS) cardinality for each setting. Table 2 shows the results of the application of HIB-SBS method and the compared algorithms on the four data sets.

HIB-SBS and Randomized algorithm achieved the best classification performances for DLBCL data set, followed by Relief filter. Hyb-Seq then t-test give the worst classification results. With a smaller SFS cardinality, HIB-SBS is favored for this data set.

For Breast cancer data set also, Randomized algorithm followed by HIB-SBS, with a MCE difference of about 4 %, are significantly outperforming other

Table 2. MCE rates and SFS cardinalities on cancer diagnosis data sets.

		HIB-SBS	Randomized	Hyb-Seq	Relief	t-test
DLBCL	Min MCE	0.0518*	0.0519	0.1039	0.0649	0.1429
	#SFS	31	40	10*	65	55
Lymphoma	Min MCE	0.0200	0.0667	0.0000*	0.0222	0.0222
	#SFS	5 *	30	27	30	15
Breast	Min MCE	0.3067	0.2680*	0.5258	0.4433	0.5258
	#SFS	17	75	65	6	2*
CNS	Min MCE	0.2486*	0.3333	0.3167	0.3667	0.3667
	#SFS	11	17	4*	90	25

methods which give bad predictive performance. However, the HIB-SFS uses only 17 features while Randomized algorithm uses 75.

The best performance of HIB-SBS is observed on CNS data set, where it outperforms Hyb-Seq and Randomized algorithms of 5 % and 9 % respectively. For Lymphoma cancer data set, the best performance is obtained with Hyb-Seq algorithm. With a MCE difference of 2 %, HIB-SBS gives the second best performance, then comes Relief and t-test filters with similar results. For this data set also, HIB-SBS uses the smallest SFS consisting of 5 features. Thus, SFS cardinalities of HIB-SBS range between 5 features for Lymphoma and 31 features for DLBCL data set.

We report in Table 3 the execution time (in seconds) of our proposed method and the four other methods.

Table 3. Execution times on cancer diagnosis data sets.

	Running Time (in Sec)				
	HIB-SBS	Randomized	Hyb-Seq	Relief	t-test
DLBCL	683.5028	4.5951e+003	606.7253	7.9377	1.1814
Lymphoma	236.9500	2.7602e+003	537.7784	1.7639	0.7435
Breast	2.7221e+003	4.8045e+003	729.8408	64.6571	1.6092
CNS	538.6874	7.8686e+003	513.5271	5.3489	0.8487

We can see that filters especially t-test algorithm still be the fastest algorithms with smallest execution times. This is expected as filter approaches select features independently of the classifier and thus avoid the cross-validation step used in the wrapper and hybrid algorithms. However it is noticeable that our HIB-SBS achieves execution times extremely smaller than the wrapper approach which is the randomized feature selection.

6 Conclusion

There is an interesting challenge in investigating research on filter and wrapper methods for feature selection. Wrapper methods use the bias of the induction algorithm to select features and generally perform better. However, the computational expense of wrapper methods is prohibitive on large data sets. Reducing the optimal feature subset search procedure complexity while maintaining good performance of the obtained subset is necessary for the efficiency of the selection method.

In this work we proposed HIB-SBS, a new hybrid approach based on instance learning with the main goal of accelerating the feature subset selection process by reducing the number of wrapper evaluations while maintaining good performance in terms of accuracy and size of the obtained subset.

In HIB-SBS, each instance is an expert who proposes a candidate feature subset based on an instance feature weighting technique. The candidate feature subsets are then integrated in a search procedure of the optimal feature subset, where SBS and KNN compose an evaluation system of wrappers. The best feature subset search technique is done with ten-fold cross validation.

The main challenge in this approach is that it takes advantage of the small sample size of the data set to present only a few subsets of variables to be analyzed as the number of candidate subsets is the number of instances, and so the number of wrapper evaluations decreases significantly. Our method is experimentally tested and compared with state-of-the-art algorithms over four high-dimensional low sample size data sets. The results show an impressive reduction in the execution time compared to the wrapper approach and that our proposal often outperforms other methods in terms of accuracy and cardinality of the selected subset.

References

1. Guyon, I., Elisseff, A.: An introduction to variable and feature selection. J. Mach. Learn. Res. **3**, 1157–1182 (2003)
2. Kira, K., Rendell, L.: A practical approach to feature selection. In: Sleeman, D., Edwards, P. (eds.) International Conference on Machine Learning, pp. 368–377 (1992)
3. Alizadeh, A.A., Eisen, M.B., Davis, R.E., Ma, C., Lossos, I.S., Rosenwald, A., Boldrick, J.C., Sabet, H., Tran, T., Yu, X., Powell, J.I., Yang, L., Marti, G.E., Moore, T., Hudson Jr., J., Lu, L., Lewis, D.B., Tibshirani, R., Sherlock, G., Chan, W.C., Greiner, T.C., Weisenburger, D.D., Armitage, J.O., Warnke, R., Levy, R., Wilson, W., Grever, M.R., Byrd, J.C., Botstein, D., Brown, P.O., Staudt, L.M.: Distinct types of diffuse large B-cell lymphoma identified by gene expression profiling. Nature **403**(6769), 503–511 (2000)
4. Cover, T., Hart, P.: Nearest neighbor pattern classification. IEEE Trans. Inf. Theory **13**, 21–27 (1967)
5. Sun, Y., Todorovic, S., Goodison, S.: Local learning based feature selection for high dimensional data analysis. IEEE Trans. Pattern Anal. Mach. Intell. (TPAMI) **32**, 1610–1626 (2010)

6. Jain, A., Zongker, D.: Feature selection: evaluation, application, and small sample performance. IEEE Trans. Pattern Anal. Mach. Intell. (TPAMI) **19**, 153–158 (1997)
7. Brahim, A.B., Limam, M.: A stable instance based filter for feature selection in small sample size data sets. In: Luo, X., Yu, J.X., Li, Z. (eds.) ADMA 2014. LNCS, vol. 8933, pp. 334–344. Springer, Heidelberg (2014)
8. Kohavi, R., John, G.H.: Wrappers for feature subset selection. Artif. Intell. **97**, 273–324 (1997)
9. Shipp, M.A., Ross, K.N., Tamayo, P., Weng, A.P., Kutok, J.L., Aguiar, R.C., Gaasenbeek, M., Angelo, M., Reich, M., Pinkus, G.S., Ray, T.S., Koval, M.A., Last, K.W., Norton, A., Lister, T.A., Mesirov, J., Neuberg, D.S.: Diffuse large B-cell lymphoma outcome prediction by gene-expression profiling and supervised machine learning. Nat. Med. **9**, 68–74 (2000)
10. vant Veer, L.J.: Gene expression profiling predicts clinical outcome of breast cancer. Nature **415**, 530–536 (2002)
11. Pomeroy, S.L.: Prediction of central nervous system embryonal tumour outcome based on gene expression. Nature **415**, 436–442 (2002)
12. Saeys, Y., Inza, I., Larranaga, P.: A review of feature selection techniques in bioinformatics. Bioinformatics **23**, 25072517 (2007)
13. Huang, J., Cai, Y., Xu, X.: A hybrid genetic algorithm for feature selection wrapper based on mutual information. Pattern Recogn. Lett. **28**, 18251844 (2007)
14. Bermejo, P., Gmez, J.A., Puerta, J.M.: A grasp algorithm for fast hybrid (filter-wrapper) feature subset selection in high-dimensional datasets. Pattern Recogn. Lett. **32**, 701–711 (2011)

Facial Expression Recognition Using Entire Gabor Filter Matching Score Level Fusion Approach Based on Subspace Methods

Ganapatikrishna Hegde[1,2](✉), M. Seetha[3,4], and Nagaratna Hegde[5]

[1] Department of Computer Science, SDMIT, Ujire, India
[2] VTU, Belgaum, India
gphegde123@gmail.com
[3] Department of Computer Science, GNITS, Hyderabad, India
smaddala2000@yahoo.com
[4] JNTU, Hyderabad, India
[5] Department of Computer Science, VCE, Hyderabad, India

Abstract. In this study appearance based facial expression recognition is presented by extracting the Gabor magnitude feature vectors (GMFV) and Gabor Phase Congruency vectors (GPCV). Feature vector space of these two vectors dimensions are reduced and redundant information is removed using subspace methods. Both GMFV and GPCV spaces are projected with Eigen score and projected matching scores are normalized and fused. Final matching score of each subspace method are normalized using Z-score normalization and fused together using maximum rule. Dimension of entire Gabor feature vector space consumes larger area of memory and high processing time with more redundant data. To overcome this problem in this paper entire Gabor matching score level fusion (EGMSLF) approach based on subspace methods is introduced. The JAFFE database is used for experiment. Support vector machine classifier technique is used as classifier. Performance evaluation is carried out by comparing proposed approach with state of art approaches. Proposed EGMSLF approach enhances the performance of earlier methods.

Keywords: Gabor filter · Expression recognition · Computation time · Subspace · Dimension reduction · Phase congruency

1 Introduction

Facial expression is one of the most powerful, natural, and immediate means for human beings to communicate their feelings and opinions to others. Bettadapura V. [1] has shown various states of art techniques for facial expression recognition. Face guide of Ekman P. and Friesen W [2] illustrates emotional classifications. Yang-Kai Chang, Cheng-Chang Lien [3] worked on new appearance based facial expression recognition system and they said expression variation is a cognitive activity and emotional exhibition variation of human face as an immediate means of communication path and social interaction with the internal feelings, opinions through different appearances of the face.

R. Prasath et al. (Eds.): MIKE 2015, LNAI 9468, pp. 47–57, 2015.
DOI: 10.1007/978-3-319-26832-3_6

Facial expression recognition finds important applications in many fields such as psychological studies with human-computer interaction, mind training, data-driven animations, machine learning, and computer graphics communities during the last few decades as in [1, 4–10]. Expressions exhibited by the human during various situations like pain, sad, surprise, happy, confusion, disgust, angry, fear and fatigue are primary variable expressions. Neutral and sleepy are constant expressions. There are several authors and researchers defined the facial expression recognition approaches in different ways. Basically four approaches are well known such as holistic approach appearance dependent, analytical approach geometrical dependent, part approach template dependent, segmentation approach skin color dependent.

This paper mainly focuses on appearance based entire Gabor matching score level fusion approach, both for dimensional reduction of feature vector space and improving the classification accuracy. Liu, C., Wechsler, H [12] worked on linear discriminant analysis (LDA) based Gabor feature classification. Rosdiyana Samad et al. [13] developed the algorithm to identify minimum number of Gabor wavelet parameters for natural expression recognition. They used Gabor filter for face recognition and down sampled the Gabor feature vectors and dimensions of feature vector were reduced using principal component analysis (PCA). Classification using support vector machine (SVM) claims 81.7 % recognition rate for FEEDTUM database. Priya Sisodia et al. [14] have used Gabor filter bank for detecting an appearance based features. In their work few of the most representative features are selected through feature selection method and SVM was used for classification. Le Hoang Thai et al. [15] present novel approach for Canny edge detection, PCA and Artificial Neural Network (ANN). They detected the local region of the face such as an eyebrow, eye and mouth then dimension of the face feature vector is reduced using PCA method. ANN was applied for classification on JAFFE database which have 85.7 % recognition rate. Abdulrahman et al. propose a method which is implemented using Gabor wavelet transform with PCA and LBP [16]. This hybrid approach gives 90 % average recognition rate for JAFFE database. Meher et al. [17] proposes a PCA for face recognition and expression recognition. In their work, recognition rate is found to be 81.36 % for CSU dataset and 85.5 % for ATT dataset. Yi Jin, Qiu and Qi Ruan [29] they worked on combination of Gabor wavelets and supervised locality preserving projections (SLPP) for face recognition. In this paper, Gabor filter is first designed to extract the features from the whole face images, and supervised locality preserving projections was used to train the labels and dimension reduction, which is improved by two-directional PCA to eliminate redundancy among Gabor features. They achieved 82 % to 87.50 % of recognition accuracy for varying training sample for Gabor-based locality preserve projection (LPP) method. Most of the Gabor based face recognition approaches available in literature were not used phase information of Gabor filter. Gabor magnitude was only utilized and phase part was ignored. Phase feature also can be used to improve the face and expression recognition as one part of this work.

The rest of the paper is organized as follows: Sect. 2 presents brief overview of feature extraction by entire Gabor filter design. In Sect. 3, proposed approach using subspace methods is described. In Sect. 4, experimental testing and analysis of results are presented. In Sect. 5, conclusions are made.

2 Brief Overview of Entire Gabor Filter

In this section overview of face recognition based Gabor filters is introduced. Gabor wavelet transform allows description of spatial frequency structure in the image while preserving information about spatial relations which is known to be robust to some variations, of face appearances. Gabor filters are used in this study is only to extract the texture features required for expression recognition. Gabor filters are also called Gabor wavelets they represent complex band-limited filters with an optimal localization in both the spatial as well as the frequency domain. Thus, when employed for facial feature extraction, they extract multi-resolution, spatially local features of a confined frequency band. Spatial domain 2D Gabor filter can be represented as

$$\psi_{mn} = (x, y) = \frac{f_m^2}{\pi k \eta} \left((f_m^2 / k^2)\, x'^2 + (f_m^2 / \eta^2) e^{j2\pi f_m^x} \right)' \tag{1}$$

$$\text{Here } x' = x \cos\theta_n + y \sin\theta_n \text{ and } y' = -x \sin\theta_n + y \cos\theta_n$$

$$f_m = f_{\max} / 2^{(m/2)},\ \theta_n = n\pi/8$$

As can be seen from the filters definition, each Gabor filer represents a Gaussian kernel function modulated by a complex plane wave whose center frequency and orientation are given by f_m and θ_n, respectively. The parameters κ and η determine the ratio between the center frequency and the size of the Gaussian envelope and, when set to a fixed value, ensure that Gabor filters of different scales behave as scaled versions of each other. It should also be noted that with fixed values of the parameters κ and η, the scale of the given Gabor filter is uniquely defined by the value of its center frequency f_m. While different choices of the parameters determining the shape and characteristics of the filters define different families of Gabor filters, the most common parameters used for face recognition also used in this work as $\kappa = \eta = \sqrt{2}$ and $f_{max} = 0.25$. When using the Gabor filters for facial feature extraction, researchers typically construct a filter bank featuring filters of five scales and eight orientations, that is, $m = 0, 1 .. s - 1$ and $n = 0, 1, \ldots, t - 1$, where $s = 5$ and $t = 8$. An example of the real and imaginary parts of a Gabor filter is presented in Fig. 1(a) shows magnitude output of the filtering operation with the entire Gabor filter bank of 40 Gabor filters, while Fig. 1(b) shows the real parts of the Gabor filter bank commonly used for feature extraction in the field of face recognition.

Let I(x, y) stand for a grey-scale face image of size p×q pixels and, moreover, let $\psi_{m,n}(x, y)$ denote a Gabor filter given by its center frequency f_m and orientation θ_n. The feature extraction procedure can then be defined as a filtering operation of the given face image I(x,y) with the Gabor filter $\psi_{m,n}(x,y)$ of size m and orientation n, that is

$$G_{m,n}(x, y) = I(x, y) * \psi_{m,n}(x, y) \tag{2}$$

where $G_{m,n}(x,y)$ denotes the complex filtering output that can be decomposed into its real ($E_{m,n}(x,y)$ and imaginary ($O_{m,n}(x,y)$) parts as

$$E_{mn}(x, y) = re[G_{mn}(x, y)],\ O_{mn}(x, y) = img[G_{mn}(x, y)] \tag{3}$$

Based on these results, the magnitude ($A_{m,n}(x, y)$) and phase ($\varphi_{m,n}(x, y)$) responses of the filtering operation can be computed as follows.

$$A_{m,n}(x, y) = \sqrt{E_{m,n}^2(x, y) + O_{m,n}^2(x, y)}, \tag{4}$$

$$\varphi_{m,n}(x, y) = \arctan\left(\frac{O_{m,n}(x, y)}{E_{m,n}(x, y)}\right) \tag{5}$$

2.1 Gabor Magnitude Face Recognition

Entire feature vector space of face in the database is considered as a input for the subspace method for dimensional and redundancy reduction. So it is needed to construct the Gabor filter bank after deriving the Gabor magnitude and phase congruency vectors. According to literature survey most of the authors defined about filter bank consists of Gabor filter with five scales (m=0,1, ... 4) and eight orientations (n=0,1. . . . 7). The given face image set is filtered with all 40 filters from the filter bank resulting in an inflation of data dimensionality to 40 times is initialize size. The resized dimension of each image is 111 × 126 pixels. The 40 magnitude responses reside in 559990 dimensional feature space which is more excessive size for processing and storage. Thus to overcome this large dimensionality of the Gabor magnitude and phase feature vector responses linear dimension reduction methods are used to convert Gabor feature vector space into subspace. Before considering the Gabor magnitude for construction of feature vector it is down sampled using rectangular sampling grid superimposed over the image to be sampled. These down sampled values are normalized. In this experiment, rectangular sampling grid with 16 horizontal and 16 vertical lines is used with 111 × 126 pixels of image size.

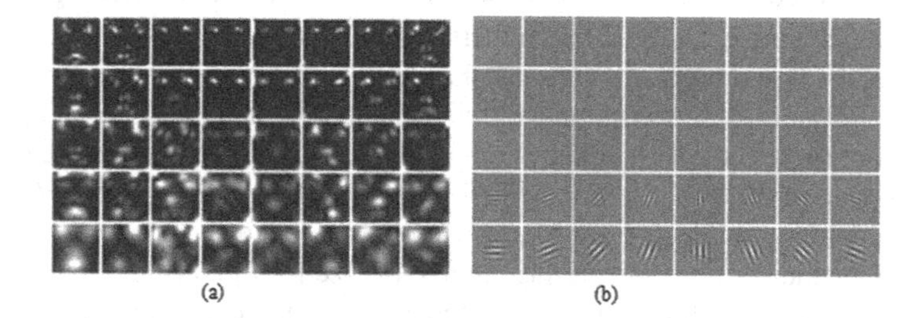

Fig. 1. a) Magnitude output (b) Real part of the Gabor filter bank

2.2 Gabor Phase Congruency Face Representation

Due to slow processing of Gabor phase with respect to spatial position of image most of the earlier Gabor based face recognition work discarded the phase information. In this paper phase congruency model is used based on face representation developed by Vitomir Struc and Nikola Pavesic [18]. For 1D signals, the phase congruency (PC(x)) is defined implicitly by the relation of the energy at a given point in the signal E(x) and sum of the Fourier amplitudes A_n as shown by Venkatesh and Owens [19].

$$E(x) = PC(x) \sum_{n} A_n \tag{6}$$

Where n denotes the number of Fourier components. Thus phase congruency at a given location of the signal x is defined as the ratio of the local energy at this location and the sum of Fourier amplitudes. Kovesi [20] extended the above concept to 2D signals by computing the phase congruency with logarithmic Gabor filters using the following equation

$$PC_{2D}(x, y) = \frac{\sum_{n=0}^{r-1} \sum_{m=0}^{p-1} A_{m,n}(x, y) \Delta\Phi_{m,n}(x, y)}{\sum_{n=0}^{r-1} \sum_{m=0}^{p-1} A_{m,n}(x, y) + \varepsilon}. \tag{7}$$

Where $A_{m,n}(x,y)$ denotes the magnitude response of the logarithmic Gabor filter at scale m and orientation n, ε represents a small constant that prevents divisions with zero, and $\Delta\Phi_{m,n}(x,y)$ stands for a phase deviation measure defined as

$$\Delta\Phi_{m,n}(x, y) = \cos(\varphi_{m,n}(x, y) - \overline{\varphi_n}(x, y)) - \left|\sin(\varphi_{m,n}(x, y) - \overline{\varphi_n}(x, y))\right| \tag{8}$$

itomir Struc and Nikola Pavesi [18], defined about oriented Gabor phase congruency model. In this work Gabor filter bank is designed by extracting the required Gabor parameters from [18].

3 Proposed Approach

Anil Jain, Karthik Nandakumar, Arun Ross, [21] they have shown that the performance of different normalization techniques and fusion rules in the context of a multimodal biometric system based on the face, fingerprint and hand geometry traits of a user. Their experimental results on a database of 100 users indicate that the application of min–max, *z*-score, and tanh normalization schemes followed by a simple sum of scores fusion method results in better recognition performance compared to other methods. However, experiments also reveal that the min–max and *z*-score normalization techniques are sensitive to outliers in the data, highlighting the need for a robust and efficient normalization procedure like the tanh normalization.

In this study four basic linear subspace methods are used to project the entire Gabor feature vectors such as PCA [23, 26], LDA [22, 24, 25], ICA and LPP respectively. In the beginning Gabor magnitude feature vectors (GMFV) and Gabor phase congruency vectors (GPCV) are extracted. These vectors are normalized. Using four linear subspace methods these normalized vectors are fused independently and formed as EGPCA, EGLDA, EGICA and EGLPP models as in Fig. 2. Final matching scores are normalized using z-score normalization technique and all the scores are fused using maximum fusion rule. Dimensionally reduced final score matrix improves the recognition rate named as EGMSLF subspace method. This method is a linear transformation of the variables into a lower dimensional space which retain maximal amount of information about the variables. In the proposed method the matching scores at the output of the individual matchers are not homogeneous. Further, the scores of the individual matchers need not be on the same numerical scale and may follow different statistical distributions. Due to these reasons, in this method score normalization is essential to transform the scores of the individual matchers into a common domain prior to combining them as in Fig. 3. Score normalization is a critical part in the design of a combination scheme for matching score level fusion.

The most commonly used score normalization technique is the z-score that is calculated using the arithmetic mean and standard deviation of the given data. This scheme can be expected to perform well if prior knowledge about the average score and the score variations of the matcher is available. Initially, estimate the mean and standard deviation of the scores from a given set of matching scores. Normalization of scores can be given by

$$S'_{ij} = \frac{S_{ij} - mean(S)}{Std(S)} \tag{9}$$

Z-scores are adaptive score normalizations that are computed in a straight forward manner. Normalized score is produced by subtracting the arithmetic mean of s from an original score and dividing this number by standard deviation *Std.*

$$NS_{EGPCA} = \frac{EGPCA_S - mean(EGPCA_S)}{Std(EGPCA_S)} \tag{10}$$

$$NS_{EGLDA} = \frac{EGLDA_S - mean(EGLDA_S)}{Std(EGLDA_S)} \tag{11}$$

$$NS_{EGICA} = \frac{EGICA_S - mean(EGICA_S)}{Std(EGICA_S)} \tag{12}$$

$$NS_{EGLPP} = \frac{EGLPP_S - mean(EGLPP_S)}{Std(EGLPP_S)} \tag{13}$$

Equations 10, 11, 12 and 13 are normalized scores of EGPCA,EGLDA,EGICA and EGLPP respectively. Proposed approach is given by

$$EGMSLF = Max[(NS_{EGPCA} + NS_{EGLDA} + NS_{EGICA} + NS_{EGLPP})/4] \tag{14}$$

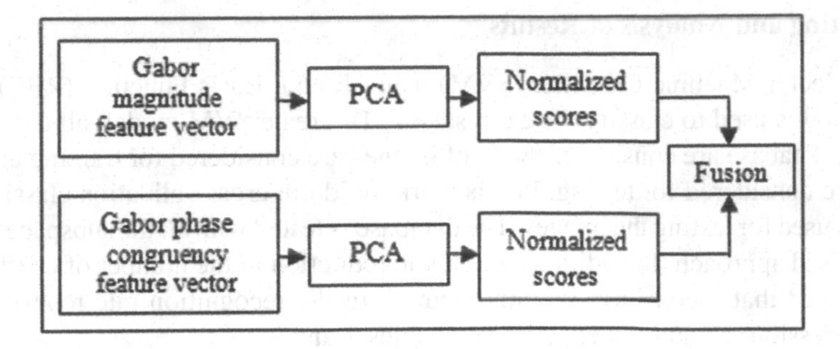

Fig. 2. Entire gabor principal component analysis (EGPCA) model

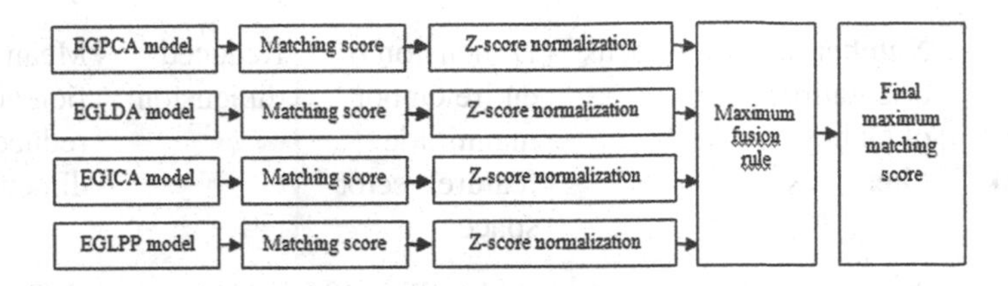

Fig. 3. Entire Gabor matching score level fusion approach (Proposed method)

4 Results and Analysis

4.1 Preprocessing

In this work, Japanese Female Facial Expression (JAFFE) database is used for experiment. This database contains 213 images of 7 facial expressions (6 basic facial expressions +1 neutral) posed by 10 Japanese female models of 256 × 256 resolution. Each image has been rated on 6 emotion adjectives by 60 Japanese subjects. All the images of this database were pre-processed to obtain pure facial expression images, which have normalized intensity, uniform size and shape. Illumination and lighting effects are also removed. The pre-processing procedure used in this work performs detecting facial feature points automatically including eyes, nose and mouth. Finally using a histogram equalization method illumination effects are removed. Figure 4 shows cropped samples of seven expressions of JAFFE database.

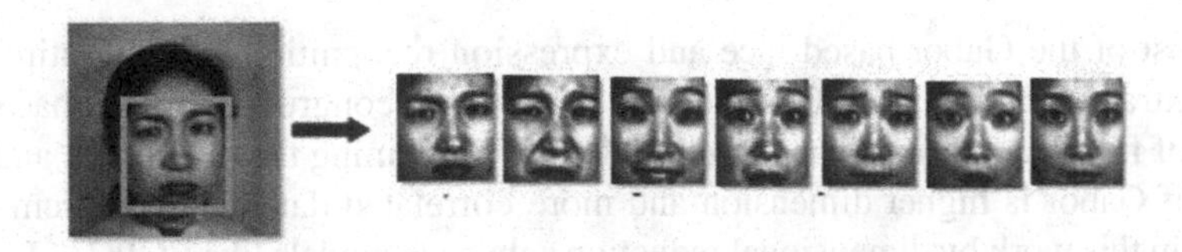

Fig. 4. Cropped samples of JAFFE database

4.2 Testing and Analysis of Results

Support Vector Machine Classifier (SVM) using Radial Basis Function (RBF) kernel [23] method is used to classify the expressions. To create SVM model, all 210 images of JAFFE database are considered. 80 % of images are considered for training and 20 % images are considered for testing. In this work, hold out cross validation classification method is used for testing the images. The database is tested with all the subspace models and proposed approach. In addition to a drastic reduction in the number of coefficients, it is observed that a considerable improvement in the recognition rate relative to the facial expression recognition experiment (Tables 1 and 2).

Table 1. Brief information about image, Gabor filter and feature vector space.

Number of scales of Gabor filter bank	Number of orientations of Gabor filter bank	Filter bank	Dimension of entire Gabor magnitude feature vector space	Reduced dimension	Mean coefficients of reduced dimension
5	8	40	111 × 126x40 =559440	1600	102.9268

Table 2. Comparison of overall expression recognition accuracy rate of subspace models

Subspace Methods	Overall accuracy rate
EGPCA	83.333 %
EGLDA	90.476 %
EGICA	85.714 %
EGLPP	88.095 %
EGMSLF(Proposed)	95.238 %
Gabor+PCA+SVM [13]	81.7 %
Gabor+LPP [29]	82 % to 87.5 %
Gabor+PCA [30]	67 % to 82 %

In most of the Gabor based face and expression recognition system entire Gabor feature extraction is not used. Complete oriented phase congruency model has several features of face recognition. The main problem of combining the magnitude and phase regions of Gabor is higher dimension and more correlated data. This problem can be resolved in this work by dimensional reduction subspace models like EGPCA, EGLDA, EGICA and EGLPP. Final matching scores are delivered by these method having less redundant coefficients values. These are normalized using Z-score normalization techniques and all the scores are fused using maximum fusion rule. 95 % of overall accuracy is achieved by proposed method. Disgust and happy expressions found to be 83.33 % and remaining 5 expressions are 100 % accuracy rates. Proposed EGMSLF approach enhances the expression classification rates compare to other subspace methods. EGMSLF approach also reduces the classification time. Comparison of expression classification accuracy rates is shown in Fig. 6 (Fig. 5).

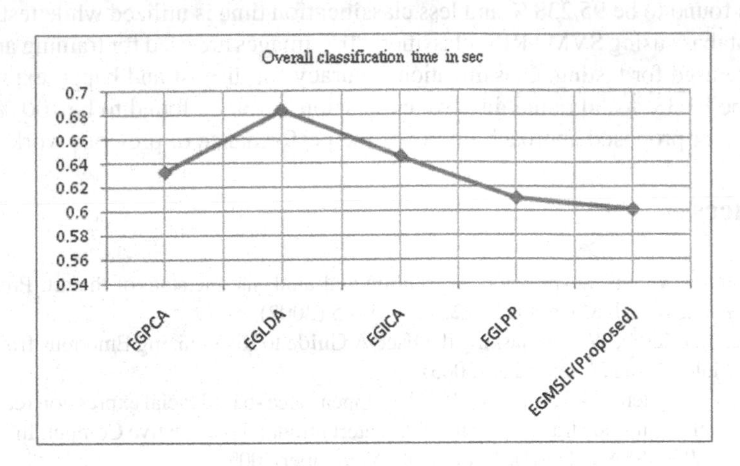

Fig. 5. Comparison of overall classification time of subspace models and proposed approach

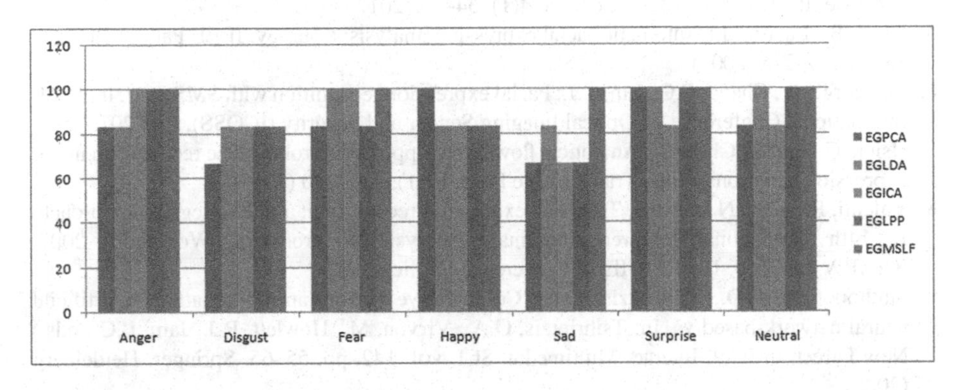

Fig. 6. Comparison of classification accuracy rates of seven expressions for subspace models

5 Conclusions

In this paper appearance based entire Gabor matching score level fusion (EGMSLF) approach based on subspace methods is presented. Most of the approaches found in literature discarded the phase part of the Gabor filter. In this work face representation for expression recognition is carried out by utilizing both Gabor magnitude and phase information. Dimension of the Gabor magnitude feature vector and Gabor phase feature vector is reduced and by removing redundant data .Using these two vectors and fusing the score values of subspace methods final matching scores are computed. Four models like EGPCA, EGLDA, EGICA and EGLPP models have been formed. Final matching scores of all these four models output is normalized using Z-score normalization and using max-rule all the normalized scores are fused. EGMSLF approach improves the performance of subspace methods. Facial expression recognition using proposed method is found to be 95.238 % and less classification time is utilized while tested with JAFFE database using SVM+RBF classifier. 80 % images are used for training and 20 % images are used for testing. Classification accuracy for disgust and happy expressions found to be 83.33 % and remaining five expression accuracy found to be 100 % which shows that for proposed approach improves the performance of previous work.

References

1. Bettadapura, V.: Face expression recognition and analysis: the state of the art. Proc. IEEE Trans. Pattern Anal. Mach. Intell. **22**, 1424–1445 (2002)
2. Ekman, P., Friesen, W.: Unmasking the Face A Guide to Recognizing Emotions from Facial Clues. Malor Books, Cambridge (2003)
3. Chang, Y.-K., Lien, C.-C., LinA, L.-Y.: New appearance -based facial expression recognition system with expression transaction matrices international. J. Innovative Comput. Inf. Control ICIC Int. 2009 ISSN **5**(11)(B), 1349–4198, November 2009
4. Wang, S., Liu, Z., Wang, Z., Wu, G.: Analyses of a multimodal spontaneous facial expression database. IEEE Trans. Affect. Comput. **4**(1), 34–46 (2012)
5. Fasel, B., Luettin, J.: Automatic facial expression analysis: a survey. IEEE Pattern Recogn. **36**(1), 259–275 (2003)
6. St.George, B., Chang, C.C., Lin, C.J.: Facial expression recognition with SMS alert. In: IEEE International Conference on Optical Imaging Sensor and Security (ICOSS), July 2013
7. Hsieh, C., Lai, S., Chen, Y.: An optical flow based approach to robust face recognition under expression variations. IEEE Trans. Image Proc. **19**(1), 233–240 (2010)
8. Valenti, R., Sebe, N., Gevers, T., Facial expression recognition: a fully integrated approach. In: 14th International Conference on Image Analysis and Processing Workshops, 2007. ICIAPW 2007, pp. 125–130. IEEE Conference Publications
9. Stathopoulou, I.-O., Tsihrintzis, G.A.: Comparative performance evaluation of artificial neural network-based vs. In: Tsihrintzis, G.A., Virvou, M., Howlett, R.J., Jain, L.C. (eds.) New Direct. in Intel. Interac. Multimedia. SCI, vol. 142, pp. 55–65. Springer, Heidelberg (2008)
10. Jafri, R., Arabnia, H.R.: A survey of face recognition techniques. J. Inf. Process. Syst. **5**(2), 69–78 (2009)
11. Suruliandi, A., Rose, R.R., Meena, K.: A combined approach using textural and geometrical features for face recognition. ICTACT J. Image Video Process. **3**(4), (2013). ISSN: 0976–9102
12. Liu, C., Wechsler, H.: Gabor feature based classification using the enhanced fisher linear discriminant model for face recognition. IEEE Trans. Image Process. **11**(4), 467–476 (2002)
13. Samad, R., Sawada, H.: Extraction of the minimum number of Gabor wavelet parameters for the recognition of natural facial expressions. Artif. Life Robot. **16**(1), 21–31 (2011). Springer
14. Sisodia, P., Verma, A., Kansal, S.: Human facial expression recognition using gabor filter bank with minimum number of feature vectors. Int. J. Appl. Inf. Syst. **5**(9), 9–13 (2013)

15. Thai, L.H., Nguyen Do Thai Nguyen, Hai, T.S.: A facial expression classification system integrating canny, principal component analysis and artificial neural network. arXiv preprint arXiv:1111.4052 (2011)
16. Abdulrahman, M., Gwadabe, T.R., Abdu, F.J., Eleyan, A.: Gabor wavelet transform based facial expression recognition using PCA and LBP. In: Signal Processing and Communications Applications Confer and Communications Applications 2014 22nd, pp. 2265–2268. IEEE (2014)
17. Goyani, M., Dhorajiya, A., Paun, R.: Performance analysis of FDA based face recognition using correlation, ANN and SVM. Int. J. Artif. Intell. Neural Netw. 108–111 (2011)
18. Vitomir Struc and Nikola Pavesic, research Article, The Complete Gabor–Fisher Classifier for Robust Face Recogntion, EURASIP Journals on Advances in Signal Processing (2010)
19. Venkatesh, S., Owens, R.: An energy feature detection scheme. In: Proceedings of the Conference on Image Processing, pp. 553–557. Singapore (1989)
20. Kovesi, P.: Image features from phase congruency. Videre: J. Comput. Vis. Res. **1**(3), 1–26 (1999)
21. Jain, A., Nandakumar, K., Ross, A.: Score normalization in multimodal biometric systems. Int. J. Pattern Recogn. **38**, 2270–2285 (2005). ELSEVIER Publishers
22. Muda, A.K., Huoy, C.Y., Hidayat, E., FajrianNur, A., Ahmad, S.: A Comparative study of feature extraction using PCA and LDA for face recognition. In: 7th International Conference on Information Assurance and Security, pp. 354–359, December 2011
23. Wang, C., Lan, L., Gu, M., Zhang, Y.: Face recognition based on principle component analysis and support vector machine. In: 3rd International Workshop on Intelligent Systems and Applications, pp. 1–4. May 2011
24. Lu, C., Jumahong, H., Liu, W.: A new rearrange modular two-dimensional LDA for face recognition. In: International Conference on Machine Learning and Cybernetics, vol. 1, pp. 361–366, July 2011
25. Ji, G.-L, Wan, J.-W., Yang, M.: Random sampling LDA incorporating feature selection for face recognition. In: International Conference on Wavelet Analysis and Pattern Recognition, pp. 180–185, July 2010
26. Shah, J.H., Sharif, M., Raza, M., Azeem, A.: A survey: linear and nonlinear pca based face recognition techniques. In: The International Arab Journal of Information Technology, vol. 10, no. 6, November 2013
27. Poh, N., Bourlai, N., Kittler, J.: Benchmarking quality-dependent and cost- sensitive score-level multimodal biometric fusion algorithms. IEEE. TIFS **4**, 849–866 (2009)
28. Liu, C., Wechsler, H.: Gabor feature based classification using the enhanced fisher linear discriminant model for face recognition. IEEE Trans. Image Process. **11**(4), 467–476 (2002)
29. Jin, Y., Ruan, Q.-Q.: Face recognition using Gabor-based improved supervised locality preserving projections computing and informatics **28**, 81–95 (2009)
30. Praseeda L.V., Dr. Sasikumar, M.: Analysis of facial expression using gabor and SVM. Int. J. Recent Trends Eng. **1**(2), 2009

Cluster Dependent Classifiers for Online Signature Verification

S. Manjunath[1], K.S. Manjunatha[2(✉)], D.S. Guru[2], and M.T. Somashekara[3]

[1] Department of Computer Science, Central University of Kerala, Kasargod 671316, India
manju_uom@yahoo.co.in

[2] Department of Studies in Computer Science, University of Mysore, Manasagangothri, Mysore 570006, Karnataka, India
kowshik.manjunath@gmail.com, dsg@compsci.uni-mysore.ac.in

[3] Department of Computer Science and Applications, Bangalore University, Bangalore 560056, India
somashekara_mt@hotmail.com

Abstract. In this paper, the applicability of notion of cluster dependent classifier for online signature verification is investigated. For every writer, by the use of a number of training samples, a representative is selected based on minimum average distance criteria (centroid) across all the samples of that writer. Later k-means clustering algorithm is employed to cluster the writers based on the chosen representatives. To select a suitable classifier for a writer, the equal error rate (EER) is estimated using each of the classifier for every writer in a cluster. The classifier which gives the lowest EER for a writer is selected to be the suitable classifier for that writer. Once the classifier for each writer in a cluster is decided, the classifier which has been selected for a maximum number of writers in that cluster is decided to be the classifier for all writers of that cluster. During verification, the authenticity of the query signature is decided using the same classifier which has been selected for the cluster to which the claimed writer belongs. In comparison with the existing works on online signature verification, which use a common classifier for all writers during verification, our work is based on the usage of a classifier which is cluster dependent. On the other hand our intuition is to recommend to use a same classifier for all and only those writers who have some common characteristics and to use different classifiers for writers of different characteristics. To demonstrate the efficacy of our model, extensive experiments are carried out on the MCYT online signature dataset (DB1) consisting signatures of 100 individuals. The outcome of the experiments being indicative of increased performance with the adaption of cluster dependent classifier seems to open up a new avenue for further investigation on a reasonably large dataset.

Keywords: Writer representative · Signature clustering · Cluster dependent classifier · Online signature verification

R. Prasath et al. (Eds.): MIKE 2015, LNAI 9468, pp. 58–69, 2015.
DOI: 10.1007/978-3-319-26832-3_7

1 Introduction

Biometric based authentication is receiving a greater attention as a replacement for password or token based authentication modes. In biometric authentication, the identify of a person can be inferred either through physical biometric traits such as finger print, palm geometry, iris, face etc., or through behavioral biometric traits such as gait, voice, signature etc., [1]. Compared to other behavioral biometrics, signature is the most widely accepted means of authentication in many countries legally. Signature verification methods can be either offline (static) or online (dynamic). In an offline mode, verification is done based on the informations extracted from the signature image while in case of online mode, verification is done considering the additional dynamic informations extracted from the acquisition devices such as PDA, pressure sensitive tablet [2].

Different online verification methods proposed in the literature are categorized as parametric and function based approaches [1]. In parametric based approaches, suitable parameters extracted from the signature trajectory are used to represent the entire signature and verification is done considering the similar parameters of a test signature and reference signatures. In function based approaches, signature is represented by means of time functions of a signature trajectory and the verification is done by comparing the corresponding time functions of a test signature and reference signatures. Parameter based approaches enjoy the advantage of compact representation and also the matching time is less. Function based approaches takes more matching time but yet result in low error rates compared to parametric based approaches. Further parametric features are categorized as local and global features [2]. Local features are extracted from sampled points of a signature trajectory while global features correspond to the entire signature. Details about the different categories of feature for online signature are available in the review papers [3, 4].

In verification, which is a two class classification problem, the test signature is assigned the label of a genuine or a forgery class by comparing the test signature with corresponding reference signatures of a writer using a suitable classifier. For online signature verification, many classifiers have been attempted by different researchers such as distance based classifier [5], HMM [6, 7], SVM [8], PNN [9], Bayesian [10], Symbolic classifier [11], Random Forest [8]. The performance of a verification system is measured in terms of two error rates namely false acceptance rate (FAR) and false rejection rate (FAR). These two errors indicate the percentage of forgery samples wrongly classified as genuine signatures and percentage of genuine signature wrongly classified as forgeries respectively. The point where these two errors are almost equal for a particular threshold is called equal error rate (EER) which is estimated using receiver operating characteristics (ROC) curve.

Instead of deciding the label of a test signature based on the decision of a single classifier, the combined decision of several classifiers are taken into consideration which leads to several fusion based approaches [12–15]. In these works it is well established that fusion based approaches outperforms the performance of a single classifier. Recently [16] proposed a novel approach named multi-domain classification where a signature is divided into different segments and for each segment, the most profitable domain of representation is detected. In the verification stage, DTW is used to evaluate the originality of each segment of the unknown signature.

In the above cited works, it is observed that the decision on acceptance or rejection of a test signature is taken by common classifier or common fusion of classifiers for all writers. The performance of a verification system depends on several factors such as features used, size and quality of training samples etc. As some writers are more consistent in signing than others, there will be variations in the training signatures of different writers which result in different distributions for different writers. Hence same classifier may not be effective for all writers.

For a verification system to be effective, it requires the usage of writer dependent characteristics such as writer dependent threshold and writer dependent classifier rather than using a common threshold and a common classifier for all writers. Most of the existing works are based on the usage of writer dependent threshold [2, 11, 12, 17]. In all these works, even though the threshold adapted is writer dependent, classifier used is same for all writers. Hence these models are referred to as writer independent models. Writer independent models are computationally efficient but they fail to consider the characteristics of individual writers. [18] proposed a hybrid model by exploiting the benefits of writer independent and writer dependent models for offline signature verification. But no attempt can be traced in the literatures on the usage of writer specific classifier for online signature verification.

Designing a verification model, with a suitable classifier for every writer is computationally expensive. Instead, we can group different writers into clusters and a suitable classifier can be designed for each cluster so that all the writers in a cluster can be trained with the same classifier. Clustering not only reduces the number of classifiers to be trained but also identify writers with a common set of discriminating features. Few attempts have been made for online signature verification based on clustering. [19] proposed an approach for online signature verification by clustering signature samples of a writer and representing each cluster in the form of interval valued symbolic feature vector. [20] proposed a cluster based approach for template creation. In these works, clustering is done to minimize the number of representatives for each writer but during verification, same classifier is used for all writers.

Considering these issues, in this work, we propose an approach for online signature verification by clustering different writers with similar characteristics and selecting a suitable classifier for each cluster. To preserve intra-class variation, a single template is created for each writer by considering the training samples which serves as a representative of the entire class. Template signatures are clustered into different groups using k-means clustering algorithm. To decide the suitability of a classifier for each writer, EER resulted from each classifier is taken into consideration and the classifier which yields lowest EER is selected as the best classifier for a particular writer. Finally the classifier that has been selected for majority of writers in each cluster is considered to be a suitable classifier for that particular cluster. During verification, the same classifier that has been selected for the cluster to which the writer belongs is used.

The paper is organized as follows. In Sect. 2 we discuss the different stages of our model. Description of the experimental frame work with description of the dataset used is given in Sect. 3. In Sect. 4, obtained results are presented. Comparative analysis of our model with other existing contemporary model is given in Sect. 5. Conclusions are drawn in Sect. 6.

2 Proposed Model

The Proposed model has four different stages. In the first stage, a single template for each writer is created. In the second stage, using the template signature of each writer, writers are clustered by employing k-means clustering. In the third stage, we estimate the performance of each classifier for each writer of a cluster. The classifier which performs better for a maximum number of writers of that cluster is selected as the classifier for all writers of that cluster. The details of classifier selected for each clusters are stored in the database. The same procedure is carried out for all other clusters also. In the fourth stage, verification is carried out where the authenticity of the test signature is decided by means of the selected classifier. The architecture of the proposed model is shown in Fig. 1.

Let there be N number of writers say $W = \{W_1, W_2, W_3, \ldots, W_N\}$ for which an online signature verification system has to be designed. Let there be n number of samples for each writer, $W_i = \{S_1^i, S_2^i, S_3^i, \ldots S_n^i\}, 1 \leq i \leq n$ and m be the number of features extracted from each sample of each writer forming a database $Sig_{Database} = (N \times n) \times m$. The samples in the database $Sig_{Database}$ can be visualized as points in a m dimensional feature space. In feature space, the writers who have similar characteristics in terms of features will be close to each other. Hence we can group all writers into k clusters and study the suitability of selecting cluster dependent classifier.

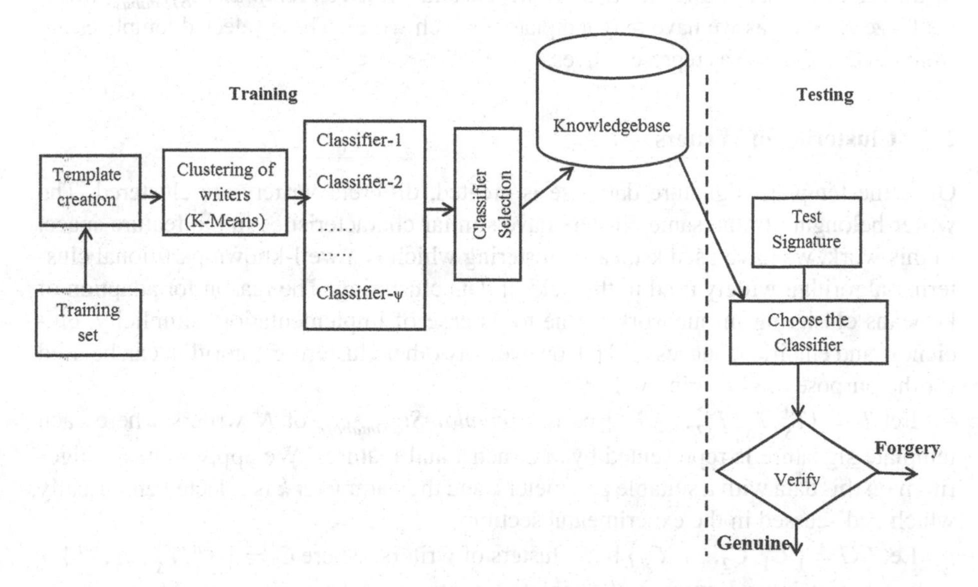

Fig. 1. Architecture of the proposed model

2.1 Template Creation

As we need to create a cluster of writers $W = \{W_1, W_2, W_{3,}\ldots, W_N\}$ and as we have n number of samples for each writer $W_i = \{S_1^i, S_2^i, S_3^i, \ldots S_n^i\}, 1 \leq i \leq n$, clustering of all writers using all n samples lead to a confusion in understanding to which cluster a writer belong as samples of same writer may belong to different clusters. Hence we intend to create a single template for each writer and then clustering different writers based on the chosen representative of each writer solves the above mentioned problem. Template signature for each writer is created considering n number of samples of each writer. The template signature for i^{th} writer W_i is created as follows.

1. Let $\{s_1, s_2, \ldots, s_n\}$ be the set of n genuine signatures of ith writer available for training purpose.
2. Calculate the pair-wise distance $dist(s_j^i, s_{j'}^i)$ from s_j^i and $s_{j'}^i$ $j = 1, 2, \ldots, n$ and $j' = 1, 2, \ldots, n, j \neq j'$
3. Compute the average distance of each signature $Avg(s_j^i) = \frac{1}{n-1} \sum_{j'=1}^{n} dist(s_j^i, s_{j'}^i)$, $j = 1, 2, \ldots, n, j \neq j'$.
4. Template signature of ith writer is selected as a signature s_j^i which has a minimum average distance to other signatures of ith writer and is given by $T_S^i = \min(Avg(S_j^i), j = 1, 2, \ldots, n$.

Similarly, we create template signature for all writers and let $T = \{T_S^i, T_S^i, T_S^3, \ldots, T_S^N\}$ be the set of template signatures of all N writers and termed as $TemplateSig_{Database}$ which is of size $N \times m$, as we have only template for each writer. These selected templates are stored in the database as representatives.

2.2 Clustering of Writers

Once the template signature database is created, different writers are clustered. The writer belonging to the same clusters have similar characteristics in the feature space. In this work, we have used k-means clustering which is a well-known partitional clustering algorithm widely used in the field of data clustering. The reason for adaption of k-means clustering in our work is due to its ease of implementation, simplicity, efficiency, and empirical success [21]. However any other clustering algorithm can be used for the purpose of clustering writers.

Let $T = \{T_S^1, T_S^2, T_S^3, \ldots, T_S^N\}$ be the $TemplateSig_{Database}$ of N writers where each template signature is represented by m dimensional features. We apply k-means algorithm on this data with a suitable parameter k and the parameter k is selected empirically which is discussed in the experimental section.

Let $CG = \{C_1, C_2, \ldots, C_k\}$ be k clusters of writers, where $C_i = \{T_S^a, T_S^b, \ldots, T_S^l\}$ is a cluster containing l number of template signatures of different users obtained by k-means clustering. As each template signature corresponds to a writer, $C_i = \{T_S^a, T_S^b, \ldots, T_S^l\}$ can be rewritten as $C_i = \{W_a^i, W_b^i, \ldots, W_l^i\}$ where, W_j^i being the

j^{th} writer of ith cluster. For each writer in each cluster, we identify the suitable classifier and select the classifier for the cluster as discussed in the following subsection.

2.3 Cluster Dependent Classifier Selection

The decision regarding the classifier suitable for a particular cluster is arrived as follows. We have clustered N writers into k clusters i.e., $CG = \{C_1, C_2, \ldots, C_k\}$ and $C_i = \{W^i_a, W^i_b, \ldots, W^i_l\}$, where W^i_j being the jth writer of ith cluster. For each writer in an individual cluster we collect all n number of samples. Out of n number of samples, n_1 samples are used for training purpose and n_2 samples are used for validation. For validation we need forgery samples also and hence we considered n_3 number of forgery samples during validation process.

Let there be ψ number of classifiers. Given a writer W^i_a of ith cluster with n number of samples and m numbers of features, we have a data matrix of size $n \times m$ for i^{th} writer. Out of $n \times m$ data matrix, $n_1 \times m$ is used as training set and trained each of the ψ classifiers. Using $n_2 \times m$ and $n_3 \times m$, False acceptance rate (FAR) and False rejection rate (FRR) are calculated and finally Equal error rate (EER) is obtained for each classifier. Calculation of FAR, FRR and EER is discussed in Sect. 3, that is for i^{th} writer we have

$$E_c = \left\{ EER^i_{C_1}, EER^i_{C_2}, EER^i_{C_3}, \ldots, EER^i_{\psi} \right\} \tag{1}$$

where $EER^i_{C_j}$ refers to EER of C^{th}_j classifier for i^{th} writer.

The experimentation is carried out with X number of trails by changing the training and validation samples. The training and validation samples are randomly selected without overlapping in each of the X trails. In each trial, the classifier with a minimum error rate is selected.

$$\text{i.e } \psi^X_{sel} = \min\{E_c\} \tag{2}$$

Let $\psi_{sel} = \{\psi^1_{sel}, \psi^2_{sel}, \psi^3_{sel}, \ldots, \psi^X_{sel}\}$ be the set of classifiers selected in X different trials, where ψ^p_{sel} is the classifier selected at p^{th} trial. In order to select the best classifier among the ψ_{sel} list, we rank the classifiers based on its frequency of selection for a particular writer as defined in (3).

$$Frequency\left(\psi_j, W^i_a\right) = \frac{Number\ of\ times\ \psi_j\ is\ selected\ for\ i^{th}\ writer}{Number\ of\ trails\ conducted} \tag{3}$$

The classifier having the highest frequency of selection say ψ^{ia}_j shall be the best classifier for the writer W^i_a of ith cluster. Similarly for all writers in the cluster, the classifier is selected using the above mentioned procedure. The classifier which is selected with the highest frequency is selected as the classifier for all writers belonging to i^{th} cluster. Similarly, for all other clusters, the classifier is selected and the selected classifier is assigned as best classifier for those writers belonging to that corresponding cluster. In the knowledgebase we store the classifier selected for each cluster and later this will be used in verifications stage. Once the classifier is decided for the entire cluster, we train the system for all writers of that cluster with the selected classifier before the verification stage.

2.4 Signature Verification

During verification, given a test signature S_{test} of a claimed writer i, the genuinity of S_{test} is verified as follows. First identify the cluster label to which the claimed writer i belongs. Once the cluster label is known, the classifier ψ^i suitable for the cluster to which the writer i belong is selected for verification. The given unknown sample S_{test} is fed to the classifier ψ^i and the claimed identity is established as a genuine or a forgery.

3 Experimental Setup

In this section, description of the database used for experimentation along with training and testing details is presented.

Database: We have evaluated the performance of our model on MCYT (DB1), a subcorpus of MCYT dataset, consisting of signatures of 100 writers. MCYT (DB1) contains 50 signature samples of 100 writers. Out of the 50 samples of each writer, 25 are genuine and the remaining 25 are skilled forgery samples. We have considered 100 global features for each writer and complete list of 100 global features can be found in the work of [12].

Experimental Setup: We conducted experimentation under two different training conditions (a) with 05 genuine signatures (b) with 20 genuine signatures. The reason for selecting 05 and 20 genuine signatures for training is to compare our model with other existing models for online signature verification. We conducted verification experiments with both skilled and random forgeries. Skilled forgery is nothing but the forgery created by professional forgers with sufficient practice and random forgery is nothing but the genuine signature of other writers. In case of testing with skilled forgery, all the remaining genuine signatures and all the 25 skilled forgeries are used for calculating FRR and FAR respectively. These two categories of testing are mentioned as Skilled_05 and Skilled_20 respectively. Similarly in case of testing with random forgery, all the remaining genuine signatures and one signature from every other writer is used for estimating FRR and FAR respectively. These two categories of testing are mentioned as Random_05 and Random_20 respectively. To fix-up the classifier for each cluster, the training set is further divided into training and validation set. Based on the error rate obtained with validation set, the classifier for each cluster is decided. For validation purpose, we have considered 50% of the available genuine signatures and equal number of random forgeries.

After creating a representative for each writer and clustering of writers as discussed in Sects. 2.1 and 2.2, we estimate the FAR, FRR and EER for each writer resulting from each of the available classifiers. For each writer, the classifier which resulted in lowest EER is considered as the suitable classifier for that writer. Finally, the classifier which has been selected for majority of writers in each cluster is decided as the suitable classifier for the entire cluster. Once the classifier for all clusters are fixed up, we test the performance of the system on the unseen test data. The resulting FAR, FRR and EER obtained with test sample is reported as the error rate of the system. We conducted experimentations with different number of clusters varying from 5 to 10. In this work, for classifier selection, we have considered 06 different classifiers namely Naïve Bayesian (NB), Nearest neighbor (NN), Support vector machine (SVM), Probabilistic neural network (PNN), Fisher Linear discriminating analysis (FLD) Principal component analysis (PCA) classifier which are widely used in the field of pattern recognition.

4 Experimental Results

We conducted 10 different trials by randomly selecting training and testing samples in each trial to measure the performance of the system. In this work we used FAR, FRR and EER as the performance measures which are generally used for measuring the performance level of a biometric system. The EER obtained for different categories of testing with the proposed model is given in Table 1.

Table 1. EER of the proposed model under varying number of clusters for different categories of testing

No. of clusters	Skilled_05	Skilled_20	Random_05	Random_20
5	13.10	1.00	9.21	0.40
6	13.00	1.10	7.60	0.40
7	12.83	1.00	7.95	0.40
8	12.60	1.20	6.52	0.40
9	12.69	1.20	8.03	0.40
10	12.75	1.20	8.13	0.40

Further to demonstrate the superiority of cluster dependent classifier over a common classifier, we also conducted experiments without any classifier selection. In this case, the same classifier has been used for all writers. Result obtained with the usage of common classifier for all writers is shown in Table 2. From Table 2, it can be deduced that, usage of a common classifier results in higher EER compared to cluster dependent classifier thereby establishing the superiority of the proposed model. In Table 2, the labels C_1, C_2, C_3, C_4, C_5 and C_6 denote the classifier NB, NN, SVM, PNN, FLD and PCA respectively. For instance, the row corresponding to C_1 denote the EER obtained when the NB classifier is used.

Table 2. EER with the usage of common classifiers for different categories of testing

Classifier Label	Skilled_05	Skilled_20	Random_05	Random_20
C_1	27.21	6.00	27.50	7.60
C_2	13.19	1.40	8.27	0.60
C_3	13.27	1.25	8.19	0.60
C_4	16.89	13.60	14.24	14.00
C_5	14.40	3.90	11.11	2.20
C_6	12.90	1.50	8.31	0.60

5 Comparative Study

In this section we present a comparative analysis of the proposed model with other existing models for online signature verification. It is well known that comparing different verification models is challenging due to change in the dataset used for experimentation, features used, training and testing size. To have a fair comparison, we have taken into consideration other models which have used MCYT (DB1). In Table 3, we compare the EER of our model with other existing models for online signature verification models. From Table 3, it is clear that, none of the individual models got lowest EER for all four categories of testing. Our model outperforms all other models in case of Skilled_20 and Random_20. Generally a writer dependent model required more training samples for capturing the characteristics of individual writer [18]. In case of skilled_05 and random_05 since only 05 signatures are available for training purpose, the error rate we achieved is high compared to other models. All other models mentioned in Table 3 are writer independent models. Even in case of Skilled_05, the EER that we achieved is lower than the EER of BASE model [15], cluster based symbolic model [19] and forgery quality estimation model with GMM [22]. In case of Random_05, the EER of our model is lower than the EER of NND classifier model [14] and BASE model [15]. In Table 3, for some categories of testing, the respective authors have not quoted the result and hence the corresponding entries are left blank.

Table 3. Comparative analysis of various online signature verification models based on EER

Model	Skilled_05	Skilled_20	Random_05	Random_20
Proposed model	12.6	1.0	6.5	0.4
Single class classifier Model [14]				
a. Parzen Window Classifier (PWC)	9.7	5.2	3.4	1.4
b. Nearest Neighbour descriptor (NND)	12.2	6.3	6.9	2.1
c. Mixture of Gaussian Description(MOGD_3)	8.9	7.3	5.4	4.3
d. Mixture of Gaussian Description (MOGD_2)	8.1	7.0	5.4	4.3
e. Support Vector Descriptor	8.9	5.4	3.8	1.6
Two class classifier model [15]				
a. BASE	17.0		8.3	
b. KHA c. Random subspace(RS) d. Random subspace with ensemble (RSB) e. Fusion of RSB+KHA	11.3 9.0 9.0 7.6		5.8 5.3 5.0 2.3	
Symbolic classifier [11]	5.8	3.8	1.9	1.7
Cluster based symbolic classifier [19]	15.4	4.2	3.6	1.2
Forgery quality estimation model [22]				
a. DTW	6.5			
b. HMM	11.5			
c. Gaussian Mixture Model (GMM)	17.7			

6 Conclusion

In this work, we have made a successful attempt on the introduction of notion of cluster dependent classifier for online signature verification. Our intuition is to recommend a suitable classifier only for those writers with similar characteristics and a different classifier for writers of different characteristics. We have clustered representatives of each writer obtained from the training sample for recommending a suitable classifier for each cluster. The outcome of the proposed model outperforms existing state of the art works reported in literature for online signature verification in achieving lowest EER for a large training size.

Acknowledgement. The authors would like to thank J.F. Aguilar and J.O. Garcia for sharing MCYT-100, a sub corpus of online signature data set and thanks to Prof. Anil K. Jain for his associated support to get the dataset.

References

1. Plamondon, R., Lorette, G.: Automatic signature verification and writer identification: the state of the art. Pattern Recogn. **2**(2), 107–131 (1989)
2. Jain, A.K., Griess, F.D., Connell, S.D.: On-line signature verification. Pattern Recogn. **35**(12), 2963–2972 (2002)
3. Impedovo, D., Pirlo, G.: Automatic signature verification: the state of the art. IEEE Trans. Syst. Man Cybern. Part C Appl. Rev. **38**(5), 609–635 (2008)
4. Zhang, Z., Wang, K., Wang, Y.: A survey of on-line signature verification. In: Sun, Z., Lai, J., Chen, X., Tan, T. (eds.) CCBR 2011. LNCS, vol. 7098, pp. 141–149. Springer, Heidelberg (2011)
5. Khan, M.K., Khan, M.A., Khan, U., Ahmad, I.: On-line signature verification by exploiting inter-feature dependencies. In: Proceedings of the ICPR, pp. 796–799 (2006)
6. Fierrez, J., Garcia, J.O., Ramos, D., Rodriguez, J.G.: HMM-based on-line signature verification: feature extraction and signature modeling. Pattern Recogn. Lett. **28**(16), 2325–2334 (2007)
7. Zou, J., Wang, Z.: Application of HMM to online signature verification based on segment differences. In: Sun, Z., Shan, S., Yang, G., Zhou, J., Wang, Y., Yin, Y. (eds.) CCBR 2013. LNCS, vol. 8232, pp. 425–432. Springer, Heidelberg (2013)
8. Parodi, M., Gómez, J.C.: Legendre polynomials based feature extraction for online signature verification. Consistency analysis of feature combinations. Pattern Recogn. **47**, 128–140 (2014)
9. Meshoul, S., Batouche, M.: A novel approach for Online signature verification using fisher based probabilistic neural network. In: IEEE International Symposium on Computers and Communications (ISCC), pp. 314–319 (2010)
10. Muramatsu, M., Kondo, M., Sasaki, M., Tachibana, S., Matsumoto, T.: A markov chain monte carlo algorithm for bayesian dynamic signature verification. IEEE Trans. Inf. Forensics Secur. **1**(1), 22–34 (2006)
11. Guru, D.S., Prakash, H.N.: Online signature verification and recognition: An approach based on Symbolic representation. IEEE Trans. Pattern Anal. Mach. Intell. **31**(6), 1059–1073 (2009)
12. Fiérrez-Aguilar, J., Nanni, L., Lopez-Peñalba, J., Ortega-Garcia, J., Maltoni, D.: An on-line signature verification system based on fusion of local and global information. In: Kanade, T., Jain, A., Ratha, N.K. (eds.) AVBPA 2005. LNCS, vol. 3546, pp. 523–532. Springer, Heidelberg (2005)
13. Nanni, L., Majorana, E., Lumini, A., Campisi, P.: Combining local, regional and global matchers for a template protected on-line signature verification system. Expert Syst. Appl. **37**(5), 3676–3684 (2010)
14. Nanni, L.: Experimental comparison of one-class classifiers for on-line signature verification. Neurocomputing **69**(7–9), 869–873 (2006)
15. Nanni, L., Lumini, A.: Advanced methods for two-class problem formulation for on-line signature verification. Neurocomputing **69**, 854–857 (2006)
16. Pirlo, G., Cuccovillo, V., Impedovo, D., Mignone, P.: On-line signature verification by multi-domain classification. In: 14th International Conference on Frontiers in Handwriting Recognition (ICFHR), pp. 67–72 (2014)
17. Fiérrez-Aguilar, J., Krawczyk, S., Ortega-Garcia, J., Jain, A.K.: Fusion of local and regional approaches for on-line signature verification. In: Li, S.Z., Sun, Z., Tan, T., Pankanti, S., Chollet, G., Zhang, D. (eds.) IWBRS 2005. LNCS, vol. 3781, pp. 188–196. Springer, Heidelberg (2005)

18. Eskander, G.S., Sabourin, R., Granger, E.: Hybrid writer-independent–writer-dependent offline signature verification system. IET Biometrics **2**(4), 169–181 (2013)
19. Guru, D.S., Prakash, H.N., Manjunath, S.: On-line signature verification: an approach based on cluster representation of global features. In: International conference on Advances in Pattern Recognition (ICAPR), pp. 209–212 (2009)
20. Liu, N., Wang, Y.: Template selection for on-line signature verification. In: 19th International Conference on Pattern Recognition (ICPR), pp. 1–4 (2008)
21. Jain, A.K.: Data clustering: 50 years beyond K-means. Pattern Recogn. Lett. **31**(8), 651–666 (2010)
22. Houmani, N., Salicetti, S.G., Dorrizi, B.: On measuring forgery quality in online signatures. Pattern Recogn. **45**(3), 1004–1018 (2012)

Classification Using Rough Random Forest

Rajhans Gondane(✉) and V. Susheela Devi

Indian Institute of Science, Bangalore, India
rajhans.gondane@gmail.com, susheela@csa.iisc.ernet.in

Abstract. The Rough random forest is a classification model based on rough set theory. The Rough random forest uses the concept of random forest and rough set theory in a single model. It combines a collection of decision trees for classification instead of depending on a single decision tree. It uses the concept of bagging and random subspace method to improve the performance of the classification model. In the rough random forest the reducts of each decision tree are chosen on the basis of boundary region condition. Each decision tree uses a different subset of patterns and features. The class label of patterns is obtained by combining the decisions of all the decision trees by majority voting. Results are reported on a number of benchmark datasets and compared with other techniques. Rough random forest is found to give better performance.

1 Introduction

In today's world, data are increasing day by day so we need a good machine learning algorithm to classify the data. We use a supervised learning approach called a decision tree which will select a set of reducts from the feature space as well as the decision rules based on the training data. In decision tree classifier the goal is to generate minimum number of decision rules which will label the test data correctly with higher accuracy. ID3 and C4.5 are the well known algorithms for constructing decision trees [12,13].

The concept of bagging is used to generate diverse ensemble classifiers [4] in random forest but the problem with applying only bagging is that the first splitting node in the decision tree remains the same [2] (even if we sample the data with replacement). So we need to add some more randomisation in the decision tree by using feature subset method.

The accuracy of random forest depends on the accuracy of its individual decision trees. In 2002, J. Wei, D. Huang, S. Wang and Z. Ma. concluded that the accuracy of decision tree constructed using rough set theory gives better results than any other precise learning approach [17]. So in this respect we use the concept of rough set theory in random forest to construct the decision trees.

The Rough random forest is an ensemble learning model which is constructed using three basic building blocks: Rough Set Theory (RST), Decision Tree and Random Forest. In this the reducts of decision trees of rough random forest are chosen based on boundary region of attributes. This means the attribute which has smallest boundary region is chosen for split.

R. Prasath et al. (Eds.): MIKE 2015, LNAI 9468, pp. 70–80, 2015.
DOI: 10.1007/978-3-319-26832-3_8

2 Background Theory

2.1 Rough Set Theory

Rough set theory [8,11,16] is a mathematical theory of data analysis introduced by Z. Pawlak in 1982. It is an extension of set theory concept which supports approximation in decision making. It deals with imprecise and uncertain information. The rough sets are characterized using the notion of Indiscernibility relation which means patterns of Equivalence class should be indiscernible and patterns to be indiscernible should follow the equivalence relation. The concept of RST is defined by the topological operation called Approximation.

In RST, the information system [8] may be defined as $I = (\mathbb{U}, \mathbb{A}, V, f)$ where $\mathbb{U}$ is a non-empty finite set of pattern, $\mathbb{A}$ is the non-empty finite set of attribute of each pattern such that $\mathbb{A} = (C \cup D)$ where C is a subset of conditional attributes and D is a subset of decision attributes, V is the range of attributes defined as $V_a(x)$: value of some pattern $x \in \mathbb{U}$ corresponding to feature $a \in \mathbb{A}$ and f is a generic function defined as: $\{f : \mathbb{U} \rightarrow V_a\}$ for every $a \in \mathbb{A}$. The basic aspects used in rough set theory are as follows:

(i) **Equivalence Relation:** The relation is said to be an equivalence relation if for any $P \in \mathbb{A}$ it satisfies the following condition:

$$\{(x, y) \in \mathbb{U}^2 \mid \forall a \in P : a(x) = a(y)\} \tag{1}$$

The P indiscernibility equivalent relation can be denoted as $[x]_P$, where $x \in \mathbb{U}$

(ii) **Lower Approximation:** A set $X \subseteq \mathbb{U}$ with respect to P (an equivalence relation) is the set of all objects which can be surely classified as X with respect to P. It can be mathematically denoted as:

$$\underline{P}X = \{x \in \mathbb{U} | [x]_P \subseteq X\} \tag{2}$$

The lower approximation can be viewed as a core part of decision making and lower approximation of two equivalence classes may not overlap.

(iii) **Upper Approximation:** A set $X \subseteq \mathbb{U}$ with respect to P (an equivalence relation) is the set of all objects which can be possibly classified as X with respect to P. It can be mathematically denoted as:

$$\bar{P}X = \{x \in \mathbb{U} | [x]_P \cap X \neq \phi\} \tag{3}$$

The Lower approximation is always a subset of Upper approximation. If some pattern is not in any of the lower approximations then it will definitely be in more than two upper approximations.

(iv) **Boundary Region:** It is a difference of Upper approximation and Lower approximation denoted as

$$\{\bar{P}X - \underline{P}X\} \tag{4}$$

If there is no element in the boundary region it means there is no roughness involved and the set is called a Crisp Set, otherwise it is a Rough set.

2.2 Decision Tree

The Decision tree classification is one of the best supervised learning techniques that are built from a set of training samples [14]. The Decision tree was constructed based on two approaches: Recursive top-down partitioning process and divide and rule for dividing the search space into several subsets.

In Decision tree, the best feature is selected as a split point so that the data in each descendent subset are purer than the data in the parent superset and finally it will classify into some classes. Each path from root to the leaf node is called a *Decision rule* and the subset of features included in the final decision tree are called *Reducts* of decision tree.

$$E(S) = -\sum_{i=1}^{n} P_i \log_2 P_i \tag{5}$$

$$Gain(S, C) = E(S) - \sum_{j=1}^{m} f_S(S_j) E(S_j) \tag{6}$$

P_i : Proportion of Label i elements in set S
$E(S)$: Entropy of Set S
n : Number of classes
m : Number of different values of conditional attribute $C = (C_1, C_2, \ldots, C_m)$
S_j : Subset of element having C_j value
$f_S(S_j)$: Frequency of subset S_j in Set S

Algorithms like ID3 and C4.5 are used for constructing decision tree [12,13]. These algorithm used Entropy (*Eq.* 5) and Information gain (*Eq.* 6) for choosing the best splitting point for the feature based on the available objects. The attribute which gives the highest gain is chosen as the attribute to split on at any node. The ID3 algorithm is based on Information theory and its main goal is to minimize the number of decision rules. The C4.5 is improved algorithm of ID3. It allows missing data by simply ignoring them i.e. the gain value is calculated by looking only at those records which have some values and during testing, the missing value of the attribute is predicted on the basis of known value of that attribute. Information gain favours the attribute having high branching factor which means it is likely to choose the attribute which will split the data into smaller fragments.

The issue with decision tree is that if the depth of the tree is very low, then it has to face the problem of *underfitting* and if the depth is more then it has to face the problem of *overfitting*. Mostly because of larger depth the decision tree becomes an over-complex classifier that does not generalize well [10]. Pruning and Ensemble learning are the two techniques used to solve this problem.

Ensemble learning uses the fact that instead of using a single classifier using a number of classifiers can give more accurate results [4]. If the ensemble of decision trees are trained based on the same training data without sampling then the ensemble model will try to find two points of different classes which are closest to each other and put a decision boundary and try to maximize the decision boundary. As shown in Fig. 1(a) just because of one outlier the position of decision boundary is affected. This issue can be solved by training a number of classifiers each with only a part of the training samples.

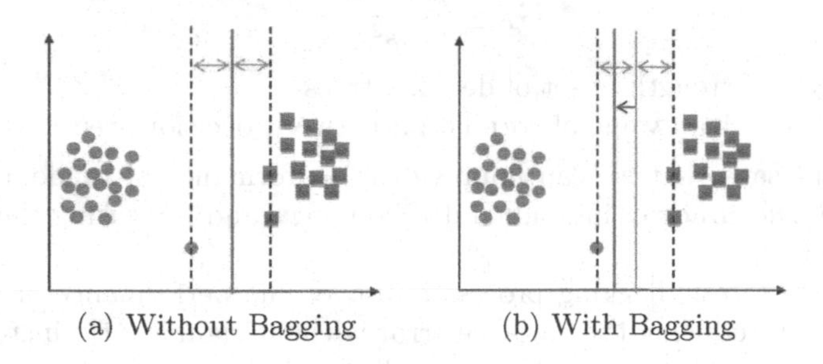

Fig. 1. Ensemble learning

In 1996, L Breiman proposed a technique called Bagging [3]. It is based on bootstrapping and aggregation. Bagging process generates many bootstrap samples from training data with replacement and trains classifiers with one bootstrap sample each and then aggregates the results of the classifiers.

In Bagging if some classifier is trained with bootstrap samples in which the outlier is not present, then that classifier will give better performance than the classifier trained based on original training data [15]. These classifiers will give better accuracy than those classifiers which are trained based on bootstrap samples with outlier. If the presence of classifiers trained without outliers is in majority then it will improve the performance of ensemble model. Thus, the ensemble model with bagging sometimes gives better performance than an individual classifier [15]. As shown in Fig. 1(b) when bagging is used in our classification model we can see that our decision boundary moved slightly towards left. This small movement makes sense in terms of outliers.

2.3 Random Forest

The Random forest is an ensemble learning technique developed by Breiman [2]. The Random forest was constructed using two concepts: subset of samples and then random subset of features.

The main idea behind random forest is to combine many decision trees constructed using subset of training samples and choosing best feature as split point among random subset of features at each split node. Suppose there are m decision trees then the sample data is divided into m subset of samples to train each decision tree. If there are d features in the dataset then the best split point is calculated from randomly selected $\sqrt{d}$ features at each split node of the tree. This extra randomness in the random forest increases the performance. After m trees are generated they vote for the most popular class for making the final decision.

In Random forest, an upper bound for the generalization error depends on the accuracy of individual decision trees and the diversity between the trees [2]. It is given by:

$$\mathbf{G}_E \leq \frac{\bar{\rho}(1 - s^2)}{s^2} \tag{7}$$

s : Strength of set of decision trees
$\bar{\rho}$: Mean value of correlation between decision trees

Equation 7 says that we can improve the performance of random forest by improving the accuracy of individual decision trees and increasing the diversity between the trees.

In Random forest, bagging process improves the performance of the single decision tree by reducing the variance error without significantly increasing the bias [1] and no pruning is performed, so all decision trees of random forest are maximal trees [18]. In 2005, M.Hamaz and D.Larocque [6] concluded that the random forest is significantly better than bagging, boosting and single decision tree and it is more accurate and more robust to noise than the other methods.

3 Related Work

In 2002, a rough set based decision tree algorithm was proposed which uses boundary region (*Eq.* 4) condition for deciding the split point for the decision tree. The algorithm was named ACRs [17]. In ACRs algorithm, the selection of an attribute for branching is chosen on the basis of smallest *boundary region* ($\bar{B}X_i - \underline{B}X_i$, where $B \in \mathbb{A}$ and X_i is a decision class) in the attribute set. In other words, we can say that we will select the attribute which has smallest set of elements in the difference of upper approximation and lower approximation. The ACRs algorithm gives higher accuracy then the entropy based algorithm under specific conditions and also reduces the number of decision rules [17].

In 2007, an algorithm for building decision tree based on dependency of attributes(ADTDA) was proposed. If there are d decision classes $X_1, X_2, \ldots, X_d$ in decision attribute D then the degree of dependency (k_B) of any conditional attribute (say B) can be calculated based on Positive region ($POS_B(D)$) as:

$$k_B = \frac{|POS_B(D)|}{|\mathbb{U}|} = \frac{|\underline{B}X_1 \cup \underline{B}X_2 \ldots \cup \underline{B}X_d|}{|\mathbb{U}|} \tag{8}$$

In ADTDA algorithm [7], the degree of dependency of decision attributes (*Eq.* 8) on condition attributes, based on rough set theory, is used as a heuristic for selecting the attribute that will best separate the samples into individual classes.

In 2009, FID3 Algorithm [5] based on Fixed information gain (*Eq.* 9) was developed. The FID3 algorithm used the concept of dependency of attributes as well as the Information Gain to calculate the Fixed information gain.

$$Gain_{fix}(B) = \sqrt{k_B \times \frac{Gain(B)}{m}} \tag{9}$$

$Gain_{fix}(B)$: Fixed information gain of attribute B
$Gain(B)$: Information gain of attribute B
m : The number of different values in attribute B

In the FID3 Algorithm we choose attributes with highest fixed information gain ($Eq.$ 9) as the splitting node, and then recursively create branches based on current splitting node until it reaches a leaf node.

4 Methodology

In Ensemble learning the performance of the model does not depend only on the combination of classifiers, but it also depends on the performance of individual classifiers. The rough-set based $ACRs-$algorithm for constructing decision trees gives better performance than the entropy and information gain based decision tree algorithms [17]. The overall idea of rough random forest is based on these two facts.

The Rough random forest proposed by us constructs rough decision trees as in [17] but the method of calculating the boundary region of features is novel as describe below. In other words, we use bagging and random feature selection to form a number of rough decision trees where the novel approach for calculating boundary region is used to construct the rough decision tree.

In 2010, a fault in calculation of boundary region of features was diagnosed by X.Z.X. Zhang, J.Z.J. Zhou, Y.H.Y. He, Y.W.Y.Wang, and B.L.B. Liu [19]. They identified that in multi-class model the boundary region of a class include the lower approximations of other classes. So it is necessary to delete the union of lower approximation of other classes from the boundary region. Let there be n decision classes in data sample and B is some conditional attribute ($B \in \mathbb{A}$) then its boundary region ($br_B(d=i)$) based on any decision class i can be defined as:

$$br_B(d=i) = \bar{B}(d=i) - \bigcup_{j=1}^{n} \underline{B}(d=j) \tag{10}$$

where, $\underline{B}(d=i)$ and $\bar{B}(d=i)$ indicates the Lower and Upper approximation of attribute B based on decision class equal to i. Instead of calculating boundary region of an attribute based on some specific class and make our classification model biased toward some class, we calculate an *Overall boundary region* for making our decision of selecting best split feature unbiased as:

$$BR_B = \mathbb{U} - \bigcup_{i=1}^{n} \underline{B}(d=i) = \bigcup_{i=1}^{n} br_B(d=i) \tag{11}$$

$\mathbb{U}$: Set of all elements
$br_B(d=i)$: Boundary region of attribute B based on decision class i
BR_B : Overall boundary region of B

Therefore, we use an Overall boundary region approach in construction of rough decision trees of the rough random forest. For illustration, let us take the

Table 1. A,B,C,D - Features d is Decision attribute

No.	A	B	C	D	*d*
1	1	2	2	1	1
2	1	2	3	2	1
3	1	2	2	3	1
4	2	2	2	1	1
5	2	3	2	2	2
6	1	3	2	1	1
7	1	2	3	1	2
8	2	3	1	2	1
9	1	2	2	2	1
10	1	1	3	2	1
11	2	1	2	2	2
12	1	1	2	3	1

same data samples as the paper on Rough Set Based Decision Tree (2002) [17] shown in Table 1. It consists of 12 different elements, 4 conditional attributes and 2 decision classes. So we will construct the decision tree by boundary region approach. Let us calculate the Overall boundary region of attributes as:

$$\underline{A}(d=1) \cup \underline{A}(d=2) = \phi, \qquad BR_A = \{1, ..., 12\}$$
$$\underline{B}(d=1) \cup \underline{B}(d=2) = \phi, \qquad BR_B = \{1, ..., 12\}$$
$$\underline{C}(d=1) \cup \underline{C}(d=2) = \{8\}, \qquad BR_C = \{1, ..7, 9, ..., 12\}$$
$$\underline{D}(d=1) \cup \underline{D}(d=2) = \{3, 12\}, \qquad BR_D = \{1, 2, 4.., ..., 11\}$$

We can see that attribute D gives less elements in its Overall boundary region which means that least implicit data is left after splitting. So attribute D is selected as the root of the decision tree as shown in Fig. 2(a). Then the other attributes are tested on branch $D = 2$ and their Overall boundary region is: $BR_A = \{8\}$, $BR_B = \{2, 5, 8, 9, 10, 11\}$, $BR_C = \{5, 9, 11\}$.

Since the Overall boundary region of attribute A is smaller than the other two attributes we will select that as our split point. We repeat the same procedure till we need to assign labels to that node. Finally we will get our complete rough decision tree as shown in Fig. 2(b), where L_1, L_2 indicate the class labels. In Rough random forest, if there are m rough decision trees then as a process of bootstrapping, the training data is divided into m bootstrap samples. Then each rough decision tree i is trained using i^{th} bootstrap sample of training data. While training the rough decision tree, the best split feature is calculated among $\sqrt{d}$ features chosen at random from d features at each spilt.

While testing, the test data is passed through each rough decision tree of rough random forest and decision class is calculated based on each decision tree result. Then the final result of the rough random forest is calculated based on

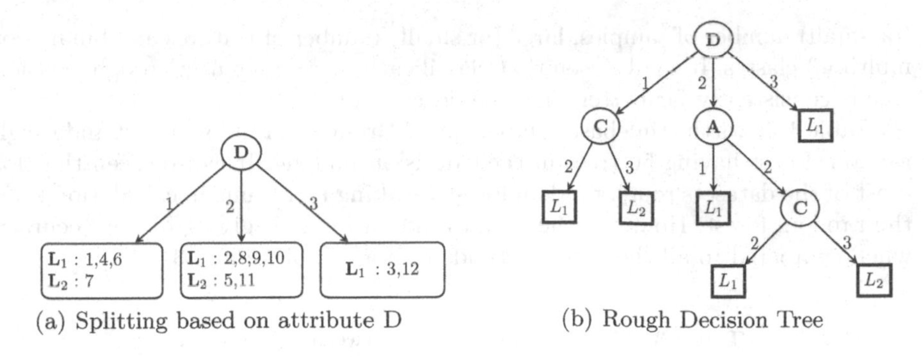

(a) Splitting based on attribute D (b) Rough Decision Tree

Fig. 2. Overall boundary region based splitting and Rough decision tree

the vote of most popular class and that final class label calculated by majority voting is considered as class label of rough random forest.

5 Implementation and Results

In Rough random forest algorithm approximately 2/3 of observations are considered as training set and the remaining 1/3 of data is said to be Out of bag (OOB). This OOB data is used for testing each decision tree. Let there be m decision trees in the rough random forest so the training data is divided into m samples each of 2/3 size of training data. In rough random forest we used only binary rough decision tree which will divide the set of elements into two parts based on split point value. The splitting value which will divide the samples approximately in equal parts is chosen as split points for continuous predictor variables in a decision tree. Due to the randomness involved in the random forest algorithm, we ran each experiment 10 times and the average and standard deviation of the accuracy obtained has been reported in the Tables 2, 3 and 4.

Table 2 compares the results of ID3 decision tree and rough decision tree on some well known datasets taken from UCI-Machine Learning Repository [9]. In our experiments, we have covered different possibility of datasets having large

Table 2. ID3 Decision Tree v/s Rough Decision Tree

Dataset	Samples	Features	Classes	ID3 Decision Tree	Rough Decision Tree
wine	178	14	3	74.78 ± 7	90.65 ± 4
wdbc	596	31	2	86.18 ± 2	91.04 ± 3
digit	5620	65	10	32.24 ± 4	74.12 ± 3
libras	360	91	15	30.13 ± 6	39.66 ± 5
letter	20000	17	26	48.54 ± 1	57.30 ± 1
MiniBooNE	130064	51	2	69.38 ± 2	82.27 ± 2

(or small) number of samples, large (or small) number of features and binary (or multiple) classes. It can be seen that classification accuracy using rough decision tree is consistently far better than the decision tree.

Table 3 shows the timebase comparison of Breiman's random forest and rough random forest having 50 trees in their decision making. It can be seen that for most of the datasets rough random forest is taking more time in calculation than the random forest. However, the rough random forest is giving higher accuracy when compared to all the other methods in both Tables 2 and 3.

Table 3. Timebase comparison between RF and RRF

Dataset	Random Forest		Rough Random Forest	
	Accuracy(*percent*)	Time (*sec.*)	Accuracy (*percent*)	Time (*sec.*)
wine	93.97 ± 2	0.4340	97.46 ± 2	0.3497
wdbc	89.37 ± 1	3.2992	92.63 ± 1	1.9908
digit	56.26 ± 4	17.488	91.88 ± 1	53.916
libras	44.05 ± 4	5.716	51.65 ± 4	6.667
letter	63.35 ± 1	185.93	64.23 ± 1	203.66
MiniBooNE	71.94 ± 1	181.09	88.12 ± 1	2103.63

One of the parameter to be determined is the number of decision trees to use. Table 4 shows the comparison of random forest and rough random forest for different number of trees. It can be seen that the performance of both random forest and rough random forest is improving with increase in number of trees and rough random forest is consistently giving better classification accuracy than the Breiman's random forest irrespective of number of trees. The classification accuracy keeps on increasing with increase in number of trees, but converges to a certain extend for a tree size of 100.

Table 4. Random Forest (RF) and Rough Random Forest (RRF)

Dataset	Number of Trees					
	20 Trees		50 Trees		100 Trees	
	RF	RRF	RF	RRF	RF	RRF
wine	92.54 ± 3	96.98 ± 2	93.97 ± 2	97.46 ± 2	94.06 ± 1	98.25 ± 1
wdbc	88.53 ± 1	92.21 ± 1	89.37 ± 1	92.63 ± 1	90.21 ± 1	93.16 ± 1
digit	51.46 ± 3	87.81 ± 4	56.26 ± 4	91.88 ± 1	59.85 ± 3	92.79 ± 1
libras	40.55 ± 5	50.63 ± 4	44.05 ± 4	51.65 ± 4	46.20 ± 5	52.64 ± 3
letter	62.19 ± 1	63.62 ± 1	63.30 ± 1	64.23 ± 1	63.77 ± 1	64.93 ± 1
MiniBooNE	70.63 ± 1	86.42 ± 1	71.94 ± 1	88.12 ± 1	72.14 ± 1	89.64 ± 1

6 Conclusion

It can be seen that with slight increase of time, the rough random forest gives better performance as compared to information gain based random forest. Especially for the sparse dataset the rough random forest gives much better classification accuracy than the random forest approach. As in the case of random forest, the classification accuracy of rough random forest consistently increases with increase in number of rough decision trees.

In future, we will use other criteria for splitting in the rough random forest. We also need to determine the optimal size of the rough random forest. It is also necessary to find the optimal number of decision trees to use. We also plan to experiment on a number of large datasets to explore how well rough random forest work on them.

References

1. Bonissone, P., Cadenas, J., Garrido, M., Dıaz-Valladares, R.: A fuzzy random forest: Fundamental for design and construction. In: Proceedings of the 12th International Conference on Information Processing and Management of Uncertainty in Knowledge-Based Systems (IPMU 2008), pp. 1231–1238 (2008)
2. Breiman, L.: Random forests. Mach. Learn. **45**(1), 5–32 (2001)
3. Breiman, L.: Bagging predictors. Mach. Learn. **24**, 123–140 (1996)
4. Dietterich, T.G.: An experimental comparison of three methods for constructing ensembles of decision trees: Bagging, boosting, and randomization. Mach. Learn. **40**(2), 139–157 (2000)
5. Ding, B., Zheng, Y., Zang, S.: A new decision tree algorithm based on rough set theory. In: 2009 Asia-Pacific Conference on Information Processing, pp. 326–329, July 2009
6. Hamza, M., Larocque, D.: An empirical comparison of ensemble methods based on classification trees. J. Stat. Comput. Simul. **75**, 629–643 (2005)
7. Huang, L., Huang, M., Guo, B., Zhuang, Z.: A new method for constructing decision tree based on rough set theory. In: 2007 IEEE International Conference on Granular Computing (GRC 2007), pp. 241–241, November 2007
8. Komorowski, J., Pawlak, Z., Polkowski, L., Skowron, A.: Rough sets: A tutorial. Rough fuzzy hybridization: A new trend in decision-making, pp. 3–98 (1999)
9. Lichman, M.: UCI machine learning repository (2013)
10. Patel, B.R., Rana, K.K.: A survey on decision tree algorithm for classification. Int. J. Eng. Dev. Res. IJEDR **2**, 1–5 (2014)
11. Pawlak, Z.: Rough sets. Int. J. Comput. Inform. Sci. **11**(5), 341–356 (1982)
12. Quinlan, R.J.: C4.5: Programs for Machine Learning. Morgan Kaufmann Publishers Inc., San Francisco (1993)
13. Quinlan, J.R.: Induction of decision trees. Mach. Learn. **1**(1), 81–106 (1986)
14. Rokach, L., Maimon, O.: Data Mining with Decision Trees: Theroy and Applications. World Scientific Publishing Co., Inc., River Edge (2008)
15. Skurichina, M., Duin, R.P.W.: Bagging, boosting and the random subspace method for linear classifiers. Pattern Anal. Appl. **5**, 121–135 (2002)
16. Suraj, Z.: An introduction to rough set theory and its applications. In: ICENCO, Cairo, Egypt (2004)

17. Wei, J., Huang, D., Wang, S., Ma, Z.: Rough set based decision tree. In: 2002 Proceedings of the 4th World Congress on Intelligent Control and Automation, vol. 1, pp. 426–431 (2002)
18. Yeh, C.C., Lin, F., Hsu, C.Y.: A hybrid KMV model, random forests and rough set theory approach for credit rating. Knowl. Based Syst. **33**, 166–172 (2012)
19. Zhang, X., Zhou, J., He, Y., Wang, Y., Liu, B.: Vibration fault diagnosis of hydroturbine generating unit based on rough 1-v-1 multiclass support vector machine. In: 2010 Sixth International Conference on Natural Computation (ICNC), vol. 2, pp. 755–759, August 2010

Extending and Tuning Heuristics for a Partial Order Causal Link Planner

Shashank Shekhar(✉) and Deepak Khemani

Department of Computer Science and Engineering,
Indian Institute of Technology Madras, Chennai, India
{sshekhar,khemani}@cse.iitm.ac.in

Abstract. Recent literature reveals that different heuristic functions perform well in different domains due to the varying nature of planning problems. This nature is characterized by the degree of interaction between subgoals and actions. We take the approach of learning the characteristics of different domains in a supervised manner. In this paper, we employ a machine learning approach to combine different, possibly inadmissible, heuristic functions in a domain dependent manner. With the renewed interest in Partial Order Causal Link (POCL) planning we also extend the heuristic functions derived from state space approaches to POCL planning. We use Artificial Neural Network (ANN) for combining these heuristics. The goal is to allow a planner to learn the parameters to combine heuristic functions in a given domain over time in a supervised manner. Our experiments demonstrate that one can discover combinations that yield better heuristic functions in different planning domains.

Keywords: Learning · ANN · POCL planning · Graphplan · Heuristic · Admissibility · PE-ANN · VHPOP · Satisficing planning

1 Introduction

The applications of heuristics in classical state space planning has received significant interest in the past. While there is no efficient universal domain independent heuristic function [34], there are many good heuristic evaluation functions with varying performances on different domains. One approach is to combine different heuristic functions in a manner that the combination parameters can be tuned by supervised learning process in individual domains. In the last decade researchers have also investigated the use of heuristics derived from state space approaches [17] in POCL planning [13,20]. The advantage of greater flexibility of non-linear plan generated by a POCL planner during the execution [15] encourages us to use POCL heuristics as we extend this work. But a partial plan has a complex structure and therefore developing a well informed heuristic function is a tedious task [32]. Our focus here is to learn to combine different, possibly inadmissible, heuristic functions. We adapt the use of neural networks to combine heuristic functions [25] to POCL planning. Our results demonstrate that

R. Prasath et al. (Eds.): MIKE 2015, LNAI 9468, pp. 81–92, 2015.
DOI: 10.1007/978-3-319-26832-3_9

the planner does indeed learn a different way of combining heuristic functions over different domains. The basic rationale behind this approach is that different heuristic functions take a different view of the current problem and arrive at different estimates of the distance to the goal. While some functions may grossly underestimate the distance in some domains, others may overestimate the distance in those domains. The approach we extend uses a set of solved examples in a domain to learn a suitable combination of such functions in that domain. For a partial plan, the notion of distance is that which estimates the work to be done to transform it into a solution plan. Here we implicitly assume that the faster plan completion process will also result in better plans.

The rest of the paper is organized as follows. In the next section we introduce POCL planning in brief. Then we describe the adaptation of neural networks aimed at arriving at high (close to the target) heuristic estimates. Following that we describe a few existing and some adapted heuristic functions for POCL planning. Finally we present our experimental results and conclude.

2 Background

POCL planning algorithm dissociates the task of finding actions that constitute a plan and the placement of those actions in the plan. Following [17,33,34] we work with grounded actions *in lieu* of partially instantiated operators, trading delayed commitment for speed. An introduction to *least commitment planning* can be found in [31]. A partial plan is a 4-tuple, $\Pi = (A, O, L, B)$, where A is a set of actions, O is a set of ordering constraints between actions, L is a set of causal links (CL) among actions and B is a set (empty in our case) of binding constraints. A CL between two actions a_i and a_j represented as $a_i \xrightarrow{p} a_j$ states that the action a_i produces a proposition p which is consumed, as a precondition, by the action a_j. An ordering between actions ($a_i \prec a_j$) signifies that a_i is scheduled before a_j in the plan. An Open Condition (OC), $\xrightarrow{p} a_j$ in POCL planning is a proposition p that is a precondition for the action a_j and for which the supporting *causal link* is absent. An unsafe link (UL) (also called as a *threat*) is a causal link $a_i \xrightarrow{p} a_j$ that can *potentially* be broken by an action a_k if it were to be scheduled in between a_i and a_j and a negative effect of a_k unifies with p. The set F of *flaws* is the set of all such OC and UL in a partial plan.

A partial plan is a solution plan when the set F of flaws is empty. We start with a null partial plan with only OCs and no threat [20]. A null partial plan has two dummy actions a_0 and a_∞ ($a_0 \prec a_\infty$). The plan refinement procedure involves selection of OCs, along with a *resolver* for each OC. The resulting partial plan may have a threat. In our implementation we try to resolve all the threats observed after the resolution of one OC [2,28] before we pick the next.

A solution plan is a partial plan with no flaws. The POCL planning is a two stage process (a) selecting the most promising partial plan and (b) selecting and resolving a flaw from it. We need good heuristic functions in both the stages. Our learned heuristics using variants of neural networks will be employed in the first stage, to select the most promising partial plan for further refinement.

3 Penalty Enhanced ANNs

The basic idea in this paper is to estimate a heuristic value from the values returned by a set of existing heuristic functions. Following the work [25] we use an artificial neural network (ANN) [14] to define the estimate in terms of inputs. Given that some of the input values may be gross overestimates, we need to train a network so as to penalize inputs that are larger than the (known) target value in each training example. The estimates have been combined using a variant of ANN called Penalty-Enhanced Artificial Neural Network (PE-ANN) [25]. Suppose an ANN with Y as target variable and x_i for $i \in \{1, 2, ..., 4\}$ as input variables. The error function $E(t)$ is defined as $E(t) = \theta(t) - Y(t)$, where $\theta(t)$ is the predicted value and $Y(t)$ is the target value. The mean-squared error (MSE) is, $MSE = \Sigma_{t \in X} E^2(t) \Big/ |X|$, where X is the training data. The MSE is a symmetric function that penalizes the error on both sides equally.

For algorithms like WA* [19], IDA* [12] or dynamically weighted A* [30], we penalize the model predictions that are higher than the target value more than the ones that are lower. This results in our predications being by and large lower that the target value, required for admissibility. The PE-ANN model addresses this concern and its error function [25] given below is biased towards underestimation,

$$E_{new}(t) = \left\{ a + \frac{1}{1 + \exp(-bE(t))} \right\} \times E(t)$$

Here a and b are constants which decide the slope of the error curve, $E_{new}(t)$ regulates the penalization criteria for $(E(t) > 0)$ and $(E(t) < 0)$, where $E(t) > 0$ is penalized more.

We adapt and use the PE-ANN model in POCL planning, in which the error function *with* and *without* regularization (L2 norm) [3] is used. This follows evidence that regularization decreases the tendency of overfitting when the size of the training set is smaller than what is ideally required. This is often the case in a planning domain (for example the *Towers of Hanoi* domain) where the number of different instances that can be solved are not many.

A brief description of the approach is as follows. For each planning problem in the training set, we have a set of n heuristic values along with the target value that is calculated using Graphplan [4]. The training data is used to learn the weights for the PE-ANN using a suitable error function that determines the weights. The error functions we use are described later in this paper. For a new (test) instance t, we input the n values from the existing heuristic functions to the trained PE-ANN model to predict the heuristic value $\theta(t)$. The heuristic value at the output node is instrumental in deciding *which* of the candidate partial plans is selected for refinement.

We give up on the least commitment strategy as we use fully grounded actions, but gain in speed-up of execution [17,33,34]. In [33,34] authors show the advantages of POCL planning with lifted actions as well. Therefore one extension could be to explore the possibility of reverting to partial plans with lifted actions. Next, we describe the detailed approach for training of neural networks.

4 Training the PE-ANN

The idea is not bound to strict *admissibility* of heuristic functions in any form (either individually or combined). In POP, a partial plan (π) is a node in a the space of partial plans, the function used by WA* for selecting a candidate is, $f(\pi) = g(\pi) + w \times h(\pi)$. Here $g(\pi)$ is the number of actions in π and $h(\pi)$ is the heuristic value that estimates the number of refinements required to resolve the remaining flaws, w is the weight factor. The heuristic value $h(\pi)$ thus plays a crucial role in the selection process of π from a set of candidate partial plans. The PE-ANN model uses the gradient descent approach in the algorithm Back-propagation to learn the weights starting from randomly initialized weights. The weight update rules and the gradient descent process are described below.

4.1 Weight Update Rules

The error function defines a surface over which gradient descent seeks the minimum. The error function employed in the PE-ANN imposes higher penalties on edges that transmit the signal from overestimating inputs. The network then learns to suppress such inputs and generate an output that is generally lower than the target value. Our PE-ANN model uses the L2-norm, the cost function is devised as follows,

$$E'(t) = \sum_{i=1}^{N} \left\{a + \frac{1}{1 + \exp\big(-bE(t)\big)}\right\} \times E(t) + \gamma \left(\sum_{m} \beta_m^2 + \sum_{m}\sum_{l} \alpha_{ml}^2\right)$$

where $E'(t)$ is error for an instance t, a and b are the parameters that are set empirically, γ is the regularization coefficient, l is the number of nodes in the input layer, m is the number of nodes in the hidden layer, and α and β are the weight of the edges emanating from the first two layers.

The weight update rules for the PE-ANN for adapted cost function with regularization for minimization problem are given below.

$$R(\theta) = \sum_{i=1}^{N} \left\{\left(a + \frac{1}{1 + \exp\big(-b(T - Y_i)\big)}\right) \times (T - Y_i)\right\}^2$$

where $R(\theta)$ is sum of Squared Error also represented as $R(\alpha, \beta)$, T is the predicted output that is $T = \beta_0 + \beta^T z$, Y_i is the target value, a and b are experimental parameters and N is the number of training instances in the training set. In the above expression T is dependent on vector z (defined above), where m^{th} element of z is, $z_m = \sigma(\alpha_{0m} + \alpha_m^T x)$. The partial derivative of $R(\theta)$ *wrt* β_m is, $\partial R(\theta)/\partial \beta_m = 2 \times E'(t) \times \big(\partial E'(t)/\partial \beta_m\big)$, and $\sigma(g)$ is a sigmoid function over g defined as, $\sigma(g) = 1/\big(1 + \exp(-g)\big)$. The partial derivative of $\sigma(G)$ *wrt* G is given as, $\partial\sigma(G)/\partial(G) = \sigma(G) \times \big(1 - \sigma(G)\big)$.

$$R(\theta) = \sum_{i=1}^{N} \left\{\big(a + \sigma\big(b(T - Y_i)\big)\big) \times (T - Y_i)\right\}^2 \tag{1}$$

In order to minimize $R(\theta)$ in Eq. 1, we are required to assess its sensitivity to each of the weights. We take the partial derivative of $R(\theta)$ with respect to each of the weight parameters to calculate the effect of changing weights. Simplification of the partial derivatives gives final weight update rules corresponding to α and β with *L2-norm* as regularization. The weight update for the weight vector β is,

$$\begin{aligned}\frac{\partial R(\theta)}{\partial \beta_m} &= 2 \times (a + \sigma(b(T - Y_i))(T - Y_i)) \\ &\quad \times \Big\{ b \times \sigma\big(b(T - Y_i)\big) \times \big[1 - \sigma\big(b(T - Y_i)\big)\big] \times z_m \\ &\quad \times (T - Y_i) + \big[a + \sigma\big(b(T - Y_i)\big)\big] \times z_m \Big\} + \gamma \times 2\beta\end{aligned}$$

Similarly, the weight update for the weight matrix α is quantified as,

$$\begin{aligned}\frac{\partial R(\theta)}{\partial \alpha_{ml}} &= 2 \times \big(a + \sigma\big(b(T - Y_i)\big) \times (T - Y_i)\big) \times \Big\{ b \times \sigma\big(b(T - Y_i)\big) \\ &\quad \times \Big(1 - \sigma\big(b(T - Y_i)\big)\Big) \times \beta_m \frac{\partial(\sigma(\alpha_m^T x))}{\partial(\alpha_{ml})} \times (T - Y_i) \\ &\quad + \big[a + \sigma(b(T - Y_i))\big] \times \beta_m \frac{\partial(\sigma(\alpha_m^T x))}{\partial(\alpha_{ml})} \Big\} + \gamma \times 2\alpha\end{aligned}$$

where $\gamma > 0$, and $\frac{\partial \sigma(\alpha_m^T x)}{\partial \alpha_{ml}} = \sigma(\alpha_m^T x)\big(1 - \sigma(\alpha_m^T x)\big)x_l$. From above equations, the gradient descent weight update rules for parameters α and β for a single training instance in r^{th} iteration are obtained and given as, $\alpha_{ml}^{r+1} = \alpha_{ml}^r - \eta_\alpha \times \frac{\partial R(\theta)}{\partial \alpha_{ml}}$ and $\beta_m^{r+1} = \beta_m^r - \eta_\beta \times \frac{\partial R(\theta)}{\partial \beta_m}$, where η_α and η_β are the learning rates that may get different values for the gradient descent rules. The $\gamma * 2\beta$ and $\gamma * 2\alpha$ factor can be removed from the update rules in order to train PE-ANN without the regularized cost function. The details of the parameters values in our experiments are given in tables for each domain.

5 Multiple Heuristics

An efficient partial order planner does a controlled search in the plan space. It attempts to minimize search (flaw resolutions) by making informed choices during the planning process [27]. A good strategy is to select the most demanding flaw and a refinement that leaves the minimum refinements to be done subsequently. There are good flaw selection techniques based on Most-Cost or Most-Work *etc.* associated with the flaws [34]. One possible way of selecting a (refined) partial plan from the list of candidates based on the minimum number of actions needed to resolve all the open conditions in it [17]. We consider the definition of $h^*(\pi)$,

Definition 1. *For given a partial plan* π, $h^*(\pi)$ *gives the minimum number of new actions required to convert a partial plan to a solution plan.*

It is desirable to have a close estimate to h^*. The heuristics used as *inputs* to the neural network to arrive at our combination heuristic are described below.

5.1 $h_{max}(\pi)$: Max Heuristic

The simplest heuristic h_{max} [5] counts the number of steps required individually for each open goal, and takes the maximum value. We adapt this heuristic for search in plan space as, $h_{max}(\pi) = \max_{\xrightarrow{q} a_j \in OC(\pi)} h_{max}(q)$ $s.t.$ a_i provides q, where the factor $h_{max}(q)$ is $min\{h_{max}(q), 1 + h_{max}(precond(a_i))\}$.

5.2 $h_{add}(\pi)$: Additive Heuristic

The additive heuristic [5–7] adds up the steps required by each individual open goal. Younes et al. adapt the additive heuristic for POCL planning [34].

5.3 $h^r_{add}(\pi)$: Positive Interaction Heuristic

Younes et al. address the positive interactions among subgoals while ignoring the negative interactions [34]. The heuristic $h^r_{add}(\pi)$ is a substitute for $h_{add}(\pi)$ as the latter has no provision for actions reuse.

The heuristics defined below are based on the reachability analysis of the partial states from initial state by Graphplan [4] structure. We treat all the OCs present in π as constituting a state S.

5.4 $h_{relax}(\pi)$: Relax Heuristic

Nguyen et al. address positive subgoal interactions using a serial planning graph for the subgoal reachability analysis [17]. We describe and use the existing heuristic called $h_{relax}(\pi)$ (a variant of FF heuristic [10]) proposed in [17].

Now we describe some of the *adapted* heuristic functions (proposed in [16] applicable in state space search) for the search in POCL planning.

5.5 $h_{set-level}(\pi)$: Set-level Heuristic

We altered and use $h_{set-level}(S)$ proposed in [11] for plan space search where its usage depends on the preprocessing step (the assumption of hypothetical state S). The altered heuristic $h_{set-level}(\pi)$ is equal to $level(S)$.

5.6 $h_{partition-2}(\pi)$: Partition-2 Heuristic

Nguyen et al. have shown that in certain cases $h_{set-level}(S)$ estimates the same numerical values for two different states [16]. To overcome this, $h_{partition-2}$ is devised and that is adapted for POCL planning as, $h_{partition-2}(\pi) = \sum_{S_i} lev(S_i)$.

5.7 $h_{adjust-sum}(\pi)$: Adjust-sum Heuristic

We adapt the idea (proposed in [16]) discussed in the previous heuristic function where we state that $level(S - p_1) \leq level(S)$, p_1 is achieved first and by the time $(S - p_1)$ is achieved, it may clear the achieved proposition p_1 because of negative interactions. It is defined as, $h_{adjust-sum}(\pi) = h_{add}(\pi) + h_{set-level}(\pi) - h_{max}(\pi)$.

Table 1. This table is divided into **two** parts. The **first** half captures the results in blocksworld domain. We finalize the parameters as, |X| = 955, a = 1.0, b = 3.0, after cross-validation $\gamma = 0.1$, $L_{rates} \in [0.6, 0)$ and 13 nodes in the hidden layer. h_{closed} indicates the feature which has its value closer to the h_{reg} (learned heuristic) and h_{val} is the value of h_{closed}. In the **second** half, we captures the results in elevator domain. Here |X| = 724, a = 1.0, b = 3.0, $\gamma = 0.01$, $L_{rates} \in [0.6, 0)$ and 10 nodes in the hidden layer. Also the similar table has shortened caption.

Ins	h_{low}	h_{high}	h_{target}	PE-ANN			PE-ANN (*L2 norm*)		
				h_{reg}	h_{closed}	h_{val}	h_{reg}	h_{closed}	h_{val}
1	3	18	9	10.05	h_{vhpop}	7	8.97	h_{vhpop}	7
2	2	10	10	11.32	h_{combo}	10	9.13	h_{combo}	10
3	6	15	12	6.25	h_{vhpop}	6	7.0	$h_{set-level}$	7
4	4	19	11	6.27	h_{max}	4	6.95	$h_{set-level}$	9
5	6	14	11	6.28	h_{vhpop}	6	6.98	$h_{set-level}$	7
1	3	24	13	5.82	h_{vhpop}	8	5.97	h_{vhpop}	8
2	5	20	17	15.80	$h_{partition-2}$	15	15.11	$h_{partition-2}$	15
3	3	20	11	5.78	h_{vhpop}	9	5.96	h_{vhpop}	9
4	3	12	12	15.27	h_{combo}	12	12.12	h_{combo}	12
5	2	8	10	5.76	h_{vhpop}	7	7.87	h_{vhpop}	7

5.8 $h_{adjust-sum2}(\pi)$: Adjust-sum2 Heuristic

This heuristic is adapted from [16] as well for search in plan space. It is defined as, $h_{adjust-sum2}(\pi) = h_{relax}(\pi) + h_{set-level}(\pi) - h_{max}(\pi)$. This heuristic performs better when we consider binary mutexes [18].

5.9 $h_{combo}(\pi)$: Combo Heuristic

Nguyen et al. show that the combo heuristics work better for a few domains. We alter the original definition and consider a variation called $h_{combo}(\pi)$ and defined as, $h_{combo}(\pi) = h_{relax}(\pi) + h_{set-level}(\pi)$ [16]. We replace $h_{add}(\pi)$ with $h_{relax}(\pi)$ as this is more accurate in most of the planning domains [16].

Even though Graphplan may sometimes give *target* values more than the *optimal* value for some problems, it has been used for the calculation of target values (h_{target}) for smaller sized problems. For some of the planning domains (*e.g. Towers of Hanoi*) we observe the lack of sufficient training instances. We do not consider those domains. Where we have sufficient instances the L2 norm is used.

6 Empirical Evaluation

All the experiments have been performed on Intel dual-core PC with 4Gb of RAM. We present the results obtained by PE-ANN *with* and *without* regularization. The results demonstrate that penalizing the overestimates more does tend

Table 2. This table shows the results in Travel domain. We divide this table into **three** major parts. The *second* and *third* parts capture the predictions by the learned models using all 9 features and the most uncorrelated 4 features, respectively. In **second** part, the parameters are |X| = 359, a = 1.0, b = 3.0, $\gamma = 0.001$, $L_{rates} \in [0.6, 0)$ and 7 nodes in the hidden layer. While in the **third** part |X| = 359, a = 1.0, b = 3.0, $\gamma = 0.001$, $L_{rates} \in [0.6, 0)$ and 7 nodes in the hidden layer with 4 features.

Ins	h_{low}	h_{high}	h_{target}	PE-ANN			PE-ANN (*L2 norm*)			PE-ANN			PE-ANN (*L2 norm*)		
				h_{reg}	h_{closed}	h_{val}	h_{reg}	h_{closed}	h_{val}	h_{reg}	h_{closed}	h_{val}	h_{reg}	h_{closed}	h_{val}
1	3	10	5	3.43	h_{max}	3	3.48	h_{max}	3	3.96	h_{max}	3	4.12	h_{max}	3
2	5	14	17	3.46	-	-	3.51	-	-	19.87	h_{combo}	14	18.21	h_{combo}	14
3	4	12	6	3.45	-	-	3.49	-	-	4.02	h_{max}	4	4.36	h_{max}	4
4	3	10	5	3.44	h_{max}	3	3.48	h_{max}	3	3.96	h_{max}	3	4.13	h_{max}	3
5	4	12	12	3.44	-	-	3.49	-	-	12.29	h_{combo}	12	11.81	h_{combo}	12

to keep the learned heuristic within the target value. The datasets used in our experiments are generated as follows. The partial plans for which the heuristic value is to be estimated are generated by our implementation of VHPOP (currently accepts only some varieties of problems). The target value corresponding each partial plan is obtained using Graphplan. Some sets of open goals cannot be part of any consistent state, and for them Graphplan cannot find a plan. For such sets, we set the target to infinity and exclude them from the training set. That is, we exclude all partial plans for which Graphplan cannot find a solution for the set of open goals, which do not constitute a consistent state.

Dataset Preparation: Our datasets are made up of estimated and actual plan costs, with integer values. The estimated costs are the input heuristic values for the PE-ANN, and the actual cost found the target value. Each row in the dataset has the heuristic values computed by the functions described in the preceding section, and the last column is the target value (h_{target}) found by Graphplan. The datasets used in training are complete, with no missing values, and took O(hour) for preparation in each domain. As mentioned earlier, problems like the *Towers of Hanoi* produce small datasets, since even moderately sized problems cannot be solved by Graphplan within a time limit of **15 min**.

Model Learning: The standard Backpropagation algorithm using gradient descent is applied to train the PE-ANN and its variant with regularization. Each dataset is shuffled because there might be multiple entries occurring in the training samples belonging to the same plan space. Shuffling the order prevents localization, and results in the edge weights and coefficients to be determined globally by the entire data [3,8]. The time required for training phase depends on the L_{rates} and number of *epochs*. Starting with higher L_{rates} it took O(minute) for learning the parameters in each domain. The number of training instances (|X|) used along with other parameters are in the captions of the tables. The overhead in our approach to planning is due to the fact that we have to train the networks for each domain, and during the planning process one needs to compute the heuristic values using all the heuristic functions that provide the input to the neural network. The main motivation for this approach is that we may not be aware of the degree of subgoal interaction in a new domain, and thus be unable to choose an appropriate heuristic function. The pay-off is that for each new domain we are able to work with heuristic functions that often yield the good underestimating values.

Now we discuss the related work done in the past. Many approaches of combining heuristics have been discussed and used earlier where the learning phase is not employed. In [9,23,24,35], the authors study and show different possible combinations of heuristics where learning is not required. These approaches have the advantage of not requiring training samples to start with, but they cannot arrive at the non linear relationships that ANNs can find. The problem arises due to the usage of Graphplan for our approach can be tackled by online learning [29] or bootstrapping [1]. Sapena et al. have tried combining different heuristics to accelerate the performance of forward partial order planner [26]. The previous efforts have shown the advantages of combining different heuristics in place of selecting one. In future, we intend to *compare* the existing approaches of combining different heuristic functions to our approach.

Performance Evaluation: The results of our experiments in three domains are presented in Tables 1 and 2, each with the data from five test instances. We generate all possible pair of $\langle\langle h_1, h_2, ..., h_9\rangle,\ h_{target}\rangle$ by solving bigger sized problems and pick randomly five instances for testing. For each test instance, the tables show the lowest and the highest heuristic estimates, h_{low} and h_{high}, from the nine heuristic functions that provide the inputs to the neural networks. Also shown is the target value h_{target} computed by the algorithm Graphplan. For each test instance the heuristic estimate generated by both the PE-ANN and the regularized PE-ANN networks is shown along with the input heuristic function that matches h_{reg} the most. In cases where h_{reg} is lower than h_{low} the corresponding heuristic function is shown by a dash ("-"). We hazard a guess that the predicted value is lower than *all* the inputs because of the non-linear nature of the combination produced by the neural networks in which the penalty for some inputs pulls the prediction below h_{low}. The objective is to show that in general the estimates generated by the networks are high values but lower than the targets, as required for admissibility.

Table 1 shows the outcomes of the experiments performed in the Blocksworld and Elevator domain. The learning rate L_{rate} is initially set high which leads to early convergence to a region close to the minimum. When gradient descent approaches the minimum the L_{rate} is reduced to enable it to converge using smaller steps along with α and β were initialized to random values. The problem instances reported in Table 1 are bigger compared to the ones used for training. All the predictions for all the instances by both the networks are closer to the target, but the regularized network estimates are closest. The predictions are even more informed than the estimates of individual features used in the experiments. The accuracy decreases if γ increases, and some uneven behavior is observed in the accuracy when $\gamma \in [90, 100]$ which is not been studied further.

Table 2 is divided into three major parts. The last two parts capture the outputs of the learned models using all 9 features and only 4 uncorrelated features as inputs. The table shows results in the Travel domain. We do not have enough training samples for this domain. In the *second* part of the table, both the algorithms perform equally worse. This may be due to the small number of training instances. The domain has high subgoal dependence as there is big difference between h_{target} and h_{high}. Predicting values lower than the h_{low} is not a desirable property. To overcome the drawback of small number of training instances, we reduce the number of features by removing correlated ones. In *third* part, only 4 features (h_{max}, h_{add}, $h_{adjust-sum2}$ and h_{combo}) are selected. We keep the same number of training instances with reduced number of features which resulting in better performance. The regularized network yields better predictions when it is a smaller network with only four inputs, probably because the number of training instances are sufficient for the smaller networks.

Our *first* observation is about the heuristic functions $h_{set-level}$, $h_{partition-2}$, $h_{adjust-sum}$, $h_{adjust-sum2}$ and h_{combo} that we have adapted and used. We observe in the datasets (not shown in this paper due to lack of space) that the extended heuristics individually perform well compare to the state of the art. Often, the extended heuristics are closer to the target values. In the testing phase, the values returned by learned functions are even closer to the target values than the individual heuristics. Though in some domains they underperform at a few stages as compared to other state-of-the-art POCL heuristics. Our *second* observation concerns is that, in all our experiments the heuristic estimates generated were high underestimating values. Though there were some instances where the values were much lower than the optimal value, for example as shown in Table 2. To address this drawback, we tried training a smaller neural network here with only four inputs instead of nine, the performance improved.

7 Summary and Future Work

We describe an attempt to learn domain specific ways to combine different heuristic values to arrive at consistently better estimates over a range of planning domains and problems. One can see from Tables 1 and 2 that different heuristic functions perform better in different domains; also in a single domain, different heuristic functions perform better on different instances of problems. This is probably due to the fact that planning by search is essentially a process of non-monotonic reasoning, in which the reasoner asserts and retracts fluents describing the (current) state of the world. This is something that a simple domain independent heuristic function is unable to make predictions about and therefore a non-linear function to combine the different estimates, as done by ANN, can be made.

Our future work is to validate this thesis experimentally using a *partial order planner*. Another possibility is to reintroduce some aspects of partially instantiated operators, to try and cut down on the space required to completely instantiate all operators for each problem. Recent work [21, 22] has shown that this can work, specially when dealing with large problems, where the grounding phase itself takes up most of the computation cycles.

References

1. Arfaee, S.J., Zilles, S., Holte, R.C.: Learning heuristic functions for large state spaces. Artif. Intell. **175**(16), 2075–2098 (2011)
2. Barrett, A., Weld, D.S.: Partial-order planning: evaluating possible efficiency gains. Artif. Intell. **67**(1), 71–112 (1994)
3. Bishop, C.M.: Pattern Recognition and Machine Learning, vol. 1. Springer, New York (2006)
4. Blum, A.L., Furst, M.L.: Fast planning through planning graph analysis. Artif. Intell. **90**(1), 281–300 (1997)
5. Bonet, B., Geffner, H.: Planning as heuristic search. Artif. Intell. **129**(1), 5–33 (2001)
6. Bonet, B., Loerincs, G., Geffner, H.: A robust and fast action selection mechanism for planning. In: AAAI/IAAI, pp. 714–719 (1997)
7. Haslum, P., Geffner, H.: Admissible heuristics for optimal planning. In: AIPS, pp. 140–149. Citeseer (2000)
8. Hastie, T., Tibshirani, R., Friedman, J., Hastie, T., Friedman, J., Tibshirani, R.: The Elements of Statistical Learning, vol. 2. Springer, New York (2009)
9. Helmert, M.: The fast downward planning system. J. Artif. Intell. Res. (JAIR) **26**, 191–246 (2006)
10. Hoffmann, J.: FF : the fast-forward planning system. AI Mag. **22**(3), 57 (2001)
11. Kambhampati, S., Parker, E., Lambrecht, E.: Understanding and extending graphplan, pp. 260–272 (1997)
12. Korf, R.E.: Depth-first iterative-deepening: an optimal admissible tree search. Artif. Intell. **27**(1), 97–109 (1985)
13. McAllester, D., Rosenblatt, D.: Systematic nonlinear planning (1991)
14. Mitchell, T.M.: Machine learning and data mining. Commun. ACM **42**(11), 30–36 (1999). http://doi.acm.org/10.1145/319382.319388
15. Muise, C., McIlraith, S.A., Beck, J.C.: Monitoring the execution of partial-order plans via regression. In: IJCAI Proceedings-International Joint Conference on Artificial Intelligence, vol. 22, p. 1975 (2011)
16. Nguyen, X., Kambhampati, S.: Extracting effective and admissible state space heuristics from the planning graph. In: AAAI/IAAI, pp. 798–805 (2000)
17. Nguyen, X., Kambhampati, S.: Reviving partial order planning. In: IJCAI, vol. 1, pp. 459–464 (2001)
18. Nigenda, R.S., Nguyen, X., Kambhampati, S.: Altalt: combining the advantages of graphplan and heuristic state search (2000)
19. Pearl, J.: Heuristics: Intelligent Search Strategies for Computer Problem Solving. Addison-Wesley, Reading (1984)
20. Penberthy, J.S., Weld, D.S.: UCPOP: A sound, complete, partial order planner for adl, pp. 103–114. Morgan Kaufmann (1992)
21. Ridder, B.: Lifted Heuristics: Towards More Scalable Planning Systems. Ph.D. thesis, King's College London (University of London) (2014)
22. Ridder, B., Fox, M.: Heuristic evaluation based on lifted relaxed planning graphs. In: Proceedings of the Twenty-Fourth International Conference on Automated Planning and Scheduling, ICAPS 2014, Portsmouth, New Hampshire, USA, 21–26 June 2014. http://www.aaai.org/ocs/index.php/ICAPS/ICAPS14/paper/view/7948

23. Röger, G., Helmert, M.: Combining heuristic estimators for satisficing planning. In: ICAPS 2009 Workshop on Heuristics for Domain-Independent Planning, pp. 43–48 (2009)
24. Röger, G., Helmert, M.: The more, the merrier: combining heuristic estimators for satisficing planning. Alternation **10**(100s), 1000s (2010)
25. Samadi, M., Felner, A., Schaeffer, J.: Learning from multiple heuristics. In: AAAI, pp. 357–362 (2008)
26. Sapena, O., Onaindıa, E., Torreno, A.: Combining heuristics to accelerate forward partial-order planning. In: Constraint Satisfaction Techniques for Planning and Scheduling, P. 25 (2014)
27. Schubert, L., Gerevini, A.: Accelerating partial order planners by improving plan and goal choices. In: Proceedings of Seventh International Conference on Tools with Artificial Intelligence, pp. 442–450. IEEE (1995)
28. Smith, D.E., Peot, M.A.: Postponing threats in partial-order planning. In: Proceedings of the Eleventh National Conference on Artificial Intelligence, pp. 500–506. AAAI Press (1993)
29. Thayer, J.T., Dionne, A.J., Ruml, W.: Learning inadmissible heuristics during search. In: ICAPS (2011)
30. Thayer, J.T., Ruml, W.: Using distance estimates in heuristic search. In: ICAPS, pp. 382–385. Citeseer (2009)
31. Weld, D.S.: An introduction to least commitment planning. AI Mag. **15**(4), 27 (1994)
32. Weld, D.S.: AAAI-10 classic paper award: systematic nonlinear planning a commentary. AI Mag. **32**(1), 101 (2011)
33. Younes, H.L., Simmons, R.G.: On the role of ground actions in refinement planning. In: AIPS, pp. 54–62 (2002)
34. Younes, H.L., Simmons, R.G.: Versatile heuristic partial order planner. J. Artif. Intell. Res. (JAIR) **20**, 405–430 (2003)
35. Zhu, L., Givan, R.: Simultaneous heuristic search for conjunctive subgoals. In: 1999 Proceedings of the National Conference on Artificial Intelligence (AAAI), vol. 20, pp. 1235–1241. AAAI Press, MIT Press, Menlo Park, Cambridge (2005)

Symbolic Representation of Text Documents Using Multiple Kernel FCM

B.S. Harish(✉), M.B. Revanasiddappa, and S.V. Aruna Kumar

Department of Information Science and Engineering,
Sri Jayachamarajendra College of Engineering, Mysuru, India
bsharish@sjce.ac.in, {revan.cr.is,arunkumarsv55}@gmail.com

Abstract. In this paper, we proposed a novel method of representing text documents based on clustering of term frequency vector. In order to cluster the term frequency vectors, we make use of Multiple Kernel Fuzzy C-Means (MKFCM). After clustering, term frequency vector of each cluster are used to form a interval valued representation (symbolic representation) by the use of mean and standard deviation. Further, interval value features are stored in knowledge base as a representative of the cluster. To corroborate the efficacy of the proposed model, we conducted extensive experimentation on standard datset like Reuters-21578 and 20 Newsgroup. We have compared our classification accuracy achieved by the Symbolic classifier with the other existing Naive Bayes classifier, KNN classifier and SVM classifier. The experimental result reveals that the classification accuracy achieved by using symbolic classifier is better than other three classifiers.

Keywords: Classification · Text documents · Representation · Symbolic feature · Multiple Kernel FCM

1 Introduction

Digital information on the web is increasing day by day and most of the information is in textual form. Text classification is one of the solution to provide better result in information retrieval (IR) system. Text classification (Categorization) is the process of automatically classifying text documents into predefined categories. Basically there are two approaches to classify text documents. First approach is the rule based, in this approach classification rules are defined manually and documents are classified based on rules. Second approach is machine learning approach, here classification rules or equations are found automatically using sample labeled documents. This class of approaches has much higher precision than rule based approaches. Therefore, machine learning based approaches are replacing the rule based approaches for text classification [1]. A well classified corpus can facilitate document searching, filtering and navigating for both users and information retrieval tools. Text classification is used in number of interesting fields like, search engine, automatic document indexing for information

R. Prasath et al. (Eds.): MIKE 2015, LNAI 9468, pp. 93–102, 2015.
DOI: 10.1007/978-3-319-26832-3_10

retrieval system, document filtering, spam filtering used for emails, text mining, digital library, word sense [2].

The major challenges of text classification are the high dimensionality of the feature space, representation of document, similarity measures of different documents, preserving semantic relationship in a documents and sparsity. To tackle the above said problems a number of methods have been introduced in literature. Before applying machine learning techniques to classify text document, we need to transform the text documents, which are typically strings of characters. In literature, many representation schemes like Bag of Word (BOW), Vector Space Model (VSM), Universal Networking Language (UNL), N-Gram, Latent Semantic Indexing (LSI), Locality Preserving Indexing (LPI), Regularized Locality Preserving Indexing (RLPI) etc. are applied for text classification. Bag of Words (BOW) approach is most widely adopted representation model in text classification. In this approach a text is represented as a vector of word weights [3]. However, BOW representation suffers from two limitations, it tends to break terms into their constituent words and it treats synonymous words as independent features. Vector Space Model (VSM) is an algebraic model for representing text documents as vectors of identifiers, such as index terms. Major limitation of VSM is, it represents terms with very high dimensionality and the resulting vector is very sparse. Universal Networking Language (UNL) presents a document in the form of a graph, with nodes as the universal words and relations between them as links [4]. This method requires the construction of a graph for every document and hence it is unwieldy to use for an application where large numbers of documents are present.

Hotho et al., in [5] proposed an ontology representation for a document to keep the semantic relationship between the terms in a document. The ontology model preserves domain knowledge of a term demonstrated in a document. However, automatic ontology construction is a difficult task due to the lack of structured knowledge base. Cavnar., [6] used a sequence of symbols (byte, a character or a word) called N-Grams, that are extracted from a long string in a document. In the N-Gram scheme, it is very difficult to decide the number of grams to be considered for effective document representation. Another approach in [7] uses multi-word terms as vector components to represent a document. But this method requires a sophisticated automatic term extraction algorithms to extract the terms automatically from a document. Latent Semantic Indexing (LSI) is based on Singular Value Decomposition (SVD), which project the document vectors into a subspace. LSI finds the best subspace estimation to the original document space in the sense of minimizing the global reconstruction error [8]. In other words, LSI seeks to uncover the most representative features rather the most discriminative features for document representation. Therefore, LSI might not be optimal in discriminating documents with different semantics. To discover the discriminating structure of a document space, He et al., [9] proposed a Locality Preserving Indexing (LPI). LPI can have more discriminating power than LSI even though LPI is also unsupervised. An assumption behind LPI is that close inputs should have similar documents. The computational complexity of LPI is very expensive because it involves eigen-decomposition of two dense matrices. It is almost infeasible to apply LPI on very large dataset. Hence to reduce the computational complexity of LPI, Cai et al., [10] proposed Regularized Locality Preserving Indexing (RLPI). RLPI is fundamentally based on LPI. Specifically, RLPI decomposes the LPI problem as a graph embedding problem and a regularized least square problem. Such modification avoids eigen-decomposition of dense matrices and can significantly reduce both time and memory cost in computation. However, RLPI fails to preserve the intraclass variations among documents of different classes.

In text classification, clustering techniques has been used as an alternative representation scheme, which automatically groups the text documents into a list of meaningful categories. Several approaches of clustering have been proposed in literature to solve high dimensionality problem. Baker and McCallum [11] proposed Distributional Clustering method. This method cluster the words into groups based on the distribution of class labels associated with each word. Bekkerman et al., [12] proposed word-clustering representation model based on the information bottleneck method, which generates a compact and efficient representation for documents. However, distributional clustering and word-clustering methods are agglomerative in nature and results in sub-optimal word clusters, and high computational cost. To overcome these drawbacks, Dhillon et al., [13] proposed new information theoretic divisive algorithm for feature clustering. The feature clustering is a powerful alternative to feature selection for reducing the dimensionality of text document.

Fuzzy clustering which is also called as soft clustering, is used in text classification, to improve accuracy and performance. In literature many researchers worked on fuzzy clustering techniques for reducing high dimensionality of features. A fuzzy set is class of objects with a continuum of grades of membership [14]. Generally, fuzzy membership functions are defined in terms of numerical values of an underlying crisp attribute and membership ranges between 0 to 1. To reduce high dimensionality of feature vector, Anilkumarreddy et al., [15] proposed fuzzy based incremental feature clustering method. Jiang et al., [16] proposed fuzzy self constructing feature clustering (FFC) algorithm, which is an incremental clustering approach to reduce the dimensionality of the features in text classification. In this algorithm, each cluster is characterized by a membership function with statistical mean and deviation. Features that are similar to each other are grouped into the same cluster. If a word is not similar to any existing cluster, a new cluster is created. Puri, [17] proposed Fuzzy Similarity Based Concept Mining Model (FSCMM). This model mainly reduces feature dimensionality and removes ambiguity at each level to achieve high classifier performance. Term frequency vectors of each cluster are used to form a symbolic representation by the use of mean and standard deviation. Carvalho [18] proposed fuzzy C-means clustering algorithm for symbolic interval data. It aims to provide fuzzy partitions of a set of pattern clusters and a set of corresponding representatives (prototypes) for each cluster by optimizing an adequacy criterion based on suitable squared Euclidean distances between vectors of intervals.

In literature, most of the clustering based classification method uses conventional term document matrix representation. Since, the value of the term frequency different from document to document in the class, preserving these variations has been difficult. To overcome this problem, Guru et al., [19] proposed a symbolic based representation model. In this model, text documents are represented by the use of interval valued symbolic features. Harish et al., [20] extended work presented in [19] by applying adaptive FCM. In this method, to preserve the intra class variation multiple clusters are created for each class by using an adaptive FCM algorithm. Unfortunately, FCM is effective only for linear data. The important variant of FCM is kernel FCM (KFCM) and Multiple Kernel FCM (MKFCM) [21]. These variants are widely used for clustering the non linear data.

The result of KFCM depends on the selection of right kernel function. Unfortunately, for many applications, selection of right kernel function is not easy. This problem is overcome by using Multiple Kernel FCM (MKFCM) [22]. Multiple Kernel FCM is based on the KFCM, which uses the composite kernel function. MKFCM gives the flexibility in selection of kernel functions. Addition to flexibility it also offers to combine the different information from multiple heterogeneous or homogeneous sources in kernel space. Huang et al., [22] applied Multiple kernel Fuzzy clustering for text clustering. This method uses four kernel function (i.e. Euclidean distance, Cosine similarity, Jaccard coefficient, Pearson correlation coefficient) to calculate pairwise distance between two document. In this paper, we proposed a novel method for representing text documents based on clustering of term frequency vector. In order to cluster the term frequency vectors, we make use of Multiple Kernel FCM (MKFCM). After clustering, term frequency vector of each cluster is used to form an interval valued representation (symbolic representation) by the use of mean and standard deviation. Further, interval value features are stored in the knowledge base as a representative of the cluster. In document classification, the features of the test document are compared with corresponding interval value features which stored in the knowledge base. Based on the degree of belongigness, we assign the class label.

The rest of the paper is organized as follows: The proposed method is presented in Sect. 2. Details of dataset used, experimental settings and results are presented in Sect. 3. The paper is concluded along with future works in Sect. 4.

2 Proposed Method

The proposed method has two stages: (i) Multiple Kernel FCM (MKFCM) based representation and (ii) Document Classification.

2.1 Multiple Kernel FCM (MKFCM) Based Representation

In the proposed system, initially documents are represented by term document matrix. To reduce the dimensionality of term document matrix, we employ the regularized locality preserving indexing (RLPI) technique. Unfortunately, the RLPI has considerable intraclass variations. Thus, to overcome this problem, we proposed a Multiple Kernel FCM (MKFCM) clustering based representation method. In the proposed method, we capture the intraclass variations through MKFCM clustering and represent each cluster by an interval valued feature vector. The training documents are clustered using MKFCM algorithm. Let $d_1, d_2, d_3, ..., d_N$ be a set of N training documents and $F_k = f_{k2}, f_{k3}, ..., f_{km}$ be the set of m features of each document. The objective function of the MKFCM algorithm is as follows:

$$J(w, U, V) = \sum_{i=1}^{N} \sum_{c=1}^{C} u_{ic}^{m} \left\| \Phi(d_i) - \Phi(v_i) \right\|^2 \quad (1)$$

where u_{ic} is the membership value of the ith document to the cth cluster. v_c is c^{th} cluster center. Φ is an implicit nonlinear map, and

$$\left\| \Phi(d_i) - \Phi(v_c) \right\|^2 = K_L(d_i, d_i) + K_L(v_c, v_c) - 2K_L(x_i, v_c) \quad (2)$$

where $K_L(d_i, v_c)$is the composite multiple kernel function which is defined as

$$K_L(d_i, v_c) = w_1 K_1(d_i, v_c) + w_2 K_2(d_i, v_c) + w_3 K_3(d_i, v_c) + ... + w_l K_l(d_i, v_c) \quad (3)$$

$w = (w_1, w_1, w_1, ..., w_l)$ is a vector consisting of weights.
subject to $w_1 + w_2 + w_3 + ... + w_l = 1$ and $w_l \geq 0 \; \forall_l$.

In the proposed method we used four kernels to calculate the pairwise distance between document and cluster center. They are Euclidean distance, Cosine Similarity, Jaccard coefficient, Pearson correlation coefficient.

The main objective of the MKFCM is to find the combination of weights w, memberships U and cluster centers V, which minimize the objective function in Eq. 1. To obtain the membership value (u_{ic}), we solve Eq. 1 by using Lagrange multiplier. The membership becomes:

$$u_{ic} = \frac{1}{\sum_{\acute{c}=1}^{C} \left(\frac{D_{ic}^2}{D_{i\acute{c}}^2} \right)^{\frac{1}{m-1}}} \quad (4)$$

where $D_{ic}^2 = \left\| \Phi(d_i) - \Phi(v_c) \right\|^2$

The weights w is obtained by solving the Eq. 1, by applying Lagrange multiplier. The weights w becomes:

$$w_l = \frac{\frac{1}{\beta_l}}{\frac{1}{\beta_1} + \frac{1}{\beta_2} + ... + \frac{1}{\beta_L}} \quad (5)$$

where the coefficient β_l is given by:

$$\beta_l = \sum_{i=1}^{N} \sum_{c=1}^{C} u_{ic}^{m} \alpha_{icl} \quad (6)$$

where the coefficient α_{icl} is given by:

$$\alpha_{icl} = K_l(d_i, d_i) - 2 \sum_{j=1}^{N} u_{jc} K_l(d_i, d_j) + \sum_{j=1}^{N} \sum_{k=1}^{N} u_{jc} u_{kc} K_l(d_j, d_k) \quad (7)$$

The training documents are clustered using MKFCM clustering method. Now we capture the intraclass variations of each feature in the form of interval value. i.e. $[f_{ck}^-, f_{ck}^+]$. where $f_{ck}^- = \mu_{ck} - \sigma_{ck}$ and $f_{ck}^- = \mu_{ck} + \sigma_{ck}$. μ_{ck} is the mean of the kth feature of documents present in c^{th} cluster. σ_{ck} is the standard deviation of the kth feature of documents present in c^{th} cluster. The interval value represent the upper and lower limits of feature value of document cluster. Now, the reference document for a class C_c is formed by representing each feature in the form of an interval value. i.e.

$$RF_c = [f_{c1}^-, f_{c1}^+], [f_{c2}^-, f_{c2}^+], ..., [f_{cm}^-, f_{cm}^+] \tag{8}$$

This interval value features are stored in knowledge base as a representative of the cth cluster. thus, the knowledge base has N number of symbolic vectors representing clusters corresponding to a class.

Algorithm 1. Proposed Method

Data: RLPI features N documents with F_i features, set of kernel function K_l, Number of Clusters C, fuzzification degree (m) and Convergence Criteria (ε)
Result: Symbolic feature vector RF
Initialize membership matrix U
repeat
  Calculate the normalized membership value as: $u_{ic} = \frac{u_{ic}^m}{\sum_{i=1}^{N} u_{ic}}$
  Calculate the α_{icl} coefficient using Eq. 7
  Calculate the β_l coefficient using Eq. 6
  Update the weights w_l using Eq. 5
  Calculate distance as: $D_{ic}^2 = \sum_{l=1}^{L} \alpha_{icl} {w_l}^2$
  Update the membership value using Eq. 4
until $\{U(t) - U(t-1)\} < \varepsilon$;
Calculate μ_{ck} and σ_{ck} of each cluster C_c
Represent each cluster using symbolic vector RF as shown in Eq. 8

2.2 Document Classification

In proposed system, for document classification we considered a test document which is described by a set of m feature values of type crisp. The features of the test document are compared with corresponding interval value features which stored in the knowledge base. The number of feature of a test document, which fall inside the corresponding interval, is defined to be the belongigness. To decide the class label of the test document, we calculate the degree of the belongigness B_c. The degree of belongingness is calculated as follows:

$$B_c = \sum_{k=1}^{m} C\left(f_{tk}, [f_{cm}^-, f_{cm}^+]\right) \tag{9}$$

where

$$C\left(f_{tk}, \left[f_{cm}^{-}, f_{cm}^{+}\right]\right) = \begin{cases} 1 & \text{if } (f_{tm} \geq f_{cm}^{-} \text{ and } f_{tm} \leq f_{cm}^{+}) \\ 0 & \text{otherwise} \end{cases} \tag{10}$$

The feature of the test document falling into the respective feature interval of the reference class contributes a value 1 towards B_c. We compute the B_c value for all clusters of remaining classes and assign the class label to test document for which class has highest B_c.

3 Experimental Setup

3.1 Dataset

For experimentation we have used the classic Reuters 21578 collection as the benchmark dataset. Originally Reuters 21578 contains 21578 documents in 135 categories. However, in our experiment, we discarded those documents with multiple category labels, and selected the largest ten categories. For the smooth conduction of experiments we used ten largest classes in the Reuters 21578 collection with number of documents in the training and test sets as follows: earn (2877 vs 1087), trade (369 vs 119), acquisitions (1650 vs 179), interest (347 vs 131), money-fx (538 vs 179), ship (197 vs 89), grain (433 vs 149), wheat (212 vs 71), crude (389 vs 189), corn (182 vs 56). The second dataset is standard 20 Newsgroup Large. It is one of the standard benchmark dataset used by many text classification research groups. It contains 20000 documents categorized into 20 classes. For our experimentation, we have considered the term document matrix constructed for 20 Newsgroup.

3.2 Experimentation

In this section, we present the results of the experiments conducted to demonstrate the effectiveness of the proposed method on both the datasets viz., Reuters 21578 and 20 Newsgroup. We used the four kernels to calculate the pairwise distance between two documents, they are: Euclidean distance(KFC_{ed}), Cosine similarity (KFC_{cs}), Jaccard Coefficient (KFC_{jc}) and Pearson correlation coefficient (KFC_{pcc}). We set the fuzzification degree (m) to 2 and the Convergence Criteria (ε) as 0.0001. We have compared our classification accuracy achieved by the Symbolic classifier with the other existing Naive Bayes classifier (NB), KNN classifier (KNN) and SVM classifier (SVM). During experimentation, we conducted two different sets of experiments. In the first set of experiments, we used 50 % of the documents of each class of a dataset to create training set and the remaining 50 % of the documents for testing purpose. On the other hand, in the second set of experiments, the numbers of training and testing documents are in the ratio 60:40. Both experiments are repeated 5 times by choosing the training samples randomly. As measures of goodness of the proposed method, we computed percentage of classification accuracy. The average value of classification accuracy of 5 trials is presented in Tables 1 and 2. For both the experiments, we have randomly selected the training documents to create the symbolic feature vectors for each class.

It can be observed from Tables 1 and 2, our proposed method based symbolic representation achieved a better results. The reason behind this is, for given dataset we don't know which kernel will perform well. When we combine the kernels, they contribute more to the clustering and therefore, improve the results.

Table 1. Comparative analysis of the proposed method with other classifiers on Reuters-21578 dataset

		Kernels				
Classifiers	Training vs Testing	KFC_{ed}	KFC_{cs}	KFC_{jc}	KFC_{pcc}	MKFCM
NB	50:50	62.85	66.90	64.55	63.15	68.10
	60:40	63.10	67.50	65.10	64.40	69.55
KNN	50:50	63.55	68.85	64.10	63.50	69.80
	60:40	64.10	69.10	64.65	63.80	70.45
SVM	50:50	64.10	69.00	63.55	64.60	70.15
	60:40	65.55	69.85	65.10	63.55	70.65
Symbolic classifier	50:50	67.40	68.55	64.55	63.40	69.65
	60:40	68.60	68.90	65.10	63.85	71.55

Table 2. Comparative analysis of the proposed method with other classifiers on 20-Newsgroup dataset

		Kernels				
Classifiers	Training vs Testing	KFC_{ed}	KFC_{cs}	KFC_{jc}	KFC_{pcc}	MKFCM
NB	50:50	63.50	68.60	67.20	66.15	69.10
	60:40	64.25	69.35	67.15	67.00	70.20
KNN	50:50	64.50	69.10	63.45	64.25	70.25
	60:40	65.75	70.60	64.20	64.95	71.60
SVM	50:50	65.35	71.00	64.20	63.55	71.95
	60:40	66.10	71.95	64.80	63.90	72.40
Symbolic classifier	50:50	65.20	72.10	65.35	64.35	73.20
	60:40	66.60	73.45	65.60	65.90	76.85

4 Conclusion

In this paper, a new text document representation is presented. A text document is represented by the use of symbolic features. The main contribution of this paper is the introduction of Multiple Kernel FCM (MKFCM) clustering algorithm to form a interval value representation. To corroborate the efficacy of the proposed model, we conducted extensive experimentation on standard text datasets. The experimental results reveal that the symbolic representation using feature clustering techniques achieves better classification accuracy over the existing cluster based symbolic representation approaches.

In future, we are intended to introduce symbolic feature selection methods to further reduce the dimensionality of feature matrix. Our future research work will also emphasize in enhancing the ability and performance of our model by considering other parameters to capture intra class variations effectively.

References

1. Nedungadi, P., Harikumar, H., Ramesh, M.: A high performance hybrid algorithm for text classification. In: 2014 Fifth International Conference on the Applications of Digital Information and Web Technologies (ICADIWT), pp. 118–123. IEEE (2014)
2. Sebastiani, F.: Machine learning in automated text categorization. ACM Comput. Surv. (CSUR) **34**(1), 1–47 (2002)
3. Salton, G., Wong, A., Yang, C.S.: A vector space model for automatic indexing. Commun. ACM **18**(11), 613–620 (1975)
4. Choudhary, B., Bhattacharyya, P.: Text clustering using universal networking language representation. In: The Proceedings of Eleventh International World Wide Web Conference, pp. 1–7 (2002)
5. Hotho, A., Maedche, A., Staab, S.: Ontology-based text document clustering **16**, 48–54 (2002)
6. Cavnar, W.: Using an n-gram-based document representation with a vector processing retrieval model, pp. 269–269. NIST SPECIAL PUBLICATION SP (1995)
7. Milios, E., Zhang, Y., He, B., Dong, L.: Automatic term extraction and document similarity in special text corpora. In: Proceedings of the Sixth Conference of the Pacific Association for Computational Linguistics, pp. 275–284. Citeseer (2003)
8. Deerwester, S.C., Dumais, S.T., Landauer, T.K., Furnas, G.W., Harshman, R.A.: Indexing by latent semantic analysis. JAsIs **41**(6), 391–407 (1990)
9. He, X., Cai, D., Liu, H., Ma, W.Y.: Locality preserving indexing for document representation. In: Proceedings of the 27th Annual International ACM SIGIR Conference on Research and Development in Information Retrieval, pp. 96–103. ACM (2004)
10. Cai, D., He, X., Zhang, W.V., Han, J.: Regularized locality preserving indexing via spectral regression. In: Proceedings of the Sixteenth ACM Conference on Conference on Information and Knowledge Management, pp. 741–750. ACM (2007)
11. Baker, L.D., McCallum, A.K.: Distributional clustering of words for text classification. In: Proceedings of the 21st Annual International ACM SIGIR Conference on Research and Development in Information Retrieval, pp. 96–103. ACM (1998)
12. Bekkerman, R., El-Yaniv, R., Tishby, N., Winter, Y.: Distributional word clusters vs. words for text categorization. J. Mach. Learn. Res. **3**, 1183–1208 (2003)
13. Dhillon, I.S., Mallela, S., Kumar, R.: A divisive information theoretic feature clustering algorithm for text classification. J. Mach. Learn. Res. **3**, 1265–1287 (2003)
14. Zadeh, L.A.: Similarity relations and fuzzy orderings. Inf. Sci. **3**(2), 177–200 (1971)

15. Anilkumarreddy, T., Madhukumar, B., Chandrakumar, K.: Classification of text using fuzzy based incremental feature clustering algorithm. Int. J. Adv. Res. Comput. Eng. Technol. **1**(5), 313–318 (2012)
16. Jiang, J.Y., Liou, R.J., Lee, S.J.: A fuzzy self-constructing feature clustering algorithm for text classification. IEEE Trans. Knowl. Data Eng. **23**(3), 335–349 (2011)
17. Puri, S.: A fuzzy similarity based concept mining model for text classification. Int. J. Adv. Comput. Sci. Appl. **2**(11), 115–121 (2012)
18. Carvalho, F.D.A.: Fuzzy c-means clustering methods for symbolic interval data. Pattern Recogn. Lett. **28**(4), 423–437 (2007)
19. Guru, D.S., Harish, B.S., Manjunath, S.: Symbolic representation of text documents. In: Proceedings of the Third Annual ACM Bangalore Conference, pp. 1–8. ACM (2010)
20. Harish, B.S., Prasad, B., Udayasri, B.: Classification of text documents using adaptive fuzzy c-means clustering. In: Thampi, S.M., Abraham, A., Pal, S.K., Rodriguez, J.M.C. (eds.) Recent Advances in Intelligent Informatics. AISC, vol. 235, pp. 205–214. Springer, Heidelberg (2014)
21. Müller, K.R., Mika, S., Rätsch, G., Tsuda, K., Schölkopf, B.: An introduction to kernel-based learning algorithms. IEEE Trans. Neural Netw. **12**(2), 181–201 (2001)
22. Huang, H.C., Chuang, Y.Y., Chen, C.S.: Multiple kernel fuzzy clustering. IEEE Trans. Fuzzy Syst. **20**(1), 120–134 (2012)

GIST Descriptors for Sign Language Recognition: An Approach Based on Symbolic Representation

H.S. Nagendraswamy, B.M. Chethana Kumara(✉), and R. Lekha Chinmayi

Department of Studies in Computer Science, University of Mysore, Mysore 570006, India
hsnswamy@compsci.uni-mysore.ac.in, {chethanbm.research, lekha.2405}@gmail.com

Abstract. This paper presents an approach for recognizing signs made by hearing impaired people at sentence level. The signs are captured in the form of video and each frame is processed to efficiently extract sign information to model the sign and recognize instances of new test signs. Low-dimensional global "gist" descriptors are used to capture sign information from every frame of a sign video. *K*-means clustering is used to choose fixed number of frames, which are discriminative enough to distinguish between signs. Also, selection of fixed number of frames helps us to deal with unequal number of frames among the instances of same sign due to different signers and reduce the complexity of subsequent processing. Further, we exploit the concept of symbolic data analysis to effectively represent a sign. A fuzzy trapezoidal membership function is used to establish the similarity between test and a reference sign and a nearest neighbour classification technique is used to recognize the given test sign. A considerably large database of signs (UoM-ISL) is created and an extensive experimentation is conducted on this database to study the efficacy of the proposed methodology. The experimental results are found to be encouraging.

Keywords: Gist descriptor · Sign language · Symbolic representation · Video sequence

1 Introduction

Technology is becoming pervasive and is profoundly changing the modern life. It is to increase convenience, decrease fraud, remove middle men, and provide access to information in addition to provide improved conditions of life, comforts and health. However, this progress in technology is not up to that level where the sophisticated set of tools has emerged to support the disadvantaged people towards improving their daily life. The people with disabilities are really in need of the tools to become less dependent to lead an independent life with confidence and privacy.

With this motivating factor of improving the life of people with disabilities, specifically focusing on hearing and speech impaired people, research work can be taken up to come up with solutions towards an assistive interactive tool. The interactive tool is expected to make them confident and capable to interact with the society. Hearing impaired people communicate using signs by moving their hands and gesturing with their faces, which are

R. Prasath et al. (Eds.): MIKE 2015, LNAI 9468, pp. 103–114, 2015.
DOI: 10.1007/978-3-319-26832-3_11

not understood by most of us. Hence, there is a huge communication gap between such disadvantaged people and the others. The hearing people would need an assistance of another person, who can understand a sign language to help them in communicating with hearing impaired people by translating the signs. A professional interpreter is needed especially in a formal setting as in official matters and discussions to bridge the communication gap between a hearing impaired person and others. However, this kind of service not only makes the disabled person dependent on such professionals but it is also rather expensive in addition to not being practical for day-to-day communicative requirements of the people. Therefore, replacement of a trained professional by a software tool for automatic recognition and translation of a sign language into a vocal language would go a long way in helping hearing impaired people to integrate and communicate with the society and also to an extent help them to be less dependent on trained professionals. Such an automated tool would acquire gestures, analyze them, recognize them and then produce equivalent sentences in a vocal language. Similar to a vocal language, sign languages are evolved over time by addition and deletion of signs of words. Signs may be stopped being used because they may become out of fashion or they may gain negative perception. New signs get introduced due to technological and societal changes.

In view of this, since two decades, several researchers have made an attempt to standardize sign languages and a number of models have been devised for recognizing signs. Some of the techniques proposed by the research community, which gained importance due to their performance are Hidden Morkov Models (HMM) and Electromyography (EMG) segmentation [1], Localized contour sequence [17], VU-Kinect block to track 3D objects [17], Leap device based tracking [16], Sensor based glove technique [6, 12, 15, 17, 20, 24, 25, 27, 33], Size function [24], Moment based size function [13], Fourier descriptors [28], Hu moments [14], Principal axes of inertia [10], Conditional Random Field (CRF) model [22], Ichetrichef moments [6]. However, the research works reported for sign language recognition have addressed the task either at finger spelling level [2, 5, 32] or at word level [18, 24, 31]. But, signs made by the hearing impaired people are very abstract and difficult to capture in terms of finger spelling or word level. Hence, there is a need to analyze the signs at simple sentence level and devise a method to capture sign information at sentence level and to provide robust representation for recognition.

With this backdrop, we made an attempt to propose a sign language recognition technique at sentence level in this work. We used GIST descriptor to capture signers hand movements, facial expression and gesture while making a sign. From the literature survey, we also understand that the patterns can be better described in terms of symbolic data rather than crisp data [4]. It has also been proved that the Pattern Recognition methods based on symbolic data [7, 8, 10, 11, 21] have outperformed Pattern Recognition methods based on crisp data. Therefore, we explore the suitability of interval-valued type symbolic data [9, 10] for effective representation of signs incorporating intra-class variations due to different signers and due to different instances.

2 Proposed Method

The proposed sign language recognition technique exploits the concept of GIST features to extract texture information from every frame in the sign video. In order to address the problem of unequal number of frames for a sign due to different signers, it is proposed to find a fixed number of frames for each video through the concept of clustering. The *K*-means clustering algorithm is used to find the fixed number of clusters of frames from the given video. One of the frames in the cluster, which possess more similarity with all the other frames in a cluster, is considered as a representative for that cluster. Thus, every sign is characterized by a fixed number of frames, which are discriminative enough in distinguishing different signs for the purpose of recognition. Feature vectors of different instances of the same sign are aggregated to form interval-valued feature vector. Thus, every sign is represented in terms of interval-valued type symbolic feature vector to make the system more effective and robust. A fuzzy trapezoidal membership function [9, 29] is used to compute the similarity between test sign feature vector and reference sign feature vectors. A nearest neighbor classifier is used to recognize a given test sign as one among the reference sign in the knowledgebase and the associated text is displayed. The following sections elaborate the proposed methodology in detail.

2.1 GIST Feature Extraction

The GIST refers to the meaningful information that an observer can identify from a glimpse at a scene [3, 26]. The gist description usually includes the semantic label of the scene, a few objects and their surface characteristics [26], as well as the spatial layout and the semantic properties related to the function of the scene. Therefore, a model of scene gist includes a description of semantic information that human observers comprehend and infer about the scene [3, 23].

GIST descriptors are used to represent a low-dimensional image that contains enough information to identify the scene in an image. The GIST descriptors introduced in [22] can represent the dominant spatial structure of the scene from a set of perceptual dimensions. The GIST descriptors of an image can be captured by analyzing the spatial frequency and orientation. Intuitively, GIST summarizes the gradient information (Scale and Orientation) for different parts of an image, which provides a rough description (the gist) of the scene.

A GIST descriptor for a given image (frame) is computed as follows:

1. An input image is convolved with 32 Gabor filters at 4 scales and 8 orientations to produce 32 feature maps of the same size of the input image.
2. The obtained feature maps are divided into 16 regions (4×4 grids) and then the feature values within each region are aggregated.
3. The 16 aggregated values of all 32 feature maps are concatenated to produce a 512 (16×32) GIST descriptor.

Low-level features channels of GIST descriptor include intensity (0–255), orientation (0, 45, 90 and 135) and color at several sub-channels. Each Sub-channel is associated with nine pyramidal representations of filter outputs. Feature maps are produced by performing center surround operation between filter outputs at four different scales [30].

Figure 1 shows an instance of GIST feature extraction from a frame of a sign video. Figure 2 shows few example frames with their corresponding feature maps for a ***sign I want Coffee*** and Table 1 presents the GIST features extracted from these frames.

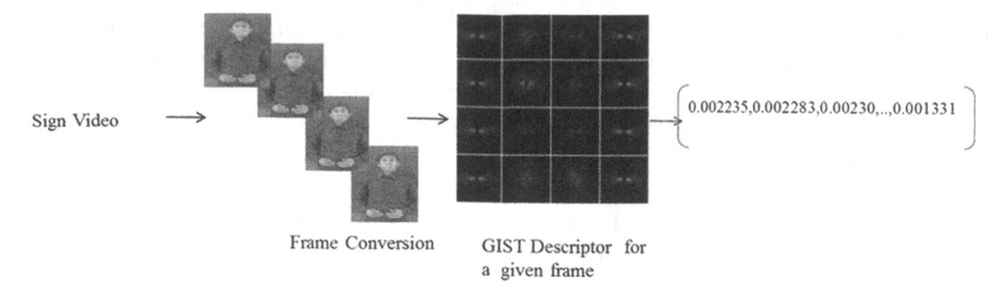

Fig. 1. GIST feature extraction from the frames of a given sign video

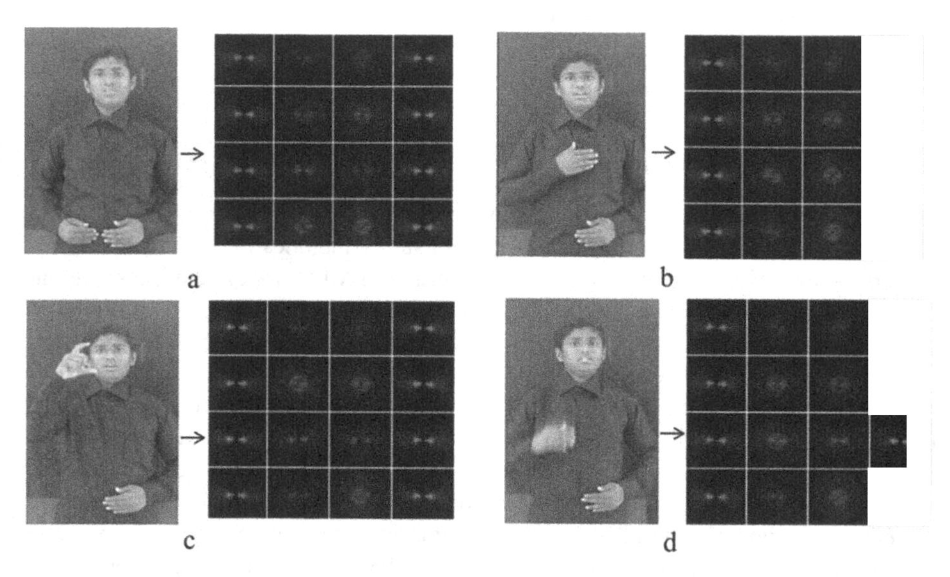

Fig. 2. (a) (b) (c) and (d) shows GIST Descriptor of four different frames of a sign ***I want Coffee***.

Table 1. GIST feature values extracted from four different frames of a sign ***I want Coffee***

Frames of sign	GIST features						
	1	2	3	...	510	511	512
f1	0.00221	0.00223	0.00227	...	0.00134	0.00132	0.00132
f2	0.00220	0.00222	0.00225	...	0.00137	0.00135	0.00134
f3	0.00217	0.00221	0.00223	...	0.00136	0.00128	0.00135
f4	0.00216	0.00218	0.00216	...	0.00134	0.00121	0.00130

2.2 Selection of Key Frames

The same sign made by the same signer at different instances of time or by different signers may contain unequal number of frames due to the speed at which signs are made and captured. Also, the adjacent frames in the sign video may not differ significantly in terms of their content. Therefore, it is more appropriate to eliminate the frames, which do not differ significantly from previous frames and to select a fixed number of frames for a given sign video, which are sufficient enough to capture sign information irrespective of a signer or instance. In order to address the problem of selection of key-frames from sign video, the concept of *K*-means clustering algorithm is studied. The value of *K* is empirically chosen after conducting several experimentations. Once the *K* number of clusters is obtained for each of the signs, each cluster of frames is further processed to select the cluster representative frame. In any cluster, the frame which possesses maximum similarity with all the frames in the cluster is considered as a key-frame for that cluster. Thus, every sign is characterized by *K*- number of key-frames.

2.3 Sign Representation

In addition to the problem of unequal number of frames for the same sign due to different signers or due to same signer at different instances of time, the instances of signs also possess some variations in terms of their content. So, it is recommended to capture such intra-class variations effectively to produce more robust representation for a sign. It is understood from the literature survey that the interval-valued type symbolic representation can be explored to provide robust representation. Therefore, in order to capture intra-class variations among the various instances of the same sign, the feature vectors of the corresponding key-frames of all the instances of the sign are aggregated. The *min* () and *max* () operations are used to find the minimum and maximum feature values to form interval-valued type feature vector. Derivation of interval-valued type symbolic feature vector to represent a sign in the knowledge base is described as follows:

Let $S = \{S_1, S_2, S_3, \ldots, S_n\}$ be the *n* number of signs considered by the system for recognition.
Let $S_i = \{s_i^1, s_i^2, s_i^3 \ldots, s_i^t\}$ be the *t* number of instances of a sign S_i made by the signers at different instances of time.
Let $\left\{F_{i1}^{(k)}, F_{i2}^{(k)}, F_{i3}^{(k)}, \ldots, F_{im}^{(k)}\right\}$ be the *m* number of frames chosen for the video of kth instances of a sign Si, where $F_{ij}^{(k)} = \left\{gf_{1j}^k, gf_{2j}^k, gf_{3j}^k, \ldots, gf_{lj}^k\right\}$ be the feature vector representing j^{th} frame of the kth instance of a sign S_i, and l is the number of features ($l = 512$).

From *t* number of instances of a sign we chose randomly ($t/2$), ($3t/2$) and ($2t/3$) number of instances respectively for (50:50), (60:40) and (40:60) percentages of training and testing. In each experiment, the feature value of the samples chosen randomly for training is aggregated to form interval-valued type symbolic feature vector, which represents the sign in the knowledgebase more effectively.

Let q be the number of training samples randomly selected for a sign S_i and

$$\begin{aligned} F_{ij}^{(1)} &= \left\{ gf_{1j}^{(1)}, gf_{2j}^{(1)}, gf_{3j}^{(1)}, \dots gf_{lj}^{(1)} \right\} \\ F_{ij}^{(2)} &= \left\{ gf_{1j}^{(2)}, gf_{2j}^{(2)}, gf_{3j}^{(2)}, \dots gf_{lj}^{(2)} \right\} \\ &\vdots \\ F_{ij}^{(q)} &= \left\{ gf_{1j}^{(q)}, gf_{2j}^{(q)}, gf_{3j}^{(q)}, \dots gf_{lj}^{(q)} \right\} \end{aligned}$$

be the feature vectors representing the jth frame of all the q training samples respectively. The minimum and maximum values for the first feature is computed as follows

$$gf_{1j}^{-} = gf_{1j}^{(Min)} = Min\left\{ gf_{1j}^{(1)}, gf_{1j}^{(2)}, gf_{1j}^{(3)}, \dots, gf_{1j}^{(q)} \right\}$$

$$gf_{1j}^{+} = gf_{1j}^{(Max)} = Max\left\{ gf_{1j}^{(1)}, gf_{1j}^{(2)}, gf_{1j}^{(3)}, \dots, gf_{1j}^{(q)} \right\}$$

Similarly, we compute the minimum and maximum values of all the features and form an interval as

$$\left[gf_{2j}^{-}, gf_{2j}^{+}\right], \left[gf_{3j}^{-}, gf_{3j}^{+}\right], \dots, \left[gf_{lj}^{-}, gf_{lj}^{+}\right].$$

Thus, the aggregated j^{th} frame of reference feature vector representing a particular sign S_i is given by

$$RF_j^i = \left\{ \left[gf_{1j}^{(i)-}, gf_{1j}^{(i)+}\right], \left[gf_{2j}^{(i)-}, gf_{2j}^{(i)+}\right], \left[gf_{3j}^{(i)-}, gf_{3j}^{(i)+}\right], \dots, \left[gf_{lj}^{(i)-}, gf_{lj}^{(i)+}\right] \right\}$$

Table 2 shows an example j^{th} frame feature vector of four instances of a particular sign ***I want Coffee*** and its interval-type representation.

Table 2. Interval-valued type feature vector representing a particular frame of a sign ***I want Coffee***.

Instances of sign	GIST features						
	1	2	3	…	510	511	512
S_1	0.00238	0.00234	0.00222	…	0.00122	0.00121	0.00123
S_2	0.00238	0.00235	0.00224	…	0.00104	0.00104	0.00122
S_3	0.00240	0.00236	0.00223	…	0.00132	0.00125	0.00122
S_4	0.00243	0.00237	0.00228	…	0.00132	0.00125	0.00127
Interval	[0.00238, 0.00243]	[0.00234, 0.00237]	[0.00222, 0.00228]	…	[0.00104, 0.00132]	[0.00104, 0.00125]	[0.00122, 0.00127]

From Table 2, we can observe that the feature values extracted from the frames of four instances of a sign "***I Want Coffee***" varies significantly. Hence, interval-valued type representation found to be more appropriate to capture such variations and to provide robust representation for a sign.

2.4 Matching and Recognition

In order to recognize a given test sign made by the signer, the video sequence of a test sign is processed to obtain frames, and the features are extracted from each frame as discussed in the previous section. The extracted features are organized in a sequence to represent the test sign. Since the test sign involves only one instance, the test sign is represented in the form of a crisp feature vector.

The task of recognition is accomplished by comparing the test sign feature vector with all the reference sign feature vectors stored in the knowledgebase. A similarity value is computed through this process and the reference sign, which possess maximum similarity value with the test sign is considered and the text associated with this sign is displayed.

The similarity measure [19] is used for the purpose of comparing reference sign feature vector with the test sign feature vector as follows

Let $TF_j = \{tf_{1j}, tf_{2j}, tf_{3j}, \ldots, tf_{lj}\}$ be the crisp feature vector representing the jth frame of a test sign and let $RF_j = \left\{\left[gf_{1j}^-, gf_{1j}^+\right], \left[gf_{2j}^-, gf_{2j}^+\right], \left[gf_{3j}^-, gf_{3j}^+\right], \ldots, \left[gf_{lj}^-, gf_{lj}^+\right]\right\}$ be the interval-valued type symbolic feature vector representing the jth frame of a reference sign.

Similarity between the test and reference sign with respect to d^{th} feature of the jth frame is computed as

$$SIM\left(tf_{dj}, [gf_{dj}^-, gf_{dj}^+]\right) = \begin{cases} 1 & if\ (gf_{dj}^+ \geq tf_{dj} \geq gf_{dj}^-) \\ \frac{(tf_{dj} - r_1)}{(gf_{dj}^- - r_r)} & if\ (gf_{dj}^- \geq tf_{dj} \geq r_1) \\ 0 & if\ ((tf_{dj} > r_2) || (tf_{dj} < r_1)) \\ \frac{(r_2 - tf_{dj})}{(r_2 - gf_{dj}^+)} & if\ (r_2 \geq tf_{dj} \geq gf_{dj}^+) \end{cases} \quad (1)$$

$\delta = gf_{dj}^+ - gf_{dj}^-$, $r_1 = gf_{dj}^- - \delta$, $r_2 = gf_{dj}^+ + \delta$, If the value of tf_{dj} lies within the interval $[gf_{dj}^-, gf_{dj}^+]$ then the similarity value is 1, otherwise the similarity value depends on the extent to which the value tf_{dj} is closer to either the lower bound r_1 or the upper bound r_2. If the value of tf_{dj} does not lie within the interval $[r_1, r_2]$ then the similarity value is zero. The similarity between reference and test sign with respect to the jth frame is given by

$$SIM\left(RF_j, TF_j\right) = \sum_{d=1}^{l} SIM(tf_{dj}, [gf_{dj}^-, gf_{dj}^+]) \quad (2)$$

The total similarity between reference and test sign due to all the frames is computed as

$$SIM\,(RF, TF) = \sum_{J=1}^{L} SIM(RF_J, TF_J) \tag{3}$$

Where L is the number of frames used to represent the reference and test sign, which is 50. The nearest neighbor classification technique is used to recognize the given test sign as one among the known sign stored in the sign knowledge base.

3 Experimentation

We have conducted experiments on our own sign language dataset to validate the feasibility of the proposed methodology. The dataset describes the sentences used by hearing impaired people in their day to day life. We have considered the videos of signs, which are signed by the communication impaired students of different schools of Mysore zone. The dataset contains 680 (11.3 h) sign videos of 17 different signs expressed by four different students with ten instances.

Several experiments are conducted for different percentages of training and testing (60:40, 50:50 and 40:60). We have also repeated the experiments for 50 trials with random sets of training and testing samples. In each trail, performance of the system is measured in terms of F-measure and the average F-measure of all the trials is recorded.

Table 3 gives classwise performance measures in terms of F-measure for respective percentages of training and testing. Figure 3 shows the confusion matrices obtained for 60:40, 50:50 and 40:60 % of training and testing samples respectively for a random run. Table 4 gives the average F-measure for different percentages of training and testing samples. We can observe from the Table 3 that F-measure for different percentages of training and testing samples of class 12 is Zero and it is fully merged with class 1. The signs belonging to class 1 and class 12 are similar in terms of their signing action and hence, it is difficult to differentiate these signs. Therefore, there is a misclassification between these two signs, which can be observed from Table 3 as well as confusion matrices in Fig. 3.

The right number of frames considered for representing a sign plays an important role in improving the performance of the recognition system. In order to study the significance of number of key-frames chosen for a sign through K-means clustering, we have conducted experiments for various values of K (30, 40, …, 90) and we have observed that for 50 frames, we have achieved 78.71 average F-measure, which is significantly high compared to the number of frames 30 and 40. But, when we increase the number of frames to 60, 70, 80 and 90, we have not observed much improvement in the recognition rate. Thus, in our experiments, we have considered 50 frames for every sign. The results obtained are tabulated in Table 5.

Table 3. Classwise performance measures in terms of F-measure obtained by the proposed methodology for one of the random experiments.

Class Index	Training: Testing			Class Index	Training: Testing		
	60:40	50:50	40:60		60:40	50:50	40:60
1	0.64	0.64	0.64	10	0.97	1.00	0.93
2	1.00	1.00	1.00	11	1.00	0.97	0.96
3	1.00	0.97	0.96	12	0.00	0.00	0.00
4	0.97	1.00	0.96	13	0.97	0.97	0.96
5	1.00	0.97	0.96	14	0.97	1.00	0.96
6	0.92	1.00	0.90	15	1.00	0.97	0.94
7	0.92	1.00	1.00	16	0.97	1.00	1.00
8	1.00	0.97	1.00	17	0.90	0.90	0.89
9	0.97	0.95	0.96				

Table 4. Overall recognition performance of the proposed method interms of F-measure.

Sign language data set	Ratio of training and testing	Recognition rate
		Average for overall classes
UoM sign language data	60:40	92.47 ± 0.44
	50:50	92.00 ± 0.44
	40:60	91.34 ± 0.52

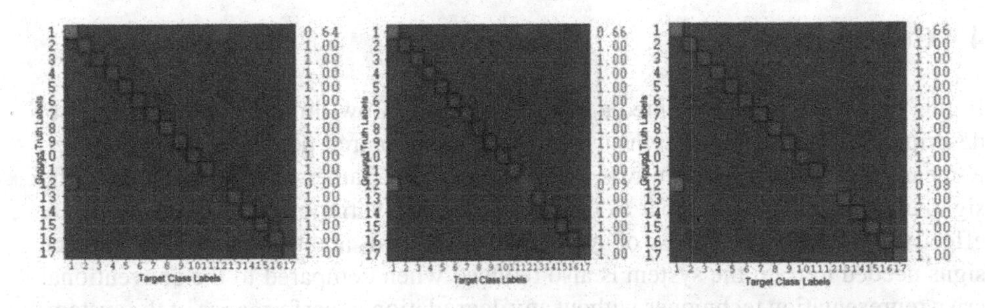

Fig. 3. Confusion matrix for (60:40), (50:50) and (40:60) training and testing samples respectively.

In order to demonstrate the superiority of the proposed method (GIST + Symbolic representation + *K*-NN) with other classifiers, we also conducted experiments using classifiers like SVM (GIST + Crisp representation + SVM) and *K*-NN (GIST + Crisp representation + *K*-NN). The results obtained are presented in Table 6. We can observe from the table that the proposed method (GIST + Symbolic representation + *K*-NN) outperforms the other methods.

Table 5. Average F-measure of the proposed method for different number of frames of sign videos.

No. of frames	Average F-measure (%)
30	72.72
40	73.91
50	**78.71**
60	78.80
70	78.87
80	78.83
90	78.98

Table 6. Results obtained by the different classifiers

Different classifiers	Recognition rate
GIST + Crisp representation + SVM	39.46
GIST + Crisp representation + *K*-NN	59.18
GIST + Symbolic representation + *K*-NN (**Proposed Method**)	**92.47**

4 Conclusion

In this paper, we have presented a method based on Low-dimensional global "gist" descriptors for recognizing signs used by hearing impaired people at sentence level. *K*-means clustering algorithm was used to select key-frames and to characterize the signs in terms of fixed number of frames. Intra-class variations among the signs are effectively captured by the use of symbolic data analysis and the number of reference signs needed to train the system is also reduced when compared to the conventional crisp representation techniques without any degradation in performance of the system. A considerably large database of signs of Indian Sign Language UoM-ISL is created and extensive experiments are conducted on this data set. Experimental results obtained by the proposed sign language recognition system are more encouraging for the data set considered. However, scalability of the proposed system needs to be tested. Minor variations in signing actions but conveying different messages may lead to less between-class variations, which results in misclassification. Hence, cognition of various features in different domain need be addressed to resolve such issues.

Acknowledgement. We would like to thank the students and the teaching staff of Sai Ranga Residential Boy's School for Hearing Impaired, Mysore, and N K Ganpaiah Rotary School for physically challenged, Sakaleshpura, Hassan, Karnataka, INDIA, their immense support in the process of UoM-ISL Sign language dataset creation.

References

1. Al-Ahdal, M., Tahir, N.: Review in sign language recognition systems. In: IEEE Symposium on Computers and Informatics (ISCI), pp. 52–57 (2012)
2. Ghotkar, A.S., Kharate, G.K.: Study of vision based hand gesture recognition using Indian sign language. IJSS Intell. Syst. **7**(1), 96–115 (2014)
3. Oliva, A., Torralba, A.: Building the gist of a scene: the role of global image features in recognition. In: Martinez-Conde, S., Macknik, S.L., Martinez, L.M., Alonso, J.-M., Tse, P.U. (eds.) Progress in Brain Research, vol. 155 (2006). ISSN 0079-6123
4. Bock, H.H., Diday, E.: Analysis of Symbolic Data. Springer, Berlin (2000)
5. Dahmani, D., Larabi, S.: User-independent system for sign language finger spelling recognition. J. Vis. Commun. Image Represent. **25**(5), 1240–1250 (2014)
6. Gourley, C.: Neural network utilizing posture input for sign language recognition. Technical report Computer Vision and Robotics Research Laboratory, University of Tenessee Knoxville, November 1994
7. Gowda, K.C., Diday, E.: Symbolic clustering using a new dissimilarity measure. Pattern Recogn. **24**(6), 567–578 (1991)
8. Gowda, K.C., Ravi, T.V.: Divisive clustering of symbolic objects using the concepts of both similarity and dissimilarity. Pattern Recogn. **28**(8), 1277–1282 (1995)
9. Guru, D.S., Prakash, H.N.: Online signature verification and recognition: an approach based on symbolic representation. IEEE Trans. Pattern Anal. Mach. Intell. **31**(6), 1059–1073 (2009)
10. Guru, D.S., Nagendraswamy, H.S.: Clustering of interval-valued symbolic patterns based on mutual similarity value and the concept of *k*-mutual nearest neighborhood. In: Narayanan, P.J., Nayar, S.K., Shum, H.-Y. (eds.) ACCV 2006. LNCS, vol. 3852, pp. 234–243. Springer, Heidelberg (2006)
11. Guru, D.S., Kiranagi, B.B., Nagabhushan, P.: Multivalued type proximity measure and concept of mutual similarity value useful for clustering symbolic patterns. Pattern Recogn. Lett. **25**(10), 1203–1213 (2004)
12. Handouyahia, M., Zion, D., Wang, S.: Sign language recognition using moment based size functions. In: Vision Interface 1999, Trois-Rivieres, Canada, 19–21 May 1999
13. Handouyahia, M.: Sign Language Recognition using moment based size functions. MSc en Informatique demathematique et d informations, universite de sherbrooke, sherbrooke (1998)
14. Hu, M.: Visual pattern recognition by moment invariants. IRE Trans Inf. Theory **8**, 179 (1962)
15. Kang, S., Nam, M., Rhee, P.: Colour based hand and finger detection technology for user interaction. In: International Conference on Convergence and Hybrid Information Technology, pp. 229–236 (2008)
16. Karthick, P., Prathiba, N., Rekha, V.B., Thanalaxmi, S.: Transforming Indian sign language into text using leap motion. Int. J. Innovative Res. Sci. Eng. Technol. (An ISO 3297: 2007 Certified Organization) **3**(4), 10906–10910 (2014)

17. Kong, W.W., Ranganath, S.: Towards subject independent continuous sign language recognition: a segment and merge approach. Pattern Recogn. **47**, 1294–1308 (2014)
18. Liwicki, S., Everingham, M.: Automatic recognition of finger spelled words in British sign language. In: IEEE Computer Society Conference on Computer Vision and Pattern Recognition Workshops, CVPR Workshops 2009. IEEE (2009)
19. MohanKumar, H.P., Nagendraswamy, H.S.: Change energy image for gait recognition: an approach based on symbolic representation. Int. J. Image Graphics Signal Proc. (IJIGSP) **6**(4), 1–8 (2014)
20. Murakami, K., Taguchi, H.: Gesture recognition using recurrent neural network. In: Actes de CHI 1991 Workshop on User Interface by Hand Gesture, pp 237–242. ACM (1991)
21. Nagendraswamy, H.S., Naresh, Y.G.: Representation and classification of medicinal plants: a symbolic approach based on fuzzy inference technique. In: Proceedings of the Second International Conference on Soft Computing for Problem Solving (SocProS 2012), 28–30 December 2012
22. Oliva, A., Torralba, A.: Modeling the shape of the scene: a holistic representation of the spatial envelope. Int. J. Comput. Vis. **42**(3), 145–175 (2001)
23. Oliva, A., Schyns, P.: Coarse blobs or fine edges? Evidence that information diagnosticity changes the perception of complex visual stimuli. Cogn. Psychol. **34**, 72–107 (1997)
24. Holden, E.-J., Lee, G., Owens, R.: Australian sign language recognition. Mach. Vis. Appl. **16**(5), 312–320 (2005)
25. Harling, P.A.: Gesture input using neural networks. Technical report, University of York, UK (1993)
26. Potter, M.C.: Meaning in visual search. Science **187**(4180), 965–966 (1975)
27. Rensink, R.A.: The dynamic representation of scenes. Vis. Cogn. **7**, 17–42 (2000)
28. Haralick, R.M., Shaipro, L.G.: Local invariant feature detectors: a survey. In: Computer and Robot Vision, vol. 2. Addison-Wesley Publishing Company, Boston (1993)
29. Ross, T.J.: Fuzzy Logic with Engineering Applications. Wiley, New York (2009)
30. Siagian, C., Itti, L.: Comparison of gist models in rapid scene categorization tasks. In: Proceedings of Vision Science Society Annual Meeting (VSS 2008), May 2008
31. Starner, T., Weaver, J., Pentland, A.: Real-time american sign language recognition using desk and wearable computer based video. IEEE Trans. Pattern Anal. Mach. Intell. **20**(12), 1371–1375 (1998)
32. Suraj, M.G., Guru, D.S.: Secondary diagonal FLD for fingerspelling Recognition. In: International Conference on Computing: Theory and Applications, International Conference on Computing: Theory and Applications, ICCTA 2007, pp. 693–697, (2007). doi:10.1109/ICCTA.2007
33. Takahashi, T., Kishino, F.: Hand gesture coding based on experiments using a hand gesture interface device. SIGCHI Bull. **23**, 67–74 (1991)

A Graph Processing Based Approach for Automatic Detection of Semantic Inconsistency Between BPMN Process Model and SBVR Rules

Akanksha Mishra[1] and Ashish Sureka[2](✉)

[1] Indraprastha Institute of Information Technology-Delhi (IIIT-D), New Delhi, India
akanksha1361@iiitd.ac.in

[2] Software Analytics Research Lab (SARL), New Delhi, India
ashish@iiitd.ac.in

Abstract. Business Process Modeling Notation (BPMN) is a technique for graphically drawing and illustrating business processes in diagramtic form. Semantic of Business Vocabulary and Business Rules (SBVR) is a declarative language used to define business vocabulary, rules and policy. Several times inconsistencies occur between BPMN and SBVR as they are independently maintained. Our aim is to investigate techniques for automatically detecting inconsistencies between business process and rules. We present a method for inconsistency detection (between BPMN and SBVR) based on converting SBVR rules to graphical representation and apply sub graph-isomorphism to detect instances of inconsistencies between BPMN and SBVR models. We propose a multi-step process framework for identification of instances of inconsistencies between the two models. We first generate an XML of BPMN diagram and apply parsing and tag extraction. We then apply Stanford NLP Parser to generate parse tree of rules. The detailed information about the parse tree is stored in the form of Typed Dependency which represent grammatical relation between words of a sentence. We utilize the grammatical relation extract triplet (actor-action-object) of a sentence. We find node-induced sub-graph of all possible length of nodes of a graph and apply VF2 Algorithm to detect instances of inconsistency between sub graphs. Finally, we evaluate the proposed research framework by conducting experiments on synthetic dataset to validate the accuracy and effectiveness of our approach.

Keywords: Business Process Modeling · Business rule modeling · Inconsistency detection · Business process intelligence · Graph matching algorithms

1 Research Motivation and Aim

Business process diagrams (such as BPMN) and the business rules (such as SBVR) are widely used in organizations to design and architect an IT system.

R. Prasath et al. (Eds.): MIKE 2015, LNAI 9468, pp. 115–129, 2015.
DOI: 10.1007/978-3-319-26832-3_12

The quality of process diagram and business rule is of up most importance to the organization - particularly their consistency. The notion of consistency refers to the absence of modeling faults within a model. As these are human created artifacts, there is a possibility of inconsistency between the two developed model for the same scenario. Inconsistency comprises of syntactic inconsistency and semantic inconsistency. BPMN and SBVR models are developed using standard editors and hence there is no possibility of syntactic error within a model (the editor automatically checks for syntactic inconsistencies). However, it might be possible that the developed IT system does not show an expected behavior. In this work, our focus is on the detection of instances where the developed IT system does not produce an expected behavior due to semantic inconsistencies between the BPMN and SBVR models resulting in an incorrect implementation.

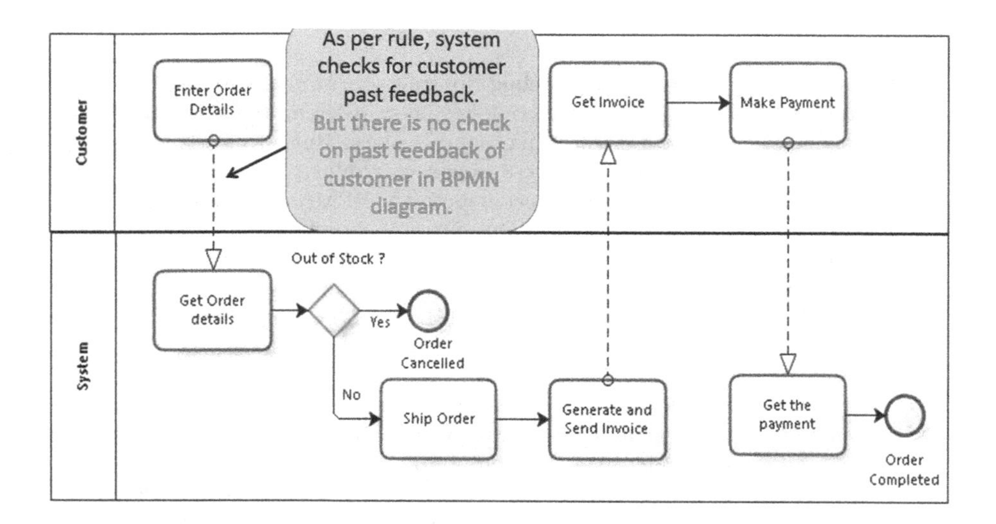

Fig. 1. An instance of BPMN process model for order fulfillment process

We came up with the real life scenarios that demonstrates the instances of semantic inconsistency between BPMN Process Model and SBVR Rules. We demonstrate the instance of semantic inconsistency between BPMN Process Model defined for the Order Fulfillment Process and SBVR rule specified for the same scenario. Figure 1 shows the BPMN Process Model for the Order Fulfillment Process. BPMN Process Model depicts the complete flow of activities that are involved in the order fulfillment process. Figure 2 shows the subset of SBVR rules for the same scenario. In SBVR rule, it is mentioned that it is necessary to check the previous record of the customer before the acceptance of order. But there is no mention about checking how the relationship of customer is with the system. We studied several real-world BPMN models from BPMR org website[1], concept draw[2] and business process incubator[3].

The research aim of the work presented in the paper is the following:

1. To automate the process of detecting semantic inconsistency violation across BPMN Process Model and SBVR rule model.
2. To extract Triplet from a English Statement efficiently and accurately and develop a system to detect faults within models using graph-matching based approaches.
3. To conducted experiment on synthetic dataset based on several real world scenarios for the purpose of validating the proposed approach (performance evaluation).

- Term: system
- Term: customer
- Term: order
- Fact type: system receives order
- Fact type: system cancels order
- Fact type: system checks for customer past feedback
- Fact type: system has bad feedback about customer
- Fact type: system checks for order availability if customer has good feedback
- Rule: It is obligatory that the system receives the order and checks for the customer past feedback.
- Rule: It is necessary that the system checks for the order availability if the customer has good feedback.
- Rule: It is possible that the system has bad feedback about the customer and cancels the order.

Fig. 2. Subset of SBVR rule model for order fulfillment process

2 Related Work and Novel Research Contributions

We conduct a literature review of papers closely related to the work presented in this paper. Integration of process model and rule model has attracted the attention of several researchers. One of the initial research related to this field is in the year 1991 by Krogstie et al. [7]. Muehlen et al. [16] proposed a framework relevant to academics and industry but there was a lack of validation in the framework. Sheen et al. [14] investigated automatic way to transform business rules to business process and analysed the impact of mode driven technologies. Habich et al. [6] proposed an SBVR annotated approach for the modeling of rule and process together. Skersys et al. [13] mapped meta models of BPMN

[1] http://www.bpmn.org/.
[2] http://bit.ly/1ViwCPa.
[3] https://www.businessprocessincubator.com.

with SBVR and related different elements of BPMN with different elements of SBVR. Sharma et al. [11] reviewed business rule meta-model and evolved process template from the same. Skersys et al. [12] proposed a method to extract business vocabularies from BPMN process model and they confirmed that it is accurately possible for semi transformation if the process model is developed following defined modeling practices. However, in case of fully transformation there can be affect on extraction efficiency and success rate. Mickeviciute et al. [8] observed that business process model should be designed in line with the business rules so that it is possible to transform business process models to business rules.

Rusu et al. [10] proposed an approach for the extraction of subject-predicate-object triplet from a sentence using four open-source parsers. Dali et al. [5] proposed machine learning approach to extract subject-predicate-object triplets from English sentences. Zin Thu Muint et al. in [9] observed that other techniques developed are using one or the other parser affecting processing time for each sentences. They proposed an approach with the help of domain specific ontology and machine readable dictionary WordNet.

Ullmann [15] proposed the matching algorithm for the graph isomorphism and sub graph isomorphism. This algorithm is based on backtracking procedure to reduce space search with an effective look ahead function. Another algorithm related to sub graph isomorphism is proposed by Cordella et al. [3]. They introduced a unique representation for matching process known as state space representation. Cordella et al. [4] proposed another work about the performance of the graph matching algorithm popularly known as VF Algorithm [3]. Cordella et al. [1,2] proposed an improved version of graph matching algorithm to solve isomorphism problems on attribute relational graphs. They have focused on the semantic information of the graph and tried to organize search space in such a way to reduce memory requirements for different sized graphs. This algorithm is popularly known as VF2 Algorithm developed to improving the VF Algorithm [3] by reducing spatial complexity to O(N) in best and worst case. In context to existing work and to the best of our knowledge, the study presented in this thesis makes the following novel contributions:

1. There have been substabtial research work done on integration of BPMN and SBVR but none of them have focused on the model quality characteristic i.e. consistency. To the best of our knowledge, the work presented in this paper is the first focusing on the notion of consistency between the BPMN Process Model and SBVR Rule Model (using graph matching based approaches).
2. While there has been work done in the area of triplet (actor-action-object) extraction, our work is the first that used grammatical relations between words stored in the form of Typed Dependency to extract the actor, action and object of a sentence.
3. While there has been work done in the integration of BPMN and SBVR, but there is no dataset available in combination of both. In our work, we have proposed a synthetic dataset for the research problem in integration of BPMN Process Model and SBVR Rule Model.

4. We propose a novel research framework to detect the instances of semantic inconsistency between BPMN Process Model and SBVR Rule Model. We demonstrate the effectiveness of our approach, discuss the strengths and limitations of our tool based on conducting experiments on synthetic dataset.

3 Research Framework and Solution Approach

Figure 3 shows the high level architecture diagram of the proposed solution approach. The proposed approach is a multi-step process primarily consisting 3 phases: conversion of BPMN diagram to a graph, conversion of SBVR rule set to graph and apply graph isomorphism algorithm on the two generated graph.

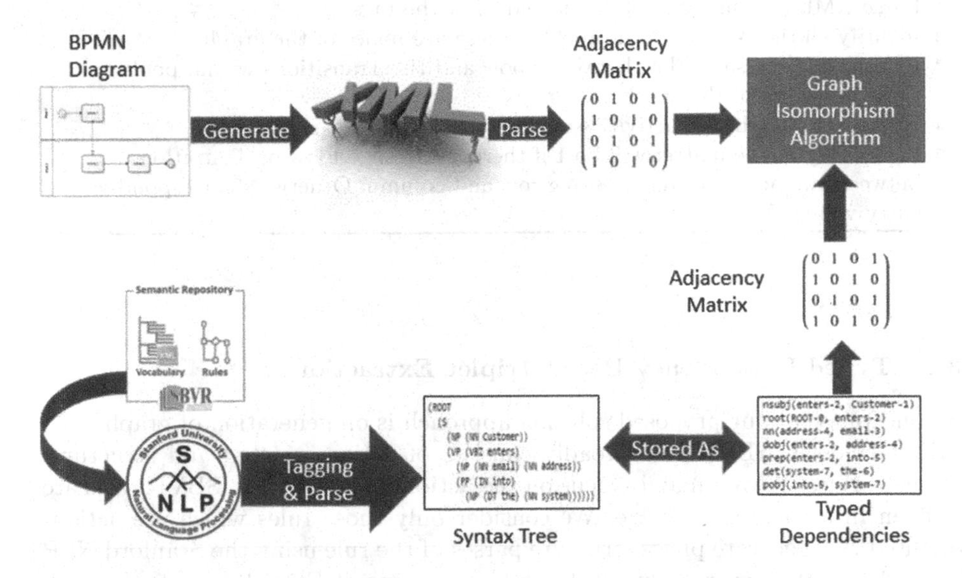

Fig. 3. Research framework

3.1 XML Based BPMN Graph

First phase involves the generation of XML representation of the BPMN diagram using one of the popular modeling tool for BPMN namely Bizagi Modeler[4]. We parse generated XML file and extract values of tags namely Activity, Message Flow and Transitions. We make use of these extracted values to identify nodes and edges of the graph. Activity are considered as nodes of the graph. And, Message Flow and Transitions are considered as edges of the graph.

The pseudo-code for XML based BPMN Graph algorithm is shown in Algorithm 1. The input to the Algorithm is an XML file containing complete

[4] http://www.bizagi.com/.

information about BPMN diagram. This XML file comprises of various tags like Pools, Lanes, Activity, Message Flow, Associations, Artifacts, WorkFlow-Processes, Transition, and many more. Each tag comprises of several attributes like Name, Id, From (optional) and To (optional). We focus only on the tags namely Activity, Message Flow and Transition. The algorithm returns nodes and edges of the graph in the form of adjacency matrix which is the representation of the graph.

Algorithm 1. XML based Adjacency Matrix

Data: XML of BPMN Diagram
Result: Adjacency Matrix
1 Parse XML file and Extract Name and Id of the tags
2 Identify all the Activity in a process - these are nodes of the graph
3 Identify the Message Flow between pools and the Transitions within pools - these are edges of the graph
4 Adjacency Matrix has activities as the nodes of the matrix
5 Each entry in the matrix will be 1 if there is Message Flow or Transitions between two activities representing row and column. Otherwise corresponding entry will be 0.

3.2 Typed Dependency Based Triplet Extraction

Second phase of our proposed solution approach is on generation of graph from SBVR rule set. SBVR rule broadly consists of action-oriented and structure-oriented rules. We are mainly focus on the action-oriented rules. SBVR rules are written in a natural language. We consider only those rules which are action-oriented. We generate phase structure parses of the rule using the Stanford NLP parser[5]. We then extract typed dependency parses of English sentences which represent the dependency between individual words of a sentence. Also, a typed dependency parse labels these dependencies with the grammatical relations. We extract triplet (Subject-Verb-Object) from a rule using the mentioned grammatical relation. Identification of rule as action-oriented or structure-oriented is done on the basis of whether the verb used in a rule is transitive or intransitive. This phase of our solution approach consists of further three sub phases:

1. Triplet Extraction from sentences based on Typed Dependency
2. Classification of rule as action-oriented or structure-oriented based on Triplet Extracted
3. Generation of Graph using action-oriented rules.

First sub-phase of second phase involves the generation of parse tree of rule. The information of parse tree is stored in the form of Typed Dependency which

[5] http://nlp.stanford.edu:8080/parser/.

represents grammatical relations between words of a sentence. The pseudo-code for Triplet Extraction from a rule based on Typed Dependency of Stanford NLP Parser is shown in Algorithm 2. The input to the algorithm is the file comprising of rules on a scenario. The algorithm returns the triplet (Actor-Action-Object) of each rule. Figure 4 depicts how the Typed Dependency are generated for an English Sentence. As shown in the Fig. 4, there is a rule written in Natural Language. We used Stanford Parser, a Natural Language Parser API to tag English Sentence, then generate a parse tree and that parse tree information is stored in the form of Typed Dependency which represent grammatical relations between two words of a English Sentence.

Second sub-phase of second phase involves the classification of a rule as action-oriented or structure-oriented on the basis of the verb used in a rule is transitive or intransitive. The pseudo-code for the classification of rule is shown in Algorithm 3. The input to the algorithm is the triplet of a rule. The algorithm returns triplet of rules which are action-oriented.

Third sub-phase of second phase involves the generation of a graph considering action-oriented rules. Actor and Object are considered as nodes of the graph. Action represent the edge from the actor to the object. We are making a list of edges in the form of A,B where A and B represents Actor or Object. Thus, we obtained adjacency list of SBVR rule-set which is the representation of the graph.

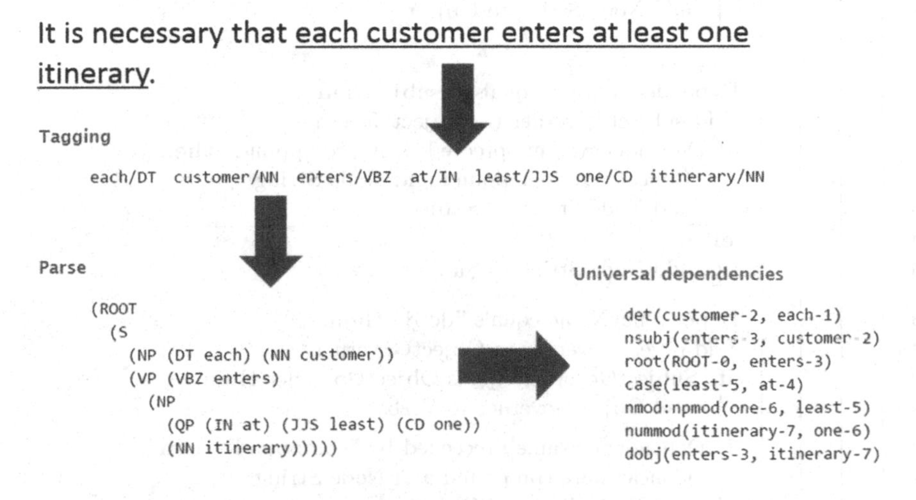

Fig. 4. Generation of typed dependency for a english sentence

3.3 Subgraph Graph Algorithm: VF2 Algorithm

Third and final phase of our solution approach focuses on the use of graph isomorphism algorithm to compare two graphs generated by the initial two phases. VF2 Algorithm is one of the popular and most widely used sub graph-graph

Algorithm 2. Typed Dependency based Triplet Extraction

```
   Data: Rule written in Natural Language
   Result: Triplet Extraction from Sentences
1  foreach Rule in File do
2      foreach Object in List do
3          if DependencyName not equals "mark" then
4              if DependencyName equals "root" then
5                  add Node String to Verb
6              if DependencyName equals "nsubj" or "nsubjpass" then
7                  if DependencyName preceeded by "compound" then
8                      Concatenate compound and Node String
9                      add Node String to Subject
10                 else
11                     add Node String to Subject
12             if DependencyName equals "dobj" then
13                 if DependencyName preceeded by "compound" then
14                     Concatenate compound and Node String
15                     add Node String to Object
16                 else
17                     add Node String to Object
18         else
19             if DependencyName equals "nsubj" then
20                 add subject governer to SubjectGoverner
21                 if DependencyName preceeded by "compound" then
22                     Concatenate compound and Node String
23                     add Node String to Subject
24                 else
25                     add Node String to Subject
26             if DependencyName equals "dobj" then
27                 add object governer to ObjectGoverner
28                 if SubjectGoverner equals ObjectGoverner then
29                     add ObjectGoverner to Verb
30                 if DependencyName preceeded by "compound" then
31                     Concatenate compound and Node String
32                     add Node String to Object
33                 else
34                     add Node String to Object
```

Algorithm 3. Triplet based Rule Classification

```
   Data: Actor-Action-Object of a rule
   Result: Classification of a rule
 1 Read a file
 2 foreach Rule in File do
 3     Create an arraylist for an Actor, an Object and an Action
 4     if Actor != Null and Verb != NULL and Object != NULL then
 5         add Actor, Verb and Object to arraylist
 6         Transitive Verb
 7         Rule participate in Business Process
 8     else
 9         Intransitive Verb
10         Rule do not participate in Business Process
```

isomorphism algorithm that is used to compare sub graph of a graph with the another graph to identify whether they are isomorphic or sub isomorphic to each other. We generate all possible sub graph of both the graphs and then comparing them to identify if the subset of business process diagram is isomorphic or sub isomorphic to the subset of business rule. We then identify which part of BPMN diagram is isomorphic to which part of SBVR rule.

The pseudo-code of VF2 Matching Algorithm [1] is shown in Algorithm 4. The input to the algorithm is the initial state and intermediate state. Algorithm returns the mapping between two graphs.

Algorithm 4. PROCEDURE Match(s)

```
   Data: an intermediate state s; the initial state s0 has M(s0) = φ
   Result: the mappings between the two graphs
 1 if M(s) covers all nodes of G2 then
 2     OUTPUT M(s)
 3 else
 4     Compute the set P(s) of the pairs candidate for inclusion in M(s)
 5     foreach p in P(s) do
 6         if the feasibility rules succeed for the inclusion of p in M(s) then
 7             Compute the state s' obtained by adding p to M(s)
 8             CALL Match(s')

 9     Restore Data Strucutres
10 END PROCEDURE Match
```

The pseudo-code for finding sub isomorphic pairs of mapping of nodes of one sub graph to another sub graph is shown in Algorithm 5. The input to the algorithm is the two adjacency lists generated by the intial two phases. Algorithm returns the isomorphic mapping pairs of sub graphs.

Algorithm 5. Subgraph-Subgraph Mapping Algorithm

Data: Two Adjacency Lists
Result: mapping between two graph which are isomorphic
1 Create a Graph $G1$ by reading file containing adjacency list of first graph
2 Create a Graph $G2$ by reading file containing adjacency list of second graph
3 Find node-induced subgraph of all possible combination of length of nodes of $G1$
4 **foreach** *subgraph* G' *of* $G1$ **do**
5 Find subgraph of all possible combination of length of nodes of $G2$
6 **foreach** *subgraph* G'' *of* $G2$ **do**
7 Perform Matching between subgraph G' and G''
8 **if** *(G',G'') are isomorphic* **then**
9 Print Mapping between (G',G'')

4 Performance Evaluation and Results

We create an experimental test-bed which constitutes small synthetic dataset depicting multiple scenarios covering various aspect of daily life. We cover almost all sectors ranging from business to education, financial to health care and many more. We design dataset for both the consistent and inconsistent scenarios. Table 1 shows the synthetic dataset for the research problem in integration of BPMN Process Model and SBVR Rule Model. Table 1 lists all the different types of scenario depicting consistent or inconsistent ones.

Table 1. Collection of synthetic data set of consistent and inconsistent scenarios

SNo.	Scenario	Consistent Scenario
1	Travel Reservation	No
2	Cab Booking	Yes
3	Employee Reimbursement Request	Yes
4	Order Fulfillment	No

4.1 Consistent Scenario: Cab Booking Result

We perform experiment on cab booking scenario to detect whether it is possible to identify consistency or inconsistency between BPMN and SBVR. This scenario depicts the sequences of activities that take place during the cab booking process. It shows the communication between customer, travel agent and cab driver. Booking is initiated by the customer by requesting for a cab. Travel agent checks the availability of the cab and inform the availability status to the customer. If the cab is available and customer confirms the cab booking, then travel agent assigns the cab driver for the customer. Cab driver picks up the

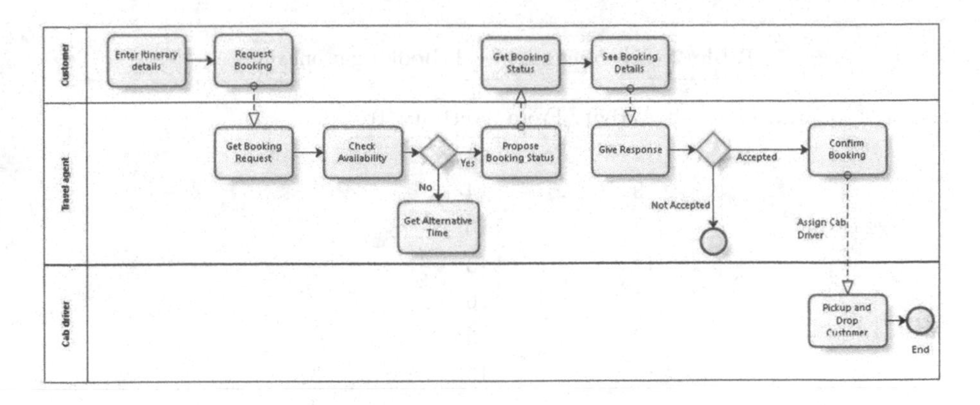

Fig. 5. BPMN diagram for cab booking

customer and drops him or her at scheduled destination. We designed BPMN diagram considering defined modeling practices using popular BPMN Editor[6]. Figure 5 depicts the BPMN process model for the cab booking scenario. We use Bizagi Process Modeler to generate XPDL of the BPMN diagram. We extract tags namely Activity, Message Flow and Transition. We assign numbering to Activities. Again, we assigned numbering to the list of message flow and transition as shown in Table 2. We construct graph for the BPMN diagram using a graph visualization software, Graphviz[7]. The input to the Graphviz is in the form of adjacency list. Figure 6 shows the graph generated for BPMN diagram.

We develop SBVR Business Vocabulary and Business Rules for the cab booking scenario using SBVR Lab 2.0[8]. We extract triplet (Actor-Action-Object) from a English Sentence using Stanford NLP Parser. We consider only those triplets which participate in the business process. Extracted triplets are shown in the Table 3. We construct graph for the SBVR rules using a graph visualization software, Graphviz. The input to the Graphviz is the extracted triplet. Actor and Object form the nodes of the graph. Action represent the relationship between actor and object of the English sentence. Figure 7 depicts the graph generated for the SBVR rules.

We apply VF2 sub-graph isomorphism on the two graphs generated in the Figs. 6 and 7. We use Python NetworkX Library[9] for the subgraph isomorphism algorithm. We identified subgraphs of BPMN graph of size of length of SBVR graph or less than that. We identified that if we take length of sub graph as of 2 less than number of nodes of SBVR Graph then we get two sub isomorphic graphs.

---

[6] http://www.bizagi.com/.
[7] http://www.graphviz.org/.
[8] http://www.sbvr.co/.
[9] https://networkx.github.io/.

Table 2. List of edges of cab booking scenario

Activity From	Activity To
12	14
3	4
4	10
11	13
5	6
2	5
14	15
8	3
6	7
7	9
10	11
1	2

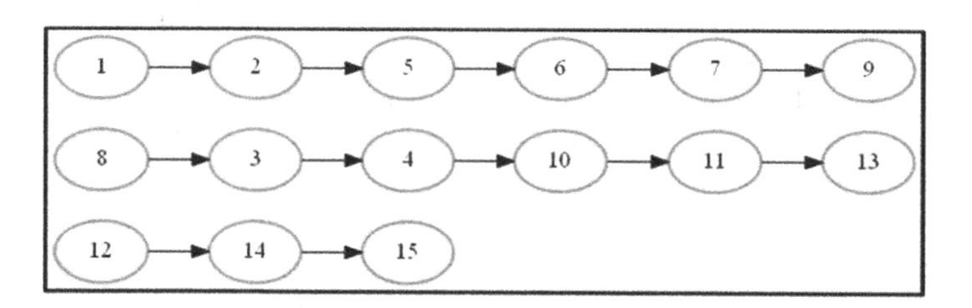

Fig. 6. BPMN graph for cab booking

Table 3. List of triplet extracted from business rules for cab booking

Actor	Action	Object
Customer	Enters	Itinerary details
Customer	Request	Cab booking
Travel agent	Checks	Availability
Travel agent	Provides	Information
Customer	Confirms	Cab booking
Customer	Cancels	Cab booking
Travel agent	Assigns	Cab driver
Customer	Confirms	Cab booking
Cab driver	Picks	Customer

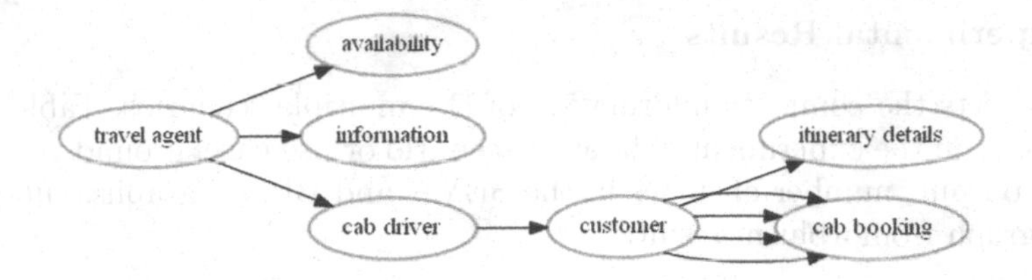

Fig. 7. SBVR graph for cab booking

4.2 Inconsistent Scenario: Order Fulfillment

We perform another experiment on an inconsistent scenario of order fulfillment. We designed BPMN diagram for the process of taking an order to the product delivery in the organization. We designed this BPMN diagram and generated XPDL of the BPMN diagram using one of the popular process modeler, Bizagi Process Modeler. We extracted 'id' attribute of the activity tag and they will be nodes of the BPMN graph. Also, we extracted 'from' and 'to' attribute of the MessageFlow and Transition tags and these will be edges of the BPMN Graph. As these edges will be directional from 'from' attribute to 'to' attribute of Message Flow and Transition tags. Activity will be numbered and accordingly message flow and transition will be numbered.

We designed SBVR Business Vocabulary and Business Rules for the Order Fulfillment process using SBVR Lab 2.0[10]. Business Vocabulary and Rules are written in Natural Language. We then parsed business rules using Stanford NLP parser. We extracted triplet (Subject-Verb-Object) of the business rules and categorized rules as action-oriented or structure-oriented. We pruned multiple edges of the BPMN and SBVR graph. We also considered both the graphs as undirected graph. We then applied VF2 subgraph isomorphism algorithm on the resultant BPMN graph and SBVR graph. We obtained two isomorphic graphs of size two less than length of SBVR graph.

Table 4. Collection of synthetic data set of consistent and inconsistent scenarios

Scenario	Expected	BPMN Nodes(G1)	SBVR Nodes(G2)	Subgraph of BPMN(G3)	subisomorphic (G2,G3)
Travel Reservation	Inconsistent	17	10	7	True
Cab Booking	Consistent	15	7	5	True
Employee Reimbursement	Consistent	13	5	3	True
Order Fulfillment	Inconsistent	10	7	4	True

[10] http://www.sbvr.co/.

4.3 Experimental Results

Table 4 depicts the complete information of the multiple scenarios. Table 4 provides details on the experimental dataset: scenario or use-case, ground-truth and expected output, number of nodes in the SBVR and BPMN graphs, sub-graph and sub-graph isomorphism result.

5 Conclusion

We present a novel approach for detection of instances of semantic inconsistency between BPMN Process Model and SBVR Rule Model developed for an IT system in an organization. The key components of the proposed framework are: extraction of tags from generated XML of BPMN diagram, triplet (actor-action-object) extraction of action-oriented rules from SBVR rule-set using grammatical relations between words represented by Typed Dependency of parse tree using Stanford NLP parser, and find node-induced sub graph of all possible length of nodes of a graph and apply VF2 Algorithm to detect instances of semantic inconsistency between sub graphs. We create a synthetic dataset comprises of multiple number of consistent and inconsistent scenarios. We conducted a series of experiments on synthetic dataset to demonstrate that the proposed approach is effective.

References

1. Cordella, L.P., Foggia, P., Sansone, C., Vento, M.: An improved algorithm for matching large graphs. In: 3rd IAPR-TC15 Workshop on Graph-based Representations in Pattern Recognition, Cuen, pp. 149–159 (2001)
2. Cordella, L., Foggia, P., Sansone, C., Vento, M.: A (sub)graph isomorphism algorithm for matching large graphs. Pattern Anal. Mach. Intell. IEEE Trans. **26**(10), 1367–1372 (2004)
3. Cordella, L.P., Foggia, P., Sansone, C., Vento, M.: Subgraph transformations for the inexact matching of attributed relational graphs. Springer, Heidelberg (1998)
4. Cordella, L.P., Foggia, P., Sansone, C., Vento, M.: Performance evaluation of the vf graph matching algorithm. In: Proceedings of the International Conference on Image Analysis and Processing, 1999, pp. 1172–1177. IEEE (1999)
5. Dali, L., Fortuna, B.: Triplet extraction from sentences using svm. In: Proceedings of SiKDD (2008)
6. Habich, D., Richly, S., Demuth, B., Gietl, F., Spilke, J., Lehner, W., Assmann, U.: Joining business rules and business processes. In: Proceedings of IT (2010)
7. Krogstie, J., McBrien, P., Owens, R., Seltveit, A.H.: Information systems development using a combination of process and rule based approaches. In: Andersen, R., Bubenko, J.A., Solvberg, A. (eds.) Advanced Information Systems Engineering. LNCS, vol. 498, pp. 319–335. Springer, Heidelberg (1991)
8. Mickeviciute, E., Nemuraite, L., Butleris, R.: Applying SBVR business vocabulary and business rules for creating BPMN process models. In: Abramowicz, W., Kokkinaki, A. (eds.) BIS 2014 Workshops. LNBIP, vol. 183, pp. 105–116. Springer, Heidelberg (2014)

9. Myint, Z.T.T., Win, K.K.: Triple patterns extraction for accessing data on ontology. Int. J. Future Comput. Commun. **3**(1), 40 (2014)
10. Rusu, D., Dali, L., Fortuna, B., Grobelnik, M., Mladenic, D.: Triplet extraction from sentences. In: Proceedings of the 10th International Multiconference Information Society-IS, pp. 8–12 (2007)
11. Sharma, D.K., Prakash, N., Sharma, H., Singh, D.: Automatic construction of process template from business rule. In: 2014 Seventh International Conference on Contemporary Computing (IC3), pp. 419–424. IEEE (2014)
12. Skersys, T., Kapocius, K., Butleris, R., Danikauskas, T.: Extracting business vocabularies from business process models: Sbvr and bpnm standards-based approach. Comput. Sci. Inf. Syst. **11**(4), 1515–1535 (2014)
13. Skersys, T., Tutkute, L., Butleris, R., Butkiene, R.: Extending bpmn business process model with sbvr business vocabulary and rules. Inf. Technol. Control **41**(4), 356–367 (2012)
14. Steen, B., Pires, L.F., Iacob, M.E.: Automatic generation of optimal business processes from business rules. In: 2010 14th IEEE International Enterprise Distributed Object Computing Conference Workshops (EDOCW), pp. 117–126. IEEE (2010)
15. Ullmann, J.R.: An algorithm for subgraph isomorphism. J. ACM **23**, 31–42 (1976)
16. Zur Muehlen, M., Indulska, M., Kittel, K.: Towards integrated modeling of business processes and business rules. In: ACIS 2008 Proceedings, p. 108 (2008)

An Improved Intrusion Detection System Based on a Two Stage Alarm Correlation to Identify Outliers and False Alerts

Fatma Hachmi[1(✉)] and Mohamed Limam[2]

[1] ISG, University of Tunis, Tunis, Tunisia
hachmi.fatma@gmail.com
[2] University of Tunis, Tunisia and Dhofar university, Salalah, Oman

Abstract. To ensure the protection of computer networks from attacks, an intrusion detection system (IDS) should be included in the security architecture. Despite the detection of intrusions is the ultimate goal, IDSs generate a huge amount of false alerts which cannot be properly managed by the administrator, along with some noisy alerts or outliers. Many research works were conducted to improve IDS accuracy by reducing the rate of false alerts and eliminating outliers. In this paper, we propose a two-stage process to detect false alerts and outliers. In the first stage, we remove outliers from the set of meta-alerts using the best outliers detection method after evaluating the most cited ones in the literature. In the last stage, we propose a binary classification algorithm to classify meta-alerts whether as false alerts or real attacks. Experimental results show that our proposed process outperforms concurrent methods by considerably reducing the rate of false alerts and outliers.

Keywords: Clustering · Binary classification · False positives · Intrusion detection systems · Outliers

1 Introduction

The ultimate goal of computer security is to protect networks against criminal activities such as violation of privacy, corruption of data and access to unauthorized information. In fact, computers are in need for powerful security technologies to secure the information system and to prevent hackers from destroying it. In fact, intrusion detection systems (IDS)s are considered as essential components for the protection of computer networks. Therefore, their accuracy depends on their ability to detect real threats on the network and to alarm the administrator about them. Despite the major role of an IDS as a component of the security infrastructure, it is still far from perfection since it tends to generate a lot of noisy alerts and a high rate of false alerts. A false alerts is defined as a signal triggered by an IDS reporting an attack but in reality it is just a normal network traffic. An outlier is defined as a noisy and inconsistent data point characterized by its dissimilarity from other observations in a given data set.

R. Prasath et al. (Eds.): MIKE 2015, LNAI 9468, pp. 130–139, 2015.
DOI: 10.1007/978-3-319-26832-3_13

To enhance the accuracy of IDSs, we propose a two-stage process to eliminate outliers and to reduce the rate of false alerts. First, we clean the set of alerts by removing noisy meta-alerts or outliers. In fact, we evaluate the most cited outliers detection methods and then we select the best one to be integrated in our proposed process. Four outliers detection approaches are commonly used in the literature: the clustering-based, the density-based, the distance-based and the distribution-based approaches. In the second stage, we begin by clustering the set of cleaned alerts into a set of meta-alerts based on several attributes extracted from the alert database. Then, a binary classification algorithm (BCA) is proposed to identify false positives (FPs). The remainder of this paper is organized as follows. Section 2 gives an overview of the related works. Section 3 describes the proposed method for outliers detection and FPs reduction. Experimental results and performance comparisons are given in Sect. 4. Finally, conclusion and future work are given in Sect. 5.

2 Related Works

Many research works were conducted to reduce the rate of false alerts generated by intrusion detection systems. Reference [1] introduced a new alert correlation technique to extract attack strategies based on the causal relationship between alarms. This technique is based on two approaches: Multilayer perceptron (MLP) and support vector machine (SVM). In the experimental study they used DARPA 2000 to test their proposed technique. In fact MLP and SVM algorithm require a training set to build a model that will be used to predict the right decision for a new observation. Using elementary alerts from the training set is not valuable since one event may produce multiple alert signature. In order to improve the accuracy of an IDS, [2] proposed a two-stage alarm correlation and filtering technique. The first stage aims to classify the generated alerts based on the similarity of some attributes to form partitions of alerts using SOM with k-means algorithm. The second one aims to classify the meta-alerts created in the first stage into two clusters: true alarms and false ones. The binary classification is based on the collection of seven features extracted from the set of meta-alerts using SOM with k-means to cluster the input set. In the experimental study, they used DARPA 1999 to test their proposed technique. Unfortunately the use of SOM with k-means in the second stage is not efficient since the administrator should manually determine which cluster contains the true alarms by examining the two attributes: alarm frequency and time interval. Reference [3] introduced a network detection technique based on neural networks. First the detected network traffic is clustered using SOM. The partitions are displayed to the administrator so he can recognize attacks clusters. Then, MLP algorithm is applied to the output of SOM to efficiently model the network traffic. In the experimental study they used DARPA 1999 training data set along with three types of attacks to test their proposed technique: denial of service attacks, portscans attacks, and distributed denial of service attacks. Unfortunately, this technique requires a well experienced network administrator to recognize the attacks. Moreover, this technique represents a real challenge if other attacks are added. Reference [4] developed a network detector based on SOM algorithm called NSOM. It allows the classification of real time network traffic. Once the detected alerts are normalized and the classification features are extracted, alarms that represent a normal behaviour will be grouped in one or more clusters and attacks will be placed outside. In fact, results show that NSOM allows the classification of normal traffic and attacks graphically and dynamically. But if there are different types of attacks NSOM will face a serious problem since it has to cluster each type of attack distinctly. Reference [5] introduced the decision support classification (DSC) alert classification. It collects the alerts generated in an attack-free environment. So, all alarms are considered as FPs in this environment and the recorded patterns in this case define the normal behavior and are called patterns of FPs. Then, DSC removes FPs based on these patterns. Reference [6] used a knowledge-based evaluation for the proposed post-processor for IDS alarms. This system uses background information concerning the hosts available in the network and generates a score for each alarm based on the exploited vulnerability. This score measures the importance of each alert. Then, based on the value of score threshold, a binary classifier groups the alert as real attacks or FPs. Reference [7] proposed a correlation framework that reduces the number of processed alerts in the first phases by removing the inconsistent and false alerts. Reference [8] proposed a Beysian network model for classifying the alerts generated by IDSs as attacks or false alarms. Reference [9] used a Genetic Fuzzy Systems within a pairwise learning framework to improve IDSs. Reference [10] developed a novel approach called the cluster center and nearest neighbor (CANN) where two distance measures are computed then summed.

3 The Proposed Method

This work aims to improve the accuracy of IDSs by eliminating inconsistent alerts and false ones. To achieve this goal, we propose a two-stage process that begins with a cleaning the set of alerts to remove outliers. Once only consistent alerts remain, a clustering step is performed by applying k-means algorithm. Then, we propose a BCA that aims to identify FPs by comparing the similarity of meta-alerts with a labeled training set used as a classification model.

3.1 Outliers Detection

Outlier detection methods aim to clean databases from unusual objects. This helps building consistent data sets that can be used to extract knowledge in different domains. Four outliers detection approaches are commonly used in the literature. Distribution-based approach introduced by [11], consists of developing statistical models from a given data set for the normal behavior and then perform statistical tests to decide if an observation belongs to this model or not. This approach assumes that each data set has a distribution. [12] proposed the active outlier method (AO). AO invokes a selective sampling mechanism which is based on active learning. However, this approach is not appropriate in multidimensional scenarios since they are univariate in nature and a prior knowledge about the data distribution is needed. Also, the construction of a probabilistic model based on empirical data is a difficult computational task and the chosen sample is not guaranteed to match the distribution law.

The Clustering-based approach introduced by [13], considers clusters of small sizes as clustered outliers. But if the separation between clusters is large enough, then all clusters are considered as outliers. So, the clustering approach by itself is not sufficient to detect outliers efficiently.

The distance-based approach introduced by [14], considers an object O in a data set as an outlier if there are less than M object, within the distance d from O. The major limitation of this approach is that it is difficult to set the values of M and d. An extended method based on the distance of an observation O from its kth nearest neighbor (KNN) is proposed by [15]. It sorts the top k vectors based on the distance between it and its KNN. Besides, [16] propose an algorithm that computes the outlier factor of each object as the sum of distances from its KNN. For large data sets, KNN-based methods are very time-consuming. Reference [17] introduced neighborhood approximation properties. Reference [18] proposed a new method called neighborhood outlier detection (NED) to detect outliers based on neighborhood rough set. But, the efficiency of this method depends on the appropriate selection of the neighborhood parameters.

The density-based approach originally proposed by [19], gives to each object a factor called the local outlier factor (LOF) to measure the degree of an object being an outlier.

To remove outliers from the set of meta-alerts, we use four methods, one from each approach, AO, k-means, NED and LOF. We evaluate them on different data sets and then based on the overall results of the proposed process, we select the best method in the context of intrusion detection.

3.2 False Positives Reduction

In this final stage, our interest is to reduce the rate of false alarms generated by IDSs. As an input set, we use the cleaned set of alerts which is generated by the previous stage. Then, we reduce the huge number of alerts by clustering the testing sets and finally a BCA is applied to identify the set of FPs.

The Clustering Step. The clustering step aims to reduce the number of generated alerts by correlating the similar ones together. The correlation is based on the similarity-based technique. It is based on maximizing the degree of similarity between objects in the same cluster and minimizing it between clusters. Therefore, the classification is based on the similarity of some selected attributes. Four attributes are used. In fact, IP addresses represent the identity of the hacker in the network and the timestamp define the time frame of a given event while the protocol defines its nature.

- Source IP address
- Destination IP address
- Timestamp
- Protocol

Before clustering the alerts, we normalize all the attributes to ensure having reliable clustering results. As a clustering algorithm, we propose to use k-means algorithm since it is a simple unsupervised learning method. K-means defines k centroids, one for each cluster and then clusters all data into the pre-defined k clusters. The grouping is done by computing the sum of squared Euclidean distances from the mean of each cluster. Moreover, k-means is very appropriate to ensure having reliable results from the next stages of our proposed process since it aims to maximize the distance between the clusters and to minimize the dispersion within them.

Since the detection of FPs requires wide knowledge about the network traffic and expertise in the domain, we use a training set to generate a model for the network traffic. But, the generated alerts are not valuable in the creation of the model since a single event may produce multiple alerts. So, the aforementioned attributes are used to cluster the training set. First, we split the training set on two big pre-clusters, the first one clusters the false alarms and the second one groups real attacks. Second, inside each pre-cluster we apply k-means algorithm to group similar alerts together. Therefore, the training set is transformed from a set of elementary insignificant alerts into a set of labeled clusters.

The Binary Classification Algorithm. The binary classification aims to identify FPs from the set of meta-alerts created by the clustering step. The attributes list is extracted from each consistent cluster. Four attributes are judged useful for the binary classification, namely

- Number of alerts in each cluster
- Signature type
- Protocol number
- Alert priority

Based on these extracted attributes, we propose a BCA to classify each meta-alert from the testing set whether as true alert or a false one. In fact, there is two labeled training clusters. The first one F includes the false meta-alerts and the second one T includes the true meta-alerts. Based on the aforementioned attributes, we compute the Euclidean distance between each meta-alert (MA) from the testing set and the centroid C_1 and C_2 of the two clusters F and T respectively. If the distance between MA and C_1 (Dist(MA_i,C_1)) is lower than MA and C_2 (Dist(MA_i,C_2)), then the probability that MA is a false alerts is high. To ensure that MA is correctly classified as a false alert, we propose to test its similarity with each false meta-alert (MT) inside F. Therefore, if the distance between MA and MT is lower than the maximum distance between false meta-alerts inside F, then MA is a false meta-alert otherwise it is a real attack. The proposed algorithm is detailed as follows.

Algorithm 1. Proposed classification algorithm

Input: meta-alerts of the testing set, training clusters
Output: False alarms

```
Begin
N is the number of meta-alerts
F is the set of training false positives
T is the set of training true alerts
FN : number of clusters inside F
TN : number of clusters inside T
C1 : centroid of the set F
C2 : centroid of the set T
For each meta-alert (MA) i from N
    If Dist(MA_i,C1) ⪯ Dist(MA_i,C2)
        For each meta-alert (MT) j from FN
            If Dist(MA_i,MT_j) ⪯ AVG(Dist(MT_j in F)
            insert MA_i in the cluster F
            FN=FN+1
        End For
        C1 is the new centroid of cluster F
    End IF
End For
End
```

4 Experimental Results

To test the efficiency of the proposed technique, we used a public data set named DARPA 1999, commonly used for the evaluation of computer network sensors. Our experiments are based on the off-line evaluation sets:

- As a training set, we use the first and third weeks of the training data which are attacks free and the second week of the training data which contains a selected subset of attacks. The primary purpose of this training set is the detection of false alerts from the testing set.
- A selected sample from the fourth week is used as our first testing data set in order to evaluate our proposed process.
- A selected sample of the fifth week is used as our second testing data set.

4.1 The First Stage

To remove outliers from the clusters generated by the first stage, we use four methods namely, AO, K-means, NED and LOF.

Tables 1 and 2 give the number of outliers generated by each method for the first and second testing sets respectively.

We create four different subsets for each testing set. In the first one, we remove the outliers detected by LOF and then we apply BCA to evaluate its

Table 1. Number of outliers in testing set 1

Outliers detection methods	Detection rate
LOF	81
k-means	66
AO	52
NED	79

Table 2. Number of outliers in testing set 2

Outliers detection methods	Detection rate
LOF	72
k-means	66
AO	44
NED	81

performance with LOF. The second one includes the consistent alerts after eliminating the outliers detected by NED. In the third one we remove the noisy alerts identified by AO and the final one contains the alerts after removing the outliers defined by K-means. For each testing set, we evaluate the performance of the proposed BCA based on those four subsets and the best outlier detection method is used for our proposed process. The overall process is also compared to the methods SOM with k-means and DSC proposed by [2,5] respectively.

4.2 The Second Stage

The Clustering Step Evaluation. In this step, we follow the approach used by [3] to select the appropriate k value because we don't know the exact number of clusters. The system performs 500 randomised trials and the best classifications are selected based on the minimal sum of squared errors and the highest frequency.

To ensure having the best clustering solution, we test the different values of k generated from the randomized trials until we get the optimal partitioning of the data. The latter is based on several validity measures to test the quality of the partitions. The first validity measure, separation index (SI), determines the average number of data and the square of the minimum distances of the cluster centers. Indeed, a small value of SI indicates an optimal portioning. The second validity measure, Dunns index (DI), is used to identify whether clusters are well separated and compact or not. A big value of DI implies a good clustering. The third validity measure, Xie and Benis index (XB), quantifies the ratio of total variation within cluster and the separation of clusters. To have an optimal number of partitions, the value of XB index should be minimized.

Table 3 values of SI, DI, and XB for different values of k using the two testing sets. As illustrated, we notice that the best solution is when k is equal to 85

Table 3. Clustering evaluation of testing set 1

Testing set	Pairs of clusters	SI	DI	XB
1	k=66	5.6814e-004	0.0149	4.2338
1	k=85	4.5002e-004	0.0274	3.6083
2	k=79	5.8351e-004	0.0236	4.7062
2	k=77	3.6143e-004	0.0285	3.1397

for the first testing set. However, for the second testing set the best clustering solution is provided when k is equal to 77.

The BCA Evaluation. The effectiveness of our proposed technique is evaluated using true positives rate (TPR) which represent the false meta-alerts successfully classified as FPs. TPR is given by:

$$TPR = \frac{number\ of\ detected\ false\ alerts}{Real\ number\ of\ false\ alerts} \tag{1}$$

Table 4. TPR for testing set 1

	TPR
LOF, K-means and BCA	72.5
k-means, Kmeans and BCA	68.9
AO, K-means and BCA	54.8
NED, K-means and BCA	88.7
SOM with K-means	79.3
DSC	75.3

Rates are given in percentages

Table 4 illustrates the TPR for the first testing set respectively. As shown the combination of NED, K-means and BCA outperforms concurrent methods for FPs reduction since it generates better results than SOM with K-means and DSC.

In fact, we notice that NED is the best outliers detection approach among the others in the context of intrusion detection.

As illustrated in Table 5 which summarizes TPR for the second testing set, the process AO, k-means and BCA has the lowest TPR for the two testing sets.

In fact, we deduce that AO is not efficient for outliers detection in the context of intrusion detection. Our proposed process NED, K-means and BCA outperforms concurrent methods since it generates the best TPR. In addition, it outperforms the method SOM with k-means and DSC for the second testing set.

Table 5. TPR for testing set 2

	TPR
LOF, K-means and BCA	74.2
k-means, Kmeans and BCA	66.3
AO, K-means and BCA	60.2
NED, K-means and BCA	79.1
SOM with K-means	75
DSC	70.3
Rates are given in percentages	

4.3 Time Performance Evaluation

To evaluate the time performance of our proposed process, we compare its running times with competitor methods using the aforementioned testing sets. Table 6 shows the experimental results of running times for the three FPs detection methods. It is clear that the running time of our method is only a little higher than SOM with k-means. However, DSC has the highest running time among all other methods.

Table 6. Experimental results of running times

Methods	Testing set 1	Testing set 2
NED, K-means and BCA	0.22	0.28
DSC	0.35	0.40
SOM with K-means	0.20	0.25

5 Conclusion

An IDS is an essential part of any security package since it ensures the detection of intrusive activities if the information system has been hacked. However, an IDS tends to generate large databases where the majority of detected alerts are false alarms along with many outliers. In this work we propose a two-stage alarm correlation technique to improve the accuracy of an IDS. The aim of the first stage is to remove outliers from the set of alerts and the second one begins by a clustering step to reduce the cardinality of the testing set and ends by the identification of FPs. As our technique is tested using off-line data sets, it will be of interest to extend this work to study alarm correlation for multiple sensors.

References

1. Zhu, B., Ghorbani, A.: Alert correlation for extracting attack strategies. Int. J, Netw. Secur. **3**(3), 244–258 (2006)

2. Tjhai, C., Furnell, M., Papadaki, M., Clarck, L.: A preliminary two-stage alarm correlation and filtering system using som neural network and k-means algorithm. Comput. Secur. **29**, 712–723 (2010)
3. Bievens, A., Palagiri, C., Szymanski, B., Embrechts, M.: Network-based intrusion detection using neural networks. Intell. Eng. Syst. Artif. Neural Netw. **12**, 579–584 (2002)
4. Labib, K., Vemuri, R.: Nsom: A real time network-based intrusion detection system using self-organizing map. In: Networks Security (2002)
5. Zhang, Y., Huang, S., Wang, Y.: Ids alert classification model construction using decision support techniques. In: International Conference on Computer Science and Electronics Engineering, pp. 301–305 (2012)
6. Gupta, D., Joshi, P.S., Bhattacharjee, A.K., Mundada, R.S.: Ids alerts classification using knowledge-based evaluation. In: International Conference on Communication Systems and Networks, pp. 1–8 (2012)
7. Elshoush, H.-T., Osman, I.-M.: An improved framework for intrusion alert correlation. In: WCE12: Proceedings of the 2012 World Congress on Engineering, pp. 1–6 (2012)
8. Benferhat, S., Boudjelida, A., Tabia, K., Drias, H.: An intrusion detection and alert correlation approach based on revising probabilistic classifiers using expert knowledge. Int. J. Appl. Intell. **38**(4), 520–540 (2013)
9. Elhag, S., Fernandez, A., Bawakid, A., Alshomrani, S., Herrera, F.: On the combination of genetic fuzzy systems and pairwise learning for improving detection rates on intrusion detection systems. Expert Syst. Appl. **42**, 193–202 (2015)
10. Lin, W.-C., Ke, S.-W., Tsai, C.-F.: Cann: An intrusion detection system based on combining cluster centers and nearest neighbors. Knowl. Based Syst. **78**, 13–21 (2015)
11. Rousseeuw, P.J., Leroy, A.M.: Robust regression and outlier detection. John Wiley & Sons, New York (1987)
12. Abe, N., Zadrozny, B., Langford, J.: Outlier detection by active learning. In: Proceedings of the 12th ACM SIGKDD International Conference on Knowledge Discovery and Data Mining, pp. 504–509. ACM Press, New York, NY, USA (2006)
13. Jain, A.K., Murty, M.N., Flynn, P.J.: Data clustering: A review. ACM Comput. Surv. **31**(3), 264–323 (1999)
14. Knorr, E.M., Ng, R.T.: Algorithms for mining distance-based outliers in large datasets. In: Proceedings of the 24th International Conference on Very Large Databases, New York, NY, pp. 392–403 (1998)
15. Ramaswamy, S., Rastogi, R., Kyuseok, S.: Efficient algorithms for mining outliers from large data sets. In: Proceedings of the ACM SIDMOD International Conference on Management of Data, pp. 211–222 (2000)
16. Angiulli, F., Pizzuti, C.: Fast outlier detection in high dimensional spaces. In: Elomaa, T., Mannila, H., Toivonen, H. (eds.) PKDD 2002. LNCS (LNAI), vol. 2431, pp. 15–27. Springer, Heidelberg (2002)
17. Wu, W.Z., Zhang, W.X.: Neighborhood operator systems and approximations. Inf. Sci. **144**, 201–217 (2002)
18. Chen, Y.M., Miao, D.Q., Zhang, H.Y.: Neighborhood outlier detection. Expert Syst. Appl. **37**(12), 8745–8749 (2010)
19. Breunig, M.M., Kriegel, H.P., Ng, R.T., Sander, J.: Lof: Identifying densitybased local outliers. In: Proceedings of the 2000 ACM SIGMOD International Conference on Management of Data, Dallas, pp. 93–104 (2000)

A Geometric Viewpoint of the Selection of the Regularization Parameter in Some Support Vector Machines

Nandyala Hemachandra(✉) and Puja Sahu

Indian Institute of Technology Bombay, Mumbai, India
{nh,puja.sahu}@iitb.ac.in

Abstract. The regularization parameter of support vector machines is intended to improve their generalization performance. Since the feasible region of binary class support vector machines with finite dimensional feature space is a polytope, we note that classifiers at vertices of this unbounded polytope correspond to certain ranges of the regularization parameter. This reduces the search for a suitable regularization parameter to a search of (finite number of) vertices of this polytope. We propose an algorithm that identifies neighbouring vertices of a given vertex and thereby identifies the classifiers corresponding to the set of vertices of this polytope. A classifier can then be chosen from them based on a suitable test error criterion. We illustrate our results with an example which demonstrates that this path can be complicated. A portion of the path is sandwiched between two finite intervals of path, each generated by separate sets of vertices and edges.

Keywords: Support vector machines · Regularization path · Polytopes · Neighbouring vertices · Prediction error · Parameter tuning · Linear programming

1 Introduction

A classical learning problem is that of binary classification wherein the learner is trained on a given data set (training set) and predicts the class of a new data point. Let the n point training set be $\{(\mathbf{x}_i, y_i)\}_{i=1}^{n}$, where $\mathbf{x}_i \in \mathbb{R}^m$ is a vector of m features and $y_i \in \{-1, +1\}$ is the label of $\mathbf{x}_i$, $i \in \{1, \cdots, n\}$. We consider the class of linear classifiers, $(\mathbf{w}, b)$, with $\mathbf{w} \in \mathbb{R}^m$ and $b \in \mathbb{R}$. The classifier predicts the class of data point $\mathbf{x}$ as -1 if $\mathbf{w} \cdot \mathbf{x} + b < 0$ and predicts the class as $+1$ otherwise, i.e., the predicted class for $\mathbf{x}$ is $\text{sign}(\mathbf{w} \cdot \mathbf{x} + b)$. Such classifiers are called linear Support Vector Machines (SVMs).

Among finite dimensional models for binary class prediction, the class of polynomial kernels form an important class. These are quite popular in natural language processing (NLP) because fast linear SVM methods can be applied to the polynomially mapped data and can achieve accuracy close to that of using highly nonlinear kernels [2].

R. Prasath et al. (Eds.): MIKE 2015, LNAI 9468, pp. 140–149, 2015.
DOI: 10.1007/978-3-319-26832-3_14

The standard soft-margin SVM optimization problem (SVM QP), for a given $\lambda > 0$ [6] is:

$$\begin{aligned} \min_{\mathbf{w},b,\boldsymbol{\xi}} \ & \lambda||\mathbf{w}|| + \sum_{i=1}^{n} \xi_i \\ \text{s. t. } & y_i(\mathbf{w} \cdot \mathbf{x}_i + b) \geq 1 - \xi_i \quad \forall i \in \{1 \ldots n\} \\ & \xi_i \geq 0 \quad \forall i \in \{1 \ldots n\} \end{aligned} \tag{1}$$

The objective function is a sum of the regularization penalty term with regularization parameter $\lambda \in \mathbb{R}^+$ and the classification error (measured from the margins, $\mathbf{w} \cdot \mathbf{x} + b = \pm 1$, of the classifier) as captured by $\{\xi_i\}_{i=1}^{n}$.We are working with linearly inseparable data. Therefore, $\xi_i > 0$ for at least one $i \in \{1, \ldots, n\}$.

The purpose of the regularization parameter λ is to improve the generalization error of the SVM. It is known that a proper choice is needed; see, for example, Figure 4 of [7]. The purpose of this paper is to investigate this choice in fairly basic SVMs by considering the polyhedral nature of the feasible region of the above SVM QP.

The main results of this paper are summarized as follows: We characterize, to the best of our knowledge for the first time, the polytope, P, associated with the feasible space of (1), in terms of its vertices and give an algorithm that lists all its vertices. We notice that, starting off from a vertex, the path is generated by vertices and edges (one-dimensional facets) as well as facets of higher dimensions. The regularization parameter, λ, for any classifier can be identified by linear programs; and for classifiers corresponding to vertices, this is an interval. The SVMs are generally assessed in terms of their performance on $0-1$ loss criterion. We find that the vertex classifiers dominate other boundary classifiers on a single test point using this $0-1$ loss function. This means that for the SVMs that we consider, a suitable choice of λ as a design parameter can be replaced by a search among the finite but large number of the vertices of P.

Different approaches have been employed to select an optimal regularization parameter, λ, for the SVM QP. The task of tracing an entire regularization path was pioneered by [7]. The sets E, L and R of [7] in the feature space $(\mathbf{w}, b)$ correspond to a vertex v of the polytope P in the lifted space in $(\mathbf{w}, b, \boldsymbol{\xi})$. Another approach [3] considers finding the optimal parameters for SVM based classifiers with kernel functions that could be infinite dimensional, where λ is included in the parameter vector. Bounds on the test error are obtained, based on the leave-one-out testing scheme and these are differentiable with respect to the parameter vector. A gradient based scheme is proposed for finding optimal parameters. Apart from tracing the path, various other aspects have been studied regarding the design of the SVMs, such as the feature selection problem [3,10].

Note that, to trace the path, [7] use the dual optimization program to the SVM QP to study the trajectories of the primal and dual variables as a function of the regularization parameter; whereas the polytope considered in this paper resides in the primal space itself. As a consequence, we need to search among a finite, albeit a large, set of vertices. And unlike [3], we restrict our analysis to the case of finite dimensional kernels, which can be handled using fast algorithms.

2 The Polytope of the Feasible Region of SVMs

First we notice that the feasible region is unbounded; hence it admits a Minkowski decomposition into a base polytope, P, and a recession cone. We want to concentrate on characterizing P in terms of its vertices and more importantly, the role of these vertices in the regularization path of the SVM.

Theorem 1. *For a given $\lambda \geq 0$, the optimal point (the classifier for the SVM) lies on the boundary of the polytope P.*

Proof. Consider the unconstrained problem with the same objective function of SVM QP: $\lambda||\mathbf{w}|| + \sum_i \xi_i$. This optimization is separable into two optimization problems: $\min_{\mathbb{R}^m} \lambda||\mathbf{w}||$ and $\min_{\mathbb{R}^n} \sum_i \xi_i$. While the first one has the optimal value zero at $\mathbf{w} = 0$, the second one is unbounded. Hence the unconstrained problem has an unbounded value, whereas the SVM QP has a finite non-negative optimal value. The SVM QP is a convex minimization problem and hence its finite optimal solution will lie on the boundary of its feasible region, the polytope P. □

Theorem 2. *A classifier on the vertex of the polytope dominates a boundary classifier, i.e., a classifier corresponding to an edge or a facet, on $0-1$ loss function.*

Proof. Consider two classifiers $(\mathbf{w}^1, b^1)$ and $(\mathbf{w}^2, b^2)$ on the λ-path, lying on two different vertices of the polytope. Suppose, for a test data $\hat{\mathbf{x}}$, we have $\text{sign}(\mathbf{w}^1 \cdot \hat{\mathbf{x}} + b^1) = +1$ and $\text{sign}(\mathbf{w}^2 \cdot \hat{\mathbf{x}} + b^2) = -1$. A classifier $(\tilde{\mathbf{w}}, \tilde{b})$ on the related edge can be identified as $(\tilde{\mathbf{w}}, \tilde{b}) = \alpha(\mathbf{w}^1, b^1) + (1-\alpha)(\mathbf{w}^2, b^2)$ for some $\alpha \in (0, 1)$. We can see that

$$\text{sign}(\tilde{\mathbf{w}} \cdot \hat{\mathbf{x}} + \tilde{b}) = \begin{cases} 1 & \text{if } \alpha \geq \alpha_0, \\ -1 & \text{if } \alpha < \alpha_0, \end{cases}$$

where $\alpha_0 \in (0, 1)$ is the normalized distance of $\hat{\mathbf{x}}$ from $\mathbf{w}^1, b^1$. Thus, $(\tilde{\mathbf{w}}, \tilde{b})$ can be dominated on the grounds of 0–1 loss function by one of the vertices depending on the true label of $\hat{\mathbf{x}}$. Dominance over the classifiers belonging to a facet of the polytope can be shown using similar arguments.

Remark 1. The above result was shown for a single test point. When we have a collection of the points, we will have a 'dominating set' of vertices, which may or may not lie on the λ-path.

2.1 Characterization of the Vertices in Terms of Active Constraints

We recall the following (see [1,9], etc.): For a polytope in $\mathbb{R}^k$, a vertex is a point of zero dimension. A vertex in $\mathbb{R}^k$ can be identified as a solution of k linearly independent linear equations. A vertex is an extreme point of the polytope, and can not be obtained as a convex combination of any two distinct points. An edge is a facet of dimension one and is a convex combination of two vertices of the polytope. A facet of dimension two is a convex combination of three vertices and so forth. We define a vertex classifier as a classifier corresponding to a vertex on the polytope of the feasible region of the standard SVM model. The edge and facet classifiers are defined in a similar fashion. Henceforth, these notations will be used in the rest of the paper.

The dimension of $(\mathbf{w}, b, \boldsymbol{\xi})$ is $(m+1+n)$ and we have n linear inequalities with n positively constrained variables, $\boldsymbol{\xi} = (\xi_1, \ldots, \xi_n)$. (We make the assumption that $(m+1) < n$). Rewriting them as n equalities with positive slack variables s_i, we get a set of $2n$ linear constraints whose intersection gives us a polytope as the feasible region. Hence, if at least $(m+n+1)$ of these $2n$ constraints are active and are linearly independent, the resulting unique solution is a vertex.

So, with s_i, we have the following:

$$y_i(\mathbf{w} \cdot \mathbf{x}_i + b) - 1 = s_i - \xi_i \quad \forall i \in \{1 \ldots n\}, \tag{2}$$

where $s_i, \xi_i \geq 0 \; \forall i \in \{1 \ldots n\}$.

At a vertex v, then, for a given $i \in \{1, \cdots, n\}$, if $\xi_i - s_i \neq 0$, then only one of ξ_i or s_i is non-zero. It can be noted that,

$$s_i - \xi_i = \begin{cases} (-\infty, -2) & \text{if } \mathbf{x}_i \text{ is misclassified by } (\mathbf{w}, b) \text{and outside the margin} \\ -2 & \text{if } \mathbf{x}_i \text{ is misclassified by } (\mathbf{w}, b) \text{ and on the margin} \\ (-2, -1) & \text{if } \mathbf{x}_i \text{ is misclassified by } (\mathbf{w}, b) \text{ and within the margin} \\ -1 & \text{if } \mathbf{x}_i \text{ is correctly classified by } (\mathbf{w}, b) \text{ and on the classifier} \\ (-1, 0) & \text{if } \mathbf{x}_i \text{ is correctly classified by } (\mathbf{w}, b) \text{ and within the margin} \\ 0 & \text{if } \mathbf{x}_i \text{ is correctly classified by } (\mathbf{w}, b) \text{ and on the margin} \\ & \text{i.e., } \mathbf{x}_i \text{ is a support vector} \\ (0, \infty) & \text{if } \mathbf{x}_i \text{ is correctly classified by } (\mathbf{w}, b) \text{ and outside the margin.} \end{cases}$$

We can identify three different categories of classifiers based on the above values of $(s_i - \xi_i)$, as mentioned in the following theorem:

Theorem 3. *There are three types of linear SVM classifiers for the case of binary classification problem:-*

(i) The classifiers for which the points can be within, on or outside the margin. Thus, $(s_i - \xi_i) \in (-\infty, \infty) \; \forall i \in \{1, \ldots, n\}$ for such classifiers.
(ii) The $(\mathbf{0}, 1)$ and $(\mathbf{0}, -1)$ classifiers, for which ξ_i is either 0 or 2 and $(s_i - \xi_i) \in \{0, -2\} \; \forall i \in \{1, \ldots, n\}$.
(iii) The classifiers for which all the points are within or on the margins. For such classifiers $(s_i - \xi_i) \in (-2, 0) \; \forall i \in \{1, \ldots, n\}$.

We have another important characterization of a vertex of the polytope in terms of support vectors of the classifier.

Theorem 4. *A vertex $v = (\mathbf{w}, b, \boldsymbol{\xi})$ of the polytope P has at least one correctly classified point on its margin, also known as the support vector. Therefore, for $v \in P$, we have*

$$|\{i \in \{1, \ldots, n\} | \xi_i = s_i = 0\}| \geq 1. \tag{3}$$

Proof. For the type (ii) classifiers, the above is trivially true using (2). For the type (i) and type (iii) classifiers, using the definition of a vertex of a polytope, at least $(m+n+1)$ of the $2n$ constraints in (1) have to be active. Thus, for any set of at least $(m+1+n)$ active constraints, I^*

$$\exists \text{ at least one } i \in I^* \text{ such that } y_i(\mathbf{w} \cdot \mathbf{x}_i + b) = 1 - \xi_i$$
$$\text{and,} \quad \xi_i = 0. \qquad \square$$

Rewriting the constraints in (1) in a matrix form, we have: $A \cdot (\mathbf{w}, b, \boldsymbol{\xi}) \geq b$, where A corresponds to the coefficient matrix of the two sets of constraints of the SVM QP and $b = \begin{pmatrix} \mathbf{1} \\ \mathbf{0} \end{pmatrix}$.

Let us denote by $I^*(v)$, the set of (indices of) active constraints at vertex v. So, given a set of active constraints, $I^*(v)$, we can find the corresponding vertex $v = (\mathbf{w}, b, \boldsymbol{\xi})$ by solving the following equation:

$$A[I^*(v),] \cdot (\mathbf{w}, b, \boldsymbol{\xi}) = b[I^*(v)], \tag{4}$$

where $A[I^*(v),]$ is a sub-matrix of A with rows corresponding to $I^*(v)$. Such a vertex corresponds to a basic solution of the feasible region. It is a feasible vertex if and only if it satisfies (1). Equivalently, it is not a feasible point if s_i is strictly negative. A feasible vertex corresponds to a basic feasible solution [1].

Given an active set of constraints, $I^*(v)$, Algorithm VERTEX(*Active*)), as described in the technical report [8], computes a vertex v corresponding to these active constraint set, if it exists and is feasible.

2.2 Neighbours of a Vertex of the Polytope, *P*

Given that a vertex is characterized by the set of constraints, $I^*(v)$, that are active at that point, we can find a neighbouring vertex $\tilde{v}$ by changing $I^*(v)$ in the following way:

Replace an active constraint by the one that is currently inactive at v. The constraint $i \in I^*(v)$ to leave the active set is the one such that $\xi_i = s_i = 0$. The existence of such a constraint $i \in I^*(v)$ is guaranteed by Theorem 3. The incoming constraint $j \in I^*(v)$ is chosen so that $\{j \mid (s_j > 0 \,\&\, \xi_j = 0) \text{ or } (\xi_j > 0 \,\&\, s_j = 0)\}$ at v. And at the neighbour $\tilde{v}$, we set $\xi_j = s_j = 0$ to ensure a support vector for $\tilde{v}$.

If the solution to (4) with these new active constraints is feasible, then it is a valid neighbour of v. Note that if the given vertex v is degenerate, then, the above change in active constraint set $I^*(v)$ can lead to another degenerate vertex and hence not a neighbouring vertex. Such degenerate vertices need to be ignored in the list of neighbours of v.

Such a careful updating of the set of neighbouring vertices avoids potential cycling while listing the set of all vertices of the polytope P. The set of all such neighbours of given v is denoted by $N(v)$ and can be found as in Algorithm NEIGHBOUR(v), described in our technical report [8].

2.3 Vertices of the Polytope, P

We observe that for $\lambda = 0$, the SVM QP is a linear program and its optimal solution by simplex type algorithms will be at a vertex, say v_0, and the set of active constraints, $I^*(v_0)$, can be easily obtained. Intializing with this vertex v_0 and its active set $I^*(v_0)$, Algorithm 1 VERTEX SEARCH finds the set of all vertices of polytope P using Algorithm NEIGHBOUR(v) [8] as required, which in turn calls procedure VERTEX(*Active*) [8] with $I^*(v)$.

Algorithm 1 . VERTEX SEARCH(P)

1: Solve the SVM QP at (1) with $\lambda = 0$. Let this optimal classifier be $(\mathbf{w}^0, b^0, \boldsymbol{\xi}^0)$.
   This corresponds to a vertex, say $v_0 \in P$
2: Set $Current \leftarrow \{v_0\}, N(P) \leftarrow \phi$
3: **while** ($Current \neq \phi$) **do**
4: $\quad Next \leftarrow \phi$
5: $\quad$ **for all** $v \in Current$ **do**
6: $\quad\quad Next \leftarrow Next \cup$ NEIGHBOUR(v)
7: $\quad$ **end for**
8: $\quad N(P) \leftarrow N(P) \cup Current$
9: $\quad Current \leftarrow Next \setminus N(P)$
10: **end while**
11: **return** $N(P)$

---

3 The Regularization Path

As the optimal classifiers for SVM QP are on the boundary of the polytope, by Theorem 1, the set of classifiers given by the regularization path is a subset of the set of vertices and related edges of the polytope of feasible region. Since the classifier is chosen by 0–1 loss function, using this in SVM design phase itself, one can argue that vertex classifiers on the path dominate those at the related edges.

However, for some set of test points, the dominating vertex classifier (as in Theorem 2) may or may not be on the path (see the example in Sect. 4). In the following discussion, we focus on the classifiers at vertices that generate some portions of the regularization path.

Before describing the procedure to identify the vertices on the path traced by the parameter λ, we mention a few results which will be used by this procedure.

Using the fact that, at optimality, the gradient of the objective function in a convex setting needs to be a member of the normal cone at that point and the KKT system gives an algebraic representation of this geometric phenomena, we have the following:

Theorem 5. *The bounds $[\lambda_l, \lambda_u]$ on the range of values of the regularization parameter λ for which a given classifier $(\mathbf{w}, b, \boldsymbol{\xi}) \in P$ is optimal for SVM QP at (1), can be obtained as solutions to the following two linear programs, respectively:*

$$\begin{aligned} \lambda_l &= \min_{\lambda \geq 0} \lambda \\ over & \quad S(\mathbf{w}, b, \boldsymbol{\xi}) \end{aligned} \tag{5}$$

$$\begin{aligned} \lambda_u &= \max_{\lambda \geq 0} \lambda \\ over & \quad S(\mathbf{w}, b, \boldsymbol{\xi}) \end{aligned} \tag{6}$$

where $S(\mathbf{w}, b, \boldsymbol{\xi}) = \{(\lambda, \alpha_1, \ldots, \alpha_n)\}$ *such that*

$$\begin{aligned} \sum_{i=1}^{n} \alpha_i y_i \mathbf{x}_i &= 2\lambda \mathbf{w} \\ \sum_{i=1}^{n} \alpha_i y_i &= 0 \\ \alpha_i \left[y_i(\mathbf{w} \cdot \mathbf{x}_i + b) - \xi_\mathbf{i}) \right] &= 0 \quad \forall i \in \{1, \ldots, n\} \\ (1 - \alpha_i)\xi_i &= 0 \quad \forall i \in \{1, \ldots, n\} \\ 0 \leq \alpha_i &\leq 1 \quad \forall i \in \{1, \ldots, n\} \end{aligned}$$

The next result says that a portion of the λ-path of a given SVM is partitioned by intervals corresponding to some of the vertices and edges of the polytope P. Also, we can see that the λ value for an edge classifier is a harmonic mean of the bounds on the λ interval of the related vertices. Please refer to our technical report [8] for details of the proof.

Theorem 6. *(i) For a classifier* $(\mathbf{w}, b)$ *which is a vertex* $v := (\mathbf{w}, b, \boldsymbol{\xi})$ *on the polytope* P*, the range* $[\lambda_l, \lambda_u]$ *of* λ *values, for which* v *is optimal, is an interval in* $\mathbb{R}$*. In fact, this range is always finite since the gradient of the objective is never parallel to the generators of the normal cone.*

(ii) For a classifier on an edge or a facet of P*, the feasible* λ *value is a singleton, i.e.,* $\lambda_l = \lambda_u$*. Specifically, for an edge point classifier,* e_{v_1,v_2}*, lying between two on-the-path vertices* v_1 *and* v_2 *such that* $\lambda_u(v_1) < \lambda_l(v_2)$*, we have*

$$\lambda(e_{v_1,v_2}) = \frac{\lambda_u(v_1)\lambda_l(v_2)}{\beta\lambda_l(v_2) + (1-\beta)\lambda_u(v_1)}. \tag{7}$$

(iii) A portion of the regularization path corresponding to vertex-edge boundary of P *can be decomposed into intervals corresponding to vertices on the path and the edges between them. This is so because, for an edge point classifier,* e_{v_1,v_2}*, as described above, we have*

$$\lambda(e_{v_1,v_2}) \in (\lambda_u(v_1), \lambda_l(v_2)). \tag{8}$$

To trace the regularization path, we solve the SVM QP for $\lambda = 0$ which is a linear program. This gives us a classifier corresponding to a vertex in P, say v_0. The range of λ for v_0 can be obtained via the solutions to the linear programs: (5) and (6). We know that λ traces a continuous path along the boundary of P, so the next vertex on the path will be a neighbour of v_0, found using the procedure NEIGHBOUR(v) [8]. Many of the neighbouring vertices are not optimal classifiers for any value of λ and hence, the LPs (5) and (6) become infeasible at such neighbouring vertices. We will have one such neighbour for which there exists an interval of λ and hence it becomes the next vertex on the λ-path. Then we search amongst the neighbours of the current vertex, to find the next vertex on the path. This procedure continues iteratively till all the neighbours of the current vertex become infeasible for the path. Such a vertex corresponds to the last but one vertex on the path.

Yet, there can be instances, as we will show in our example, where the path does not retain continuity along the vertex-edge boundary of the polytope. This happens when none of the neighbours of the current vertex are optimal for any value of λ. This forces us to search exhaustively for next generation neighbours which are optimal for some value of the parameter λ.

4 An Illustrative Example

The purpose of this example is to illustrate two aspects of the regularization path: a contiguous portion of the path composed of vertices and edges of the polytope, and another portion on facets of two or more dimensions. Interestingly, this portion is sandwiched between intervals generated by some vertices and edges. We tabulate the λ intervals for the vertices of P on the λ-path for a binary classification SVM model. We consider a training set with 50 points drawn from two bivariate normal distributions. The two classes have means $(0, 0)$ and $(1, 0)$ and same covariance matrix $(0.5, 0; 0, 0.5)$. A list of 15002 valid vertices with an index was generated using `rcdd` package [5] in R programming language.

The vertices are arranged so that the lower and upper bounds on the λ intervals are in an increasing order (Table 1). It was observed that each vertex on this path is a neighbour of the previous vertex on the list (except the first vertex, v_0). The portion of the path that occurs between vertices 9158 and 10470 corresponds to classifiers on the facets, since none of the first generation neighbours of vertex 9158 are optimal for any value of λ. The next optimal vertex is 10470 which may be obtained as a fifth generation neighbour to the vertex 9158.

Using Theorem 6, we note that only 15 of these 15002 vertices are on the λ-path. The set of first 13 vertices correspond to the first portion of the path, followed by a portion on the facets of two or higher dimensions. The last two vertices and the edge involving them correspond to the next segment of the path. As mentioned above, there are no more optimal vertices on the path.

We have plotted the test error of these classifiers in Fig. 1, computed on 5 test data sets of 15 points with the same distribution as the training data set. From Fig. 1, we can also see that the classifiers corresponding to the facets are dominated by the vertex classifiers in terms of test error. Hence, it is sufficient to consider the vertices corresponding to low values of parameter λ. The test error is lower for the last but one vertex on the path for the given test sets.

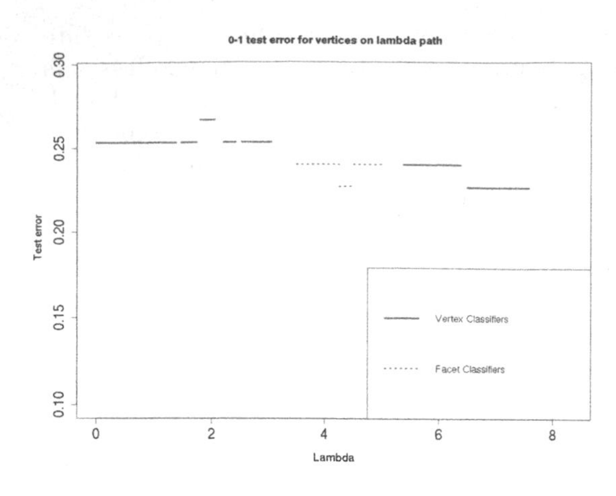

Fig. 1. Test error (averaged over 5 test data sets) for the λ-path for bivariate normal data with 50 points

Table 1. (λ_l, λ_u) for vertices on λ-path for bivariate normal data with 50 points

Vertex Index	$\lambda_l(v)$	$\lambda_u(v)$
7274 (v_0)	0	0.002712
7325	0.002776	0.059187
7327	0.059275	0.339041
7326	0.339110	0.522892
7328	0.523507	0.602838
7426	0.610561	0.745904
7420	0.749820	0.818734
7421	0.851777	1.191843
7424	1.198099	1.370998
7425	1.456195	1.723649
7352	1.785567	2.047029
9038	2.206312	2.428423
9158	2.535443	3.053205
-	3.5	3.5
-	4.25	4.25
10470	5.372196	6.368380
10471	6.486113	7.576469

5 Discussion

The polyhedral structure of the feasible space of the standard SVM optimization model allows us to trace the λ-path on a subset of vertices of the base polytope. It was observed that for initial values, the λ-path comprises of vertices and related edges. We have examples where the path has classifiers that are on facets of the SVM polytope, P. We have restricted ourselves to the subset of vertices that dominate the whole polytope of feasible classifiers on 0–1 loss. The vertices and their neighbours can be identified via suitable active constraint sets. We noticed in our limited computational exercise, that the tracing of the λ-path has encountered numerical instabilities; such problems are also reported by [7].

Some aspects that naturally need attention include the need to come up with a scheme to pick that neighbour of a vertex which generates the adjacent interval on the λ-path, if such an interval exists. Perhaps, a more broader and important aspect is to be able to restrict ourselves to a suitable subset of vertices, which may or may not be on the λ-path, but have a promising test error.

A leave-one-out scheme [3] can be employed for testing the design of the classifier. Such leave-one-out training sets can be viewed as suitable perturbations of a given training set, and corresponding robust classification problems can be formalized. We can ensure the stability of the SVM algorithm by establishing such equivalence with a robust optimization formulation [13].

Besides the test error, some other measures such as bias, variance and the margin [4] of the classifier can be used for design. An analysis of error decomposition of the learning algorithms such as decision trees, k-NNs, etc. is done by [4], where the main consideration is for the algorithms that are consistent in the use of the same loss function for training as well as testing. This analysis was further taken up by [12] to the case of SVMs, where training makes use of hinge loss and testing is based on $0-1$ loss.

Some statistical properties about the risk of the classifier [11,14] can be explored to improve the efficiency of our algorithms. These results may help us pick 'good' vertex classifiers, for example, via the λ intervals corresponding to good generalizations guarantees.

References

1. Bertsimas, D., Tsitsiklis, J.N.: Introduction to linear optimization. Athena Scientifc Belmont, MA (1997)
2. Chang, Y.-W., Hsieh, C.-J., Chang, K.-W., Ringgaard, M., Lin, C.-J.: Training and testing low-degree polynomial data mappings via linear SVM. J. Mach. Learn. Res. **11**, 1471–1490 (2010)
3. Chapelle, O., Vapnik, V., Bousquet, O., Mukherjee, S.: Choosing multiple parameters for support vector machines. Mach. Learn. **46**(1), 131–159 (2002)
4. Domingos, P.: A unified bias-variance decomposition. In: Proceedings of 17th International Conference on Machine Learning, pp. 231–238. Morgan Kaufmann, Stanford CA (2000)
5. Geyer, C.J., Meeden, G.D.: Incorporates code from cddlib written by Komei Fukuda. rcdd: Computational Geometry, R package version 1.1-9 (2015)
6. Hastie, T., Tibshirani, R., Friedman, J.H.: The Elements of Statistical Learning: Data Mining, Inference, and Prediction. Springer series in statistics. Springer, Heidelberg (2001)
7. Hastie, T., Rosset, S., Tibshirani, R., Zhu, J.: The entire regularization path for the support vector machine. J. Mach. Learn. Res. **5**, 1391–1415 (2004)
8. Hemachandra, N., Sahu, P.: A geometric viewpoint of the selection of the regularization parameter in some support vector machines. Technical report, IE & OR, IIT Bombay, Mumbai, September 2015. http://www.ieor.iitb.ac.in/files/SVMpath_TechReport.pdf, September 30, 2015
9. Hiriart-Urruty, J.-B., Lemaréchal, C.: Fundamentals of Convex Analysis. Grundlehren Text Editions. Springer, Heidelberg (2004)
10. Jawanpuria, P., Varma, M., Nath, S.: On p-norm path following in multiple kernel learning for non-linear feature selection. In: Proceedings of the 31st International Conference on Machine Learning, pp. 118–126 (2014)
11. Ingo Steinwart and Andreas Christmann. Support vector machines. Springer Science & Business Media (2008)
12. Valentini, G., Dietterich, T.G.: Bias-variance analysis of support vector machines for the development of svm-based ensemble methods. J. Mach. Learn. Res. **5**, 725–775 (2004)
13. Huan, X., Caramanis, C., Mannor, S.: Robustness and regularization of support vector machines. J. Mach. Learn. Res. **10**, 1485–1510 (2009)
14. Zhang, T.: Statistical behavior and consistency of classification methods based on convex risk minimization. Ann. Stat. **32**, 56–85 (2004)

Discovering Communities in Heterogeneous Social Networks Based on Non-negative Tensor Factorization and Cluster Ensemble Approach

Ankita Verma(✉) and K.K. Bharadwaj

School of Computer and Systems Sciences,
Jawaharlal Nehru University, New Delhi 110067, India
{vermaankita333,kbharadwaj}@gmail.com

Abstract. Identification of the appropriate community structure in social networks is an arduous task. The intricacy of the problem increases with the heterogeneity of multiple types of objects and relationships involved in the analysis of the network. Traditional approaches for community detection focus on the networks comprising of content features and linkage information of the set of single type of entities. However, rich social media networks are usually heterogeneous in nature with multiple types of relationships existing between different types of entities. Cognizant to these requirements, we develop a model for community detection in Heterogeneous Social Networks (HSNs) employing non-negative tensor factorization method and cluster ensemble approach. Extensive experiments are performed on 20Newsgroup dataset which establish the effectiveness and efficiency of our scheme.

Keywords: Heterogeneous social networks (HSNs) · Community detection · Social network analysis · Non-negative tensor factorization (NTF) · Cluster ensemble

1 Introduction

Social network analysis has been an area of inexorable interest for the researchers in recent years. Analysis of the social networks not only unveils the latent social structures but also explores different interactions among social actors. A social network is a social structure, community, or society made of nodes which are generally accounts, individuals or organizations. It indicates the ways in which they are connected through various social familiarities, affiliations and relationships. Social networks can be categorized into two types: (a) Homogeneous Social Networks and (b) Heterogeneous Social Networks as shown in Fig. 1. Homogeneous Social Network comprises of a set of entities linked together with the links representing single type of relationship. Heterogeneous Social Network consists of different type of relationships between entities. For a more generic case, it can also have multiple types of entities [3].

The heterogeneous network involves a multitude of relations and types of entities. Generally, it consists of three types of information, attributes of the entities in the data, homogeneous relationship between same type of entities and heterogeneous relationship between different types of entities [15]. Figure 2 specifies three different structures

R. Prasath et al. (Eds.): MIKE 2015, LNAI 9468, pp. 150–160, 2015.
DOI: 10.1007/978-3-319-26832-3_15

of multi-type relational data. Figure 2(a) specifies a simple bi-type relation with one heterogeneous relation R12 such as word-document data. Figure 2(b) refers to the multi-type relational data with two heterogeneous relations R12 and R23, one homogeneous relation R1. Figure 2(c) denotes a general multi-type relational data with homogeneous and heterogeneous relations as well as attributes F5 of type-5 entity. The third category of structure comprising of homogeneous, heterogeneous relations as well as attributes of entities is the most generic one.

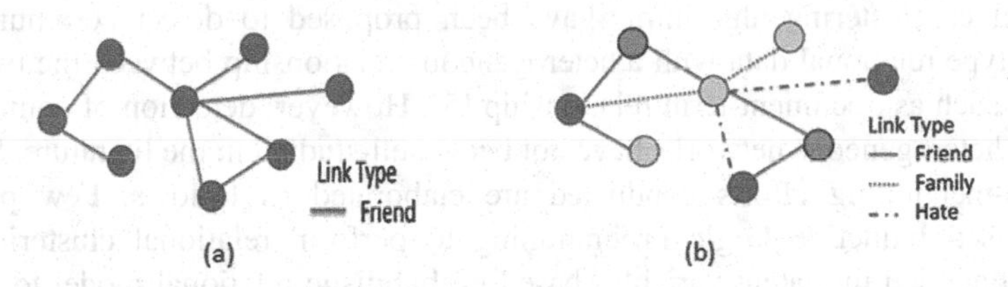

Fig. 1. (a) Homogeneous network with single type of entity and relation (b) Heterogeneous relation with multiple types of relations and entities [3]

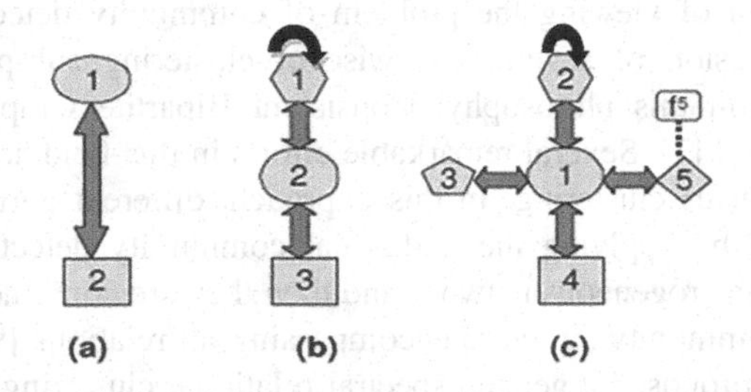

Fig. 2. Examples of structure of heterogeneous network [14]

Community detection is an important tool for social network analysis. A community is a densely connected subset of the nodes that is sparsely connected with the rest of the network. In this article, we propose a novel scheme for discovering communities in the heterogeneous networks. At the core of the scheme, we employ tensor factorization as a relational learning tool for leveraging the relational information embedded in the data. Our approach is based on Canonical Decomposition also known as Parallel Factorization with non-negativity constraint imposed on it. Traditional clustering methods are applied on the result of the factorization step to obtain an ensemble of clustering results. Finally, cluster ensemble is combined to achieve the desired community structure.

The organization of the rest of the article is as follows: Sect. 2 elaborates some of the state-of-the-art techniques for discovering communities in the heterogeneous network. The detailed description of our proposed scheme for community detection is provided in the Sect. 3. Experimental study and the obtained results are mentioned in the Sect. 4. Finally, Sect. 5 concludes the article with some research directions.

2 Related Work

Community discovery is not a naïve tool for network analysis; it has gained extensive attention of research community for several years. Although, the focus of the research is kept restricted to the homogeneous networks only but community discovery in heterogeneous networks emerged as the challenging field in the recent years. Detection of community structure in heterogeneous social network requires different insight and approach as compared to the discovery of communities in traditional setting.

Several co-clustering algorithms have been proposed to detect communities in simple bi-type relational data with a heterogeneous relationship between the two types of entities such as document-term relationship [5]. However, detection of communities in general heterogeneous networks have not been well-studied in the literature. Some of the worth-mentioning efforts conducted are elaborated as follows. Few proposed methods used Inductive Logic Programming to perform relational clustering [10]. Getoor [8] applied the latent variables based probabilistic relational model to perform clustering. A framework named ReCom (Reinforcement Clustering of Multi-type Interrelated data objects) discovers communities in heterogeneous Web data using mutual reinforcement learning [22].

Another approach of viewing the problem of community detection in heterogeneous network as fusion of several pair-wise co-clustering sub-problems was put forward and based on this philosophy, Consistent Bipartite Graph Co-Partitioning (CBGC) was proposed [7]. Several remarkable efforts in this field are based on cluster ensemble and consensus clustering. In this approach, different partitioned set of the objects are obtained by applying the traditional community detection algorithms to each relation of the heterogeneous network and then they are combined across multiple relations into one community structure encompassing all relations [9, 17, 19].

Long et al. [14] proposed a general spectral relational clustering model applicable to heterogeneous networks with various structures. Some of the researchers also focused on the multi-view clustering [13, 24]. Non-negative matrix factorization has been successfully applied for heterogeneous data clustering [4]. Tang et al. [21] present a unified view of community discovery based on an integration technique through structural features embedded in the relational data. A fuzzy approach to cluster heterogeneous network involving different types of entities and relationships clusters different types of objects simultaneously [16]. Wu et al. [23] have given different weights to relations reflecting the intrinsic importance of the relation in the multi-relational network. Another interesting line of research is the evolution of communities in dynamic multi-mode networks [20]. Thus, we can conclude that there has been a lack of research to be conducted for general heterogeneous social networks.

3 The Proposed Framework for Discovering Communities in Heterogeneous Social Networks (HSNs)

Our overall proposed framework is divided into three major steps as illustrated below:

- **Step 1:** The heterogeneous social network is modeled as a three way tensor X whose first and second mode depict the set of multiple types of entities in the network and third mode of the tensor corresponds to the multiple relationships existing between them.
- **Step 2:** On the tensor representing the given network, we apply non-negative tensor factorization to reveal hidden structural features of the entities. In the work presented in this article, we have used Candecomp/Parafac (CP) factorization to factorize X to A, B, C factor matrices.
- **Step 3:** Given factor matrices A and B, we can apply any traditional clustering method to obtain two different clustering results corresponding to each of them. These two different results are then combined together to achieve final community structure.

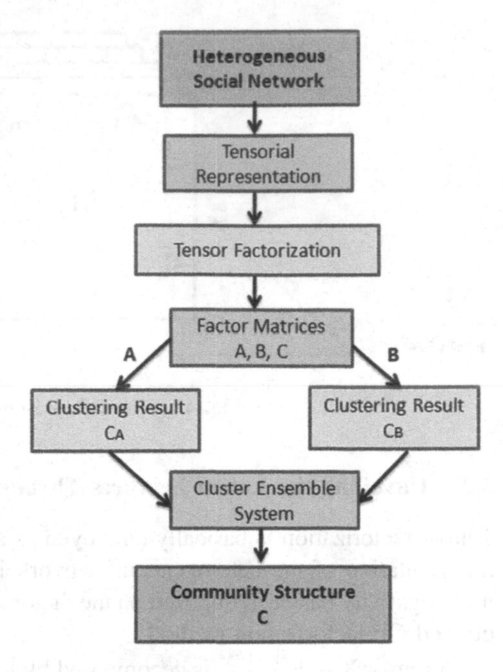

Fig. 3. The framework of our proposed scheme

Figure 3 depicts our scheme diagrammatically. In the following sub-sections, we have explained each step of the scheme in detail.

3.1 Model Formulation

For modeling heterogeneous networks involving multitude of relations and types of entities, a three-way tensor is employed. In this three-way tensor, two modes are identically formed by the concatenated entities of different types and the third mode holds the different types of the relations in the domain [18]. For a heterogeneous network with N number of total entities and M types of relationships between them, a tensor $X \in \mathbb{R}^{N \times N \times M}$ is created, whose entries are filled as given in Eq. (1).

$$\mathrm{X(i,j,k)} = \begin{cases} 1, & \textit{if there exist } \mathrm{k_{th}} \textit{ relationship between } \mathrm{i_{th}} \\ & \textit{entity as subject and } \mathrm{j_{th}} \textit{ entity as object} \\ 0, & \textit{for unknown or non} - \textit{existing relation} \end{cases} \tag{1}$$

If any relation is weighted, then the value 1 can be replaced with the corresponding weight of the relation. As for an example, given a network with n types of entities represented by sets $E_1, E_2, E_3 \ldots, E_n$ and m types of relationships between then represented by $R_1, R_2, \ldots, R_m$, the tensor X created is shown by Fig. 4.

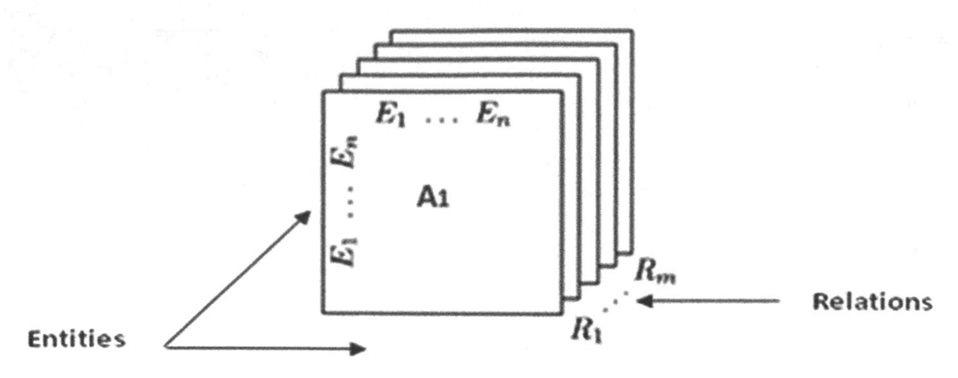

Fig. 4. Tensor Representation of the HSN

3.2 Unveiling the Latent Features Through Tensor Factorization

Tensor factorization is basically employed as a tool for relational learning. The tensor representation of the heterogeneous network is non-negative; therefore, constraint of non-negativity has been imposed on the factorization. In our proposed scheme, we have utilized CP factorization method.

A tensor $X \in \mathbb{R}^{N \times N \times M}$ is decomposed by R-Rank PARAFAC into three component matrices $A, B \in \mathbb{R}^{N \times R}$, $C \in \mathbb{R}^{M \times R}$ and R principal factors (pf) in descending order. Formally, the tensor X can always be expressed as a sum of rank-1 tensors in the form

$$X = \sum_{j=1}^{R} \lambda_j \mathbf{a}_j \mathrm{o} \mathbf{b}_j \mathrm{o} \mathbf{c}_j + E \tag{2}$$

where a_j, b_j, c_j are column vectors of factor matrices, λ_j denotes jth principal factor, $E \in \mathbb{R}^{N \times N \times M}$ is the tensor representing error, o outer product. Figure 5 shows the pictorial representation of the decomposition method. The set of vectors $\{a_1, a_2, \ldots, a_R\}$ can be written as matrix A where each vector forms the column of the matrix. Similarly, matrices B and C can also be formed. So, after the decomposition process X can be represented by $\lambda; A, B, C$. The matrix A depicts the association of subject and pf, B represents the association of object and pf, and C shows the relationship and pf association [6].

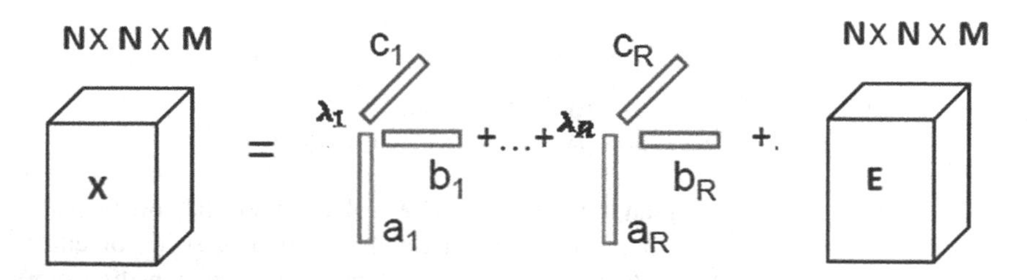

Fig. 5. Pictorial representation of the Candecomp/Parafac Factorization

3.3 Cluster Ensemble Approach for Discovering Communities

The next step that follows is the discovery of communities in the network from the latent features revealed by the factorization step. This task is achieved using cluster ensemble approach. A cluster ensemble system involves two steps: the first step takes the dataset as input and produces ensemble of different partitions of the set of data objects, the second step takes this cluster ensemble and combine different results into one clustering solution as the final outcome.

For our proposed scheme, the factors A and B are matrices with R columns, each one corresponding to one extracted component. The rows of A correspond to network entities which participate in different relationships as the subjects. Similarly, the rows of B correspond to entities participating as the objects. The entries of A and B give the membership weight of entities to the different components. Thus, A and B provide the community structure of the network. On each of the factor matrix, we can apply any similarity based community detection method to obtain partition of the set of entities. For an instance, if the community structure of the type i entity is required, then matrices A_i and B_i are created with the rows of A and B corresponding to those entities which are of n type. Given input matrices A_i, B_i and number of desired cluster K, a cluster ensemble of clustering solutions $C_{emsemble} = \{C_A, C_B\}$ is obtained. This cluster ensemble is then combined into single consensus clustering result by using Cluster-based Similarity Partitioning Algorithm (CSPA) [19].

CPSA induces a pair-wise similarity measure between objects based on the clustering result. For each partition, the similarity between two objects is 1 if they belong to same cluster and 0 otherwise. We construct binary membership indicator matrices $H_A, H_B \in \{0,1\}^{N \times K}$ corresponding to C_A and C_B respectively. The similarity matrix can be computed as:

$$S = \frac{1}{2}\left[H_A(H_A)^T + H_B(H_B)^T\right] = \frac{1}{2}\left[H^*(H^*)^T\right] \tag{3}$$

where, $H^* = [H_A, H_B]$. Based on this similarity matrix between nodes, we can apply similarity based community detection methods to find out final community structure.

4 Experimental Evaluation

In this section, we have performed several experiments to evaluate the effectiveness of our proposed scheme. The performance of our scheme is also compared with some of the state-of-the-art methods of community detection in HSNs. The description of the dataset and the evaluation metrics used in our study are given as follows:

4.1 Dataset Description

The dataset used in this experimental study is 20Newsgroup data [11] which contains about 18828 documents from 20 different categories. There are three types of entities (document, term and category) linked together with two types of relationships as:

- **Document-Term:** This relation specifies the number of times a word appear in a particular document.
- **Document-Category:** This relation indicates the category a particular document belongs to.

We have created three small datasets TM1, TM2, TM3 from the aforementioned dataset, whose taxonomical structure is enlisted in Table 1. The datasets are generated by randomly selecting 100 documents from each of the chosen subtopics. For e.g. dataset TM1 consists of two communities of documents C1 and C2; C1 is the *rec.sport* cluster with two categories *baseball* and *hockey*, C2 is the talk.politics cluster with categories *guns* and *mideast*.

Table 1. Structure of the 20Newsgroup datasets.

Data sets	No. of documents	Description
TM1	400	C1: {rec.sport.basebal, rec.sport.hockey} C2: {talk.politics.guns, talk.politics.mideast}
TM2	700	C1: {comp.graphics, comp.windows.x} C2: {rec.sport.baseball, rec.sport.hockey} C3: {sci.electronics, sci.med, sci.space}
TM3	800	C1: {comp.sys.ibm.pc.hardware, comp.sys.mac.hardware} C2: {rec.autos, rec.motorcycles} C3: {sci.crypt, sci.med} C4: {talk.politics.guns, talk.politics.misc}

4.2 Evaluation Metrics

As the ground truth information is available with the given dataset, so to measure the performance of the community discovery method, we used three supervised metrics: Normalized Mutual Information (NMI), Pair-wise F-score (PWF) and Entropy (E).

Given true community structure of the network as $C = \{C_1, C_2, C_3, \ldots, C_K\}$ and the community structure achieved by the algorithm is $C' = \{C'_1, C'_2, C'_3, \ldots, C'_K\}$. For each C'_i, its entropy can be measured as:

$$\text{entropy}\left(C'_i\right) = \sum\nolimits_{j=1}^{K} \Pr_i\left(C_j\right) \log_2 \Pr_i\left(C_j\right) \tag{4}$$

where $\Pr_i\left(C_j\right)$ is the proportion of C_j data points in C'_i. So the total entropy of C' is given in Eq. 5. Lesser the value of entropy, better the community structure is.

$$\text{entropy}_{\text{total}} = \sum\nolimits_{i=1}^{K} \frac{n_i}{n} \text{entropy}\left(C'_i\right) \tag{5}$$

where n is the total number of entities and n_i is the number of entities in C'_i.

The Normalized Mutual Information (NMI) [19] between C and C′ is given as:

$$\mathrm{NMI}(C, C') = \frac{\sum_{i=1}^{K} \sum_{j=1}^{K} n_i^j \log\left(\frac{n.n_i^j}{n_i.n_j}\right)}{\sqrt{\left(\sum_{i=1}^{K} n_i \log \frac{n_i}{n}\right)\left(\sum_{j=1}^{K} n_j \log \frac{n_j}{n}\right)}} \quad (6)$$

where n_i is the number of entities in C_i', n_j is the number of entities in C_j and n_i^j is the number of common entities in C_i' and C_j. Higher the NMI, closer will be the partition to the ground truth.

Another metric that we have used is pair-wise F-score. Let T denotes the set of node-pairs having the same label; S denotes the set of node-pairs assigned to the same community. The pair-wise F-score can be computed as:

$$\begin{aligned} \mathrm{precision} &= \frac{|S \cap T|}{|S|} \quad \mathrm{recall} = \frac{|S \cap T|}{|T|} \\ \mathrm{PWF} &= \frac{2 \times \mathrm{precision} \times \mathrm{recall}}{\mathrm{precision} + \mathrm{recall}} \end{aligned} \quad (7)$$

4.3 Performance of the Proposed Community Discovery Framework

In our framework any traditional similarity based clustering algorithm can be plugged in for community discovery task. But for this experimental setup, we have used K-means algorithm for clustering. The rank of factorization is an important aspect on which the quality of reconstruction achieved depends. In our experiments, different ranks of factorization are used for different datasets. Rank of factorization of TM1, TM2 and TM3 is 10, 40 and 60 respectively.

We have compared the performance of our proposed scheme (NTF_CE) with the three baseline approaches as follows:

- Mutual Reinforcement K-means (MRK) based on the idea of mutual reinforcement clustering mentioned in Sect. 2.
- K-means algorithm applied to only factor matrix A (AK).
- K-means algorithm applied to only factor matrix B (BK).

We investigate the overall performance of our proposed scheme for above mentioned metrics. Table 2 shows the NMI, entropy value and pair-wise F-score of all the algorithms based on all three datasets. We observe that our proposed algorithm has shown considerably better performance than mutual reinforcement K-means. It has also seen that the performance of the community structure obtained by combining clustering results of factor matrices A and B yields better results than community structure obtained by any of individual matrix. The recall and precision values of different algorithms based on all datasets are also evaluated. Figure 6 shows the plotted values for both precision and recall.

Table 2. Performance of various algorithms on 20Newsgroup datasets

Algorithms	TM1			TM2			TM3		
	NMI	Entropy	PWF	NMI	Entropy	PWF	NMI	Entropy	PWF
MRK	0.4556	0.6887	0.6652	0.4775	0.8571	0.5774	0.5560	0.9876	0.5268
AK	0.6441	0.3635	0.8380	0.1286	1.5460	0.4861	0.2087	1.7756	0.3980
BK	0.6252	0.3835	0.8264	0.6998	0.5936	0.7113	0.5345	1	0.5243
NTF_CE	0.9091	0.1103	0.9655	0.7472	0.3936	0.7667	0.6067	0.9444	0.6648

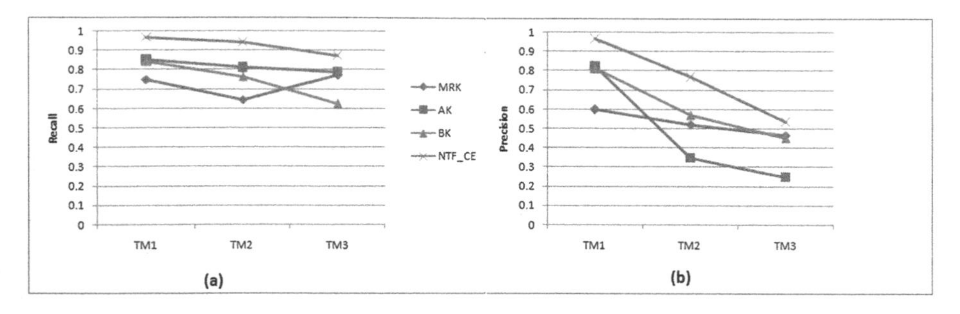

Fig. 6. Recall for different algorithms for datasets TM1, TM2, TM3 (b) Precision for different algorithms for datasets TM1, TM2, TM3

5 Conclusion and Future Work

The objective of our proposed work is to model detection of communities in heterogeneous social networks. The basic idea of the proposed scheme is to employ tensor factorization as a learning tool for exploiting relational information from the network in order to reveal latent features of the entities. Once the latent features of the network are revealed, we applied cluster ensemble approach to find out final community structure. We have performed extensive experimental study to prove the efficacy of our proposed framework.

As a future work, exploration of dynamics and temporal aspects of detected communities can be considered [12, 20]. The notions of trust, distrust, reputation and social- psychological theories find its successful application in social network analysis [1, 2]. We will try to incorporate these notions in community discovery process.

References

1. Anand, D., Bharadwaj, K.K.: Pruning trust–distrust network via reliability and risk estimates for quality recommendations. Soc. Netw. Anal. Min. **3**(1), 65–84 (2012)
2. Bharadwaj, K.K., Al-Shamri, M.Y.H.: Fuzzy computational models for trust and reputation systems. Electron. Commer. Res. Appl. **8**(1), 37–47 (2009)

3. Cai, D., Shao, Z., He, X., Yan, X., Han, J.: Mining hidden community in heterogeneous social networks. In: Proceedings of 3rd International Workshop on Link discovery, pp. 58–65 (2005)
4. Chen, Y., Wang, L., Dong, M.: Non-negative matrix factorization for semi-supervised heterogeneous data co-clustering. IEEE Trans. Knowl. Data Eng. **22**(10), 1459–1474 (2010)
5. Dhillon, I.S.: Co-clustering documents and words using bipartite spectral graph partitioning. In: Proceedings of 7th ACM SIGKDD International Conference Knowledge Discovery Data Mining, pp. 269–274 (2001)
6. Franz, T., Schultz, A., Sizov, S., Staab, S.: TripleRank: ranking semantic web data by tensor decomposition. In: Bernstein, A., Karger, D.R., Heath, T., Feigenbaum, L., Maynard, D., M, E., T, Krishnaprasad (eds.) ISWC 2009. LNCS, vol. 5823, pp. 213–228. Springer, Heidelberg (2009)
7. Gao, B., Liu, T.Y., Zheng, X., Cheng, Q.S., Ma, W.Y.: Consistent bipartite graph co-partitioning for star-structured high-order heterogeneous data co-clustering. In: Proceedings of ACM KDD, pp. 41–50 (2005)
8. Getoor, L.: An introduction to probabilistic graphical models for relational data. IEEE Data Eng. Bull. **29**(1), 32–39 (2006)
9. Goder, A., Filkov, V.: Consensus clustering algorithms: comparison and refinement. In: Proceedings of 10th Workshop on Algorithm Engineering and Experiments, pp. 109–117 (2008)
10. Holder, L.B., Cook, D.J.: Graph-based relational learning: current and future directions. ACM SIGKDD Explorations Newsletter **5**(1), 90–93 (2003)
11. Lang, K.: News weeder: learning to filter netnews. In: Proceedings of International Conference on Machine Learning, pp. 331–339 (1995)
12. Lin, Y.R., Chi, Y., Zhu, S., Sundaram, H., Tseng, B.L.: Analyzing communities and their evolutions in dynamic social networks. ACM Transactions on Knowledge Discovery from Data **3**(2), 1–31 (2009)
13. Long, B., Philip, S.Y., Zhang, Z.: A general model for multiple view unsupervised learning. In: Proceedings of SIAM International Conference on Data Mining, pp. 822–833 (2008)
14. Long, B., Zhang, Z.M., Wu, X., Yu, P.S.: Spectral clustering for multi-type relational data. In: Proceedings of 23rd International Conference on Machine learning, pp. 585–592 (2006)
15. Long, B., Zhang, Z.M., Yu, P.S.: A probabilistic framework for relational clustering. In: Proceedings of 13th ACM SIGKDD International Conference on Knowledge Discovery Data Mining, pp. 470–479 (2007)
16. Mei, J.P., Chen, L.: A fuzzy approach for multi-type relational data clustering. IEEE Trans. Fuzzy Syst. **20**(2), 358–371 (2012)
17. Nguyen, N., Caruana, R.: Consensus clusterings. In: Proceedings of 7th IEEE International Conference on Data Mining, pp. 607–612 (2007)
18. Nickel, M., Tresp, V., Kriegel, H.P.: A three-way model for collective learning on multi-relational data. In: Proceedings of 28th International Conference on Machine Learning, pp. 809–816 (2011)
19. Strehl, A., Ghosh, J.: Cluster ensembles—a knowledge reuse framework for combining multiple partitions. J. Mach. Learn. Res. **3**, 583–617 (2003)
20. Tang, L., Liu, H., Zhang, J., Nazeri, Z.: Community evolution in dynamic multi-mode networks. In: Proceedings of 14th ACM SIGKDD International Conference on Knowledge Discovery Data Mining, pp. 677–685 (2008)
21. Tang, L., Wang, X., Liu, H.: Community detection via heterogeneous interaction analysis. Data Min. Knowl. Disc. **25**(1), 1–33 (2011)

22. Wang, J., Zeng, H., Chen, Z., Lu, H., Tao, L., Ma, W.Y.: ReCoM: reinforcement clustering of multi-type interrelated data objects. In: Proceedings of 26th Annual International ACM SIGIR Conference on Research and Development in Information Retrieval, pp. 274–281 (2003)
23. Wu, Z., Yin, W., Cao, J., Xu, G., Cuzzocrea, A.: Community detection in multi-relational social networks. In: Proceedings of International Conference on Web Information Systems Engineering, pp. 43–56 (2013)
24. Zhou, D., Burges, C.J.: Spectral clustering and transductive learning with multiple views. In: Proceedings of 24th International Conference on Machine learning, pp. 1159–1166 (2007)

On the Impact of Post-clustering Phase in Multi-way Spectral Partitioning

R. Jothi(✉), Sraban Kumar Mohanty, and Aparajita Ojha

Indian Institute of Information Technology, Design and Manufacturing Jabalpur,
Jabalpur, Madhya Pradesh, India
{r.jothi,sraban,aojha}@iiitdmj.ac.in
http://www.iiitdmj.ac.in

Abstract. Spectral clustering is one of the most popular modern graph clustering techniques in machine learning. By using the eigenvalue analysis, spectral methods partition the given set of points into number of disjoint groups. Spectral methods are very useful in determining non-convex shaped clusters, identifying such clusters is not trivial for many traditional clustering methods including hierarchical and partitional methods. Spectral clustering may be carried out either as recursive bi-partitioning using fiedler vector (second eigenvector) or as muti-way partitioning using first k eigenvectors, where k is the number of clusters. Although spectral methods are widely discussed, there has been a little attention on which post-clustering algorithm (for eg. K-means) should be used in multi-way spectral partitioning. This motivated us to carry out an experimental study on the influence of post-clustering phase in spectral methods. We consider three clustering algorithms namely K-means, average linkage and FCM. Our study shows that the results of multi-way spectral partitioning strongly depends on the post-clustering algorithm.

Keywords: Clustering · Spectral partitioning · Muti-way partitioning

1 Introduction

Clustering is the process of discovering natural grouping of objects so that objects within the same cluster are similar and objects from different clusters are dissimilar according to certain similarity measure. Clustering algorithms include wide range applications such as pattern recognition, bio-informatics, social network analysis and satellite image segmentation. Various methods for clustering are broadly classified into hierarchical and partitional methods [1].

The Hierarchical clustering methods build a set of nested clusters from the individual points by progressively merging or splitting the clusters. The well known examples of hierarchical clustering methods are BIRCH, CURE and CHAMELEON. Hierarchical clustering methods require pre-computation of dissimilarity matrix of $O(n^2)$ complexity which limits the application of these algorithms to very large datasets [2].

R. Prasath et al. (Eds.): MIKE 2015, LNAI 9468, pp. 161–169, 2015.
DOI: 10.1007/978-3-319-26832-3_16

Partitioning algorithms divide the points into k groups, where k is the number of clusters to be specified by the user in advance. K-means is widely used partitional clustering method due to its simplicity and linear time complexity. But the incorrect choice of initial centers may degrade the performance of the K-means algorithm [1,2]. Moreover, K-means may not perform well when the clusters are not defined by regular geometric curves [3].

Detecting clusters of non-convex shapes is becoming a research interest in recent years. Spectral methods have been widely used for this purpose as they can detect non-convex shaped clusters, which is not trivial for traditional clustering algorithms such as K-means and hierarchical clustering. By using the eigenvalues of the Laplacian matrix, which is the difference between degree and adjacency (weight) matrices of the data similarity matrix, these methods partition the given dataset into disjoint groups [4]. Spectral clustering may be carried out as either recursive bi-partitioning [5] or multiway partitioning [6].

Although spectral clustering has rich literature, to the best of our knowledge there has been a little attention on which post-clustering algorithm (for eg. K-means) should be used in multi-way spectral partitioning. This motivated us to carry out an experimental study to observe the impact of post-clustering phase in spectral methods. We consider three clustering algorithms namely K-means, Average Linkage (AL) and Fuzzy C-means (FCM) [1,2]. This study shows that the results of multi-way spectral partitioning strongly depends on the post-clustering algorithm.

The rest of the paper is organized as follows. The description of spectral clustering algorithm is given in Sect. 2. The experimental analysis is shown in Sect. 3. Discussion on the results is given in Sect. 4. The conclusion and future scope are given in Sect. 5.

2 Spectral Clustering Algorithm

Spectral clustering is a popular clustering method which makes use of the eigenvectors of the data similarity graph to partition the dataset. Lots of work has been carried out on spectral clustering and the notable references can be seen in [5–8]. The important prepossessing step in spectral clustering is graph construction from the given set of points $X = \{x_1, x_2, \cdots, x_n\}$. The choice of similarity graph that is used to model the given dataset has a great impact on the results of spectral methods. Different similarity graphs have been studied in the literature and the most commonly used graph construction methods are as follows [9]:

Definition 1 [ε-neighborhood Graph]: Given a set of points (vertices), ε-neighborhood graph connects the points whose pairwise distance is smaller than ε, where ε is user specified parameter.

Definition 2 [K-Nearest Neighbor Graph]: A pair of points x_i and x_j are said to be connected in KNN- graph, if x_j belongs to the set of K-nearest neighbors of x_i.

Definition 3 [Fully-connected Graph]: All those points with non-zero similarity are connected with each other, where similarity is defined using the gaussian similarity function $s(x_i, x_j) = exp(- \parallel x_i - x_j \parallel^2 /(2\sigma^2))$. Here, the parameter σ is used to control the width of the neighborhood [9].

Once similarity graph is constructed, next step is to obtain Laplacian matrix of the similarity graph as defined below [9].

Unnormalized form: $L = D - A.$

Normalized form: $L_{sym} = D^{-1/2} L D^{-1/2}$

where A is the adjacency matrix (weight matrix) of the similarity graph with each $A(i,j)$ denoting weight of the edge (x_i, x_j); D is the degree matrix which is a diagonal matrix with each $D(i,i)$ representing the total weight of edges incident on vertex x_i. Finally, the first k eigenvectors of the Laplacian matrix are stacked as columns in a $n \times k$ matrix U, where each row $u_i, 1 \leq i \leq n$ is treated as a data point in the projected space. This new representation is called as the transformed space and the clusters are identified by applying a clustering algorithm such as K-means on the matrix U. The spectral clustering algorithm is summarized in Algorithm 1 .

Algorithm 1 *Spectral Multi-way Partitioning Algorithm [9].*

Input: *Dataset X, Number of clusters k.*
Output: *k clusters.*

1 *Construct the similarity graph $G = (V, E)$ of the given dataset X.*
2 *Let A be the adjacency matrix of G, where each $A(i,j) = 1$ if $(x_i, x_j) \epsilon E$, else $A(i,j) = 0$*
3 *Find the degree matrix D of G, where each $D(i,j) = 0, i \neq j$ and $D(i,i) = \sum_{j=1}^{n} A(i,j)$.*
4 *Compute Laplacian Matrix $L = D - A$.*
5 *Compute the first k eigenvectors $v_1, v_2, \cdots, v_k$.*
6 *Form a $n \times k$ matrix U, where each eigenvector $v_i, 1 \leq i \leq k$ is stacked as a column in U.*
7 *Treat each row u_i of U, $1 \leq i \leq n$ as a data point.*
8 *Apply a post-clustering algorithm such as K-means to partition the points $(u_i)_{i=1,2,\cdots,n}$ into k clusters.*

K-means algorithm is widely used algorithm in spectral multi-way partitioning [9]. To the best of our knowledge there is no theoretical study on the choice of clustering algorithm to be applied on the projected space. Consider a sample dataset with two clusters as shown in Fig. 1. The KNN graph with

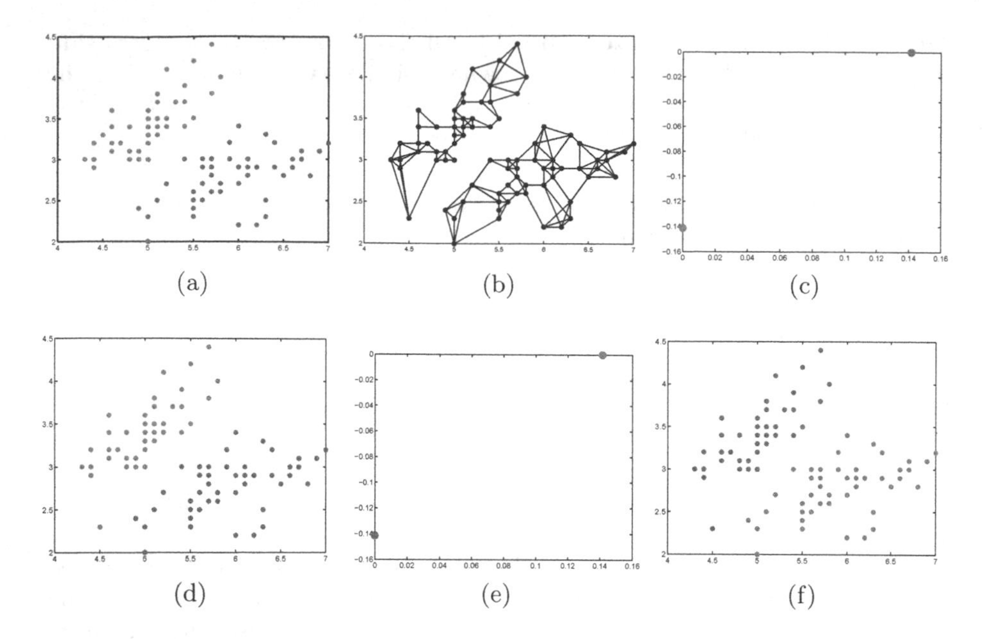

Fig. 1. Spectral muti-way partitioning with identical results from different post-clustering algorithms : (a) A given dataset with two clusters. (b) KNNG graph (k=5) (c) Applying K-means as a multi-way partitioning algorithm on the projected space. (d) Mapping of results in (c) to original space. (e) Applying average linkage as a multi-way partitioning algorithm on the projected space. (f) Mapping of results in (e) to original space.

$K = 5$ is shown in Fig. 1b. Here the clusters are clearly separated. And hence spectral clustering produces same partitioning on both the versions i.e. spectral muti-way partitioning with K-means and Average Linkage (AL). But if there is a lot of overlapping between two clusters as shown in Fig. 2, then we may obtain different results from muti-way partitioning with K-means and AL. It is clear from the figures Figs. 1 and 2 that the performance of the spectral clustering depends on the post-clustering algorithm used in the k-way partitioning.

3 Experimental Study

The motivation of this experimental study is to observe the impact of post-clustering phase in spectral multi-way partitioning. Hence we apply three well known clustering algorithms such as K-means, Average Linkage (AL) and FCM during the post-clustering phase and accordingly the algorithms are referred as SC-Kmeans, SC-AL and SC-FCM respectively. The results of spectral clustering is compared against standard versions of K-means, AL and FCM. The graph construction technique that we used here is mutual K-Nearest Neighborhood graph with K=5.

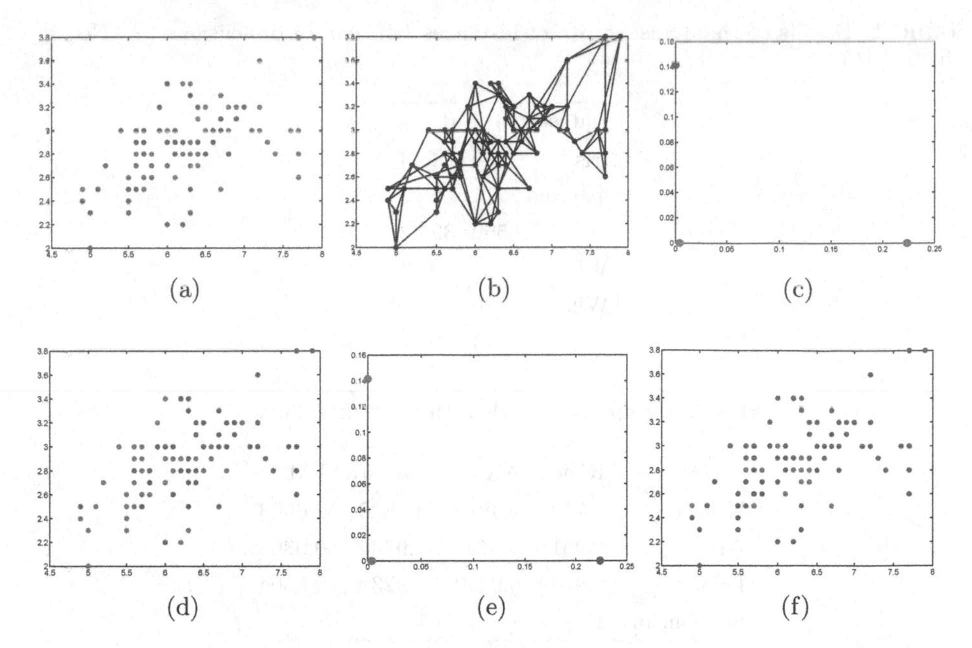

Fig. 2. Spectral muti-way partitioning with identical results from different post-clustering algorithms : (a) A given dataset consisting of two clusters with overlapping. (b) KNNG graph (k=5) (c) Applying K-means as a multi-way partitioning algorithm on the projected space. (d) Mapping of results in (c) to original space. (e) Applying average linkage as a multi-way partitioning algorithm on the projected space. (f) Mapping of results in (e) to original space.

The experimental study considers six real world datasets Iris, Thyroid, Breast Cancer Wisconsin Diagnostic (WDBC), Breast Cancer Wisconsin (WBC), Wine and Glass dataset. All the above datasets are taken from UCI machine learning repository [10]. The details of the datasets are given in Table 1. The comparison criteria used for evaluation are external cluster validity indices such as Rand index, Adjusted Rand index (ARand), Jaccard coefficient and Fowlkes and Mallows (FM) [11]. The ideal partitioning is achieved when these indices attain 1.

Iris dataset consists of 3 classes (Iris setosa, Iris virginica and Iris versicolor) of 50 samples each, where each class refers to a type of iris plant. The evaluation results on Iris dataset is shown in Table 2. It is observed that both SC-Kmeans and SC-FCM outperforms the other algorithms. Among the spectral methods, the SC-AL version performs slightly lower than SC-Kmeans and SC-FCM. It can also be seen that FCM and SC-AL algorithms obtain same partitioning results as indicated by the all the validity indices.

Table 3 demonstrates the performance of the various algorithms on Thyroid dataset which contains 215 samples of 3 classes. The results show that FCM performs relatively better than others. SC algorithms achieves closer results to that of FCM. Among the spectral variants SC-Kmeans and SC-FCM obtains slightly improved results than SC-AL.

Table 1. Details of the datasets: No. of instances (n), No. of dimensions (d), No. of clusters (k).

Dataset	n	d	k
Iris	150	4	3
Thyroid	215	5	3
WDBC	569	32	2
WBC	683	9	2
Wine	178	13	3
Glass	214	9	7

Table 2. Comparison of algorithms on Iris dataset.

Algorithm	Rand	ARand	Jaccard	FM
K-means	0.7197	0.4289	0.4820	0.6613
AL	0.9911	0.9799	0.9734	0.9865
FCM	0.9739	0.9410	0.9239	0.9604
SC-Kmeans	1	1	1	1
SC-AL	0.9739	0.9410	0.92399	0.9604
SC-FCM	1	1	1	1

Table 3. Comparison of algorithms on Thyroid dataset.

Algorithm	Rand	ARand	Jaccard	FM
K-means	0.7965	0.5832	0.7146	0.8410
AL	0.5794	0.1114	0.5543	0.7422
FCM	0.8329	0.6592	0.7536	0.8645
SC-Kmeans	0.8193	0.6307	0.7387	0.8557
SC-AL	0.7639	0.5142	0.6841	0.8227
SC-FCM	0.8193	0.6307	0.7387	0.8557

Breast Cancer Wisconsin (Diagnostic) (WDBC) dataset consist of 569 instances of cell nucleus of breast cancer tumors. Table 4 illustrates the performance of various algorithms on WDBC dataset. K-means, FCM and SC-AL obtains similar partitioning as indicated by the quality measures. Similar to the results on iris and thyroid, SC-AL performs slightly poorer than other SC versions. It is also worthwhile to note that SC-Kmeans outperforms other algorithms.

Wisconsin Breast Cancer Database (WBC) consists of 699 breast cancer samples which fall into two groups Benign and Malignant. The results of the various algorithms on WBC is shown in Table 5. In this dataset, K-means algorithm outperforms the others. Among the spectral variants, SC-FCM seems to perform better than others.

Table 4. Comparison of algorithms on WDBC dataset.

Algorithm	Rand	ARand	Jaccard	FM
K-means	0.7504	0.4914	0.6499	0.7915
AL	0.5521	0.0523	0.5314	0.7215
FCM	0.7504	0.4914	0.6499	0.7915
SC-Kmeans	0.8055	0.6103	0.6862	0.8140
SC-AL	0.7504	0.4914	0.6499	0.7915
SC-FCM	0.7973	0.5931	0.6786	0.8086

Table 5. Comparison of algorithms on WBC dataset.

Algorithm	Rand	ARand	Jaccard	FM
K-means	0.9099	0.8182	0.8481	0.9179
AL	0.8794	0.7558	0.8043	0.8918
FCM	0.9047	0.8076	0.8406	0.9135
SC-Kmeans	0.6809	0.3402	0.5895	0.7473
SC-AL	0.5475	0.0144	0.5425	0.7347
SC-FCM	0.7882	0.5674	0.6927	0.8205

Table 6. Comparison of algorithms on Wine dataset.

Algorithm	Rand	ARand	Jaccard	FM
K-means	0.7187	0.3711	0.4120	0.5835
AL	0.6262	0.2926	0.4247	0.6192
FCM	0.7105	0.3539	0.4013	0.5728
SC-Kmeans	0.5776	0.0682	0.2433	0.3915
SC-AL	0.6016	0.1676	0.3192	0.4877
SC-FCM	0.6016	0.1676	0.3192	0.4877

Wine dataset contains 178 samples collected for chemical analysis to determine the wine quality. The results shown in Table 6 report that K-means algorithm outperforms the others. It is also seen that spectral variants SC-AL and SC-FCM produces similar results. The results shown in Table 7 report the results of different algorithms on glass dataset.

Figure 3 shows the comparison of various algorithms according to Adjusted Rand index (ARand). Table 8 shows the best performer on each of the datasets according to different quality measures. It is clear that the results of spectral clustering (k-way partitioning) depend on the choice of post-processing phase.

Table 7. Comparison of algorithms on Glass dataset.

Algorithm	Rand	ARand	Jaccard	FM
K-means	0.7932	0.3919	0.3473	0.5282
AL	0.6275	0.3103	0.3872	0.6045
FCM	0.7915	0.3758	0.3304	0.5136
SC-Kmeans	0.7610	0.3087	0.2947	0.4631
SC-AL	0.7708	0.3484	0.3263	0.4978
SC-FCM	0.7540	0.3126	0.3070	0.4732

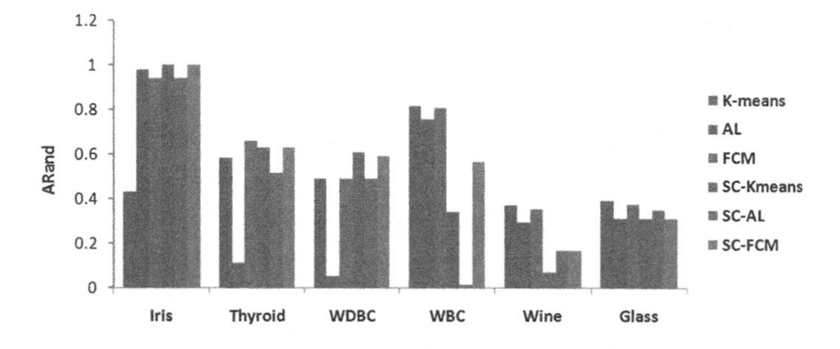

Fig. 3. Comparison of different algorithms according to Adjusted Rand index (ARnad).

Table 8. The best performing algorithm on each of the datasets.

Dataset	Rand	ARand	Jaccard	FM
Iris	SC-Kmeans, SC-FCM	SC-Kmeans, SC-FCM	SC-Kmeans, SC-FCM	SC-Kmeans, SC-FCM
Thyroid	FCM	FCM	FCM	FCM
WDBC	SC-Kmeans	SC-Kmeans	SC-Kmeans	SC-Kmeans
WBC	FCM	FCM	FCM	FCM
Wine	K-means	K-means	AL	AL
Glass	K-means	K-means	AL	AL

4 Discussion

Spectral clustering has gained popularity for detecting clusters of irregular shapes as it does not assume the shape of clusters. By the change of representation of the points from the original space into the new transformed space, spectral clustering outperforms the traditional algorithms such as K-means and hierarchical methods [9]. If the clusters are well separated, the eigenvectors of the Laplacian matrix L are piecewise constant [9]. Here all the data points belonging to a cluster map to the same point u_i in the new transformed space. In such a case, applying a clustering algorithm such as K-means can easily extract the intended clusters. But for the dataset with highly overlapping clusters as shown in Fig. 2a, even the spectral methods may fail to extract the correct clusters as the eigenvectors of the Laplacian matrix may not be piecewise constant. And hence the problem of detecting arbitrary shaped clusters faced by the traditional algorithms is also inherited in the spectral methods.

Through the experimental analysis, we summarize the following observations. First, there is a variation in the performance of the spectral methods depending on the choice of post-clustering method used. Secondly, the experimental results suggest to explore different classical algorithms in the post-processing phase as SC-AL and SC-FCM are able to achieve relatively better cluster accuracy as compared to SC-K-means. Third, although spectral methods have proven to perform better than the standard versions of the algorithms, they may not provide improved results on all the datasets. This is again confirming the fact that no universal clustering algorithm exist which can detect correct clusters on all the datasets.

5 Conclusion

This paper carried out an experimental study on spectral multi-way partitioning with respect to the choice of algorithm used in the post-clustering phase. From the experimental analysis on six real world datasets, we conclude that the results of spectral clustering depend on the choice of clustering method utilized in multi-way partitioning. As a future work we carry out an extensive analysis of different clustering algorithms in spectral partitioning context.

References

1. Jain, A.K., Murty, M.N., Flynn, P.J.: Data clustering: a review. ACM Comput. Surv. (CSUR) **31**(3), 264–323 (1999)
2. Xu, R., Wunsch, D., et al.: Survey of clustering algorithms. IEEE Trans. Neural Netw. **16**(3), 645–678 (2005)
3. Jain, A.K.: Data Clustering: User's Dilemma. In: Perner, P. (ed.) MLDM 2007. LNCS (LNAI), vol. 4571, pp. 1–1. Springer, Heidelberg (2007)
4. Schaeffer, S.E.: Graph clustering. Comput. Sci. Rev. **1**(1), 27–64 (2007)
5. Shi, J., Malik, J.: Normalized cuts and image segmentation. IEEE Trans. Pattern Anal. Mach. Intell. **22**(8), 888–905 (2000)
6. Ng, A.Y., Jordan, M.I., Weiss, Y., et al.: On spectral clustering: Analysis and an algorithm. Adv. Neural Inf. Process. Syst. **2**, 849–856 (2002)
7. Verma, D., Meila, M.: A comparison of spectral clustering algorithms. Technical report (2003)
8. Jordan, F., Bach, F.: Learning spectral clustering. Adv. Neural Inf. Process. Systems **16**, 305–312 (2004)
9. Von Luxburg, U.: A tutorial on spectral clustering. Stat. Comput. **17**(4), 395–416 (2007)
10. Repository, U.M.L. http://archive.ics.uci.edu/ml
11. Halkidi, M., Batistakis, Y., Vazirgiannis, M.: On clustering validation techniques. J. Intell. Inf. Syst. **17**(2), 107–145 (2001)

BSO-CLARA: Bees Swarm Optimization for Clustering LARge Applications

Yasmin Aboubi(✉), Habiba Drias, and Nadjet Kamel

LRIA, USTHB, BP 32 El Alia, Bab Ezzouar Algiers, Algeria
{yasminaboubi,nadjet.kamel}@gmail.com,
h_drias@hotmail.fr
http://www.lria.usthb.dz

Abstract. Clustering is an essential data mining tool for analyzing big data. In this article, an overview of literature methods is undertaken. Following this study, a new algorithm called BSO-CLARA is proposed for clustering large data sets. It is based on bee behavior and k-medoids partitioning. Criteria like effectiveness, eficiency, scalability and control of noise and outliers are discussed for the new method and compared to those of the previous techniques. Experimental results show that BSO-CLARA is more effective and more efficient than PAM, CLARA and CLARANS, the well-known partitioning algorithms but also CLAM, a recent algorithm found in the literature.

Keywords: Clustering algorithms · Bee Swarm Optimization (BSO) · Metaheuristics · Medoids

1 Introduction

One of the most important tasks for data mining is data clustering because of its numerous applications for a large spectrum of domains. Several methods have been developed for adapted situations but the challenge remains to find more effective and more efficient algorithms that respond to the majority of data mining criteria such as effectiveness, efficiency and scalability. In this paper, we restricted our study to partitioning algorithms. For that aim, a study of existing methods was undertaken in order to better master respectively the techniques with their advantages and drawbacks. More precisely, the classic algorithms k-means [2], PAM (Partitioning Around Medoids) [1], CLARA (Clustering LARge Applications) [1] and CLARANS (Clustering Large Application RANdomized Search) [5] were explored. In addition, recent approaches based on metaheuristics were also investigated. The most interesting one called CLAM (Clustering Large Applications using Metaheuristics) [3] is presented in details. For each studied method, we identify its benefits and inconvenient in order to be able to come up with a new method gathering the most benefits and eliminating the most disadvantages. The algorithm we proposed is called BSO-CLARA (BSO for Clustering LARge Applications). It combines BSO (Bees Swarm Optimization)

R. Prasath et al. (Eds.): MIKE 2015, LNAI 9468, pp. 170–183, 2015.
DOI: 10.1007/978-3-319-26832-3_17

metaheuristic with PAM. BSO is used to explore the solutions space and PAM helps to build the clusters. The new algorithm is presented and discussed through its complexity, benefits and drawbacks. Extensive experiments were performed to validate its effectiveness and efficiency. The results are exposed and analysed in comparison with all the studied algorithms found in the literature.

This paper is organized as follows. Section 2 describes the clustering problem based on partitions. Section 3 provides an overview of the classic and recent partitioning algorithms found in the literature. Section 4 outlines BSO-CLARA the proposed algorithm. Section 5 presents the results of the experiments that were conducted for the validation of the proposal.

2 An Overview on Partitioning Methods

Partitioning n objects over k groups is an old and fundamental problem, which is NP-hard. The objective is to put together the objects having similar characteristics in one cluster. We need therefore to define a similarity measure between the objects. An effective partitioning is such that the similarity between objects of the same cluster is maximal and the similarity between objects of different clusters is minimal. In order to facilitate these measures, clusters are represented by centers. Concretely, the problem is to minimize the sum of the dissimilarities between each object and the center of the cluster to which it belongs. This quantity, expressed by Formula (1), represents the absolute error, also called inertia of the partitioning. E is the sum of the absolute error for all objects in the data set, p is a given object in cluster C_j and o_j is the center or representative object of C_j, k being the number of clusters.

$$E = \sum_{j=1}^{k} \sum_{p \in C_j} d(p, o_j) \tag{1}$$

The data mining community has recently deployed a lot of efforts on developing fast algorithms for partitioning very large data sets. The most known techniques are k-means, k-medoids, CLARA and CLARANS. In this section, we present their respective algorithm as well as their benefits and disadvantages.

2.1 k-Means

k-means is the most widely used clustering algorithm because it is simple to implement. The algorithm is outlined in Algorithm 1. It starts by drawing at random k centers then it assigns each objet to a cluster according to its distance with the cluster center. The means is calculated for each cluster once all the objects are inserted in the clusters and becomes the new center. This process is repeated until no changes occur in the centers. This algorithm is known to be not effective enough because of the representation of the center as the means of the objects residing inside the cluster. However it is efficient as it consists of one loop inside which, we dispatch the objects over the clusters and calculate the means with a simple formula.

Algorithm 1. k-means

```
Input:
   k the number of clusters;
   D a set of n objects;
Output:
  a set of k clusters;
begin
  1. Select k initial centers arbitrarily from D;
  2. repeat until no changes in clusters centers
      2.1. for each object do
            2.1.1. calculate its distance from
                       the center of each cluster;
            2.1.2. assign the object to the cluster
                      with the nearest center;
   2.2. compute the means of each cluster
                     and update its center;
end.
```

2.2 PAM

k-medoids also called PAM (Partition Around Medoids) was designed by Kaufman and Roussew in 1987 [1] to palliate to the k-means drawbacks. It shares the same algorithmic structure with k-means but uses as cluster representative, its medoid. It was the first k-medoids algorithm introduced in the literature [4]. The fact to substitute the means by the medoid makes the algorithm more effective because the medoid position in the cluster is central. Therefore, the error value of Formula (1) is less than that calculated by the means. Another benefit is the insensitivity to noise and outlier. Algorithm 2 outlines k-medoids.

The complexity of the algorithm is $O(k * (n - k)^2)$. For large values of n and k, such computation becomes very costly. The algorithm is then less efficient than k-means.

2.3 CLARA (Clustering LARge Application)

CLARA designed by Kaufmann and Rousseeuw in 1990, is an improvement of PAM to handle large data sets. It decreases the overall quadratic complexity and time requirements into linear in total number of objects [8]. It runs PAM on multiple random samples, instead of the whole dataset. If the samples are sufficiently random and numerous, the medoids of the samples approximate the medoids of the dataset. Experiments reported in [1] indicate that five samples of size $(40 + 2k)$ give satisfactory results. CLARA is described in Algorithm 3.

Algorithm 2. k-medoids

```
Input:
   k the number of clusters;
   D a set of n objects;
Output:
   a set of k clusters;
begin
   1. Select k initial medoids arbitrarily from D;
   2. repeat until no changes in clusters medoids
   2.1. for each object do
        2.1.1. calculate its distance from the
               medoid of each cluster;
        2.1.2. assign the object to the cluster with
               the nearest medoid;
   2.2. for each medoid m do
        2.2.1. min = sum of the distances from m
               to the other objects;
        2.2.2. for each non-medoid object o do
               2.2.2.1. calculate the sum of the
                        distances From o to the others;
               2.2.2.2. if the sum is less than min
                        then swap o and o and update min;
end
```

Algorithm 3. CLARA

```
Input:
   k the number of clusters;
   D a set of n objects;
Output:
   a set of k clusters
begin
   1. For i = 1 to 5, repeat the following steps:
      1.1. Draw a sample of (40+2k) objects randomly from the
           entire data set and call Algorithm PAM to find
           k medoids of the sample
      1.2. For each object Oj in the entire data set, determine
           which of the k medoids is the most similar and assign it to
           the corresponding cluster;
```

```
      1.3. Calculate the average dissimilarity of the clustering
           obtained in the previous step. If this value is
           less than the current minimum, use this value as the
           current minimum and retain the k medoids found in Step 2
           as the best set of medoids obtained so far;
   2. Return to Step 1 to start the next iteration;
end.
```

It has been shown to produce relatively good quality solutions in a reasonable computation time for large data sets but it is less effective as it considers samples and not the entire datasets.

2.4 CLARANS (Clustering Large Application RANdomized Search)

CLARANS (Ng and Han 1994) is proposed in order to improve the effectiveness in comparison to CLARA. It aims at using randomized search to facilitate the clustering of a large number of objects by searching the entire dataset. It uses a sampling technique to reduce the search space and the sampling is conducted dynamically for each iteration of the search procedure.

Algorithm 4. CLARANS

```
Input:
   k the number of clusters;
   D a set of n objects;
   parameters numlocal and maxneighbor;
Output:
   a set of k clusters;
begin
   1. Initialize i to 1, and mincost to a large number;
   2. Set current to an arbitrary node in G(n,k);
   3. Set j to 1;
   4. Consider a random neighbor S of current
      and based on Eq. (1), calculate the cost differential
      of the two nodes;
   5. If s has a lower cost then set current to S,
      and go to Step 3;
   6. Otherwise, increment j by 1; If j=maxneighbor then
      go to Step 4;
```

```
    7. Otherwise, when j >maxneighbor, compare the cost
       of current with mincost. If the former is
       less than mincost, set mincost to the cost
       of current and set bestnode to current;
    8. Increment i by 1; If  i >numlocal, output bestnode,
       Otherwise,  go to Step 2;
end
```

Conceptually, the clustering process can be viewed as a search through a graph $G_{n,k}$, where each vertex is a collection of k medoids $O_{m1}, \ldots, O_{mk}$. Two nodes $S_1 = O_{m1}, \ldots, O_{mk}$ and $S_2 = O_{w1}, ..., O_{wk}$ are neighbors (that is, connected by an eadge in the graph) if their sets differ by only one object $|S1 \cap S2| = k - 1$. It is easy to see that each node has $k(n - k)$ neighbors. Each node can be assigned a cost that is defined by the total dissimilarity between every object and the medoids of its cluster. At each step, PAM examines all of the neighbors of the current node in its search for a minimum cost solution. CLARANS is outlined in Algorithm 4.

The observed results show that dynamic sampling used in CLARANS is more effective than the method used in CLARA in large application and more efficient than the swap phase of PAM in small application [6]. It outperforms then all the previous partitioning methods described previously.

These algorithms were implemented in order to compare the effectiveness and the efficiency of our proposed method to them. The results are presented in Sect. 5.

2.5 Recent Related Works

Among the very recent efforts, many algorithms using metaheuristics were designed to address the clustering problem for large data sets. Tsai et al. in [9], presented an interesting review of eight kinds of recent metaheuristics that were developed for clustering during the last decade. In another study [10], shohdohji et al. proposed a new algorithm for clustering a large amount of dataset based on Ant Colony Optimization (ACO). The Ant Colony Clustering (ACC) algorithm uses artificial ants cooperating together to solve a problem of clustering. In [3], Nguyen and Rayward-Smith proposed another algorithm for clustering large application called CLAM. The latter is a hybridization of two metaheuristics that are Variable Neighborhood Search (VNS) and Tabou Search. In the experimental results, CLAM is shown to be more effective than CLARANS in most situations. The only case where CLARANS outperforms CLAM is when both algorithms are set to perform a very small number of moves in the search space. For this reason, CLAM is used as a reference to compare our proposed method.

3 Bees Swarm Optimization (BSO)

The Bees Swarm optimization (BSO) metaheuristic proposed in [7], is inspired by the bees behavior. It simulates a swarm of bees cooperating together to solve a problem. First, the scout bee is represented by an artificial bee named BeeInit. It initializes the solution of reference drawn at random and from which a group of solutions of the search space are determined via a diversification strategy. This set of solutions is called SearchArea. Then, every bee will consider a solution from SearchArea as a starting point to perform a local search. After accomplishing its search, every bee communicates the best visited solution by storing it in a table named *Dance*. The best global solution saved in this table will become the new solution of reference for the next iteration. In order to avoid cycles, the reference solution is stored every time in a taboo list. The algorithm stops when the optimal solution is found or the maximum number of iterations is reached. The framework of this process is outlined in Algorithm 5.

Algorithm 5. BSO

```
Input:
   a search space;
output:
   the best solution;
begin
   1. Sref := BeeInit drawn randomly;
   2. while stop condition is not reached do
      2.1. insert Sref in TL the Tabu List;
      2.2. determine SearchArea from Sref;
      2.3. assign for each bee a solution of SearchArea;
      2.4. for each bee do
           2.4.1. search in the corresponding area;
           2.4.2. insert result in Dance;
      2.5. choose the best solution from Dance and assign it to Sref;
end
```

---

4 BSO-CLARA Algorithm

In this section, we propose an algorithm based on Bees Swarm Optimization for Clustering LARge Application called BSO-CLARA. To adapt BSO for Clustering large datasets, we have to define the following components: the artificial world where the bees live, the fitness function, the distance between two neighbors and the degree of the diversity. BSO-CLARA is described in Algorithm 6.

4.1 Artificial World Where the Bees Lives

The search space is represented by a graph denoted $G_{n,k}$ (n is the number of objects in the dataset and k is the number of clusters). Each node is represented by a set of k objects (a collection of k-medoids). The graph is highly inter-connected; two nodes are neighbors if and only if the cardinality of their intersection is equal to $(k-1)$ that is, if they differ by only one object. Each node has then $k(n-k)$ neighbors.

4.2 The Fitness Function

The fitness function is defined to be the total dissimilarity between every object and the medoids of its cluster. It is defined in (1).

4.3 The Distance Between Two Neighbors

The cost differential between two neighbors S_1 and S_2 is given by TC_{ij} in Eq. (2) where O_i, O_j are the difference between S_1 and S_2 (i.e. O_i and O_j not in $S_1 \cap S_2$, but $O_i \in S_1$ and $O_j \in S_2$)[5].

$$TC_{i,random} = \sum_i C_{jirandom} \tag{2}$$

4.4 The Degree of the Diversity

The degree of diversity offered by a solution S is measured as the minimum distance that separates it from the other solutions not found in tabu list. It is expressed as:

$diversity(S) = min$ { $d(S,T_1)$, $d(S,T_2)$, ...,$d(S,T_m)$ / T_i not in Tabu list and m being the number of bees}

The distance between two solutions S and T_i is given by Eqs. 3 and 4.

$$d(S,T_i) = \sum_{j=1}^{k} S_j \oplus T_{ij} \tag{3}$$

$$a \oplus b = \begin{cases} 0 & \text{if } a{=}b \\ 1 & \text{otherwise .} \end{cases} \tag{4}$$

4.5 Complexity

For each iteration, m solutions are exploited and each one explores $(k(n-k))$ neighbors. So, the complexity of BSO-CLARA algorithm is:
$O(MaxIter * m * k(n-k))$.

Algorithm 6. BSO-CLARA

```
Input:
   k the number of clusters;
   D a set of n objects;
   MaxIter, maxNeighbors, m the number of bees;
Output:
   a set of k clusters;
begin
   1. i=0;
      TabuList := empty;
      let sref  be the set of medoids of BeeInit drawn at random;
      s*=sref;
      Cost(s*);
   2. While (i<MaxIter)
      2.1. TabuList := TabuList + sref;
      2.2. build m solutions not in TabuList distant from sref;
      2.3. affect to each bee a solution (a set of k medoids);
      2.4 for each bee do
          2.4.1. Search();
          2.4.2. store the best Set Of Medoids in Dance table;
      2.5. Sref := best Set Of Medoids in Dance table;
      2.6. If( cost(sref) < cost(s*)) then s*=sref;
      end while
   3. Return s*;
end
```

5 Experimental Results

BSO-CLARA was implemented in java NetBeans 7.1, on a personal computer with an Intel®Core(TM) i7-2600S CPU (2.80 GHz) with 8 GB RAM. The numerical tests have been performed on two benchmarks available from the UCI machine learning repository.

5.1 Description of the Benchmarks

The first dataset called CONCRETE [12] includes 1030 instances, each with 10 attributes, among them eight input attributes, which are cement, blast furnace slag, fly ash, water, superplasticizer, coarse aggregate, fine aggregate and age. We run BSO-CLARA algorithm to cluster the dataset instances into 10 clusters.

The second dataset is the Wisconsin Breast Cancer database [11]. The benchmark contains 699 instances, each object characterized by nine attributes, which are clump thickness, uniformity of cell size, uniformity of cell shape, marginal adhesion, single epithelialcell size, bare nuclei, bland chromatin, normal nucleoli and mitoses.

The instances belong to one of the two possible classes: benign or malignant.

Table 1. Paramater seting

Number of bees	Neighbors	Cost
5	5 %	306,1
	10 %	298,3
	20 %	293,5
	50 %	291,6
	100 %	288,9
10	5 %	297,3
	10 %	293,7
	20 %	291,6
	50 %	288,9
	100 %	288,9
15	5 %	294,9
	10 %	291,6
	20 %	288,9
	50 %	288,9
	100 %	288,9
20	5 %	293,5
	10 %	291,6
	20 %	288,9
	50 %	288,9
	100 %	288,9
25	5 %	291,6
	10 %	288,9
	20 %	288,9
	50 %	288,9
	100 %	288,9
30	5 %	288,9
	10 %	288,9
	20 %	288,9
	50 %	288,9
	100 %	288,9

5.2 Experimentations and Results

The first series of experiments use the data set Wisconsin. Table 1 results aim at setting the empirical parameters. We fix the number of *MaxIter* at 5, then we vary the number of bees from 5 to 30 by step of 5 and the number of neighbors at 5 %, 10 %, 20 %, 50 % and 100 % of $k(n-k)$.

The average cost depends on the number of neighbors and the number of Bees. Figures 1 and 2 illustrate the performance of BSO-CLARA with respect to these parameters. We notice that the fitness decreases with the increase number of bees. The same remark is observed for the number of neighbors.

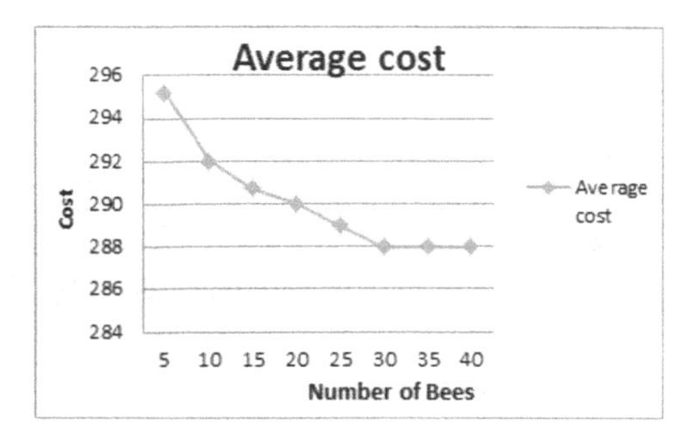

Fig. 1. Number of bees vs. cost

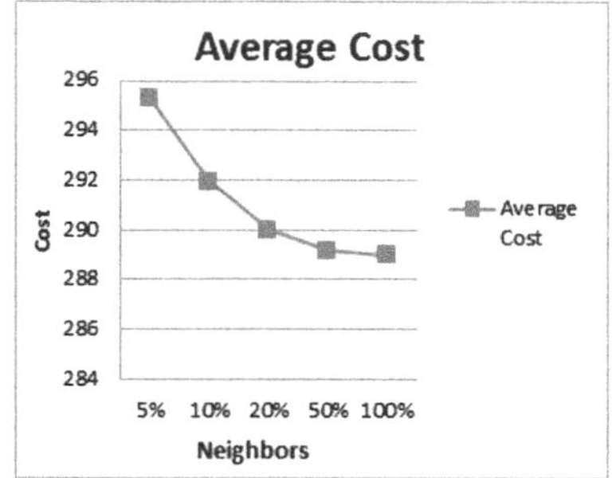

Fig. 2. Neighbors vs. cost

In order to evaluate BSO-CLARA in practice, we compare its performance with that of different k-medoids clustering techniques, using the datasets mentionned previously that is, CONCRETE and Wisconsin Breast Cancer. Five algorithms are used for these experiments: BSO-CLARA, CLARANS, CLAM,

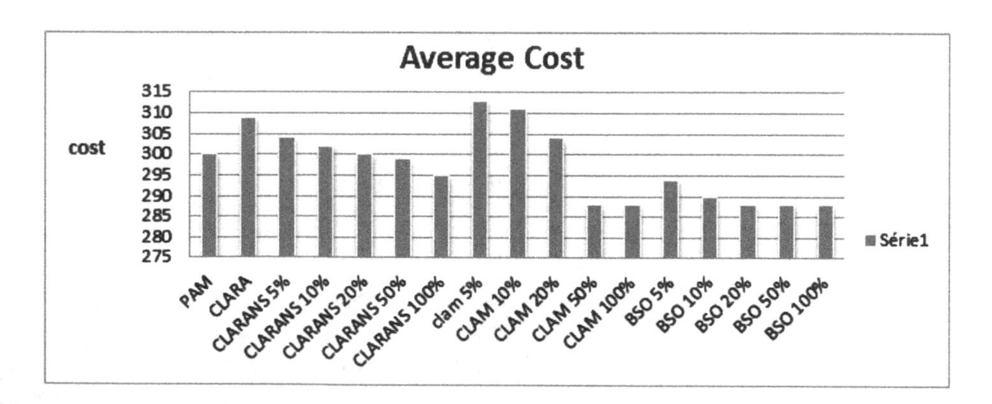

Fig. 3. Comparing different k-medoids clustering algorithms using Wisconsin dataset

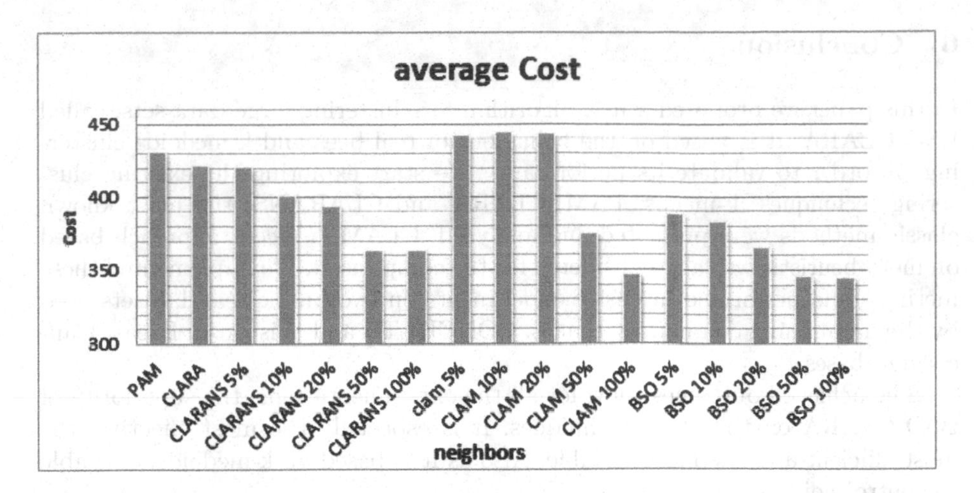

Fig. 4. Comparing different k-medoids clustering algorithms using CONCRETE dataset

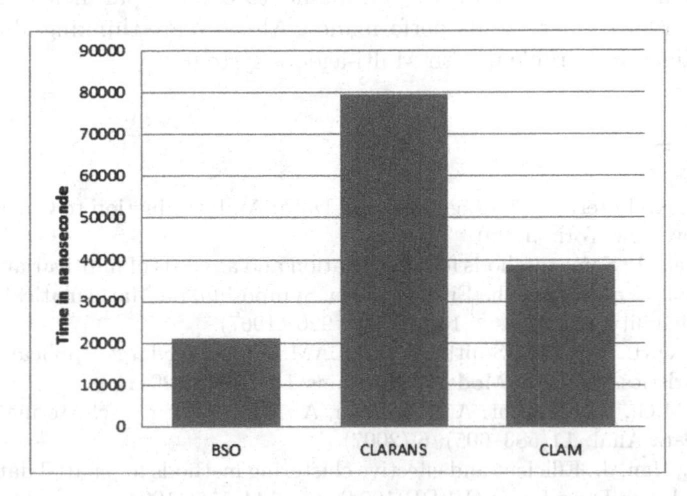

Fig. 5. Comparing the runtime for the best k-medoids algorithms using the Wisconsin dataset

CLARA and PAM. BSO-CLARA, CLARANS and CLAM are tested with different number of neighbours: 5 %, 10 %, 50 % and 100 % of *k(n-k)*.

Figures 3, 4 and 5 show the performance yielded by these methods. We remark that BSO-CLARA outperforms the four other techniques. On the Wisconsin dataset, BSO-CLARA with 20 % of neighours performs as CLAM and CLARANS with 100 % neighbours. Also on CONCRETE dataset, BSO with 50 % of neighbours outperforms all the other techniques.

Figure 5 depicts the optimal runtime for the best techniques. It is clear that BSO-CLARA is the most efficient for the Wisconsin dataset.

6 Conclusion

In this paper we proposed a new algorithm for clustering large data sets, called BSO-CLARA. It is based on the behaviour of real bees and k-medoids clustering. In order to validate its performance, we start exploring the existing clustering techniques. k-means, PAM, CLARA and CLARANS, the most known classic methods were presented and analyzed. CLAM, a recent approach based on meta-heuristic was also considered for the comparison. We implemented these methods and performed intensive experiments on two public real datasets used by the datamining community that is, CONCRETE and Wisconsin Breast Cancer databases.

The achieved outcomes show for all the experimentations, the superiority of BSO-CLARA on the other techniques. It is especially the most effective, the most efficient and the most scalable. Also as it is based on k-medoids, it is able to control noise and outliers.

For the time being, we are investigating other recent metaheuristics for the same problem. For the near future, we intend to explore parallelism for BSO-CLARA in order to increase its performance. Also, we are thinking about modelling the clustering problem with Multi-agents systems.

References

1. Leonard, K., Peter, J.: Finding Groups in Data: An Introduction to Cluster Analysis. Wiley, New York (1990)
2. MacQueen, J.: Some methods for classification and analysis of multivariate observations. In: Proceedings of the Fifth Berkeley Symposium on Mathematical Statistics and Probability, vol. 1, issue 14, pp. 921–926 (1967)
3. Nguyen, Q.H., Rayward-Smith, V.J.: CLAM: clustering large applications using metaheuristics. J. Math. Model. Algorithms **10**, 57–78 (2011)
4. Omran, M.G., Engelbrecht, A.P., Salman, A.: An overview of clustering methods. Intell. Data Anal. **11**(583–605), 6 (2007)
5. Ng, R.T., Han, J.: Efficient and effective clustering methods for spatial data mining. In: Very Large Data Bases (VLDB 1994), pp. 144–155 (1994)
6. Ng, R.T., Han, J.: Clarans: a method for clustering objects for spatial data mining. IEEE Trans. Knowl. Data Eng. **14**(5), 1003–1016 (2002)
7. Sadeg, S., Drias, H., Yahi, S.: Cooperative bees swarm for solving the maximum weighted satisfiability problem. In: Cabestany, J., Prieto, A.G., Sandoval, F. (eds.) IWANN 2005. LNCS, vol. 3512, pp. 318–325. Springer, Heidelberg (2005)
8. Shirkhorshidi, A.S., Aghabozorgi, S., Wah, T.Y., Herawan, T.: Big data clustering: a review. In: Murgante, B., et al. (eds.) ICCSA 2014, Part V. LNCS, vol. 8583, pp. 707–720. Springer, Heidelberg (2014)
9. Tsai, C.-W., Huang, W.-C., Chiang, M.-C.: Recent development of metaheuristics for clustering. In: Park, J.J.J.H., Adeli, H., Park, N., Woungang, I. (eds.) Mobile, Ubiquitous, and Intelligent Computing. LNEE, vol. 274, pp. 629–636. Springer, Heidelberg (2014). http://dblp.uni-trier.de/db/conf/music/music2013.htmlTsaiHC13a

10. Tsutomu, S., Fumihiko, Y., Yoshiaki, T.: A new algorithm based on metaheuristics for data clustering. Zhejiang Univ. Sci. A **12**, 921–926 (2010)
11. WIlliam H, W.: UCI Repository of Machine Learning Databases. University of California, Irvinc (1992)
12. Yeh, I.C.: UCI Repository of Machine Learning Databases. University of California, Irvine (2007)

ECHSA: An Energy-Efficient Cluster-Head Selection Algorithm in Wireless Sensor Networks

Bibudhendu Pati[1], Joy Lal Sarkar[1],
Chhabi Rani Panigrahi[1](✉), and Mayank Tiwary[2]

[1] Department of Computer Science and Engineering,
C.V Raman College of Engineering, Bhubaneswar, India
{patibibudhendu,panigrahichhabi}@gmail.com,
joy35032@rediffmail.com

[2] Department of Information Technology,
C.V Raman College of Engineering, Bhubaneswar, India
mayanktiwari09@gmail.com

Abstract. In Wireless Sensor Networks (WSNs) a key issue is the limited battery power of sensor nodes. To increase the network lifetime is a great challenge where different nodes have different energy labels. To work with this challenge we propose an Energy-Efficient Cluster-Head Selection Algorithm (ECHSA) based on Nash Equilibrium (NE) decision of game theory where, each cluster in the network acts as a player and each player chooses his best strategy followed by other players. We have also compared ECHSA with existing protocols. The simulation results show increase in performance of our proposed approach as compared to the existing approaches.

Keywords: WSNs · Game theory · Energy consumption · Residual energy

1 Introduction

To work with high data rate applications, WSNs need sufficient energy labels of sensor nodes. But, sensor nodes suffer from limited amount of energy [5] and the nodes become dead when their energy labels is zero [2]. The energy consumption of sensor nodes is one of the challenging factors in WSNs because once a node is dead it is very difficult to replace battery especially in risk areas [7,10]. There are many situations where sensor nodes consume their energy. For example, to sense the information, transmitting and receiving packets sensor nodes consume their energy [4]. In WSNs, clusters are formed by a group of sensor nodes and one Cluster-Head(CH) is selected from each cluster based on the highest energy labels of nodes [15]. The nodes in each cluster communicate with their CH and different information from different nodes are aggregated by the corresponding CH including it's own information and is forwarded to the sink node via another CH. In case of single-hop, the nodes which are farthest from the CH consume

R. Prasath et al. (Eds.): MIKE 2015, LNAI 9468, pp. 184–193, 2015.
DOI: 10.1007/978-3-319-26832-3_18

more energy but in case of multi-hop the nodes which are closer to the CH consume more energy due to relaying. In this work, we propose a CH selection algorithm named as ECHSA based on NE of game theory [9]. In ECHSA, each cluster is considered as a player and from each cluster a CH is selected based on NE decision.

The rest of the paper is organized as follows: Sect. 2 describes the related work. Section 3 presents the energy consumption model. Section 4 describes proposed ECHSA along with the analysis. Section 5 presents the results obtained along with the analysis of results. Finally, we conclude the paper in Sect. 6.

2 Related Work

In WSNs, every sensor node suffers from running out of energy due to various activities such as during transmission and receive etc. which decreases the performance of the network [1]. There are several protocols proposed in the literature to increase the energy efficiency requirements in WSNs [1]. In [4], authors proposed a Stable Election Protocol (SEP) called as Deterministic-SEP (D-SEP) which works in a distributed environment which is very useful for electing CH. In [1], authors proposed a Balanced Energy-Efficient Grouping (BEEG) protocol which divides the nodes into various groups based the initial energy of nodes.

In [11], authors proposed a Markov model for sensor networks where a sensor maintains two operational modes: *sleep* and *active* mode to save energy. A sensor node consumes less energy when it is in sleep mode as compared to the active mode. In [3], authors proposed Cluster Head Relay (CHR) routing protocol which uses two kinds of sensors called Low-end sensors (L-sensor) and powerful High-end sensor(H-sensor). A L-sensor works in a multi-hop fashion whereas a H-sensor aggregates the data and forwards these to the sink node. In [13], authors proposed an Adaptive Fidelity Energy-Conserving Algorithm (AFECA) which works with duty cycling scheme. In AFECA, the sleep nodes wake up when they get Route REQuest (RREQ) packets. In [16], authors proposed Distributed Energy-Efficient Clustering (DEEC) algorithm that chooses CH based on the residual energy of nodes. In [12], authors proposed an Unequal Cluster-based Routing (UCR) protocol in WSNs which determines the process of cluster formation by assuming network-wide announcements.

3 Proposed Energy Consumption Model

In this section, we present the energy consumption model for CH selection in WSNs. The energy consumed due to various activities are described as follows:

- *Energy consumption due to sensing by a sensor.* Sensing a system is one of the important characteristics of sensors which links the sensor nodes to the physical world [14]. The energy consumption due to sensing by a sensor node n_i is denoted as $E_{sensor-sensing}(n_i)$ and is computed by using Eq. 1.

$$E_{sensor-sensing}(n_i) = K_{msg} V_s I_{sensing} t_{sensing} \tag{1}$$

Where, $t_{sensing}$ is the time duration for node sensing including CH, $I_{sensing}$ is the total amount of current required for sensingK_{msg} bit packet and V_s is the supply voltage.

- *Energy consumption due to logging by a sensor.* For reading K_{msg} bit packet and to write into the memory sensor losses it's energy [14]. The total energy consumed due to sensor logging by a node n_i is denoted as $E_{sensor-logging}(n_i)$ and is computed by using Eq. 2.

$$E_{sensor-logging}(n_i) = \frac{K_{msg}V_s}{8}(I_w t_w + I_r t_r) \tag{2}$$

Where, I_w and I_r are current for writing and reading of one byte data at the time duration t_w and t_r respectively.

- *Energy consumption for transmitting and receiving data unit.* The energy consumption for transmitting K_{msg} bits message by a node n_i is denoted as $E_{trans}(n_i)$ and is computed by using Eq. 3.

$$E_{trans}(n_i) = \begin{cases} E_{elec}.K_{msg} + l_{amp}.K_{msg}.d_2; & \text{When } d \leq d_0 \\ E_{elec}.K_{msg} + l_{fs}.K_{msg}.d_4; & \text{When } d > d_0 \end{cases} \tag{3}$$

In Eq. 3, E_{elec} is the energy consumption for transmitting and receiving data unit. In Eq. 3, l_{amp} and l_{fs} are the amplifier parameters of transmission corresponding to the multi-path fading model and free space model respectively and d represents the distance between any two nodes and d_0 is the threshold distance.

Energy consumption for receiving K_{msg} bits message by a node n_i is denoted as $E_{recv}(n_i)$ and is computed by using Eq. 4.

$$E_{recv}(n_i) = E_{elec}.K_{msg} \tag{4}$$

- *Energy consumption due to transition from sleep to active state.* Each sensor node consumes energy during transition from sleep to active state. The energy consumed by m number of nodes which make transition from sleep to active state is denoted as $E_{slp-actv}$ and is computed using Eq. 5.

$$E_{slp-actv} = me_{sa} \tag{5}$$

For a single node n_i, the energy consumption due to transition from sleep to active state is denoted as $E_{slp-actv}(n_i)$ and is computed by using Eq. 6.

$$E_{slp-actv}(n_i) = e_{sa} \tag{6}$$

Where, e_{sa} is the energy consumption by the sensor nodes when nodes are transfer to the sleep to active state, while the cost of transition from active state to sleep state can be neglected [11]. So, the total energy consumed by a sensor node n_i denoted by $E_{total}(n_i)$ is computed by using Eq. 7.

$$\begin{aligned} E_{total}(n_i) =& E_{sensor-sensing}(n_i) + E_{sensor-logging}(n_i) \\ &+ E_{trans}(n_i) + E_{recv}(n_i) + E_{slp-actv}(n_i) \end{aligned} \tag{7}$$

Let $E_{initial}(n_i)$ be the initial energy of a node n_i then the residual energy is denoted as $E_{residual}(n_i)$ and is computed by using Eq. 8.

$$E_{residual}(n_i) = E_{initial}(n_i) - E_{total}(n_i) \tag{8}$$

4 Energy-Efficient Cluster-Head Selection Algorithm (ECHSA)

A non-cooperative game [8] G consists of four tuples vector $T = \{P, S, R, A\}$. Where, P is the set of n players, S is the set of n number of strategies corresponding to each player, R is the set of residual energy of n players, and A is the finite set of actions of n players. The ECHSA is given in Algorithm 1. In ECHSA, we assume that there are n number of players and each player corresponds to a cluster and there are k number of nodes in each cluster.

Each player has q number of strategies denoted as S (steps $2-4$). We assume that each player acts as a cluster as in steps $7-9$. Each residual energy $E_{residual}$ is set as a strategy for each player p_i (step 9). We then check whether the action of one player(x_i) is greater than or equal to the action of another player(x_{-i}) from the action profile x^* which is the condition of NE. After getting NE we

Algorithm 1. The ECHSA Algorithm

```
1: procedure ECHSA(P, S, nash)
2:     P ← {p_1, p_2, p_3, ..., p_n}
3:     N ← {n_1, n_2, n_3, ..., n_k}
4:     S ← {s_1, s_2, s_3, ..., s_q}
5:     nash ← false
6:     CH ← NULL
7:     for each player p_i ∈ P do
8:         for each node n_k ∈ N do
9:             Compute E_residual
10:            s_i ← E_residual
11:    for ∀ p_i ∈ P and ∀ s_i ∈ S do
12:        if (s_i(x_i, x) ≥ s_i(x_-i, x)) then
13:            nash ← true
14:            CH ← nash
15:        else nash ← false
16:    if (nash > 1) then
17:        for each player p_i ∈ P do
18:            for each s_i ∈ nash do
19:                select s_max
20:                CH ← s_max
```

then select it as a CH (steps 11 – 14). If there are multiple NE, each player chooses its best strategy among all NEs denoted as s_{max} and selects it as a CH (steps 16 – 20). Each player in the game has different number of strategies and each player selects his best strategy according to his payoffs. Here, we assume that the residual energy of different nodes are different.

4.1 Best Response Function to Find Nash Equilibrium

Nash equilibrium mainly works with two components:
(a) At first, each player believes in the other player's action and they must take the actions based on the model of rational choices.
(b) During the game, each player believes that the actions which is taken by another player must be correct.

To find the NE of a game we used *best response function* [9] where, every player maintains only a few actions by checking each action profile. We can define NE using *best response function* and is given as in Eq. 9.

The action profile x_* is called NE when there exists a best response function.

$$x_p{}^* \; is \; in \; G_p \; is \; in \; G_p(x_{-p}{}^*) \; for \; every \; player \; p \tag{9}$$

From each list x_{-p} of other player's action, every player in P maintains one best response. Here, for each player p and each list x_{-p} of the other player's action, denote the single member of $G_p\,(x_{-p}\,)$ by $b_p\,(x_{-p}\,)$ (that is, $G_p(x_{-p}) = \{b_p(x_{-p}\}$) then from Eq. 9:

$$x_p{}^* = b_p(x_{-p}{}^*) for \; every \; player \; p, \tag{10}$$

Figure 1 shows a NE for 2-players game. In Fig. 1, Player-1 and Player-2 have three strategies such as (B, M, T) and (L, C, R) respectively. The value-pair indicates that the residual energy of nodes that is the strategy of two players. For example, a pair (2.14, 1.12) indicates that when Player-1 chooses the strategy M then Player-2 choose the strategy L and vice versa. In this game, when Player-1 chooses T, Player-2 has three strategies (L,C,R). But, Player-2 will choose his best payoff and hence it chooses L. Similarly, when Player-1 chooses M or B

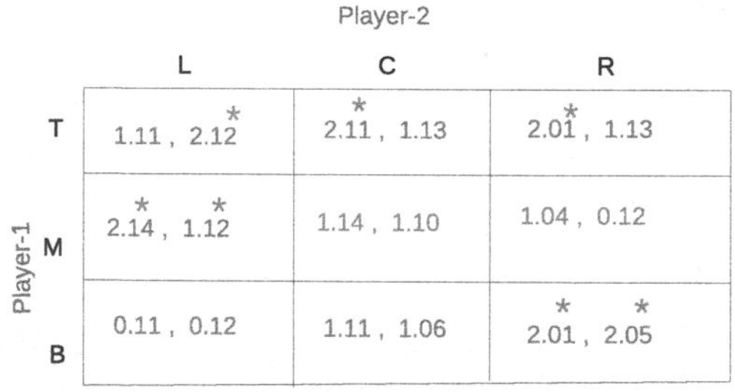

Player-1 \ Player-2	L	C	R
T	1.11 , 2.12*	2.11* , 1.13	2.01* , 1.13
M	2.14* , 1.12*	1.14 , 1.10	1.04 , 0.12
B	0.11 , 0.12	1.11 , 1.06	2.01* , 2.05*

Fig. 1. NE for 2-player game.

then Player-2 chooses L and R respectively. Moreover, when Player-2 chooses the strategy L, Player-1 has three strategies (T,M,B). But, Player-1 also wants to choose his best payoff and so Player-1 chooses M. But, there may be the case arises when Player-2 chooses R, Player-1 chooses T and B because both are having the same payoffs. The star symbol (*) in Fig. 1 indicates the best payoff of each player. In this game, (M,L) and (B,R) are the NE because at that point both players choose their best strategy.

In Fig. 2, there are n number of players and the big circle indicates best payoff of each player. When Player-1 selects one strategy, the other players have different strategies.

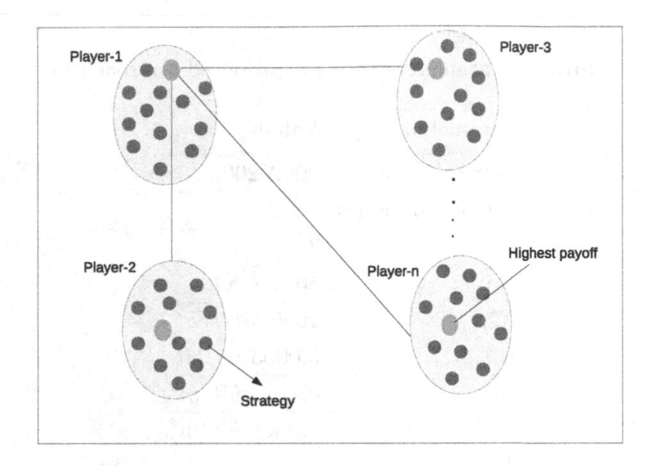

Fig. 2. n-player game.

4.2 Cluster Formation

For distribution of CH in every region, there is a statistical distribution of sensor nodes in a given space [6,7]. According to ECHSA, after selection of CHs, each CH broadcasts its announcement packets within a radius suppose βr_i. Where, β is the system parameter. But, it is important to confirm that atleast one broadcast packet is received by all non-CHs in the radius βr_i. Each CH should be confirmed that every non-CH receives this packet using a technique as in [6]. Based on the announcement packets from multiple CHs, each non-CH selects their CH and sends its information. CH then aggregates the data and forwards to the sink node via another CH.

5 Results and Analysis

We implemented ECHSA in MATLAB (R2013a). We compute residual energy by taking the random values of different parameters for each node as given in Table 1. We compute residual energy of different clusters in each round for different strategies. We have also compared ECHSA with existing protocols that is UCR [12], DEEC [16], and BEEG [1]. To show the nodes which have high energy that affects network lifetime, we deploy the nodes having high energy ranging from 10, 20, 30, ..., 90 in a cluster.

The graph as shown in Fig. 3 shows the comparison of average number of nodes when first node dies(N_{dead}) with respect to the number of rounds(Nh) in ECHSA and other existing protocols such as BEEG, DEEC, and UCR. Where in case of ECHSA 90 nodes are dies after 1706 rounds whereas in case of UCR, DEEC, BEEG same number of nodes dies after 1590, 980, 667 rounds respectively. The graph as shown in Fig. 4 shows the average number of nodes alive (N_a) over round for all four methods as mentioned above. The results as shown

Table 1. Parameters used for proposed approach

Parameters	Values
Network Size	$200X200m^2$
Number of nodes	100
$E_{initial}$	3.8 j
E_{elec}	50 mj/bit
K_{msg}	2000 bit/sec
$E_{slp-actv}$	0.0003 nj/bit
l_{amp}	0.1 $nj/bit/m^2$
l_{fs}	0.2 $nj/bit/m^4$
d	$\leq 10m$
e_{sa}	0.0003 nj/bit

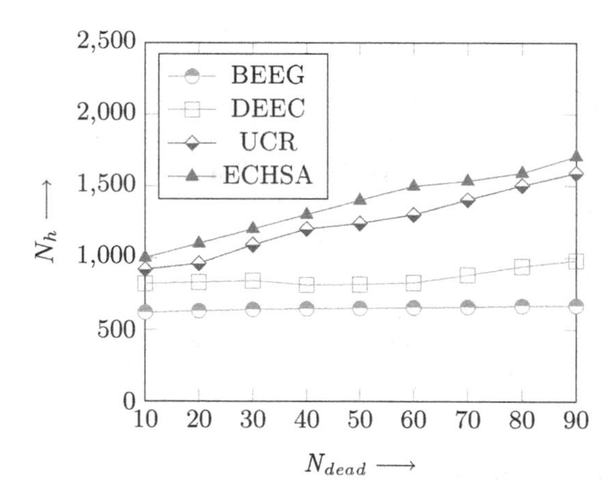

Fig. 3. Average number of rounds off when first node dies with respect to number of varying high energy nodes

in Fig. 3 and Fig. 4 indicate that ECHSA performs better as compared to the existing methods. Figure 4 also indicates that the number of live nodes is higher in ECHSA as compared to the existing approaches which results in increase in the network lifetime.

Figure 5 shows the comparison of average residual energy labels of different protocols with ECHSA. The graph as shown in Fig. 5 indicates that after 3000 rounds the nodes are almost dead in case of DEEC, BEEG but after 3500 rounds the value of $E_{residual}$ is 1.0 in case of UCR whereas the value of $E_{residual}$ is 2.1 after 3500 rounds in case of ECHSA. The simulation results indicate that ECHSA provides better performance as compared to the existing approaches. This may be due the fact that ECHSA computes residual energy of nodes in every round based on NE decision and a NE always results in an optimal solution.

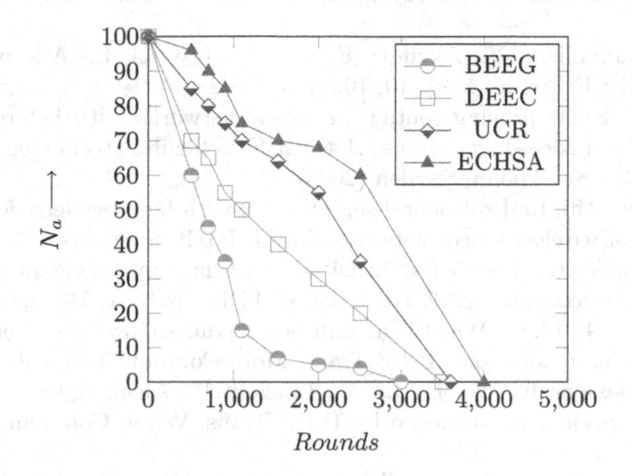

Fig. 4. Average number of nodes alive over rounds

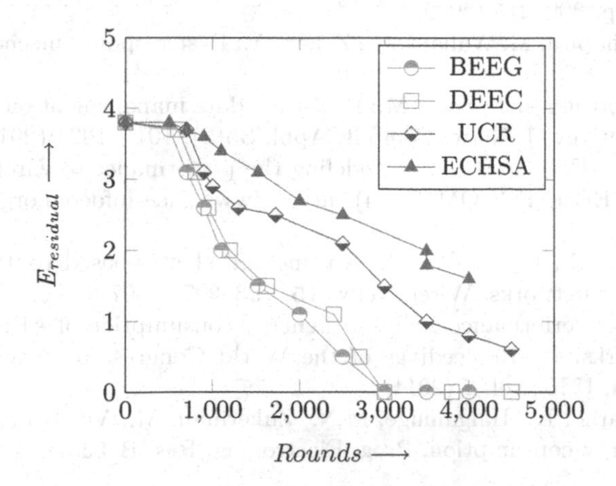

Fig. 5. Average residual energy labels

6 Conclusion

In this work, an energy efficient algorithm named as ECHSA is proposed based on a NE decision. Simulation is conducted to evaluate the performance of ECHSA. The results obtained indicate that the proposed ECHSA can achieve better performance as compared to the existing algorithms. In future, we will try to use this game theoretic approach for real-time query scheduling in heterogeneous WSNs.

References

1. Chang, L., Liaw, J.J., Chu, H.-C.: Improving lifetime in heterogeneous wireless sensor networks with the energy grouping protocol. ICIC Int. **8**(9), 6037–6047 (2001)
2. Sankarasubramaniam, Y., Akyildiz, F., Su, W., Cayirci, E.: A survey on sensor networks. JIEEE Comm. Mag. **40**, 102–114 (2002)
3. Du, X., Lin, F.: Improving routing in sensor networks with heterogeneous sensor nodes. In: Proceedings of the 61st IEEE Vehicular Technology Conference, pp. 2528–2532. Stockholm, Sweden (2005)
4. Bala, M., Awasthi, L.: Proficient d-sep protocol with heterogeneity for maximizing the lifetime of wireless sensor networks. Int. J. Intell. Syst. Appl. **7**, 1–15 (2012)
5. Sene, M., Diallo, O., Rodrigues, J.: Distributed data management techniques for wireless sensor networks. IEEE Trans. Parallel Distrib. Syst. **26**(2), 604–623 (2013)
6. Ye, M., Chen, G., Li, C., Wu, J.: On multihop distances in wireless sensor networks with random node locations. IEEE Trans. Mob. Comput. **9**(4), 540–552 (2010)
7. Vural, S., Moessner, K., Wei, D., Jin, Y., Tafazolli, R.: An energy-efficient clustering solution for wireless sensor networks. IEEE Trans. Wirel. Commun. **10**(1), 3973–3983 (2011)
8. Widgerand, J., Grosu, D.: Parallel computation of nash equilibria in n-player games. In: International Conference on Computational Science and Engineering, Vancouver, pp. 209–215 (2009)
9. Nisan, N., Schapira, M., Valiant, G., Zohar, A.: Best-response mechanisms. In: ICS (2011)
10. Diallo, O., Rodrigues, J., Sene, M.: Real-time date management on wireless sensor networks: a survey. J. Netw. Comput. Appl. **35**(3), 1013–1021 (2012). Elsevier
11. Chiasserini, C.-F., Garetto, M.: Modeling the performance of wireless sensor networks. In: IEEE INFOCOM (2004). http://www.ieee-infocom.org/2004/Papers/06_1.PDF
12. Ye, M., Chen, G., Li, C., Wu, J.: An unequal cluster-based routing protocol in wireless sensor networks. Wirel. Netw. **15**, 193–207 (2007)
13. Chung, Y. W.: Performance analysis of energy consumption of AFECA in wireless sensor networks. In: Proceedings of the World Congress on Engineering, WCE 2011, London, U.K., vol. 2 (2011)
14. Ramamohanarao, K., Halgamuge, M.N., Zukerman, M., Vu, H.L.: An estimation of sensor energy consumption. Prog. Electromag. Res. B **12**(4), 259–295 (2009)

15. Pati, B., Das, H., Panigrahi, C.R., Sarkar, J.L.: A novel approach for real-time data management in wireless sensor networks. In: Proceedings of 3rd International Conference on Advanced Computing, Networking and Informatics, pp. 599–607. IEEE Computer Society Press (2015)
16. Zhu, Q., Quing, L., Wang, M.: Design of a distributed energy-efficient clustering algorithm for heterogeneous wireless sensor networks. Comput. Commun. **29**, 2230–2237 (2006)

Optimal Core Point Detection Using Multi-scale Principal Component Analysis

T. Kathirvalavakumar[1](✉) and K.S. Jeyalakshmi[2]

[1] Research Center in Computer Science, V.H.N.S.N. College (Autonomous), Virudhunagar 626 001, India
kathirvalavakumar@yahoo.com

[2] Department of Computer Science, N.M.S.S.Vellaichamy Nadar College (Autonomous), Madurai 625 019, India
jeyal2007@gmail.com

Abstract. Core point plays a vital role in fingerprint matching and classification. The fingerprint images may be of poor quality because of sensor type and user's body condition. To detect the core point in noisy and poor quality fingerprint images, we have estimated the dominant orientation field based on principal component analysis and multi-scale pyramid decomposition to produce correct orientation field. The proposed work detects the optimal upper and lower core points using shape analysis of orientation field and binary candidate region images in fingerprints. Experiments are carried out on FVC databases and it is found that the proposed algorithm has high accuracy in locating exact core points.

Keywords: Fingerprint image · Core point · Orientation field · Principal component analysis · Multi-scale pyramid decomposition

1 Introduction

Now a days in the electronically connected world, identity fraud is increased. Hence the focus on biometrics as a form of identification is evolving. Fingerprints are more used in law enforcement agencies because of its uniqueness, reliability and ease of use. Core point is a reference point used to identify uniqueness among different samples. Core point detection is used to recognize the fingerprint even though there are different poses of same finger [1]. Since singular points (core and delta point), are unique landmarks of fingerprint, which are also called as level 1 features of a fingerprint, they are used as reference points for fingerprint matching and classification [2,3].

The core point has played an important role in most fingerprint identification techniques [4] and is widely used in fingerprint classification [5–7] and fingerprint matching [8–10]. The typical core point detection methods are based on

T. Kathirvalavakumar–This work is Funded by University Grants Commission Major Research Project, New Delhi, INDIA.

R. Prasath et al. (Eds.): MIKE 2015, LNAI 9468, pp. 194–203, 2015.
DOI: 10.1007/978-3-319-26832-3_19

Poincare Index (PI) analysis. Even though the PI analysis is simple and robust to image rotation compared with other methods, their performance tends to degrade while the image quality is poor. Yongming et al. [11] have proposed a method of singular point localization by successive approximation method and PI algorithm. Huang et al. [12] have presented a shrinking and expanding algorithm (SEA) based on a scale-pyramid model with fault line analysis to locate singular points. Ignatenko et al. [13] have detected the core point based on the integration of a directional vector field over a closed contour. Wrobel et al. [14] have proposed a new method based on IPAN99 algorithm. The drawback of the proposed methods in [11–14] is that the input fingerprint images should be enhanced but it is the time consuming task. Weng et al. [15] have proposed a method that combines modified PI and muti-resolution analysis for identifying singular points. Bo et al. [16] have calculated PI only on region of interest (ROI). Iwasokun et al. [17] have proposed a modified PI to detect singular points but it is failed with very poor quality images. Fei [18] has presented a novel method that applies corner detection algorithm on orientation image and has high accuracy only in upper core point detection. Weiwei et al. [19] have presented an algorithm that uses Multi-Resolution Direction Field and High-Resolution Direction Field to search and locate the positions of fingerprint singular points. Julasayvake et al. [20] have proposed a work that combines Direction of curvature technique (DC) and Geometry of region (GR). DC technique has been used in coarse core point determination whilst GR technique is used in the fine finding of the core point. This modified algorithm fails in locating the core point of the fingerprint with arc structure and double loop structure. Akram et al. [21] have proposed a method that uses GR for isolating ROI, applies PI on ROI. Kundu et al. [22] have proposed a method that uses shape template masks to detect the reference points using some soft computing tool like fuzzy set theory. Rahimi et al. [23] and Porwik et al. [24] have been used directional mask to detect singular points. LingLing et al. [25] have proposed a work that uses the concept of normal lines of gradient of double orientation field for singular points detection. Awad et al. [26] have proposed a method that uses complex filter to identify singular points. Bahgat et al. [27] have proposed work that uses a mask to locate the core point simply from the ridge orientation map and detects the core point at the end of the discontinuous line appearing in the orientation map presented by a gray-scale. Jiang et al. [1] have introduced reference point detection based on hierarchical analysis of orientation coherence. Most of the core point detection algorithms can efficiently detect the core point when the image quality is fine. When the image quality is poor, the core point detection rate is decreased [28].

In the proposed method, orientation field is estimated based on PCA and multi-scale pyramid decomposition. After the homogeneous zone division process BCRI is constructed and by shape analysis core point is identified. Proposed method identifies core point even though image is noisy and its quality is poor.

2 Segmentation

Segmentation is the process of separation of foreground image from background image i.e., separation of ridge & valley area from unrecoverable area, non ridge, and non valley area. Image is segmented by generating mask using 'erode' and 'open' morphological operations. The segmentation procedure [29] follows:

Step 1: Re-size the input image.
Step 2: Erode the re sized image with the structuring element of size w × w.
Step 3: Enhance the contrast of the eroded image by power law transformation.
Step 4: Convert the enhanced image into binary image based on the threshold obtained from Otsu method.
Step 5: Generate the mask by negating the binary image.
Step 6: Perform morphological operation 'open' on the mask to remove islands.
Step 7: Use the generated mask on the input image and get the segmented image.

3 Orientation Field Estimation and Smoothening

Orientation is the direction of flow of ridges over the point. Estimation of orientation plays a vital role in detecting accurate core point position. In this method, orientation field is estimated using multi-scale PCA proposed by Feng and Milanfar [30]. The method follows:

Step 1: Compute gradient of the input image.
Step 2: Construct gradient pyramid with n layers.
Step 3: Form local blocks with 1 pixel overlapped on each layer.
Step 4: Estimate local orientation on each block using PCA for all layers.

In the resultant orientation image O, orientation of the block O(i,j) is an unoriented direction lying in $[0,\pi)$ [31], where i and j represent block position. The orientation image needs to be smoothened using Gaussian low pass filter to remove the noise, if it exists.

4 Homogeneous Zones Division

Core point is always at the point where different homogeneous areas in the quantized orientation image are met. In the proposed work, smoothened orientation image is quantized into 8 homogeneous zones, $\theta_k, (1 \leq k \leq 8)$, the orientation value of each position in the zone lies in the range $\left[\frac{(k-1)}{8}\pi - \omega_0 \ \frac{(k-1)}{8}\pi + \omega_0\right]$ where $\omega_0 = \frac{\pi}{16}$.

5 Binary Candidate Region Image Construction

The region of the ridges having the core point has the shape, similar to symbol cap ($\cap$) in the upper core point and/or cup ($\cup$) in the lower core point [32]. The orientation values of the neighboring blocks around the core point are to be in non decreasing values from 0 to π which occur in clockwise direction. $BCRI_U$ for cap shaped candidate region is constructed by subtracting the orientation of every block from its right block and assign value 1 or 0 based on the Eq. (1) [32].

$$BCRI_U(\theta_L) = \begin{cases} 1 & if -\frac{3\pi}{4} \leq (\theta_L - \theta_R) \leq -\frac{\pi}{4} \; and \; \theta_L \neq 0° and \; \theta_R \neq 0° \\ & or \\ & -\frac{3\pi}{4} \leq (\theta_L - \theta_R) \leq -\frac{\pi}{2} \; and \; \theta_L = 0° \\ & or \\ & \frac{\pi}{4} \leq (\theta_L - \theta_R) \leq \frac{\pi}{2} \; and \; \theta_R = 0° \\ 0 & otherwise \end{cases} \quad (1)$$

Similarly, $BCRI_L$ for cup shaped candidate region is constructed using the Eq. (2) [32].

$$BCRI_L(\theta_L) = \begin{cases} 1 & if \frac{\pi}{4} \leq (\theta_L - \theta_R) \leq -3\frac{\pi}{4} \; and \; \theta_L \neq 0° \; and \; \theta_R \neq 0° \\ & or \\ & -\frac{\pi}{2} \leq (\theta_L - \theta_R) \leq -\frac{\pi}{4} \; and \; \theta_L = 0° \\ & or \\ & \frac{\pi}{2} \leq (\theta_L - \theta_R) \leq 3\frac{\pi}{4} \; and \; \theta_R = 0° \\ 0 & otherwise \end{cases} \quad (2)$$

6 Core Point Identification

Candidate region is connected when adjacent vertical positions in $BCRI_U$ have the value 1. Upper core point is located by finding the orientation values of 8 neighboring blocks around the bottom most of each connected candidate region is checked with the following conditions.

Condition 1: Orientation values are in non decreasing order vary from 0 to π in clockwise direction.

Condition 2: At least 4 different homogeneous zones are there.

If the conditions satisfy then the corresponding candidate region is the location of the upper core point. If more than one candidate region satisfy the conditions then the location of the upper core point is calculated by averaging x coordinates and y coordinates. Similarly lower core point is identified by applying the above said procedure by considering the top most candidate region instead of bottom most of the connected candidate region.

Procedure

Step 1: Separate ridge & valley area from non-valley area, non-ridge, and unrecoverable area using Sect. 2.
Step 2: Estimate orientation field using Sect. 3.
Step 3: Smoothen the orientation field using Gaussian low pass filter.

Step 4: Quantize the smoothened orientation image into 8 homogeneous zones.
Step 5: Construct $BCRI_U$ and $BCRI_L$ using (1) and (2) correspondingly.
Step 6: Identify connected candidate region.
(a) Locate bottommost block of connected candidate region.
(b) Check orientation values of 8 neighbors are in non-decreasing order in clock wise direction.
(c) Check 4 unique homogeneous zones are in neighboring blocks.
(d) If (b) & (c) satisfies, the candidate region is upper core point location.
Step 7: If more than one upper core points are identified then average the coordinates of the located blocks to locate exact upper core point.
Step 8: Perform steps 6 and 7 to find lower core point, but in (a) locate topmost block of connected candidate region instead of bottommost block.

7 Results and Discussion

The performance of the proposed algorithm is tested with the help of the public fingerprint databases FVC 2000 DB1_B [33], DB3_B [33], FVC 2002 DB1_B [34] and FVC 2004 DB1_B [35]. It has totally 320 fingerprint images. Each database consists of 8 different poses of 10 fingers. The sensor type, image size and resolution of each database are given in Table 1. The proposed algorithm is implemented in MATLAB 2013a.

Table 1. Characteristics of Fingerprint Databases

Database	Sensor type	Image size	Images	Resolution
FVC 2000 DB1_B	Low-cost Optical Sensor	300×300	80	500 dpi
FVC 2000 DB3_B	Optical Sensor	448×478	80	500 dpi
FVC 2002 DB1_B	Optical Sensor	388×374	80	500 dpi
FVC 2004 DB1_B	Optical Sensor	640×480	80	500 dpi

The measures Core detection rate and False alarm rate (FAR) are used to find the accuracy of the core point detection methods in literature.

- Core Detection Rate: ratio of detection by algorithm and by manual.
- False Alarm Rate: ratio of false core point detection by the algorithm and core point detection by manual.

To assess the effectiveness of the proposed work, the performance of the proposed algorithm is compared with the method proposed by Karu et al. [2], and Rosa [28] and is shown in Table 2. Rosa [28] uses orthogonal gradient magnitude of fingerprint orientation field to detect core point whereas Karu et al. [2] have used Poincare index analysis. Both methods have not discriminated between the upper and lower core points but the proposed method discriminates. Table 2 shows proposed method detects more core points and less false core point than [2]

Table 2. Performance Comparison

Method Name	# Correctly detected core point	# False core point detected	Core point detection rate	FAR
Method of Karu et al. [2]	269	51	84.06	15.94
Method of Rosa [28]	297	23	92.81	7.19
Proposed Method	**301**	**19**	**94.06**	**5.94**

and [28]. Hence the rate of core point detection is greater and rate of FAR is lesser than [2] and [28]. The accuracy of the detected core point is measured using the Euclidean distance error which is the distance between manually located core point position $P_m(x_o, y_m)$ and algorithm detected core point position $P_d(x_d, y_d)$, and is defined as:

$$d_e = \sqrt{|x_m - x_d|^2 + |y_m - y_d|^2} \tag{3}$$

If the distance error is more than 20 pixels then the detected core point is false. Table 3 shows the detected core points for different thresholds of pixels in distance error and reveals that the proposed algorithm has detected the core points more accurately.

Table 3. Distance Error for different threshold

	Upper core point		Lower core point	
Distance Error (d_e) pixels	# Core points	Ratio	# Core points	Ratio
$\leq$ 10	308	97.2	46	83.6
> 10 & $\leq$ 20	8	2.5	3	5.5
> 20	1	0.03	6	10.9
Not Detected	3	-	9	-

The standard deviation of the distance error is calculated using Eq. (4). Table 4 shows minimum, mean, maximum and standard deviation of the distance error. The proposed algorithm has detected core point position with the mean error 5.486 and standard deviation error 4.199.

$$\sigma = \sqrt{\frac{1}{N}\sum_{i=1}^{N}\left|d_e(i) - \sum_{i=1}^{N} d_e(i)\right|^2} \tag{4}$$

During the experiment it has been identified that most of the missed core points are around the border. Also it has been identified that few core points are missed when the number of neighbors of different direction zones varying from 0 to π

Table 4. Statistics on Distance Error

Type of Core point	Min	Mean	Max	Standard Deviation
Upper	0	5.015	20.6	2.718
Lower	0	8.189	44.5	8.306
Upper & Lower	0	5.486	44.5	4.199

Table 5. Processing time of each step in the proposed method

PROCESS	TIME (s)
Segmentation	0.038394724
Estimation of Orientation	3.366057063
Homogeneous Zone Division	0.0000459248
Construction of $BCRI_U$	0.0000389749
Construction of $BCRI_L$	0.00003679
Upper Core point detection	0.001304191
Lower Core point detection	0.027383

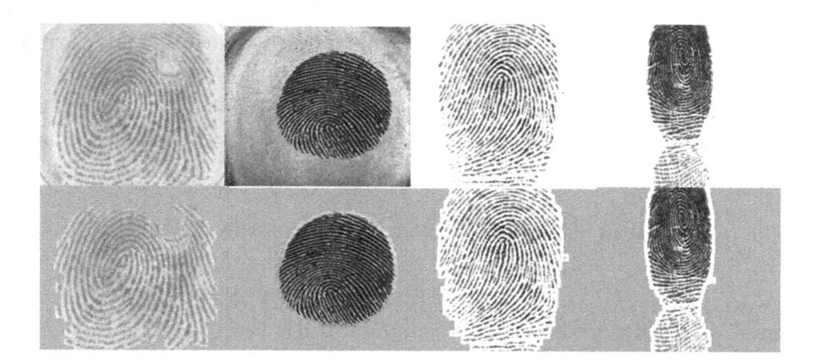

Fig. 1. Fingerprint images before segmentation (top row) and after segmentation (bottom row)

is 3. In a condition 2, if we select number of unique homogeneous zones as 3 instead of 4, more false core points have been detected. Hence, number of unique homogeneous zones around core point position is fixed in condition 2 as 4.

The processing time of each step in the proposed method is shown in Table 5. Estimation of orientation field consumes more time than the other steps in the proposed method. The proposed method spends more time in finding the correct estimation of orientation field that leads to accurate detection of core point quickly.

In the proposed method, input fingerprint images are resized and segmented. Top row of Fig. 1 shows the fingerprint images before segmentation and bottom row of corresponding location shows after segmentation. Orientation image is calculated, smoothened and then divided into homogeneous zones. Figure 2 explains the way how the core point is detected in the fingerprints with left and right loop. Figure 3 shows the core point detection process in a double loop and whorl type fingerprints. Figure 4 shows the False/Missed core point.

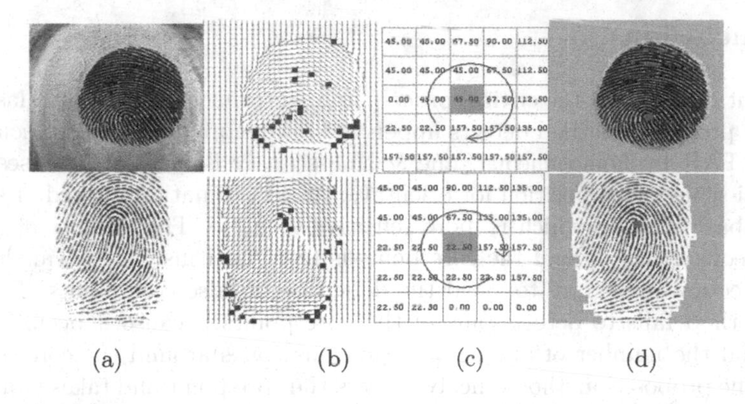

Fig. 2. (a) Fingerprint Image, (b) $BCRI_U$ superimposed on orientation image, (c) Neighboring Blocks around upper core point, (d) detected core point

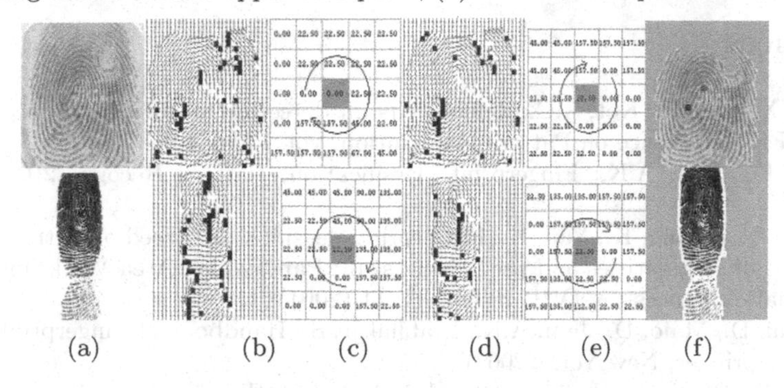

Fig. 3. (a) Fingerprint Image, (b) $BCRI_U$ superimposed on orientation image, (c) Neighboring Blocks around upper core point, (d) $BCRI_L$ superimposed on orientation image, (e) Neighboring Blocks around lower core point, (f) Detected core point

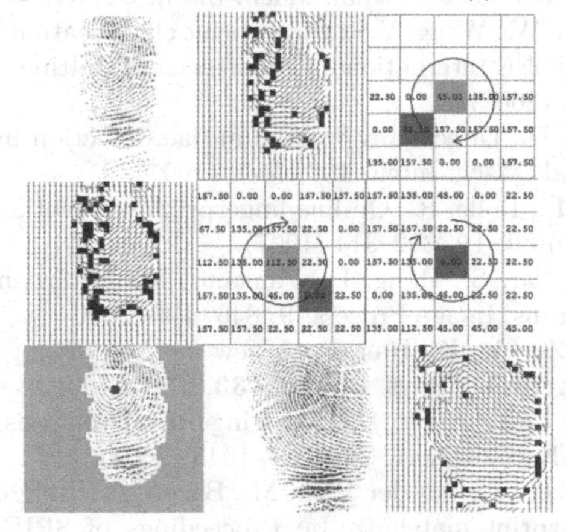

Fig. 4. Fingerprint Image, $BCRI_U$ superimposed on orientation image, Neighboring Blocks around upper core point, Detected false core point & Missed core point

8 Conclusion

Core point detection is the critical process in fingerprint matching and classification. The proposed method detects upper and lower core points more accurately with low FAR by homogeneous zone & shape analysis of BCRI. It uses PCA and multi-scale decomposition for orientation field estimation instead of typical gradient base method which is more sensitive to noise. The number of unique homogeneous zones around the core point is fixed as 4 instead of 3 in the core point detection procedure to avoid the detection of false core points. The proposed method fails to detect some of the core points because it occurs at the border and the number of unique homogeneous zones around the core point is only 3. The proposed method quickly detects the core point and takes more time for orientation field estimation.

References

1. Jiang, X., Liu, M., Kot, A.C.: Reference point detection for fingerprint recognition. In: IEEE Conference on Pattern Recognition, vol. 1, pp. 540–543 (2004)
2. Karu, K., Jain, A.K.: Fingerprint classification. Pattern Recogn. **29**, 389–404 (1996)
3. Zhang, Q., Huang, K., Yan, H.: Fingerprint classification based on extraction and analysis of singularities and pseudoridges. In: Pan-Sydney Area Workshop Visual Information Processing (VIP 2001), vol. 11 (2001)
4. Maltoni, D., Maio, D., Jain, A.K., Prabhakar, S.: Handbook of Fingerprint Recognition. Springer, New York (2003)
5. Wang, S., Wang, Y.: Fingerprint enhancement in the singular point area. IEEE Signal Process. Lett. **11**, 16–19 (2004)
6. Jain, A.K., Prabhakar, S., Hong, L.: A multichannel approach to fingerprint classification. IEEE Trans. Pattern Anal. Mach. Intell. **21**, 348–359 (1999)
7. Wang, S., Zhang, W.W., Wang, Y.S.: Fingerprint classification by directional fields. In: Proceedings of IEEE International Conference on Multimodal Interfaces (ICMI 2002), pp. 395–398 (2002)
8. Maio, D., Maltoni, D.: Direct gray-scale minutiae detection in fingerprints. IEEE Trans. Pattern Anal. Mach. Intell. **19**, 27–40 (1997)
9. Jain, A.K., Hong, L., Bolle, R.: On-line fingerprint verification. IEEE Trans. Pattern Anal. Mach. Intell. **19**, 302–314 (1997)
10. Jain, A.K., Prabhakar, S., Hong, L., Pankanti, S.: Filterbank-based fingerprint matching. IEEE Trans. Image Process. **9**, 846–859 (2000)
11. Yang, Y., Zulong, Z., Lin, K., Han, F.: A new method of singular points accurate localization for fingerprint. Phys. Procedia **33**, 67–74 (2012)
12. Huang, C.Y., Liu, L.M., Hung, D.C.D.: Fingerprint analysis and singular point detection. Pattern Recogn. Lett. **28**, 1937–1945 (2007)
13. Ignatenko, T., Kalker, T., van der Veen, M., Bazen, A.: Reference point detection for improved fingerprint matching. In: Proceedings of SPIE-IS & T Electronic Imaging, pp. 1–9 (2006)
14. Wrobel, K., Doroz, R.: New Method for finding a reference point in fingerprint images with the use of the IPAN99 algorithm. J. Med. Inform. Technol. **13**, 59–63 (2009)

15. Weng, D., Yin, Y., Yang, D.: Singular points detection based on multi-resolution in fingerprint images. Neurocomputing **74**, 3376–3388 (2011)
16. Bo, J., Ping, T.H., Lan, X.M.: Fingerprint singular point detection algorithm by poincar index. WSEAS Trans. Syst. **7**, 1453–1462 (2008)
17. Iwasokun, G.B., Akinyokun, O.C.: Fingerprint singular point detection based on modified poincare index method. Int. J. Signal Process. Image Process. Pattern Recogn. **7**, 259–272 (2014)
18. Fei, S., Peng, S., Bo-tao, W., An-ni, C.: Fingerprint singular points extraction based on the properties of orientation model. J. China Univ. Posts Telecommun. **18**, 98–104 (2011)
19. Weiwei, Z., Wang, Y.: Singular point detection in fingerprint image. In: Proceedings of the 5th Asian Conference on Computer Vision (2002)
20. Julasayvake, A., Choomchuay, S.: An algorithm for fingerprint core point detection. In: IEEE - International Symposium on Signal Processing and its Applications, pp. 1–4 (2007)
21. Akram, M.U., Tariq, A., Nasir, S., Khanam, A.: Core point detection using improved segmentation and orientation. In: 6th ACS/IEEE International Conference on Computer Systems and Applications, pp. 637–644 (2008)
22. Kundu, M.K., Maiti, A.K.: Accurate localizations of reference points in a fingerprint image. In: Kuznetsov, S.O., Mandal, D.P., Kundu, M.K., Pal, S.K. (eds.) PReMI 2011. LNCS, vol. 6744, pp. 293–298. Springer, Heidelberg (2011)
23. Rahimi, M.R., Pakbaznia, E., Kasaei, S.: An adaptive approach to singular point detection in fingerprint images. Int. J. Electron. Commun. AEUE **58**, 367–370 (2004)
24. Porwik, P., Wieclaw, L.: A new approach to reference point location in fingerprint recognition. IEICE Electron. Express **1**, 1–7 (2004)
25. Fan, L.L., Wang, S., Guo, T.D.: Global and local information combined to detect singular points in fingerprint images. Sci. China Inf. Sci. **55**, 1–13 (2012)
26. Awad, A.I., Baba, K.: Singular point detection for efficient fingerprint classification. Int. J. New Comput. Architectures Appl. (IJNCAA) **2**, 1–7 (2012). The Society of Digital Information and Wireless Communications
27. Bahgat, G.A., Khalil, A.H., Kader, N.S.A., Mashali, S.: Fast and accurate algorithm for core point detection in fingerprint images. Egypt. Inform. J. **14**, 15–25 (2013)
28. Rosa, L.: Core Point Detection Using Orthogonal Gradient Magnitudes of Fingerprint Orientation Field. http://www.advancedsourcecode.com/fingerprint.asp
29. Wu, C., Tulyakov, S., Govindaraju, V.: Robust point-based feature fingerprint segmentation algorithm. In: Lee, S.-W., Li, S.Z. (eds.) ICB 2007. LNCS, vol. 4642, pp. 1095–1103. Springer, Heidelberg (2007)
30. Feng, X.G., Milanfar, P.: Multiscale principal components analysis for image local orientation estimation. In: IEEE - Signals, Systems and Computers, vol. 1, pp. 478–482 (2002)
31. Maltoni, D., Maio, D., Jain, A.K., Prabhakar, S.: Handbook of Fingerprint Recognition, p. 103. Springer, Heidelberg (2009)
32. Park, C.H., Lee, J.J., Smith, M.J.T., Park, K.H.: Singular point detection by shape analysis of directional fields in fingerprints. Pattern Recogn. **39**, 839–855 (2006)
33. http://bias.csr.unibo.it/fvc2000/download.asp
34. http://bias.csr.unibo.it/fvc2002/download.asp
35. http://bias.csr.unibo.it/fvc2004/download.asp

Recognition of Semigraph Representation of Alphabets Using Edge Based Hybrid Neural Network

R.B. Gnana Jothi(✉) and S.M. Meena Rani

V.V.Vanniaperumal College for Women, Virudhunagar 626001, Tamilnadu, India
gnanajothi_pcs@rediffmail.com, smmeenarani@gmail.com

Abstract. Graph structured data are classified by connectionist models such as Graph neural network (GNN), recursive neural network. These models are based on the label of the nodes of the graph. An attempt has been made to consider the network based on edges. If a graph structured data is represented as semigraph, the number of edges will be reduced leading to a reduction in the number of networks in GNN. In this paper uppercase English alphabets represented as graphs are recognized using edge based hybrid neural network by viewing the graphs as semigraph. Experimental results show that the edge based hybrid neural network is able to identify all the graphs of alphabets correctly and outperforms edge based GNN.

Keywords: Alphabet recognition · Semigraph · Graph structured data · Recursive neural network · Recurrent network · Feedforward network

1 Introduction

Semigraph and Hypergraph are generalisation of graphs which have more than two nodes in an edge. The nodes in an edge of a semigraph follow some order while the hypergraphs have no such order. Semigraphs connect more than two nodes in an edge. The number of edges in a semigraph is less than the number of edges in the corresponding graph. As a result of it, characteristics of a graph related to edges can be studied by considering the corresponding semigraph. Semigraph introduced by Sampathkumar [12] are better model than graph in applications where there is a need to connect several points by an edge [4]. The concept of domination in bipartite semigraphs is introduced by Swaminathan and Venkatakrishnan [15]. They have also stated that road networks can be modelled by semigraphs and traffic routing and traffic density in junctions can be studied through domination in semigraphs. Jeya Bharathi et al., [8] have associated the graph splicing scheme of Freund [5] with semigraphs.

Alphabets can be recognized using neural networks. Reetika Verma and Rupinder Kaur [11] have used neural network algorithm and SURF feature extraction to recognize English alphabets. Mithun Biswas and Ranjan Parekh

R. Prasath et al. (Eds.): MIKE 2015, LNAI 9468, pp. 204–215, 2015.
DOI: 10.1007/978-3-319-26832-3_20

[10] have recognized the English characters based on dynamic window using Artificial neural network. Umesh Kumar and Poonam Dabas [14] have proposed a work based on Hopfield network and Backpropagation network for Pattern recognition. They have proved that the success rate for this combination of networks is high compared with backpropagation alone.

The idea of taking non graphical but graphic objects and looking at them from the point of view of graph theory is not new [3]. Many problems such as web page ranking, face detection, mutagenesis problem, etc., are considered as graphical objects and solved by GNN [13]. GNN which is capable of processing general type of graphs both cyclic and acyclic, implements a function that maps a graph G and one of its nodes into an k-dimensional Euclidean space. The ranking of nodes in an attack graph is an important step towards analysing network security. Liang et al., [9] have modelled GNN for the task of ranking attack graphs. They have shown that GNN is able to learn rank function from examples and is able to generalize the function to unseen possibly noisy data. Bandinelli et al., [2] have introduced Layered GNN and have tested it on subgraph matching problem and some other benchmark problems. Yong et al., [16] have trained GNN to encode and process a relatively large set of XML formatted documents for mining.

Character recognition is one of the main criteria in coding and decoding. Any new way of character recognition is certainly an additional advantage in this field. This motivated us to use a new structure namely semigraph in character recognition. The English uppercase letters represented as semigraphs are identified by edge based GNN [6]. Hybrid neural network for classification of graph structured data based on nodes is introduced by Gnana Jothi and Meena Rani [7]. In this paper, the English uppercase letters are represented by graphs. These graphs are treated as semigraphs and are recognized by hybrid edge based neural network. Experimental results show that the hybrid edge based neural network outperforms edge based GNN.

2 Semigraph Representation for Alphabets

A semigraph G is a pair (V,X), where V is a non empty set whose elements are called vertices(nodes) of G, and X is a set of n-tuples, called edges of G, of distinct vertices, for various $n \geq 2$ satisfying the following conditions.

1. Any two edges have atmost one vertex in common.
2. Two edges $(u_1, u_2, ...u_n)$ and $(v_1, v_2, ...v_m)$ are considered to be equal iff
   (a) $m = n$ and
   (b) either $u_i = v_i$ or $u_i = v_{n-i+1}$ for $1 \leq i \leq n$.

Semigraph is a natural generalization of graph and it resembles graph when drawn in plane. In a semigraph each edge is an n-tuple. According to our convenience we can choose this n. Edges of a graph are combined together to form a single edge in a semigraph. Viewing a graph as a semigraph leads to reduction in the number of edges and thereby reducing the number of networks.

Let $E = \{v_1, v_2, ...v_n\}$ be an edge of a semigraph G. Then v_1 and v_n are the end vertices of E and v_i, $2 \leq i \leq n-1$ are the middle vertices of E. In the semigraph representation, an end vertex which is not a middle vertex is represented by thick dot. If a vertex is a middle vertex of a single edge it is denoted by small circle. If a middle vertex of an edge E is an end vertex of an edge E' a small tangent is drawn to the circle at the end of the edge E'. A graph

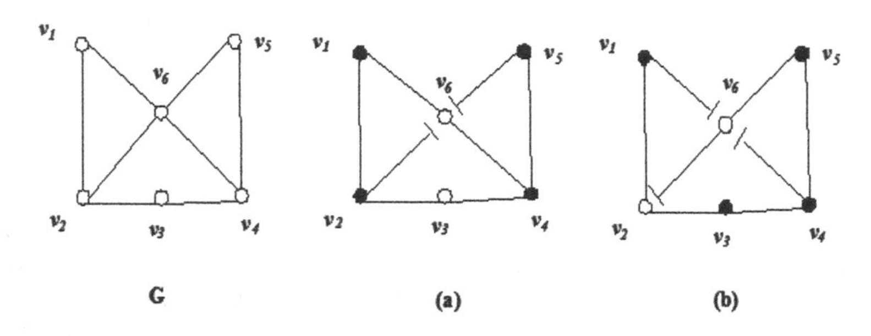

Fig. 1. Semigraph

G and its semigraph representation in two ways are given in Fig. 1. In Figure 1(a) edges are $\{v_1, v_2\}$, $\{v_2, v_3, v_4\}$, $\{v_4, v_5\}$, $\{v_5, v_6\}$, $\{v_2, v_6\}$, $\{v_1, v_6, v_4\}$. In Fig. 1(b) edges are $\{v_1, v_2, v_3\}$, $\{v_3, v_4\}$, $\{v_4, v_5\}$, $\{v_5, v_6, v_2\}$, $\{v_1, v_6\}$, $\{v_4, v_6\}$.

The uppercase English alphabets can be viewed as graphs. The alphabets A, E, F, H, I, J, K, L, M, N, T, V, W, X, Y, Z are formed by only straight lines. They are represented as graphs by introducing nodes at the end points and at points where two or more lines meet (Fig. 2).

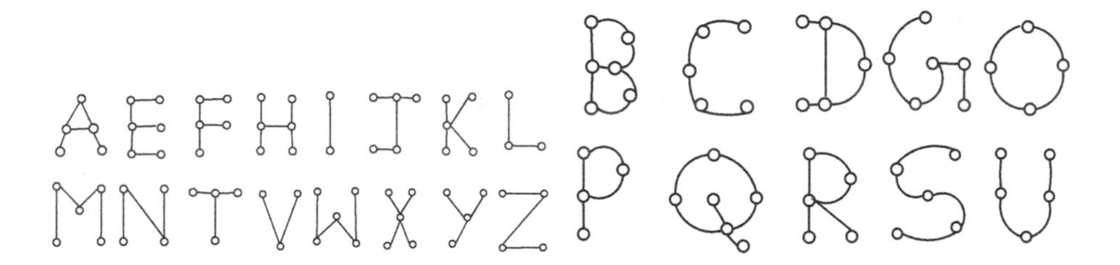

Fig. 2. Alphabets with straight edges **Fig. 3.** Alphabets with curved edges

But alphabets B, C, D, G, O, P, Q, R, S, U are formed with curves and straight lines. They are viewed as graphs by introducing nodes at end points, at point of intersection of lines and curves, and at points where the sign of the slope of the curve changes (Fig. 3).

The semigraph of alphabets are considered to have edges with two vertices and/or three vertices only for convenience. In the semigraph of alphabets, a

Fig. 4. Semigraph representation of alphabets

straight line with two endpoints is taken as edge e and the line with three nodes is taken as edge e'. In the curved portion, as edges of semigraphs can have atmost one node in common, the edges are viewed as e or e' restricting the total number of edges in all the semigraphs to be atmost 4. Figure 4 gives the semigraph representation of alphabets.

3 Edge Based Hybrid Neural Network

Consider a graph G=(V,E) where V is the set of vertices called nodes and E is the collection of arcs connecting two nodes of V. Let ne[e] denote the set of edges adjacent to an edge e in E. Let co[e] represent the set of vertices common to e and the neighbouring edges of e. The label attached to node n and to an edge e are given by $l_n \varepsilon R^c$ and $l_e \varepsilon R^d$ respectively.

In the hybrid network based on edges, the input layer of the input network consists of label and state vector of the edge, label and state vector of the neighbouring edges, characteristics of the node at which neighbouring edges are incident and a bias; hidden layer consists of h neurons and the output layer consists of s neurons which is the state vector of the edge e. The state vector x_e is calculated using a function f_w given by

$$x_e = f_w(l_e, x_e, l_{ne[e]}, x_{ne[e]}, l_{co[e]}, bias) \quad (1)$$

where l_e, x_e represent the label and state vector of the edge e; $l_{ne[e]}$, $x_{ne[e]}$ represent the label and state vectors of the neighbouring edges of e; for $e_1 \varepsilon co[e]$, label

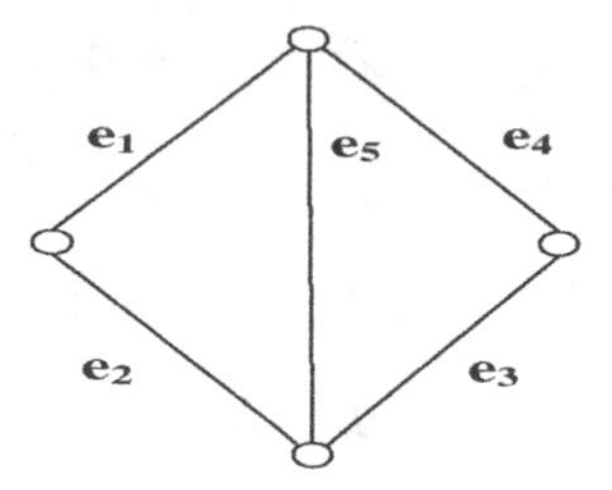

Fig. 5. Graph G

$l_{(e,e_1)} \varepsilon R^m$ represents the characteristics of the node on which the neighbouring edges e and e_1 are incident. $f_w : R^{(d+s)+|ne[e]|(d+s+m)+1} \rightarrow R^s$ is the function that implements the input network for an edge e.

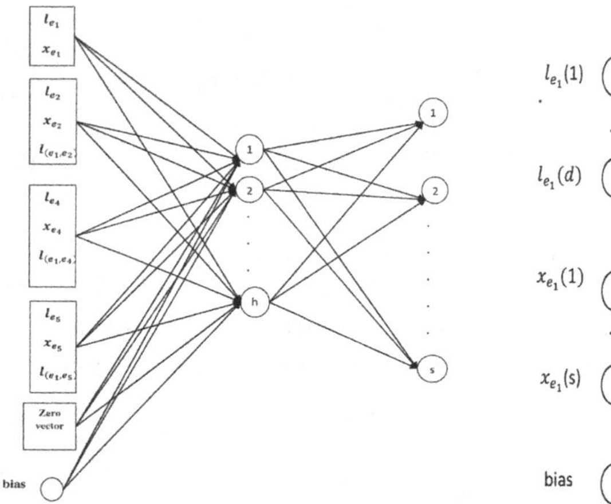

Fig. 6. Input network of edge e_1

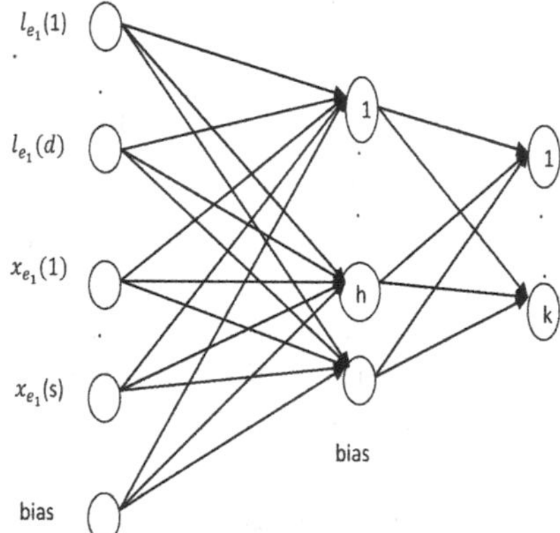

Fig. 7. Output network of edge e_1

Figure 5 represents a graph G and Fig. 6 represents the input network of the edge e_1 of the graph G. The state vectors (i.e.) output of input networks are stabilized using recurrent architecture. Let x_e(t) denote the tth iteration of x_e. Thus the states are computed by the following iteration as,

$$x_e(t+1) = f_w(l_e, x_e(t), l_{ne[e]}, x_{ne[e]}(t), l_{co[e]}, bias) \quad (2)$$

The output corresponding to each edge of a graph is produced by a single hidden layer FNN called output network with the stabilized state of the edge generated by the input network and its label in the input layer. For each edge e, the output o_e is computed by the output function

$$o_e = g_w(x_e, l_e) \quad (3)$$

$g_w : R^{s+d} \rightarrow R^k$ is the function that implements the output network. Figure 7 represents the output network of edge e_1. Both f_w and g_w involve FNN calculation.

Let X, L denote the vectors constructed by stacking all the states and all labels respectively of the graph. Equation (1) can be written as

$$X = F_w(X, L) \tag{4}$$

where F_w is the global input function. The global output function is

$$O = G_w(X, L_E) \tag{5}$$

Where O, L_E are the vectors obtained by stacking all the outputs and all the edge labels. Proposed hybrid neural network based on edges is given in Fig. 8. Construction of proposed neural network is based on graph with maximum number of edges and input layer of input network is based on edge with maximum neighbours in the graphs. Equations (4) and (5) are well defined if state updating has a unique solution and this depends on the properties of the input network. The input function f_w is implemented by a Multilayered Perceptron with recurrent architecture. The recurrent network is guaranteed to get a convergent representation by defining conditions on the weights that guarantee convergence.

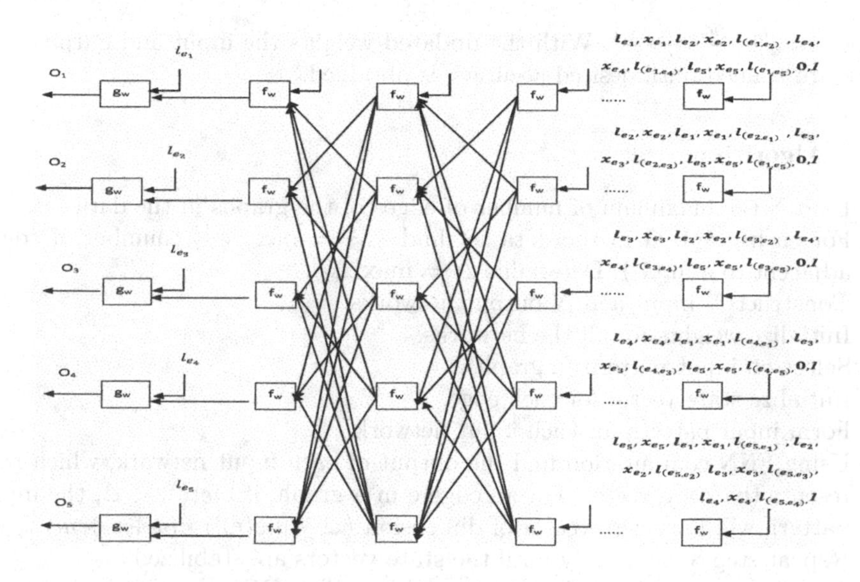

Fig. 8. Proposed hybrid neural network based on edges

4 Training Procedure for Hybrid Edge Based Neural Network

After representing all English alphabets as semigraph, input and output network are to be constructed for each edge. State vectors and weight of the networks of all edges are to be initialized. The output of each input network called state vector is computed using feedforward network computations and the network is trained using backpropagation algorithm. State vector of each edge is to be computed recursively until it is stabilized. The stabilized state vector and the label of the edge are taken as input for the output network. The sum of squared error

$$e_w = \sum_e (t_e - o_e)^2 \tag{6}$$

where t_e is the desired target of edge e, is calculated after calculating the output of each output network. Weights of output networks and then input networks are to be updated using error gradient method of the backpropagation algorithm. Based on the error e_w, change of weight for the output network is calculated as

$$\frac{\partial e_w}{\partial w} = \frac{\partial e_w}{\partial O} \frac{\partial G_w(X, L_E)}{\partial w} \tag{7}$$

where $O = G_w(X, L_E)$. Change of weight formula for input network is

$$\frac{\partial e_w}{\partial w} = b \frac{\partial F_w}{\partial w}(X, L) \tag{8}$$

where $b = \frac{\partial e_w}{\partial O} \frac{\partial G_w(X, L_E)}{\partial x}$. With the updated weights the input and output network are trained until desired accuracy is obtained.

4.1 Algorithm

1. Find N the maximum of number of edges of the graphs in the data set.
2. For each graph G in the data set find $\triangle_G = \max_{e \epsilon X(G)}\{$number of edges adjacent to e in G $\}$. Determine $\triangle = \max \triangle_G$.
3. Construct N input and N output networks.
4. Initialize weights for all the networks.
5. Select an input pattern (a graph).
6. Initialize state vector for each edge.
7. Form input pattern for each input network.
8. Using FNN computation find the output of each input network, which represents the state vector. For an edge e in a graph, if $|ne(e)| < \Delta$, the input pattern will have zero vector of dimension $(\Delta - |ne(e)|) * (d + s + m)$.
9. Repeat step 8 recursively until the state vectors are stabilized.
10. Calculate output for each output network using FNN computation.
11. Find the error for each output network using (6).
12. Update the weights of input and output network correspondingly based on error gradient method using (7) and (8)
13. Repeat steps 5 to 12 for all patterns.
14. Repeat steps 5 to 13 until Mean squared error of the validation data set satisfies the termination condition.

5 Results and Discussion

To recognize English alphabets through edge based network, each alphabet is viewed first as a graph and then as semigraph with edges e or e' as specified in Sect. 3. The edges are given label of dimension 1. The edges are labeled according to the type of edge, which is a line or curve. If the type of edge is a line, then it may be a horizontal line, vertical line or slant line. The label considered for an edge which is horizontal is 0.01, for a vertical edge it is 0.02, for an edge formed as slant line is 0.03 and 0.04 for a curve. In order to make the edge labels distinct a small noise with mean 0 and standard deviation 1 is added to the labels of the edges.

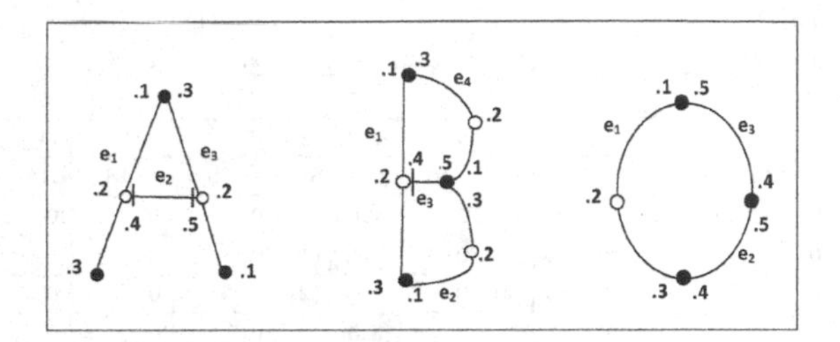

Fig. 9. Semigraph representation of A,B,O with labels

The semigraph representation of alphabets have two types of edges e and e'. Edges of type e has two nodes which are given labels 0.4 and 0.5. Edges of type e' has three vertices which are given labels $0.1, 0.2$ and 0.3. The label of a common node $l_{(u,v)}$ at which an edge u and neighbouring edge v is incident is of dimension 3. The first two components are the labels of the common node in the two edges, and the third component is the angle between u and v at the common node. The angle characteristic is represented as -0.18 if the angle is less than 180^0 and 0.18 if it is greater than or equal to 180^0. In Fig. 9, in the semigraph representation of alphabet A, $l_{[e_1,e_2]}$ is $(0.2, 0.4, -0.18)$ and $l_{[e_1,e_3]}$ is $(0.1, 0.3, -0.18)$.

The networks are modelled to identify English uppercase alphabets. The output for these 26 alphabets are distinguished by considering the binary equivalent of the number that represents the alphabets position in the set of English alphabets. Five output neurons are considered in the output network as the binary representation needs 5 binary digits. Since a semigraph is to be identified as a particular alphabet all edges in a particular alphabet are given the same target.

6 Results of Edge Based Hybrid Neural Network

Semigraph representation of all alphabets are formed with a maximum of four edges. Four input networks and four output networks are constructed here. Table 1 shows the input pattern corresponding to the alphabet A formed by

Table 1. Input pattern for the semigraph of alphabets A,B,O

Alphabet	A				B				O			
Input for	e_1	e_2	e_3	e_4	e_1	e_2	e_3	e_4	e_1	e_2	e_3	e_4
l_e	.03	.01	.03	0	.02	.04	.01	.04	.04	.04	.04	0
x_e	$x_{e_1}(1)$	$x_{e_2}(1)$	$x_{e_3}(1)$	0	$x_{e_1}(1)$	$x_{e_2}(1)$	$x_{e_3}(1)$	$x_{e_4}(1)$	$x_{e_1}(1)$	$x_{e_2}(1)$	$x_{e_3}(1)$	0
	$x_{e_1}(2)$	$x_{e_2}(2)$	$x_{e_3}(2)$	0	$x_{e_1}(2)$	$x_{e_2}(2)$	$x_{e_3}(2)$	$x_{e_4}(2)$	$x_{e_1}(2)$	$x_{e_2}(2)$	$x_{e_3}(2)$	0
$l_{ne[e]}$	.01	.03	.03	0	.04	.02	.02	.02	.04	.04	.04	0
$x_{ne[e]}$	$x_{e_2}(1)$	$x_{e_1}(1)$	$x_{e_1}(1)$	0	$x_{e_2}(1)$	$x_{e_1}(1)$	$x_{e_1}(1)$	$x_{e_1}(1)$	$x_{e_2}(1)$	$x_{e_1}(1)$	$x_{e_1}(1)$	0
	$x_{e_2}(2)$	$x_{e_1}(2)$	$x_{e_1}(2)$	0	$x_{e_2}(2)$	$x_{e_1}(2)$	$x_{e_1}(2)$	$x_{e_1}(2)$	$x_{e_2}(2)$	$x_{e_1}(2)$	$x_{e_1}(2)$	0
$l_{co[e]}$	.2	.4	.3	0	.3	.1	.4	.3	.3	.4	.5	0
	.4	.2	.1	0	.1	.3	.2	.1	.4	.3	.1	0
	−.18	−.18	.18	0	.18	−.18	−.18	.18	.18	−.18	.18	0
$l_{ne[e]}$	.03	.03	.01	0	.01	.01	.04	.01	.04	.04	.04	0
$x_{ne[e]}$	$x_{e_3}(1)$	$x_{e_3}(1)$	$x_{e_2}(1)$	0	$x_{e_3}(1)$	$x_{e_3}(1)$	$x_{e_2}(1)$	$x_{e_3}(1)$	$x_{e_3}(1)$	$x_{e_3}(1)$	$x_{e_2}(1)$	0
	$x_{e_3}(2)$	$x_{e_3}(2)$	$x_{e_2}(2)$	0	$x_{e_3}(2)$	$x_{e_3}(2)$	$x_{e_2}(2)$	$x_{e_3}(2)$	$x_{e_3}(2)$	$x_{e_3}(2)$	$x_{e_2}(2)$	0
$l_{co[e]}$	.1	.5	.2	0	.2	.3	.5	.1	.1	.5	.4	0
	.3	.2	.5	0	.4	.5	.3	.5	.5	.4	.5	0
	−.18	−.18	−.18	0	−.18	.18	−.18	−.18	−.18	.18	−.18	0
$l_{ne[e]}$	0	0	0	0	.04	0	.04	0	0	0	0	0
$x_{ne[e]}$	0	0	0	0	$x_{e_4}(1)$	0	$x_{e_4}(1)$	0	0	0	0	0
	0	0	0	0	$x_{e_4}(2)$	0	$x_{e_4}(2)$	0	0	0	0	0
$l_{co[e]}$	0	0	0	0	.1	0	.5	0	0	0	0	0
	0	0	0	0	.3	0	.1	0	0	0	0	0
	0	0	0	0	−.18	0	.18	0	0	0	0	0
Bias	1	1	1	1	1	1	1	1	1	1	1	1

straight lines, alphabet B formed by line and curves, and alphabet O formed by curves.

The weights of input networks are initialized with random numbers from $(0,1)$ and output network with random numbers from $(-1,1)$. Convergence of the state vector is guaranteed by considering the weights of the input networks [1] to satisfy the condition

$$\sum_i \sum_j w_{ij}^2 < \frac{1}{\max_i f(net_i)}$$

Each edge is adjacent to a maximum of three edges. The dimension of the state vector of each edge is taken to be 2 and it is initialized with zero vector. Input layer of input network has $3*(2+1+3)+(2+1)+1(=22)$ neurons. Both input and output network have sigmoidal activation function. The number of hidden neurons in the two FNN's are different. The number of hidden neurons of the input networks is considered to be 5. There was no significant change in convergence when the number is changed. The number of hidden neurons in the output network increased as the number of alphabets are increased for realization by the edge based hybrid neural network. The first 20 alphabets are recognized with only 13 hidden neurons. As the alphabet graphs from 21st alphabet to the 26th alphabet are similar to the alphabet graphs within 20 alphabets, the

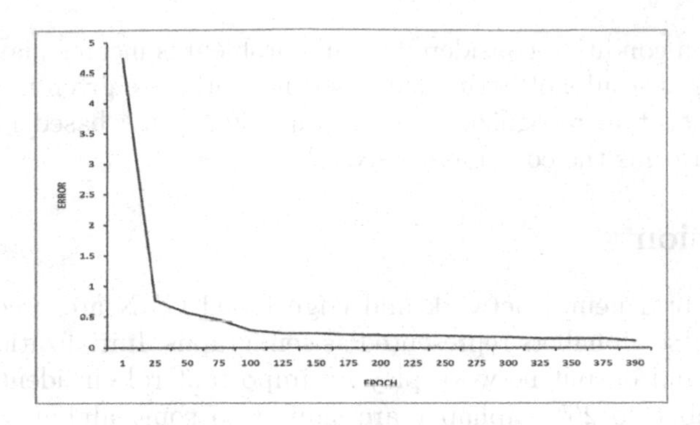

Fig. 10. Learning curve for edge based hybrid neural network

Table 2. Time taken to identify the alphabets.

Alphabets	# hidden neurons in output network		Time (sec.)		Epoch	
	Edge based hybrid neural network	Edge based GNN	Edge based hybrid neural network	Edge based GNN	Edge based hybrid neural network	Edge based GNN
First 5	5	5	20	23	43	32
First 10	7	8	108	260	126	377
First 15	11	12	175	2209	112	493
First 20	13	16	489	4198	223	962
First 26	37	37	3269	5382	891	688

number of hidden neurons in the output network increased much and it took 37 hidden neurons for the output network to identify all the 26 alphabets correctly. The experiment is carried out for first 5 alphabets, 10 alphabets, 15 alphabets, 20 alphabets and 26 alphabets for 10 different runs and the results are averaged.

Figure 10 shows the learning curve for the problem of recognizing all the Uppercase English alphabets by edge based hybrid neural network. The learning rate and momentum for the two FNN's are also different. For input network the learning rate and momentum are 0.01, 0.05 respectively and for output network they are 0.009 and 0.1 respectively. The termination condition considered in this problem is mean squared error

$$\frac{1}{n}\sum_{e}(t_e - o_e)^2 = 0.1$$

when the first 10 alphabets are considered and it is 0.09 when all the English alphabets are considered, where n is the number of alphabets considered. The learning rate for both the FNN is 0.01 and momentum up to first 20 alphabets is 0.05 and 0.1 for all the 26 alphabets. The value of μ is considered as 0.001.

The termination condition considered in this problem is mean squared error 0.1. The experimental results of both edge based networks are given in Table 2. The results signify that in recognizing the 26 alphabets edge based hybrid neural network outperforms the edge based GNN.

7 Conclusion

Edge based hybrid neural network and edge based GNN are used to identify English uppercase alphabets represented as semigraphs. Initialization of weights in both input and output network play an important role in identification. As the 21st alphabet to 26th alphabet are similar to some alphabet in the first 20 English alphabets, more hidden neurons are needed in the output network to recognize all alphabets than first 20 alphabets. The length and position of the edges are not considered for recognizing the alphabets. It implies that the recognition does not depend upon the size and orientation of the alphabets. It has been observed from the experimental results that edge base neural network outperforms edge based GNN in terms of execution time, number of epochs and number of hidden neurons in the output network.

In the process of training the edge based neural networks to recognize the semigraphs representing the alphabets, matrices corresponding to those semigraphs are used. Whenever a secret message is to be sent, each alphabet can be coded into its corresponding matrix and decoding can be done via the network. This work opens many novel methods in coding and decoding.

References

1. Aribowo, A., Lukas, S., Handy: Hand written alphabet recognition using hamming network. Seminar Nasional Aplikasi Teknologi Informasi, G-1 - G-5 (2007)
2. Bandinelli, N., Bianchini, B., Scarselli, F.: Learning long-term dependencies using layered graph neural networks. In: International Joint Conference on Neural Networks, pp. 1–8 (2010)
3. Bondy, J.A.: The graph theory of the greek alphabet. In: Alavi, Y., Lick, D.R., White, A.T. (eds.) Graph Theory and Applications. Lecture Notes in Mathematics, vol. 303, pp. 43–54. Springer, Heidelberg (1972)
4. Deshpande, C.M., Gaidhani, Y.S.: About adjacency matrix of semigraphs. Int. J. Appl. Phy. Math. **2**, 250–252 (2012)
5. Freund, R.: Splicing system of graphs. In: Proceedings of the First International Symposium on Intelligence in Neural and Biological Systems, p. 189. IEEE computer society, Washington, DC (1995)
6. Jothi, R.B.G., Rani, S.M.M.: Edge based graph neural network to recognize semigraph representation of english alphabets. In: Prasath, R., Kathirvalavakumar, T. (eds.) MIKE 2013. LNCS, vol. 8284, pp. 402–412. Springer, Heidelberg (2013)
7. Gnana Jothi, R.B., MeenaRani, S.M.: Hybrid neural network for classification of graph structured data. Int. J. Mach. Learn. Cybern. **6**, 465–474 (2015)
8. Jeya Bharathi, S., Padmashree, J., Sinthanai Selvi, S., Thiagarajan, K.: Semigraph structure in DNA splicing system. In: Sixth International conference on Bio-inspired Computing- Theories and Applications, pp. 27–29. IEEE Conference Publication (2011)

9. Lu, L., Safavi-Naini, R., Hagenbuchner, M., Susilo, W., Horton, J., Yong, S.L., Tsoi, A.C.: Ranking attack graphs with graph neural networks. In: Bao, F., Li, H., Wang, G. (eds.) ISPEC 2009. LNCS, vol. 5451, pp. 345–359. Springer, Heidelberg (2009)
10. Biswas, M., Parekh, R.: Character recognition using dynamic windows. Int. J. Comput. Appl. **41**, 47–52 (2012)
11. Verma, R., Kaur, R.: An efficient technique for character recognition using neural network and SURF feature extraction. Int. J. Comput. Sci. Inf. Technol. **5**, 1995–1997 (2014)
12. Sampathkumar, E.: Semigraphs and their applications. Report submitted to DST (Department of Science and Technology), India, May 2000
13. Scarselli, F., Gori, M., Tsoi, A.C., Hagenbuchner, M., Monfardini, G.: The graph neural network model. IEEE Trans. Neural Netw. **20**, 61–80 (2009)
14. Kumar, U., Dabas, P.: Recognizing numeric, alphabets and special chracters by pattern recognition using neural network. Int. J. Eng. Res. Technol. **3**, 1879–1883 (2014)
15. Venkatakrishnan, Y.B., Swaminathan, V.: Bipartite theory of semigraphs. WSEAS Trans. Math. **11**, 1–9 (2012)
16. Yong, S.L., Hagenbuchner, M., Tsoi, A.C., Scarselli, F., Gori, M.: Document mining using graph neural network. In: Fuhr, N., Lalmas, M., Trotman, A. (eds.) INEX 2006. LNCS, vol. 4518, pp. 458–472. Springer, Heidelberg (2007)

Small Eigenvalue Based Skew Estimation of Handwritten Devanagari Words

D.S. Guru[1], Mahamad Suhil[1(✉)], M. Ravikumar[1], and S. Manjunath[2]

[1] Department of Studies in Computer Science, Manasagangothri, Mysore, India
dsg@compsci.uni-mysore.ac.in, {mahamad45,ravi2142}@yahoo.co.in
[2] Department of Computer Science, Central University of Kerala, Kasaragod, Kerala, India
manju_uom@yahoo.co.in

Abstract. In this work, a novel technique for estimating skew angle in handwritten Devanagari words is proposed. Orientation of the Shirorekha, a horizontal line present at the top and touching all the characters of a word, is used as a clue to identify its skew. The method exploits the eigenvalue analysis of the covariance matrix formed by the edge pixels of the word image over a small connected region of support for the purpose extracting straight line segments. The line segments thus obtained are grouped according to their orientations. The orientation of a line segment is computed as a function of angles of its associated edge pixels. The angle of an edge pixel is identified using the eigenvector corresponding to the small eigenvalue associated with it. The line segments of each group are processed to locate a longest connected line which is decided to be the Shirorekha. The method is very fast when compared to Hough transform based line detection approach in addition to being robust to noise. Performance of the method is studied on a dataset consisting of 400 word images extracted from handwritten Devanagari documents especially of Hindi language with arbitrary orientations ranging from −45° to 45° under different scaling.

Keywords: Small eigenvalue · Region of support · Line detection · Skew estimation · Handwritten analysis

1 Introduction

Today, there is a great amount of digital content available over the World Wide Web. However, there are still plenty of paper based hard documents present in many places including government documents and ancient records which contain crucial information. Effective conversion of such content needs efficient techniques in converting a scanned document into its editable equivalent. Document processing has thus become a major area of research today because of the vast number of challenges present in converting such scanned documents into editable form. A good amount of work has already been done in converting printed English documents into editable form since past three decades. However, a major challenge faced by the researchers today is handling handwritten documents of various scripts. A number of applications including document

R. Prasath et al. (Eds.): MIKE 2015, LNAI 9468, pp. 216–225, 2015.
DOI: 10.1007/978-3-319-26832-3_21

transcription, automatic mail routing, and machine processing of forms, cheques and faxes etc., require processing of handwritten documents.

In India, Devanagari is one of the major scripts being in use in government offices after English. So, there is an urgent need for the development of an OCR for the Devanagari documents especially for handwritten documents. The major challenge in designing an OCR system for handwritten documents of any script is skew correction. Skew in handwritten documents starts with the user during writing itself and hence can be of any orientation at any location on a document. This is the main difference between a handwritten and a printed document when speaking of the skew present in them. Most of the times, in a printed document skew occur due to errors in scanning and hence normally the skew will be in only one direction throughout the document. But, in handwritten documents like Devanagari, there can be any amount of skew present in any location of the document. So, the methods developed for skew detection in printed documents perform very poor on the handwritten documents. One more important aspect that needs attention is, in handwritten documents the skew may be present at the word level and hence the document level or line level skew estimation techniques will fail to address the issue effectively. Hence, an efficient and effective skew estimation technique at word level is very important for the success of any handwritten OCR system which is the motivation behind this work.

In this paper, we address the problem of skew estimation and correction in handwritten Devanagari word images. One special property of Devanagari words is the 'Shirorekha' – a horizontal line present at the top and touching all the characters of a word. The proposed method exploits this special property to detect the skew present in handwritten Devanagari words using a robust and fast line detection algorithm proposed in [1] based on small eigenvalue analysis over a small connected region of support.

The rest of the paper is organized as follows. Section 2 presents the survey of the literature on skew estimation. The proposed small eigenvalue based skew estimation technique is presented in Sect. 3. In Sect. 4, the experimentation and results are discussed along with the dataset used. Conclusions are given in Sect. 5 followed by references.

2 Literature Survey

The methods available in the literature on skew estimation can be broadly classified into three categories based on whether they work for printed documents or for handwritten documents or for both. The techniques can further be divided into three groups based on whether they work at document level, line level or word level. Most of the works available in the literature are for printed documents. As our work focuses on handwritten documents, we focus our survey only on the techniques proposed for handwritten documents and also on those which are for both handwritten and printed documents.

In literature, there are many works available on skew estimation of documents which we preferred to segregate into Devanagari and Non-Devanagari as our work proposes model for Devanagari scripts. Most of the Non-Devanagari are the works proposed for English language. Some of the important works have been discussed here. A skew angle estimation technique is proposed in [2] using horizontal histogram and the Wigner-Ville distribution (WVD). They avoid the calculation of the histograms for every angle and subsequent application of WVD for every skew angle to reduce the computational cost. Authors claim that the accuracy of the estimation procedure does not depend on the resolution level, the presence of borders, the number of columns, the presence of graphics, or the size or the amount of different fonts. An approach using WVD for skew estimation of both printed and handwritten documents is presented in [3] based on the document's horizontal projection profile. The maximum intensity of the WVD of the horizontal histogram is used as the measure for skew angle estimation. In the case of printed document, entire document is processed at once whereas, for handwritten documents, block level processing is done to handle the difficulties effectively. In [4], a Radon transform based projection profile technique for skew estimation and correction in words of handwritten documents is proposed. The authors claim that the method is robust to noise and it is fast compared to other state of the art techniques. A method based on distance transform of the binarized documents is proposed for skew detection in documents is proposed in [5]. The method computes the dominant orientation of the gradient of the distance transform to estimate the skew of the document. The method is applicable to documents of printed English, Chinese handwritten as well as cursive scripts and is robust to variations in the text properties. A new technique called 'Viskew' is proposed in [6] to estimate the skew of text lines in documents. A skeleton is created using the transition map to generate an axis line which is then used to estimate skew angle. The approach was applied to a set of 150 images of documents with a total amount of more than 1,700 text lines. A method for skew estimation in unconstrained handwritten Kannada script is proposed based on mixture models [7]. The mixture of Gaussians is learnt by the Expectation-Maximization (EM) algorithm and skew angle is estimated using the cluster angle of individual words. A moment-based method is proposed in [8] for the estimation of the skew at word level in handwritten documents by using the group segmentation method. In [9], a method for estimation of skew angle of digitized printed Persian document images is proposed by using morphological operations and connected component analysis. Another approach makes use regression analysis and morphological operations for skew estimation in printed English documents [10]. A method using projection profile analysis and regression fit is proposed in [11] with a dynamic and static threshold to estimate the skew of a printed English documents.

To the best of our knowledge, only three works in the literature of skew estimation are devoted towards the Devanagari scripts. Skew estimation of the Devanagari word images using heuristic properties is proposed [12]. However, the model is too complex as it needs to scan the whole image every time, many parameters need to be estimated such as scan direction and further the image has to be noise free during scanning which may not be possible always. In [13], the authors have proposed a method for skew estimation in multilingual documents. The documents used are with multiple skews. Initially, morphological connected component analysis is applied for segmenting words and skew of each word is estimated by fitting a minimum confining ellipse. Word level skew is identified using the major-axis orientation followed by the block orientation by K-means clustering. According to the authors, the estimation may go wrong if the lines touch or close to each other. Moreover, the failure rate is high for Hindi scripts as some words may be elongated in the direction of minor axis. In [14], a method for skew estimation and correction of Devanagari word images using Hough transform is proposed. The authors argue that the method has given 97 % of accuracy for a dataset of Hindi and Marathi word images. However, different parameters involved in the method and how to fix them up has not been clearly explained. Also, as Hough transform based line detection is too complex and the overall time needed for computation of skew of all words will be very high which affects the efficiency of the method.

From the literature it can be observed that, skew estimation of handwritten documents is very challenging when compared to that of the printed documents. The techniques will either have less accuracy or the computational cost will be high. Also, skew detection at word level is very much necessary for the handwritten documents. Moreover, there are very less number of works reported in the literature of skew estimation on Devanagari documents especially of handwritten type. A major challenge for skew estimation of handwritten Devanagari documents is that, estimation of skew need to be done up to individual words since every word may have its own skew. Besides, conventional skew estimation techniques do not work on Devanagari documents since they carry entirely different structural properties when compared to other scripts such as English. Hence, efficient techniques which can detect the skew of a Devanagari word image very quickly are demanding so that the overall time required for skew estimation and correction of an entire document is reduced. In this direction, we propose a novel skew estimation technique for words in Devanagari documents exploiting the property of small eigenvalue in preserving geometrical property of a set of points [1].

3 Proposed Method

3.1 Overview

Given is a handwritten Devanagari word image with arbitrary orientation, it is initially skeletonized using morphological operation to obtain the edge image. The binary edge image thus obtained is passed through the small eigenvalue based line detection algorithm to extract linear segments as discussed in [1]. The line segments obtained are then clustered based on their orientations computed using the Eigen vectors corresponding to the small Eigen values of the pixels associated with the linear segments. From every cluster, the small line segments which are close to one another (decided by a threshold) are connected to form a single long line. The longest line obtained among all the clusters is decided to be the Shirorekha of the given word and its orientation is computed using its endpoints. The major stages involved in the proposed method are shown in Fig. 1.

3.2 Extracting Line Segments [1]

Let I_{edg} be the edge image of the input image I of size $M \times N$ (Fig. 2(a)) with n edge pixels obtained using morphological skeletonization as shown in Fig. 2(b). Consider a mask W of size $k \times k$ for some odd integer $k > 1$. Let $e_1, e_2, \ldots, e_n$ be the edge pixels in I_{edg}. For every edge pixel e_i, place the window W with e_i being the center of it as shown in Fig. 2(c), compute a set of edge pixels e_j connected to e_i and covered by W. The set of all e_js' is called as the family of e_i denoted by F_i. Let $F_1, F_2, \ldots, F_n$ be the families of

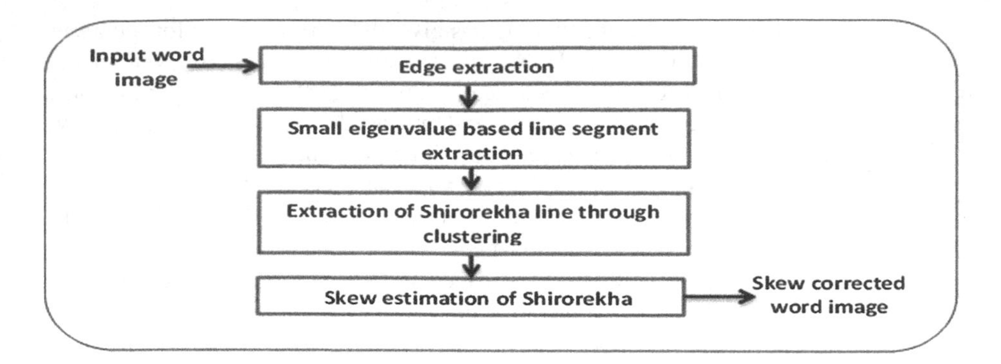

Fig. 1. Architecture of the proposed skew estimation method

the edge pixels e_1, e_2, …, e_n respectively. Let λ_1, λ_2, …, λ_n be the small eigenvalues computed for the covariance matrices of the edge pixels associated with families F_1, F_2, …, F_n respectively. In the output image, $I^{|}_{edg}$, change the values of the pixels which are members of F_i to λ_i. While doing this, it is not surprising that an edge pixel which is a member of two or more families will have more than one λ associated with it. This situation is resolved by replacing the value of that edge pixel to the minimum of all the λs' associated with it. An edge pixel which has a membership to only one family, though has λ value equal to zero, is associated with a very large value so that it can later be discarded by treating it as a noise pixel. The output image $I^{|}_{edg}$ thus generated is called as small eigenvalue image. In the second step of the process, thresholding is applied to $I^{|}_{edg}$ to filter all those pixels which are not linear edge pixels. So, the final output image I_{out} will be a binary image where the pixels which carry the small eigenvalue less than a pre-defined threshold t in $I^{|}_{edg}$ will carry a high value and rest will be made as low including non-edge and noise pixels.

3.3 Identifying the Line Segment Corresponding to Shirorekha, Skew Estimation and Correction

The output image I_{out} obtained by the small eigenvalue based line detector is further processed to identify the line segment corresponding to Shirorekha. The pixels of I_{out} are now termed as line-pixels in our discussion as each of them belongs to one or the other line in the given image. Normally, the length of Shirorekha will be very high when compared to any other line segment in any Devanagari word. Thus a line with maximum length among all the lines detected in I_{out} is decided to be the Shirorekha line. But, during the line detection process the pixels around the junction of two or more lines will be removed which lead to many breaks in the Shirorekha line. Also, due to the staircase effect, some of the line-pixels will get high eigenvalues and hence get filtered during thresholding process. This will further create some breaks in the Shirorekha line. So, K-means clustering is used to group the line-pixels of a particular range of angles together. The angle of each line-pixel in I_{out} is computed using the eigenvectors corresponding to the small eigenvalue of that line-pixel. Once the line-pixels are clustered, it is easy to identify the lines which are spatially closer to each other with similar orientation. In the following paragraphs, the process of orientation computation of each line-pixel, K-means clustering to cluster the line-pixels, joining of disjoint line segments and selection of longest line segment are presented.

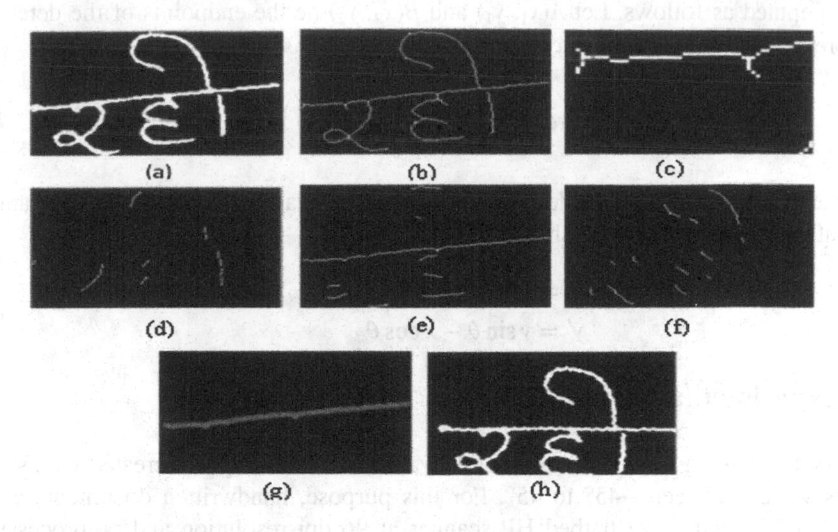

Fig. 2. (a) Input image, (b) Edge image, (c) Region of support with window of size k = 7, (d), (e) and (f) Clusters of line segments (g) Longest line extracted as Shirorekha (h) Skew corrected image

Once all the linear components present in the given word image are identified, the next job is to identify the lines which contribute in the process of skew estimation. Hence, the orientation of every line-pixel present in I_{out} need to be determined. This is done by the use of eigenvector of the corresponding small eigenvalue selected for that line-pixel. i.e., the angle of a line-pixel p_i in I_{out} with small eigenvalue λ_i and eigenvector corresponding to λ_i being $V_i = [x_i\ y_i]$ is computed as follows,

$$\theta_i = \arctan\left(\frac{y_i}{x_i}\right) \tag{1}$$

So, during the computation of small eigenvalue of a pixel its corresponding eigenvector is preserved to be able to use for the estimation of the orientation of the pixel.

Let $p_1, \ldots, p_m$ be the line-pixels present in I_{out} with $\theta_1, \ldots, \theta_m$ being their corresponding angles. K-means clustering is employed to this set of angles to identify the clusters of line-pixels. Figure 2(d)–(f) show clusters of line-pixels obtained by K-means clustering. Every cluster of line-pixels thus obtained is processed separately to connect the disjoint lines using their spatial positions. That is, if L_1 and L_2 are the two connected line components with same orientation θ and if the number of pixels between the endpoints of L_1 and L_2 in the direction θ is less than a predefined threshold c then the two line components L_1 and L_2 are connected to form a single longer line segment L.

After the broken line segments are joined, the line which is a longest among all the other lines due to all clusters is extracted as shown in Fig. 2(g) and it is decided to be the Shirorekha line of the given word image.

The skew of this line is the angle of inclination of the line with respect to the origin and computed as follows. Let $A(x_1, y_1)$ and $B(x_2, y_2)$ be the endpoints of the detected Shirorekha line, then the angle of this line is computed as in Eq. (2) below.

$$\theta_{skew} = \arctan\left(\frac{y_2 - y_1}{x_2 - x_1}\right) \times \left(\frac{180}{\Pi}\right) \quad (2)$$

The skew correction is then made by rotating the word image by the obtained skew angle θ. That is, every pixel (x, y) is shifted to (x', y') where,

$$\begin{aligned} x' &= x\cos\theta + y\sin\theta \\ y' &= y\sin\theta - x\cos\theta \end{aligned} \quad (3)$$

4 Experimentation and Results

A dataset consisting of 400 handwritten Devanagari word images is created with skew angles varied between −45° to 45°. For this purpose, handwritten documents were collected, scanned using flatbed HP scanner at 96 dpi resolution and preprocessing techniques such as binarization and noise removal were carried out and the words were segmented. The maximum size (in terms of Width × Height) of the created images is 312 × 100 with 57 × 82 being the minimum size. Some of the example images in the dataset can be seen in the Fig. 3.

The parameters are set empirically as follows. Size of the window for mask processing, W, is varied from 5 × 5 to 9 × 9 in steps of one and set to 7 × 7. Small eigenvalue threshold, t, is varied from 0.01 to 0.1 in steps of 0.01 and best results were obtained for 0.08. Similarly, the number of clusters for K-means clustering, K is set to 4 by experimenting with values from 3 to 8.

To evaluate the performance of the proposed model, the skew of each word is estimated manually by drawing a line on each word and the orientation of each line is stored. The stored orientation of the line of each word is compared with the angle produced using our model. Average Relative Error (ARE) and Accuracy are used as measures to compare the skew angle obtained by the proposed model with that of the ground truth as given in Eqs. (4) and (5) respectively.

$$Average\ Relative\ Error = mean\left(\left|\frac{\theta_{acutal} - \theta_{obtained}}{\theta_{actual}}\right|\right) \quad (4)$$

$$Accuracy = \frac{no\ of\ correctly\ estimated\ words}{Total\ no\ of\ words\ in\ the\ dataset} \quad (5)$$

Fig. 3. Samples from the dataset of handwritten Devanagari images

We have also implemented the Hough transform based line detection to detect the Shirorekha followed by the skew estimation. Table 1 shows results obtained by the Hough transform based method and the proposed method for some example images. It can be seen that, the proposed method outperforms the Hough transform based method in almost all the cases.

Quantitatively, when the ARE and Accuracy of both the methods are computed for the entire dataset, the ARE and Accuracy obtained by the proposed method happened to be 0.31 and 92 % respectively. Whereas, the same measures for the Hough transformed based approach were 0.36 and 90.31 % respectively. So, it is guaranteed by the experimentation that the proposed small eigenvalue based Shirorekha detection works well when compared to the Hough transform based technique. Also, as it is argued in [1], the complexity of the Hough transform based line detection is very high when compared to that of small eigenvalue based line detection.

Table 1. Sample results obtained by the proposed method along with that of Hough transform based approach

Original Image and Skew	Hough Transform based		Proposed Small Eigenvalue based Method	
	Image with Shirorekha Detected and Estimated Skew	Relative Error	Image with Shirorekha Detected and Estimated Skew	Relative Error
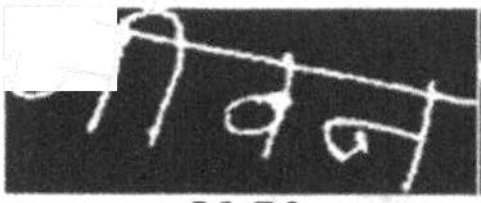 -13.76	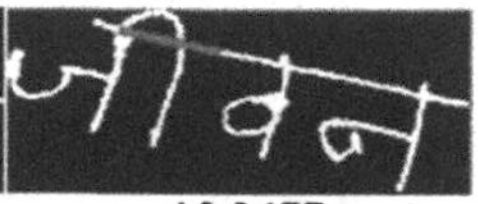 -12.8477	0.066	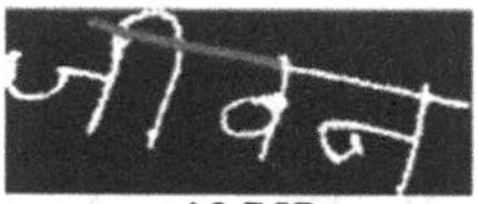 -12.757	0.072
 13.214	 12.3179	0.068	 12.095	0.085
 8.535	 0	1	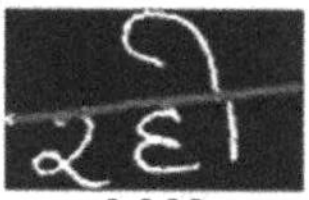 9.008	0.055
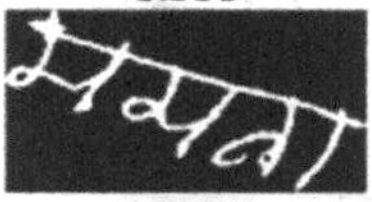 -17.612	 -17.9044	0.017	 -17.789	0.01
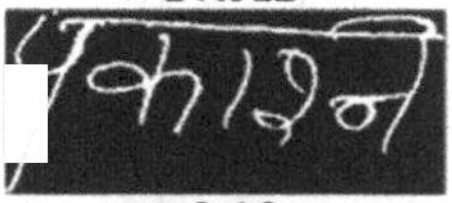 -3.46	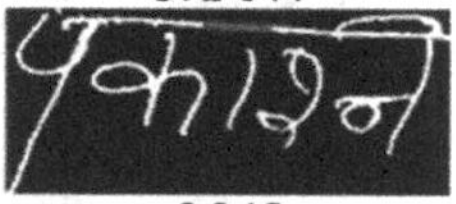 -2.245	0.351	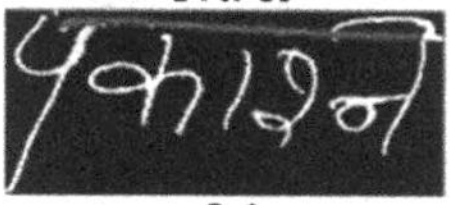 -3.1	0.104
 13.91	 13.172	0.053	 13.59	0.023

5 Conclusions

Estimation of skew of a handwritten document of Devanagari script essentially requires the skew estimation at word level. Shirorekha of a Devanagari word image can be effectively utilized as a clue to identify the skew of a word. In this paper, a novel method for the estimation of skew in Devanagari handwritten words is proposed. The method has exploited the concept of small eigenvalue analysis for the purpose of extraction of Shirorekha of a given Devanagari word image. The proposed method is experimented with a dataset of 400 handwritten Hindi word images extracted from handwritten documents. The results show that the method outperforms the Hough transform based approach.

Acknowledgements. The second author of this paper acknowledges the financial support rendered by the University of Mysore under UPE grants for the High Performance Computing laboratory.

References

1. Guru, D.S., Shekar, B.H., Nagabhushan, P.: A simple and robust line detection algorithm based on small eigenvalue analysis. Pattern Recogn. Lett. **25**, 1–13 (2004)
2. Kavallieratou, E., Fakotakis, N., Kokkinakis, G.: An unconstrained handwriting recognition system. IJDAR **4**(4), 226–242 (2002)
3. Kavallieratou, E., Fakotakis, N., Kokkinakis, G.: Skew angle estimation for printed and handwritten documents using the Wigner-Ville distribution. J. Image Vis. Comput. **20**, 813–824 (2002)
4. Cao, Y., Wang, S., Li, H.: Skew detection and correction in document images based on straight-line fitting. J. Pattern Recogn. Lett. **24**(12), 1871–1879 (2003)
5. Yosef, N.B., Hagbi, K.K., Dinstein, I.: Fast and accurate skew estimation based on distance transform. In: Proceedings of DAS 2008, pp. 402–407 (2008)
6. Mello, C.A.B., Sánchez, Á., Cavalcanti, G.D.C.: Multiple line skew estimation of handwritten images of documents based on a visual perception approach. In: Real, P., Diaz-Pernil, D., Molina-Abril, H., Berciano, A., Kropatsch, W. (eds.) CAIP 2011, Part II. LNCS, vol. 6855, pp. 138–145. Springer, Heidelberg (2011)
7. Aradhya, V.N.M., Naveena, C., Niranjan, S.K.: Skew estimation for unconstrained handwritten documents. In: Abraham, A., Mauri, J.L., Buford, J.F., Suzuki, J., Thampi, S.M. (eds.) ACC 2011, Part III. CCIS, vol. 192, pp. 297–303. Springer, Heidelberg (2011)
8. Brodić, D., Milivojević, Z.: Estimation of the handwritten text skew based on binary moments. Radioengineering **21**(1), 162–169 (2012)
9. Ashkan, M.Y., Guru, D.S., Punitha, P.: Skew estimation in Persian documents: a novel approach. In: Proceeding of International Conference CGiV 2006, pp. 64–70 (2006)
10. Guru, D.S., Punitha, P., Mahesh, S.: Skew estimation in digitized documents: a novel approach. In: ICVGIP (2004)
11. Shivakumara, P., Guru, D. S., Kumar, G.H., Nagabhushan, P.: Skew estimation of binary document images using static and dynamic thresholds useful for document image mosaicing. In: Proceedings of WITSA – 2003. Jamia Milia Islamia, New Delhi (2003)
12. Kapoor, R., Bagai, D., Kamal, T.S.: Skew angle detection of a cursive handwritten Devanagari script character image. J. Indian Inst. Sci. **82**, 161–175 (2002)
13. Guru, D.S., Ravikumar, M., Manjunath, S.: Multiple skew estimation in multilingual handwritten documents. IJCSI **10**(5–2), 65–69 (2013)
14. Tripati, A.J., Ravindra, S.H.: Skew detection and correction of Devanagari script using Hough transform. Procedia Comput. Sci. **45**, 305–311 (2015)

Recognizing Handwritten Arabic Numerals Using Partitioning Approach and KNN Algorithm

T. Kathirvalavakumar(✉) and R. Palaniappan

Research Center in Computer Science, V.H.N.S.N College (Autonomous), Virudhunagar 626 001, India
kathirvalavakumar@vhnsnc.edu.in, svrpalani@yahoo.co.in

Abstract. A method has been proposed to classify handwritten Arabic numerals in its compressed form using partitioning approach and K-Nearest Neighbour (KNN) algorithm. Handwritten numerals are represented in a matrix form. Compressing the matrix representation by merging adjacent pair of rows, by OR-ing the bits in corresponding positions, reduces its size in half. Considering each row as a partitioned portion, clusters are formed for each partition of a digit separately. Leaders of clusters of partitions are used to recognize the patterns by Divide and Conquer approach and KNN algorithm. Experimental results show that the proposed method recognize the patterns accurately.

Keywords: Handwritten numerals · Divide and conquer · Cluster · Leader algorithm · K-Nearest Neighbour · Classification

1 Introduction

Now-a-days more attention is needed in processing large data as it needs more memory and computations. More methods have been designed to minimize data size, computation and recognition [9,15]. classification of handwritten numerals need more memory and computation as Handwritten numerals data size is large.

Ravindra Babu et al. [15] have used run length encoded binary data for classification. This method focus on minimizing storage and computation on data. This method representing data in compressed form and classifying the compressed data without decompression. This method performs clustering and classification just using distance method. They have shown that this method works well for handwritten numerals. Saif and Naqvi [16] have proposed a near minimum sparse pattern coding based scheme for binary image compression. They have applied their scheme on co-ordinate representation of rectangular regions via a number of matrices. Such representations allow for efficiently coding these vertices and hence compress and hence compress the image significantly. Kumar et al. [8] have

T. Kathirvalavakumar—The work of T. Kathirvalavakumar is supported by University Grants Commission for Major Research Project, Government of India.

R. Prasath et al. (Eds.): MIKE 2015, LNAI 9468, pp. 226–234, 2015.
DOI: 10.1007/978-3-319-26832-3_22

presented a scheme for offline handwritten Gurmukhi character recognition based on diagonal features and transition features using KNN classifier. Diagonal and transition features of a character have been computed based on distribution of points on the bitmap image of character.

Basappa et al. [2] have proposed the projection distance metrics methods for numeral recognition and general regression neural networks for the classification of the unconstrained handwritten Kannada numeral character images. Noor et al. [11] have proposed a system to recognize Arabic numerals using fourier descriptors as the main classifier feature set and to improve the recognition accuracy a simple structure based classifier is added as a supplementary classifier. Moro et al. [10] have dealt with an optical character recognition system for handwritten Gujarathi numbers. They have added the values at level horizontal, vertical and two diagonals and form profile vector. Neural network is used to classify the patterns. DevinderSingh and Khera [4] have proposed a system to recognize digits using backpropagation neural network. They have segmented image, binary digital image is skeleton to reduce the width of digit into just a single line. Co-ordinate bounding box, area, centroid, eccentricity, equiv-diameter features are used by backpropagation neural networks as a classifier to recognize digits. Surinta et al. [18] have proposed a novel handwritten character recognition method for isolated handwritten Bur gale digits. A contour angular technique is introduced for the patterns. Dhandra et al. [6] have used zone based features to recognize handwritten and printed mixed kannada digits. KNN and SVM have been used for classification. Garg and Ahuja [7] have proposed a modified Hough transformation technique and four view profile projection technique have been used to extract the features of numerals. SVM and MLP have been used for classification. Rajiv Kumar and Ravulakolla [14] have presented a study on the performance of transformed domain features in Devnagari digit recognition. The recognition performance is measured from features obtained in direct pixel value, fourier transform, discrete cosine transform, gaussian pyramid, laplacian pyramid, wavelet transform and curvelet transform using classification schemes namely feedforward, function fitting, pattern recognition, cascade neural networks and K-Nearest Neighbours. They have observed that Gaussian pyramid based feature with KNN classifier yielded good accuracy.

An efficient hierarchical clustering algorithm is proposed by Vijaya et al. [19] for effective clustering and prototype selection for pattern classification. This method uses incremental clustering principles to generate a hierarchical structure for finding the subgroups/subclusters with each cluster. They have presented Leader-Sub-leader an extension of the leader algorithm.

Al-Omari and Al-Jarrah [1] have presented a system to recognize handwritten numerals using Probabilistic neural network. It involves feature vector based on centre of gravity and a set of vectors to the boundary points of the digit object. Dhandra et al. [5] have proposed an approach for recognizing Kannada, Telugu and Devanagari handwritten numerals using Probabilistic neural network. Pal et al. [12] have proposed a method for recognizing the numerals of Devanagari, Bangala, Telugu, Oriya, Kanada, and Tamil by the directional features computed from the blocks of a bounding box of numerals. Shivanand Rumma et al. [17]

have proposed a method for recognizing handwritten Kannada numerals by Radial basis function. Kumar et al. [8] have proposed unconstrained offline handwritten numeral recognition system using local and global features of profile of numeral image, majority voting scheme and neural network. Rajashekararadhya and Vanaja Ranjan [13] have proposed a zone based feature extraction method for recognizing handwritten numerals. This method involves character centroid and average distance of pixel present in the zone and the centroid of the character. Nearest neighbor, feedforward neural network and support vectors are also used in this work.

In the proposed work, Handwritten numerals are represented in matrix form with binary values. Logical OR operation is applied on adjacent pixels in a vertical direction. Cluster the bits in each row of a digit with Leader algorithm. Using divide and conquer technique, KNN algorithm and majority voting algorithms patterns are classified. Rest of the chapter is organized as follows. Chapter 2 describes compression, chap. 3 explains clustering with leader, Chap. 4 describes classification, Chap. 5 gives proposed procedure, and Chap. 6 discusses experimental results followed with conclusion section.

2 Compression

Patterns of each digit are represented in matrix form with binary values. Consider each row of a matrix as a pattern. Apply logical OR operation on bits of same columns of adjacent pair of rows of a matrix without row overlapping. Resultant matrix is with half the size of the original matrix as the number of row is reduced by half. The logical OR operation does not omit any of its 'on' character as the characteristic of arabic numerals is continuous in its structure. So the shape of the digit does not change because of logical OR operation but shrink in its size vertically. The original image and compressed image are shown in Figs. 1 and 2.

3 Clustering with Leader

Compressed digit is in a matrix form. The matrix is partitioned into groups equivalent to number of rows of the matrix. Bits of each row are member of the group. Bits in each row of a digit are clustered based on distance measure. By considering all patterns of a particular digit clusters are formed for each partition of those digits separately. Similarly for every digit clusters are formed. Clustering technique with leader concept [19] is used to group meaningful patterns so as to improve classification accuracy with minimum input-output operations. In this method [19], first pattern is treated as a cluster leader. Remaining patterns are compared with the leaders of existing clusters and is assigned to member of a cluster when leader is with minimum distance. If the distance between pattern and the leader is greater than predefined distance then the pattern is a leader of a new cluster. Distance between pattern is computed by the Manhattan formula as follows:

$$Manhattandistance = |X - Y| \tag{1}$$

```
0 0 1 1 1 1 1 1 1 1 0 0
0 0 1 1 1 1 1 1 1 1 0 0
0 0 1 1 1 0 0 0 1 1 0 0
0 1 1 1 0 0 0 1 1 1 0 0
1 1 1 0 0 0 0 1 1 0 0 0
1 1 1 0 0 1 1 1 0 0 0 0
1 1 1 0 0 1 1 1 0 0 0 0
1 1 1 1 1 1 1 1 1 1 1 0
0 1 1 1 1 1 1 1 0 1 1 0
0 0 1 1 1 1 0 1 0 1 1 0
0 0 1 1 1 0 0 0 0 0 1 0
0 0 1 1 1 1 1 1 1 1 1 0
0 0 1 1 0 0 0 0 1 1 1 0
0 0 1 1 1 1 1 1 1 1 1 0
0 0 1 1 1 1 1 1 1 1 0 0
0 0 0 0 1 1 1 1 0 0 0 0
```

Fig. 1. Handwritten digit 8 in matrix form

4 Classification

Patterns to be classified is represented in matrix form. Compress the pattern by doing logical OR operation on the bits of same column of adjacent row. Patterns on each row is considered and find distance of the pattern with the leaders of corresponding row of all digits. Distance measures are kept with its digit. Calculated distance measures of each partition are sorted separately. K minimum distance measures are selected from each partition. As the merging of the separated partition form a digit, selected minimum distance of first partition is added with the selected K distances of next partitions in all combinations. These added values are added with the distance measures of next partitions in all combinations. This is repeated for all remaining partitions. The obtained final added values are sorted. K minimum values are selected from that and a digit which are having more distance values is a class of the pattern which need classification.

5 Proposed Procedure

Every handwritten digit to be viewed in a matrix form with binary value. The problem domain with handwritten numerals are divided into training and testing numerals. Consider all numerals in a training pattern for clustering. Consider every pair of row of a digit without overlapping. Matrix to be compressed to its half size by applying logical OR operation on bits occur in columns of selected pair of rows. Consider each row of a compressed matrix as a pattern. Include the digit of a matrix on each pattern as a target value. Group all handwritten

0 0 1 1 1 1 1 1 1 1 0 0
0 1 1 1 1 0 0 1 1 1 0 0
1 1 1 0 0 1 1 1 1 0 0 0
1 1 1 1 1 1 1 1 1 1 1 0
0 1 1 1 1 1 1 1 0 1 1 0
0 0 1 1 1 1 1 1 1 1 1 0
0 0 1 1 1 1 1 1 1 1 1 0
0 0 1 1 1 1 1 1 1 1 0 0

Fig. 2. Handwritten digit 8 in compressed form

numerals to be grouped based on its digit value. Using Leader algorithm [19] clusters to be formed for particular row of all numerals of a digit in a group separately. Every cluster leader to be tagged with its target value. By forming cluster leaders for each row of a compressed matrix the numerals are partitioned into many based on the size of the rows of the matrix. This procedure is Divide method. Now the leaders of compressed matrix of digits represent the numerals. The numerals in a testing group to be classified using KNN and merge method. Every numeral of a testing group to be represented in a matrix form. Compress the matrix using logical OR operation procedure which is applied on training numerals. Every row of the compressed matrix to be considered as a pattern and Classification procedure to be applied on the patterns of a matrix with voters algorithm. It involves KNN with merge procedure.

5.1 Training Algorithm

step 1. Convert 192 bits of numerals into 16 × 12 size and treat 193rd bit as target value for the 16 rows of a digit.

Step 2. Repeat step 1 for all training patterns of the problem.

step 3. Apply logical OR operation on bits of each column of adjacent two rows without row overlapping. Now 16 × 12 becomes 8 × 12

step 4. Now treat every resultant pattern as with 8 partitions

- TRAINING

step 5. Now form clusters for every partition of each digit separately
i.e.) cluster for partition 1 of digit $0, 1, 2, 3, ..., 9$
i.e.) cluster for partition 2 of digit $0, 1, 2, 3, ..., 9$
i.e.) cluster for partition 3 of digit $0, 1, 2, 3, ..., 9$
....
....
....
i.e.) cluster for partition 8 of digit $0, 1, 2, 3, ..., 9$

- TESTING

step 6. Read a test pattern with 192 bits of numerals and convert into 16×12. Apply logical OR as in Training procedure. Resultant obtained matrix is with size 8×12. Bits in each row to be considered as a pattern of a partitioned compressed matrix

step 7. Compare the partition of the test pattern with the leaders of the corresponding partition of the training patterns using Manhattan distance measure

step 8. Merge first two adjacent partition values by adding every value in one partition with the values in adjacent partitions all combinations

step 9. Sort the resultant values

step 10. Select first K values

step 11. Add these selected values with next adjacent partition as in step $9, 10, \&11$

step 12. Now repeat the procedure for all partitions

step 13. Apply voters algorithm on the final selected values which will give the class of the Test pattern

6 Experimental Results and Discussion

The proposed method is applied on Handwritten digit data [9] having 667 patterns per digit. Totally 6670 patterns, each with 193 bits, are used for training and 3330 patterns are used for testing. The last bit of the training patterns represent target class of the patterns. The experiment is carried out using MathLab software in the Intel Quad core system.

Every pattern is converted into 16 number of patterns with single digit as target value. After compression by logical OR operation, 16 patterns become 8 patterns. clusters are formed among patterns of each partition of a digit separately. When threshold for distance measure is considered as 2, total number of clusters formed is 2216, but when threshold is considered as 3 total number of clusters formed is 1063 which is displayed in Table 1. Table 2 shows number of clusters formed for each partition when threshold is 3. When KNN algorithm is applied on these cluster leaders, classification accuracy not differ much for the threshold 2 and 3. The obtained classification accuracy is 98.7 % when threshold =3. Ravindra Babu et al. [15] have used Run length encoding for the patterns and apply KNN method for classification on the encoded patterns and got 92.47 % accuracy. Vijaya et al. [19] have used leader algorithm and KNN method for classification but got 97.34 %. Monu et al. [9] have generated synthetic patterns and KNN algorithm and obtained 96.28 %. The proposed method give good accuracy than the results of the existing literatures which is shown in Table 3. This method uses simple logical OR operation to

Table 1. Number of cluster Leaders for different Thresholds

Digit	# of clusters when threshold =3	#of clusters when threshold =2
0	121	246
1	25	56
2	164	259
3	109	228
4	120	193
5	130	277
6	112	218
7	85	215
8	89	283
9	108	241

Table 2. Number of cluster leaders on each partition when Threshold=3

Digit	part. 1	part. 2	part. 3	part. 4	part. 5	part 6	part. 7	part. 8
0	7	14	21	17	18	19	17	8
1	3	4	3	3	2	2	4	4
2	9	17	21	21	19	25	25	27
3	10	16	15	9	17	16	17	9
4	10	17	20	19	18	15	13	8
5	14	14	16	14	18	20	23	11
6	5	7	12	19	20	20	21	8
7	11	15	17	11	11	7	6	7
8	9	14	20	13	13	7	7	6
9	12	17	18	18	13	13	8	9

Table 3. Comparison of classification accuracy

Method	Accuracy(%)
Babu et al. [15]	92.47
Monu et al. [9]	96.28
Vijaya et al. [19]	97.34
Proposed Work	98.7

compress the patterns without affecting its characteristics. Also as this proposed method uses leader algorithm, only cluster leaders of the compressed patterns are used for processing. It reduces the computation time for processing. When memory requirement is considered number of bits in original 6670 patterns are 1280640. As only 1063 clusters are enough for processing, these cluster leaders are enough to store it in a memory. It has totally 12756 bits for storing. Babu et al. work needs 139054 bits for storing the patterns.

7 Conclusion

Handwritten characters are compressed by logical OR operation, leaders of clusters are selected from the formed clusters of each partition of every digit. Only those leaders are used for classifying the digit by divide and conquer technique and KNN algorithm. Novelty of this work is compressing the given digits by logical OR operation and is used for forming cluster for individual row of each digit separately. Experimental results show that the proposed method classify the training patterns accurately when the data are in the compressed form.

References

1. Al-Omari, F.A., Al-Jarrah, O.: Handwritten indian numerals recognition system using probabilistic neural networks. Adv. Eng. Inf. **18**, 9–16 (2004)
2. Kodada, B.B., Shivakumar, K.M.: Unconstrained handwritten kannada numeral recognition. Int. J. Inf. Electron. Eng. **3**, 230–232 (2013)
3. Desai, A.A.: Gujarati handwritten numeral optical character recognition through neural network. Pattern Recog. **43**, 2582–2589 (2010)
4. Singh, D., Khehra, B.S.: Digit recognition system using back propagation neural network. Int. J. Comput. Sci. Commun. **2**, 197–205 (2011)
5. Dhandra, B.V., Benne, R.G., Hangarge, M.: Kannada, telugu, and devanagari handwritten numeral recognition with probabilistic neural network: a novel approach. IJCA Spec. Issue Recent Trends in Image Process. Patten Recog., pp. 83–88 (2010)
6. Dhandra, B.V., Mukarambi, G., Hangarge, M.: Zone based features for handwritten and printed mixed kannada digits recognition. In: Proceedings of the International Conference on VLSI, Communication and Instrumentation. Int. J. Comput. Appl., pp. 5–11 (2011)
7. Garg, M., Ahuja, D.: A novel approach to recognize the off-line handwritten numerals using MLP and SVM classifiers. Int. J. Comput. Sci. Eng. Technol. **4**, 953–958 (2013)
8. Kumar, R., Goyal, M.K., Ahmed, P., Kumar, A.: Unconstrained handwritten numeral recognition using majority voting classifier. In: IEEE International Conference on Parallel Distributed and Grid Computing(PDGC), pp. 284–289 (2012)
9. Agrawal, M., Gupta, N., Shreelekshmi, R., Murty, M.N.: Efficient pattern synthesis for nearest neighbour classifier. Pattern Recog. **38**, 2200–2203 (2005)
10. Kamal, M., Fakir, M., El Kessab, B.D., Bouikhalene, B., Daoui, C.: Gujarati handwritten numeral optical character through neural network and skeletonization. Jurnal Sistem Komputer **3**, 40–49 (2013)

11. Noor, S.M., Mohammed, I.A., George, L.E.: Handwritten arabic (indian) numerals recognition using fourier descriptor and structure base classifier. J. Al-Nahrain Univ. **14**, 215–224 (2011)
12. Pal, U., Sharma, N., Wakabayashi, T., Kimura, F.: Handwritten numeral recognition of six popular indian scripts. In: Ninth International Conference on Document Analysis and Recognition, ICDAR, vol. 2, pp. pp. 749–753 (2007)
13. Rajashekararadhya, S.V., Ranjan, P.V.: Handwritten numeral/mixed numerals recognition of south-indian scripts: the zone based feature extraction method. J. Theor. Appl. Inf. Technol. **7**, 63–79 (2009)
14. Kumar, R., Ravulakollu, K.K.: Offline handwritten devnagari digit recognition. ARPN J. Eng. Appl. Sci. **9**, 109–115 (2014)
15. Babu, T.R., Murty, M.N., Agrawal, V.K.: Clasification of run length encoded binary data. Pattern Recog. **40**, 321–323 (2007)
16. Zahir, S., Naqvi, M.: A near minimum sparse pattern coding based scheme for binary image compression. In: Proceedings of IEEE International Conference on Image Processing, pp 289–292 (2005)
17. Rumma, S., Vishweshwarayya, C.H., Bhuvaneshwari, B.D.: Handwritten kannada numeral recognition using radial basis function. Int. J. Comput. Appl. **98**, 18–20 (2014)
18. Surinta, O., Schomaker, L., Wiering, M.: A comparison of feature and pixel-based methods for recognizing handwritten bangla digits. In: 12th International Conference on Document Analysis and Recognition, pp. 165–169 (2013)
19. Vijaya, P.A., Murty, M.N., Subramanian, D.K.: Leaders-subleaders: an efficient hierarchical clustering algorithm for large data sets. Pattern Recog. Lett. **25**, 505–513 (2004)

Fuzzy Based Support System for Melanoma Diagnosis

Anand Gupta[1], Devendra Tiwari[1](✉), Siddharth Agarwal[2], and Monal Jain[1]

[1] Department of Computer Engineering, Netaji Subhas Institute of Technology, New Delhi 110078, India
omaranand@nsitonline.in, {dev.gl.tiwari,monal94}@gmail.com
[2] Department of Instrumentation and Control Engineering, Netaji Subhas Institute of Technology, New Delhi 110078, India
siddharth691@gmail.com

Abstract. Early detection of Melanoma (skin cancer) and its classes (Malignant, Atypical, Common Nevus) is always beneficial for patients. Till now researchers have designed many Computer Aided Detection (CAD) systems which have focused on providing binary results (i.e. either presence or absence of any class of melanoma). As these systems do not provide relative extent of lesions belonging to each class, they usually lack decision support for dermatologists (in case of suspiciousness of a lesion) and complete reliability for routine clinical use. To overcome these problems, a two stage framework is proposed incorporating a new fuzzy membership function based on Lagrange Interpolation Curve Fitting method. This framework returns analogue values for a lesion which represents its relative extent in a particular class (helpful in recreating suspiciousness), hence having a greater degree of acceptability among dermatologists. A two stage CAD framework proposed here uses PH^2 dermoscopic image dataset as input. In the first stage pre-processing, segmentation and feature extraction is performed while in the next stage fuzzy membership values for the three classes are calculated using Gaussian, Bell and the proposed membership function. A comparative study is done on the basis of sensitivity and specificity for the three membership functions.

1 Introduction

Melanoma is a form of skin cancer, common among Northern and North-Western Europeans having primary cause as Ultraviolet radiation. Common Nevus (typical mole), Atypical Melanoma (also known as Dysplastic Nevus) and Malignant Melanoma are different classes of skin cancer in order of increasing severity. Malignant melanoma which is a severe case of skin cancer is a cause of majority of deaths (75 %) related to skin cancer. Malignant Melanoma can be prevented if it is diagnosed in its early stage. Many signs have been identified by researchers for early detection of melanoma which are categorized as ABCD-rule of dermoscopy [12]. Dermatologist having vast experience can only accurately diagnose melanoma using these signs. Therefore it has been a major area of research

R. Prasath et al. (Eds.): MIKE 2015, LNAI 9468, pp. 235–246, 2015.
DOI: 10.1007/978-3-319-26832-3_23

Table 1. Possible cases and their course of action

Membership value for class to which lesion belongs	Membership values for other classes	Case	Future course of action
Very high	Very low	True +ve	No suspiciousness, Medication
High but comparable	Low but comparable	True +ve	Suspiciousness recreated, Recommendation to pathologist
Low but comparable	High but comparable	False -ve	Suspiciousness recreated, Recommendation to pathologist

* All membership values are relative.

in Computer Vision and Machine Learning to develop Computer Aided Detection (CAD) system that detects melanoma accurately. These CAD systems are designed to output binary results for the dermatologists i.e. either presence or absence of any class of melanoma. These CAD systems have little acceptance among dermatologists [4] as they don't help in decision making process. Thus, they are not extensively used in hospitals today. An effort has been made here to design a CAD framework that uses fuzzy membership functions assigning membership values to each lesion. Fuzzy membership values describe extent of lesions belonging to a particular class. With the help of fuzzy membership values this framework helps the dermatologists by recreating the suspiciousness that he encounters during the process of melanoma diagnosis. G R Day et al. [3] have discussed the importance of suspiciousness in their paper. Following example illustrates the importance of membership values in decision making process.

Suppose a skin cancer patient X (having malignant melanoma) visits a dermatologist Y. After observing the lesion superficially, dermatologist Y finds the lesion suspicious. So he tests the lesion under the available CAD system. If the CAD system returns a binary result that the lesion does not have malignant melanoma (false negative result) then he would not recommend any further tests or medication relying on the CAD system completely. It could have fatal effects on the health of patient X. To overcome this problem if a CAD system returns membership values of the lesion for all three classes then the dermatologist would observe the three values (the extent of lesion belonging to each class). Table 1 enlists the possible cases for which this framework may benefit and their future course of actions.

On the basis of the observation and comparison of the three membership values Y can now take a well informed decision. During further visits of X, Y can monitor the effect of medication prescribed by keeping a track of extent of melanoma in the lesion.

Next section briefly tabulates the prior work being done in this domain.

1.1 Related Work

Korotkov K et al. in the review paper [6] have given a detailed survey of the work done in the analysis of Pigmented Skin Lesions (PSLs) using CAD systems. An automated skin lesion segmentation algorithm is developed for PH^2 [8] dataset in [1]. Later color and shape features are used to classify melanoma. Stanley et al. [11] have used relative color histogram approach over training set of images to diagnose melanoma. Hansen et al. [5] used features like asymmetry, edge-abruptness, color distribution, etc. from ABCD-rule to design a probabilistic framework for detection of melanoma. The use of fuzzy based classification systems in dermoscopic images are not fully explored as compared to other systems. One of the initial use of fuzzy based classification system on dermoscopic images is done in [10] to distinguish benign skin lesions from malignant. Relative color histogram technique is used on fuzzy set with trapezoidal membership functions to classify melanoma lesion. Patwardhan et al. [9] have focused on improving the sensitivity and specificity of melanoma diagnosis by using both crisp and fuzzy based membership partitioning. They have proposed multi-dimensional Gaussian and Bell fuzzy membership functions and ADWAT crisp classification method on wavelet transform based bi-modal channel energy features on epi-illumination images. Since Gaussian and Bell Membership functions have been used in fuzzy based Melanoma diagnosis, therefore these functions are used for comparison with our work.

Inspite of many advancements, these CAD systems are not yet fully accepted by dermatologists. Dreiseitl et al. [4] have conducted a sudy involving 52 volunteer dermatologists to investigate reaction of physicians when faced with decision support system, whereas G R Day et al. [3] have suggested a method to identify problematic skin lesions and to reproduce the suspiciousness that dermatologists experience.

1.2 Motivation

Following are the issues that motivated us to propose a new framework for such problems:

- Study by Dreiseitl et al. [4] indicates unwillingness of physicians in using current computer aided detection systems as these systems are not providing the necessary decision support required. Thus a system is required to provide support in the form of suspiciousness that physicians encounters during the diagnosis. This property is not available in the currently available systems.
- Due to low correlation among features belonging to ABCD-rule; Gaussian and Bell membership functions provide very small membership values that may return accurate binary results but do not signify the extent of lesion belonging to a particular class.
- It is observed during the experiments that Gaussian and Bell membership functions have low sensitivity and specificity for PH^2 dataset on features belonging to Asymmetry, Border and Color.

1.3 Contribution

This paper has the following two contributions to overcome drawbacks of current CAD systems:

- Instead of giving binary results this framework gives analogue fuzzy values thus aiding the dermatologist in decision making process by recreating the suspiciousness.
- A new method for determination of fuzzy membership functions based on Lagrange Interpolation Curve Fitting technique for melanoma diagnosis for features with low correlation.

1.4 Problem Definition

To evaluate and compare analogue values obtained from Gaussian, Bell and our proposed fuzzy membership function based on Lagrange Curve Interpolation method using PH^2 - a dermoscopic image dataset.

1.5 Organization of the Paper

Section 2 of the paper gives a detailed description of both stages of the framework. Section 3 describes the various experiments performed on the system and the discussions related to the results obtained. Section 4 concludes the paper and mentions the related future works.

2 Framework

The framework proposed for melanoma diagnosis is divided into two stages. First stage caters to the image processing requirements on training and test dataset, while in the second stage fuzzy membership values are calculated on test dataset using three membership functions- Gaussian, Bell and the proposed

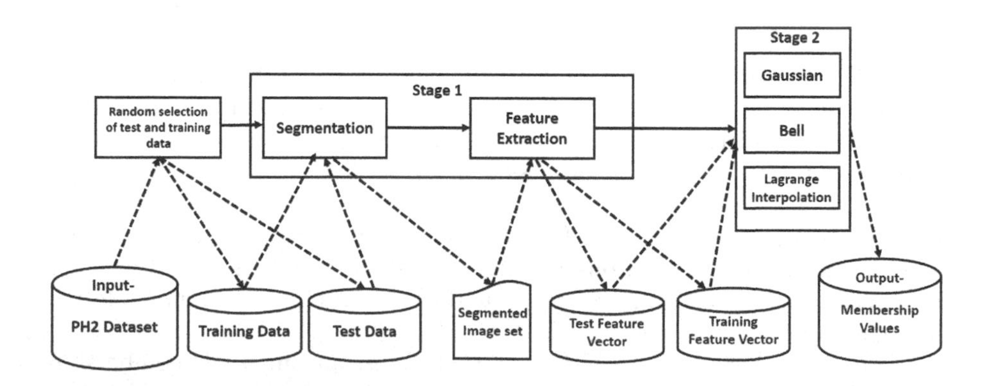

Fig. 1. Work flow model of the framework

membership function using Lagrange Interpolation Curve Fitting method. Initially PH^2 dataset is randomly divided into training dataset and test dataset. Figure 1 shows the work flow model of the framework. Detailed description of each stage is given below.

2.1 Stage One

In stage one, training dataset and test dataset are given as input. Then segmentation is performed followed by feature extraction. Segmentation is described in the next sub-section.

Segmentation. In this framework the segmentation algorithm developed by Abuzaghleh et al. [1] for PH2 dataset is used. Segmentation process also incorporates pre-processing and post-processing. Various steps involved are shown in Fig. 2. Image is first re-sized to a scale of 0.6 and changed to double precision. RGB image obtained is then converted to grayscale form using weighted sum of R, G and B components (0.2989*R + 0.5870*G + 0.1140*B). For better performance of segmentation algorithm, hair are removed using dull razor algorithm [7]. A two dimensional low-pass Gaussian filter of size 9×9 is then applied on the image. Otsu's method is used in next step to calculate a threshold for binary segmentation. White corners of the image are removed by masking the image using mask

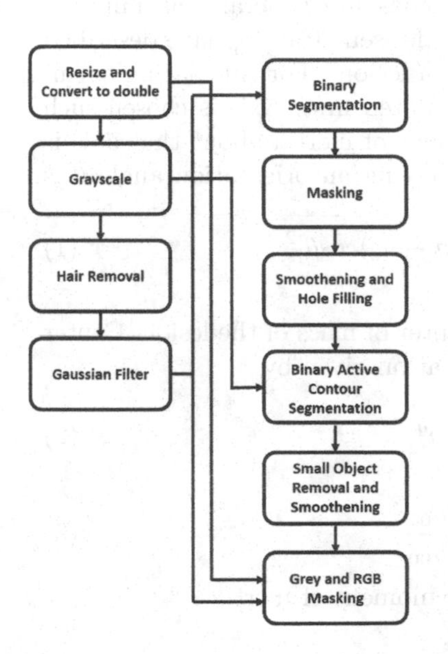

Fig. 2. Flow chart for pre-processing, segmentation and post-processing process belonging to stage one of the framework.

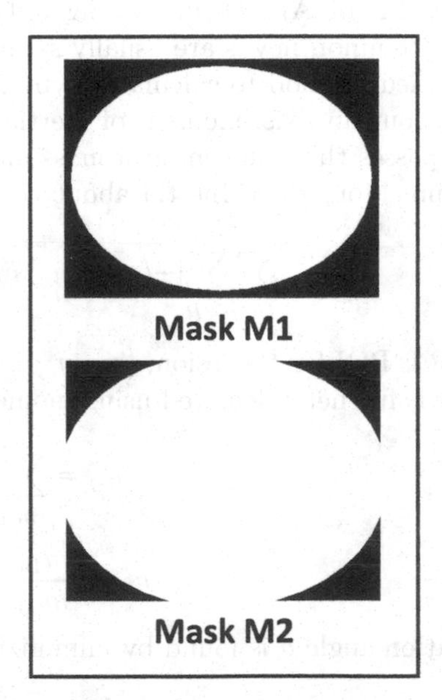

Fig. 3. Masks M1 and M2 used in segmentation process

M1 shown in Fig. 3. Next step is to smoothen the image mask which is done by first creating disk shaped element and then applying morphological close operation followed by morphological open operation. Then holes are filled using morphological reconstruction operation. Grayscale image obtained is segmented using binary active contour method. Thereafter area opening algorithm is applied on the segmented image to remove small objects of size fewer than 50 pixels. Finally after smoothening, morphological open and close operations are again applied, to obtain gray and RGB region of interest (ROI). It is done by masking with mask M2 shown in Fig. 3. Thus at the end of segmentation process of stage one we obtain gray and RGB ROI of the lesion as shown in Fig. 4.

Feature Extraction. After the segmentation process is completed gray and RGB ROIs are obtained which are fed into feature extraction process to obtain the required set of features. In this work 9 feature sets are extracted for each class of melanoma and used as the basis for classification of a lesion. Features are descriptors that provide necessary information in classifying algorithm for distinguishing a lesion among different classes. Selection of features is of fundamental importance in designing any CAD system. Therefore 9 features belonging to the ABCD-rule are used for melanoma classification. The 9 features are described as follows.

Asymmetry. The first two features used are asymmetry. Asymmetry falls under category 'A' of ABCD-rule. Malignant melanomas are generally asymmetric whereas common nevus are usually symmetric. Hansen et al. [5] have described the detailed method to calculate asymmetry in a lesion . For calculating asymmetry about an axis, moment of inertia is calculated first. Axis is chosen such that it passes through center of mass and moment of inertia about that axis is minimum. Moment of Inertia about this axis having an orientation angle θ is given by

$$\sum_{(m,n)\epsilon R}\sum[-(m-m_c)sin\theta+(n-n_c)cos\theta]^2 \tag{1}$$

where R is ROI for the lesion, (m_c, n_c) is the center of mass of the lesion. Center of mass is further calculated using moment equation given by

$$m_{pq}=\sum_{(m,n)\epsilon R}\sum m^p n^q \tag{2}$$

$$m_c=\frac{m_{10}}{m_{00}}, n_c=\frac{m_{01}}{m_{00}} \tag{3}$$

Orientation angle θ is found by minimizing the moment of inertia

$$\theta=\frac{1}{2}\tan^{-1}[\frac{2m_{11}^c}{m_{20}^c-m_{02}^c}] \tag{4}$$

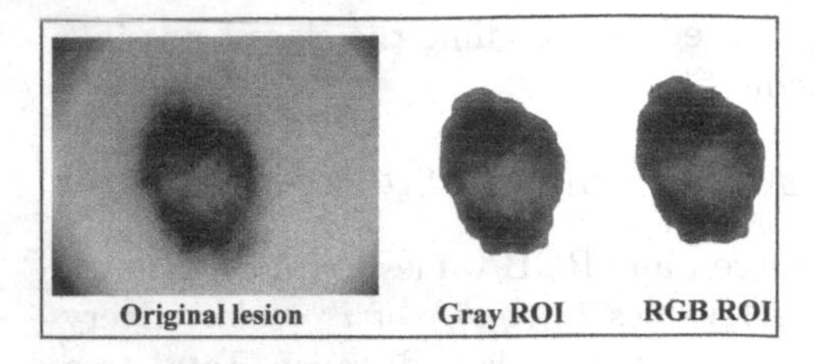

Fig. 4. Original lesion and segmented ROIs for an atypical melanoma lesion (Color figure online)

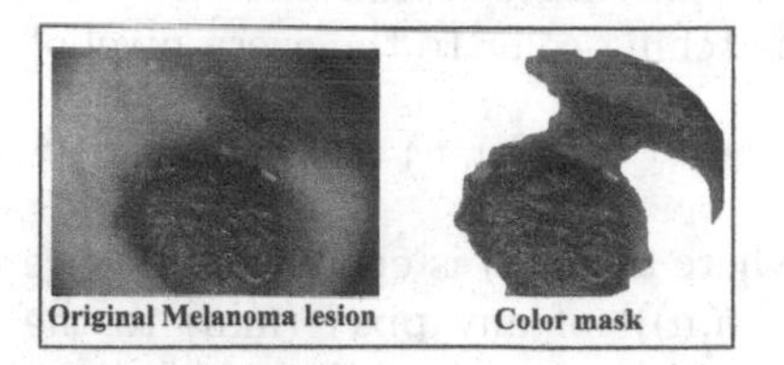

Fig. 5. Original lesion and color mask of malignant melanoma with color feature value of 0.5001 (Color figure online)

where m^c_{pq} is centralized moment given by equation

$$m^c_{pq} = \sum_{(m,n)\epsilon R}\sum (m - m_c)^p (n - n_c)^q \tag{5}$$

The axis obtained is the principal axis and asymmetry is obtained by folding the two halves of the lesion about this axis and then measuring the non-overlapping area of the lesion. Feature 1 is obtained by calculating the ratio of this non-overlapping area over the total area of lesion for major axis. Another axis called minor axis is orthogonal to the major axis. Feature 2 is obtained by measuring the asymmetry about this minor axis.

Border Descriptors. Belonging to the category 'B' (Border Descriptors) of the ABCD rule, 6 features used are - Lesion Area, Lesion Perimeter, Convex Hull Area, Convex Hull Perimeter, Edge Abruptness and Variance of Edge Abruptness along skin lesion. Lesion area and perimeter are calculated by calculating the number of pixels inside and on the lesion boundary respectively. Malignant melanoma has the largest area and perimeter than other classes as it is the more advanced form of melanoma. Convex Hull of a lesion is the smallest convex polygon containing the lesion completely inside it. The convex hull for each image is calculated and its area and perimeter are used as features [2]. Edge abruptness is used to detect the change of pigmentation between the lesion and the surrounding skin. Malignant melanoma is indicated by the presence of a sharp abrupt edge. To calculate edge abruptness [5], the gradient of equally weighted components of image ((R+B+G)/3) is found and only the gradient values of the edges around the lesion boundary is stored. The mean and variance of all the edges act as our next two features.

Color Features. The next feature used is Color belonging to category 'C' of ABCD-rule. Presence of wide variety of colors in lesion indicate malignancy. Malignant melanoma is rich in blue, black and white colors. The ratio of total area of the lesion that matches these colors to the grand total area of the lesion is calculated. Euclidean distance measure is used for assigning the closest possible matching color to the every pixel of lesion [5].

$$d_i^2(m,n) = [r(m,n) - r_i]^2 + [g(m,n) - g_i]^2 + [b(m,n) - b_i]^2 \quad (6)$$

where $d_i(m,n)$ is euclidean distance between any RGB values [(r(m,n), g(m,n), b(m,n))] of any pixel (m,n) to the RGB values (r_i, g_i, b_i) of ith color. Every pixel is assigned a color which has the minimum euclidean distance. Total area of desired melanoma colors blue, black and white are then calculated from the color mask to find the required color feature. Figure 5 shows a color mask for a malignant melanoma lesion.

2.2 Stage Two

In the second stage, three membership functions are used namely Gaussian, Bell and a proposed membership function using Lagrange Interpolation Curve Fitting method to evaluate fuzzy membership values of a test lesion for each class. Training and test feature vectors obtained from previous stage are used as input.

Membership Function Using Lagrange Interpolation Curve Fitting. Figure 6 gives the steps involved in determination of membership function. The first step in this process is grouped frequency distribution. Grouped frequency distribution of each training feature vector F_i^c for each class c - malignant melanoma or atypical melanoma or common nevus is created (where i is the particular feature). Number of groups in each frequency distribution is kept constant at 5. Groups are created between the minimum and maximum value of the feature vector. Height of grouped frequency distribution is normalized between 0 and 1. After obtaining normalized grouped frequency distribution a sample data consisting of x component and y component is obtained. X component is the particular training feature vector and Y component is corresponding height of frequency distribution. This sample data is fed to Lagrange interpolation curve fitting method and the corresponding curves for each feature of each class are obtained. After bounding these curves y = $f_i^c(x)$ along x axis between minimum and maximum of feature vector and along y axis between 0 and 1 we get membership curves for each feature of each class i.e. 9 membership curves for 3 classes thus total of 27 curves. Combined membership value of μ^c is defined for any lesion having feature vector t by calculating average of $\mu_i^c(t_i)$ as

$$\mu^c(t) = \frac{\sum\limits_{i=1to9} \mu_i^c(t_i)}{9} \quad (7)$$

where t_i is any i^{th} feature of a feature vector t obtained from any test image belonging to an unknown class.

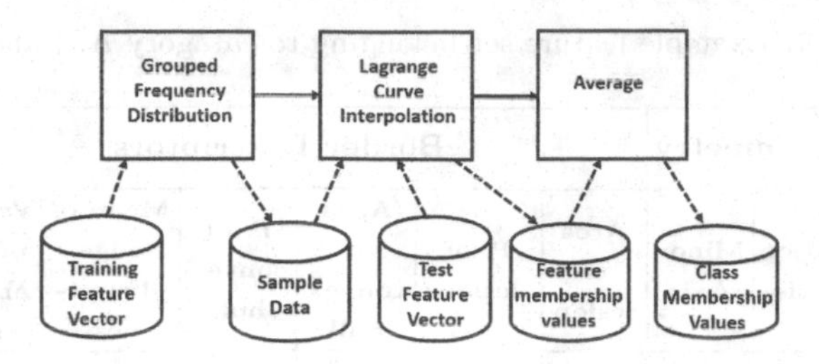

Fig. 6. Steps involved in Lagrange interpolation curve fitting method to evaluate membership values

Multivariate Gaussian and Bell Membership Functions. Patwardhan et al. [9] have used multivariate Gaussian and Bell membership function on features obtained from epi-illumination and multi-spectral trans-illumination images for classification of melanoma. Multivariate Gaussian and Bell membership function described in [9] are used here on 9 features to compare the results obtained from membership function using Lagrange interpolation curve fitting method. Value of parameter representing sharpness of Bell curve (N_c) is taken as 3 chosen experimentally for best possible accuracy. Next section briefly describes the experiments performed and discusses the results so obtained.

3 Experimentation

The implementation of the framework explained in previous section is carried out on a machine having Intel® Core^{TM} i5-3230M Processor 2.60 GHz, 8 GB RAM, Windows 8.1 Pro (x64) in MATLAB.

3.1 Dataset Description

PH^2 dataset [8] used for experiment is acquired and managed by Dermatology Service of Hospital Pedro Hispano (Matosinhos, Portugal). These images are acquired under similar conditions using Tuebinger Mole Analyzer system having magnification of 20x. There are total 200 image lesions in this dataset, comprising 40 from Malignant Melanoma and 80 lesions each from Common Nevus and Atypical Melanoma category. Training and test dataset are obtained by randomly partitioning the PH^2 dataset in the ratio 75:25 respectively for three classes.

3.2 Results and Discussions

Table 2 shows three images belonging to each class and their respective feature values. It is evident from Table 2 that all the features used are able to distinguish each class of Melanoma. Feature values are either increasing or decreasing which is a reflection of the severity of Melanoma classes e.g. Area of Lesion is 77267 for Malignant Melanoma, 12355 for Atypical Melanoma and 5918 only for Common Nevus.

Table 2. Example feature set belonging to catagory A,B and C

Class	Lesion Image	Asymmetry		Border Descriptors						Color
		Major Axis	Minor Axis	Area of lesion	P. of lesion	Area of convex hull	P. of convex hull	Mean of edge abrupt-ness	Variance of edge abrupt-ness	MM Colors
MM		0.2059	0.2847	77267	1783.4	106057	1226.3	0.3907	0.0064	0.5942
AM		0.2625	0.2703	12355	487.98	14194	441.6	0.4866	0.0103	0.0967
CN		0.0924	0.0678	5918	276.918	6059	280.3	0.6394	0.0015	0

MM-Malignant Melanoma; AM-Atypical Melanoma; CN-Common Nevus; P-Perimeter

Table 3. Fuzzy membership values for test image.

Lesion Image	Gaussian			Bell			Lagrange		
	MM	AM	CN	MM	AM	CN	MM	AM	CN
	$1.2962\text{x}\ 10^{-15}$	$4.0011\text{x}\ 10^{-28}$	2.1480×10^{-289}	6.7144×10^{-5}	1.3678×10^{-6}	4.5366×10^{-10}	0.7018	0.4007	0.1241
	4.4876×10^{-15}	2.0152×10^{-11}	1.7237×10^{-11}	9.2396×10^{-5}	4.3747×10^{-4}	9.8712×10^{-5}	0.3123	0.8071	0.6781
	3.5753×10^{-15}	4.3249×10^{-11}	1.0283×10^{-8}	8.6927×10^{-5}	6.3306×10^{-4}	0.0014	0.2649	0.3086	0.9791

MM-Malignant Melanoma; AM-Atypical Melanoma; CN-Common Nevus

Table 3 shows fuzzy membership values for the same images as used in Table 2. It can be observed here that membership values calculated using Gaussian and Bell membership function are very low. Reason behind such low membership values is low correlation among the feature values represented by K_c (covariance matrix). For such features, membership function using Lagrange Interpolation curve fitting method has higher membership values which better indicate degree of extent the lesion belongs to a particular class. Following points give relevant examples from the dataset to highlight the cases mentioned in Table 1.

- Lagrange membership values for lesion images in Table 3 relate to first case of Table 1 where lesions have higher membership value in the class, which it belongs to.
- In the event of false negative result (as explained in third case of Table 1) shown in Fig. 7, if binary results are given as output, dermatologist has no option but to either use or ignore the system, whereas in proposed framework by observing analogue membership values dermatologist can further recommend tests.

Table 4. Sensitivity and specificity of three membership functions

Membership functions	Sensitivity			Specificity		
	MM	AM	CN	MM	AM	CN
Gaussian	80	35	94.11	97.29	88.88	60
Bell	90	65	52.94	83.78	74.07	90
Lagrange	90	80	88.23	94.59	88.88	90

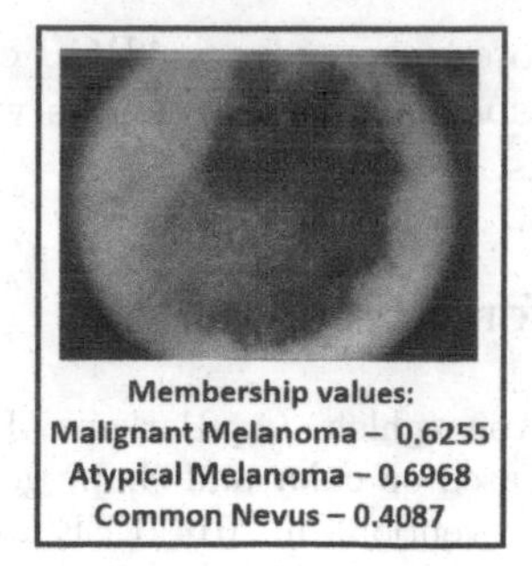

Fig. 7. Malignant melanoma image lesion for false negative result using Lagrange

- This framework fails for the case when fuzzy membership value is relatively very high for the class to which the lesion does not belong. Sensitivity and specificity should be improved in future works such that these cases are avoided.

Sensitivity and Specificity for Gaussian, Bell and Lagrange membership functions are shown in Table 4. It is observed that Lagrange Interpolation curve fitting method has higher sensitivity for Malignant Melanoma and Atypical Melanoma than those obtained from Gaussian membership functions. Lagrange membership function has also higher specificity for Common Nevus class. It further has higher sensitivity than Bell membership function for Atypical Melanoma and Common Nevus class whereas its specificity exceeds that of Bell membership function for Malignant Melanoma and Atypical Melanoma class.

4 Conclusion

This paper proposes a two stage framework which take PH^2 dermoscopic image dataset as input. In the first stage it calculates feature vector by applying segmentation and feature extraction techniques based on ABCD feature categories. The obtained feature vectors are then used by second stage to calculate membership values using Gaussian, Bell and the proposed method using Lagrange Interpolation. It is observed that Gaussian and Bell Membership Function gives very small membership values when considering features from ABCD feature categories. These membership values may help in providing binary result but are not suitable for fuzzy based decision support system proposed in this paper. To overcome this problem a fuzzy membership function using Lagrange Interpolation is proposed. It provides distinctive values that can identify lesion with its Melanoma class. The obtained values also indicate the extent to which the lesion belongs to Melanoma classes. These analogue values can help in a better decision support for medical practitioner by incorporating suspiciousness in the result, and that is the actual motivation behind this work. However, combination of more features from ABCD classes may be used for improvement in sensitivity and specificity, and authors would like to work in this direction in the future works.

References

1. Abuzaghleh, O., Barkana, B.D., Faezipour, M.: Automated skin lesion analysis based on color and shape geometry feature set for melanoma early detection and prevention. In: 2014 IEEE Long Island Systems, Applications and Technology Conference (LISAT), pp. 1–6. IEEE (2014)
2. Celebi, M.E., Aslandogan, Y.A.: Content-based image retrieval incorporating models of human perception. In: Proceedings of the International Conference on Information Technology: Coding and Computing, ITCC 2004, vol. 2, pp. 241–245. IEEE (2004)
3. Day, G., Barbour, R.: Automated skin lesion screening-a new approach. Melanoma Res. **11**(1), 31–35 (2001)
4. Dreiseitl, S., Binder, M.: Do physicians value decision support? a look at the effect of decision support systems on physician opinion. Artif. Intell. Med. **33**(1), 25–30 (2005)
5. Hintz-Madsen, M., Hansen, L.K., Larsen, J., Drzewiecki, K.T.: A Probabilistic Neural Network Framework for Detection of Malignant Melanoma. Danmarks Tekniske Universitet, København (1999)
6. Korotkov, K., Garcia, R.: Computerized analysis of pigmented skin lesions: a review. Artif. Intell. Med. **56**(2), 69–90 (2012)
7. Lee, T., Ng, V., Gallagher, R., Coldman, A., McLean, D.: Dullrazor: a software approach to hair removal from images. Comput. Biol. Med. **27**(6), 533–543 (1997)
8. Mendonça, T., Ferreira, P.M., Marques, J.S., Marcal, A.R., Rozeira, J.: Ph 2-a dermoscopic image database for research and benchmarking. In: 2013 35th Annual International Conference of the IEEE Engineering in Medicine and Biology Society (EMBC), pp. 5437–5440. IEEE (2013)
9. Patwardhan, S.V., Dai, S., Dhawan, A.P.: Multi-spectral image analysis and classification of melanoma using fuzzy membership based partitions. Comput. Med. Imaging Graph. **29**(4), 287–296 (2005)
10. Stanley, R.J., Moss, R.H., Van Stoecker, W., Aggarwal, C.: A fuzzy-based histogram analysis technique for skin lesion discrimination in dermatology clinical images. Comput. Med. Imaging Graph. **27**(5), 387–396 (2003)
11. Stanley, R.J., Stoecker, W.V., Moss, R.H.: A relative color approach to color discrimination for malignant melanoma detection in dermoscopy images. Skin Res. Technol. **13**(1), 62–72 (2007)
12. Stolz, W., Riemann, A., Cognetta, A., Pillet, L., Abmayr, W., Holzel, D., Bilek, P., Nachbar, F., Landthaler, M.: Abcd rule of dermatoscopy-a new practical method for early recognition of malignant-melanoma. Eur. J. Dermatol. **4**(7), 521–527 (1994)

KD-Tree Approach in Sketch Based Image Retrieval

Y.H. Sharath Kumar(✉) and N. Pavithra

Department of Information Science and Engineering,
Maharaja Institute of Technology,
Belawadi, Srirangapatna Tq, Mandya, Karnataka, India
{sharathyhk, pavia02}@gmail.com

Abstract. In this work, we developed a model for representation and indexing of objects for given input query sketch. In some applications, where the database is supposed to be very large, the retrieval process typically has an unacceptably long response time. A solution to speed up the retrieval process is to design an indexing model prior to retrieval. In this work, we study the suitability of Kd-tree indexing mechanism for sketch based retrieval system based on shape descriptors like Scale invariant feature transform(SIFT), Histogram of Gradients (HOG), Edge orientation histograms (EOH) and Shape context (SC). To corroborate the efficacy of the proposed method, an experiment was conducted on Caltech-101 dataset. And we collected about 200 sketches from 20 users. Experimental results reveal that indexing prior to identification is faster than conventional identification method.

Keywords: Region merging · SIFT · EOH · SC · HOG · Kd-tree

1 Introduction

A Sketch is a rapidly executed freehand drawing that may serve a number of purposes, it might record something that the artist sees, it might record or develop an idea for later use or it might be used as a quick way of graphically demonstrating an image, idea or principle. It is an excellent way to quickly explore concepts. A sketch is a natural form of user interactions. Sketch interfaces have recently attracted a lot of attention in the area of computer vision. Still many researchers have been spent lot of effort on shape-to-image matching problems.

Present–day CAD systems and vector–based drawing applications provide powerful tools to create and edit vector graphics in many domains, such as architecture, mechanics, automobile industry or mould industry (for technical drawings) and presentation charts, clip-art drawings or illustration graphics (for more general drawings). Even though reusing past drawings is common practice in such domains, there are almost no developed mechanisms to support this activity in an automated manner. Thus, it becomes important to develop new systems to support automatic classification and retrieval of vector drawings based on their contents, rather than relying solely on textual annotations or metadata for such purposes.

R. Prasath et al. (Eds.): MIKE 2015, LNAI 9468, pp. 247–258, 2015.
DOI: 10.1007/978-3-319-26832-3_24

2 Related Work

Here we discuss some of the work on sketch based image retrieval. Rui et al. [1], developed a image retrieval system for free-hand sketches using Gradient Field Histogram of gradients (GF-HOG) descriptor. GF-HOG explains invariant image descriptor by encapsulating local spatial structure in the sketches and facilitate task of developing efficient codebook for effective retrieval. Manuel [2], presented an efficient sketch based retrieval system using indexing. The topological and geometrical features are extracted and sketches are indexed based on the topological relationships between objects. The NB-tree is used for finding an object or image which is stored in multidimensional database. Huet et al. [3], discussed an method for content based image retrieval system for user written drawings. Sketches are described using histogram of attributes and retrieved based on similar occurrences. Leung and Chen [4], proposed sketch based retrieval system works on free form in literature survey [4] hand drawings. The geometrical relationship between multiple strokes is explored for matching. Mathias et al. [5], discussed method for searching an image in a database of over one million images using similar structure and also by analyzing gradient orientations. Then, best matching images are clustered based on dominant color distributions. Annie and Govindan [6], presented a new novel approach for sketch based image retrieval using counter let edge detection. For identifying edges in different direction the counter let transformation are used. The histogram descriptors are used to compute the similarity between the query sketch and images. Suyu and Karthik [7], proposed a search method with multi-class probability estimates for 3D sketches. Given a free-hand user sketch, combination of classifiers are in literature survey [7] used to estimate the likelihood of the sketch belonging to each predefined category. Rong et al. [8], presented a new approach to allow efficient retrieval of complex vector drawings by content using simple sketches as queries. In this work, a new hierarchical topology is proposed, which allows comparing complex drawings to simple queries. The graph spectrum is used for topology description.

From the above literature review, the sketch based image retrieval accuracy and high performance is still a great challenge today, most of the query-by-sketch methods suffer from the scalability issue due to the lack of efficient indexing mechanism. The image retrieval performance is poor when the system contains large dataset. Many authors have used single descriptor for image retrieval but still many drawbacks could be viewed from them and also they could not achieve better result. In this work we are using various descriptors with effective indexing mechanism for high performance. This paper is organized as follows. The detail of the proposed method is given in Sect. 2. In Sect. 3, the process of sketch indexing is presented. In Sect. 4 the details on experimental settings and performance analysis along with the results are summarized. Finally conclusion is drawn in Sect. 5.

3 Proposed Model

The objects of all classes are segmented using region merging segmentation. From a segmented image the features like Scale Invariant Feature Transform, Histogram of Gradients, Edge Orientation Histograms and shape context are extracted and stored in a database. However storing feature in a database in an efficient manner is required such that a list of possible objects selected for matching process should be very minimum. Hence, there is a need of backend tool called indexing mechanism which stores the data in some predefined manner so that during matching only a few potential objects can be considered. The multi-dimensional feature vectors obtained from sketch image are indexed using the Kd-tree. Top matches will be retrieved from the Kd-tree. The block diagram of the proposed sketch retrieval system is given in Fig. 1.

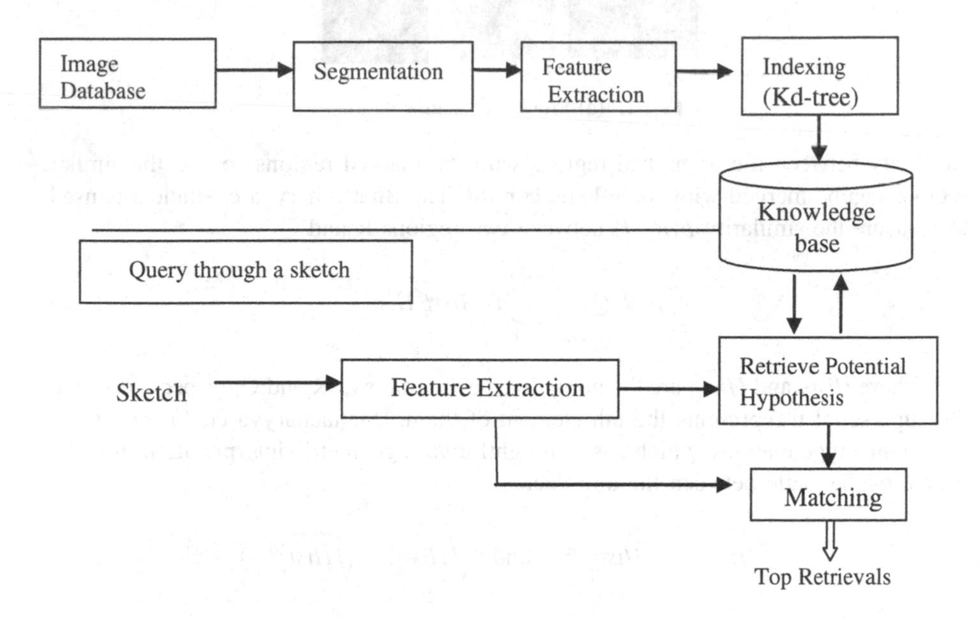

Fig. 1. Block diagram of the proposed sketch indexing model

3.1 Segmentation

For object segmentation, we enhance the work done by the Ning et al. in literature survey [9] which deals with less number of images. In our method, an initial segmentation is required to partition the image into homogeneous regions for merging. For initial segmentation we used Quick shift segmentation [10]. The major advantage of using Quick shift segmentation is greatly reduces the number of image primitives compared to the pixel representation and also it preserves boundary of different objects in the image. Therefore, quick shift has become a popular preprocessing for many computer vision applications.

After initial segmentation, many small regions are available (see in Fig. 2). The color histograms are used as descriptors to represent the regions as the initially segmented small regions of the desired object often vary a lot in size and shape, while the colors of different regions from the same object will have high similarity. The RGB color space is used to compute the color histogram. Each color channel is quantized into 16 levels and then the histogram of each region is calculated in the feature space of $16 \times 16 \times 16 = 4096$ bins. The key issue in region merging is how to determine the

Fig. 2. (a) Process of segmentation

similarity between the unmarked regions with the marked regions so that the similar regions can be merged with some logic control. The Bhattacharyya coefficient is used to measure the similarity $\rho(R, Q)$ between two regions R and Q.

$$\rho(R, Q) = \sum_{u=1}^{4096} \sqrt{Hist_R^u . Hist_Q^u} \tag{1.1}$$

Where $Hist_R$ and $Hist_Q$ are the normalized histograms of R and Q, respectively, and the superscript u represents the uth element of them. Bhattacharyya coefficient ρ is a divergence-type measure which has a straightforward geometric interpretation. It is the cosine of the angle between the unit vectors

$$(\sqrt{Hist_R^1} \ldots \sqrt{Hist_R^{4096}})^T \text{ and } (\sqrt{Hist_Q^1} \ldots \sqrt{Hist_Q^{4096}})^T \tag{1.2}$$

Higher the Bhattacharyya coefficient between R and Q, larger is the similarity between them.

In object segmentation, the central part image region with the size 10 × 10 is called as object marker region and the boundary of the image is called as background marker region. We use light blue markers to mark the object while using dark blue markers to represent the background shown in Fig. 2. After object marking, each region will be labeled as one of three kinds of regions: the marker object region, the marker background region and the non-marker region. To extract the object contour, non-marker region must be assigned to either object region or background region. The region merging method starts from the initial marker regions and all the non-marker regions will be gradually labeled as either object region or background region. Figure 2, shows the results of object segmentation on images. In figure (a) Input image (b) foreground and background marked (c) initial segmentation (d) regions with marker (e) contour extraction (f) segmented object.

3.2 Feature Descriptors

Different features are chosen to describe different properties of sketches. As a sketch can be described using only shape features we choose well known descriptors viz,

Scale Invariant Feature Transform (SIFT), Histogram of Gradients (HOG), Edge Orientation Descriptors (EOH) and Shape context. The following section describes the use of these features separately.

Histogram of Oriented Gradients (HOG). Histogram of oriented gradients [11] can be used as feature descriptors for the purpose of sketch based object retrieval, where the occurrences of gradient orientation in localized parts of a sketch image play important roles. The basic idea behind HOG is that the appearances and shapes of local regions within a object image can be well described by the distribution of intensity gradients as the votes for dominant edge directions. Such feature descriptor can be obtained by first dividing the image into small contiguous regions of equal size, called cells, then collecting a histogram of gradient directions for the pixels within each cell, and finally combining all these histograms.

Edge Orientation Histograms (EOH). Edge orientation histograms (Freeman and Roth, [12]; Levi and Weiss, [13]) are also interesting for our work, since objects often present strong edges. They rely on the richness of edge information and maintain invariance properties to global illumination changes. The basic idea is to build a histogram with the directions of the gradients of the edges (borders or contours). It is possible to detect edges in a image but it in this work we are interested in the detection of the angles. The sobel operators could give an idea of the strength of the gradient in five particular directions. The convolution against each of this mask produces a matrix of the same size of the original image indicating the gradient (strength) of the edge in any particular direction. It is possible to count the maximum gradient in the final matrix and use that to complete a histogram.

Scale Invariant Feature Transform (SIFT) Descriptors. The first stage of the SIFT algorithm finds the coordinates of key points in a certain scale and assign an orientation to each one of them. The results of this guarantee invariance to image location, scale and rotation. Later, a descriptor is computed for each key point. This descriptor must be highly distinctive and partially robust to other variations such as illumination and 3D viewpoint. To create a descriptor, Lowe [14] proposed an array of 4×4 histograms of 8 bins. These histograms are calculated from the values of orientation and magnitude of the gradient in a region of 16×16 pixels around the point so that each histogram is formed from a sub-region of 4×4. The descriptor vector is a result of the concatenation of these histograms. Since there are $4 \times 4 = 16$ histograms of 8 bins each, the resulting vector is of size 128. This vector is normalized in order to achieve invariance to illumination changes. The distinctiveness of these descriptors allows us to use a simple algorithm to compare the collected set of feature vectors from one image to another in order to find correspondences between feature points in each image.

Shape-Context. The Shape-Context descriptor describes the shape of sketches. Given a set of points (usually edges) in an image, the Shape-Context descriptor [15] can be used for describing the relationship between these points. This is done by creating log-polar histograms of the points relative to a reference point. In most applications the Shape-Context is sensitive to scale, but scale-invariance can be achieved by using multiple radii for the descriptor. Typically, the Shape-Context is computed from edges in the image, so it relies on robust edge detection.

4 Sketch Indexing

In the proposed indexing model, the multi-dimensional feature vectors obtained from images of each individual are indexed separately using Kd-tree. The Kd-tree is one of the most prominent multidimensional space partitioning data structures (Dixit et al., [16]) for organizing points in a k-dimensional space and it is a useful data structure for searching based on a multidimensional key. The construction algorithm of Kd-tree is very similar to the planar case. At the root, we split the set of points into two subsets of roughly the same size by a hyper-plane perpendicular to the x1-axis. In other words, at the root, the point set is partitioned based on the first coordinate of the points. At the children nodes of the root the partition is based on second coordinate and so on, until depth of $k - 1$ at which partition occurs based on last coordinate where k is the dimension of the feature space. After depth k, again, partitioning is based on first coordinate. The recursion stops only when one point is left, which is then stored at the leaf. Because a k-dimensional Kd-tree for a set of n points is a binary tree with n leaves, it uses O(n) storage with construction time being O(n log(n)). In addition to this, in Kd-tree there is no overlapping between nodes [17]. Kd-tree is an appropriate data structure for object retrieval system particularly in the analysis of execution of range search algorithm and it decreases the search time as it is supporting the range search with a good pruning. When query feature vector of multidimensional is given, range search is invoked using Kd-tree to retrieve top matches that lie within a certain distance (threshold) from the query. These top matches are subsequently used for object retrieval.

5 Database

In this section for experimentation, we apply our method to the Caltech 101 object categorization [12] database. Since we would like to test how a general technique is so we picked some of the object classes from Caltech-101. The dataset consists of 400 images divided into 20 classes. Figure 3 shows some of samples in each class. In addition to images, we also collected 200 sketches from 20 users from which some are

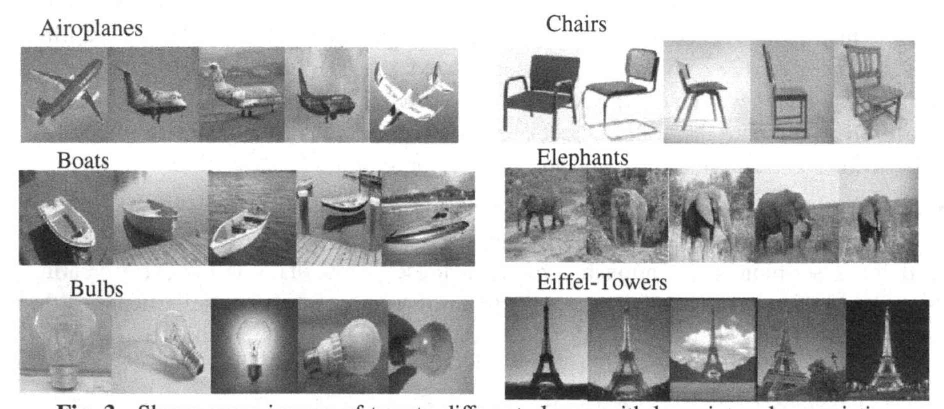

Fig. 3. Shows some images of twenty different classes with large intra class variation

collected by drawing on papers and some are painted using MS paint application. Figure 4 shows some of the sketches that are used for experimentation.

Fig. 4. Shows sketch images used for experimentation

Table 1. Average retrieval accuracy obtained for the proposed indexing scheme for top ten images by considering individual features

Classes	Descriptors			
	SIFT	HOG	EDH	SC
Eiffel - tower	5.2	5.6	4.2	3.1
Elephants	4.7	8.3	5.2	3.9
Flowers	6.1	2.7	5.1	4.5
Homes	2.5	6.3	2.4	3.7
Mug	7.4	4.9	4.2	4.4
Revolvers	4.1	6.4	4.3	5.3
Butterflies	4.2	2.0	4.6	3.4
Buses	2.8	7.8	3.7	3.8
Bulbs	8.5	3.1	8.2	5.4
Boats	3.6	2.7	5.3	4.9
Airplanes	5.5	2.2	7.8	6.6
Chairs	8.0	6.7	7.3	3.2
Tables	3.6	3.6	5.6	4.7
Wrenches	3.2	6.0	4.8	5.0
Guitars	3.3	6.2	5.7	4.7
Monitors	7.8	4.8	7.8	5.5
Mouses	5.7	4.6	6.0	4.4
Pyramids	2.3	7.2	2.2	2.0
Umbrella	3.8	3.5	4.1	3.8
Faces	3.2	2.9	3.4	4.7

Table 2. Average retrieval accuracy obtained for the proposed indexing scheme for top twenty images by considering individual features

Classes	Descriptors			
	SIFT	HOG	EDH	SC
Eiffel - tower	3.9	2.7	2.9	2.5
Elephants	4.1	3.8	2.3	4.0
Flowers	2.8	4.9	2.3	2.6
Homes	2.7	4.3	3.9	3.2
Mug	3.3	4.5	2.7	3.1
Revolvers	5.3	3.0	3.8	3.3
Butterflies	3.6	4.4	3.5	4.5
Buses	4.6	5.0	3.9	3.5
Bulbs	2.4	4.7	2.7	3.2
Boats	2.6	2.9	2.0	3.3
Airplanes	3.7	2.5	2.5	4.7
Chairs	4.7	1.4	2.1	3.3
Tables	3.0	4.7	2.4	4.9
Wrenches	3.1	2.9	2.6	2.3
Guitars	3.1	4.6	2.0	3.6
Monitors	4.2	4.6	2.9	3.0
Mouses	2.4	3.3	2.3	4.7
Pyramids	3.7	4.1	3.1	3.2
Umbrella	2.2	3.3	3.6	2.6
Faces	4.0	4.2	4.6	3.3

Table 3. Average retrieval accuracy obtained for the proposed indexing scheme for top ten images by considering combination two features

Classes	Descriptors combinations					
	HOG + EOH	HOG + SIFT	SOG + SC	EOH + SIFT	EOH + SC	SIFT + SC
Eiffel - tower	5.8	6.2	4.7	6.8	6.6	5.5
Elephants	6.6	6.7	5.9	9.1	6.1	6.8
Flowers	5.3	6.6	6.0	6.5	5.6	5.5
Homes	5.2	3.3	3.2	6.7	5.5	5.1
Mug	6.6	6.0	6.4	5.7	5.4	5.8
Revolvers	6.9	5.2	5.2	7.3	7.3	5.6
Butterflies	5.4	5.1	5.2	5.0	4.4	5.4
Buses	5.6	4.7	4.7	6.2	6.7	4.1
Bulbs	6.4	8.9	7.2	7.0	4.1	7.6
Boats	4.8	6.4	5.7	5.0	4.6	6.3
Airplanes	4.0	6.6	6.7	5.8	6.0	8.6
Chairs	8.2	8.5	6.7	7.8	5.7	5.9
Tables	4.7	6.3	5.8	5.4	5.2	6.7
Wrenches	5.4	5.8	5.3	7.5	6.3	5.4
Guitars	5.1	6.6	5.7	7.2	7.5	6.5
Monitors	7.4	8.6	7.6	6.5	6.2	7.8
Mouses	6.8	6.7	6.9	6.4	5.9	7.8
Pyramids	6.5	3.8	3.2	7.1	5.4	3.1
Umbrella	4.8	6.0	4.4	5.7	4.4	5.3
Faces	3.9	4.8	5.8	4.6	5.3	5.4

6 Experimentation

In this section, we present the details of an experimentation conducted to demonstrate our proposed model on Caltech-101 dataset. Given a sketch as a query, the first ten and twenty similar images are retrieved by comparing with Caltech-101 dataset. We study average retrieval accuracy for all 20 classes by considering top ten and twenty retrieved images. Tables 1 and 2 shows the average retrieval accuracy for top ten and twenty images by considering individual features like SIFT, HOG, EOH and SC for each object. Tables 3 and 4 shows the average retrieval accuracy for top ten and twenty images by considering combination of two features SIFT + HOG, SIFT + EOH, HOG + EOH, HOG + SC, EOH + SC, SIFT + SC for each object. Tables 5 and 6 shows the average retrieval accuracy for top ten and twenty images by considering three combination of SIFT + HOG + EOH, SIFT + HOG + SC, SIFT + EOH + SC, HOG + EOH + SC for each object. Tables 7 and 8 shows the average retrieval accuracy for top ten and twenty images by considering final combination of features SIFT + HOG + EOH + SC. From above tables we observed that the final fusion of various descriptors (SIFT + HOG + EOH + SC) achieves maximum accuracy when compare to individual descriptor.

Table 4. Average retrieval accuracy obtained for the proposed indexing scheme for top twenty images by considering combination two features

Classes	Descriptors combinations					
	HOG + EOH	HOG + SIFT	SOG + SC	EOH + SIFT	EOH + SC	SIFT + SC
Eiffel - tower	4.4	4.6	6.8	6.1	4.5	3.9
Elephants	7.0	4.8	4.3	3.4	4.8	5.1
Flowers	5.3	4.1	5.4	4.6	4.6	5.0
Homes	3.1	4.3	4.4	4.4	4.8	4.1
Mug	3.5	6.0	4.1	5.1	5.6	5.2
Revolvers	4.2	5.0	4.6	4.7	5.9	6.2
Butterflies	5.5	6.7	4.6	5.1	6.8	4.3
Buses	6.9	4.2	4.6	4.0	4.5	3.9
Bulbs	5.8	5.0	5.6	5.5	6.1	6.5
Boats	4.7	4.3	4.4	4.3	6.1	4.2
Airplanes	5.0	4.5	5.1	4.9	4.5	4.1
Chairs	4.7	4.1	5.4	4.9	4.5	5.1
Tables	3.4	3.5	5.7	4.5	6.8	4.9
Wrenches	6.4	4.8	5.7	4.2	5.3	6.6
Guitars	4.6	5.0	5.3	5.4	3.5	6.9
Monitors	4.8	4.3	5.3	6.6	5.3	6.9
Mouses	4.4	4.1	4.8	6.2	6.4	5.2
Pyramids	5.0	4.9	5.5	5.2	5.1	5.6
Umbrella	4.9	4.4	6.7	6.5	6.7	6.5
Faces	4.8	4.5	5.0	7.0	5.0	6.4

Table 5. Average retrieval accuracy obtained for the proposed indexing scheme for top ten images by considering combination three features

Classes	Descriptors combinations			
	SIFT + HOG + EOH	SIFT + HOG + SC	SIFT + EOH + SC	HOG + EOH + SC
Eiffel - tower	6.9	8.0	7.1	7.5
Elephants	6.5	6.1	6.4	6.1
Flowers	6.1	7.2	7.2	6.5
Homes	8.0	6.8	7.4	6.1
Mug	6.0	6.9	6.5	6.1
Revolvers	7.7	6.4	7.4	7.5
Butterflies	6.4	6.3	6.5	6.3
Buses	7.7	7.1	7.5	7.6
Bulbs	7.6	7.8	7.4	7.7
Boats	7.0	7.9	7.1	7.2
Airplanes	6.2	7.1	6.5	7.0
Chairs	7.0	6.9	7.1	6.8
Tables	7.5	7.6	7.4	7.1
Wrenches	7.7	6.5	7.5	6.9
Guitars	7.7	6.5	7.0	6.4
Monitors	6.6	6.4	6.5	6.7
Mouses	7.1	6.2	7.4	6.5
Pyramids	8.0	6.9	7.0	6.8
Umbrella	8.2	6.1	7.6	7.2
Faces	7.1	5.6	7.0	6.0

Table 6. Average retrieval accuracy obtained for the proposed indexing scheme for top twenty images by considering combination three features

Classes	Descriptors combinations			
	SIFT + HOG + EOH	SIFT + HOG + SC	SIFT + EOH + SC	HOG + EOH + SC
Eiffel - tower	7.4	7.2	7.0	7.1
Elephants	7.8	8.0	7.7	7.6
Flowers	7.0	7.4	7.8	7.1
Homes	6.4	5.8	6.9	6.8
Mug	7.0	6.5	7.1	7.5
Revolvers	7.9	7.8	8.0	8.2
Butterflies	6.6	6.6	6.6	6.7
Buses	6.6	7.3	6.8	7.0
Bulbs	7.8	7.7	7.6	7.8
Boats	5.2	6.5	6.6	6.2
Airplanes	5.6	6.3	6.1	6.4
Chairs	8.9	8.1	8.1	8.2
Tables	5.4	6.9	6.8	5.5
Wrenches	6.2	7.7	6.3	7.1
Guitars	7.4	6.2	7.1	6.3
Monitors	8.8	6.5	7.1	7.2
Mouses	7.9	7.3	7.6	7.7
Pyramids	5.6	7.6	6.0	7.4
Umbrella	6.5	6.0	6.2	6.9
Faces	5.2	7.2	6.2	7.1

Table 7. Average retrieval accuracy obtained for the proposed indexing scheme for top ten images by considering combination of all the features

Classes	Descriptors combinations
	HOG + EOH + SC + SIFT
Eiffel - tower	8.7
Elephants	8.3
Flowers	7.8
Homes	7.1
Mug	7.2
Revolvers	8.3
Butterflies	7.8
Buses	7.5
Bulbs	7.9
Boats	7.1
Airplanes	6.8
Chairs	8.5
Tables	7.1
Wrenches	7.3
Guitars	7.5
Monitors	8.1
Mouses	7.9
Pyramids	7.6
Umbrella	7.0
Faces	7.2

Table 8. Average retrieval accuracy obtained for the proposed indexing scheme for top twenty images by considering combination of all the features

Classes	Descriptors combinations
	HOG + EQH + SC + SIFT
Eiffel - tower	7.4
Elephants	6.8
Flowers	6.3
Homes	8.6
Mug	6.9
Revolvers	7.9
Butterflies	7.8
Buses	7.5
Bulbs	6.1
Boats	6.8
Airplanes	6.5
Chairs	7.5
Tables	7.0
Wrenches	7.6
Guitars	7.9
Monitors	8.1
Mouses	8.1
Pyramids	8.6
Umbrella	7.6
Faces	6.8

7 Conclusion

In this work, we proposed Kd-tree based indexing approach to index a object dataset for given input sketch. In the proposed method we represent each object by shape descriptors of SIFT, HOG, EOH and SC. Experimentations are conducted on sub classes of Caltech-101 dataset and sketch based retrieval system to assess the advantage of using indexing technology. From Experimentation we can understand that the combination of all the features descriptors achieves a good accuracy in terms of performance with indexing approach.

References

1. Rui, H., Mark, B., John, C.: Gradient field descriptor for sketch based retrieval and localization. In: Proceedings of IEEE Transaction International Conference on Image Processing, pp. 1025–1028 (2010)
2. Manuel João Caneira Monteiro da Fonseca, Sketch-based retrieval in large sets of drawings, PhD Thesis, IST/UTL, July 19, 2004

3. Hute, B., Gennarino, G., Nicky, K., Bernard, M.R.: Skeletons for retrieval in patent drawings. In: Proceedings of the IEEE International Conference on Image Processing (ICIP 2001), Thessaloniki, Greece, vol. 2, pp. 737–740 (2001)
4. Leung, W.H., Chen, T.: User-independent retrieval of free-form hand-drawn sketches. In: Proceedings of the IEEE International Conference on Acoustics Speech and Signal Processing (ICASSP 2002), Orlando, Florida, USA, vol. 2, pp. 2029–2032. IEEE Press (2002)
5. Mathias, E., Kristian, H., Tamy, B., Marc, A.: An evaluation of descriptors for large- scale image retrieval from sketched feature lines. In: International Conference on Computer Graphics and Interactive Techniques, computer Graphics, no. 5, pp. 482–498 (2010)
6. Annie, J.J., Govindan, V.K.: Edge histogram for sketch based image retrieval using contourlet edge detection. In: International Conference on Electrical Engineering and Computer Science, pp. 272–276 (2012)
7. Suyu, H., Karthik, R.: Classifier combination for sketch-based 3D part retrieval. In: International Conference on Computer Graphics and Interactive Techniques, Computers and Graphics, no. 31, pp. 598–609 (2007)
8. Rong, Z., Liuli, C., Liqing, Z.: Sketch-based image retrieval on a large scale database. In: MM 2012, Nara, Japan, 29 October–2 November 2012
9. Ning, J., Zhang, L., Zhang, D., Wu, C.: Interactive image segmentation by maximal similarity based region merging. Pattern Recogn. **43**(2), 445–456 (2010)
10. Vedaldi, A., Soatto, S.: Quick shift and kernel methods for mode seeking. In: Proceedings of the European Conference on Computer Vision, vol. 4, pp. 705–718 (2008)
11. Dalal, N., Triggs, B.: Histograms of oriented gradients for human detection. In: IEEE Computer Society Conference on Computer Vision and Pattern Recognition, vol. 1, pp. 886–893 (2005)
12. Freeman, W.T., Roth, M.: Orientation histograms for hand gesture recognition. In: International Workshop on Automatic Face- and Gesture- Recognition, pp. 296–301. IEEE Computer Society (1995)
13. Levi, K., Weiss, Y.: Learning object detection from a small number of examples: the importance of good features. In: Proceedings of the IEEE Conference on CVPR, pp. 53–60 (2004)
14. Lowe, D.G.: Distinctive image features from scale-invariant keypoints. Int. J. Comput. Vis. **60**, 91–110 (2004)
15. Belongie, S., Malik, J.: Shape matching and object recognition using shape contexts. IEEE Trans. Pattern Anal. Mach. Intell. **24**(24), 509–522 (2002)
16. Dixit, V., Singh, D., Raj, P., Swathi, M., Gupta, P.: Kd-tree based fingerprint identification system. In: Proceedings of the Second International Conference on Anti-counterfeiting, Security and Identification, pp. 5–10 (2008)
17. Samet, H.: The Design and Analysis of Spatial Data Structures. Addison-Wesley, Boston (1994)

Benchmarking Gradient Magnitude Techniques for Image Segmentation Using CBIR

K. Mahantesh[1], V.N. Manjunath Aradhya[2(✉)], and B.V. Sandesh Kumar[1]

[1] Department of ECE, Sri Jagadguru Balagangadhara Institute of Technology, Bangalore, India
{mahantesh.sjbit,bvsk39}@gmail.com

[2] Department of MCA, Sri Jayachamarajendra College of Engineering, Mysore, India
aradhya.mysore@gmail.com

Abstract. As image segmentation has become a definite prerequisite in many of the image processing and computer vision applications, an effort towards evaluating such segmentation techniques is indeed found very less in literature. In this paper, we carried out a comprehensive evaluation of five different gradient magnitude (GM) based image segmentation techniques using CBIR (Content Based Image Retrieval). Firstly, boundary probabilities are detected using the gradient magnitude based techniques such as - Canny edge detection (pbCanny), Second moment matrix (pb2MM), Multi-scale second moment matrix (pb2MM2), Gradient magnitude (pbGM) and Multi-scale gradient magnitude (pbGM2). Further, Ridgelets are applied to these boundaries to extract radial energy information exhibiting linear properties and PCA to reduce the dimensionality of these features. Finally, probabilistic neural network (PNN) classifiers are used to classify and observe the performance of gradient magnitude techniques in classification process. We observed the performance of these algorithms on the most challenging and popular image datasets namely Corel-1K, Caltech-101, and Caltech-256.

Keywords: Image segmentation · Gradient magnitude · pbCanny · pb2MM · pb2MM2 · pbGM · pbGM2 · CBIR · Ridgelets · PCA · Neural networks · PNN

1 Introduction

Generating perceptually meaningful regions is an exigent task; development of an efficient image segmentation algorithm for images with complex background and multiple objects is becoming more challenging, comparison and evaluation of these algorithms is even more challenging. Segmentation subdivides an image into its constituent regions, and level of details to which subdivision applied depends on the problem being solved. Most of the segmentation algorithms are designed based on two basic issues -discontinuity and similarity. Discontinuity deals with the abrupt changes in intensity (e.g. edges). Similarity property is based on region partition with predefined criteria such as thresholding & region

R. Prasath et al. (Eds.): MIKE 2015, LNAI 9468, pp. 259–268, 2015.
DOI: 10.1007/978-3-319-26832-3_25

merging/splitting. Some of the contemporary algorithms can be seen in [3,4]. Masks are created to detect intensity discontinuities, in [3], the authors has reviewed various masks and their performances on edge discontinuities. Hough Transform is a practical method for defining shapes through global pixel linking and curve detection methods. Segmentation techniques based on thresholding, connected components, contours, watersheds are best examples for identifying similarities [4].

Soccer image analysis was carried out by choosing different set of color components belonging to different color spaces. Adaptive hybrid color space was introduced by sequentially selecting learned features [5]. Color data exploited using dichromaticity and reflection properties of objects made segmentation less sensitive toward shadow, shading and highlights [6]. The product of saliency and repeatedness gives co-saliency and substitutes in defining global energy to undergo regularization of k-means in learning bag of words toward co-segmentation of multiple images [7]. Fuzzy membership functions of 9 chromatic colors for Hue component (from HSV colorspace model) are calculated from 12 colors palette generated after quantization. The degrees of these members are added to obtain fuzzy histograms and hence DCs can be identified to represent dominant colors in an image [8]. Quantitative evaluation of four different segmentation algorithms: Spectral Embedding Min-Cut, Normalized Cuts, Local Variation, Mean Shift was carried on images from Berkeley Segmentation Database (BSD). Precision and recall measures used as a standard metrics to measure the relativity between segmented image and human segmented boundaries created benchmark for recent and future avenues [9].

Due to the unavailability of generalized evaluation metrics and limited standard datasets along with its ground truth, many of the major segmentation algorithms remain to be validated. Evaluation of segmentation techniques can be either qualitative or quantitative. However, qualitative evaluation of many state-of-the-art segmentation techniques fails due to its intuitive property of human perceptions to identify objects, image contours and its boundaries [1]. The goal of this paper is to use five different algorithms to detect boundaries by finding the probability of a boundary's existence and to draw quantitative evaluations using CBIR technique.

This paper is organized as follows: Sect. 2 briefs GM-based segmentation techniques selected for comparison. Section 3 explains CBIR technique used to extract features and probabilistic neural network classifier for classification. Section 4 demonstrates the experimental results including the performance evaluation and comparison of segmentation techniques. Finally, concluded in Sect. 5.

2 Gradient Magnitude Based Segmentation Techniques

In this paper, we use five different algorithms to detect boundaries by finding the output images that show us the probability of a boundary's existence. The methods featured are finding the boundary probability through the following algorithms: Canny edge detection (pbCanny), Second moment matrix (pb2MM), Multi-scale second moment matrix (pb2MM2), Gradient magnitude (pbGM), Multi-scale gradient magnitude (pbGM2). Brief overviews on the aforesaid techniques are discussed in the following subsections.

2.1 PbGM

The gradient is a vector consisting of certain magnitude and direction. let $\nabla f = \begin{pmatrix} \delta f/\delta x \\ \delta f/\delta y \end{pmatrix}$ be the gradient function of $'f'$, where magnitude and direction can be calculated using the following equations.

$$Magnitude(\nabla f) = \sqrt{(\delta f/\delta x)^2 + (\delta f/\delta y)^2} = \sqrt{M_x^2 + M_y^2} \tag{1}$$

$$Direction(\nabla f) = atan2(M_y/M_x) \tag{2}$$

To increase the speed of computation, the magnitude of gradient can also be approximated using $magn(\nabla f) = |M_x| + |M_y|$. The strength of the edge is given by the magnitude of gradient and the direction of edge is always perpendicular to the direction of gradient.

2.2 PbGM2

In pbGM2, different scales are used to analyze and classify image features. Gradient magnitudes can be defined using the properties of multi-scale gradient watershed regions. Probabilities of these boundaries corresponding to the definite edges of objects are easily identified. Hence, multi-scale analysis imposes scale-based hierarchy on watersheds associated to identify the boundaries. These hierarchies are further used to label boundaries obtained due to watershed giving valuable and in-depth properties of multi-scale edges. Sub-trees of these regions are selected based on hierarchical parameters to extract visually sensible edges of an object [10].

2.3 PbCanny

Pb canny [12], introduced by John Canny outperforms many of newer algorithms due to its inherent property of removing noise and preserving edge features before finding edges. Few important steps are mentioned below:

1. Convolve given image $f(r,c)$ with a Gaussian function to get smooth image $f^*(r,c)$. i.e. $f^*(r,c) = f(r,c) * G(r,c)$.
2. Apply first order gradient to find edge strength, magnitude and its direction.
3. Apply non maximal threshold and find edges with local maxima of gradient magnitude.
4. Find local maxima to find the edges.
5. Apply thinning to broad ridges and retain the points w.r.t the largest local change.

2.4 Pb2MM

As pbCanny fails to detect boundaries/edges in textured regions due to the presence of subtle intensity variations, one can think of analyzing gradients at multiple orientations that helps in exploiting multiple incident edges within the textured regions. pb2MM (Second moment matrix) is one of the best approach in finding solutions for such cases and are derived from the gradient of an image. It identifies the predominant directions in the gradient of neighboring pixel and also identifies the angles to its coherent directions. Boundaries are detected by observing the spectrum of eigen values from the derived spatially averaged second moment matrix [2].

2.5 pb2MM2

The Multi-scale second moment matrix of a function Iis in contrast to other one-parameter scale-space features an image descriptor that is defined over two scale parameters. One scale parameter, referred to as local scale **t**, is needed for determining the amount of pre-smoothening when computing the image gradient $(\nabla(I))$ $(x;t)$. Another scale parameter, referred to as integration scale **s**, is needed for specifying the spatial extent of the window function that determines the weights for the region in space over which the components of the outer product of the gradient by itself $(\nabla I)(\nabla I)^T$ are accumulated [11].

3 Content Based Image Retrieval (CBIR)

CBIR is a technique which uses visual features such as color, shape, texture and spatial layout to retrieve similar images from the large image dataset. In this paper, we consider CBIR as an evaluation technique to evaluate the performance of segmentation techniques. CBIR can be broadly classified into feature extraction and classification. In our experiment, we make use of ridgelet transform to extract features and probabilistic neural network for classification.

3.1 Ridgelet PCA

Being inspired from the fact of representing line singularities in higher dimensional multi-resolution feature bands and to handle higher dimensional intermittency ridgelets are proposed in our earlier work mentioned. Ridgelets can be obtained by applying 1-D wavelet on slices of 2-D radon images. Ridgelet & PCA are applied to the segmented images which give the inherent benefit of spectral analysis in developing robust and invariant geometrical features with structural information in reduced feature space. Detailed explanation can be found in [13].

3.2 Probabilistic Neural Network (PNN)

Features obtained in the previous section are used as input to the PNN classifier. Due to its excellent generalization performance, SVM (Support Vector Machines) is most promising classifier in machine learning. However, SVM's are slow and still remains to be a bottleneck for large datasets and multi-class classification. We have made use of PNN, since it is based on concepts used for conventional pattern recognition problems [14].

PNN Architecture is similar to that of supervised learning Architecture, but PNN does not carry weights in its hidden layer. Each node of hidden layer acts as weights of an example vector. The hidden node activation is defined as the product of example vector $'E'$ input feature vector $'F'$ given as $h_i = E_i X F$. The class output activations are carried out using the following equation:

$$C_j = \frac{\sum_{i=1}^{n} e^{\frac{(h_i - 1)}{\gamma^2}}}{N} \quad (3)$$

Where $'N'$ is example vectors belonging to class $'C'$, $'h_i'$ are hidden node activation and $'\gamma'$ is smoothing factor.

4 Experimental Results and Performance Analysis

We have conducted extensive experiments in order to observe the behaviour of segmentation algorithms and to study their theoretical claims. All our experiments are carried out on a PC machine with an intel core2duo 2.20 GHz processor and 3 GB RAM under Matlab 10.0 programming platform. Figure 1 shows the original and segmented results obtained by applying pbGM, pbGM2, pb2MM, pb2MM2, pbCanny techniques respectively. Performance measures such as correlation, Hausdorff distance, Jaccard & Dice coefficients, root mean square error (RMSE) mentioned in [15], fails to precisely locate the boundaries and evaluate it with its ground truth. One of the main constraints in using these metrics is that it requires the ground truth / manually segmented images for evaluation purpose.

We demonstrated and evaluated these segmentation techniques by applying CBIR technique on segmented edge images using three most popular and very large datasets namely Corel-1K [16], Caltech-101 [22] & Caltech-256 [21] extensively used for retrieval and classification. The corresponding benchmarking procedures for training and testing and evaluation results are discussed in the following sections.

In order to test the efficacy of the system, several experimental procedures are mentioned in the literature. We considered identical conditions in order to conduct experiments on each image category. Each category datasets are divided into train and test datasets. First 'N' numbers of training images are drawn orderly as labelled set and remaining images in dataset are considered as test images. We then learn models based on the proposed approaches and evaluated their performances on the test images.

4.1 Corel 1-K

In this section, we reported the recognition accuracies obtained due to proposed methods using highly optimum number of training samples i.e. 30 training images

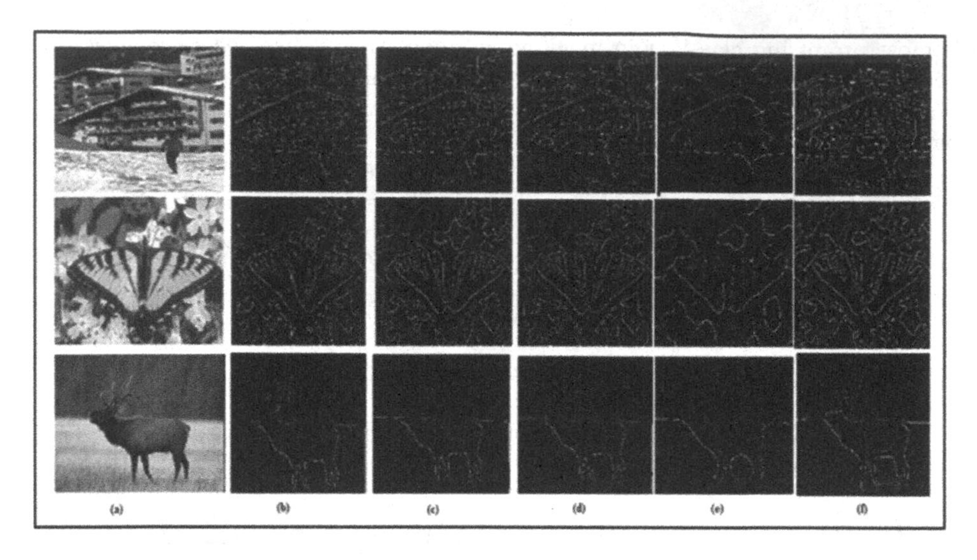

Fig. 1. (a): Sample images. (b) - (f): Resultant segmented images obtained using pbGM, pbGM2, pb2MM, pb2MM2, pbCanny respectively.

per category. Table 1 gives the performance analysis of the proposed & existing methods for Corel-1K dataset.

pbCanny with Ridgelet PCA and PNN has outperformed the Kernel & 2D Markov models capturing the semantic context [16], WBCH model (Wavelet Based Color Histogram) in which texture and color features are extracted using Haar wavelet transformation and color histogram (HSV color space) exhibiting robustness toward translation and scaling of objects [17] and showed better results compared to Statistical quantized histogram texture features such as

Table 1. Performance analysis of the proposed & existing methods for Corel-1K.

Methods	Classification rates(%)
Lu et al. [16]	77.9
Manimala et al. [17]	81
Malik et al. [18]	80
Ridgelet PCA	76
pbGM + Ridgelet PCA	84
pbGM2 + Ridgelet PCA	85
pb2MM + Ridgelet PCA	82
pb2MM2 + Ridgelet PCA	83
pbCanny + Ridgelet PCA	86

mean, standard deviation, skewness, kurtosis, energy, entropy and smoothness extracted from the quantized DCT coefficients (1 DC and 3 AC coefficients) [18].

4.2 Caltech-101 &Caltech-256

As per the observations made from the literature, we found that common experimental set up was considered for both Caltech-101 & Caltech-256 datasets. Following common benchmarking experimental procedure mentioned in [19,20, 24,25]. Each experiment is repeated two times by varying the training samples into 15 and 30 images per category. The remaining images are tested to obtain an average of per class recognition in each stage for each of the classifiers. Due to its very high computation cost and large memory requirement to store the stable feature parameters, we have to restrict training case to maximum of 30 training images per category for both Caltech-101 & Caltech-256 datasets.

Table 2 contains a summary of the performances of proposed methods and their assessment with few well known techniques identified from the literature.

Table 2. Classification rates(%) of the Proposed & Existing Methods for Caltech-101.

Methods	15 Train	30 Train
Serre et al. [19]	35	42
Holub et al. [20]	37	43
Rigamonti et al. [21]	-	45
Berg et al. [22]	45	-
Mutch & Lowe [23]	33	41
Ridgelet PCA	33.6	42
pbGM + Ridgelet PCA	34	42.6
pbGM2 + Ridgelet PCA	35	43
pb2MM + Ridgelet PCA	32	39
pb2MM2 + Ridgelet PCA	33	40
pbCanny + Ridgelet PCA	36	44

Table 3 summarizes the classification rates obtained on Caltech-256 dataset and gives performance evaluation of proposed models against some of the most popular and leading techniques.

Following are the observations made from Tables 2 and 3:

1. The performance of pbCanny is well above the Fisher score features extracted using 3 different detectors and classifying them using SVM [20].
2. Serre constructed C2 features combining simple features S1 and complex features C1. S1 responses generated due to Gabor filters and C1 generated due to shift and size of the object very similar to the primary visual cortex showed deprived results [19].

Table 3. Classification rates(%) of the Proposed & Existing Methods for Caltech-256.

Methods	15 Train	30 Train
Van et al. [24]	-	27
Jianchao et al. [25]	25	29
Ridgelet PCA	14	19
pbGM + Ridgelet PCA	16	21
pbGM2 + Ridgelet PCA	17	22
pb2MM + Ridgelet PCA	15	19
pb2MM2 + Ridgelet PCA	18	20
pbCanny + Ridgelet PCA	20	24

3. Relevance sparsity model found its best recognition rate of 43.8 % due to the investigations carried out on sparsity constraints by convolving linear filters within an image [21].
4. Berg et al. [22], randomly picked 30 images from each category and splitting them into 15 for train and remaining 15 for test. The correctness rate of 45 % obtained using geometric blur features, seems to be less competitive when compared to proposed methods coresponding to 15 training image per category.
5. pbCanny with ridgelet found to be very adequate in storing stable parameter compared to the kernel codebook approach where Gaussian distributions are extracted as features using SIFT descriptors and are further smoothened using kernels to generate codebook [24].
6. pbCanny also leads Linear SPM technique which is an extension of SVM using SPM kernel approach in combination with sparse codes which has been identified as one of the most successful techniques in image classification [25].

5 Conclusion

Manually segmenting images for large datasets in order to generate ground truth segmented images for evaluation purpose is obviously a cumbersome task. So, in this paper we introduce CBIR technique to evaluate the performances of GM-based segmentation techniques on large datasets. Canny segmentation and Gradient magnitude at multiple scales showed good tendencies towards accurate image retrieval compared to the other segmentation methods. Ridgelet transforms gained competence by locating positions of lines and reducing those line features to point like features with different set of orientations confining linear singularities along with its length is first of its kind in the literature used for multi-class classification. We complement our results with the efficiency of pbCanny segmentation which proved to be the best in creating benchmarking segmentation database [26].

References

1. Han-Hui, H., Chi-Yu, L., Jin-Jang, L.: Saliency-directed color image segmentation using modified particle swarm optimization. Sig. Process. **92**, 1–18 (2012)
2. Martin, D.R., Fowlkes, C.C., Malik, J.: Learning to detect natural image boundaries using local brightness, color, and texture cues. IEEE Trans. Pattern Anal. Mach. Intell. **26**(1), 530–549 (2004)
3. Fram, J.R., Deutsch, E.S.: On the quantitative evaluation of edge detection schemes and their comparison with human performances. IEEE Trans. Comput. **24**, 616–628 (1975)
4. Woods, R.E., Gonzalez, R.C.: Digital image processing. Prentice Hall, Upper Saddle River (2002)
5. Jack-Gerard, P., Nicolas, V., Ludovic, M.: Color image segmentation by pixel classification in an adapted hybrid color space - application to soccer image analysis. Comput. Vis. Image Underst. **90**, 190–216 (2003)
6. Guimei, Z., Lu, W., Jun, C., Jun, M.: Edge and corner detection. Opt. Laser Technol. **45**, 756–762 (2013)
7. Shang-Hong, L., Kai-Yueh, C., Tyng-Luh, L.: From co-saliency to co-segmentation: An efficient and fully unsupervised energy minimization model. IEEE - Comput. Vis. Pattern Recogn. **26**, 2129–2136 (2011)
8. Manuel, J., Fonseca, J., Amante, C.: Fuzzy color space segmentation to identify the same dominant colors as users. In: 'DMS' Knowledge Systems Institute, pp. 48–53 (2012)
9. Allan, D., Jepson, F., Estrada, J.: Benchmarking image segmentation. Int. J. Comput. Vis. **85**, 167–181 (2009)
10. Fowlkes, C., Malik, J., Arbelaez, P., Maire, M.: Contour detection and hierarchical image segmentation. IEEE. PAMI. **33**, 898–916 (2011)
11. Fowlkes, C., Malik, J., Arbelaez, P., Maire, M.: Using contours to detect and localize junctions in natural images. In: CVPR, pp. 1–8. IEEE (2008)
12. Stella Yu, X.: Segmentation induced by scale invariance. IEEE. Comput. Vis. Pattern Recogn. **1**, 444–451 (2005)
13. Mahantesh, K., Manjunath Aradhya, V.N.: An exploration of ridgelet transform to handle higher dimensional intermittency for object categorization in large image datasets. In: International Conference on Applied Information and Communications Technology (ICAICT). Procedia Technology, pp. 515–521 (2014)
14. Specht, D.F.: Probabilistic neural networks. Neural Netw. **3**, 109–118 (1990)
15. Mahantesh, K., Manjunath Aradhya, V.N.: An impact of complex hybrid color space in image segmentation. In: International Symposium on Intelligent Informatics (ISI) vol. 235, pp. 73–82 (2013)
16. Lu, Z., Ip, H.H.: Image categorization by learning with context and consistency. In: IEEE CVPR, pp. 2719–2726 (2009)
17. Manimala, S., Hemachandran, K.: Content based image retrieval using color and texture. Sig. Image Process. Int. J. (SIPIJ) **3**(1), 39–57 (2012)
18. Baharum, B., Fazal, M.: Analysis of distance metrics in content based image retrieval using statistical quantized histogram texture features in the dct domain. J. King Saud Univ. Comput. Inf. Sci. **25**(2), 207–218 (2013)
19. Wolf, L., Serre, T., Poggio, T.: Object recognition with features inspired by visual cortex. IEEE-CVPR **2**, 994–1000 (2005)
20. Welling, M., Holub, A., Perona, P.: Exploiting unlabelled data for hybrid object classification. In: NIPS Workshop on Inter-Class Transfer, Whistler (2005)

21. German, G., Engin, T., Fethallah, B., Roberto, R., Vincent, L.: On the relevance of sparsity for image classification. Comput. Vis. Image Underst. **125**, 115–127 (2014)
22. Berg, T.L., Berg, A.C., Malik, J.: Shape matching and object recognition using low distortion correspondence. In: IEEE CVPR, vol. 1, pp. 26–33 (2005)
23. Jim, M., David, G.L.: Muticlass object recognition with sparse, localized features. In: IEEE CVPR. vol. 1, pp. 11–18 (2006)
24. van Gemert, J.C., Geusebroek, J.-M., Veenman, C.J., Smeulders, A.W.M.: Kernel codebooks for scene categorization. In: Forsyth, D., Torr, P., Zisserman, A. (eds.) ECCV 2008, Part III. LNCS, vol. 5304, pp. 696–709. Springer, Heidelberg (2008)
25. Yihong, G., Thomas Huang, J., Kai, Y.: Linear spatial pyramid matching using sparse coding for image classification. In: IEEE-CVPR, pp. 1794–1801 (2009)
26. Allan, D.J., Francisco, J.E.: Benchmarking image segmentation algorithms. Int. J. Comput. Vis. **85**, 167–181 (2009)

Automated Nuclear Pleomorphism Scoring in Breast Cancer Histopathology Images Using Deep Neural Networks

P. Maqlin[1(✉)], Robinson Thamburaj[1], Joy John Mammen[2], and Marie Theresa Manipadam[3]

[1] Department of Mathematics, Madras Christian College, Chennai, India
maq.linparamanandam@yahoo.com
[2] Department of Transfusion Medicine and Immunohaematology, Christian Medical College, Vellore, India
[3] Department of Pathology, Christian Medical College, Vellore, India

Abstract. Scoring the size/shape variations of cancer nuclei (nuclear pleomorphism) in breast cancer histopathology images is a critical prognostic marker in breast cancer grading and has been subject to a considerable amount of observer variability and subjectivity issues. In spite of a decade long histopathology image analysis research, automated assessment of nuclear pleomorphism remains challenging due to the complex visual appearance and huge variability of cancer nuclei.This study proposes a practical application of the deep belief based deep neural network (DBN-DNN) model to determine the nuclear pleomorphism score of breast cancer tissue. The DBN-DNN network is trained to classify a breast cancer histology image into one of the three groups: score 1, score 2 and score 3 nuclear pleomorphism by learning the mean and standard deviation of morphological and texture features of the entire nuclei population contained in a breast histology image. The model was trained for features from automatically-segmented nuclei from 80 breast cancer histopathology images selected from publicly available MITOS-ATYPIA dataset. The classification accuracy of the model on the training and testing datasets was found to be 96 % and 90 % respectively.

Keywords: Nuclear pleomorphism · Deep neural networks · Breast cancer

1 Introduction

Breast cancer grading can be described as the microscopic examination of breast biopsy tissue slides to determine how different the tumour appears on comparison with normal breast tissue in order to ascertain the aggressiveness of the disease. The Nottingham (Elston-Ellis) modification of the Scar -Bloom-Richardson grading protocol (NGS) (El-ston and Ellis, 2002) has been used worldwide for breast cancer grading [9]. The NGS protocol is based on the evaluation of three

R. Prasath et al. (Eds.): MIKE 2015, LNAI 9468, pp. 269–276, 2015.
DOI: 10.1007/978-3-319-26832-3_26

morphological features: (a) degree of tubule or gland formation in the tumor, (b) nuclear pleomorphism (size and shape variation of cancer nuclei compared to normal nuclei), and (c) mitotic nuclei count. Figure 1 depicts the NGS protocol. Studies prove that es-timating nuclear pleomorphism is the most subjective [3] and least reproducible feature compared to the other two [2,4] and [12]. These issues have led to tremendous research in morphological analysis of cancer nuclei via many image analysis and machine learning techniques. A summary of the state-of-the-art work in this area has been discussed in Sect. 2.

Semiquantitative method for assessing histological grade in breast cancer

Feature	Score
Tubule gland formation	
Majority of Tumor >75%	1
Moderate degree 10% - 75%	2
Little or none <10%	3
Nuclear Pleomorphism	
Small, regular, uniform cells	1
Modrate increase in size and variability	2
Marked variation	3
Mitotic count	
Depends on field area. Example for 0.152 mm^2	
5-6	1
6-10	2
>11	3

The score of the three features are added to determine the grade of the breast cacner histopathology slide:
Grade1 : well differentiatied (3-5 points)
Grade2 : moderately differentiated (6-7 points)
Grade3 : poorly differentiated (8-9 points)

Fig. 1. Nottingham (Elston-Ellis) modification of the Scarff -Bloom-Richardson grading protocol [9]

This paper applies a recent inference of Restricted Boltzmann machine (RBM) in Deep Neural Network (DNN) model, suggested in [8], called DBN-DNN to automatically determine the nuclear pleomorphism score of breast cancer histopathology images. A DBN-DNN model consists of two main steps: pre-training and fine tuning.

Pre-training: Each layer of the DBN-DNN model is an independently trained RBM. An RBM is a generative artificial neural network that learns the probability distribution over its set of inputs. In this work the RBM with Bernoulli's distribution is used. First the RBM is trained in an unsupervised way using a contra divergence algorithm (CD). The states of the hidden nodes inferred during this training are used as the inputs to the next level of RBM. Thus a sequence of RBMs are trained and stacked to form a DBN. Then the DNN is constructed by adding a final layer. The parameters of the trained DBN are used to initialize the DNN. The DNN thus constructed can be called as DBN-DNN

Fine-Tuning: The constructed DBN-DNN is fine-tuned in a supervised manner using a classical back-propagation algorithm.

A DBN-DNN construction is illustrated in Fig. 2.

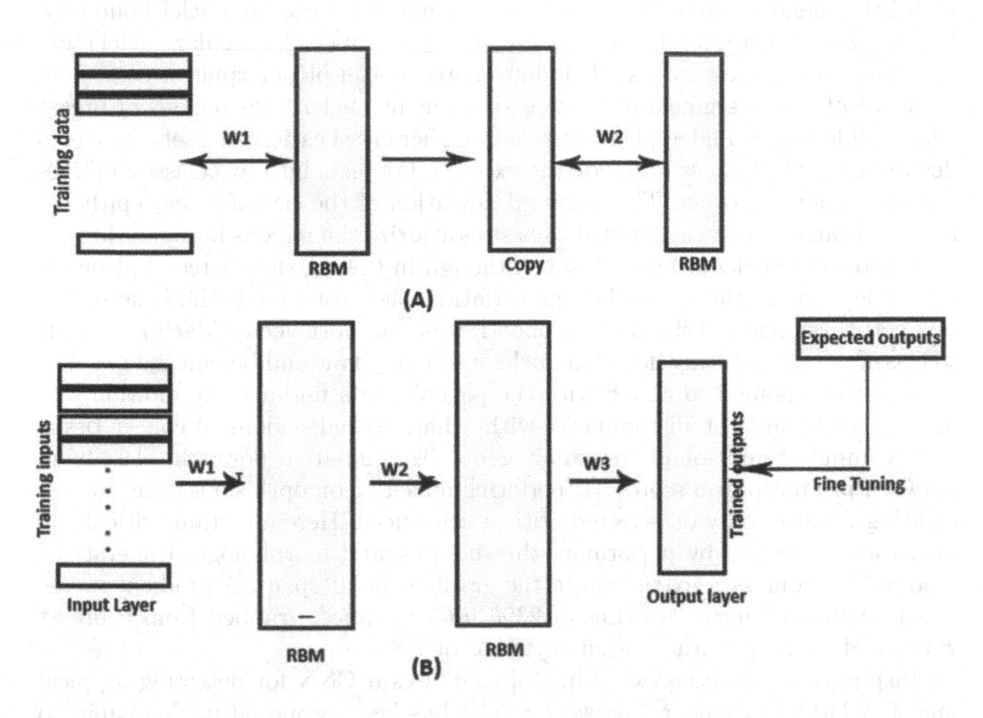

Fig. 2. (A) DBN formed by stacking of pre-trained RBMs and (B) DBN-DNN

In this study a set of 20 aggregate size and shape features of the nuclei population of a histopathology image is used to train the DBN-DNN model in determining the nuclear pleomorphism score of the image. The feature extraction and the training of the DBN-DNN model on these features is explained in Sect. 3.

The remainder of the paper is organized into six sections. Related works are discussed in Sect. 2. Section 3 gives a brief overview of the dataset. Section 4 and 5 present the proposed methodology and the experimental results respectively followed by the concluding remarks in Sect. 6.

2 Related Works

Accurate nuclei detection is the first and most important step for automated nuclei pleo-morphism scoring. Interested authors may find a detailed review of the state-of-the-art nuclei detection, segmentation and classi cation methods discussed in [5]. These methods are based on techniques such as thresholding, watershed, morpholog-ical operations, active contours, active-shape models, graph based segmentation and neural networks The review states that in-spite of the immense research activity in the area an accurate detection and grading of breast cancer nuclei remains challenging due to complex appearance of high grade cancer cells with-heterogeneous characteristics of cancer nuclei - irregular nuclei boundary, high variation in chromatin distribution and appearance of vesicular nuclei etc.

One of the earliest works [11] in automated nuclei pleomorphism, presents a sequence of image segmentation steps to segment nuclei from regions of breast whole- slide images and applies the standard sher classi cation on a set of features derived from the images to divide the extracted objects into two classes: epithelial cell nuclei and other. The standard deviation of the size of (area) epithelial nuclei within an area was claimed to best categorize the regions between three di erent scores of nuclei pleomoprhism. Although in theory, the degree of pleomorphism depends on the size and shape variation, the proposed method claimed to be e ective using nuclei size factor alone. [1] used Support Vector Machine (SVM) to classify au-tomatically detected nuclei into malignant and benign groups. The classi cation claimed to match with the pathologist's findings. In addition, the median nuclei area of all the nuclei within handpicked regions of cancer tissue were computed and found to exhibit good discriminative power in classifying between pleomorphism scores. [7] performs nuclei pleomoprhism scoring by calculating features only on selected critical cell nuclei. Herein certain critical cell nuclei are extracted by performing thresholding and morphological operations followed by boundary extraction in the gradient polar space. A gradient model classi ed the cell nuclei features of 2396 image frames, grabbed from 6 breast whole slide images, with an accuracy error of 7.84.

Inspired by the recent work in [10] that uses a CNN for detecting atypical nuclei, a DBN-DNN neural network model has been proposed in this paper to learn the size, shape and texture features of the nuclei population in the breast cancer. This enables a fast and accurate scoring.

3 Dataset

The dataset consists of 80 H&E stained breast cancer histopathology images of various nuclear pleomorphism scores, randomly selected from the MITOS-ATYPIA dataset. Each image is a frame selected from a whole slide image at x 20 magnification captured by an Aperio Scanscope XT scanner. The images are assigned scores for nuclear pleomoprhism, by three different pathologists, ranging from 1 to 3. The score assigned to each frame is a discrete number between 1 and 3. The overall score of the image is the majority among the three pathologists. Figure 3 shows three images of differrent nuclear pleomorphism scores.

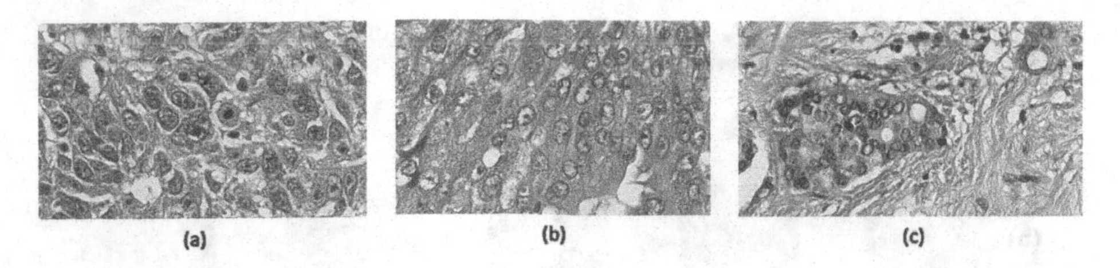

Fig. 3. Breast cancer histopathology sections exhibiting nuclei pleomorphism (a) Score 3, (b) Score 2 and (c) Score 1

4 Methodology

The nuclei pleomorphism scoring framework consists of three main modules 1. Nuclei detection. 2. Feature set preparation 3. DBN training and classification. Figure 4 depicts a block digram of the proposed methodology. Matlab R2013b is the computational tool used to develop the method.

4.1 Nuclei Detection

The boundaries of individual nuclei are automatically detected using convex grouping technique described in our earlier work [6]. Figure 5 shows sub images of detected nuclei from various grades of cancer.

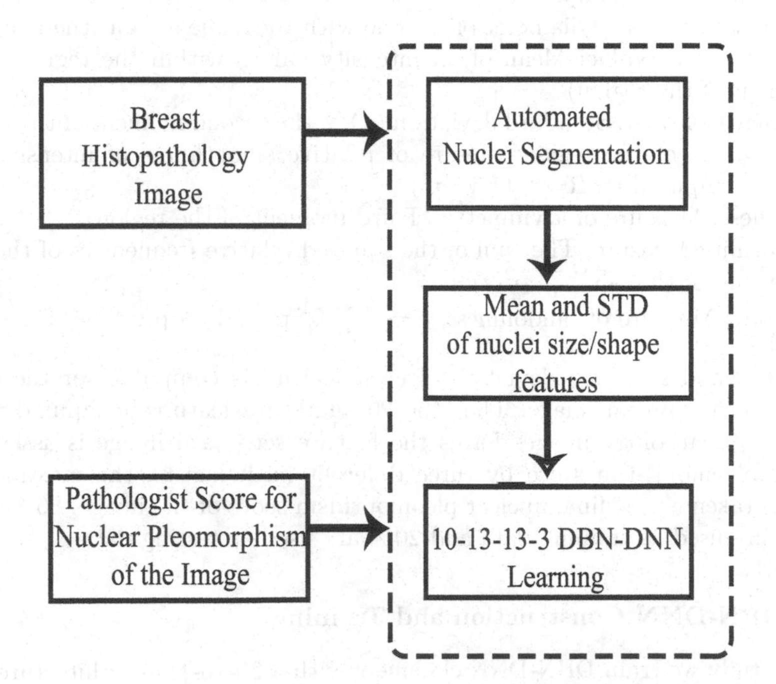

Fig. 4. Block diagram of the proposed methodology

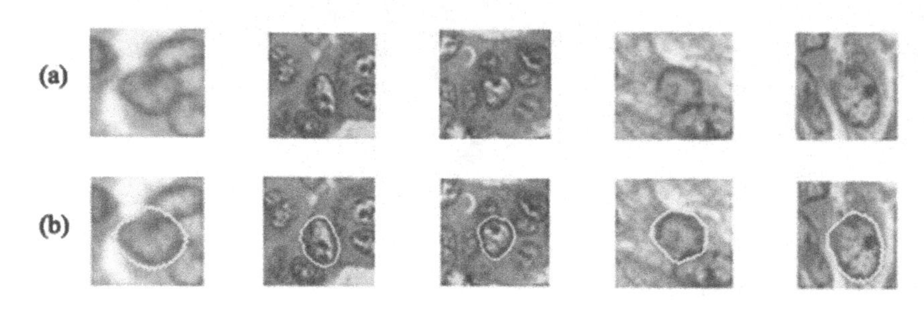

Fig. 5. (A) Original sub images of nuclei (B) Resultant images showing detected nuclei

4.2 Feature Extraction

The nuclear pleomorphism score of a breast cancer image is assigned based on the degree of size and shape variations of cancer nuclei from normal ones. Listed below are the 10 attributes extracted from the segmented nuclei of histopathology images.

- Area: Total number of pixels within the nuclei boundary.
- Solidity: Proportion of pixels in the convex hull that are also inside the nuclei boundary.
- Eccentricity: The eccentricity of the ellipse that has the same second-moments as the nuclei boundary.
- Equivdiameter: The diameter of a circle with the same area as the nuclei.
- Average Gray Value: Mean of all intensity values within the region. (First moment of the region)
- Average Contrast: Standard deviation (σ) of the second moment of the region.
- Smoothness of the region: Mesure of relative smoothness of intensity in a region computed by $R= 1-1/(1+\sigma^2)$
- Skewness: Measure of asymmetry (Third moment of the region).
- Uniformity Measure: The sum of the squared relative frequencies of the gray scale values. $U = -\sum_{i=0}^{L-1} p^2(z)$
- Entropy: Measure of randomness. $e = -\sum_{i=0}^{L-1} p(z_i) \log_2 p(z_i)$

The mean and standard deviation of each feature is computed for the nuclei population within an image. Thus the 20 aggregate features computed for 80 breast histopathology images forms the feature set. Each image is assigned a nuclear pleomoprhism score by three different pathologists, the maximum of which is taken as the final nuclear pleomoprhism score of the image. 80 % of the images are used as training data and 20 % are used as testing data.

4.3 DBN-DNN Construction and Training

In this study we train DBN-DNN classifiers with a 20-13-13-3 architecture. The DBN-DNN takes 20 inputs(mean and standard deviation values of the features of nuclei population in a single image). Two layers of RBM are then stacked, each with 13 hidden units. The output layer consists of three output states to determine the score of nuclear pleomorphism. The DBN-DNN was trained with inputs from training dataset. During the training process the stacked RBMS are pre-trained independently in an unsupervised way from one another using the contra divergence algorithm. The states of the hidden units in one RBM are used as the inputs for the next RBM. The final outputs inferred by the stacked RBM are fine-tuned in a supervised manner using the classical back propagation algorithm.

5 Results

The test results of the DBN-DNN classifier for each image in the testing dataset have been compared against the score given by the pathologists. On training the classifier with a training set of features from 80 images and testing it for a set of 20 images, the classification root mean square error and error rate in classifying images into their respective pleomorphism scores is shown in Table 1. The first and second rows in the table show the root mean square error and classification error rate computed on training and testing the DBN-DNN classifier. The error rate is likely to decrease with an increase in the amount of training data.

Table 1. DBN DNN classifier performance compared to pathologist

Classification Task	Root Mean Square Error	Error Rate
Training	0.0328	0.0401
Testing	0.2407	0.1066

6 Conclusion

An automated method to determine the nuclear pleomorphism score of a breast histopathology image has been presented here. The novelty of the approach lies in the application of a recent inference of RBM in neural networks, called DBN-DNN which helps in a faster and accurate method of nuclear pleomorphism scoring. Pre-training of stacked RBMs by unsupervised learning, the outputs of which are fine-tuned by supervised back propagation, enables quick and accurate classification. As part of the next phase of work, training of the classifier on detected nuclei image pixels, instead of the extracted features and comparison of this method with conventional neural network models is to be done.

Acknowledgement. The first two authors would like to acknowledge the Department of Science and Technology (DST), Govt. of India for providing computing facility at the institution through the FIST programme. The authors would also like to thank the organizers of the MITOS-ATYPIA 2014 contest for their consent to use images from their dataset.

References

1. Cosatto, E., Miller, M., Graf, H.P., Meyer, J.S.: Grading nuclear pleomorphism on histological micrographs. In: 19th International Conference on Pattern Recognition, ICPR 2008, pp. 1–4 December 2008
2. Dunne, B., Going, J.J.: Scoring nuclear pleomorphism in breast cancer. Histopathology **39**(3), 259–265 (2001)
3. Elston, C.W., Ellis, I.O.: Pathological prognostic factors in breast cancer. i. the value of histological grade in breast cancer: experience from a large study with long-term follow-up. Histopathology **41**(3A), 151–152 (2002). discussion 152–3
4. Frierson, Jr., H.F., Wolber, R.A., Berean, K.W., Franquemont, D.W., Gaffey, M.J., Boyd, J.C., Wilbur, D.C.: Interobserver reproducibility of the nottingham modification of the bloom and richardson histologic grading scheme for infiltrating ductal carcinoma. Am. J. Clin. Pathol. **103**(2), 195–198 (1995)
5. Irshad, H., Veillard, A., Roux, L., Racoceanu, D.: Methods for nuclei detection, segmentation, and classification in digital histopathology: A review of - current status and future potential. Biomed. Eng. IEEE Rev. **7**, 97–114 (2014)
6. Paramanandam, M., Thamburaj, R., Manipadam, M.T., Nagar, A.K.: Boundary extraction for imperfectly segmented nuclei in breast histopathology images – a convex edge grouping approach. In: Barneva, R.P., Brimkov, V.E., Šlapal, J. (eds.) IWCIA 2014. LNCS, vol. 8466, pp. 250–261. Springer, Heidelberg (2014)
7. Dalle, J.R., Li, H., Huang, C.H., Leow, W.K., Racoceanu, D., Putti, T.C.: Nuclear pleomorphism scoring by selective cell nuclei detection
8. Tanaka, M., Okutomi, M.: A novel inference of a restricted boltzmann machine. In: Proceedings of the 2014 22Nd International Conference on Pattern Recognition, ICPR 2014, pp. 1526–1531. IEEE Computer Society, Washington, DC, USA (2014)
9. Tavassoli, F.A., Devilee, P.: International Agency for Research on Cancer, and World Health Organization. Pathology and Genetics of Tumours of the Breast and Female Genital Organs. Who/IARC Classification of Tumours. IARC Press, Lyon (2003)
10. Wang, H., Cruz-Roa, A., Basavanhally, A., Gilmore, H., Shih, N., Feldman, M., Tomaszewski, J., Gonzalez, F., Madabhushi, A.: Mitosis detection in breast cancer pathology images by combining handcrafted and convolutional neural network features. J. Med. Imaging **1**(3), 034003 (2014)
11. Watson, S.K., Watson, G.H.: Assessment of nuclear pleomorphism by image analysis, WO Patent App. PCT/GB2004/000, 400, August 26 2004
12. William, H., Nick, W., Olvi, L., Mangasarian, L.: Importance of nuclear morphology in breast cancer prognosis. Clin. Cancer Res. **5**(11), 3542–3548 (1999)

Hybrid Source Modeling Method Utilizing Optimal Residual Frames for HMM-based Speech Synthesis

N.P. Narendra(✉) and K. Sreenivasa Rao

School of Information Technology, Indian Institute of Technology Kharagpur, Kharagpur 721302, India
narendrasince1987@gmail.com, ksrao@iitkgp.ac.in

Abstract. This paper proposes a new hybrid source modeling method for improving the quality of HMM-based speech synthesis. The proposed method is an extension of recently proposed source model based on optimal residual frame [1]. The source or excitation signal is first decomposed into a number of pitch-synchronous residual frames. Unique variations are observed in the pitch-synchronous residual frames present at the beginning, middle and end regions of excitation signal of a phone. Based on the observation, one optimal residual frame is extracted from each of the beginning, middle and end regions of excitation signal of a phone. The optimal residual frames extracted from every region of excitation signal are separately grouped in the form of decision tree. During synthesis, for every phone, three optimal residual frames are selected from three decision trees based on target and concatenation costs. Using three optimal residual frames, the excitation signal of a phone is constructed. The proposed hybrid source model is used for synthesizing speech under HTS framework. Subjective evaluation results indicate that the proposed source model is better the two existing source modeling methods.

Keywords: HMM-based speech synthesis · Hybrid source modeling · Excitation signal · Optimal residual frame

1 Introduction

Hidden Markov model (HMM)-based speech synthesis has attained popularity over past few years due to its flexibility, smoothness and small foot print. Most of the current text-to-speech (TTS) conversion systems utilize HMM-based speech synthesizer. HMM-based speech synthesis systems (HTS) are based on source-filter theory. Here, filter refers to the time-varying resonators formed in the vocal-tract system and source refers to the excitation signal formed by vibration of vocal folds. In order to produce good quality speech, both vocal-tract and source signal should be modeled accurately. In this paper, the quality of synthesized speech is improved through efficient representation and modeling of source signal.

Recently, several studies have been conducted to develop a better source or excitation modeling method for improving the quality of HTS. One of the earlier

R. Prasath et al. (Eds.): MIKE 2015, LNAI 9468, pp. 277–286, 2015.
DOI: 10.1007/978-3-319-26832-3_27

approaches of modeling excitation signal was based on mixed excitation algorithm [2]. Later Zen et al. [3] used the speech transformation and representation using adaptive interpolation of weighted spectrum (STRAIGHT) vocoding method [4] for HMM-based speech synthesis system. In [5], mixed excitation is constructed by state-dependent filtering of pulse trains and white noise sequences. During training, filters and pulse trains are jointly optimized through a procedure which resembles analysis-by-synthesis speech coding algorithms. In [6], the excitation signal is constructed by modifying a single natural instance of glottal flow pulse according to the source parameters generated from HMMs.

Instead of using the parameters derived from statistical models, a hybrid approach is proposed which utilizes both the parameters and the real instances of residual segments for generating the excitation signal [7–9]. In [7], the source signal was generated by selecting suitable residual frames from the codebook based on target residual specification. Raitio et al. [8] generated excitation signal by selecting suitable glottal source pulses from the database based on the target and concatenation costs. In [9], the excitation signal is generated as a combination of small fraction real residual frame around glottal closure instant and noise component represented in terms of amplitude envelope and energy. In [1], optimal residual frame estimated from every phone are efficiently grouped in the form of decision tree. During synthesis, the excitation signal of a phone is generated by selecting suitable optimal residual frame from the leaf of decision tree.

In this paper, a hybrid source model is proposed by extending the recently proposed source modeling method based on optimal residual frame [1]. Time-varying characteristics of excitation signal are captured by extracting one optimal residual frame in each of the beginning, middle and end regions of a phone. Three optimal residual frames extracted all phones are grouped separately in three decision trees. During synthesis, suitable optimal residual frames are selected from the leaves of decision trees based on target and concatenation costs. Using optimal residual frames, excitation signal of a phone is constructed. In this paper, the terms *source*, *excitation* and *residual* are used interchangeably. This paper is organized as follows. Section 2 describes the proposed hybrid source model. Different steps in speech synthesis using the proposed method are explained in Sect. 3. Subjective evaluation of proposed method is provided in Sect. 4. Section 5 provides the conclusion of the present work.

2 Proposed Hybrid Source Model

The proposed hybrid source model utilizes natural residual segments for generation of excitation signal. The proposed method is an extension of recently proposed source model [1] which utilizes single optimal residual frame estimated from every phone for generation of excitation signal. In this work, time-domain variations are better captured by extracting optimal residual frames in three regions of excitation signal of a phone. The flow diagram indicating the sequence of steps in the proposed hybrid source model is shown in Fig. 1. Initially, energy is extracted from every frame of excitation signal. The excitation signal is decomposed into a number of pitch-synchronous residual frames (explained in Sect. 2.1). The excitation signal of every phone is divided into three equal regions, namely, beginning, middle and end. From the residual frames extracted from every region, single optimal residual frame is estimated (described in Sect. 2.2). The residual frame which has the lowest mean acoustic distance with all other residual frames in a particular region is considered as the optimal residual frame. To compute acoustic distance, acoustic features are extracted from every residual frame (explained in Sect. 2.2). The optimal residual frames extracted from every region are clustered separately in the form of decision tree (explained in Sect. 2.3). In order to cluster the residual frames, positional and contextual features are extracted from every phone label. Three decision trees are developed corresponding to the optimal residual frames in three regions.

At the time of synthesis, depending on the input phone, three optimal residual frames are selected from the leaves of three decision trees based on target and concatenation costs (described in Sect. 2.4). Using three optimal residual frames, beginning, middle and end regions of excitation signal of phone is constructed by pitch-synchronous overlap add (PSOLA). The steps involved in the proposed hybrid source model are explained in the following section.

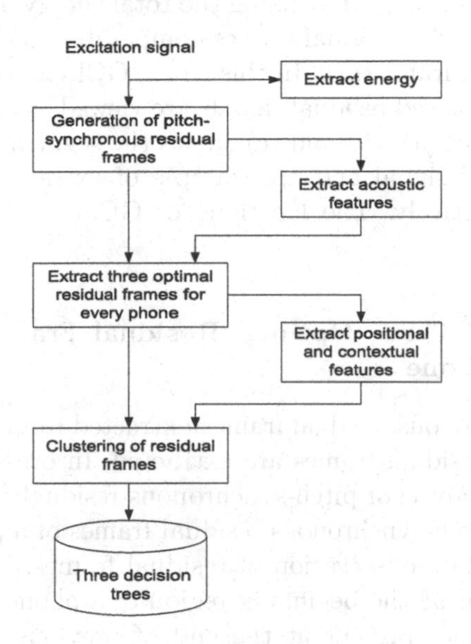

Fig. 1. Flowchart indicating the sequence of steps in proposed hybrid source model

2.1 Generation of Pitch-Synchronous Residual Frames

The pitch-synchronous residual frames are extracted from the excitation signal of a phone using the knowledge of glottal closure instants (GCIs). Using GCIs, the boundaries of pitch cycles are marked on the excitation signal. GCIs are estimated from the speech signal using zero-frequency filtering (ZFF) method [10].

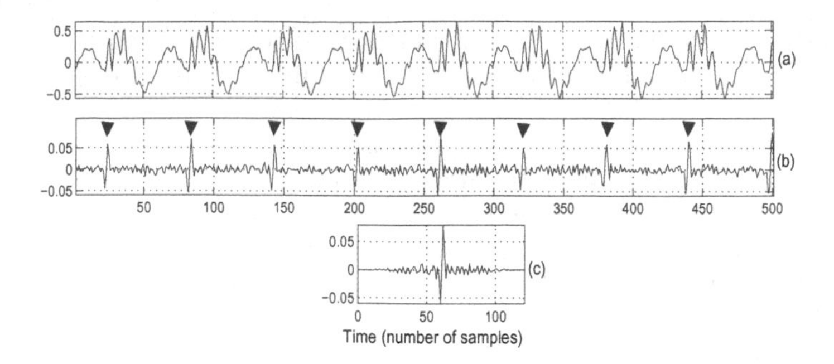

Fig. 2. (a) Speech signal, (b) Residual signal and (c) Pitch-synchronous residual frame. The locations of GCIs are shown by downward arrows (▼).

Using GCI positions as anchor points, two-pitch period long residual signals are extracted and they are Hanning windowed. The extracted residual signals are normalized both in pitch period and energy. Pitch normalization is performed by upsampling the residual frames to the lowest pitch period of the speaker. The energy normalization is achieved by fixing the total energy of residual frame to 1. These operations make the residual frames comparable so that they can be analyzed under a common framework. In this work, GCI centered two-pitch period long and Hanning windowed residual signals are viewed as the pitch-synchronous residual frames. Figure 2(a), (b) and (c) shows the segment of speech signal, its corresponding residual signal and an example of extracted pitch-synchronous residual frame, respectively. The locations of GCIs are shown by downward arrows (▼) in Fig. 2(b).

2.2 Estimation of Three Optimal Residual Frames from Every Phone

From the pitch-synchronous residual frames extracted from the excitation signal of a phone, optimal residual frames are extracted. In order to extract optimal residual frames, an analysis of pitch-synchronous residual frames is carried out. Generally, adjacent pitch-synchronous residual frames of a phone exhibit strong correlation [11]. On close observation of residual frames, it is noticed that the residual frames present at the beginning region of a phone have some influence from the residual frames present at the end of previous phone. The residual frames present at the middle region of a phone are relatively stable and their shapes are almost identical. The residual frames present at the end of a phone have some influence from the residual frames present at the beginning of next phone. As residual frames are varying in different regions of excitation signal, by representing each of the beginning, middle and end regions of excitation signal with an optimal residual frame can result in capturing the time-varying characteristics of excitation signal. The optimal residual frame closely represents all other residual frames in a particular region of excitation signal.

To efficiently extract optimal residual frames in beginning, middle and end regions of excitation signal, every residual frame is represented by a set of new acoustic features. The new acoustic features are as follows: (1) energies in six parts of residual frame, (2) ratio of maximum to minimum peak amplitude of residual frame, (3) pulse spread of residual frame, and (4) linear Predictor Coefficients (LPC). The acoustic features are proposed based on the previous studies performed on the analysis of variation residual frames for different phonetic classes [12]. Single instance of residual frame indicating different acoustic features considered for finding the optimal residual frame is shown in Fig. 3. The residual frame is divided into six equal parts (S1 to S6). In every part of the residual frame, energy is computed. This way of energy computation captures the distribution of energy in the residual frame. The ratio of maximum (Amax) to minimum (Amin) peak indicates the orientation of residual frame, whether it is oriented towards positive or negative direction. The pulse spread is the time interval (in number of samples) where the peak of HE of residual frame drops to certain threshold α (the value of α is chosen to be 20 %). 10-dimensional LPCs represent the spectral envelope of the residual frame.

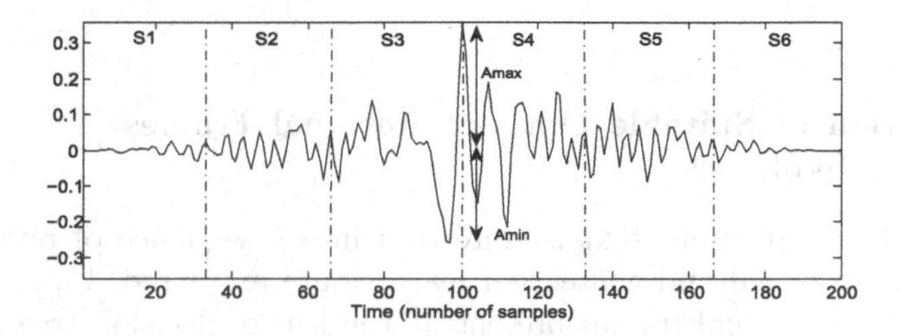

Fig. 3. Single instance of residual frame indicating different acoustic features considered for finding the optimal residual frame

In order to find the optimal residual frame, mean acoustic distance between a given residual frame with all other residual frames in a particular region of phone is computed. Acoustic distance between two residual frames is computed as the weighted Mahalanobis distance of acoustic features. Mean of all acoustic distances is calculated between a given residual frame and all other residual frames. Similarly mean acoustic distances of all residual frames in a particular region of phone are computed. The residual frame which has the lowest mean acoustic distance is considered as the optimal residual frame of a particular region of phone. Using this procedure, optimal residual frames are estimated from three regions of a phone. Similarly, optimal residual frames are estimated from all phones present in the database.

2.3 Developing Decision Trees Using Optimal Residual Frames

During synthesis, three optimal residual frames extracted from all phones are available for excitation signal generation. The optimal residual frame which best matches with the given target phone specification is selected from the database. In order to accomplish this, the residual frames extracted from all phones are grouped into number of clusters. All the units present in the cluster have similar acoustic characteristics. For the given target unit specification, appropriate cluster is selected which consists of a small set of candidate residual frames. Suitable residual frame is selected from the cluster based on target and concatenation costs.

To cluster the residual frames, acoustic distance between residual frames is computed using features proposed in Sect. 2.2. The acoustic distance is the Mahalanobis distance of features of two residual frames. Using this acoustic measure, the acoustic impurity of cluster of residual frames is computed which is the mean acoustic distance of all the residual frames present in the cluster. A decision tree is constructed by splitting the cluster of residual frames into subclusters based on the questions related to positional and contextual features of phone. The questions that best minimize the acoustic impurity of subclusters are asked at each node of decision tree. Minimum number of units present in the cluster is specified to be 10. By considering the optimal residual frames extracted from three regions of all phones present in the database, three decision trees are developed.

2.4 Selection of Suitable Optimal Residual Frames During Synthesis

During synthesis, an input text is converted into a sequence of phones. From each phone, positional and contextual features are extracted. Using these features, cluster of residual frames present at the leaf of decision tree is selected by answering the questions at each node. To select the most appropriate unit from the cluster of residual frames or candidate units, target cost computation is carried out. Features used as the subcosts of target cost are position of phrase (PP), position of word (PW), position of syllable (PS), phone identity (PI), previous phone identity (PPI) and next phone identity (NPI). PP, PW and PS are categorized as begin, middle and end. If target unit specification matches with that of candidate unit, then the subcost is assigned as zero otherwise it is one. Target cost is computed as the weighted sum of subcosts.

To ensure smooth variation residual frames present in three regions of phone and at adjacent phones, concatenation cost is computed. Acoustic features used to describe the residual frames (described in Sect. 2.2) are used as the subcosts for concatenation cost. Concatenation cost is computed as the weighted sum of difference between the features of two candidate units. Sequence of candidate units which has the lowest sum of target and concatenation costs are chosen as the best sequence of optimal residual frames of the utterance. The weights for target and concatenation subcosts are chosen manually based on informal listening tests.

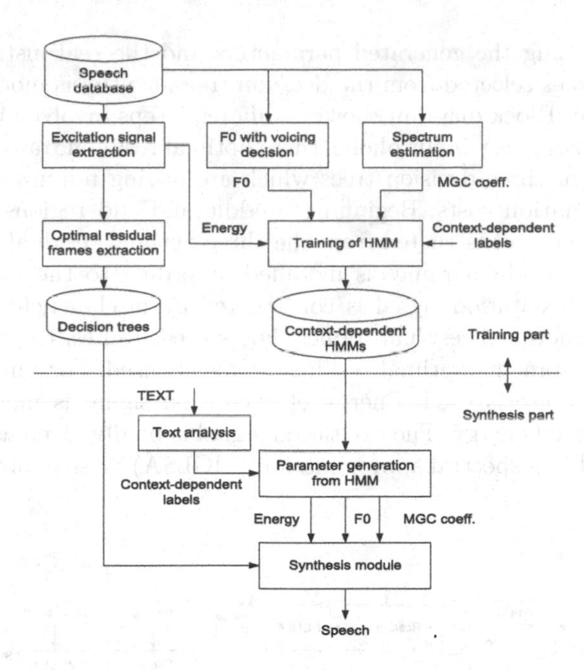

Fig. 4. Overall block diagram of HTS with the proposed source model

3 Speech Synthesis Using the Proposed Hybrid Source Model

The proposed hybrid source model is used for synthesizing the speech under HTS framework. In this work, publicly available HTS toolkit [13] is used for implementing the HTS. Overall block diagram of HTS with the proposed hybrid source model is shown in Fig. 4. During training, spectrum and F0 with voicing decision are extracted from the speech utterances present in the database. The spectrum estimation consists of extracting MGC coefficients (34th order) with the parameter values $\alpha = 0.42$ ($Fs = 16$ KHz) and $\gamma = -1/3$. Log F0 is obtained using the recently proposed voicing decision and F0 estimation method based on the instant of significant excitation [14]. Root mean square (RMS) energy is extracted from every frame of excitation signal. MGC, F0 and energy parameters are extracted for a frame size of 25 ms with frame shift of 5 ms, and these parameters are modeled by HMMs in a unified framework. From the excitation signal of every phone, three optimal residual frames are extracted from beginning, middle and end regions. The optimal residual frames are systematically arranged in three decision trees.

During synthesis, text analysis module converts an input text into a sequence of context-dependent phoneme labels. According to the label sequence, a sentence HMM is constructed by concatenating context-dependent HMMs. From the sentence HMM, MGC coefficients, log F0 including voicing decisions and energy are generated. Positional and contextual features are extracted from the target phones. Using the generated parameters and the real instances of optimal residual frames selected from the decision trees, synthesis module generates the speech signal. Block diagram showing different steps involved in synthesis is shown in Fig. 5. For every input phone, three optimal residual frames are selected from the leaves of three decision trees which are having minimum sum of target and concatenation costs. Beginning, middle and end regions of excitation signal of a phone are constructed from the three optimal residual frames. Pitch of three optimal residual frames is modified according to the target pitch by resampling. The excitation signal is constructed by pitch-synchronous overlap adding of pitch modified residual frames. For voiced frames, the excitation signal constructed from the optimal residual frames is used. For unvoiced frames, white Gaussian noise is used. Energy of excitation signal is modified according to the generated energy. The excitation signal is finally given as input to the Mel-Generalized Log spectral approximation (MGLSA) filter to obtain synthetic speech signal.

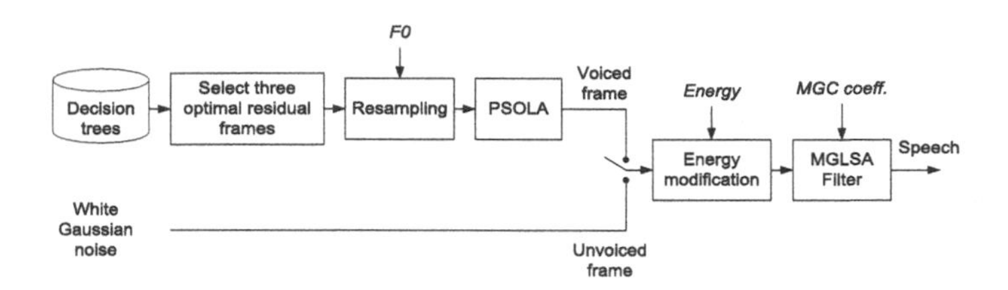

Fig. 5. Block diagram showing different stages in synthesis. Parameters generated from HMMs are shown in italics.

4 Subjective Evaluation

The proposed method is evaluated using one female (SLT) and male speaker (AWB) from CMU Arctic speech database. The training set of each of the speaker consists of about 1100 phonetically balanced English utterances. The duration of training set is about 56 and 79 min for SLT and AWB speakers, respectively. For evaluation, 20 sentences which were not part of training data are used. Subjective evaluation is conducted with 20 research scholars in the age group of 23–35 years. The subjects have sufficient speech knowledge for proper assessment of the speech signals. The quality of synthesized speech from the proposed method is compared with two source modeling methods, namely, pulse-HTS and SORF-HTS. In pulse-HTS, sequence of pulses positioned according to the generated pitch is used as the excitation signal. In single optimal residual frame (SORF)-HTS, the excitation signal of a phone is represented by a single optimal residual frame [1]. Before evaluation, the energies of synthesized speech signals are normalized to the same level.

Subjective evaluation of HTS is performed using two measures, namely, comparative mean opinion scores (CMOS) and preference tests. In CMOS test, subjects were asked to listen to two versions (i.e., one from the proposed method and

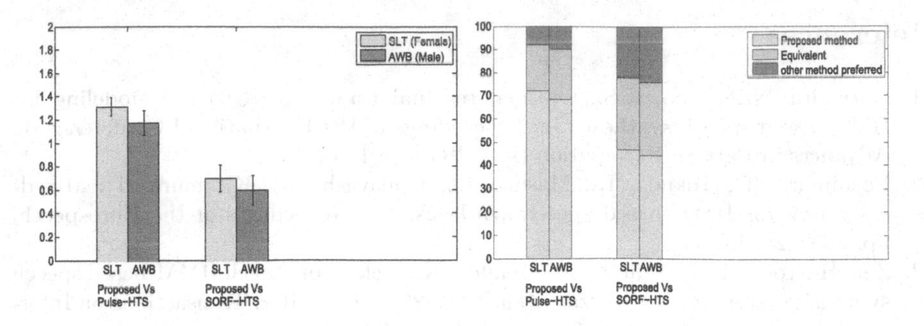

Fig. 6. (a) CMOS with 95 % confidence intervals and (b) preference scores obtained by comparing the proposed method with two existing methods.

other from the existing methods) and to provide a score according to their overall preference on a 7-point scale (-3 to +3). Two versions of synthesized speech were randomly shuffled and played to subjects. A positive score indicates that the proposed method is preferred over other method and negative score implies the opposite. Figure 6(a) provides the CMOS scores with 95 % confidence intervals. CMOS scores of both female and male speakers are varying from 0.6 to 1.4 which indicates that the proposed method is better than the existing methods. CMOS scores of female speaker are relatively higher compared to male speaker. CMOS score of proposed method Vs SORF-HTS is around 0.6. This indicates that the speech synthesized with the excitation signal generated using three optimal residual frames is better than the speech synthesized with the excitation signal generated using single optimal residual frame. Using three optimal residual frames, time-varying characteristics of excitation signal are preserved. In preference test, subjects were asked to give a preference between pair of synthesized speech utterances. Subjects were given the option either to prefer one of the synthesized speech utterances or to prefer both as equal. Preference scores are shown in Fig. 6(b). For both female and male speakers, subjects preferred the proposed method compared to other two methods.

5 Conclusion

This paper proposed a hybrid source model for improving the quality of HMM-based speech synthesis. By analyzing the characteristics of excitation signal, three optimal residual frames are extracted from beginning, middle and end regions of every phone. Three optimal residual frames extracted from all phones are systematically grouped in the form of three decision trees. During synthesis, depending on the input phone, suitable residual frames are selected from leaves of three decision trees based on target and concatenation costs. Both CMOS and preference scores indicated that the proposed source modeling method is better than two existing source modeling methods.

References

1. Narendra, N.P., Rao, K.S.: Optimal residual frame based source modeling for HMM-based speech synthesis. In: Proceedings of the International Conference on Advances in Pattern Recognition (ICAPR), pp. 1–5 (2015)
2. Yoshimura, T., Tokuda, K., Masuko, T., Kobayashi, T., Kitamura, T.: Mixed-excitation for HMM-based speech synthesis. In: Proceedings of the Eurospeech, pp. 2259–2262 (2001)
3. Zen, H., Toda, T., Nakamura, M., Tokuda, K.: Details of Nitech HMM-based speech synthesis system for the Blizzard Challenge 2005. In: IEICE Transactions on Information and Systems, vol. E90-D, pp. 325–333 (2007)
4. Kawahara, H., Masuda-Katsuse, I., de Cheveigne, A.: Restructuring speech representations using a pitch-adaptive time-frequency smoothing and an instantaneous-frequency-based F0 extraction: possible role of a repetitive structure in sounds. Speech Commun. **27**, 187–207 (1999)
5. Maia, R., Toda, T., Zen, H., Nankaku, Y., Tokuda, K.: An excitation model for HMM-based speech synthesis based on residual modeling. In: Proceedings of the Speech Synthesis Workshop 6 (ISCA SW6) (2007)
6. Raitio, T., Suni, A., Yamagishi, J., Pulakka, H., Nurminen, J., Vainio, M., Alku, P.: HMM-based speech synthesis utilizing glottal inverse filtering. IEEE Trans. Audio, Speech, Lang. Process. **19**(1), 153–165 (2011)
7. Drugman, T., Moinet, A., Dutoit, T., Wilfart, G.: Using a pitch-synchrounous residual codebook for hybrid HMM/frame selection speech synthesis. In: Proceedings of the International Conference on Acoustics, Speech and Signal Processing, (ICASSP), pp. 3793–3796 (2009)
8. Raitio, T., Suni, A., Pulakka, H., Vainio, M., Alku, P.: Utilizing glottal source pulse library for generating improved excitation signal for HMM-based speech synthesis. In: Proceedings of the International Conference on Acoustics, Speech and Signal Processing, (ICASSP), pp. 4564–4567 (2011)
9. Cabral, J.P.: Uniform concatenative excitation model for synthesising speech without voiced/unvoiced classification. In: Proceedings of the Interspeech, pp. 1082–1086 (2013)
10. Murty, K.S.R., Yegnanarayana, B.: Epoch extraction from speech signals. IEEE Trans. Audio, Speech Lang. Process. **16**(8), 1602–1613 (2008)
11. Yumoto, E., Gould, W., Baer, T.: Harmonics-to-noise ratio as an index of the degree of hoarseness. J. Acoust. Soc. Am. **71**(6), 1544–1550 (1982)
12. Narendra, N.P., Rao, K.S.: Time-domain deterministic plus noise model based hybrid source modeling for HMM-based speech synthesis. In: Speech Communciation, 2015 (Under review)
13. HMM-based speech synthesis system (HTS). http://hts.sp.nitech.ac.jp/
14. Narendra, N.P., Rao, K.S.: Robust voicing detection and F0 estimation for HMM-based speech synthesis. Circ. Syst. Sig. Process. **34**(8), 2597–2619 (2015)

Significance of Emotionally Significant Regions of Speech for Emotive to Neutral Conversion

Hari Krishna Vydana(✉), V.V. Vidyadhara Raju, Suryakanth V. Gangashetty, and Anil Kumar Vuppala

Speech and Vision Lab, International Institute of Information Technology, Hyderabad, India
{hari.vydana,vishnu.raju}@research.iiit.ac.in, {svg,anil.vuppala}@iiit.ac.in

Abstract. Most of the speech processing applications suffer from a degradation in performance when operated in emotional environments. The degradation in performance is mostly due to a mismatch between developing and operating environments. Model adaptation and feature adaptation schemes have been employed to adapt speech systems developed in neutral environments to emotional environments. In this study, we have considered only anger emotion in emotional environments. In this work, we have studied the signal level conversion from anger emotion to neutral emotion. Emotion in human speech is concentrated over a small region in the entire utterance. The regions of speech that are highly influenced by the emotive state of the speaker is are considered as emotionally significant regions of an utterance. Physiological constraints of human speech production mechanism are explored to detect the emotionally significant regions of an utterance. Variation of various prosody parameters (Pitch, duration and energy) based on their position in the sentences is analyzed to obtain the modification factors. Speech signal in the emotionally significant regions is modified using the corresponding modification factor to generate the neutral version of the anger speech. Speech samples from Indian Institute of Technology Kharagpur Simulated Emotion Speech Corpus (IITKGP-SESC) are used in this study. A subjective listening test is performed for evaluating the effectiveness of the proposed conversion.

Keywords: Emotionally significant regions · Emotion recognition · Automatic speech recognition · Physiological constraints · Emotional environments · Emotive to neutral conversion · Adaptation scheme

1 Introduction

Human speech is implicitly embedded with a wide range of paralinguistic information. The paralinguistic information comprises of emotion, voice quality, pathological state etc. [26]. Among the paralinguistic information handling emotive speech has a great deal of practical importance. Performance of most of the

R. Prasath et al. (Eds.): MIKE 2015, LNAI 9468, pp. 287–296, 2015.
DOI: 10.1007/978-3-319-26832-3_28

speech processing applications (automatic speech recognition (ASR), speaker identification and audio search) degrades while operating in real life environments as they have to operate on emotive speech [5,15]. Human computer interaction through speech can be carried out with ease when machines are provided with an ability to operate on emotive speech [3,25]. Most of the systems are developed using the data from neutral environments and operated in emotional environments, speaker may not be producing the neutral speech and demanding such type of signal from the user constrains him/her and reduces the ease of operation. A mismatch between developing and operating environments is inevitable, at the same time developing a model using the data from all different emotions may not be affordable most of the times. As human emotions subtle, complex and slide from one state from another. Most of the real life emotions are mixed in most cases it is more likely to have a mismatch between developing and operating environments. So a proper adaption scheme is a pre-requisite to operate speech processing applications. The major motivation for this work comes from the desire to develop speech systems that are more adaptable to real life emotional environments. The main goal of this work is to provide sufficient knowledge for speech processing applications to carryout the tasks efficiently in emotional environments.

Emotive speech is produced when human is in a different state from the normal state [4]. The features extracted from emotive speech are highly influenced by the emotive state of the speaker. Imparted variation in speech due to emotion can be compensated at different levels i.e., signal level, feature level and model level. In this work, we mainly focus on the signal level compensation to the emotional variations imparted in speech. Most of the works in literature have focused on feature and model level compensations to variations imparted in speech owing to emotional state of the speaker, but the signal level compensation can also be an effective approach that can supplement the feature and model level approaches.

In literature a few works have focused on developing emotion independent speech systems. Emotion independent speaker verification systems are developed in [6,13,14]. Acoustic variation imparted by emotional state of the speaker is studied in [19] and a significant shift in the vowel triangles and their position in f2/f1 dimensional space is observed, which is reported as a reason for the degradation in the performance of ASR in emotional environments. To compensate the mismatch between developed acoustic model and operating environments maximum likelihood linear regression (MLLR) and maximum a-posteriori estimation (MAP), adaptation algorithms are used to adapt the neutral ASR to emotive speech utterances and a significant improvement is observed [20]. Static and dynamic adaptation strategies are studied for adapting ASR to emotional environments [15]. In a static adaptation neutral ASR is adapted using the emotional data before the testing process. In dynamic adaptation scheme a fast emotion recognition system is employed to detect the emotion of the spoken utterance and a fast adaptation scheme is selected based on the detected emotion. As mismatch in the environments is better handled in dynamic adaptation strategy a better performance is noted.

In contrast to the above methods, a neural network based feature transformation is performed in [9], to map the spectral features of emotive speech data to neutral speech data. Most of the earlier methods have concentrated on model adaptation strategies to handle emotional environments. As human emotions are complex and the emotional state of the speaker slides from one emotion to another, so even after the adaptation the occurrence of a mismatch in environments is more likely. This motivated us to study a signal level adaptation which can deal with this highly varying emotional environments. As a primary study towards the broad goal, we concentrate on generating the neutral speech from its anger version. Signal level adaptation is a less popular concept and not many works have focused on signal level adaptation. Some works in the literature have focused on expressive speech generation from the neutral version by modifying the prosody accordingly [22–24]. In this study, we mainly focus on converting the expressive speech to its neural version. Prosody parameters pitch, duration and energy are modified accordingly for converting anger sentence to its neutral state. The emotive gestures produced in an emotive utterance are confined to a shorter duration and they usually don't span over the entire utterance [3]. During this study, the regions of speech in an utterance that are highly influenced by the emotive state of the speaker are computed and they are termed as emotionally significant regions. Prosody parameters of emotionally significant regions of speech are modified accordingly to generate the neutral version of emotive speech.

The rest of the paper is structured as follows: Sect. 2 discusses about the database used during the study. The technique used to detect emotionally significant regions of an utterance is presented in Sect. 3. Section 4 analyses of various prosody parameters to generate the neutral version of an emotional utterance. Section 5 describes the proposed algorithm for converting anger speech to its neutral version. Section 6 describes the evaluation of the proposed algorithm by subjective listening tests. Conclusion and future scope of this work is described in Sect. 6.

2 Database

Speech data from Indian Institute of Technology-Simulated Emotion speech corpus (IITKGP-SESC) [8] is used during the course of the present study. The speech corpus is recorded from professional artists in All India Radio(AIR), Vijayawada, India. Ten professional artists (5 male and 5 female) within the age group of 25–40 years with a professional experience of 8–12 years have taken part in the preparation of the database. The database comprises of speech data in eight emotions (Anger, Happy, Fear, Neutral, Sarcastic, Surprise, Disgust and sad). A total of 10 sessions are conducted and in each session fifteen Telugu sentences are expressed in neutral and 7 other emotions. The entire database comprises of 1200 utterances (15 sentences $\times$ 8 emotions$\times$ 10 artists $\times$ 10 sessions). Speech data is collected with a sampling rate of 16 KHz and a bit rate of 16 bits/samples. During the course of the study, expressive speech samples comprises of anger speech only.

3 Emotionally Significant Regions of Speech

Emotive speech is produced by the speaker when he/she is in a different state from the normal state [4]. Earlier studies [3,7,12,17,21,23] have shown that the emotive state of a speaker is concentrated in a short duration of utterance. The smallest unit that can carry the emotional intent (i.e., information related to emotion) is a word [2]. Physiological constraints of human speech dictate the lung pressure for utterances in a normal state (non-emotive state). Lung pressure may vary along the utterance to give lexical stress but the variance in the lung pressure is less than 30 % from the mean lung pressure over the entire sentence or utterance. When the speaker is in a normal state (non-emotive state) [16]. The lung pressure is observed to get manifested as glottal flow at vocal folds [1]. Strength of the excitation is employed as an acoustic correlate to effectively capture the variations in glottal flow along the utterance [11]. During this work, we locate the regions whose strength of the excitation fluctuates above and below 30 % from the mean strength of the excitation of the entire sentence, as emotionally significant regions of an utterance. Prosody parameters of speech signal in these regions is modified accordingly to generate neutral version from the emotive utterance.

3.1 Detecting Emotionally Significant Regions of an Utterance

In this work, strength of the excitation is computed from zero frequency filter (ZFF) based method [10]. Slope of the zero frequency filtered signal at epoch gives the strength of the excitation at that epoch.

3.2 Algorithm to Detect Emotionally Significant Regions of an Utterance

- Compute strength of excitation (SoE) at an epoch as mentioned in [10] and repeat the same value till the next epoch.
- A 20 ms mean smoothing is performed on the obtained step signal to generate a smooth contour and is termed as the strength of excitation (SoE) curve.
- SoE_{avg} is the average value of strength of excitation over the entire sentence.
- The regions of speech whose strength of excitation fluctuates 30 % above and below the SoE_{avg} are considered as emotionally significant regions of that utterance.

Proposed method for detecting emotionally significant regions of an utterance is shown in Fig. 1. Anger speech utterance is shown in Fig. 1(a). Figure 1(b) shows the zero frequency filtered signal computed using the ZFF method. Strength of the excitation at every epoch computed using the algorithm described in [10] is shown in Fig. 1(c). Strength of the excitation computed in the above step is mean smoothed using a frame size of 20 ms is illustrated in Fig. 1(d). Detected emotionally significant regions of an utterance is shown in Fig. 1(e). Emotionally significant regions of an utterance computed from the above algorithm are

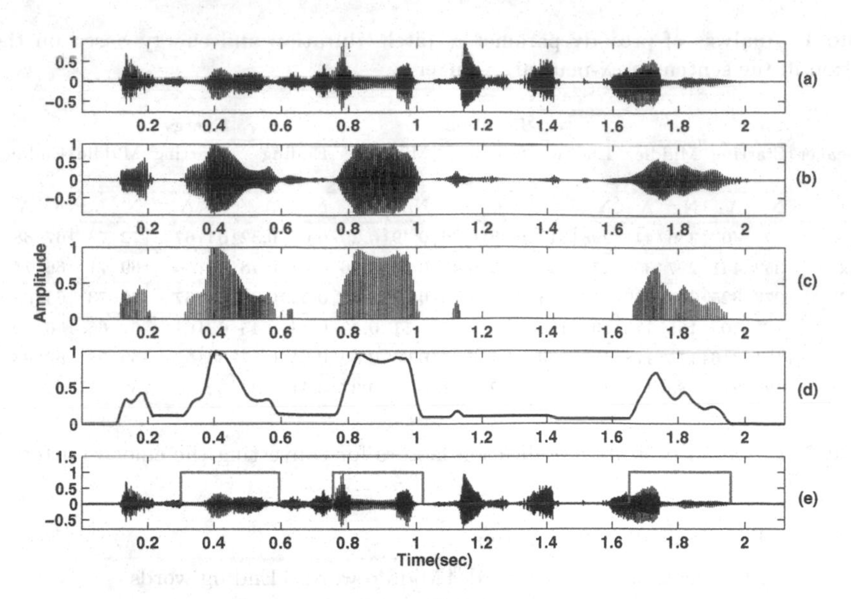

Fig. 1. Proposed method to detect emotionally significant regions of an utterance. (a) Speech signal, (b) ZFF signal, (c) strength of the excitation at every epoch, (d) strength of the excitation mean smoothed using a frame size of 20 ms and (e) detected emotionally significant regions of speech.

considered as the regions in an utterance, where the emotional intent of the speaker is significantly high. Prosody of the signal in these regions is modified accordingly to generate a neutral version of the emotive utterance.

4 Analysis of Various Prosody Parameters to Generate the Neutral Version of an Emotive Utterance

When a sentence with same lexical content is uttered in various emotive states, the variation is perceived due to non lexical content of that sentence (i.e., Prosody) [24]. In this study, we confine our observations to study the variations imparted by the anger emotion and convert it to its neutral version. During this study, three basic prosody parameters (i.e., Pitch, duration and energy) are analyzed to study the variation imparted in prosody due to the emotive state of the speaker. The influence of prosody varies differently based on the position of a word in a sentence. Variation imparted by the emotion on various parts of a sentence is studied by segmenting sentence in to three regions i.e., starting, middle and ending regions of a sentence. The speech signal between two pause segments is considered as a word (i.e., acoustic word). The first two words of a sentence are considered as starting words of a sentence, last two words are considered as ending words of a sentence and the remaining words are considered as the middle words of the sentence. To illustrate the analysis carried out

Table 1. Analysis of prosody parameters pitch, duration and energy based on their position in the sentences. N-neutral, A-anger

Speaker ID	Pitch						Duration						Energy					
	Starting		Middle		Ending		Starting		Middle		Ending		Starting		Middle		Ending	
	N	A	N	A	N	A	N	A	N	A	N	A	N	A	N	A	N	A
F-1	371	407	336	441	298	312	0.45	0.36	0.49	0.29	0.65	0.32	64	67	72	78	67	68
F-2	352	441	297	359	274	328	0.69	0.41	0.52	0.36	0.66	0.46	66	66	69	71	66	74
F-3	277	325	235	289	213	258	0.66	0.50	0.47	0.37	0.53	0.39	62	67	70	73	64	69
M-1	185	203	165	173	160	168	0.9	0.67	0.51	0.43	0.64	0.45	63	61	67	66	56	59
M-2	181	216	175	178	173	190	0.71	0.54	0.63	0.50	0.65	0.57	58	56	59	58	62	63
M-3	190	219	183	187	138	150	0.52	0.31	0.47	0.37	0.51	0.37	52	56	59	61	54	57

Table 2. Proposed prosody modification factors for converting the emotive utterance to neutral.

Female modification factors			
Modification factors	Starting words of a sentence	Middle words of a sentence	Ending words of a sentence
Pitch	0.82	0.9	0.84
Duration	1.36	1.27	1.45
Energy	1.14	0.94	1.3
Male modification factors			
Modification factors	Starting words of a sentence	Middle words of a sentence	Ending words of a sentence
Pitch	0.90	0.97	1.04
Duration	1.41	1.11	0.98
Energy	1.03	0.98	1.01

on various prosody parameters based on the position of a word in the sentence, data from six sentences (3 male and 3 female) are shown in the Table 1.

In Table 1, column 1 specifies the speaker ID. The variations in pitch, duration and energy for neutral and anger emotions in different specified positions of a sentence (i.e., starting middle and ending words of a sentence) are shown in Columns 2–7, 8–13 and 14–19. In Table 1, row 1 specifies the prosody parameters considered during the analysis, row 2 specifies the position of the word in the sentence. In row 3 of Table 1, 'N' refers to neutral and 'A' refers to anger emotion. The average pitch, duration and energy of a word in neutral sentence is normalized by using the corresponding values from anger sentence. To study the variation imparted in prosody due to the emotive state of the speaker. The obtained factors are called normalized factors and normalized factors >1 indicates that the specific prosody parameter is increasing compared to anger utterance and factors less than one indicates a decrease from anger. Major trend in the factors is observed (i.e.,>1 or <1) is computed. The average of all the values in that corresponding trend is considered as the modification factor. Modification factors obtained from above analysis is tabulated in Table 2. During this study, Modification factors for male and female speakers are obtained separately as presented in Table 2.

5 Proposed Method for Converting Emotive Speech to Neutral Speech

The algorithm to convert the emotive utterance to its neutral version is as follows.

- Compute the emotionally significant regions of an utterance using the algorithm mentioned in Sect. 3.1
- Select the modification factors from the Table 2 based on the position of the emotionally significant region of an utterance i.e., starting, middle or ending words of a sentence.
- Prosody of the speech signal in the detected emotionally significant regions is modified using the selected modification factors.
- A PSOLA [18] based prosody modification method is employed to modify pitch and duration.
- A simple amplitude scaling is used for energy modification.

The neutral utterance generated by modifying the prosody in emotionally significant regions of an utterance is termed as synthesized neutral utterance. The performance of the proposed conversion is evaluated by a subjective listening test. During this study, prosodic parameters (i.e., pitch, duration and energy) of anger sentence are modified using the PSOLA method [18]. The parameters of the neutral sentence are modified using the following expression
Neutral parameters = Anger parameters × modification factor
where parameters $\in \{Pitch, Duration, Energy\}$

6 Evaluation of the Proposed Method

To evaluate the performance of the proposed method a subjective listening test is performed. Evaluation is done by 25 human subjects comprising of students working in speech processing and other research areas in the institute. Actuarial recordings of neutral and anger files are played and later 50 synthesized neutral utterances (25 male and 25 female) are given and an opinion score for a degree of 5 is collected. Performance of the proposed method is compared against gross level modification method and tabulated in Table 4. Pitch and duration of the anger sentence is modified globally by using PSOLA [18] based prosody modification method. Scores of evaluation are taken based on the scale mentioned in Table 3. Mean opinion scores of the above evaluation are given in Table 4.

Table 3. Ranking used in perceptual test to judge the similarity of the synthesized neutral emotion with the natural neutral emotion.

Rating	Speech quality	Description for evaluating the converted neutral sentences
1	Very poor	Sounds exactly like anger
2	Poor	Sounds slightly different from anger
3	Good	sounds different from neutral
4	Very good	Sounds similar to neutral
5	Excellent	Sounds exactly like neutral

Table 4. Mean opinion scores (MOS) obtained to judge the similarity between the neutral utterance and the synthesized neutral utterance.

MOS scores	Conversion methods	
	Gross level	Proposed modification
	2.2	3.3

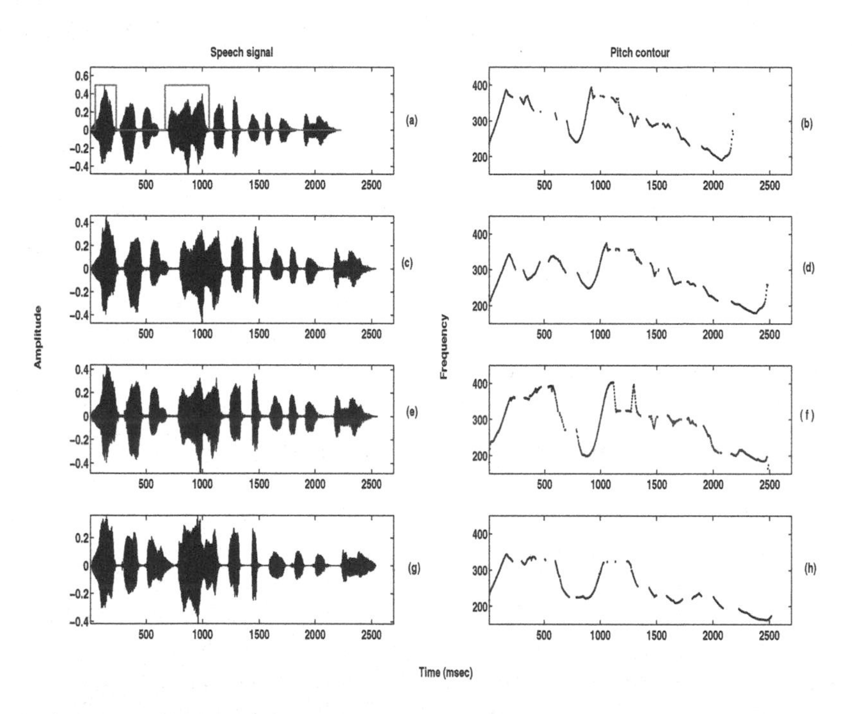

Fig. 2. Anger to neutral conversion for female speaker. In the figure, speech waveform, pitch contour of anger sentence is shown in (a & b), synthesized neutral by using gross level, and proposed conversion method is shown in (c & d), (e & f), respectively, and speech waveform, pitch contour natural neutral speech are shown in (g & h).

From Fig. 2(b), it can be noted that the pitch contour obtained by the gross level modification (Fig. 2(b)) still holds the gross pattern of natural anger sentence (Fig. 2(a)), but in a different pitch range. In the proposed method, prosody parameters of speech segments in emotionally significant regions are modified by selecting the modification factors specific to their positions in the sentence. The pitch contour obtained by converting the emotive sentence to neutral in the proposed method has the gross pattern similar to the natural neutral utterance. The similar trend can also be observed in Table 4, i.e., the MOS of the listening test described in Sect. 6.

7 Conclusion and Future Scope

During this work, we have generated a neutral version of an anger utterance. Exploring the physiological constraints of human speech production mechanism, the regions of speech that are highly influenced by the emotional state of the speaker is computed. These regions are termed as emotionally significant regions of an utterance. The variation imparted in prosody due to emotional state (anger) of the speaker is analyzed. Modification factors are computed with relevance to position of a word in a sentence is computed. Only the speech segments in emotionally significant regions are modified using the corresponding modification factors to generate the neutral version of an anger utterance. A subjective listening test is carried out to analyze effectiveness of the proposed modification is compared with the gross level modification. A significant improvement in the mean opinion score is observed in the proposed conversion method. Similar work can be extended to all the real life emotions. The effectiveness of the synthesized neutral speech is to be studied as a supplementary adaptation scheme to operate speech systems in emotive environments as a future task.

References

1. Alku, P.: Glottal wave analysis with pitch synchronous iterative adaptive inverse filtering. Speech Commun. **11**(2), 109–118 (1992)
2. Batliner, A., Steidl, S., Seppi, D., Schuller, B.: Segmenting into adequate units for automatic recognition of emotion-related episodes: a speech-based approach. Adv. Hum. Comput. Interact. **2010**, 3 (2010)
3. Cowie, R., Douglas-Cowie, E., Tsapatsoulis, N., Votsis, G., Kollias, S., Fellenz, W., Taylor, J.G.: Emotion recognition in human-computer interaction. Sig. Process. Mag. IEEE **18**(1), 32–80 (2001)
4. Gangamohan, P., Kadiri, S.R., Yegnanarayana, B.: Analysis of emotional speech at subsegmental level. In: INTERSPEECH, pp. 1916–1920 (2013)
5. Hansen, J.H., Bou-Ghazale, S.E., Sarikaya, R., Pellom, B.: Getting started with susas: a speech under simulated and actual stress database. In: Eurospeech, vol. 97, pp. 1743–1746 (1997)
6. Hansen, J.H., Womack, B.D.: Feature analysis and neural network-based classification of speech under stress. IEEE Trans. Speech Audio Process. **4**(4), 307–313 (1996)
7. Kadiri, S.R., Gangamohan, P., Yegnanarayana, B.: Discriminating neutral and emotional speech using neural networks. ICON (2014)

8. Koolagudi, S.G., Maity, S., Kumar, V.A., Chakrabarti, S., Rao, K.S.: IITKGP-SESC: speech database for emotion analysis. In: Ranka, S., Aluru, S., Buyya, R., Chung, Y.-C., Dua, S., Grama, A., Gupta, S.K.S., Kumar, R., Phoha, V.V. (eds.) IC3 2009. CCIS, vol. 40, pp. 485–492. Springer, Heidelberg (2009)
9. Krothapalli, S.R., Yadav, J., Sarkar, S., Koolagudi, S.G., Vuppala, A.K.: Neural network based feature transformation for emotion independent speaker identification. Int. J. Speech Technol. **15**(3), 335–349 (2012)
10. Murty, K.S.R., Yegnanarayana, B.: Epoch extraction from speech signals. IEEE Trans. Speech Audio Lang. Process. **16**(8), 1602–1613 (2008)
11. Murty, K.: Significance of excitation source information for speech analysis. Ph.D. thesis, Department of Computer Science and Engineering, Indian Institute of Technology Madras (2009)
12. Ortony, A., Clore, G.L., Collins, A.: The cognitive structure of emotions. Cambridge University Press, Cambridge (1990)
13. Raja, G.S., Dandapat, S.: Speaker recognition under stressed condition. Int. J. Speech Technol. **13**(3), 141–161 (2010)
14. Reynolds, D.A., Quatieri, T.F., Dunn, R.B.: Speaker verification using adapted gaussian mixture models. Digital Signal Process. **10**(1), 19–41 (2000)
15. Schuller, B., Stadermann, J., Rigoll, G.: Affect-robust speech recognition by dynamic emotional adaptation. In: Proceedings of Speech Prosody. Citeseer (2006)
16. Stevens, K.N.: Acoustic Phonetics, vol. 30. MIT press, Cambridge (2000)
17. Tao, J., Kang, Y., Li, A.: Prosody conversion from neutral speech to emotional speech. IEEE Trans. Audio Speech Lang. Process. **14**(4), 1145–1154 (2006)
18. Valbret, H., Moulines, E., Tubach, J.P.: Voice transformation using psola technique. In: 1992 IEEE International Conference on Acoustics, Speech, and Signal Processing ICASSP-92, vol. 1, pp. 145–148. IEEE (1992)
19. Vlasenko, B., Philippou-Hübner, D., Prylipko, D., Böck, R., Siegert, I., Wendemuth, A.: Vowels formants analysis allows straightforward detection of high arousal emotions. In: 2011 IEEE International Conference on Multimedia and Expo (ICME), pp. 1–6. IEEE (2011)
20. Vlasenko, B., Prylipko, D., Wendemuth, A.: Towards robust spontaneous speech recognition with emotional speech adapted acoustic models. In: Poster and Demo Track of the 35th German Conference on Artificial Intelligence, KI-2012, pp. 103–107. Citeseer, Saarbrucken, Germany (2012)
21. Vlasenko, B., Wendemuth, A.: Location of an emotionally neutral region in valence-arousal space: two-class vs. three-class cross corpora emotion recognition evaluations. In: 2014 IEEE International Conference on Multimedia and Expo (ICME), pp. 1–6. IEEE (2014)
22. Vuppala, A.K., Kadiri, S.R.: Neutral to anger speech conversion using non-uniform duration modification. In: 2014 9th International Conference on Industrial and Information Systems (ICIIS), pp. 1–4. IEEE (2014)
23. Vuppala, A.K., Limmayya, J., Raghavendra, G.: Neutral speech to anger speech conversion using prosody modification. In: Prasath, R., Kathirvalavakumar, T. (eds.) MIKE 2013. LNCS, vol. 8284, pp. 383–390. Springer, Heidelberg (2013)
24. Vydana, H.K., Kadiri, S.R., Vuppala, A.K.: Vowel-based non-uniform prosody modification for emotion conversion. Circuits Syst. Signal Process. **34**, 1–21 (2015)
25. Vydana, H.K., Kumar, P.P., Krishna, K., Vuppala, A.K.: Improved emotion recognition using GMM-UBMs. In: 2015 International Conference on Signal Processing And Communication Engineering Systems (SPACES), pp. 53–57. IEEE (2015)
26. Yang, B., Lugger, M.: Emotion recognition from speech signals using new harmony features. Signal Process. **90**(5), 1415–1423 (2010)

Spoken Document Retrieval: Sub-sequence DTW Framework and Variants

Akshay Khatwani[1], Komala Pawar[1], Sushma Hegde[1], Sudha Rao[1], Adithya Seshasayee[2], and V. Ramasubramanian[1(✉)]

[1] PES Institute of Technology - Bangalore South Campus (PESIT-BSC), Bangalore, India
{akshaykhatwani,komalapawar15,hegdesushma6,sudhalrao1993}@gmail.com, v.ramasubramanian@pes.edu

[2] University of California, San Diego, USA
aseshasa@eng.ucsd.edu

Abstract. We address the problem of spoken document retrieval (alternately termed content-based audio-search and retrieval), which involves searching a large spoken document or database for a specific spoken query. We formulate the search within the sub-sequence DTW (SS-DTW) framework proposed earlier in literature, adapted here to work on acoustic feature representation of the database and spoken query term. Further, we propose several variants within this framework, such as (i) path-length based score normalization, (ii) clustered quantization of acoustic feature vectors for fast search and retrieval with invariant performances and, (iii) phonetic representation of the database and spoken query term, derived from ground-truth annotation as well as HMM based continuous phoneme recognition. We characterize the performance of the proposed framework, algorithms and variants in terms of ROC curves, EER and time-complexity and present results using the TIMIT database with annotated spoken sentences from 400 speakers.

Keywords: Spoken document retrieval · Audio-search · Sub-sequence DTW · Quantized search · Phonetic audio-search

1 Introduction

The problem of 'spoken document retrieval' can be defined as searching a large speech database for one or more occurrences of a spoken query term (a word or a phrase, that is typically very short when compared to the database being searched). This problem can be seen as an 'audio-search' problem, with a wider applicability. There is a growing volume of audio-visual content in various domains, e.g. world-wide web (You-tube), multi-media (TV, movie-content, smart-phone services), broadcast content management, on-line lecture material,

A. Seshasayee—Author carried out this work as Research Associate at PESIT-BSC, Bangalore.

R. Prasath et al. (Eds.): MIKE 2015, LNAI 9468, pp. 297–311, 2015.
DOI: 10.1007/978-3-319-26832-3_29

meeting capture, voice-mail browsing, telephone conversations, forensic search on surveillance data, analytics on call-center dialogs, sports gisting, audio-guided audio-visual data search etc. This necessitates effective mechanisms of accessing and organizing such data in a manner that makes it possible to handle large and complex volumes of audio-visual data for various purposes that are unique and idiosyncratic to its particular domain [1].

In this paper, we consider the 'sub-sequence DTW' algorithm [2] (SS-DTW) for spoken document retrieval and first introduce its basic formulation. We then examine its applicability to the audio-search problem on a large speech database, namely TIMIT [3], propose several variants of the framework, namely, path-length based score normalization, VQ-indexing and phoneme-indexing, and examine their relative performances through ROC curves and EERs.

2 Overview

Here, we provide an overview of the overall framework employed in this paper for spoken-document retrieval, centered on the sub-sequence DTW (SS-DTW) based search, highlighting its components and the variants proposed here.

Figure 1 shows the block diagram of the entire system. This has two main parts, namely, off-line database indexing and the on-line query retrieval. Off-line indexing involves converting the raw speech database (a large database, being searched) into a sequence of indices which can be further searched for a given query, also represented by the same set of indices. On-line indexing involves retrieving all the best matches of a given query from the indexed database.

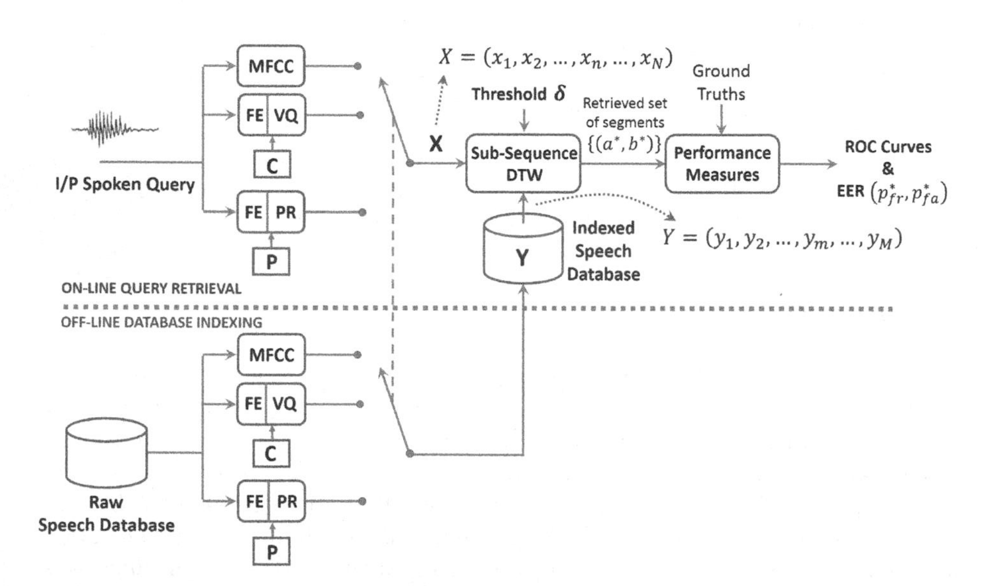

Fig. 1. Schematic of the SS-DTW framework with three indexing variants

In the following, we describe each of the components making up the framework and system shown in Fig. 1 forming the central part of this paper.

2.1 Database

Here, we describe the database and queries employed in this paper, to characterize the different variants of the SS-DTW algorithm used to perform audio-search for a spoken document retrieval task. We have used the TIMIT database [3] for this work. This is a phonetically annotated (segmented and labeled) English database, spoken by 630 speakers, each speaking 10 sentences, of which 2 sentences are common across all speakers. These two sentences are labeled 'sa1' and 'sa2'. Specifically, in our work, in order to examine various issues related to the performance of the algorithm, we have focused on the 'sa1' sentence 'She had your dark suit in greasy wash water all year' made of 11 words. We used the 'sa1' from 400 different speakers, to make up the database sequence Y to be searched, i.e., each sentence is converted to each of the three indexing representation described below in Sect. 2.2 (e.g. MFCC feature vector sequence) and as shown in the lower part of Fig. 1 constituting the 'off-line database indexing' part, forming 400 such sequences, together treated as Y. We used 'sa1' sentences spoken by 30 speakers, all different from the 400 speakers comprising Y, to provide query words X that can then be searched and retrieved from Y, where any X so chosen is guaranteed to occur 400 times in all of Y. Note that each of the query words is subjected to the same indexing as the database, and as shown in the upper part of Fig. 1 constituting 'on-line query retrieval'. This thus serves the purpose of (i) making sure the inter-speaker variability is present in such a search, which is likely to be the practical scenarios (as outlined in Sect. 1), and (ii) a large number of 'multiple' occurrences of each query is also ensured, so as to provide statistically significant measures for (False-alarm, False-rejection) based ROC curves to characterize the performance of the algorithm.

2.2 Indexing Variants

Indexing is a primary mechanism which controls various aspects of a search and retrieval system, such as the time resolution and granularity at which the data is represented, the search efficiency in terms of accuracy of retrieval, search complexity, off-line indexing complexity, storage etc. We consider an implicit indexing scheme, by way of retaining the database in the form of acoustic feature vector representation, and two explicit indexing schemes, where such an acoustic feature vector representation is further transformed to a cluster index (by means of vector quantization) or a phone index (by means of HMM based continuous phoneme recognition). In each of the indexing scheme, it should be noted that both the database (being searched) as well as the query should be transformed to the same indexing representation. Alternative strategies, namely, cross-index or cross-granular search can be considered too, but are not part of this paper.

As shown in Fig. 1, we consider three types of indexing:

1. **MFCC (mel-frequency-cepstral-coefficents):** Here, the input speech is subjected to a short-time spectral analysis and feature extraction, yielding MFCC feature vectors [4] for every 20 ms frames. Such a representation of the database and the query actually does not correspond to an explicit indexing scenario, since such a representation is the most basic and minimal means of accessing the speech information, via acoustic feature representation. In this paper, we use MFCCs of dimension 12 (without the energy co-efficient) obtained using 20 ms frame sizes, without overlap.
2. **Vector quantization (VQ):** Here, the input speech is first represented by suitable spectral features (via a front-end feature-extraction, shown as FE in Fig. 1, which could be the MFCCs themselves), followed by a vector quantization of each frame (i.e., the feature vector corresponding to a frame) using a VQ codebook $\mathbf{C}$, made of K code vectors (also MFCC vectors). This corresponds to an explicit indexing step, where each frame is represented by an index drawn from the VQ codebook $\mathbf{C}$. By this, we aim to examine the effectiveness of the granularity of the VQ indexing, by means of controlling K, the number of indices used to quantize the MFCC feature vectors. In this paper, we obtain codebooks $\mathbf{C}$ of sizes $K = 4, 8, 16, 32, 64, 256, 1024, 4096$ from the 400 sa1 sentence database using standard K-means algorithm with random initialization.
3. **Phonetic representation:** Here, the input speech is first represented by suitable spectral features (as in the VQ case above, which could be MFCCs) via a front-end feature-extraction (FE), followed by a phoneme-recognition (PR) using phoneme models ($\mathbf{P}$), i.e., acoustic models, such as the hidden Markov model (HMM) for each of the phones in the language (or database). The phoneme models are derived by HMM training (Baum-Welch training) from the phone data in the database; the phoneme-recognition (PR) is done by Viterbi decoding the feature-vector sequence of the database (in off-line indexing) and the query speech (in on-line retrieval) to yield a sequence of phoneme indices that make up the underlying speech.

   In this paper, we use a phone-set of 61 phonemes making up the entire TIMIT database. Each phone is represented by a context-independent HMM with 3 emitting states, trained from the 400 'sa1' sentences database, each of which is further converted to a phone sequence by a continuous phoneme recognition using the trained HMM acoustic-models, using Viterbi decoding. The accuracy of the phoneme recognition is crucial to the subsequent accuracy of the search-retrieval - i.e., higher the phoneme recognition accuracy, higher the retrieval accuracy. The phoneme recognition accuracy is measured in terms of word-error-rate (%WER) obtained as in a conventional automatic speech recognition for words (but here, applied on phone sequences) using string-edit (or Levenshtein distance) to measure the substitutions, insertions and deletions, used in the calculation of the %WER. Lower WER correspond to higher phoneme recognition, and the WER is implicitly controlled by the number of templates per phone used in the training of the phone HMMs (varying from 50 to 200), and the number of mixtures per state (in the Gaussian mixture models, or GMMs, forming the state-observation densities), varying from 1 to 5.

Note that the objective of examining indexed-representation (VQ or PR) as above, are twofold: (i) use of a VQ or phoneme-recognition based indexing allows reducing the complexity of search and retrieval by SS-DTW in a stage termed 'cost-matrix computation', by merely loading the values of the cost-matrix by a look-up from a $K \times K$ distance-matrix of inter-index distances of the VQ code-book of size K or the phoneme indices of size 61, (ii) to explore whether a coarser granularity representation (i.e., coarser than the fine acoustic representation in terms of MFCC vector at every 20 ms resolution) of the speech database and query can be adequate, both in terms of retrieval performance and complexity, and possibly even more optimal in the sense of being able to offer better retrieval performance, owing to such indexed representations smoothing out the high spectral variability inherent in a MFCC-kind of feature-level representation and thereby retain the minimally 'essential' information required for retrieval (e.g. phonetic content of the speech).

2.3 Sub-sequence DTW

The other components in Fig. 1 mainly consist of the sub-sequence DTW algorithm, which takes inputs as the query sequence $X = (x_1, x_2, \ldots, x_n, \ldots, x_N)$, the indexed database $Y = (y_1, y_2, \ldots, y_m, \ldots, y_M)$ and a search parameter threshold δ, and retrieves all instances (i.e., sub-sequences or speech segments) marked by a set of (begin, end) frames $\{(a^*, b^*)\}$ in the database sequence Y, that match within a cost of δ between X and each retrieved segment $Y(a^*, b^*) = (y_{a^*}, y_{a^*+1}, \ldots, y_{b^*})$ in terms of dynamic time-warping distance. This algorithm is described in detail in Sect. 3.

2.4 Performance Characterization

Such a set of retrieved best matches, for a given δ is further evaluated with the actual ground truths of the query present in Y to derive the retrieval performance measure of number of false rejections (i.e., ground truth segments not detected) and number of false alarms (i.e., non ground truth segments detected), further represented as the (probability of false rejects, probability of false alarms), denoted as $(p_{fr}(\delta), p_{fa}(\delta))$ for a given δ. Such $(p_{fr}(\delta), p_{fa}(\delta))$ obtained by varying δ over the range of matching scores possible, yields the ROC (receiver operating characteristics) curve, which helps characterize the overall performance of the retrieval system, as well as the EER point (i.e., the equal-error-rate) corresponding to the EER-threshold δ^* for which $p_{fr}(\delta^*) = p_{fa}(\delta^*)$. These completely characterize the performance of the retrieval system, and can further help in comparing different variants of the system, such as are proposed here (for e.g. the different indexing representations, path normalization, score normalization etc.) and also to derive an optimal non-EER performance by choosing appropriate 'operating points' corresponding to any desired $(p_{fr}(\delta), p_{fa}(\delta))$ along the ROC curve. More details of these measures with reference to how δ is used in the SS-DTW algorithm are given in Sect. 5.

3 Sub-sequence DTW

In this section, we describe in detail the sub-sequence DTW (dynamic time warping) formulation and algorithm [2], which we use here for performing an 'audio search' given a query in the form of a speech fragment (e.g. a word or phrase). The sub-sequence DTW algorithm is tailored to find one or more occurrences of a short sequence of features (e.g., a sequence of MFCC feature vectors derived from speech signal [4]) in a relatively much longer sequence of features (e.g., also a sequence of MFCC feature vectors). Let $X = (x_1, x_2, \ldots, x_n, \ldots, x_N)$ be the short feature sequence of length N and let $Y = (y_1, y_2, \ldots, y_m, \ldots, y_M)$ be the longer feature sequence of length M, with $M >> N$. In the context of audio-search such as we require here, the sequence X corresponds to the 'spoken query' and the sequence Y corresponds to the 'speech database' that needs to be searched to find and retrieve the potentially multiple occurrences of the query. The basic SS-DTW algorithm finds a sub-sequence $Y(a^*, b^*) = (y_{a^*}, y_{a^*+1} \ldots, y_{b^*})$, with $1 \leq a^* \leq b^* \leq M$ which is the best match to the query sequence X in the sense of minimizing the dynamic-time warping (DTW) distance between X and every possible sub-sequence in Y. This can be given as,

$$(a^*, b^*) = \arg \min_{(a,b):1 \leq a \leq b \leq M} (DTW(X, Y(a:b))) \tag{1}$$

where, $DTW(X, Y(a:b))$ refers to the optimal time-aligned distance (or cost) between X and any arbitrary sub-sequence $Y(a:b)$ in Y.

The above solution in Eq. (1) is obtained by a dynamic programming (DP) based procedure which finds the optimal path that aligns X to the optimal sub-sequence $Y(a^*, b^*)$ as given below in Steps A and B. Figure 2 shows the trellis space where these steps are computed, namely, the initialization, recursions, and the termination step involving the storing of the $\Delta(b)$ profile.

A. Forward-pass steps

A1. Initialization

A1.1 Compute cost-matrix (CM) as appropriate for the different representation of the database and query (i.e., indexing variants - MFCC, VQ and PR)

$$d(x_n, y_m), \quad n = 1, \ldots, N, \quad m = 1, \ldots, M \tag{2}$$

A1.2 Initialize accumulated cost matrix $D(n, m)$

$$D(n, 1) = \sum_{k=1}^{n} d(x_k, y_1), \qquad n = 1, \ldots, N \tag{3}$$

$$D(1, m) = d(x_1, y_m), \qquad m = 1, \ldots, M \tag{4}$$

A2. Recursion (for $m = 2, \ldots, M$ and $n = 2, \ldots, N$)

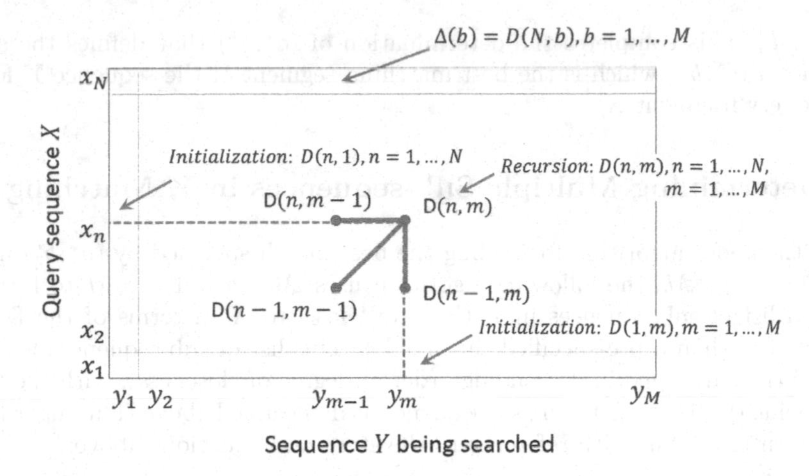

Fig. 2. Schematic of the search trellis and recursions in the SS-DTW framework

$$D(n,m) = \min\{D(n,m-1), D(n-1,m-1), D(n-1,m)\} + d(x_n, y_m) \quad (5)$$

$$\psi(n,m) = \arg \min_{(i,j)\in\{(n-1,m),(n-1,m-1),(n,m-1)\}} D(i,j) \quad (6)$$

Note here that the above back-pointer in Eq. (6) stores the co-ordinate tuple (i, j) which minimizes $D(i, j)$ from among the three possible 'candidate' predecessor co-ordinates, as specified by a given local continuity constraint [4].

A3. Termination (Store $D(N, b), b = 1, \ldots, M$)

$$\Delta(b) = D(N,b), \qquad b = 1, \ldots, M \quad (7)$$

B. Steps for determining (a^*, b^*)

B1. Finding $b^* \in [1 : M]$

$$b^* = \arg \min_{b\in[1,M]} \Delta(b) \quad (8)$$

B2. Finding a^* by backtracking

First find the optimal path p^* ending at b^* (as obtained in the forward left-to-right pass of Steps A1, A2, A3 and B1 above) by backtracking using the back-pointer function $\psi(n, m)$ in Eq. (6), as follows:

Let the optimal path be a sequence of co-ordinates of the matrix $D(n, m)$, given by $p^* = (p_1, p_2, \ldots, p_l, \ldots, p_L)$, derived by backtracking from the last co-ordinate in the path $p_L = (N, b^*)$ as,

$$p_{l-1} = \psi(p_l), l = L, \ldots, 2 \quad (9)$$

Once this optimal warping path p^* is obtained, the start of the sub-sequence $a^* \in [1 : M]$ is obtained as the maximal index such that $p_l = (a^*, 1)$ for some

$l \in [1 : L]$. This completes the determination of (a^*, b^*) that defines the subsequence $Y(a^*, b^*)$ which is the best matching segment of the sequence Y for a given query fragment X.

4 Determining Multiple Sub-sequences in Y Matching X

Given the above algorithm for finding the best match specified by (a^*, b^*) from $\Delta(b), b = 1, \ldots, M$, the following algorithm uses $\Delta(b), b = 1, \ldots, M$ to further locate a list of subsequences in Y that are 'close' to X in terms of the DTW distance less than a pre-specified threshold δ. This list of sub-sequences is rank ordered (i.e., arranged in descending order of degree of closeness), with the first entry being the 'best' matching subsequence as determined above. This algorithm is as given in [2], but with reference to the steps and equations above:

Input: $X = (x_1, x_2, \ldots, x_n, \ldots, x_N)$ query sequence
$Y = (y_1, y_2, \ldots, y_m, \ldots, y_M)$ database sequence
Output: Ranked list of all subsequences of Y that have a DTW distance to X below a threshold δ.

1. Initialize the ranked list to be the empty list.
2. Compute the accumulated cost matrix $D(n, m), m = 1, \ldots, M, n = 1, \ldots, N$ as in Eqs. (3), (4) and (5).
3. Determine the distance function $\Delta(b), b = 1, \ldots, M$ as in Eq. (7).
4. Determine the minimum $b^* \in [1 : M]$ as in Eq. (8) from the current $\Delta(b)$ (that could be in the modified form from Step (8) below in the iteration).
5. If $\Delta(b^*) > \delta$ then terminate the procedure.
6. Compute the corresponding start index $a^* \in [1 : M]$ of the optimal warping path ending at b^* using **Step B2**.
7. Extend the ranked list by the subsequence $Y(a^*, b^*)$.
8. Set $\Delta(b) = \infty$ for all b within a suitable neighborhood of b^*
9. Continue with Step (4).

Note that the above algorithm retrieves a ranked list of R 'hits' in the form of $\{(a_i^*, b_i^*), i = 1, \ldots, R\}$, such that the corresponding sub-sequences (or speech segments) in Y, given by $Y(a_i^* : b_i^*), i = 1, \ldots, R$ are the R-best solutions of Eq. (1), with their corresponding DTW aligned distances with the query X given by $DTW(X, Y(a_i^* : b_i^*))$ in Eq. (1) are in increasing order for $i = 1, \ldots, R$.

The above steps are illustrated in Fig. 3 with an actual example of searching a long speech sentence 'She had your dark suit in greasy wash water all year' (which is the so-called sa1 sentence in the TIMIT database, described earlier in Sect. 2.1) constituting the sequence Y. The query X is the word 'greasy' spoken by a speaker different from the speaker of the database sentence. Both sequences are MFCC feature vector sequence. Here, the top panel shows the local distance matrix $d(x_n, y_m), n = 1, \ldots, N, m = 1, \ldots, M$. From this, the potential warping path - indicating the match between the query 'greasy' and the corresponding occurrence of this word in the sequence Y being searched - can be visualized as the low distance valley (blue region from x-axis 73rd frame

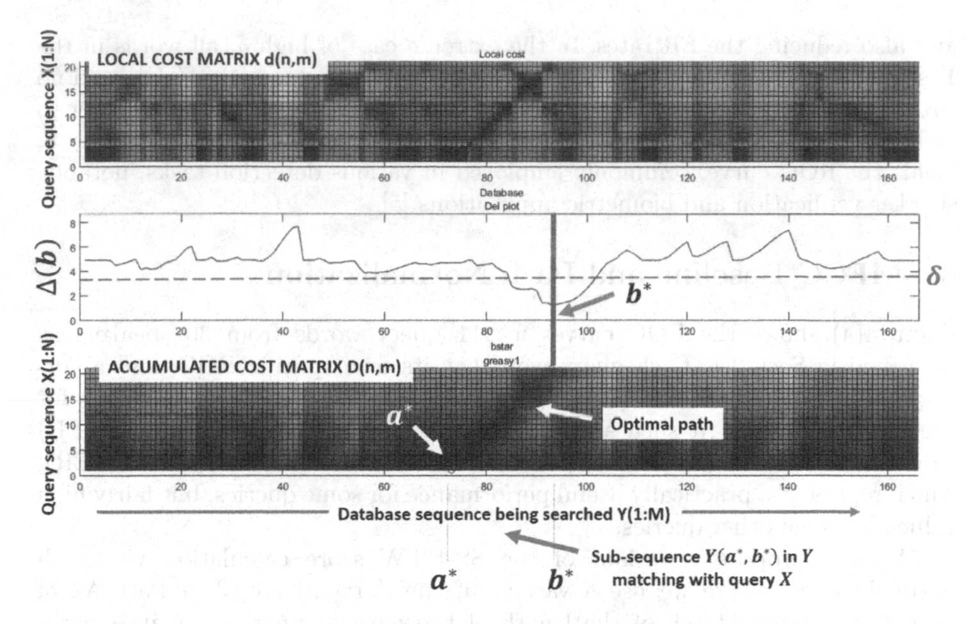

Fig. 3. SS-DTW search on speech MFCC sequence and associated parameters (Color figure online)

to 93rd frame). The bottom panel shows the accumulated distance matrix $D(n, m), n = 1, \ldots, N, m = 1, \ldots, M$ as derived by Eqs. (3), (4) and (5). Here again, the optimal warping path corresponding to the matching between X and $Y(73 : 93)$ can be visualized. The middle panel shows the $\Delta(b), b = 1, \ldots, M$ profile as derived in Eq. (7) and which forms the basis for further determining (a^*, b^*) as in Steps. B1 and B2. The location of b^* at frame number 93 can be clearly noted from this panel (i.e., the frame at which $\Delta(b)$ is minimum). Once b^* is determined, Step B2 obtains a^* and this completes the search for the one single occurrence of the word 'greasy'. Multiple occurrences can be further derived by following steps 1–9 above.

5 Performance Measures

We now present the performance measure we have used to characterize the algorithm's effectiveness - this is essentially the Receiver Operating Characteristics (ROC), made of False-alarm and False-reject probabilities. In order to derive the ROC characteristics, we note that the SS-DTW algorithm needs to have a controlling parameter that allows trading off of the False-alarm (FA) and False-reject (FR) probabilities. We propose to use the threshold δ in the algorithm (for detecting multiple occurrences of the query word) as such a controlling parameter. In order to understand this, consider the role that δ plays in reporting the ranked-list in Sect. 4. It can be easily seen that lower the δ, lower is the FA rate and higher is the FR rate, i.e., we get no false 'dips' in the $\Delta(b)$ profile below the 'low δ'. As we increase δ from some lower limit to a upper limit, the incidences of 'dips' in $\Delta(b)$ (minimas) below δ increases, thereby increasing the FA rates, and also reducing the FR rates. In the extreme case of high δ, all words in the Y sequence could be reported as 'query', thereby leading to a 100 % FA, and no FR at all. This is precisely the behavior we expect in a controlling parameter to yield the trade-off between FA and FR. Plotting FR vs FA, parameterized by δ, yields the ROC curve, commonly employed in various detection tasks, notably, speaker-verification and biometric applications [5].

6 MFCC Baseline and Path Normalization

Figure 4(a) shows the ROC curves for 11 query words from 30 speakers as described in Sect. 2.1. Each query word has its own specific ROC, and a corresponding Equal-error-rate (EER) point with an associated δ, termed δ^i_{EER} for the ith query word. At such an EER point, it can be seen that the (FA, FR) values are in the range of 10–40 % over the query words (4 words $<$ 20 % EER), which represent a practically useful performance for some queries, but fairly high values for some other queries.

We also propose a variant of the SS-DTW score calculation via 'path normalization', where we use a variant of the 'forward steps' in Part A2 of Sect. 3, by keeping track of the length of the path at any (n, m), in a path-length variable $L(n, m)$ which is initialized to $L(n, 1) = n, n = 1, \ldots, N$ and $L(1, m) = 1, m = 1, \ldots, M$ and updated (along with $D(n, m)$ in Eq. (5)), as $L(n, m) = L(n, m) + 1$ and finally normalizing the score in Step. A3 (Termination), as $\Delta(b) = D(N, b)/L(N, b), b = 1, \ldots, M$. We expect such a path-length normalization of the score to become invariant and hence insensitive to the length of the query, thereby yielding $\Delta(b)$ with uniform dynamic range across queries, and hence provide better ROCs. Figure 4(b) shows the ROCs for this path-length normalized case, and it can be seen that several query words now have lowered ROCs (6 words $<$ 20 % EER).

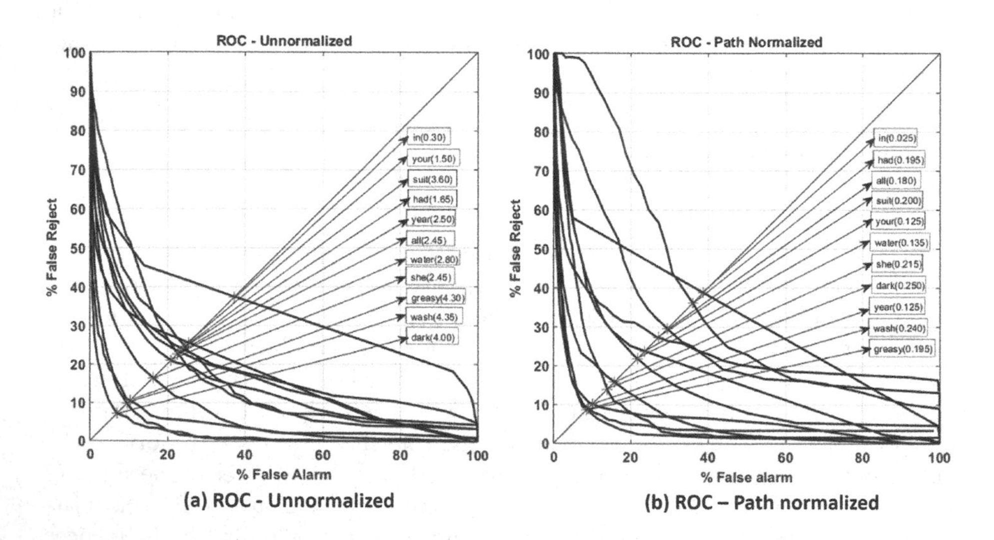

Fig. 4. ROCs for MFCC: (a) unnormalized and (b) path-length normalized cases

We also examined 'score-normalization', by making the dynamic range of $\Delta(b)$ to be in the range of $(0, 1)$ by normalizing it over each of the 'sa1' database sentence, so that the retrieval sees a uniform variability in $\Delta(b)$ across database speakers. A combination of the above path-length based normalization and score-normalization was also considered. EER results of these are reported in Fig. 5(b).

7 MFCC and VQ-indexed Performances

As described in Sect. 2.2, we now present results for the MFCC baseline and VQ-indexed search and retrieval. First, we show in Fig. 5(a), the ROCs for VQ-indexing with the VQ codebook size varying as $K = 4, 8, 16, 32, 64, 256, 1024$ and 4096, along with the ROC for MFCC as a baseline comparison for the word 'greasy'. Figure 5(a) shows the ROCs for these cases, and the following can be noted: (i) the performance of VQ-indexing is practically same as that of MFCC for most K (16-4096), with all these ROCs bunching together in the range of (10–15 % EERs), but with the interesting possibility that all VQ codebooks of sizes other than $K = 4, 8$ can offer even better performance than MFCC (thick red dash-dot curve). This may sound counter-intuitive at first, but as already pointed out in Sect. 2.2, is easily understood considering that such codebook sizes roughly corresponding to the number of phonemes in the language/database, actually preserve the essential phonetic information needed for accurate match and retrieval and inherently eliminates the high spectral variability present in a short-time (20 ms) framesize MFCCs, which can lead to noisy matches. Hence, it can be concluded that even small codebook sizes of $K = 16, 32, 64$ are adequate to derive a good performance, even while offering complexity reductions, as will be discussed further. However, very small codebook sizes $K = 4, 8$ quantize the data severely, and results in loss of spectal resolution and information and leads to increased confusability between the query and non-query parts of the database, and consequently leads to poor ROCs with higher EERs than MFCCs.

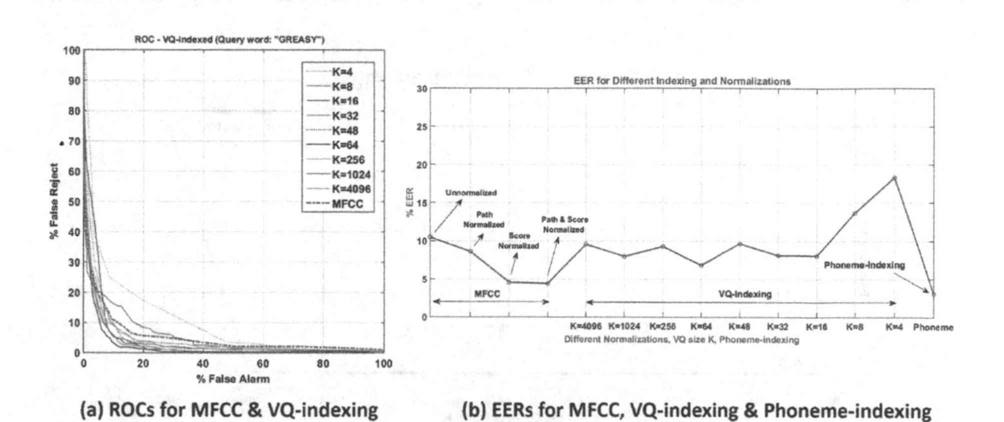

Fig. 5. ROCs for VQ-indexing: (a) ROCs for varying K and MFCC, (b) EERs for different cases - MFCC, VQ, Phoneme-indexed (Color figure online)

Figure 5(b) shows the EER corresponding to Fig. 5(a), for various cases: (i) MFCCs (unnormalized, path normalized, score normalized and both path & score normalized), (ii) VQ-indexing for various K, and (iii) Phoneme-indexing (to be discussed in next section). It can be noted that for the MFCCs, there is a progressive performance improvement across normalization, with the score-normalized scenario being the best. Other observations made in the preceding paragraph can be noted in this EER plot too.

As indicated in Sect. 2.2, the other prime motivation to use a VQ-indexing is to reduce the complexity of on-line search and retrieval. This is made possible by noting that, the individual frames of query X and Y, namely, x_n and y_m are now quantized and indexed by a code vector index of the VQ codebook $\mathbf{C}$ for any given codebook size K. By this, it becomes possible to perform the 'cost-matrix filling' operation in the initialization step of SS-DTW in Eq. (2), using a look-up table in the form of an off-line pre-computed 'distance matrix' of size $K \times K$ having the inter-code vector distances in the VQ codebook $\mathbf{C}$ of size K. This can lead to significant reduction in on-line computational complexity of SS-DTW, considering that Step. A1.1 (filling the cost-matrix) is the most computation intensive step among the Steps A and B in SS-DTW (Sect. 3). Note that the cost of further steps in SS-DTW (as in Steps A.2, A.3 and B) remains the same as for the MFCC (i.e., without VQ-indexing).

We show in Fig. 6, the relative complexity (in secs) between the MFCC-based-search and VQ-indexing of the two steps (i) on-line cost-matrix $d(x_n, y_m)$ computation and (ii) finding the set of multiple matching intervals $\{(a^*, b^*)\}$ using Steps 1–9 in Sect. 4. It can be noted that the cost-matrix filling cost is practically constant, and very low, for all K in VQ-indexing, considering that the filling cost is independent of the size of the look-up distance matrix. On the other hand, the cost-matrix computation for MFCC is relatively very high (about 20 times higher) than the cost of VQ-index based filling. Moreover, it can also be noted that the computational cost of finding the multiple matching intervals is very low (and the same) for both VQ-indexing and MFCC-representation,

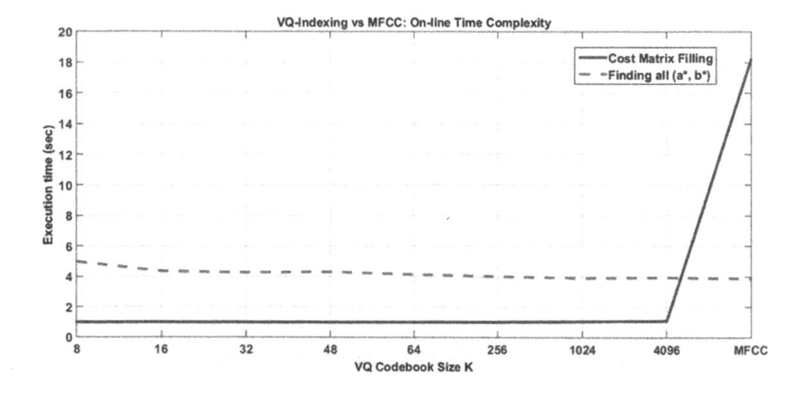

Fig. 6. On-line complexity of MFCC and VQ-indexing

considering that these involve only recursions and inexpensive comparison operations, and happen subsequent to the cost-matrix filling. Thus, we conclude that VQ-indexing is advantageous over MFCC-representation, with respect to considerations of both performance and computational complexity.

8 Phoneme-Indexed Performances

As discussed in Sect. 2.2, phoneme-indexing realizes a coarse granularity in indexing even while retaining the essential information needed for efficient retrieval. The various phone-model (HMM) training scenarios were already described in Sect. 2.2. Here, we first show in Fig. 7, the ROCs for 9 different short phrase queries formed from different 3 contiguous words in the 'sa1' sentences of TIMIT. Figure 7(a) shows the ROCs using phoneme-indexing obtained from the ground-truth, thereby not incurring any phoneme-recognition error. The remarkably low EERs (in the range of 0–10 %) and good ROCs can be noted. Figure 7(b) shows the ROCs with actual phoneme-recognition (for one of the training scenarios, namely, with 200 templates/phone and 5 mixtures/state) with a word-error-rate (WER) of 34 %. It can be noted that even such a reasonably high WER does not effect the ROC (or EER) performance of the final retrieval, in the sense that the EERs in Fig. 7(b) are in the range of 10–15 %, being only ∼5–10 % higher than those in Fig. 7(a). This shows the inherent robustness of the retrieval algorithm to errors in phoneme-transcription of the database, which makes the phoneme-indexing approach a viable alternative. Note also that the EER of the 'phoneme-indexing' case (for word 'greasy') shown in Fig. 5(b) is the lowest among all variants of MFCC-representation and VQ-indexing.

In order to isolate the effect of WER on the final ROC performance (and associated EER), we show in Fig. 8, the ROCs of 4 cases: (i) ground truth phoneme indexing and (ii) three cases of phoneme-indexing with actual phoneme recognition using HMMs trained under different training conditions, namely,

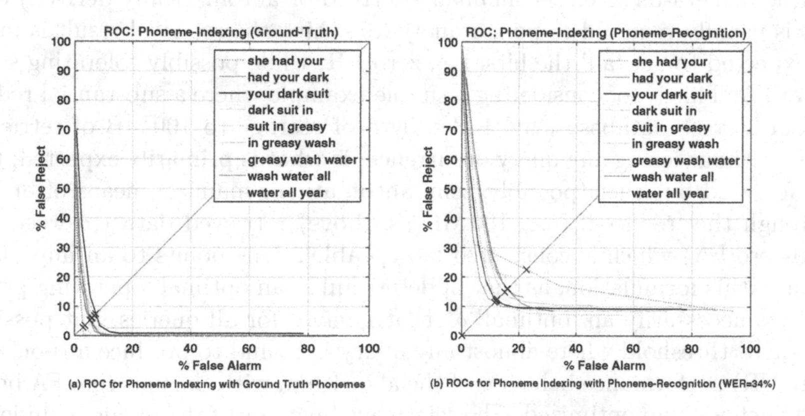

Fig. 7. ROCs for Phoneme-indexing: (a) Ground truth phoneme indexing, (b) With phoneme-recognition (34 % WER)

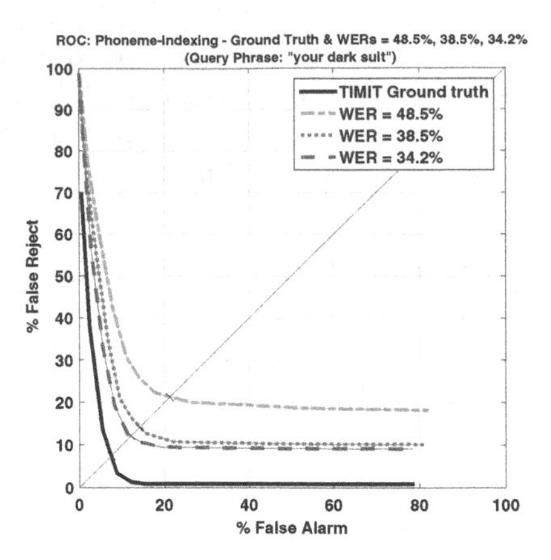

Fig. 8. Phoneme-indexing ROCs: ground truth and phoneme-recognition with 3 WERs

number of training occurrences/phoneme (T) and number of mixtures/state (Q), yielding %WERs such as 48.5 % (T=100, Q=1), 38.5 % (T=100, Q = 3) and 34.2 % (T= 200, Q = 5). Clearly, as %WER decreases, the performance reaches the ground-truth baseline, and more importantly, the relative robustness of the system, over the range of WER (ground-truth, with WER = 0 % to WERs of 34 % to 48 %) can be noted. The implication of such a robustness is further highlighted in Sect. 9.

9 Discussion

An interesting scenario arises in the case of audio search-retrieval problems in practical cases. Considering that the need for such a solution is felt in searching through very large databases (e.g. in broadcast archive management) of untagged (i.e. without meta-data, either manually derived or automatically derived) content, it is usually required to post a query for which the retrieval result is minimally expected to get 'all' the hits, i.e., zero FR, while possibly tolerating some extent of FA. This is so, considering a simple example where a substantial reduction from a large database (say, 1 Petabyte of storage to 100 MB of retrieved content) without losing any query occurrence, is what is primarily expected; this can be followed by other (possibly, semi-automatic or manual) means of browsing through the 'reduced' (e.g. 100 MB as above) retrieved data (to weed out the non-zero FA) which is considered 'acceptable'. This points to an important problem in this formulation, as that of determining an optimal 'operating point' that is not necessarily an 'optimal' δ^* that is good for all queries, but possibly a 'non-EER' threshold where almost any query is bound to produce a (non-zero FA, zero FR) performance behavior of the algorithm, with the non-zero FA being carefully defined and optimized. The relative robustness of the phoneme-indexed performance (in Sect. 8) even with moderately high word error rates (e.g. approx. 30 %) to yield good ROCs and EERs in the range of 10–15 % has high practical significance and implications to realize useful 'operating-points' which offer such reductions in retrieved data with associated (low FR, acceptable non-zero FA), e.g. (<1 %, ∼10 %) for ground-truth phoneme-indexing and (∼10 %, <20 %) for actual phoneme-recognition WER 34 %, as in Fig. 8.

10 Conclusions

We have addressed the problem of spoken document retrieval as an audio-search and proposed several variants of the sub-sequence DTW framework and studied their applicability to audio-search and effectiveness on a large speech database.

References

1. Divakaran, A.: Multimedia Content Analysis: Theory and Applications. Springer, New York (2009)
2. Muller, M.: Dynamic time warping. In: Muller, M. (ed.) Information Retrieval for Music and Motion, Chap. 4, pp. 69–84. Springer, Heidelberg (2007)
3. Fisher, W.M., Doddington, G.R., George, R., Goudie-Marshall, K.M.: The DARPA speech recognition research database: specifications and status. In: Proceedings of DARPA Workshop on Speech Recognition, pp. 93–99 (1986). https://catalog.ldc.upenn.edu/LDC93S1
4. Rabiner, L.R., Juang, B.H.: Fundamentals of Speech Recognition. Prentice Hall, Upper Saddle River (1993)
5. Rosenberg, A.E., Bimbot, F., Parthasarathy, S.: Overview of speaker recognition. In: Benesty, J., Sondhi, M.M., Huang, Y. (eds.) Handbook of Speech Processing, Chap. 36, pp. 725–741. Springer, Berlin (2008)

Improved Language Identification in Presence of Speech Coding

Ravi Kumar Vuddagiri[1], Hari Krishna Vydana[1](✉), Jiteesh Varma Bhupathiraju[2], Suryakanth V. Gangashetty[1], and Anil Kumar Vuppala[1]

[1] Speech and Vision Lab, International Institute of Information Technology, Hyderabad, India
{ravikumar.v,hari.vydana}@research.iiit.ac.in, {svg,anil.vuppala}@iiit.ac.in
[2] SRKR Engineering College, Hyderabad, India
jiteesh10@gmail.com

Abstract. Automatically identifying the language being spoken from speech plays a vital role in operating multilingual speech processing applications. A rapid growth in the use of mobile communication devices has inflicted the necessity of operating all speech processing applications in mobile environments. Degradation in the performance of any speech processing applications is majorly due to varying background environments, speech coding and transmission errors. In this work, we focus on developing a language identification system robust to degradations in coding environments in Indian scenario. Spectral features (MFCC) extracted from high sonority regions of speech are used for language identification. Sonorant regions of speech are the regions of speech that are perceptually loud, carry a clear pitch. The quality of coded speech in high sonority region is high compared to less sonorant regions. Spectral features (MFCC) extracted from high sonority regions of speech are used for language identification. In this work, GMM-UBM based modelling technique is employed to develop an language identification (LID) system. Present study is carried out on IITKGP-MLILSC speech database.

Keywords: Automatic language identification · Mobile environments · Speech coders · Sonority regions · Glottal closure region · GMM · GMM-UBM

1 Introduction

Automatic language identification (LID) refers to the task of identifying the language being spoken from the speech by an anonymous speaker. A wide range of multilingual services like voice operated airport information query systems, voice controlled assistant systems. Linguistic constraints of the spoken language can be used as an extra information to improve the performance of automatic speech

R. Prasath et al. (Eds.): MIKE 2015, LNAI 9468, pp. 312–322, 2015.
DOI: 10.1007/978-3-319-26832-3_30

recognition system, if the language being spoken implicitly from the speech. Some demanding applications like automatically routing a telephone call from a foreigner to the human operator fluent in that specific language needs an LID system to be operated in mobile environments. Complete automation of customer care services and customer feedback services demands a robust LID system to be operable in mobile environments. Due to multi-lingual culture a language identification (LID) system is highly demanded in India. In this work, we mainly focus on developing an LID system in mobile environments for Indian scenario. Most of the Indian languages are derived from the Sanskrit language (Devanagari script) and have overlapping set of phonemes, this makes the task of Indian language identification even more challenging.

The major issues that arise in mobile environments are degradation of speech quality due to coding and transmission errors. Present work is aimed at analyzing the performance of LID system based on speech coders. The effects of speech coding on speaker and language recognition systems are studied in [7]. A decrease in the performance of the recognition is noted with a decrease in the coding rate. The role of an implicit language identification system and various approaches to develop an implicit language identification system are discussed in [5]. Use of prosody features for building a language identification (LID) system is studied in [3,8]. Significance of steady vowel and transition regions for language identification is studied in [6] and almost same performance is observed in both the regions. Acoustic features used in most of the previous studies (spectral and prosody features) are strongly influenced by varying background environments and coding effects on speech. This gives motivation towards developing a system that is robust to degradations in mobile environment. Most of the above LID systems are designed with GMM as their modeling techniques, but use of GMM-UBM for the task of language identification in Indian scenario investigated in this paper. In the proposed approach features from high sonority regions of speech are used for the task of language identification. High sonority regions are the regions that are perceptually loud and, the regions where the spectral energy is mostly concentrated. The quality of coded speech is superior in high sonority regions compared to low sonority regions and the effect of background environments is less in this region. The proposed method uses the language specific information in these regions for the task of language identification.

The remaining parts of the paper is organized as follows: Details of baseline system GMM-UBM modeling technique is discussed in Sect. 2. In Sect. 3 the performance of LID system for coded speech is explained. Section 4 describes the proposed method for LID system in speech coding. In Sect. 5, the evaluation of the proposed method is presented. Conclusion and future scope are discussed in Sect. 6.

2 Baseline Method for Language Identification

In the present work, language identification (LID) is carried out on Indian Institute of Technology Kharagpur - Multi Lingual Indian Language Speech Corpus

(IITKGP-MLILSC) [2]. Database comprises of 27 regional languages collected from the radio broadcasts and television talk shows. Each language contains a minimum of ten speakers including both male and female. A minimum of one hour data is collected for each language. From each speaker, 5–10 minutes of data is recorded at 16 kHz sampling rate and 16 bits per sample. Noise samples from Noisex database are employed in the present study. From the entire available dataset, speech data from two speakers (1 male and 1 female) are used for testing the LID system and the data from the remaining speakers is used for training the LID system.

In this work spectral features namely MFCC's are employed to develop language identification (LID) systems. In this work, spectral vector is obtained by block processing the whole speech using a 20 ms window with an overlap of 10 ms. From every 20 ms speech spectral features are computed using 24 filter bands. The spectral vector represented by X is given by

$$X = [x_1, x_2, x_3 \cdots, x_k, \cdots, x_n] \tag{1}$$

where k is the frame index and x_k represents N dimensional MFCC's from the kth frame. After transforming the input speech into spectral vectors these vectors are used to develop a language model by training the Gaussian mixture models. A language specific distribution from the spectral vector is computed using Gaussian mixture modelling (GMM). During the training phase spectral vectors from input speech are used to develop a separate language model λ for each language i.

During the evaluation of the developed LID system spectral vector from the testing speech sample i.e., Y is given to all the language models i.e., $[\lambda_i | i = 1, 2, 3, \cdots, L]$, where L is the total number of languages. The LID system computes the posteriori probability for the spectral vector obtained from the testing speech sample, to identify the language model that is most likely to produce the feature vector similar to testing spectral vector (Y). The posteriori probability is given by

$$\hat{i} = \arg\max \sum_{1 \leq i \leq L} \Pr(\lambda_i | Y) \tag{2}$$

The language with the highest posteriori probability ($\hat{i}$) is assumed as language of the spoken speech sample.

2.1 GMM with a Universal Background Model

Universal background model is an improvement in the Gaussian mixture modelling technique [9]. The method is initially to select a trained model and determine the likelihood ratio of testing speech sample with the trained model and the universal background model (UBM). Given a segment of speech Y and hypothesized language of the speaker is i and the task of language identification (LID) system is to detect the whether Y has the language i. $P(Y|i)$ is the likelihood that speech segment Y has the hypothesized language and $P(Y|\hat{i})$ is the likelihood that speech segment Y does not have the hypothesized language.

$$Likelihood\, ratio = \frac{P(Y|i)}{P(Y|\hat{i})} \tag{3}$$

where $P(Y|i)$, $P(Y|\hat{i})$ are refereed as the likelihood values of the hypothesis for the given speech segment. The basic goal of language identification (LID) system is to determine the values of these likelihoods $P(Y|i)$, $P(Y|\hat{i})$. Mathematically i is represented by a model denoted by λ_i, i which characterizes the of hypothesized language in feature space X. The alternative hypothesis is denoted by $\hat{\lambda_i}$.

$$log - likelihood\, ratio : LLR(x) = logp(Y|\lambda_i) - logp(Y|\hat{\lambda_i}) \tag{4}$$

Mathematically the developed UBM model is equivalent to $\hat{\lambda_i}$. In the Gaussian mixture modelling with a universal background model (GMM-UBM) based approach parameters of λ_i are estimated by adapting the UBM to language class i.

In the GMM-UBM system, a universal background model is trained by pooling the data from all the languages. The language model is obtained by updating the well trained parameters in the UBM by adaptation. Loosely speaking, Universal Background Model (UBM) is considered to cover the language independent broad class of speech sounds. During the adaptation language dependent tuning of these well trained parameters is carried out using the training data [9]. As most of the Indian languages are derived from the Sanskrit language (Devanagari script) and have overlapping set of phonemes, a tighter coupling between the universal background model and language model is obtained by GMM-UBM based approach. The performance of the coupled approaches is shown to be better than the conventional decoupled approaches (i.e., GMM based approaches) [9].

Table 1. Comparing the performance of the language identification system by varying length of testing speech sample and number of mixture components.

Performance of GMM-UBM based baseline language identification system			
No of components	Length of testing speech sample		
	3 s	5 s	10 s
64	52	53	55
128	55	55	56
256	58	59	61
512	66	69	70
1024	68	68	69

In the present work, performance of language identification system is analyzed in speaker independent case only i.e., training the language models and testing them with test speech sample is done with data from different speakers. For analyzing influence of length testing speech sample on the performance of LID, performance of LID is computed for testing speech sample with various lengths such as 3 s, 5 s and 10 s. UBM models with varying number of mixture components (i.e., 64 to 1024) are developed using 90–120 minutes of speech data. Multiple LID systems are developed by varying the number of mixture components from 64 to 1024 to analyze the influence of number of mixture components on the performance of LID. The performance of (LID) is computed for 75 different test cases from the testing dataset and average of all the test cases is reported in the Table 1.

The performance of baseline language identification system based on GMM-UBM is given in Table 1. Column 1 is the number of mixture components used in building an LID system. Performance of LID for testing speech samples of various lengths is given in column 2–4. A significant improvement in the performance of the system is noted with an increase in number of mixture components from 64 to 1024. Though there is a slightest improvement with mixture components above 1024. The language identification (LID) system developed with 512 mixture components and 3 s testing speech sample duration is used for further analysis in the present study. A slight increase in performance is noted with an increase in length of a testing speech sample from 3 s to 10 s.

3 Language Identification for Coded Speech

The quality of speech in the mobile environment is not only degraded by varying background environments but also due to the process of coding the speech. Speech coders are employed to compress the speech and effectively utilize the bandwidth of a wireless communication system. An ideal speech coder should compress the speech without much degradation in the quality of speech. But from studies of [7], it is evident that there is a degradation in quality of speech with a decrease in the coding rate. The main aim of this work is to study the influence of speech coding on language identification system. Performance of the LID system is measured in both matching and mismatching coding environments. The baseline system from Sect. 2 is used for analyzing the performance of coded speech. In the present work, CELP (FS-1016), MELP (TI 2.4 kbps), GSM full rate (ETSI 06.10) and AMR (ITU-T G.722.2) coders are used to analyze the performance of the LID system. The operational details of these coders are discussed in the following Subsections.

3.1 Adaptive Multi-rate (ITU-T G.722.2)

The AMR (Adaptive Multi-Rate) is a standard speech coding algorithm operating at eight bit rates in the range of 4.75 to 12.2 kbps. AMR coder is well known for its link robustness. Majority of 2G mobile telecommunications use AMR as a standard coding scheme. It is also used as a coding standard in 2.5G and 3G wireless systems. Recently it was also included in the CableLabs and PacketCable 2.0.

3.2 Codebook Exited Linear Prediction (CELP FS-1016)

The concept of linear predictive coding (LPC) is used in CELP coding. Current speech sample is estimated by the linear combination of past samples in LPC technique. CELP encoder and decoder maintain a fixed codebook of various excitation signals. CELP encoder synthesizes speech with all the available codebook vectors and identifies the excitation signal that is best suited to represent the present speech frame. The index of the best suitable excitation and system parameters of that frame is sent to decoder which are then used by decoder to resynthesizes speech. In this work, CELP FS-1016 is used and it operates at a bit rate of 4.8 kbps.

3.3 Mixed Exited Linear Prediction (MELP)

To improve the quality of speech and accurately capture the underlying dynamics of speech, sophisticated details from the speech production mechanism are explored to compute additional parameters. A mixed excitation computed by combing the periodic pulses and filtered noise to make the speech sound natural. This mixed excitation is used by the synthesis to synthesize the speech. MELP coding is widely used in applications like military, satellite and secure voice applications. In the present study, MELP TI 2.4 kbps is used and it is operated at a bit rate of 2.4 kbps.

3.4 Global System for Mobile (GSM 06.10) Full Rate Coder

A uniform 13 bit PCM coded signal sampled at 8 kHz is given as an input to the GSM full rate coder. GSM full rate coder uses regular pulse excitation and long-term prediction (RPE-LTP) techniques for speech coding. A frame based processing is performed on a 20 ms frame of PCM based input signal to generate a GSM coded signal.

Table 2. Comparing the performance of baseline system in various coding environments.

Coders	PCM training	Matched training
PCM (Clean)	66	66
AMR	51	53
GSM	46	49
CELP	39	42
MELP	43	46

Performance of LID system in various coding environments is shown in Table 2. Column 1 gives the coders used for analysis. Column 2 gives the performance of LID in mismatched environments i.e., training phase is carried out using PCM coded speech and testing speech sample from one of the three speech coders. Column 3 gives the performance of LID system in matched environments i.e., speech from the same coder is used in both training and testing phases. From the results of Table 2, it is evident that the performance of LID is better in matched environments compared to mismatched environments. In spite of having low bit rate MELP coder gives better performance than CELP owing to the use of mixed excitation in MELP coder. AMR coder gives superior performance in matched environments among all the three coders used in present work. An average improvement of 3 % is observed in case of matched coding environments compared to unmatched case.

4 Proposed Approach for Language Identification in Speech Coding

From the result of Table 2, it is evident that the performance of LID is degraded by varying the use of speech coders. Though the performance of language identification system is better in matched environments we are not sure that the coding environment of testing speech sample matches with coding environment that models are developed. The issue is more complex as various wireless devices used in communication operate with various speech coders. To overcome this problem, we have explored the regions in speech that are less affected by coding process. Since all the coders use the standard techniques like LPC for speech coding. It is well known fact that quality of coded speech is superior in voiced region compared to unvoiced region and within the voiced region the efficiency of speech coding is superior in high sonority regions. So features from these regions in speech are used in the proposed approach for the task of language identification. With this motivation high sonority regions in speech are detected and the language specific information in these regions is used for the task of language identification.

The proposed approach for language identification in mobile environments has the following sequence of steps:

Training Phase:

1. Consider a speech signal from the training data set.
2. Detect high sonority regions of speech using the method presented in Subsect. 4.1.
3. The speech present in these high sonority regions is used for developing the language models.

Testing Phase:

1. Compute high sonority regions for the testing speech sample.
2. The speech present in these regions is used for testing the performance of the language identification (LID) system.

The following Subsections describe the proposed approach for detection of high sonority regions.

4.1 Detection of High Sonority Regions in Speech

Sonority of human speech is mainly due to resonances in the obstruction free vocal tract. Sonority of a resonant sound depends on the sharpness of the resonances. Spectral energy is concentrated around the formant frequencies due to sharper resonances. Concentration of spectral energy is high in the voiced region and within the voiced region spectral energy is dominant in glottal closure phase. In this work, spectral energy at formants in the glottal closure region is used as an acoustic correlate for detection of high sonority regions [10].

The approach used for detection of high sonority regions in speech contains the following sequence of steps:

- Epoch locations are computed using the ZFF method [4].
- Starting from the epoch consider a segment with speech samples containing 30 % of the glottal cycle. The 30 % of the glottal cycle ensures that the chosen speech segment is within the glottal closure phase.
- The segment of speech signal is filtered using a half Hanning window of length less than the pitch period.
- Spectral energy at formants is computed using group delay function based method [1].
- Compute the sum of the N largest peaks in spectral energy every epoch and attribute this sum as the spectral energy of formants at that epoch. Where N is the number of formants considered. During the present study, we assume that the first three formants contribute major part of spectral energy.
- Consider the same spectral energy for all the samples till the next epoch.
- The above steps are repeated at all the epochs.
- The generated contour is subjected to mean-smoothing using a 50 ms window to generate smooth contour called Formant Contour (FC). Formant Contour (FC) is shown in Fig. 1(b).

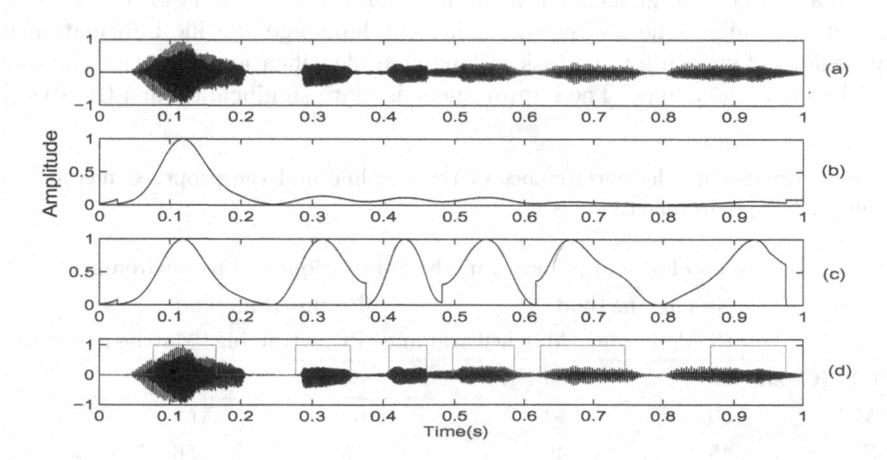

Fig. 1. Detecting high sonority regions from the speech signal. (a) Speech signal. (b) Formant contour (FC). (c) Enhanced formant contour. (d) Emphasized high sonority regions in speech.

- The evidences from the Formant contour are enhanced by computing their slope using the first order difference (FOD) and a slope counting algorithm is used to eliminate spurious peaks from evidences.
- The differenced evidence signal is then amplitude normalized between the two consecutive positive to negative zero crossings. The enhanced and normalized evidence is shown in Fig. 1(c). these are identified by peak picking algorithm.
- Each peak in the evidence is of amplitude 1. So a threshold of 0.45 is used to detect high sonority regions in speech. Detected high sonority regions in speech are indicated in Fig. 1(d).
- The region of speech with FC higher than the threshold is selected and speech in these regions is used for the task of language identification (LID) in mobile environments.

5 Evaluation of the Proposed Method

In the present study, performance of the proposed approach is analyzed by computing the performance of the language identification system in various coding environments. The performance of the proposed method in various coding environments is given by Table 3. For effective comparison, the performance of the baseline language identification system is presented along with the proposed method in Table 3.

Performance of the proposed method in various coding environments is given in Table 3. Various speech coders used in the analysis of the present approach are given in column 1. Performance of the baseline system and the proposed method under various coding environments is given in column 2–3 and column 4–5 respectively. Due to the use of high sonority regions of coded speech for the task of language identification, there is a significant improvement in degraded performance of the language identification system. As the quality of coded speech is superior in high sonority regions, using the language specific information in that regions of speech for the task of language identification gives the improvement in the performance. The improvement is more significant when the speech

Table 3. Comparing the performance of the baseline and the proposed methods in various coding environments.

Performance of baseline and proposed methods in various coding environments				
Coders	Baseline method		Proposed method	
	PCM training	Matched training	PCM training	Matched training
PCM (Clean)	66	66	66	66
AMR	50	53	60	62
GSM	45	46	55	56
CELP	40	42	49	52
MELP	43	46	52	54

data in these high sonority regions is used during the training and testing phases of the language identification (LID) system (i.e., matched training) as shown in column 5. Language identification system is developed using the PCM speech data and when sonority regions of speech from various coding environments are used for testing the language identification system an improvement of 10 % to 11 % is noted. This improvement is even more significant when language identification system is developed using speech data from high sonority regions.

6 Conclusion and Future Scope

In this paper, we have analyzed the performance of LID systems in mobile environments. During the analysis, performance of LID systems in coding environments is studied. Performance of LID system is superior in the case of matched environments compared to mismatched environments, but attaining the state of matched environment is a complex issue. To overcome this problem, we have explored speech specific knowledge to detect the regions of speech that are robust to the degradations in mobile environments. In this work, High sonority regions are considered as the regions that are robust to these degradations. High sonority regions are the regions that are perceptually loud and the regions where the spectral energy is mostly concentrated. Owing to high SNR levels, these regions are robust to background noise and due to high spectral energy concentration the quality of speech from speech coders is high in these regions. Use of features from high sonority regions in speech for language identification can effectively reduce the decline in performance of language identification system in mobile environments. In this work, we have explored GMM-UBM based modelling for developing an Indian language identification system. Performance of the baseline system can be improved by the use of more sophisticated language specific features from speech. Performance of LID systems in mobile environments can be further improved by using more sophisticated speech coders which does not degrade the quality of speech. The task of improving the performance of the baseline system would be considered for future study.

References

1. Joseph, M.A., Guruprasad, S., Yegnanarayana, B.: Extracting formants from short segments of speech using group delay functions. In: Proceedings of Interspeech, pp. 1009–1012 (2006)
2. Maity, S., Vuppala, A.K., Rao, K.S., Nandi, D.: IITKGP-MLILSC speech database for language identification. In: 2012 National Conference on Communications (NCC), pp. 1–5. IEEE (2012)
3. Mary, L., Yegnanarayana, B.: Extraction and representation of prosodic features for language and speaker recognition. Speech Commun. **50**(10), 782–796 (2008)
4. Murty, K.S.R., Yegnanarayana, B.: Epoch extraction from speech signals. IEEE Trans. Speech Audio Lang. Process. **16**(8), 1602–1613 (2008)
5. Nagarajan, T.: Implicit systems for spoken language identification. Ph.D. thesis, Indian Institute of Technology, Madras (2004)

6. Nandi, D., Dutta, A.K., Rao, K.S.: Significance of cv transition and steady vowel regions for language identification. In: 2014 Seventh International Conference on Contemporary Computing (IC3), pp. 513–517. IEEE (2014)
7. Quatieri, T.F., Singer, E., Dunn, R.B., Reynolds, D.A., Campbell, J.P.: Speaker and language recognition using speech codec parameters. Technical report, DTIC Document (1999)
8. Rao, K.S., Maity, S., Reddy, V.R.: Pitch synchronous and glottal closure based speech analysis for language recognition. Int. J. Speech Technol. **16**(4), 413–430 (2013)
9. Reynolds, D.A., Quatieri, T.F., Dunn, R.B.: Speaker verification using adapted gaussian mixture models. Digital Sig. Process. **10**(1), 19–41 (2000)
10. Vydana, H.K., Mounica.K, Vuppala, A.K.: Improved syllable nuclei detection using formant energy in glottal closure regions. In: International Conference on Devices, Circuits and Communications (Accepted). IEEE (2014)

SHIM: A Novel Influence Maximization Algorithm for Targeted Marketing

Abhishek Gupta[1(✉)] and Tushar Gupta[2]

[1] Indian Institute of Technology Roorkee, Roorkee 247667, India
{abhishekgupta76,tushargupta98}@gmail.com
[2] Wipro Technologies, Hyderabad 500032, India

Abstract. Influence maximization is the problem of finding a set of k users in a social network, such that by targeting these k users one can maximize the spread of influence in the network. Recently a new type of social network has come into existence on platforms like Zomato and Yelp, where people can publish reviews of local businesses like restaurants, hotels, salons etc. Such social network can help owners of local businesses in making intelligent business decisions through the use of Targeted Marketing.

In this paper we present Spread Heuristic based Influence Maximization (SHIM) algorithm, our novel algorithm, which uses a heuristic approach that maximizes the influence spread every time a node is added to the set of influential nodes. In our work, we also introduce a new method to find information-propagation probability based on attributes of the user. We test the proposed algorithm on academic dataset of Yelp, and a comprehensive performance study shows that SHIM algorithm achieves greater Influence Spread than several other algorithms.

Keywords: Influence maximization · Top k influential users · Local businesses · Targeted marketing

1 Introduction

Websites such as Yelp and Zomato allow people to publish reviews about local businesses, which help other people on the website to make an informed decision. The social network of such people holds great value for local businesses to market their product. As local business owners usually don't have budget for marketing their product/service to thousands of people living nearby, they adopt cheaper methods of publicity, like Facebook pages, to market their product. However, studies in recent years have shown that businesses on Facebook and Twitter are unsuccessful in reaching out to people. A recent study by Forrester [6] stated that businesses on Facebook and Twitter reach only 2 % of their fans and only 0.07 % of followers actually interact with their posts. For example, if a Restaurant has 10,000 fans on Facebook, then it is able to reach out to only 200 fans and out of these only 7 fans interact with their post. This makes it difficult for local business owners to market their product online.

R. Prasath et al. (Eds.): MIKE 2015, LNAI 9468, pp. 323–333, 2015.
DOI: 10.1007/978-3-319-26832-3_31

In this paper, we explore the possibility of helping local businesses to market their product by the use of *Targeted Marketing*. Targeted Marketing is a method to market product exclusively to a set of people, in the hope that this set of people will publicize the product in their network through word-of-mouth publicity. For example, a Chinese restaurant's potential customers are those people who visit other restaurants near that Chinese restaurant. To market the product among those people, the Chinese restaurant owner needs to attract influential individuals among those people.

This poses the problem of choosing a set of individuals such that the influence spread is maximum in network, also formally known as Influence Maximization problem. It is the problem of finding a set of K users (seeds) in a social network, such that the number of users that are influenced by these seeds is maximum. Influence Maximization problem requires a directed graph G, having *Information-Propagation* probabilities over all edges of G. Many algorithms have been proposed for Influence Maximization. However, a key challenge in solving this problem for real world networks, such as Social Networks, is to find Information-Propagation probabilities over the edges of Social Network.

In our approach, we present a novel technique to assign edge weight to an edge in social networks like Yelp, Zomato. This method is useful to convert the initially unweighted network into a weighted network. Then we applied our novel SHIM algorithm which uses the heuristic approach of maximizing the influence spread every time a node is added to the set of influential nodes. The SHIM algorithm gives better efficiency than the current algorithms.

The remainder of the paper is organized into sections. Section 2 formulates the problem and presents related work. Section 3 describes SHIM algorithm and our method of calculating Information-Propagation probability over an edge. Section 4 presents experimental results and Sect. 5 concludes the paper.

2 Problem Formulation and Related Work

2.1 Problem Formulation

Given a social network $G = (V,E)$ where $|V| = n$ nodes and $|E| = m$ edges, the Influence Maximization problem aims to find a set of k nodes such that influence spread is maximized as per a diffusion model. For every edge $(i,j) \in E$, $p(i,j)$ represents the probability of influence propagated from i to j on the edge.

Diffusion Models are models that present a rationale of how current adopters (active nodes) and potential adopters (inactive nodes) interact [7]. A number of models have been proposed to quantify the influence spread. However, the two major information diffusion models are Linear Threshold model and Independent Cascade model [8–10]. In this paper we have adopted the *Independent Cascade* (IC) model as diffusion model, where each node in G is classified as either active or inactive. An inactive node can be converted to active node only if a neighbor node tries to activate it. Given a seed set $Q \subset V$, let $Q_t \subset V$ be a set of node that is activated at time t, with $Q_0 = Q$ and $Q_t \cap Q_{t-1} = \phi$. At round $t+1$, each node $i \in Q_t$ tries to activate its neighbor in $j \in V \setminus \bigcup_{0 \leq i \leq t} Q_i$ with probability $p(i,j)$.

The influence spread of Q is denoted by $\sigma(Q)$, which represents the expected number of activated nodes when Q is seed set.

2.2 Related Work

The problem of Influence Maximization was first studied by Richardson et al. [2]. Kempe et al. [1] formally presented Influence Maximization as an optimization problem and proved it to be a NP-hard problem. They also gave two models of information diffusion, the Independent Cascade Model and the Linear Threshold model. They presented GreedyIC algorithm [1], the first algorithm for Influence Maximization, which guarantees the spread to be within *(1-1/e)* of the optimal influence spread.

However, GreedyIC algorithm is inefficient as it finds the influence spread by running Monte Carlo simulations. Chen et al. [3] had proposed NewGreedyIC and DegreeDiscount algorithm for influence maximization, which were better than the GreedyIC algorithm. NewGreedyIC is an improved version of GreedyIC algorithm, which randomly removes edges that won't contribute to diffusion from the original graph. This gives a smaller graph to run Monte Carlo simulations on, and so takes lesser time. However, it still uses Monte Carlo simulations which is time consuming.

The Degree Discount algorithm solves the problem by using degree discount heuristics. *Single Discount* algorithm discounts the degree of each node by removing the neighbors that are already in active set [3]. Similiarly *Degree Discount* algorithm calculates discount on degree in more detail. The Degree Discount algorithm performs better than Single Discount algorithm.

However, none of the previous studies take into account the overlapping part of influence spread, due to which total influence spread will be lesser than sum of their individual influence spreads. Consider Fig. 1, which describes an example where the influence spread of node A, B and C is 20, 15 and 12 respectively. The number of nodes overlapping in influence spread of A and B is 7, which makes total influence spread of node A and B equal to (20+15−7), 28. To find top-2 influential nodes in network, the Degree Discount algorithm will include A and B in the set as it includes k nodes with most degree. However, the total influence spread of node A and C is 20+12 = 32. SHIM, our novel algorithm, takes this into account and adds only that node as k^{th} node which maximizes difference between influence spread of already selected *k-1* nodes and that of k nodes after addition.

3 Proposed Approach

In this section we describe our method to find top K influential users in a social network through SHIM algorithm. The proposed method of finding information propagation probability over an edge is discussed in Section A, where we also explain our method of calculation of edge weight in the graph using attributes of a user. The proposed SHIM algorithm is discussed in section B.

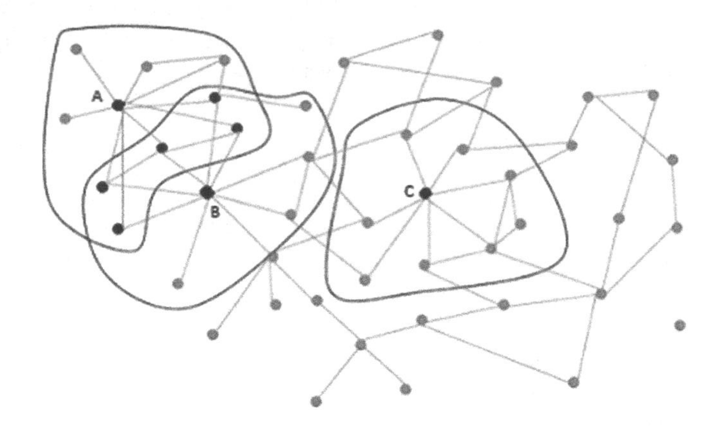

Fig. 1. Influence spread of nodes A, B and C in the network

3.1 Edge Weight and Information-Propagation Probability Calculation

First, we created a social network from Yelp's dataset [5], which is in form of user metadata such as name, review count, average stars and the user's friend list. Each node in the network represents a user and an edge is formed between two nodes if both users are friends. In following subsections we describe our method of calculating edge weight and Information-Propagation probability over an edge. Let's consider two user in the network, *UserX* and *UserY*, which are represented as nodes n_x and n_y in the graph.

(1) Edge Weight: We assign edge weight to each edge based on the similarity between UserX and UserY. To measure similarity we have developed two parameters ($w_{\text{similarFriends}}$ and $w_{\text{similarReviews}}$), which show the strength of relationship between the two users (nodes). $w_{\text{similarFriends}}$ expresses the similarity in the friend circle of UserX and UserY. It is calculated by taking the ratio of mutual friends of n_x and n_y to the total distinct friends of n_x and n_y, as denoted in Eq. (1). The reduction by 2 in denominator in Eq. (1) is because both n_x and n_y are in each other's friend list.

$$w_{similarFriends} = \frac{|n_x \bigcap n_y|}{|n_x \bigcup n_y| - 2} \tag{1}$$

$w_{\text{similarReviews}}$ expresses the similarity in opinion of UserX and UserY. X_{pos} is the set of businesses that UserX has rated positively, and X_{neg} is the set of businesses that UserX has rated negatively. The ratings are given on a scale of 1–5, where 1 is the worst and 5 is the best rating. We have considered a review with rating of 3 or below as a negative review, and one with 4 or above as positive review.

XY_{same} is the set of businesses that both User X and Y rated similarly, that is, with resonant views. It is calculated as count of reviews where for same business both users had either a negative opinion or a positive opinion, as described in Eq. (2).

$$XY_{same} = \{X_{pos} \bigcap Y_{pos}\} \bigcup \{X_{neg} \bigcap Y_{neg}\} \tag{2}$$

XY_{diff} is the set of businesses that they rated differently, that is, with dissonant views. It is calculated as count of reviews where for same business one user had a negative opinion and the other had a positive opinion, as described in Eq. (3).

$$XY_{diff} = \{X_{pos} \bigcap Y_{neg}\} \bigcup \{X_{neg} \bigcap Y_{pos}\} \tag{3}$$

The $w_{\text{similarReviews}}$ is calculated by taking the ratio of consonant views over total of consonant and dissonant views as described in Eq. (4).

$$w_{similarReviews} = \frac{|XY_{same}|}{|XY_{same}| + |XY_{diff}|} \tag{4}$$

To calculate the weight w of an edge we took average of both the parameters $w_{\text{similarFriends}}$ and $w_{\text{similarReviews}}$, as described in Eq. (5).

$$w = \frac{w_{similarFriends} + w_{similarReviews}}{2} \tag{5}$$

This gives a weighted graph, where edge weight is calculated based on similarity of nodes on that edge. To calculate influence spread in the graph, we further assign information-propagation probability to each edge in the graph.

(2) Information-Propagation probabilities: Information-Propagation probability represents the rate at which a user influences neighboring user. To calculate it we developed two parameters, *popularity(X)* and *cluster(X)*, which denote popularity of user and the clustering value of user in the network respectively.

Popularity of a User X, popularity(X), is calculated by taking product of two attributes, *reviewCount* and *averageStars*, as described in Eq. (6). The *reviewCount* is the number of reviews User X has written and *averageStars* is the average of ratings of all the reviews that User X has written.

$$popularity(X) = reviewCount * averageStars \tag{6}$$

Clustering value of node X, cluster(X), is calculated by taking ratio of sum of degree of all neighbors of node X to degree of node X, as described in Eq. (7). Here W is a node from set of nodes in neighborhood of node X represented as *Neb(X)*, and *Deg(W)* represents the degree of node W.

$$cluster(X) = \frac{\sum_{W \in Neb(X)} D(W)}{Deg(X) * (Deg(X) - 1)} \tag{7}$$

We used five number summary method [12] to remove outliers from values of popularity(X) and cluster(X). We classified the set of observations into Maximum (max), Upper Quartile (Q3), Median, Lower Quartile (Q1) and Minimum (min). The Q3 separates largest 25 % of observation from the remaining 75 %. Similarly, Q1 separates lowest 25 % of observation from the remaining 75 %. For our experiments we used observations that lied in Q3, Median and Q1 range. We then normalized the values by min-max normalization, bringing values in range [0,1]. The influence of a node, *influence(X)*, is then calculated by taking average of popularity and clustering value, as described in Eq. (8).

$$influence(X) = \frac{popularity(X) + cluster(X)}{2} \tag{8}$$

The *linkStrength(X, Y)* represents strength of a directed edge from node X to Y, and it is calculated by taking an average of influence of both X and Y, as described in Eq. (9). It is derived on the basis that people follow influential people's recommendations and moreover, influential people are also influenced by their neighbors.

$$linkStrength(X, Y) = \frac{influence(X) + influence(Y)}{2} \tag{9}$$

Information-Propagation probability is a measure of how much node Y is influenced by node X. Information-Propagation probability for a edge X to Y is ratio of linkStrength(X,Y) to sum of linkStrength of all neighbors of Y, as in Eq. (10). The rationale behind it is that the contribution of information that Y receives from X depends on their linkStrength(X,Y). For example, if node Y has 3 neighbors A,B and C with linkStrength(A,Y) = 0.25, linkStrength(B,Y) = 0.25 and linkStrenght(C,Y) = 0.5, then Y receives half of the information from C and one fourth from A and B each. The value always lies between [0,1].

$$propagationProb(X, Y) = \frac{linkStrength(X, Y)}{\sum_{W \in Neb(Y)} linkStrength(W, Y)} \tag{10}$$

3.2 SHIM Algorithm

In this section, we present a new approach to solve the problem of influence maximization based on the heuristics that tries to maximize the overall spread each time it selects a node. SHIM algorithm iteratively finds a node and adds it to the set S of top-K influential nodes. While adding k^{th} node to set S, it finds the node that maximizes the difference between spread of already selected *k-1* nodes and spread of set S after adding that k^{th} node. Algorithm 1 presents a pseudocode of the SHIM algorithm.

It takes a graph G = *(V,E,W,P)*, where edge-weight *(W)* and Information-Propagation probability *(P)* is assigned to each edge in the graph. Algorithm also takes K as an argument and returns a set of K nodes that maximize the influence. The set S stores top-K influential nodes and set A stores the nodes that have been influenced (i.e. active nodes).

Algorithm 1. SHIM Algorithm

```
   Input  : Graph G=(V,E,W,P) and K
   Output: set of K influential nodes
1  initialize S = ϕ and active node set A = ϕ
2  for i = 1 to K do
3      for each vertex v ∈ V \A do
4          Cov_v = findCoverage(G,v,I)
           Cov_v = Cov v \A
5      end
6      v_max = argmax_(v ∈ V \A){Cov_v}
7      S = S ∪ {v_max}
8      A = A ∪ {Cov_vmax}
9  end
```

Initially, set A and set S are empty. Steps 3–5 calculate the coverage of each node in the network that has not yet been activated. Step 4 finds the set of nodes are activated by node v, by calculating *findCoverage(G,v,I)*. The *findCoverage* algorithm takes a graph G and a node that is initially active as input and finds the coverage of that node using Independent Cascade Model. Steps 7–9 pick the node that activated the maximum number of non-active nodes in the network and add it to the top-K influential node set. Also the nodes activated by it are added in $Cov_{v_{max}}$, which represents coverage set of active nodes.

4 Experiment

We carried out experiments on Yelp's Dataset [5]. The dataset comprises of reviews and user metadata in 4 cities across Canada and USA. It contains 42,153 businesses, 252,898 users and 1,125,458 reviews. The social network of users created from this data consists of approximately 1 Million edges.

4.1 Dataset Description

We used Users and Reviews table, and both tables have a set of attribute as described:

(1) Users:

```
{
user id: (encrypted user id),
review count: (review count),
average stars: (floating point average, like 4.31),
friends: [(friend user ids)],
}
```

(2) Reviews:
{

business id: (encrypted business id),
user id: (encrypted user id),
stars: (star rating),
text: (review text),

}

We constructed the network, $G = (V,E,W,P)$, from the tables through the approach described in Sect. 3.1. In graph G, each node represents an individual user in the data and each edge in the G represents some kind of relationship between the two users. We use n to denote the number of vertexes and m to denote the number of edges.

4.2 Experiment Results

We benchmarked SHIM algorithm against various other Influence Maximization algorithms listed below:

- *NewGreedyIC*: It is the improved version of general greedy algorithm for influence maximization problem proposed for the IC model.
- *DegreeDiscountIC*: It is the improved version of single degree discount algorithm proposed for IC model.
- *SingleDiscount*: It is the simplest form of discount algorithm. It gives the discount in the degree measure of a node based on the number of nodes selected in top-K nodes from its neighborhood.
- *Random*: Random heuristic is the basic approach. It randomly finds K nodes from the network, one at a time and returns them as top-K influential node.

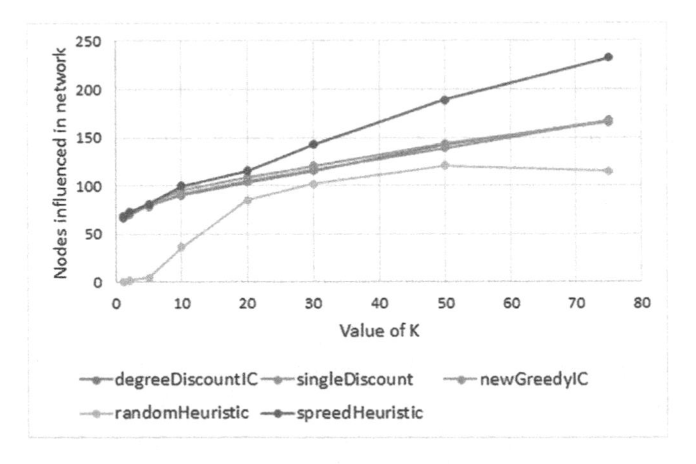

Fig. 2. Influence spread on graph G with n = 1617, E = 2058 (Color figure online)

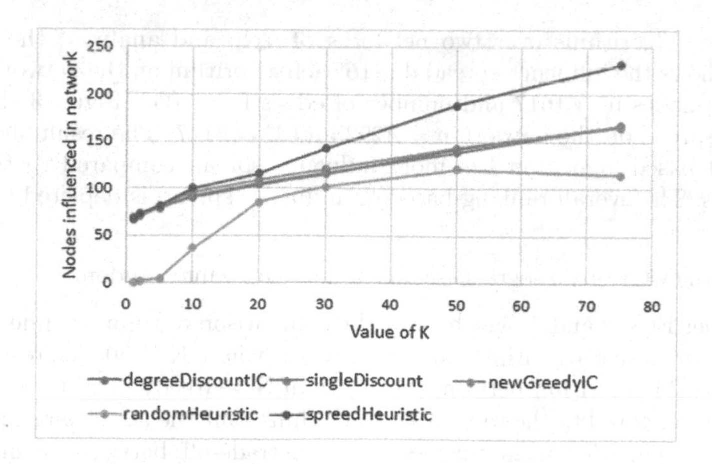

Fig. 3. Influence spread on graph G with n = 4292, E = 8147 (Color figure online)

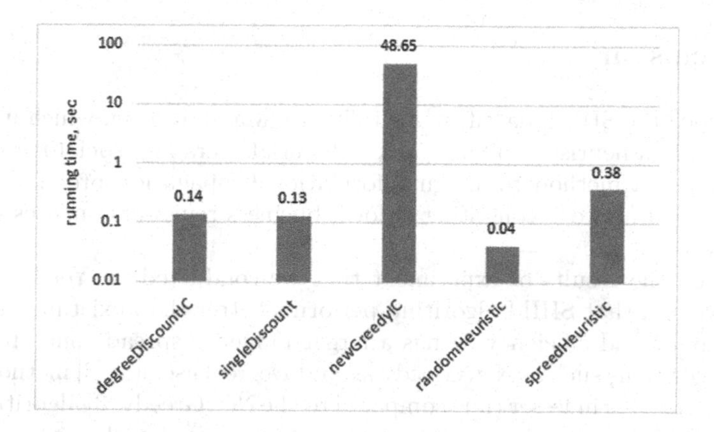

Fig. 4. Running time on graph G with n = 1617, E = 2058

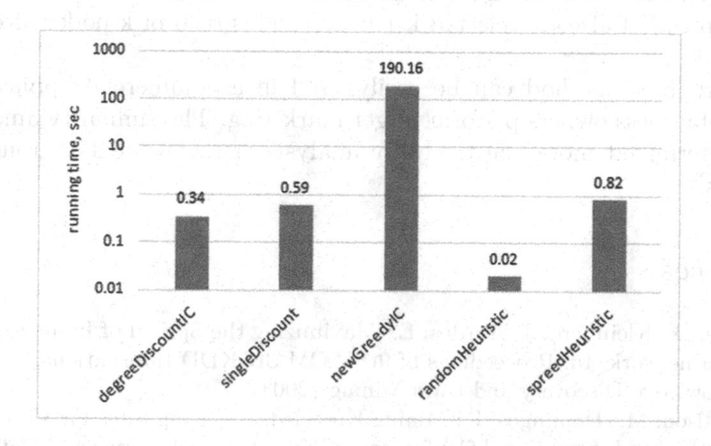

Fig. 5. Running time on graph G with n = 4292, E = 8147

We ran the algorithms over two networks of Yelp and analyzed their result. Figure 2 shows the influence spread due to each algorithm on the network, where number of nodes n = 1617 and number of edges E = 2058. Figure 3 shows the influence spread on the network n = 4292 and E = 8147. The result shows that the SHIM based algorithm has more influence spread compared to the other algorithms. The overall ranking based on influence spread is depicted by Figs. 2 and 3 as:

SHIM>newGreedy>degreeDiscount>singleDiscount>random.

Consider Figs. 4 and 5, which show the comparison of running time of SHIM algorithm wrt other algorithms on two networks where K = 50.The result shows that the SHIM algorithm performs better than NewGreedy algorithm but takes more time compared to the rest of the algorithms. But the additional time taken is marginal. The algorithm thus serves as a trade-off between accuracy and running time.

5 Conclusion

We developed the SHIM algorithm for Influence Maximization, which uses influence spread as a heuristic to find top k influential users in a social network. We also developed a method to assign Information Propagation probability to an edge in a social network consisting of local business reviewers on sites like Yelp and Zomato.

Based on the results of experiment that we conducted on Yelp's academic dataset, we find that SHIM algorithm performs better than existing algorithms w.r.t. accuracy and efficiency. It has a larger influence spread compared to the existing algorithms such as NewGreedy [3] and DegreeDiscount [3] method. Additionally, it executes in lesser time compared to the NewGreedy [3] algorithm. This is because SHIM algorithm maximizes the overall spread each time it selects a node. This is done by adding a node as kth node if it maximizes difference between spread of already selected k-1 nodes and spread of k nodes after inclusion.

The proposed method can be easily used in a commercial application to help local business owners perform target marketing. The similarity among user can be determined more effectively by analysis of review texts through NLP techniques.

References

1. Kempe, D., Kleinberg, J., Tardos, E.: Maximizing the Spread of Influence through a social network. In: Proceedings of 9th ACM SIGKDD International Conference on Knowledge Discovery and Data Mining (2003)
2. Richardson, M., Domingos, P.: Mining Knowledge-Sharing Sites for Viral Marketing. In: Eighth International Conference on Knowledge Discovery and Data Mining (2002)

3. Chen, W., Wang, Y., Yang, S.: Efficient influence maximization in social networks. In: Proceedings of the 15th ACM SIGKDD International Conference on Knowledge Discovery and Data Mining. ACM (2009)
4. Kempe, D., Kleinberg, J., Tardos, E.: Maximizing the spread of influence through a social network. In: Proceedings of the Ninth ACM SIGKDD International Conference on Knowledge Discovery and Data Mining. ACM (2003)
5. Yelp Dataset. https://www.yelp.com/datasetchallenge/dataset
6. Elliot, N.: Nate Elliott's Blog. Blogs.forrester.com. N.p., 2015. Web, 14 July 2015
7. Bass Diffusion Model. Wikipedia. Wikimedia Foundation
8. Fan, X., Li, V.O.K.: Hierarchy-based algorithm for the influence maximization problem in social networks (2013)
9. Morris, S.: Contagion. Rev. Econ. Stud. **67**(1), 57–78 (2000)
10. Granovetter, M.: Threshold models of collective behavior. Am. J. Sociol. **83**(6), 1420–1443 (1978)
11. Zhang, H., Dinh, T.N., Thai, T.: Maximizing the spread of positive influence in online social networks. In: International Conference on Distributed Computing Systems
12. Hoaglin, D.C., Mosteller, F., Tukey, J.W.: Understanding Robust and Exploratory Data Analysis, vol. 3. Wiley, New York (1983)

An Optimal Path Planning for Multiple Mobile Robots Using AIS and GA: A Hybrid Approach

Mohit Ranjan Panda[1(✉)], Rojalina Priyadarshini[2], and Saroj Pradhan[3]

[1] Department of Computer Science and Engineering, C.V. Raman College of Engineering, Bhubaneswar, India
mohit1146@gmail.com

[2] Department of Information Technology, C.V. Raman College of Engineering, Bhubaneswar, India

[3] Department of Mechanical Engineering, CET, Bhubaneswar, India

Abstract. Design of proficient control algorithms for mobile robot navigation in an unknown and changing environment, with obstacles and walls is a complicated task. The objective for building the intelligent planner is to plan actions for multiple mobile robots to coordinate with others and to achieve the global goal by avoiding static and dynamic obstacles. This paper demonstrates a hybrid method of two optimization techniques that are Artificial Immune System (AIS) and Genetic Algorithm (GA). The capability of overcoming the shortcomings of individual algorithms without losing their advantage makes the hybrid techniques superior to the stand-alone ones. The main objective behind this is to improvise the result of a path planning approach than done on AIS and GA separately. The hybridization includes two phases; in first enhancing the local searching ability by AIS and secondly to add stochasticity, instead of choosing random population, the last generation of AIS will be accepted as input to the next process of GA in the hybrid AIS-GA. From the result and observations, it can be inferred that the proposed algorithm is able to efficiently explore the unknown environment by learning from past behavior towards reaching the target. The result obtained from the hybrid algorithm is compared over AIS and GA and found to be more efficient in terms of convergence speed and the time taken to reach at the target, making it a promising approach for solving the mobile robot path planning problem.

Keywords: Mobile robot · Obstacle avoidance · Navigation · Hybrid · AIS-GA

1 Introduction

Autonomous mobile robots are intelligent agents which can execute desired tasks in various (known and unknown) environments without continuous human interference and guidance. Many types of robots are autonomous to some degree. Enabling a robot to adapt to different environment like land, underwater, air, underground or space is a challenging and vital domain in robotics research [1]. A complete autonomous robot in the real world has the capability to (a) obtain information about the environment. (b) Move from one

R. Prasath et al. (Eds.): MIKE 2015, LNAI 9468, pp. 334–346, 2015.
DOI: 10.1007/978-3-319-26832-3_32

definite point to another definite point, without human navigation guidance and assistance. (c) Avoid conditions that are unsafe and dangerous to people, property or itself. (d) Recover from failure and repair it without outside assistance. (e) Learn or gain new capabilities without outside assistance. (f) Prepare tactics and policies based on the surroundings. (g) Adapt to environment during working process [5]. This paper focuses on mobile robot path planning in an unfamiliar and composite static environment. Path planning is normally done in an offline way by taking existing knowledge about the environment. The best path is defined to be the path having minimum cost which is constituted by the shortest and collision free path. GA has been explored greatly and is used vigorously to plan a path for the robot. GA is a search strategy basing on evolutionary models [3]. They are proven to solve some of the intractable problems efficiently. The major benefits of GA are that one can change the optimization criteria for the path without modifying the whole algorithm. GA is being applied in robotics to control the motion of mobile robots in an environment containing several number of static as well as dynamic obstacles [2]. The failure of GA is due to local minima or loops encountered by path planner algorithm. AIS can also be used to optimize mobile robot path planning. In AIS the Clone Selection principle is the whole process of antigen recognition, cell creation and segregation into memory cells. Some artificial immune algorithms have been developed imitating the clone selection theory [7]. The computer scientists are instigated by the skill and capabilities of the natural immune system in some way to develop models that are applicable to some real life problems which mimic various properties of immune system [8]. Focusing on artificial immune system algorithms like, clone selection, negative selection, and the idiotypic immune system can be outlined for powerful cognition of the complex environment around the robots to distinguish between targets, and surrounding obstacles and also it exists as a problem of immature convergence. In this paper, we have hybridize AIS with GA for finding an optimal path of a mobile robot by proposing an algorithm that is designed to overcome the disadvantages mentioned above when they are being used independently. AIS-GA is a hybrid memetic algorithm based on the cited couple of techniques can be applied to produce an optimum result and proved to outperform than AIS and GA [14]. Since, it is asserted that, AIS-GA is an improvement over AIS and GA; the results are compared with that obtained from AIS and GA being used separately. The remainder of this paper is organized as follows. The next section discusses related works for autonomous mobile robot path planning. Section 2 presents the problem formulation. A discussion of AIS and GA as individual approach is contained in Sects. 3 and 4. In Sect. 5, proposed hybrid model has been discussedd in detail. Simulation results, real-time experiments and comparative outcomes are presented in Sect. 6 followed by a conclusion in Sect. 7.

2 Related Work

When GA is applied in path planning problem the first step towards this is a random generation of populations containing all possible alternative paths. Dozier et al. (1997) from NASA proposed a hybrid planner which uses a visibility-based repair methodology along with an evolutionary approach. A genetic based path planning algorithm was introduced by Wang et al. [16] where populations are created for obstacles-free paths as well as with obstacles which were supposed to be invalid. These sorts of invalid paths were used for evaluating the penalty function, later. Sedaghat et al. (2011) present GA based path planning algorithm with modified objective functions. Kala [3] cited Co-operation amongst the individual robots' evolutionary algorithms ensures generation of overall optimal paths. This piece of work tries to improvise the overall efficiency of GA by considering the base knowledge and specific genetic operators.

AIS can also be seen to address the same path finding problems by many researchers. An Immunity Clone strategy algorithm containing Immunity Monoclonal Strategy Algorithms (IMSA) and Immunity Polyclonal Strategy Algorithm (IPSA) is presented by Rouchen et al. (2003). ICS is exploited to generate solutions for multi-objective optimization task. Further Cutello et al. (2006) improvised the method by substituting the binary string representation with the real coded input patterns. The use of adapting ability of the immune system was exploited and was used by P.K. Das et al. (2010) for the robot to reach destiny successfully through a best chosen path and also with minimal rotation angle efficiency. Lixea et al. (2013) in their work have developed a method which randomly generates antibodies there by formulating the system more analogous to the actual biological process.

3 Problem Formulation

The formulation of the problem is to find out the next location of the robot from its present locations situating in its respective workspace by avoiding the collision that may occur due to other robots and static obstacles in its way towards reaching the destination. A robot can start building a path up to the selected node [5] if it aligns itself with a collision free path. If same angle of rotation is needed for turning left or right of rotation of the robot around the z-axis, the tie is illogically broken.

Let (x_j, y_j) be the present position of j-th robot at time t, $(x_j^{'}, y_j^{'})$ be the next position of the same robot at time $(t + \Delta t, v_j)$be the current velocity of j-th robot.

So, the expression for the next position $(x_j^{'}, y_j^{'})$ can be derived from the Fig. 1 as follow (Fig. 2)

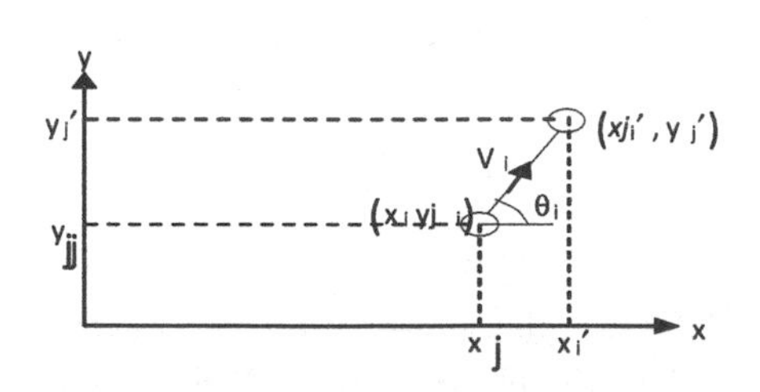

Fig. 1. Current and next position of j-th robot

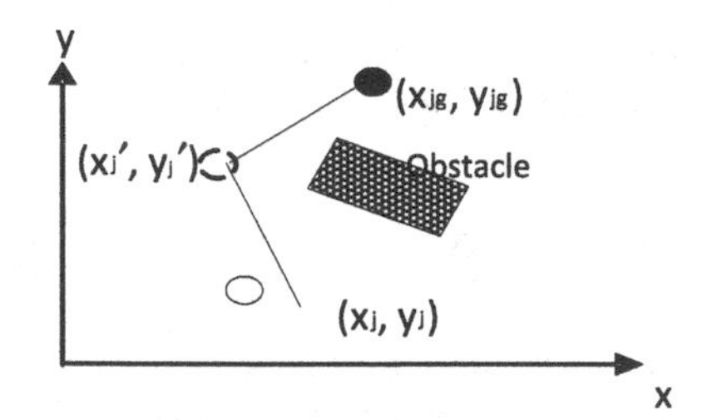

Fig. 2. Selection of next position (x_j', y_j') from current position (x_j, y_j) to avoid collision with obstacle.

$$x_j^{'} = x_j + v_j \cos \theta_j \Delta t \tag{1}$$

$$y_j^{'} = y_j + v_j \sin \theta_j \Delta t \tag{2}$$

When $\Delta t = 1$, the above set of equations reduces to

$$x_j^{\prime} = x_j + v_j \cos \theta_j \tag{3}$$

$$y_j^{'} = y_j + v_j \sin \theta_j \tag{4}$$

We are now forming a constraint that minimizes the total path length without touching the obstacle. The objective function F that determines the length of the trajectory for *'n'* number of robots,

$$F = \sum_{j=1}^{n} \left\{ \sqrt{((x_j - x_j^{\prime})^2 + (y_j - y_j^{\prime})^2)} + \sqrt{((x_j^{\prime} - x_{jg})^2 + (y_j^{\prime} - y_{jg})^2)} \right\} \tag{5}$$

Substituting $x_j^{\prime}$ and $y_j^{\prime}$ from the expressions (3) and (4) in expression (5), we obtain,

$$F = \sum_{j=1}^{n} \left\{ v_j + \sqrt{\{(x_j + v_j \cos \theta_j - x_{jg})^2 + (y_j + v_j \sin \theta_j - y_{jg})^2\}} \right\} \tag{6}$$

Let d_{ij} be the distance between *i-th* and *j-th* robots' current positions, then the constraint that these robots need not touch each other by $d_{i'j'} - 2r \geq \varepsilon$, where r denotes the radius of the robots and ε (>0) denotes a small threshold.

We now represent the multi-robot path-planning as an optimization problem [5]. The optimization problem includes the following objective function,

$$F = \sum_{i=1}^{n} \left\{ v_i + \sqrt{((x_i + v_i \cos \theta_i - x_{ig})^2 + (y_i + v_i \sin \theta_i - y_{ig})^2)} \right\} + f_{dp} \sum_{\substack{i,j=1 \\ i \neq j}}^{n \times (n-1)} \{\min(0, (d_{i,j} - 2r))\}^2 \tag{7}$$

Where f_{dp} (>0) and f_{st}(>0) denote scale factors to the second and third terms in the right hand side of the expression (7). The realization of the optimization of the final objective function using different technique will be discussed in the subsequent chapters.

4 AIS Based Path Planning

AIS is also called as adaptive systems, inspired by biological immunology and multiple immune functions, converted to mathematical models which are applied to various problem solving [6]. It is also applicable for problem like mobile robot navigation. AIS displays some important properties like feature extraction, recognition, diversity, learning, adaptation, self-organization and robustness. It is adaptable in the sense that it is having the learning capability to recognize and respond to new virus and retain a memory about those viruses for future reference and further action. This adaptivity is possible only by the dynamic functioning of the AIS, which made it enable to discard useless components and to improve on existing system [8].

4.1 Path Planning Algorithm

The current environment is defined as the Antibody of the mobile robot, including the distance between the robot and its destination [5]. An antigen is the distance between robot with each obstacle and also the position of a particular robot with respect to other robots. Each robot has the sensors to measure the distance of it from the obstacles. The position of the Robot towards its destination with an estimation of the distance corresponding to the shortest path from each step towards the Goal.

The immune networks consist of two groups. One part is between the mobile robot and the obstacle, b_i^O. The other part is between the mobile robot and the target of the immune network b_i^g.

The antibody is b_i defined as follows [5]:

$$b_i = (1 - \lambda_i)b_i^0 + b_i^g \tag{8}$$

Where λ_i is a ratio between antibody b_i^O and antibody b_i^g. The antibody with the highest α_i is selected. λ_i is given by

$$\lambda_i = \begin{bmatrix} \frac{\mathrm{d_o}}{\mathrm{d_o} + \mathrm{d_g}}, & \mathrm{d_o} > \mathrm{d_g} \\ \frac{\mathrm{d_g}}{\mathrm{d_o} + \mathrm{d_g}}, & \mathrm{d_g} < \mathrm{d_o} \\ 1, & \mathrm{d_o} = \mathrm{d_g} \end{bmatrix} \tag{9}$$

Where d_0 and d_g are the distance of the obstacles to the mobile robot and the goal of the mobile robot. The obstacle antibody b_i^0 and the goal antibody b_i^g in an immune network are calculated as:

$$b_i^0 = \left(\frac{\sum_{i=1}^{n} m_{ji}^0 b_i^0}{n} - \frac{\sum_{i=1}^{n} m_{ji}^0}{n} + m_i^0 - k_i^0 \right) b_i^0 \tag{10}$$

$$b_i^g = \left(\frac{\sum_{i=1}^{n} m_{ji}^g b_i^g}{n} - \frac{\sum_{i=1}^{n} m_{ji}^g b_i^g}{n} + m_i^g - k_i^g \right) b_i^g \tag{11}$$

Where n is the number of antibodies in the environment. The degree of stimulation by other antibodies is represented by the first term on the right hand side. The degree of suppression by other antibodies is represented by the second term. The external input from the antigens is represented by the third term. The natural death ratio is the fourth term. m_{ji}^{0} is the obstacle matching ratio and m_{ji}^{g} is the goal matching ratio.

$$\alpha_0 = \frac{D - d_0}{D} \tag{12}$$

$$m_{ji}^{0} = \begin{bmatrix} \frac{m_{ji}^{o}}{(1-\alpha_o)}, & d_0 > d_{set} \\ \frac{m_{ji}^{o}}{\alpha_o}, & d_0 < d_{set} \end{bmatrix} \tag{13}$$

$$\alpha_g = \frac{D - d_g}{D}$$

$$m_{ji}^{g} = \begin{bmatrix} \frac{m_{ji}^{g}}{(1-\alpha_o)}, & d_g > d_{set} \\ \frac{m_{ji}^{g}}{\alpha_o}, & d_g < d_{set} \end{bmatrix} \tag{14}$$

Where D is the maximum size of the maze. d_0 is the distance between the robot and the obstacle. d_{set} is the radius of the robot necessary to avoid the obstacle.

4.2 Algorithm

Initialization: Robot start position and Goal position

1. If ($Start_{position} = Goal_{position}$)
   Then goto 5
2. Check eight antibodies have finished with calculation
   If finished
   Then
   i. Select the antibody with Maximum b_i using equation (1)
   Else
   ii. goto 2
3. Check for collision
   If possibility of collision occurs
   Then
   (i)go to (4)
   Else
   (ii) Robots take action and go to (1)
4. Avoid Obstacle: Re-compute b_i of antibodies and robot comes back to previous state, then go to (2)
5. Update the parameters according to fitness of antibody.
6. STOP

5 GA Based Path Planning

In this section, we discuss our investigation on GA which helps a mobile robot to find a near optimal path between its initial positions to goal position in a cluttered environment. GAs is a effective technique to solve multi criteria based Optimization task [3]. Here we apply this technique on robotics, to determine optimal path for Robots.

GA proposes the evolutionary chromosome structure to find a path for mobile robot by avoiding multiple obstacles [4]. The GA uses fitness function, to teach robot's movements, was evaluated to achieve the desired output without changing the environment. The important property of this method is the use of dynamic chromosomes' structures and a modified crossover operator can also called as analogous crossover. Every chromosome under this population represents a different robot trajectory. Here the purpose of using GA is to minimize the deviation between the desired path and the actual path.

5.1 Proposed Algorithms

Following steps are considered to use GA to solve the path planning problem. These are (i) Convert the maze to a grid graph as because here the robot will move in a step-wise fashion on the proposed grid. (ii) The starting and ending point of the robot need to be specified, between which the path need to be established. (iii) The obstacles are static and their position on the grid need to be specified.

5.2 Initialization

Initially the population is created with population size which is predefined. The population consists of multiple numbers of chromosomes, which individually represents a solution for the problem of path finding. Here each solution is actually a path between the initial and final position on the search space. The representation of initial population can be as follows:

$$Population_{initial} = \langle q1, q2, \ldots, qn \rangle$$

In general, each structure q_i is simply an string of integer values of length L, in. So q_i represents a vector of node numbers in the grid. Generally, GAs individuals comprises of any point value between the initial and final point. So, the individuals generated is in the form of: *<b1, b2,...,bl>*.

5.3 Fitness Function and Evaluation

To generate the complete solution, the fitness value will be computed as:

$$q_i = dt_0 + dt_1 + \ldots + dt_m$$

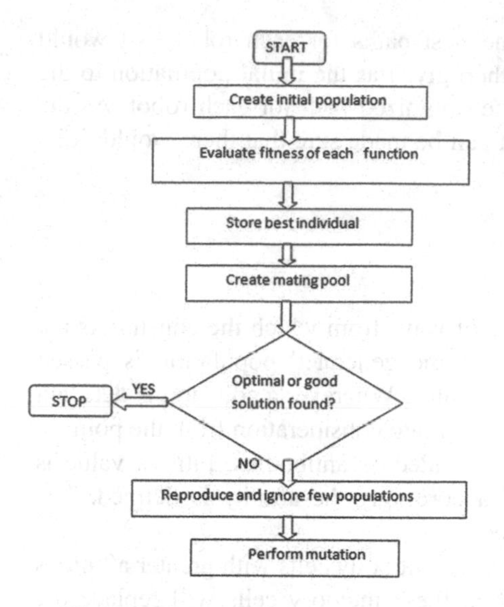

Fig. 3. Flow chart of genetic algorithm

Here, dt_0 = the distance between b_{start}, $b1$, dt_1 = The distance between $b1$, $b2$, dt_2 = The distance between $b2$, $b3$, dt_m = The distance between $b1$, b_{end} (Fig. 3).

The individual chromosome's fitness value can be computed by knowing the coordinate of each point of an individual (i.e., row and column for each b_i from a lookup table). With this, the distance between any two points in the search space can be computed (i.e., environment of the robot).

Assume there are two points in the search space $R_{current}$, R_{next}.

Following equation allow us to compute the distance between the two points:

$$D = |R_{(next,col)} - \mathbf{R}_{(current,col)}| + |\mathbf{R}_{(next,row)} - \mathbf{R}_{(current,row)}|$$

6 Hybridization

It can be observed from the above two processes that GA gives the best optimal path [13] to a robot and fails to detect collisions among the other robots. Whereas, AIS does provide a collision free path to every robot, but however, fails to optimize the path. It fails to lessen the time complexity which every robot takes to complete the journey. This indeed creates a scope for combining these two techniques, AIS with GA, to get collision free and optimized path.

6.1 Proposed Hybridization Algorithm

The hybrid algorithm is a synergy of both AIS and GA. The model is constituted by considering two phases working on AIS and GA for first and second phases respectively [12]. In other words the output of the immune phase is given to the input of Genetic phase. Here we first let the whole problem go through the Immune phase and the output of this immune phase is then passed as the initial population to the genetic phase [12]. In the immune phase, at each step, i.e., each time an antigen is detected (or the collision) proliferation and differentiation takes place. Thus, generating more clones of the antibodies. At the end, the cells with best affinity are chosen as memory cells.These memory cells specify the best paths for each robot that would cause no collisions. These memory cells are then given as the initial population to the genetic phase. Genetic phase, then extracts the optimized path for each robot. As the memory cells contain collision free paths so it can be made sure that there wouldn't be any collisions thereafter.

6.2 Generate an Initial Population

Initial population specifies a particular number of ways from which the constraints are removed. Each component (or the solution) of the generated population is passed through a fitness function to calculate its fitness value. Wherever a collision is detected (or might occur), a set of alternative paths are taken into consideration from the point of a possible collision. These paths are otherwise called as antibodies. Fitness value is then again calculated for each antibody and accordingly the affinity is defined. The antibodies with best affinities are then selected.

As the simulation reaches to the final stage, the antibody cells with greater affinities are considered to be as the memory cells. And these memory cells will replace the antibody cells with least affinities. So that in the next simulation if a similar antigen (collision detection) comes across, a decision will be made based on the memory cells, it has and will react with faster rate than the previous one. These memory cells, thus form a generation and this generation is then forwarded to the genetic phase.

6.3 Followed by the Genetic Phase of the AIS-GA

The generation formed in the immune phase, as mentioned are to be given as the initial population to the genetic algorithm. For genetic algorithms too, there are various fitness functions. Each individual (paths) is considered for the reproduction process. Once the crossover and mutation processes are over, the fitness values of each offspr ing or the newly generated string (paths) will be calculated. Depending on the fitness values the best path or the optimal paths for each robot is chosen (Fig. 4).

6.4 Rastrigin's Test Function

The Rastrigin function [12] is a non-convex function. The performance of optimization algorithms can be measured by this test function. This is an generalized example of non-linear multimodal function. Function definition:

$$f(x) = 10.n + \sum\nolimits_{i=1}^{n} (x_i^2 - 10 * \cos(2 * \pi * x_i))$$

$Xi \in [-5.12, 5.12]$ It has a global minimum at $x = 0$, where $f(x) = 0$.

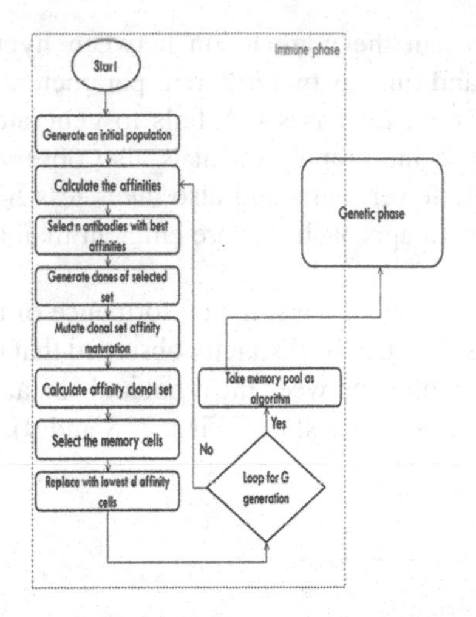

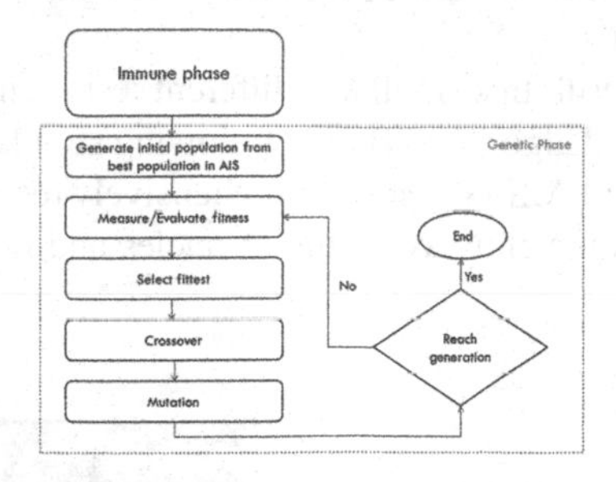

Fig. 4. (a) AIS phase of the hybrid algorithm and (b) GA phase of the hybrid algorithm

Fig. 5. The hybrid AIS-GA compared with AIS and GA by taking a path length into account

7 Simulation Result and Discussion

C- language is used to test the hybrid AIS-GA algorithm to prove its effectiveness. The result obtained from the hybrid AIS-GA algorithm was compared with the output generated, after applying the Genetic Algorithm and Artificial Immune System individually and the observation is reflected through graphs. All the simulations were performed by using a computer with processor Intel core i5processor that works with a frequency of clock of 2.3 GHZ, 4 GB of RAM with Windows 8 operating system.

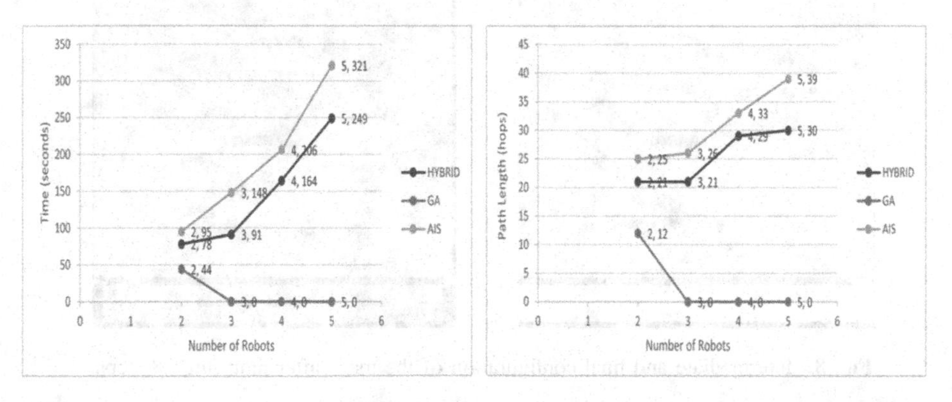

Fig. 6. The hybrid AIS-GA compared with AIS and GA by taking time consumption into account

These graphs given in Figs. 5 and 6 represent the comparison between hybrid AIS-GA, AIS and GA by taking a path length and time as two different parameters. It has been observed that, when the number of Robot increases GA fails to generate a path for those Robots, which represented by 0 in the graph. And it is also observed from the graphs that the hybrid approach travels fewer paths and also takes less time than that of AIS. From this we can infer the Hybrid approach is more efficient than the other two.

With this simulator different test scenarios are created to test the performance of the hybrid algorithm. By the following simulation result, it can be distinctly observed that the hybrid AIS-GA can comprehensively resolve the conflict between the robot and obstacles and efficiently avoid the obstacles and reach the goal successfully (Figs. 7, 8 and 9).

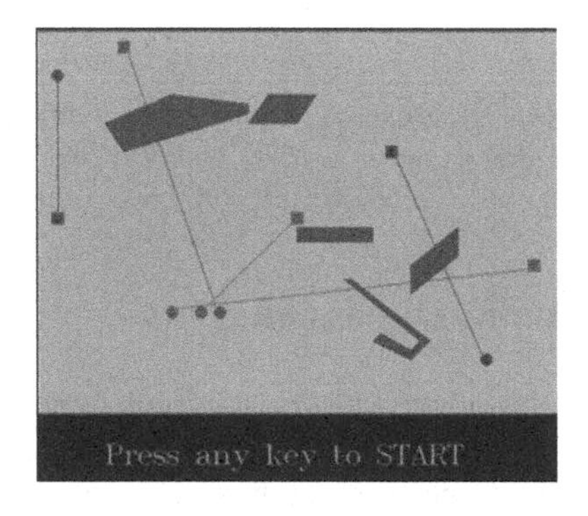

Fig. 7. Initial configuration of the maze with 5 obstacles and 5 robots

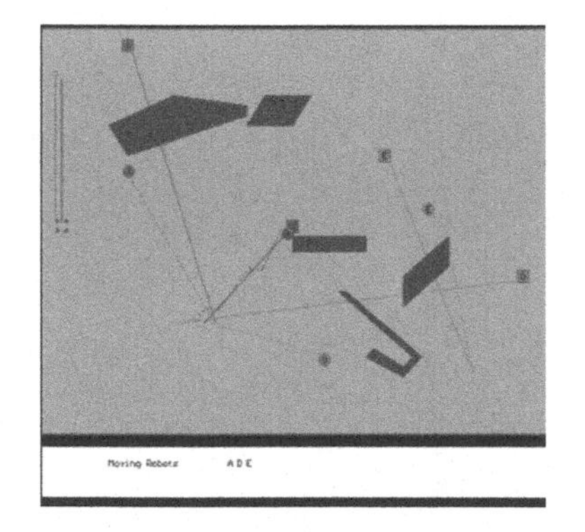

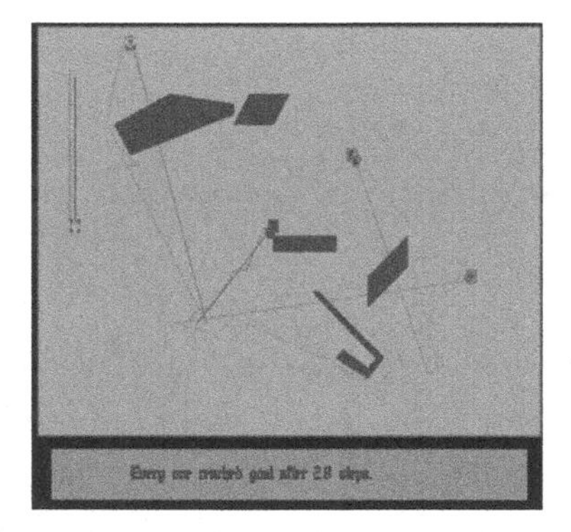

Fig. 8. Intermediate and final configuration of the maze after requiring 28 steps.

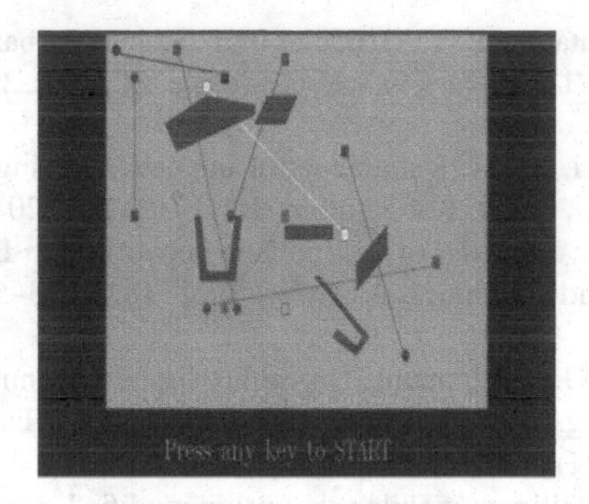

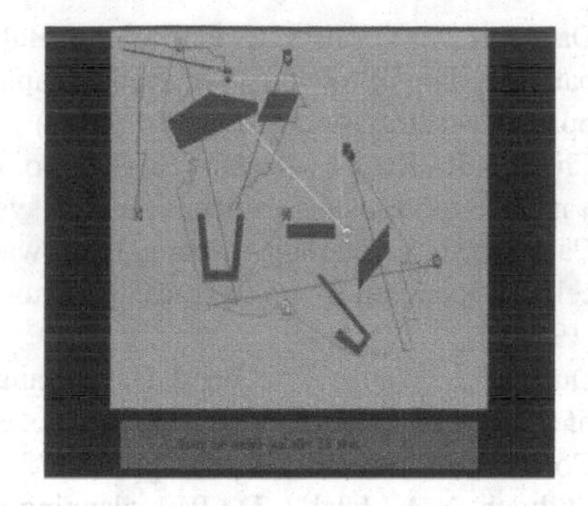

Fig. 9. Initial and final configuration of the maze with nine Robot after requiring 44 steps.

8 Conclusion and Future Work

From the study it is concluded that the AIS technique with a blend of genetic Algorithm gives a better way for navigation of multiple mobile robots. This hybrid AIS-GA algorithm overcomes the immature convergence problem of artificial immune network and local minima problem of genetic algorithm with increasing diversity of antibodies which tend to the same extreme value in solution space. Different simulated test scenarios are conducted to test the performance of the proposed algorithm. In this work, using C program we implemented the hybrid approach for better understanding of navigation of multiple mobile robots like hop count, path length, time consumption, etc. Simulations results validate the efficiency and effectiveness of the robot path planning architecture using hybrid method. But still from simulation results, it has been found that at certain points two robots are reaching at a particular coordinate from where it is difficult for them to decide which robot should proceed further. Here some decision making algorithms can be implemented for smooth path planning. Under this research work, the hybrid algorithm efficiently enables the robots to avoid collision with static obstacles. However, further development in technique may be required towards the avoidance of moving obstacles along with static obstacles.

References

1. Ismail, A.L.-T., Sheta, A., Al-Weshah, M.: A mobile robot path planning using genetic algorithm in static environment. J. Comput. Sci. **4**(4), 341–344 (2008). ISSN 1549-3636 © Science Publications
2. Nearchou, A.: Path planning of a mobile robot using genetic heuristics. Robotica **16**, 575–588 (1998)
3. Kala, R.: Multi-robot path planning using co-evolutionary genetic programming. Expert Syst. Appl. **39**, 3817–3831 (2012)
4. Xiao, J., Michalewicz, Z.: An evolutionary computation approach to robot planning and navigation. In: Hirota, K., Fukuda, T. (eds.) Soft Computing in Mechatronics, pp. 117–128. Springer, Heidelberg (2010)

5. Das, P.K., Pradhan, S.K., Patro, S.N., Balabantaray, B.K.: Artificial immune system based path planning of mobile robot. Soft Computing Techniques in Vision Science. SCI, vol. 395, pp. 1195–1202. Springer, Berlin (2010)
6. Challoo, R., Rao, P., Ozcelik, S., Challoo, L., Li, S.: Navigation control and path mapping of a mobile robot using artificial immune systems. Int. J. Rob. Autom. **1**(1), 712–718 (2011)
7. Carneiro, J., Coutinho, A., Faro, J., Stewart, J.: A model of the immune network with B-T cell co-operation I- prototypical structures and dynamics. J. Theor. Biol. **182**, 513–529 (1996)
8. Duan, Q.J., Wang, R.X., Feng, H.S., Wang, L.G.: An immunity algorithm for path planning of the autonomous mobile robot. In: Proceedings of the IEEE 8th International Multi Topic Conference, Lahore, Pakistan. pp. 69–73 (2004)
9. Shiltagh, N.A., Jalal, L.D.: Path planning of intelligent mobile robot using modified genetic algorithm. Int. J. Soft Comput. Eng. (IJSCE) **3**(2), 31–36 (2013). ISSN: 2231-2307
10. Deng, L., Ma, X., Gu, J., Li, Y.: Mobile robot path planning using polyclonal-based artificial immune network. J. Control Sci. Eng. **2013**(416715), 13 (2013)
11. Ali, M.O., Kohl, S.P., Chong, K.H., Yap, D.F.W.: Hybrid artificial immune system-genetic algorithm optimization based on mathematical test functions. In: Proceedings IEEE Conference on Research and Development (SCOReD 2010), Putrajaya, Malaysia, pp. 256–261. (December 2010)
12. Zinflou, A., Gagne, C., Gravel, M.: GISMOO: a new hybrid genetic/immune strategy for multiple-objective optimization. J. Comput. Oper. Res. **39**(9), 1951–1968 (2012)
13. Bhaduri, A.: A mobile robot path planning using genetic artificial immune network algorithm. In: World Congress on Nature and Biologically Inspired Computing, 2009. NaBIC 2009, pp. 1536–1539, 978-1-4244-5612-IEEE (2009)
14. Sugihara, K., Smith, J.: Genetic algorithms for adaptive motion planning of an autonomous mobile robot. In: Proceedings of the IEEE International Symposium on Computational Intelligence in Robotics and Automation, Monterey, California, pp. 138–143 (1997)
15. Nagib, G., Gharieb, W.: Path planning for a mobile robot using genetic algorithms. In: Proceedings of the International Conference on Electrical, Electronic and Computer Engineering (ICEEC 2004), Cairo, Egypt, pp. 185–189 (2004)
16. Wang, Y., Mulvaney, D., Sillitoe, I.: Genetic-based mobile robot path planning using vertex heuristics. In: Proceedings of the IEEE International Conference on Cybernetics and Intelligent Systems, Bangkok, pp. 463–468 (2006)

Metaheuristic Optimization Using Sentence Level Semantics for Extractive Document Summarization

P.S. Premjith[1], Ansamma John[1(✉)], and M. Wilscy[2]

[1] Department of CSE, TKM College of Engineering, Kollam, India
{premjith1190,ansamma.john}@gmail.com

[2] Department of CSE, University of Kerala, Kariavattom Campus, Trivandrum, India

Abstract. Multi document summarization is the process of automatic creation of a summary of one or more text documents. We developed a multi-document summarization system which generate an extractive generic summary with maximum relevance and minimum redundancy. To achieve this, four features associated with sentences, that can influence the summarization process are extracted. It is difficult to find the appropriate weights corresponding to the features, which leads to good results. We propose a metaheuristic optimization based on solution population with multiple objective functions. The objective functions used takes care of both the statistical and semantic aspects of the documents. Our population based optimization converges rapidly to produce candidate sentences for summary. Evaluation of the proposed system is performed on DUC 2002 dataset using ROGUE tool kit. Experimental results shows that our system outperforms the state of the art works in terms of Recall and Precision.

Keywords: Multi document summarization · Latent semantic analysis · Metaheuristic optimization · DUC · ROUGE

1 Introduction

The association of population with online world and its instantaneous growth resulted in sowing the seeds of information at a rapid rate. Internet is overwhelmed with massive amount of electronic documents especially in text category. A lot of articles are published about a single topic while many of them are redundant information. So the need of the hour is a text summarization system. Humans use a summary for surface level understanding of a topic or familiarization of a matter contained in the original text, not for an in-depth understanding. If the topic is a more familiar one he or she may peek for the diverse concepts about the topic present in the document. So we need an efficient automatic text summarization to tackle the problem.

Automatic multi document summarization (AMDS) is a system generated summary of a set of documents centered on a common theme. A broad classification of AMDS gives two categories, abstractive and extractive [9,10]. Abstractive mimics the human way of summarization by conceptual understanding of the

R. Prasath et al. (Eds.): MIKE 2015, LNAI 9468, pp. 347–358, 2015.
DOI: 10.1007/978-3-319-26832-3_33

documents and producing condensed information in a short piece of text using newly created sentences. On the other hand extractive deals with selection of sentences from documents to produce summary. Majority of the research works are carried out in extractive summarization, mainly to improve the results of existing systems. Another major classification of summarization where active research is conducted is query based and generic summarization, in the direction of extraction based systems. Query based summarization provides the summary according to the query presented by the user to the summarization system [2]. Generic summary gives condensed nontrivial information about the contents present in the given document [5]. Summarization techniques used in the aforementioned systems can be either supervised or unsupervised [4]. Supervised methods rely on building a model for summarization and training with large sets of documents along with their human generated summaries. In contrast unsupervised methods relay on heuristic formulations and make use of information only from target document set to produce a summary.

Recently many researchers started viewing summarization task as an optimization problem where one or more objective functions are formulated for optimization. Any one of the latest optimization methods such as evolutionary, simulated annealing, Genetic Algorithm, differential evolution, Particle swarm optimization or Ant colony optimization method is applied to solve the optimization problem. Such algorithms are usually slow in nature compared to traditional straightforward approaches, at the same time it tries to come up with fair results most of the time.

Our objective is to develop an extractive summarization system by selecting salient sentences from the input documents by considering summarization as an optimization problem. Four features associated with the sentences in the source documents are extracted. Sentences are ranked based on the random weights assigned to the features. Most appropriate summary is generated by the creation of possible solutions (population) iteratively from ranked sentences and optimizing the candidate solutions based on metaheuristic optimization approach. ROUGE tool kit is used for the performance analysis of our system using DUC 2002 data set. It is observed that evaluation results are good in terms of Recall and Precision values.

The rest of the paper is organized as follows: Sect. 2 discusses related works in text summarization, Sect. 3 deals with the proposed summarization system with feature extraction and metaheuristic optimization, Sect. 4 explains the result and analysis of our work and Sect. 5 conclusion.

2 Related Works

Large number of extractive summarization systems are developed over years. Sentence ranking is the most critical step of all extractive summarization system and researchers are putting lot of effort to improve the sentence ranking method to enhance the quality of the summary.

Radev et. al. [3] developed centroid based multidocument summarization system applicable for both single and multiple documents. For every sentences they extracted three features such as centroid score, positional score and over lap with first sentence. Linear combination of these features scores are used to select the salient sentences to the summary.

Josef Steinberger, Karel Jezek [18] proposed a generic summarization system by detecting semantic aspect of the text using the singular value decomposition method. It improved overall quality of the extracted summary. A new evaluation method is also developed which considers the similarity between the source document set and summary.

Leonhard Hennig [8] developed document summarizaton system based on Latent Semantic Analysis to capture the main topic of a document. Sentences are extracted to the summary based on their topic relevance and coverage. This method also satisfies the multi lingual constraint.

Rasim M.Alguliev, Ramiz M.Aliguliyev and Nijat R Isazade [1] proposed an optimization based model for generic document summarization by considering sentence to document collection, the summary to document collection and sentence to sentence relation to extract the salient sentences. Using Differential evolutionary algorithm they solved the optimization problem to come up with a good summary.

Above mentioned works performs the summarization by the analysis of statistical tf-idf measure or by the analysis of semantic features associated with the sentences, or by the application of evolutionary algorithm where optimization is performed by considering the frequency of occurrence of terms in the summary and in the documents. We developed a generic extractive summarization method by employing meta heuristic optimization approach to select sentences, based on static and semantic relevance obtained by centroid and LSA method respectively and is explained in the next section.

3 Proposed Work

In multi document summarization a generic summary provides surface level information about the salient and diverse concepts present in the documents. In this work an extractive generic summary is produced by using Feature based Sentence Scoring and metaheuristic algorithm with multiple objectives. The proposed work is decomposed into following tasks:

3.1 Document Pre-processing

Let $D = \{D_1, D_2, D_3, ..., D_d\}$, be the set of input documents presented for the summarization, in any order. Sentences from the source documentsare extracted by the sentence segmentation and tokenization process, which results in the production of sentence set $S = \{S_1, S_2, ..., S_n\}$ where each sentence $Si \in S$ is a collection of tokens. If each document D_i consist of Ni number of sentences then $n = \sum_{i=1}^{d} N_i$ Documents in D may contain information which are not relevant from summarization view point. Therefore a preprocessing step is performed to extract the required textual units and represent them in a convenient format for the further processing of our work. The sentences may contain words that do not contribute any meaningful information. In general,those words are called stop-words and are removed in the stop-word removal step using the stop-word collection provided by Brown corpus and a modified stop-word list created by the analysis of documents. Now the sentence set S is transformed into a stop-word removed set of sentences $S' = \{S'_1, S'_2, ..., S'_n\}$ such that $|S| = |S'| \, and \, |S_i| \geq |S'_i|$.

A word can exist in different forms of linguistics. As the size and number of documents increases, the number of words and its different forms also increases considerably which in turn increases the complexity of sentence representation. So the tokens are converted to its root form using Porter Stemming Algorithm [13, 19] thereby reducing the number of distinct tokens in the document set. After the stemming process we have a sentence set ST $=\{St_1, St_2, .., St_n\}$, each St_i consist of stemmed version of the tokens present in S'_i. Since we are processing multiple documents related to the same topic, there is a possibility of salient sentences to be repeated in different documents. So the summary produced by selecting salient sentences may lead to the redundancy in summary. Redundancy of information in documents can appear directly or in the form of subsumption, in which the information in one sentence is subsumed in another sentence [3]. From the study, we found that adopting cosine similarity in finding sentence subsumption is not an appropriate measure. Hence we use variant form of Simple Matching Coefficient scheme is applied progressively to sentences in ST results in the removal of subsumed sentences thereby reducing the dimensionality of set ST from n to n', where $n' \leq n$. Now $ST = \{St_1, St_2, .., St_{n'}\}$.

The sentences in ST have variable length and this is not an appropriate representation of sentences for the statistical processing of extractive summarization. Vector Space Model(VSM) is a popular method used for the representation of sentences in extractive summarization. If $T = \{t_1, t_2, t_3, ..., t_m\}$ represents all the distinct m terms appear in the set ST. VSM uses a bag-of-words approach in which a sentence S_i is represented as $S_i = \{tk_1, tk_2, ..., tk_m\}$. Each $tk_i \in S_i$, can be either 1 or 0 indicating the presence or absence of ith token of T in sentence S_i, respectively. After preprocessing of documents we proceed to the next stage, feature based scoring, where sentences are scored, based on some features such as tf-isf, title resemblance, aggregate similarity and sentence length.

3.2 Feature Based Sentence Scoring

This section of our work deals with various features utilized for scoring individual sentences. Examination of the existing works in the literature and by the analysis of gold standard summaries, scoring of sentence using features such as Term Frequency Inverse Sentence Frequency(TF-ISF), Aggregate cross sentence similarity, Title Similarity and Sentence length yielded a better numerical score estimation for sentence. These four features contribute heavily to the extractive generic summarization of multiple documents.

To felicitate the scoring process the sentences are transformed to suitable formats. Here we use two kinds of sentence representation one is using the frequency of terms in the sentence and the other using the TF-ISF value of a token in the document set D. TF and TF-ISF values are computed for every token in T. Now, a sentence St_i can be represented as $St_i = \{r_1, r_2, r_3, ..., r_j, ..., r_m\}$, where r_j is the TF or TF-ISF value of jth token of T if that token is present in St_i and zero otherwise. The reason behind the application of different formats for sentences is based on the studies performed on scoring patterns of feature. The features used for scoring are formulated as follows:

TF-ISF scoring (FS1): The TF-ISF scores for a sentence St_i is computed is computed as:

$$FS1(St_i) = \sum_{j=1}^{m} r_j \tag{1}$$

Aggregate Cross Sentence Similarity (FS2): It measures the similarity of a sentence with all other sentences. Here cosine similarity is used for computing the similarity between sentences and is shown in Eq. (2).

$$sim(St_i, St_j) = \frac{\sum_{k=1}^{m} St_{ik} * St_{jk}}{\sqrt{\sum_{k=1}^{m} St_{ik}} * \sqrt{\sum_{k=1}^{m} St_{ik}}} \tag{2}$$

$$FS2(St_i) = \sum_{j=1}^{n'} sim(St_i, St_j), \qquad St_i, St_j \in ST \tag{3}$$

Title Similarity (FS3): During the preprocessing stage the title tokens of the all titles of D are extracted. A list TL is created in such a way that TL stores the stemmed title tokens along with its frequency of occurrence. Now a title vector TV is constructed using term frequency scheme of representation i.e., $TV = \{tk_1, tk_2, tk_3, , tk_j, ...tk_m\}$, each tk_j is the frequency of occurrence of jth token of T in TL. Now Title similarity of a sentence is the cosine similarity between a sentence and TV which is computed as shown in Eq. (4).

$$FS3(St_i) = sim(St_i, TV)\ , \qquad St_i \in ST \tag{4}$$

Sentence Length (FS4): This feature is employed to penalize sentences having length within particular ranges, based on the average length of sentence in ST. Sentences which are too short and too long are penalized and others are rewarded as shown in Eq. (5).

$$\begin{aligned} FS4(St_i) &= (0.80 * len(St_i)/Avg \qquad if\ len(S_i) < Avg \\ &= 1.2 * Avg/len(St_i) \qquad if\ Avg \leq len(S_i) \leq 1.60 * Avg \\ &= 0.80 * Avg/len(St_i) \qquad if\ len(S_i) > 1.60 * Avg)) \end{aligned} \tag{5}$$

where $Avg = \frac{\sum_{i=i}^{n'} len(St_i)}{n'}$

The four features associated with each sentence is computed and each sentence in ST is transformed into a four dimensional vector $[fs_{i1}, fs_{i2}, fs_{i3}, fs_{i4}]$, where fs_{ij}, is the for FS_j^{th} feature score of ith sentence St_i. For n' sentences in

ST, a feature matrix F of order n'x 4 is constructed.The dimensionality reduced matrix F is used for further processing. In order to avoid the domination of features having higher magnitude over others, the feature scores are normalized using Min-Max Normalization.

For extractive summarization, in order to prioritize sentences, appropriate ranking mechanisms are required. Four feature based scores associated with each sentence is used to identify an aggregate single score for sentences in ST. Here the score matrix Y is computed as:

$$Y = F \, . \, W \tag{6}$$

$$where \;\; W^T = [w1, w2, w3, w4], \quad wi \,\epsilon\, [0,1], \quad 1 \leq i \leq 4$$

In reality, sentence scoring using weight values depends heavily on the nature of documents, knowledge and vocabulary of the author and style of writing. The process of generating summary is a subjective one, varying from individuals to individuals, finding the universal optimum weight values for features becomes very difficult.

3.3 Metaheuristic Optimization

Generally statistical measures are used to produce extractive summary but whatever be the statistical optimization measures used, it might not guarantee an optimized summary. Quality of the summary can be enhanced by considering the semantic aspects of the sentences in the document set. So objective functions are coined based on words-concepts and the centroid of document - summary relationships. In our work, two objectives are used to generate the summary which cannot be combined to a single meaningful scalar function. Therefore two optimal candidate summaries are produced and required summary is generated by the integration of candidate summaries. Since the problem is a multi-criterion optimization we moved to the concept of population based meta-heuristic optimization (MHO) with multiple objectives.

In population based algorithms, a pool of candidates for solution is generated and the optimum one is selected from it either by direct selection or by refining the candidates from the pool. Metaheuristic algorithms are stochastic algorithms with a master strategy to guide the stochastic process and thereby an optimal solution is produced. In our work, MHO techniques is fused with some characteristics and operations of Genetic Algorithm (GA) such as representation of solutions as chromosomes and generation of population by crossover. In most of the cases the size of the summary to be generated is decided by the end-user and is denoted as compression rate. Summary of required size can be generated by considering the parameter R, where R is proportional to the required summary size. The main steps of our metaheuristic optimization are described below.

Chromosome Encoding and Population Generation. In genetic algorithm chromosomes represent the candidate solutions of a problem. In our work each chromosome represents the candidate summary and each gene is an integer value that indicates the index of the sentence in document set. The size of the chromosome is determined by the value of parameter R. In our work, a chromosome is denoted as S_{chrom}, and collection of chromosomes (population) represents candidate summaries.

Population is generated by considering the diversification and intensification aspects of metaheuristic optimization. For each generation the population is created by identifying the two most appropriate parent chromosomes and subsequently the children are generated by the crossover operation on the parent chromosomes. This work maintains randomness in selecting the parent chromosomes which ensures the solution search on a global space.

The parent chromosome of each generation is constructed by scoring the sentences using (6). Quality of the parent is the decided by the weight vector and feature score. Weight vector is maintained randomly in every generation to avoid the dominance of some features over others. Indices of first high scored R sentences forms the first parent and subsequent R indices forms second one. Thus the feature based sentence ranking act as a guide for identifying the parent chromosomes. After the parent chromosomes are identified, the children are constructed by the crossover operation.

Formulation of Objective Functions. Fitness of an individual chromosome is identified by considering some objective functions. This fitness indicates the goodness of the solution and closeness of the solution to the optimal one. Summary generation can be considered as a multi criterion optimization problem where it is difficult to determine whether one solution for one criterion is better than another solution for another criterion. So we want to find out a combination of different criteria to come up with a good result. In this work two objective functions are used independently to generate two optimized candidate summaries. Following subsections deal with the formulation of two objective functions used in our work.

Objective function (f1). A summary is meaningful if (i) the relevance of the sentence in the summary with respect to the entire document set is maximum (ii) the overall concepts in the summary cover the entire concept of the document set (iii) the redundancy within the summary is minimum. So the first objective function is based on the work of Alguliev et al. [8] which is a statistical approach and is given as,

$$f1 = \frac{f_{cover}(S_{chrom})}{f_{diver}(S_{chrom})} \tag{7}$$

The function f_{cover} (S_{chrom}) is intended to capture the main contents of the input document set and is defined as

$$f_{cover}(S_{chrom}) = Sim(O, O^s). \sum_{i \epsilon S_{chrom}} Sim(O, S_i) \tag{8}$$

where O and O^S denote the mean vectors of the entire sentence set and the summary S_{chrom} , respectively. The diversity function is defined as follows:

$$f_{diver}(S_{chrom}) = \sum_{i \epsilon S_{chrom}} \sum_{j \epsilon S_{chrom}} Sim(S_i, S_j) \tag{9}$$

Lower value of $f_{diver}(S_{chrom})$ corresponds to higher novelty in the summary. Now we have the first objective function for optimization as:

$$f1 = \frac{Sim(O, O^s). \sum_{i \epsilon S_{chrom}} Sim(O, S_i)}{\sum_{i \epsilon S_{chrom}} . \sum_{j \epsilon S_{chrom}} Sim(S_i, S_j)} \tag{10}$$

Objective Function (f2). First objective function only checks the similarity between the centroid concepts in both summary and the document set, and diversity of sentences in the summary . The second objective function is introduced to maximize the semantic coverage of the sentences in candidate summaries. This function is based on Latent Semantic Analysis. Latent Semantic Analysis is an application of Singular Value Decomposition to text summarization. The basic idea of LSA is to find out the similarity in meaning of words using the word co-occurrence relationship. This consists of mainly two steps. In the first step the entire document is represented as term by sentence matrix A, where each column of the matrix A represents the TF-ISF vector of a sentence in stemmed sentence set ST. In the second step we are applying SVD on matrix A.

$$A = U \ . \ \Sigma . \ V \tag{11}$$

Here we are interested in the V matrix from the view point of optimization where V depicts the polarity of correlation between the sentences and concepts. In the SVD decomposition, V may contain negative values which also indicates the relationship between the sentence and reflections of the concepts identified by SVD. By exploiting the concept-sentence relationship from V, the score for a sentence is given by:

$$Sc_j = \sum_{i=1}^{nc} |V_{ij}| \tag{12}$$

where nc is the number of relevant concepts used for the scoring, whose value is set with value of parameter R. Now objective function f2 is formulated as follows:

$$f2 = \sum_{i \epsilon S_{chrom}} S_i \tag{13}$$

Here we are maximizing the objective function to get a summary with maximum semantic coverage.

Objective functions used here are non-commensurable functions, therefore two functions can not be combined to form a single optimized solution. Therefore two optimal candidate summaries are produced by optimizing the objective functions f1 and f2. Final summary is generated by the union of the candidate summaries. Low ranked sentences are removed from the generated summary if the summary size exceeds the requirement. Metaheuristic optimization based summarization procedure is shown in Algorithm 1.

Algorithm 1. Optimization Algorithm
Procedure: MHO (f1, f2, genMax)

1. Let $GC1, GC2$ represents the global best solutions for 2 objective functions, initialised with empty values.
2. Let $PC1, PC2$ represents the optimal solutions for 2 objective functions in every generation
3. Initialise weight vector $W^T = [w1, w2, w3, w4]$ with random, $wi \epsilon [0, 1]$.
4. Calculate scores for sentences using $Y := F.W$.
5. Construct a matrix X containing indices of sentences sorted based on the its scores in Y.
6. Construct two parent chromosomes,$p1 := X[1 : R]$ and $p2 := X[R+1 : 2R]$
7. Perform *Crossover(p1,p2)* to generate child chromosomes.
8. Find $PC1$, $PC2$ by evaluating the Objective Functions $f1, f2$.
9. If $f1(PC1) > f1(GC1)$ then $GC1 := PC1$
10. If $f2(PC2) > f2(GC2)$ then $GC2 := PC2$
11. Repeat steps 3–10 until number of generations exceeds genMax (required number of generations)
12. return $GC1$, $GC2$
13. end **procedure**

4 Evaluation

4.1 Experimental Data

Document Understanding Conference(DUC) 2002 dataset is used to conduct experiments for our system. These sets provides their model summaries of varying size such as 50 words, 100 words, 200 words and 400 words against which the summary generated by our system can be compared. The dataset contains four categories of documents, documents about a single natural disaster event(Category 1) and created within at most seven day window, documents about a single event in any domain created with at most a seven day window(Category 2), document about multiple distinct event of a single type (no limit on the time window) (Category 3)and documents about biological information mainly about a single individual(Category 4).

4.2 Evaluation Metrics

For evaluation of our system generated summaries, N-gram co-occurrence statistical measure of ROUGE toolkit is used [12]. The formulation of ROUGE-N is defined as follows :

$$ROUGE - N = \frac{\sum_{S \epsilon Summ_{ref}} \sum_{N-gram \epsilon S} Count_{match}(N - Gram)}{\sum_{S \epsilon Summ_{ref}} \sum_{N-gram \epsilon S} Count(N - Gram)} \tag{14}$$

where N is the size of the N-gram whether uni, bi or tri, $Count_{match}(N-Gram)$ is the number of N-grams occurring in both candidate and reference summaries and count(N-Gram) is the number of N-Grams in the reference summary. The updated ROUGE evaluation methods can generate three types of scores for a system generated summary such as Recall, Precision, and F-measure.

4.3 Results

Summaries for different categories of DUC 2002 dataset are generated by our system. A set of 20 topics, spanning across different categories are selected for summarization, where each topic consists of an average of 9 documents. Seven summaries for each topic, with a total of 140 summaries are generated, for evaluation purpose. The average ROUGE-1 and ROUGE-2 were computed for 140 summaries. Tables 1, 2 and 3 show the category wise results of ROUGE-1 and ROUGE-2 measurements for our metaheuristic optimization using functions f1 (MHO-f1),f2 (MHO-f2) and combined f1&f2 (MHO-f1f2). Even though maximum number of generation (genMax) for MHO algorithm was set to 50, in most cases the algorithm converges with optimal solutions in 25–30 generations, where by time complexity is reduced considerably compared to other evolutionary optimization algorithms.

The results produced by the objective functions f1, f2 and f1&f2 are analyzed individually. The function f1 implemented in our system is exactly similar to the one proposed by [1]. In the detailed experimental analysis it is observed that our system gives improved ROUGE-1 values due to the creation of parent chromosomes with high score sentences obtained by our feature based scoring and application of our metaheuristic optimization method. By considering the objective function MHO-f2, the average ROUGE-1 and ROUGE-2 measures are more promising than MHO-f1. By combining the two objective functions f1 and f2 our system gives improved results which is shown in Table 3.

Table 1. ROUGE-1 and ROUGE-2 scores obtained for different categories of data in DUC 2002 dataset using **MHO-f1**

Average ROUGE-1 score				Average ROUGE-2 score		
Category	Recall	Precision	F measure	Recall	Precision	F measure
C1	0.560103	0.559952	0.559713	0.274915	0.269828	0.272227
C2	0.521048	0.521344	0.520402	0.228708	0.225452	0.22671
C3	0.553298	0.54883	0.5508	0.287844	0.284102	0.285816
C4	0.537325	0.516335	0.526545	0.230765	0.219928	0.225188
Average	0.542944	0.536615	0.539365	0.255558	0.249828	0.252485

Table 2. ROUGE-1 and ROUGE-2 scores obtained for different categories of data in DUC 2002 dataset using **MHO-f2**

Average ROUGE-1 score				Average ROUGE-2 score		
Category	Recall	Precision	F measure	Recall	Precision	F measure
C1	0.547328	0.524598	0.535127	0.245628	0.236602	0.240782
C2	0.56404	0.535152	0.548984	0.273568	0.256814	0.264818
C3	0.576648	0.55543	0.565768	0.307176	0.292288	0.299462
C4	0.524618	0.511105	0.51772	0.20788	0.20128	0.204513
Average	0.553159	0.531571	0.541900	0.258563	0.246746	0.252394

Table 3. ROUGE-1 and ROUGE-2 scores obtained for different categories of data in DUC 2002 dataset using **MHO-f1f2**

Average ROUGE-1 score				Average ROUGE-2 score		
Category	Recall	Precision	F measure	Recall	Precision	F measure
C1	0.551928	0.530743	0.541065	0.262012	0.249062	0.255283
C2	0.560718	0.554628	0.557332	0.277028	0.272082	0.274362
C3	0.547894	0.55458	0.550214	0.287686	0.290712	0.288816
C4	0.560718	0.523845	0.541638	0.27855	0.259355	0.26857
Average	0.555315	0.540949	0.547562	0.276319	0.267803	0.271758

5 Conclusion

We have developed a feature based automatic generic extractive multi document summarization method, which generates summary with maximum relevance and minimum redundancy. In our work, summarization task is viewed as a metaheuristic optimization problem which is solved by coining two objective functions. The word-concept relation ship is also considered to improve selection of salient sentences in the summary. By the experimental evaluation of our method using ROUGE tool kit with DUC 2002 dataset, we observed that our system generates good quality summary in terms of precision and recall for ROUGE-1 and ROUGE-2 measures. This system can be further improved by extracting more appropriate features in the feature extraction phase.

References

1. Alguliev, R.M., Aliguliyev, R.M., Isazade, N.R.: Multiple documents summarization based on evolutionary optimization algorithm. Expert Syst. Appl. **40**(5), 1675–1689 (2013)
2. Dunlavy, D.M., OLeary, D.P., Conroy, J.M., Schlesinger, J.D.: Qcs: a system for querying, clustering and summarizing documents. Inf. Process. Manage. **43**(6), 1588–1605 (2007)

3. Erkan, G., Radev, D.R.: Lexrank: graph-based lexical centrality as salience in text summarization. J. Artif. Intell. Res. **22**, 457–479 (2004)
4. Fattah, M.A., Ren, F.: Ga, mr, ffnn, pnn and gmm based models for automatic text summarization. Comput. Speech Lang. **23**(1), 126–144 (2009)
5. Gong, Y., Liu, X.: Generic text summarization using relevance measure and latent semantic analysis. In: Proceedings of the 24th Annual International ACM SIGIR Conference on Research and Development in Information Retrieval, pp. 19–25. ACM (2001)
6. Gupta, A., Kaur, M., Singh, A., Goel, A., Mirkin, S.: Text summarization through entailment-based minimum vertex cover. In: Lexical and Computational Semantics (* SEM 2014), p. 75 (2014)
7. Hammouda, K.M., Kamel, M.S.: Models of distributed data clustering in peer-to-peer environments. Knowl. Inf. Syst. **38**(2), 303–329 (2014)
8. Hennig, L., Labor, D.: Topic-based multi-document summarization with probabilistic latent semantic analysis. In: RANLP, pp. 144–149 (2009)
9. Jing, H.: Sentence reduction for automatic text summarization. In: Proceedings of the Sixth Conference on Applied Natural Language Processing, pp. 310–315. Association for Computational Linguistics (2000)
10. Knight, K., Marcu, D.: Summarization beyond sentence extraction: a probabilistic approach to sentence compression. Artif. Intell. **139**(1), 91–107 (2002)
11. Ledeneva, Y., García-Hernández, R.A., Gelbukh, A.: Graph ranking on maximal frequent sequences for single extractive text summarization. In: Gelbukh, A. (ed.) CICLing 2014, Part II. LNCS, vol. 8404, pp. 466–480. Springer, Heidelberg (2014)
12. Lin, C.-Y.: Rouge: a package for automatic evaluation of summaries. In: Proceedings of the ACL-2004 Workshop on Text Summarization Branches Out, vol. 8 (2004)
13. Lovins, J.B.: Development of a Stemming Algorithm. MIT Information Processing Group, Electronic Systems Laboratory (1968)
14. Mishra, R., Bian, J., Fiszman, M., Weir, C.R., Jonnalagadda, S., Mostafa, J., Fiol, D.G.: Text summarization in the biomedical domain: a systematic review of recent research. J. Biomed. Inf. **52**, 457–467 (2014)
15. Nishino, M., Yasuda, N., Hirao, T., Minato, S., Nagata, M.: A dynamic programming algorithm for tree trimming-based text summarization. In: Proceedings of NAACL HLT, pp. 462–471 (2015)
16. Patil, A., Pharande, K., Nale, D., Agrawal, R.: Automatic text summarization. Int. J. Comput. Appl. **109**(17), 975–8887 (2015)
17. Silva, G., Ferreira, R., Lins, R.D., Cabral, L., Oliveira, H., Simske, S.J., Riss, M.: Automatic text document summarization based on machine learning. In: Proceedings of the 2015 ACM Symposium on Document Engineering, DocEng 2015, pp. 191–194 (2015)
18. Steinberger, J., Ježek, K.: Text summarization and singular value decomposition. In: Yakhno, T. (ed.) ADVIS 2004. LNCS, vol. 3261, pp. 245–254. Springer, Heidelberg (2004)
19. Van Rijsbergen, C.J., Robertson, S.E., Porter, M.F.: New models in probabilistic information retrieval. Computer Laboratory, University of Cambridge (1980)

Circulant Singular Value Decomposition Combined with a Conventional Neural Network to Improve the Hake Catches Prediction

Lida Barba[1,2(✉)], Nibaldo Rodríguez[2], and Diego Barba[3,4]

[1] Engineering Faculty, Universidad Nacional de Chimborazo, 060150 Riobamba, Ecuador
lbarba@unach.edu.ec

[2] School of Informatics Engineering, Pontificia Universidad Católica de Valparaíso, 2374631 Valparaíso, Chile

[3] Engineering Faculty, Universidad de Buenos Aires, C1063ACV Buenos Aires, Argentina

[4] Informatics and Electronic Faculty, Escuela Superior Politécnica de Chimborazo, 060155 Riobamba, Ecuador

Abstract. This paper presents the one-step ahead forecasting of time series based on Singular Value Decomposition of a circulant trajectory matrix combined with the conventional non linear prediction method. The catches of a fishery resource was used to evaluate the proposal, this is due to the great importance of this resource in the economy of a country, and its high variability presents difficulties in the forecasting; the catches of hakes from January 1963 to December 2008 along the Chilean coast (30°S–40°S) are the application data. The forecasting strategy is presented in two stages: preprocessing and prediction. In the first stage the Singular Value Decomposition of a circulant matriz (CSVD) resultant of the mapping time series is applied to extract the components, after the decomposition and grouping, the components interannual and annual were obtained. In the second stage a conventional Artificial Neural Network (ANN) is implemented to predict the extracted components. The results evaluation shows a high prediction accuracy through the strategy based on the combination CSVD-ANN. Besides, the results were compared with the conventional nonlinear prediction based on an Autoregressive Neural Network. The improvement in the prediction accuracy by using the proposed decomposition strategy was demonstrated.

1 Introduction

Forecasting based on historical time series is an important tool in fisheries planning. Autoregressive techniques have been evaluated to model the fishery behavior of different marine species as pacific sardines, anchovies, and hake [1–5]. The Chilean continental shelf is rich in fishery resources, their properly exploitation maximizes the economy of coastal towns. The Chilean hake (Merluccius gayi), has a great economical importance in this country; this specie is one of the main food in the coastal and an important exportation resource.

R. Prasath et al. (Eds.): MIKE 2015, LNAI 9468, pp. 359–369, 2015.
DOI: 10.1007/978-3-319-26832-3_34

Due to the significant fluctuations of the marine ecosystem, the stock of hake is in dependence of the environmental conditions. Considering the information of the Fishery National Service [6], it is observed that during some periods, the hake stock has grown up, whereas in other periods it has been decreasing. It is advisable to keep track of the evolution of this and other marine species, it is essential for the establishment of exploitation policies and control rules.

The fisheries time series have a complex nonlinear structure which involves the use of complex models with a big number of variables. An efficient alternative is the application of nonlinear methods, which could predict not only the observed time series, but rather the components that are incrusted in the time series [7].

In this work is proposed a new hake prediction strategy, this is based on the combination of the Circulant Singular Value Decomposition technique, and a conventional Autoregressive Neural Network. The CSVD strategy is applied to extract the components interannual and annual from the time series. The ANN is based on the Levenberg-Marquardt algorithm, and it is used to predict the extracted components. The results are compared with the conventional prediction of the observed time series based through the same Autoregressive Neural Network structure implemented with the proposed strategy.

The paper is structured as follows. Section 2 describes the Singular Value Decomposition of the circulant matrix. Section 3 presents the Prediction with the Autoregressive Neural Network. Section 4 describes the prediction accuracy metrics. Section 5 presents the Results and Discussions. The conclusions are shown in Sect. 6.

2 Components Extraction Based on Singular Value Decomposition of the Circulant Matrix

The Circulant matrices are an important class of matrices used decades ago in applications of mathematics, its first application is attributed to E. Catalan in 1846 [8]. In prediction it has not been used widely; although the SVD of the Hankel matrix were used to predict the traffic accidents in Chile [9].

The SVD of the circulant matrix is implemented to extract the componentes of low and high frequency of the time series. The process is presented in three steps: Mapping, Decomposition and Extraction.

2.1 Time Series Mapping

The mapping is the first step of the components extraction. The time series is the original sequence to be displayed $a_1, a_2, a_3, \dots, a_N$. Over the first row of the circulant matrix C, is represented the original sequence. Over the first row are applied circular successive rotations, therefore all the rows of the matrix will contain the time series. Considering that a circulant matrix is an square trajectory matrix, the sequence is:

$$C_{N\times N} = \begin{pmatrix} a_1 & a_2 & a_3 & \dots & a_N \\ a_N & a_1 & a_2 & \dots & a_{N-1} \\ \vdots & \ddots & \ddots & \ddots & \vdots \\ a_3 & \dots & a_N & a_1 & a_2 \\ a_2 & \dots & a_3 & a_N & a_1 \end{pmatrix} \quad (1)$$

The matrix C of order $N \times N$, is circulant if the next conditions are fulfilled:

- $c_{i,j} = c_{u,v}$, for $j - i \equiv v - u(modN)$.
- C is circulant if the corresponding matrix conjugated transpose $C*$ is also circulant.
- The Moore-Penrose pseudo-inverse matrix of a circulant matrix is also circulant.
- $\prod C = C \prod$, where $\prod$ is a circulant matrix of order $N \times N$ with the form: ($\prod = circ(0, 1, 0, \dots, 0)$. Hence the multiplication of circulant matrices is commutative [10]; and the product is also a circulant matrix.

2.2 Singular Value Decomposition

Given a matrix C of order $N \times N$, the circular convolution on a function $f \in \mathbb{R}^N$ is expressed by:

$$Cf = f \circledast c, \quad (2)$$

where $\circledast$ shows circulant convolution, and $c \in \mathbb{R}^N$ is the first row of C.

The main circulant matrices theorem relates to that the circulant matrix can be unitarily diagonalized using the Fourier matrix [8]:

$$F_{k,t} = e^{-j2\pi(k-1)(t-1)/N}, \qquad k, t = 1, \dots, N, \quad (3)$$

Then the matrix C can be decomposed in

$$C = \tilde{F}^* \hat{C} \tilde{F}, \quad (4)$$

where $\tilde{F} = \frac{1}{\sqrt{N}} F$, $\hat{C} = diag(\hat{c})$, and $\hat{c} = Fc$ along the diagonal.

Due to $\tilde{F}$ and $\tilde{H}$ are complex, and based on the symmetries in the Fourier transform and in $\hat{c}$, the SVD from the Discrete Fourier transform of $\hat{c}$ is represented by

$$C = UsV^T, \quad (5)$$

where U and V are the matrices that contain the left and right eigenvectors respectively; s is the eigenvalues vector, which is computed with

$$s = diag(abs(\hat{c})), \quad (6)$$

The inputs of U and V are obtained with the following process

$$U_{m,k} = \begin{cases} \frac{z_1}{N}, & k = 1, \\ \sqrt{\frac{2}{N}} \cos\left(\frac{2\pi(k-1)(m-1)}{N} + \theta_k\right), & k = 2, \ldots, \frac{N}{2}, \\ \frac{z_{N/2+1}}{\sqrt{N}} (-1)^{m-1}, & k = N/2+1, \\ \sqrt{\frac{2}{N}} \sin\left(\frac{2\pi(k-1)(m-1)}{N} + \theta_k\right), & k = \frac{N}{2}+2, \ldots, N, \end{cases} \quad (7)$$

where $z_1 = sgn(\hat{c}_1)$, and $z_{n/2+1} = sgn(\hat{c}_{n/2+1})$, and $\theta_k = \angle \hat{c}_k$.

$$V_{m,k} = \begin{cases} \frac{1}{\sqrt{N}}, & k = 1, \\ \sqrt{\frac{2}{N}} \cos\left(\frac{2\pi(k-1)(m-1)}{N}\right), & k = 2, \ldots, \frac{N}{2}, \\ \sqrt{\frac{1}{N}} (-1)^{m-1}, & k = N/2+1, \\ \sqrt{\frac{2}{N}} \sin\left(\frac{2\pi(k-1)(m-1)}{N}\right), & k = \frac{N}{2}+2, \ldots, N. \end{cases} \quad (8)$$

2.3 Components Extraction

The components are extracted from the eigenvalues and eigenvectors computed through the SVD of the circulant matrix. The extraction process is shown below

$$A_i(N \times N) = s(i) \times U(:, i) \times V(:, i)^T, \quad (9)$$

where $A_i(N \times N$ is a matrix which first column contains the ith row of the resultant matrix of components R, with $i = 1, \ldots, N$. therefore the ith component is:

$$P_i = R(i, :)^T, \quad (10)$$

The P_i is the ith extracted component, and $R(i, :)$, represents the ith row of the matrix of components R.

The components can be grouped in k components by simple addition, as follows:

$$G_j = \sum_{i=1}^{t} P_i, \qquad for\ j = 1,\ and\ t < N, \quad (11)$$

where G_j is the first grouped component. The second component is obtained with the addition of the rest of rows ($t+1$ to N).

3 Components Prediction with an Autoregressive Neural Newtork

The components extracted with the Singular Value Decomposition of the circulant matrix presented in the previous Section are used for the ANN prediction. The prediction of each component is used to find the hake stock prediction by simple addition of the ANN estimations, as follows

$$\hat{x}(n+1) = \hat{x_h}(n+1) + \hat{x_l}(n+1), \tag{12}$$

where n represents the time instant, $\hat{x}$ is the stock prediction, $\hat{x_i}$ is the estimated interannual component (G_1), and x_a (G_2) is the estimated annual component.

The Autoregressive Neural Network is applied to obtain de predicted components $\hat{x_i}$ and $\hat{x_a}$. The ANN has a common structure of a Multilayer Perceptron of three layers [11], the inputs are the lagged terms, which are contained in the regressor matrix Z, each row is a regressor vector z_c, at the hidden layer is applied the sigmoid transfer function, and at the output layer is obtained the predicted value. The ANN output is:

$$\hat{x}(n) = \sum_{j=1}^{Q} b_j h_j, \tag{13a}$$

$$h_j = \sum_{i=1}^{m} w_{ji} z_i(n), \tag{13b}$$

where $\hat{x}$ is the hake catch estimated value, n is the time instant, Q is the number of hidden nodes, b_j and w_{ji} are the linear and nonlinear weights of the ANN connections respectively, the sigmoid transfer function is computed with

$$f(x) = \frac{1}{1+e^{-x}}. \tag{14}$$

The ANN is denoted with $ANN(m, Q, 1)$, with m inputs, Q hidden nodes, and 1 output.

The ANN weights b and w are updated with the application of the learning algorithm.

3.1 Levenberg-Marquardt Learning Algorithm

The Levenberg-Marquardt algorithm outperforms the conventional Newton method and the gradient based methods in a widely variety of problems [12–14]. Levenberg-Marquardt (LM) is an optimization algorithm of high application due to the accuracy. The scalar u is a parameter used in LM to determine the behavior, if u increases the value, the algorithm works as the steepest descent algorithm with low learning rate; whereas if u decreases the value until zero, the algorithm works as the Gauss-Newton method ([15]). The weights of the ANN connections are updated with:

$$\omega_{n+1} = \omega_n + \Delta(\omega_n), \tag{15a}$$

$$\Delta(\omega_n) = -[J^T(\omega_n) \times J(\omega_n) + u_n \times I)]^{-1} J^T(\omega_n) \times e(\omega_n), \tag{15b}$$

where ω is the weight vector composed by $(w_{ji}, \ldots, b_j)$, $\Delta(\omega_n)$ is the weight increment, J is the Jacobian matrix, J^T is transposed Jacobian matrix, I is the identity matrix, e is the error vector and n is the time instant.

The fitness function used with LM is MSE (Mean Squared Error), the elements of the Jacobian matrix corresponds to the partial derivative of the fitness function regarding each weight, as follows

$$J(\omega_{j,i}) = \left[\frac{\partial e(\omega_{j,i})}{\partial \omega_{j,i}} \right], \tag{16}$$

where the order of the matrix (J) is $N \times k$; N is the sample size, and k is the total number of weights.

4 Prediction Accuracy Metrics

The prediction accuracy is evaluated with diverse criteria. The differences between the observed and predicted values are quantified by using forms more sensitive to significant over or underprediction, such as the modified Nash-Sutcliffe efficiency (E) [16], the Mean Absolute Percentage Error $(MAPE)$, and the Relative Percentage Error (RPE). Additionally is computed the Generalized Cross Validation (GCV) to calibrate the time window of the autoregressive models.

$$E = 1 - \frac{\sum_{i=1}^{n} |x_i - \hat{x}_i|}{\sum_{i=1}^{n} |x_i - \bar{x}|} \times 100, \tag{17}$$

$$MAPE = \left[\frac{1}{N_v} \sum_{i=1}^{N_v} |(x_i - \hat{x}_i)/x_i| \right] \times 100, \tag{18}$$

$$RPE = \left[\frac{(x_i - \hat{x}_i)}{x_i} \right] \times 100, \qquad i = 1 \ldots N_v, \tag{19}$$

$$GCV = \frac{RMSE}{(1 - P/N_v)^2}, \qquad RMSE = \sqrt{\frac{1}{N_v} \sum_{i=1}^{N_v} (x_i - \hat{x}_i)^2}, \tag{20}$$

where N_v is the validation sample size, x is the observed value, $\hat{x}$ is the predicted value, $\sigma^2(er)$ is the error variance, $\sigma^2(x)$ is the observed data variance, er is the error (residuals of $x_i - \hat{x}_i$), and P is the number of model parameters.

5 Results and Discussion

The data analyzed in this work were obtained from SERNAPESCA [6], and correspond to the monthly catches of hake along the Chilean cost (30°S – 40°S), from January 1963 to December 2008. The raw data are shown in Fig. 1a; from Figure is observed a complex nonlinear behavior. This sample has been divided into two groups, the 75 % of the data are the training sample, consequently the 25 % of the resting data is the validation sample.

5.1 Components Extraction

The components extraction from the time series was applied with the steps detailed in Sect. 2. The implementation of the CSVD gives as result n components, where n is the length of the observed time series. The number of components that were grouped to find the interannual component was found through the addition of the first 23 % of extracted components, whereas the second component was found with the addition of the rest of components. Figure 1b and c, show the components interannual and annual respectively.

From Fig. 1, is observed that the interannual component represents the long-term periods, while the annual component represents the short-term periods.

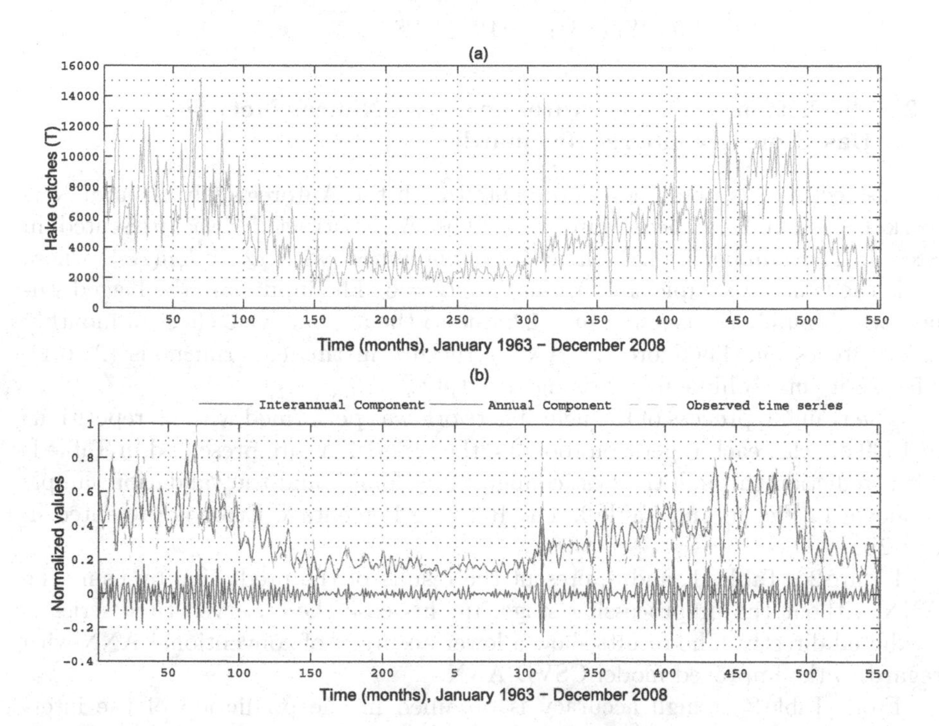

Fig. 1. (a) Observed values of hake catches (b) Components Inter-annual, annual; time series

Table 1. Results of 30 runs of total hake catches prediction based on CSVD-ANN and conventional ANN

	CSVD-ANN				Conventional ANN			
	E (%)	MAPE (%)	RMSE	RPE (±3 %)	E (%)	MAPE (%)	RMSE	RPE (±3 %)
Min	94.4	0.58	0.0025 (37.9T)	62.9	60.7	11.9	0.049 (743T)	9
Max	99.0	3.3	0.0013 (197T)	98.5	78.6	20.0	0.097 (1472T)	24
Std	0.011	0.7	0.0025 (37.9T)	9.8	0.04	1.9	0.012 (182T)	3.1
Mean	97.0	1.7	0.0074 (112T)	84.2	72.9	14.5	0.065 (986T)	15.6

Table 2. Components prediction with CSVD-ANN

	Interannual-Comp.			Annual-Comp.		
	E (%)	MAPE (%)	RMSE	E (%)	MAPE (%)	RMSE
CSVD-ANN	99.6	0.18	0.001 (15.2T)	95.1	25.7	0.0044 (66.7T)

Table 3. Hake catches prediction comparison

	CSVD-ANN	Conventional ANN
E	99.03 %	78.6 %
MAPE	0.58 %	20.0 %
RMSE	0.0025 (37.0T)	0.097 (1471.6T)
RPE	93.2 % (±3 %)	24 % (±3 %)

5.2 Prediction with the Autoregressive Neural Network Based on Levenberg-Marquardt

In this section the prediction is presented with the Autoregressive Neural Network based on Levenberg-Marquardt. The ANN structure was presented in Sect. 3. The number of hidden nodes is computed with $Q = log(N_t)$, where N_t is the training sample size, in this case $Q = 6$. The inputs are the lagged values, which number was set in $m = 22$, due to the annual cycle effect of monthly hake catches [5]. Therefore the ANN structure in this experiment is (22,6,1), with 22 inputs, 6 hidden nodes, and 1 output.

The training process of the neural network was performed with 30 repetitions of 1000 epochs; each repetition took 2 s, the repetitions are presented in Table 1. The prediction through the best configuration found, and the validation sample is shown in Fig. 2a and Table 2. The Relative Percentage Error is presented in Fig. 2b.

Figure 2c, Tables 1, and 3, present the results of the prediction by using the ANN without the preprocessing stage. In this case the observed time series is predicted directly. The results show a lower accuracy of conventional ANN with regard to the improved model CSVD-ANN.

From Table 2, a high accuracy is obtained in the prediction of the interannual component, the Interannual component prediction shows an efficiency of $E = 99.6\,\%$; whereas a lower accuracy was obtained in the prediction of the

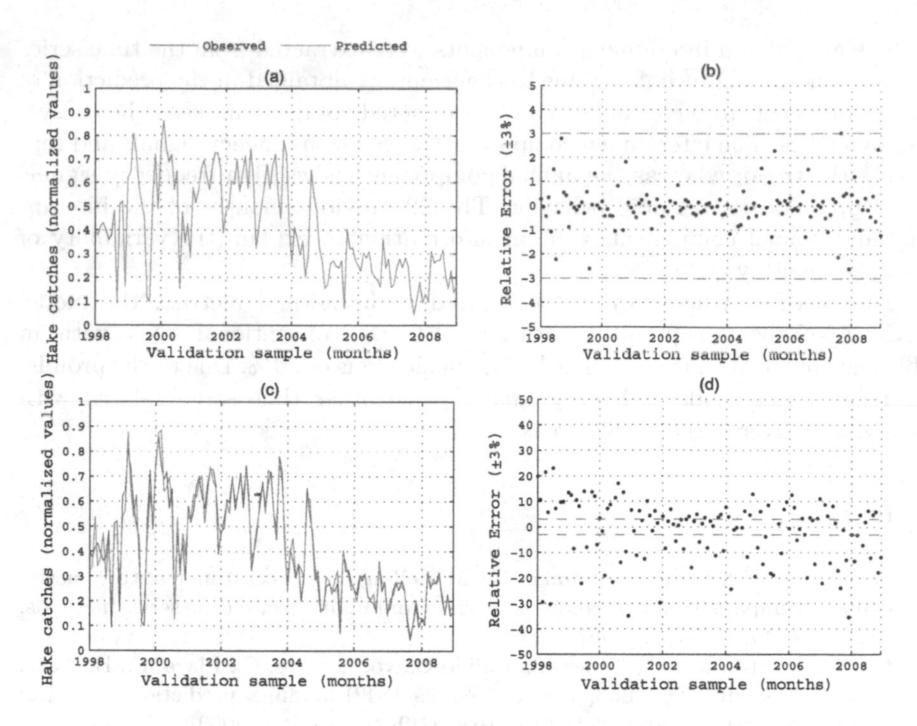

Fig. 2. (a) Prediction with CSVD-ANN (b) Relative Percentage Error (CSVD-ANN) (c) Prediction with conventional ANN (b) Relative Percentage Error (conventional ANN)

annual component, with an efficiency of $E = 95.1\,\%$. The conventional ANN was implemented without the preprocessing stage used to extract the components.

The results obtained with the proposed model CSVD-ANN and the conventional ANN are presented and compared in Fig. 2 and Table 3. The highest accuracy was reached by using the combination strategy CSVD-ANN with regard to the conventional ANN prediction. The model CSVD-ANN presents the highest efficiency $E = 99.03\,\%$, the lowest Mean Percentage Absolute Error $MAPE = 0.58\,\%$, the lowest Root Mean Squared Error $RMSE = 0.0025$ (37.9T) and the highest Relative Percentage Error, the 93.2 % of the points presents a lower RPE than $\pm 3\,\%$. The prediction through the conventional ANN shows low accuracy with $E = 78.6\,\%$, $MAPE = 20.0\,\%$, and $RMSE = 0.097$(1471.6T), and the 24 % of the predicted points shows a lower RPE than $\pm 3\,\%$.

6 Conclusions

In this paper was presented the Singular Value Decomposition of a circulant matrix to enhance the performance of a conventional ANN in one-step ahead hake catches prediction. The strategy was evaluated with the historical time series of monthly catches along the Chilean cost from year 1963 to 2008.

The annual and interannual components were extracted from the time series by using the CSVD technique; due to the accuracy obtained in the prediction of each component, it is conclude that the extracted components keep the ecosystem dynamics. The interannual component shows low frequency signals and represents the trend; whereas the annual component shows high frequency signals and represents the periodic behavior. The 22 previous months of the interannual and annual components contain information to explain the variability of the hake monthly catch.

The prediction accuracy was evaluated with residual metrics; the model CSCD-ANN shown superiority with regard to the conventional ANN, with an Efficiency of 99.03 %, the gain reached in efficiency is of 26 %. Due to the promising results, this model will be evaluated with other time series related with fishery, and diverse knowledge areas.

References

1. Stergiou, K., Christou, E., Petrakis, G.: Modelling and forecasting monthly sheries catches: comparison of regression, univariate and multivariate time series methods. Fish. Res. **29**(1), 55–95 (1997)
2. Gutiérrez-Estrada, J.C., Yánez, E., Pulido-Calvo, I., Silva, C., Plaza, F., Bórquez, C.: Pacific sardine (Sardinops sagax, Jenyns 1842) landings prediction. A neural network ecosystemic approach. Fish. Res. **100**(2), 116–125 (2009)
3. Yánez, E., Plaza, F., Gutiérrez-Estrada, J.C., Rodríguez, N., Barbieri, M., Pulido-Calvo, I., et al.: Anchovy (Engraulis ringens) and sardine (Sardinops sagax) abundance forecast of northern chile: a multivariate ecosystemic neural network approach. Prog. Oceanogr. **87**(14), 242–250 (2010)
4. Kim, J.Y., Jeong, H.C., Kim, H., Kang, S.: Forecasting the monthly abundance of anchovies in the south sea of korea using a univariate approach. Fish. Res. **161**, 293–302 (2015)
5. Rodríguez, N., Barba, L., Rubio, J.: Multiscale RBF neural network for forecasting of monthly hake catches of Southern Chile. Polibits **48**, 47–52 (2013)
6. Servicio Nacional de Pesca. https://www.sernapesca.cl/
7. Rodríguez, N., Cubillos, C., Rubio, J.M.: Multi-step-ahead forecasting model for monthly anchovy catches based on wavelet analysis. J. Appl. Math. **2014**, Article ID 798464 (2014)
8. Davis, P.J.: Circulant Matrices, 2nd edn. Chelsea Publishing, New York (1994)
9. Barba, L., Rodríguez, N., Montt, C.: Smoothing strategies combined with arima and neural networks to improve the forecasting of traffic accidents. Sci. World J. **2014**, Article ID 152375 (2014)
10. Canright, D., Chung, J., Stănică, P.: Circulant matrices and affine equivalence of monomial rotation symmetric boolean functions. Discrete Math. **338**(12), 2197–2211 (2015)
11. Freeman, J.A., Skapura, D.M.: Neural Networks, Algorithms, Applications, and Programming. Addison-Wesley, Redwood City (1991)
12. Kermani, B.G., Schiffman, S.S., Nagle, H.T.: Performance of the Levenberg-Marquardt neural network training method in electronic nose applications. Sens. Actuators B: Chem. **110**(1), 13–22 (2005)

13. Yetilmezsoy, K., Demirel, S.: Artificial neural network (ANN) approach for modeling of Pb(II) adsorption from aqueous solution by Antep pistachio (Pistacia Vera L.) shells. J. Hazard. Mater. **153**(3), 1288–1300 (2008)
14. Mukherjee, I., Routroy, S.: Comparing the performance of neural networks developed by using Levenberg Marquardt and Quasi-Newton with the gradient descent algorithm for modelling a multiple response grinding process. Expert Syst. Appl. **39**(3), 2397–2407 (2012)
15. Hagan, M., Demuth, H., Bealetitle, M.: Neural Network Design. Hagan Publishing (2002)
16. Krause, P., Boyle, D.P., Bäse, F.: Comparison of different effciency criteria for hydrological model assessment. Adv. Geosci. **5**, 89–97 (2005)

To Optimize Graph Based Power Iteration for Big Data Based on MapReduce Paradigm

Dhanapal Jayalatchumy[1(✉)] and Perumal Thambidurai[2]

[1] Department of CSE, PKIET, Karaikal, India
djlatchumy@gmail.com
[2] PKIET, Karaikal, India
ptdurai158@yahoo.com

Abstract. The next big thing in the IT world is Big Data. The values generated from storing and processing of Big Data cannot be analyzed using traditional computing techniques. The main aim of this paper is to design a scalable machine learning algorithm to scaleup and speedup clustering algorithm without losing its accuracy. Clustering using power iteration is fast and scalable. However, it requires matrix computation which makes the algorithm infeasible for Big Data. Moreover, power method converges slowly based on eigen vector. Hence, in this paper an investigation is done on convergence factor by applying a modified constraint that minimizes the computational cost by making the algorithm converge quickly. MapReduce parallel environment for Big Data is verified for the proposed algorithm using different sizes of datasets with different nodes in the cluster selecting speedup, scalability, and efficiency as the indicators. The performance of the proposed algorithm has been shown with respect to the execution time and the number of nodes. The results show that the proposed method is feasible and valid. It improves the overall performance and efficiency of the algorithm that can meet the needs of large scale processing.

Keywords: Big data · Convergence · Mapreduce · Modified constraint · Performance · Scalability · Speedup

1 Introduction

Big Data challenges data mining techniques as there is a need to handle dramatic data growth [15, 17]. To deal with massive data there is a need for powerful tools to discover knowledge. Data mining techniques are excellent knowledge discovery tools. Clustering is one among them that groups similar data [11]. PIC is one of the spectral clustering algorithms that seems to be fast, scalable and suitable for larger datasets [10, 25]. But there is a problem of memory consumption due to matrix computation. PIC computes pseudo eigen vector using power iteration. To make the algorithm more effective an Inflation based PIC is proposed which makes the algorithm converge at a faster rate. Hence, the computational time is much reduced yielding better clustering results. Motivated by the need for parallelism and effectiveness of Big Data processing, the main effort in this paper is on parallelization strategies and implementation details for minimizing computation cost [13]. The designed algorithm is made to work on

R. Prasath et al. (Eds.): MIKE 2015, LNAI 9468, pp. 370–381, 2015.
DOI: 10.1007/978-3-319-26832-3_35

MapReduce framework since scalability and performance becomes a key concern for handling mass and distributed data [1, 18].

The idea of MapReduce is to handle parallelism, fault tolerance, data distribution and node failures in a single library [4, 12, 15]. Experiments have been conducted on various datasets and the results show that the proposed algorithm performs well on this framework. It is fast, scalable and can work with Big Data. The computational cost and time complexity has been much reduced using MapReduce in the proposed algorithm. It is proved that as the number of node increases, execution time decreases, and hence performance increases. The experiment has been shown on Amazon EC2. The rest of the paper is organized as follows. Section 2 explains about the Methodology involved followed by proposed algorithm in Sect. 3. In Sect. 4, the experimental results are discussed and Sect. 5 finally concludes giving the future directions.

2 Methodology

2.1 Power Method

The power method is used to find the first eigen value and eigen vector of a matrix. Given a N*N matrix A, the eigen vector corresponding to the eigen value is identified [20, 21]. Let $\lambda_0, \lambda_1, \lambda_2 \ldots \lambda_{n-1}$ be the eigen values of A. The largest eigen value is λ_0 (i.e.) $|\lambda_0| > |\lambda_1| > |\lambda_2| \cdots\cdots\cdots\cdots |\lambda_n|$. Let there be an eigen pair (λ_j, v_j) and assume that the vectors $v_0, v_1, v_2 \ldots v_{n-1}$ be linearly independent. To start with we take an initial vector at a time 0. The vector is written as a combination of the vector that is linearly independent [14]. Hence, $x^{(0)} = \gamma_0 v_0 + \gamma_1 v_1 + \gamma_2 v_2 \cdots\cdots\cdots \gamma_{n-1} v_{n-1}$. If the first vector is written with a matrix A we create [4, 20]

$$\begin{aligned} x^{(1)} &= Ax^{(0)} \\ &= (\gamma_0 v_0 + \gamma_1 v_1 + \gamma_2 v_2 \cdots\cdots\cdots \gamma_{n-1} v_{n-1}) \\ &= \gamma_0 A v_0 + \gamma_1 A v_1 + \gamma_2 A v_2 \cdots\cdots\cdots + \gamma_{n-1} A v_n \end{aligned}$$

Since they are eigen vectors the value Av_0 becomes $\lambda_0 v_0$ and Av_1 becomes $\lambda_1 v_1 \ldots Av_{n-1}$ is $\lambda_{n-1} v_{n-1}$ and similarly others can be found. On computation we get

$$lim_{k\to\infty}(x)^k = \lim_{k\to\infty}(\gamma_0 \lambda_0 + \gamma_1 \lambda_1^k v_1 + \gamma_2 \lambda_2^k v_2 \cdots \gamma_{n-1} \lambda_{n-1}^k v_{n-1}).$$

Here $\gamma_0 \lambda_0^k v_0$ dominates other vectors. The direction of v_1 is insignificant and vector v_0 keeps on increasing. To fix the increasing value v_0 the vectors should be kept under control [2, 5, 8, 20]. So it is rewritten as

$$x^{(0)} = \gamma_0 v_0 + \gamma_1 v_1 + \gamma_2 v_2 \ldots \gamma_{n-1} v_{n-1},$$

$$x^{(1)} = \gamma_0 \frac{1}{\lambda_0} A v_0 + \ldots\ldots + \gamma_{n-1} \frac{1}{\lambda_0} A v_{n-1}$$

Finally we get

$$\begin{aligned} &= \gamma_0 v_0 + \gamma_1(0)v_1 + \cdots\cdots + \gamma_{n-1}(0)v_{n-1} \\ &= \gamma_0 v_0 \end{aligned}$$

eigen vector is finding the iteration involved and it's about its length. Thus it is derived that the process yields a vector in the direction which is largest in magnitude [14, 20]. Hence, the power method is given as $Ax = \lambda x$

$$X_{i+1} = \frac{Ax_i}{||Ax_i||} \qquad \lambda_{i+1} = Ax_i$$

2.2 Convergence Constraint

The power method does not compute matrix decomposition. Hence, it can be used for larger sparse matrix. The power method converges slowly since it finds only one eigen vector [7]. The starting vector x_0 in terms of eigenvector of A is given by $X_0 = \sum_{i=1}^{n} a^i v^i$. Then $X_k = AX_{k-1} = A^2 X_{k-2} = \cdots\cdots A^k X_0$

$$\sum_{i=1}^{n} \lambda_i^k a_i v_i = \lambda_n^k \left(a_n v_n + \sum_{i=1}^{n-1} \left(\frac{\lambda_1}{\lambda_n} \right)^k a_i v_i \right)$$

The higher the power of $\left(\frac{\lambda_1}{\lambda_n}\right)$ it goes to 0. That is it converges if λ_1 is dominant and if q^0 has a component in the direction of corresponding eigen vector x_1 [7]. It can also happen that the power method fails to converge in case if $|\lambda_1| = |\lambda_2|$. The convergence rate of power iteration depends on the dominant ratio of the operator or the matrix which is challenging when the dominance ratio is close to 1. Hence, convergence acceleration method is required to reduce the uncertainty of the results with significantly less computational cost [21]. To overcome this drawback inflated based power iteration clustering algorithm has been proposed. The advantage of this method is the improvement in the convergence rate [7]. The proposed algorithm has two advantages less computational time and fast convergence rate.

2.3 Proposed Inflated PIC

PIC is a Spectral Clustering technique that is based on eigen decomposition of n*n kernel or affinity matrix which occupies much time and space [6, 10]. PIC overcomes it by finding a pseudo eigen vector, a linear combination of eigen vectors, it uses the power iteration method to find the largest of eigen vector.

Given a dataset $X = \{x_1, x_2, \ldots, x_n\}$, a similarity function $s(x_i, x_j)$ is a function where $s(x_i, x_j) = s(x_j, x_i)$ and $s \geq 0$ if $i \neq j$, and $s = 0$ if $i = j$. A matrix $A_{ij} = s(x_i, x_j)$ is defined as an affinity matrix. The diagonal matrix is the degree matrix D associated with A and is defined [25] as $d_{ii} = \Sigma_j Aij$. A normalized affinity matrix W is defined as $D^{-1}A$. The affinity matrix is constructed to model the neighbor relations between the data points. Affinity matrix can be constructed using many techniques like the K-Nearest neighborhood, fully connected graph etc [4, 5].

For a set of data points $\{x_1, x_2 \ldots x_n\}$ where x is a d-dimensional vector the similarity between the matrices using the Euclidian distance given as [10, 24]

$$s(x_i, x_j) = \exp(-\frac{||x_i - x_j||_2^2}{2\sigma^2})$$

where σ is the scaling parameter that controls the kernel width [3, 17]. In this method power iteration is used to find the largest eigen vector [10]. It is an iterative method that starts with an arbitrary vector v^0 and iteratively updates the vector by multiplying by $Wv^{t+1}=cWv^t$ where c is the normalizing constant to keep the vector stable.

PIC on multiclass dataset causes intercollision problem [5]. Now the main aim is to replace the large N dimensional eigen value problem with a N dimensional classical problem [7]. Since it relies on matrix vector multiplication; we reduce the eigen symmetric value A, $A\psi_n = e_n\psi_n$ Where ψ_n is the nth eigen vector with eigen value e_n. The main aim is to introduce a constraint that induces an inflationary growth to find a desired extremal solution. The constraints are used to find the minimum under time without any local minima [7]. The eigen vector grows exponentially fast through a lagrangian multiplier that regulates the inflation. Nearby eigen vectors are also found. The lagrangian approach is given as

$$L = \sum\nolimits_i x_i^2 - \sum\nolimits_{ij} x_i A_{i,j} x_j + \lambda\left(\sum\nolimits_i x_i^2 - 1\right)$$

Where, λ is the Lagrangian multiplier that enforces normalization. The corresponding Euler equation is

$$\ddot{x}_i = -\sum\nolimits_j A_{i,j} x_j + \lambda(t) x_i, \text{ i = 1 .... n}$$

On integrating with time δt gives

$$p_i(t+\delta t) = p_i(t) - \sum\nolimits_j \left[A_{i,j} - \lambda(t)\delta_{ij}\right] x_j(t)\delta t$$
$$x_i(t+\delta t) = x_i(t) + p_i(t+\delta t)\delta t$$

Let λ (t) be a constant for small time step we have, $\xi_i(t) = a_i \cos(w_i t + \delta_i)$, If $e_i > \lambda$, where δ_i is a real phase shift and $w_i = \sqrt{\lambda} - e_i$ for low eigen value $e_i < \lambda$, they coordinate as $\xi_i(t) = (a_i' e^{w_i' t} + b_i' e^{w_i' t})$. The eigen modes with eigen values $e_i < \lambda(t)$ gets inflated exponentially at a time(t). However, λ(t) regulates inflation by managing the inflated and the non-inflated states [7]. The need for eigen pair and accuracy can be adjusted based on the convergence factor [25]. The algorithm for inflated PIC is shown in Algorithm 1.

Algorithm 1: Inflated Power Iteration Clustering

Input: Converted Normalized affinity matrix W

Step 1. Calculate new vector and new velocity by V_t= Power Iteration (W_t-1)

Step 2. Now find v_t from w_t,

$$v^t = \gamma W v^{t-1}$$
$$\delta^{t+1} = |v^{t+1} - v^t|$$
$$\text{where} \quad \gamma = \frac{1}{||Wv^{t+1}||}$$

Step 3. Check the acceleration $|\delta^t - \delta^{t-1}|$ and Increment the value of t.

Step 4. Inflate the vector and perform update through iteration and normalize it from getting large for the time δt

Step 5. Calculate the rate of convergence of inflation method of the ground state ξ_0 relative to the first excited state $\xi 1$ denoted as $\frac{\xi_0(t)}{\xi_n(t)}$ where ξ_0 =e_1-e_0 obtained by λ (t) = λ (t)+ ξ_{01} and λ (t)is the Lagrangian multiplier at the time t is

$$\delta t \sim \sqrt{\frac{e_1 - e_0}{e_{max} - e_0}}$$

Step 6. Repeat step 4 on incrementing the value of t until the algorithm converges using the constraint $(|\delta^t - \delta^{t-1}| \simeq 0 \ \&\& \ \frac{v_{max}-v_0}{2} \ != 0)$

Step 7. Cluster the points v_t on a k dimensional subspace using K-Means algorithm.

Output: Clusters c_1, c_2....c_k..

The rate of convergence is an important factor to control to growth of inflation. It is limited with the time step δt. If the value is large it causes the inflation to wrong modes [7]. The convergence is calculated with respect to the ground state and the first excited state which is given as $\xi_0(t)/\xi_n(t) \sim e^{\sqrt{e_1 - e_0 t}}$. The optimal rate of convergence in time δt is given $\frac{1}{\delta t} \sim \sqrt{\frac{e_{max}-e_0}{e_1-e_0}}$.

3 Algorithm Design for Inflated PIC in MapReduce

In this section a detailed process for Inflated PIC algorithm using MapReduce based on similarity measures and matrix vector multiplication is presented. The parallelism concept comes into picture for an algorithm implemented in MapReduce framework when map functions are forked for every job. These maps are run in any node under distributed environment configured under Hadoop [10, 24, 25]. The job distribution is done by the Hadoop system and datasets are required to be put in HDFS [13]. The flow of IPIC is shown in Fig. 1.

Mapper Function Phase 1- The input dataset is stored on HDFS as a sequence file <Key, Value> pairs where each is a record in the dataset [16]. The map tasks accepts the datasets and splits into <Key, Value> pair. The parallelization takes place in splitting the dataset into chunk to obtain the sub similarity matrix [10]. The mapper constructs the row sum of the sub similarity matrix and the results are send to the reduce phase.

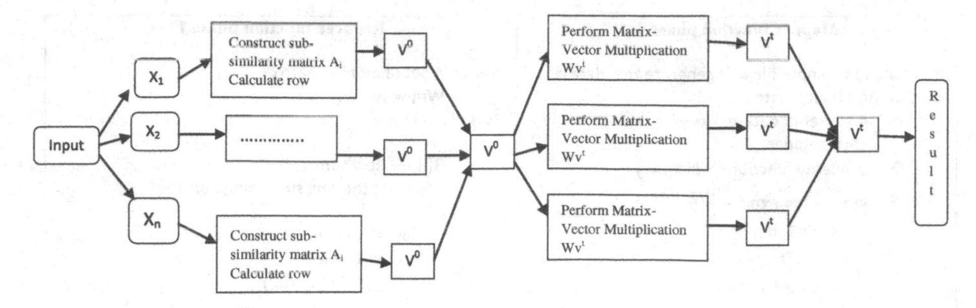

Fig. 1. MapReduce flow of inflated power iteration clustering algorithm

Reducer Function Phase 1- The input for the reducer function will be from the intermediate file [25]. The row sum from each mapper is obtained, shuffled and reduced to form the overall rowsum R. The initial vector v^0 is calculated using the norm $v^0 = \frac{R}{\|R\|}$.

Mapper Function Phase 2- The matrix vector multiplication is performed by reading the initial vector v^0 obtained during the previous phase in the mapper from HDFS [10, 14, 19]. Receive the row value and dot product with the vector to produce the element of output file. Emit (index, value) and construct the vector in the reducer.

Reducer Function Phase 2- The vector v_t obtained is normalized by its element sum and the rate of convergence is calculated [7, 25]. To calculate the rate of convergence of the inflated method consider the inflation of the ground state to the first excite state given as $\xi_0(t)/\xi_n(t) \sim e^{\sqrt{e_1 - e_0 t}}$. The optimal rate of convergence in time δt is shown [7] and the pseudo code for mapper and reducer functions are given below in Algorithm 2

$$\delta t^2 \sim \frac{e_1 - e_0}{e_{max} - e_0} \quad \text{(i.e.)} \quad \delta t \sim \sqrt{\frac{e_1 - e_0}{e_{max} - e_0}}.$$

4 Experimental Results and Discussions

Various experiments have been conducted and the proposed algorithm was tested for different data points in a d-dimensional space. [3] The performance of the algorithm has been investigated through number of iterations taken for convergence of the data points and execution time taken in seconds. Experiments was conducted on datasets like MNIST, BIRCH, IRIS, HOUSE, BREAST datasets and synthetic datasets of various sizes into 5 groups of 2mb, 10 MB, 250 MB, 500 MB and 800 MB. Speedup, efficiency and reliability are considered as evaluation indicators to test the overall performance of the algorithm based on MapReduce environment comparing it with the single machine environment [22].

Mapper function phase 1

Input: Read the input file and generate the dataset .
Output: An Affinity matrix

- A subset of datasets $x \leftarrow \{x_1, x_2, x_3 \ldots x_n\}$ in each mapper.
- Similarity function $\leftarrow s(x_i, x_j)$
- $s(x_i, x_j) = \exp\left(-\frac{\|x_i - x_j\|_2^2}{2\sigma^2}\right)$

```
for i=1 to n do
      D[i,i]=0;
   for j=1 to n do
      D[i,i]=D[i,i]+A[i,j]
   for i=1 to n do
      D[i,i]=1/ D[i,i];
loop
 for i=1 to n do
  for j= 1 to n do
    W[i,j]=D[i,j]*A[i,j]
  end
end
emit < matrix w>
```

Reducer function phase 1

Input : A set of affinity matrix
$W=\{w_1, w_2, w_3 \ldots w_n\}$
Output: Vector v^0

```
R[i] ← Row sum
Calculate the row sum using norm
{
  for i=1 to n do
   for j= 1 to n do
     R[i,j]=R[i,j]+W[i,j]
   end
  end
   {
     for (int i=0; i<n; i++)
       v^0=R/r
     end
   }
}
Output record <v^0>
```

Mapper function phase 2

Input: Initial vector v^0 obtained using the reduce phase.
Output: Matrix[Wv^t]

```
{
  for (int i=0; i<n; i++)
  {
    v^t=0;
    for(j=0;j<n; j++)
    Wv^t+=W[i][j]*v^(t-1)[j];
  }
}
emit pair(matrix, vector)
```

Reducer function phase2

Input the matrix value and apply constraint.

$$\xi_{01} \leftarrow \text{e1-e0} ; \frac{\xi_0(t)}{\xi_n(t)} \sim -e^{\sqrt{e_1 - e_{0t}}}$$

where $\xi_0 = e_1\text{-}e_0$ obtained by $\lambda(t) = \lambda(t) + \xi_{01}$ where λ (t)is the lagrangian multiplier at the time t is given as

$$\delta t \sim \sqrt{\frac{e_1 - e_0}{e_{max} - e_0}}$$

```
loop
   for i=1 to n do
    for j= 1 to n do
     V_i^t= V_i^t + Wij* V_i^(t-1)
    end
  end
```

until $(|\delta^t - \delta^{t-1}| \simeq 0 \ \&\& \ \frac{v_{max} - v_0}{2} \ != 0)$
Cluster on points <v_t > by using K means algorithm.
Output the final vectors.

Algorithm 2. Mapper/Reducer function phases

- ***Complexity of Algorithm using MapReduce***-The overall complexity of the Hadoop operation is given as $CR_D + NCR_DW_D$ where C is the number of clusters [15], N is the number of points a given mapper is responsible for clustering and RD and WD are read and write times to disk and the overall runtime of the reducer is $CR_D + MNR_DW_D$, where M is the number of mappers in the system. The complexity of the algorithm is analyzed by splitting the datasets between the available mappers. For each mapper an iteration based on the number of nodes is performed and iterated through the output from the mappers for reducer. It is calculated by the number of iterations and the dimensionality of similarity matrix to the number of mappers involved [3]. Hence, the computational complexity is reduced to a greater extent. Thus the computational complexity of proposed algorithm using MapReduce is expressed as O(k(n + 1)/No. of Mappers).

- ***Computational Time Results***-The overall complexity of the matrix vector multiplication is given as $O(n^3)$ and the computational complexity is given as $O(n^2)$ which makes it impractical for solving large scale clustering algorithms [10]. Hence, the algorithm should have a faster execution speed to reduce the complexity. The computational time for larger datasets for PIC, IPIC and K-Means on a single machine and MapReduce environments are shown in Figs. 2 and 3. IPIC is fastest among these algorithms [10, 25]. Its average computational time is less than the compared algorithms particularly when the dataset size is large. So IPIC algorithm is efficient for larger dataset and has less complexity in the MapReduce environment.
- ***Test of Speedup Ratio***- Speedup is an important measure to verify the performance. It is defined to reduce the running time. The greater the ratio of speedup, lesser is the time that the parallel processors exploit [9, 22]. This is the reason for parallelization of a sequence algorithm. The speedup is calculated using the formula [9] $S = Ts/Tp$ Where T_s is the execution time of the fastest sequential program and T_p is the execution time of the parallel program. If a parallel program is executed on (p) processor, the highest value is equal to number of processors [9]. In this system every processor needs T_s/p time of complete the job (i.e.) $S = Ts/(Tp/p) = P$. In Map Reduce environment when the algorithm is dealt with different increasing sizes datasets, the algorithm speedup ratio performance is getting better [3, 9, 22]. The reason for this is the combine operation reduces the communication cost between master and slave. Hence, we can say that the as the data size increases, speedup ratio performance will improve (i.e.) when the data node increases running time proportionally decreases.

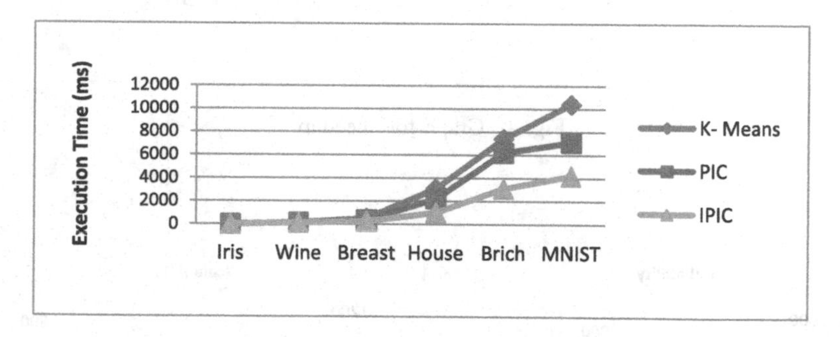

Fig. 2. Running time in single machine environment

The result of speedup ratio performance tests according to the various synthetic datasets are shown in Fig. 4. It is clearly shown that as the value of the node increases the execution time decreases with an increase in speed up.

- ***Analysis of Scalability.*** The efficiency of parallel algorithm represents the utilization of the cluster during execution. When the number of nodes and the size of processing the datasets increases, the running time changes accordingly and the

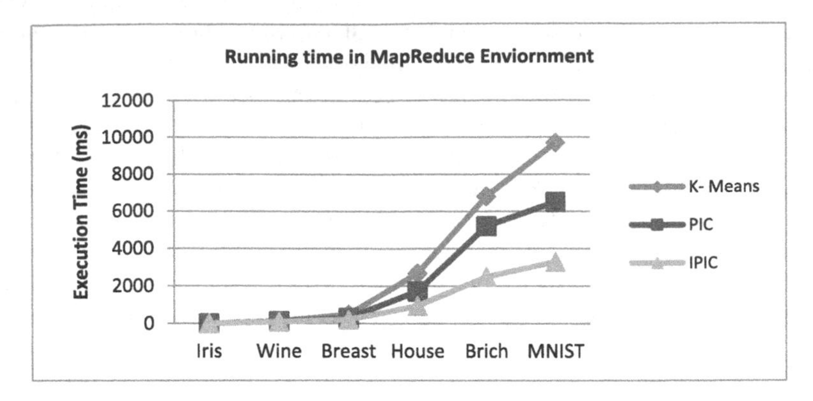

Fig. 3. Running time in MapReduce environment

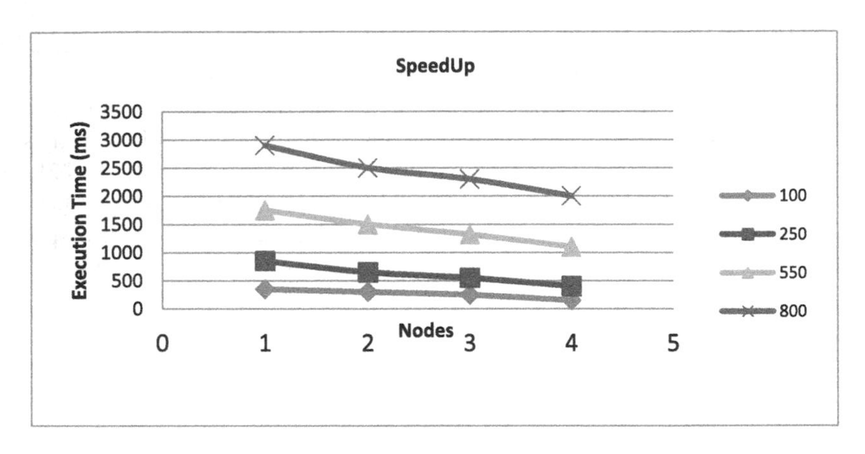

Fig. 4. Graph for speedup

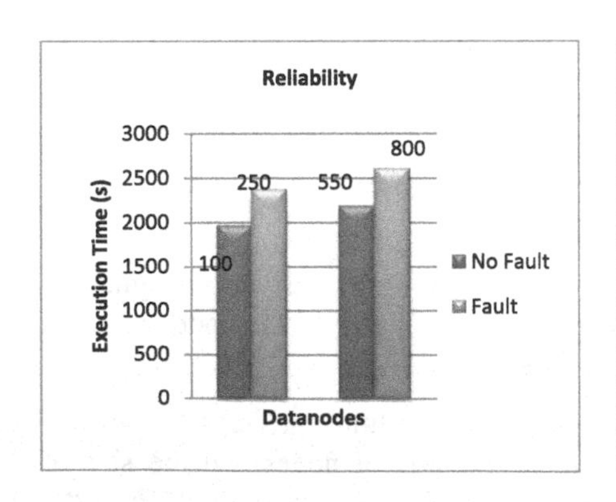

Fig. 5. Reliability of IPIC

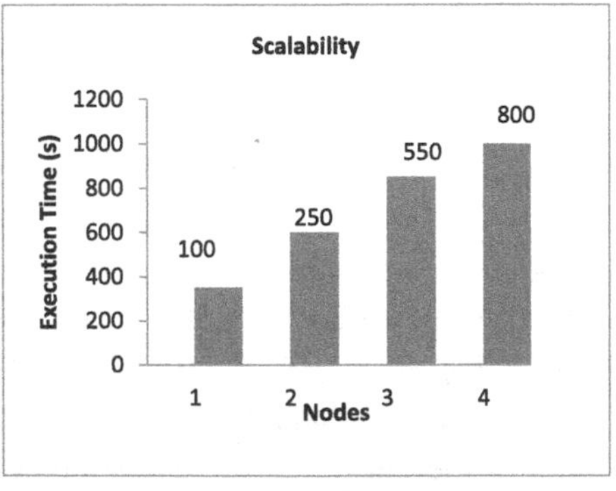

Fig. 6. Scalability of IPIC

efficiency is increased (i.e.) it has better scalability in the MapReduce environment. The formula [5, 23] to calculate efficiency is given as Efficiency (E) = S_R/p. Where S_R represents the ratio of speedup, and p represents the number of processors in the cluster. The graph for relative efficiency of the algorithm for different synthetic datasets has been shown in Fig. 6. From the figure it is seen that as the data size increases, the efficiency is greater, (i.e.) it has better scalability. Experiment results shows that the proposed algorithm has better scalability for larger datasets.

- ***Reliability.*** The algorithm is made to run in single machine and in MapReduce environment which is composed of four data nodes and it is tested for its reliability [9, 23]. In a single machine environment when the machine fails then the entire task is being dropped or fails. There is no reliability here. Whereas, in the MapReduce environment since there are more than one data nodes even if 1 data nodes shuts, the algorithm can still run using the data nodes and can produce the same result but with an increase in time. The reason is Hadoop is fault tolerant. When a fault occur other nodes performs the task assigned by the jobtracker. The task will be computed and it can be said that it has good reliability in MapReduce environment. The graph for reliability is shown in Fig. 5.

5 Conclusion

Increasing data size motivates a need to propose new algorithm and choose a new framework. In this paper an Inflated Power iteration clustering algorithm was proposed and tested in both single machine environment and Hadoop environment for real dataset and synthetic dataset. A depth analysis was done on MapReduce environment on large scale datasets. MapReduce performs well for larger data size than smaller. The algorithm was redesigned to improve the overall performance and efficiency. The computational cost and time complexity has been much reduced using MapReduce environment. We noticed that increasing the number of nodes in mapper affects the performance. The algorithm based on MapReduce shows that it has good scalability and speed when compared to single machine environment while it maintains the quality of clusters. As a future work clustering algorithm on GPU based MapReduce should be investigated to achieve even better scalability and performance.

References

1. Fashanu, A., Ale, F., Agboola, O.A., Ibidaapo Obe, O.: Performance analysis of parallel computing algorithm developed for space weather simulation. Int. J. Advancements Res. Technol. **1**(7), 2278–7763 (2012)
2. Shirkhorshidi, A.S., Aghabozorgi, S., Wah, T.Y., Herawan, T.: Big data clustering: a review. In: Murgante, B., Misra, S., Rocha, A.M.A.C., Torre, C., Rocha, J.G., Falcão, M.I., Taniar, D., Apduhan, B.O., Gervasi, O. (eds.) ICCSA 2014, Part V. LNCS, vol. 8583, pp. 707–720. Springer, Heidelberg (2014)

3. Azmoodeh, A., Hashemi, S.: To boost graph clustering based on power iteration by removing outliers. In: Herawan, T., Deris, M.M., Abawajy, J. (eds.) Proceedings of the First International Conference on Advanced Data and Information Engineering. LNEE, vol. 285, pp. 249–258. Springer, Heidelberg (2013)
4. Elsayed, A., Ismail, O., EiSharkawi, M.E.: MapReduce: state-of-the-art and research directions. Int. J. Comput. Electr. Eng. **6**(1) (2014). doi:10.7763/IJCEE.2014.v6.789
5. Buzbee, B.L.: The efficiency of parallel processing. Frontiers of Supercomputing, Los Alamos Siencee Fall 7 (1983)
6. Fowlkes, C., Belongie, S., Chung, F., Malik, J.: Spectral grouping using the Nystrom method. IEEE Trans. Pattern Anal. Mach. Intell. **26**, 214–225 (2004)
7. Heller, E.J., Kaplan, L., Pollaman, F.: Inflamatory dynamics for matrix Eigen value problems. PNAS, **105**(22), 7631–7635 (2008). doi:10.1073/pnas.0801047105
8. Xue, F.: Numerical solution of eigenvalue problems with spectral transformations. Doctor of Philosophy (2009)
9. Alecu, F.: Performance analysis of parallel algorithms. J. Appl. Quant. Methods **2**(1), 129–134 (2007)
10. Lin, F., Cohen, W.W.: Power iteration clustering. In: International Conference on Machine Learning, Haifa, Israel (2010)
11. Fahad, A., Alshatri, N., Tari, Z., Zomaya, A., Foufou, S., Bouras, A.: A survey of clustering algorithms for big data taxonomy and empirical analysis. IEEE Trans. Emerg. Top. Comput. (2014). doi:10.1109/TETC20142330519
12. Ninama, H.: Distributed data mining using message passing interface. Rev. Res. **2**(9) (2013). ISSN 2249-894X
13. Dean, J., Ghemawat, S.: MapReduce: simplified data processing on large clusters. ACM Commun. **51**(1), 107–113 (2008)
14. Lambers, J.: The eigenvalue problem: power iterations. In: MAT 610 Summer Session 2009–10
15. Yang, J., Li, X.: MapReduce based method for big data semantic clustering. In: Proceedings of the 2013 IEEE International Conference on Systems, Man, and Cybernetics (SMC 2013) (2013). ISBN 978-1-4799-0652/13. doi:10.1109/SMC.2013.480
16. Kamalraj, N., Malathi, A.: Hadoop operations management for big data clusters in telecommunication industry. Int. J. Comput. Appl. (0975-8887) **105**(12), 40–44 (2014)
17. Shim, K.: MapReduce algorithms for big data analysis. In: Proceedings of the VLDB Endowment, VLDB Endowment 21508097/12/08, vol. 5, no. 12. (2012)
18. Lancos, C.: An iteration method for the solution of Eigen value problem of linear differential and integral operators. J. Res. Nat. Bur. Stand. **48**, 255 (1959)
19. Steinbach, M., Ertöz, L., Kumar, V.: The challenges of clustering high dimensional data. In: Wille, L.T. (ed.) New Directions in Statistical Physics, Book Part IV, pp. 273–309. Springer, Heidelberg (2004). doi:10.1007/978-3-662-08968-2_16
20. Panju, M.: Iterative methods for computing eigenvalues and eigenvectors. The Waterloo Mathematics Review. University of Waterloo (2011). http://mathreview.waterlo.ca
21. Numerical methods, chapter 10.3 power method for approximating eigenvalues. www.cengage.com/resource_uploads/downloads/0618783768_138794.pdf
22. Gobil, P., Garg, D., Panchal, B.: A performance analysis of MapReduce applications on big data in cloud based Hadoop. In: ICICES2014, Chennai. IEEE (2014). ISSN 978-1-4799-3834-6/14

23. Rong, Z., Xia, D., Hang, Z.: Complex statistical analysis of big data: implementation and application of apriori and FP-growth algorithm based on MapReduce. In: 2013 IEEE 4th International Conference on Software Engineering and Service Science (ICSESS). IEEE (2013). ISSN 978-1-4673-5000-6/13. doi:10.1109/ICSESS.2013.6615467
24. Chen, W.Y., Song, Y., Bai, H., Lin, C., Chang, E.Y.: Parallel spectral clustering in distributed systems. IEEE Trans. Pattern Anal. Mach. Intell. **33**(3), 568–586 (2011)
25. Yana, W., et al.: p-PIC: parallel power iteration clustering for big data. J. Parallel Distrib. Algorithm **73**(3), 352–359 (2013)

Complex Transforms

Garimella Rama Murthy[1(✉)] and Tapio Saramaki[2]

[1] SPCRC, International Institute of Information Technology, Hyderabad, India
rammurthy@iiit.ac.in
[2] Department of Signal Processing, Tampere University of Technology, Tampere, Finland

Abstract. In this research paper, motivated by the concept of complex hypercube, a novel class of complex Hadamard matrices are proposed. Based on such class of matrices, a novel transform, called complex Hadamard transform is discussed. In the same spirit of this transform, other complex transforms such as complex Haar transform are proposed. It is expected that these novel complex transforms will find many applications. Also, the associated complex valued orthogonal functions are of theoretical interest.

Keywords: Transforms · Hadamard matrix · Complex signum function

1 Introduction

Homosapien civilizations, across the planet innovated positional number systems to conduct day-to-day life. Napier discovered the concept of logarithm that enabled him to convert the operation of multiplication of large numbers to additions of small numbers in some sense. This discovery led to the concept of "transform" in primitive form. Discoveries in algebra led to the concept of a function, more particularly periodic function. Fourier, in his studies on heat conduction problems proposed the representation of a periodic function in terms of trigonometric functions. The coefficients of so called Fourier series were computed explicitly. To extend the idea to aperiodic functions, Fourier devised the so called "Fourier transform" of real valued signals as well as complex valued signals. The concept was generalized by Laplace to more general class of functions using the so called Laplace transform.

In the case of design and analysis of linear time invariant systems, Fourier/Laplace transform enabled deriving very elegant and powerful results. Researchers naturally questioned the representability of signals that are defined on a finite support in terms of other interesting finite collection of signals. In that effort, various families of orthonormal basis functions were discovered. It has been shown that functions/signals in certain function spaces (i.e. $L^p - spaces$) can be expressed in terms of orthonormal basis functions.

With origins in solution of linear system of equations, linear algebra was established as an important subject with applications in science, engineering and other fields of human endeavor. The concepts such as vector space, linear independence, dimension, basis were provided sound logical/mathematical footing. They led to the concept of orthonormal basis vectors. With the important result that any vector can be expressed

R. Prasath et al. (Eds.): MIKE 2015, LNAI 9468, pp. 382–391, 2015.
DOI: 10.1007/978-3-319-26832-3_36

uniquely using a set of orthonormal basis vectors, researchers focused their efforts on finding various such basis vectors. Sylvester and Hadamard independently discovered matrices whose rows/columns form an orthonormal basis. Such basis vectors also lead to orthonormal basis functions on a finite support.

This research paper is organized as follows. In Sect. 2, real Hadamard matrix and real Hadamard transform are reviewed. In Sect. 3, using complex hypercube, complex Hadamard matrix is defined and the associated complex Hadamard transform is defined. In Sect. 4, other complex transforms are briefly defined. The research paper concludes in Sect. 5.

2 Real Hadamard Matrices: Real Hadamard Transform

Linear algebra as a field of human endeavor found many applications. Mathematcians conceived of finite dimensional linear operators, i.e. matrices with special structure such as Toeplitz, Hankel, Hilbert,Vandermonde matrices. In fact L-matrices were proposed in [7]. In a well defined sense, such matrices naturally arise in many applications in science, engineering and other areas. Sylvester as well as Hadamard independently conceived of one such matrix with special structure, now called the Hadamard matrix. Each of the elements of the Hadamard matrix belongs to the set $\{+1, -1\}$. Thus, the rows/columns of the Hadamard matrix (square matrix) are the corners of the symmetric, unit hypercube i.e. Mathematically, this set is specified precisely as follows:

$$S = \{\overline{X} = (x_i, x_2, x_3, \ldots \ldots, x_N) : x_i = \pm 1 \; for \; 1 \leq i \leq N\}$$

Formally, we have the following definition [4].

Definition: A Hadamard matrix of order 'm', denoted by H_m, is an m x m matrix of +1's and −1's such that

$$H_m H_m^T = m I_m$$

where I_m is the m x m identity matrix. This definition is equivalent to saying that any two rows of H_m are orthogonal.

In view of the above definition and Lemma 1, we have the following interesting result.

Lemma 1: Hadamard matrices of odd order do not exist.

Proof: Consider two vectors X, Y lying on the unit hypercube. Let the "Hamming Like distance" between them be 'd'. The inner product of X, Y is given by,

$$< X, Y > \langle X, Y \rangle = \{\text{Number of components where X, Y agree}\} - \{\text{Number of components where X, Y disagree}\}$$
$$= (N - d) - d = N - 2d.$$

Thus for the two vectors X, Y to be orthogonal, it is necessary and sufficient that "N" is an even number. Thus, Hadamard matrices of odd order do not exist. **Q.E.D**

Corollary: If the dimension is odd, there are no orthogonal vectors lying on any one of the countably many hypercubes.

Thus, we are interested in determining whether Hadamard matrices of any arbitrary EVEN ORDER exist. A partial answer to this question is well known. Specifically, it is known that Hadamard matrices of order 2^k exist for all $k \geq 0$. The so called Sylvester construction is provided below:

$$H_1 = [1]$$

$$H_2 = \begin{bmatrix} 1 & 1 \\ 1 & -1 \end{bmatrix}$$

$$H_{2^{n+1}} = \begin{bmatrix} H_{2^n} & H_{2^n} \\ H_{2^n} & -H_{2^n} \end{bmatrix}$$

- **Motivation for Real Hadamard Transform:** It is clear that the rows of real Hadamard matrix can be interpreted as orthogonal basis functions on the interval [0,1] or more generally [0,T] for a finite real number T. These orthogonal basis functions are variously called as Walsh, Hadamard, Rademacher functions. As in the case of other transforms, these basis function scan be utilized to define and study a new transform. Specifically, real Hadamard transform decomposes an arbitrary input vector into superposition of Walsh functions. It is an example of generalized class of Fourier transforms. Specifically, it can be interpreted as as being built out of size-2 discrete Fourier transforms.

- **Details of Hadamard transform:** The real Hadamard transform is a real Hadamard matrix(normalized/scaled by a normalization factor). Specifically H_m is a $2^m \times 2^m$ matrix, the Hadamard matrix (scaled by a normalization factor), that transforms 2^m real numbers x_n into 2^m real numbers X_k. The real Hadamard transform can be recursively defined in the following manner.

The 1×1 Hadamard transform H_0 is defined as 1 i.e. $H_0 = 1$. Then H_m for m > 0 is defined in the following manner:

$$H_m = \frac{1}{\sqrt{2}} \begin{pmatrix} H_{m-1} & H_{m-1} \\ H_{m-1} & H_{m-1} \end{pmatrix},$$

where the $\frac{1}{\sqrt{2}}$ is a normalization constant that is sometimes omitted.

- Hadamard transform, $\bar{v}$ of an $N \times 1$ vector $\bar{u}$ is written as $\bar{v} = \mathbf{H}\bar{u}$. (H is an $N \times N$ Hadamard matrix) and the inverse Hadamard Transform is given by

$$\bar{u} = \mathrm{H}\bar{v}.$$

- Hadamard transform is utilized in various applications such as image compression. It constitutes one among the various transforms utilized in, say image processing.

3 Novel Complex Hadamard Matrices: Complex Hadamard Transform

The In our research efforts [1–3] on complex valued neural networks, we encountered vectors whose components belong to the following set:

$$D = \{1 + j1,\ 1 - j1,\ -1 + j1,\ -1 - j1\}.$$

Definition: The finite collections of vectors whose components belong to the set D are defined to lie on the "unit complex hypercube". This definition is motivated by the definition of real valued symmetric unit hypercube). We now derive the conditions for two vectors on the complex hypercube to be unitary to each other. We provide necessary (not obvious condition) and sufficient condition.

- **Conditions on unitary vectors on complex hypercube:** Let the two complex valued vectors lying on it be denoted by X, Y. They can be represented in the following manner:

$$X = A + jB,\ Y = C + jD,\ \text{where}$$

A, B, C, D lie on the real valued unit hypercube.

- **Definition:** X and Y unitary/orthogonal when

$$X^{*}Y = 0,\ \textit{where}$$

X^{*} denotes the conjugate transpose of X.

Now we find the conditions for unitarity/orthogonality of X, Y. We need the following definitions. Let

$$d_1 \textit{ is the number of places where } A, C \textit{ differ},$$
$$d_2 \textit{ is the number of places where } B, D \textit{ differ},$$
$$d_3 \textit{ is the number of places where } A, D \textit{ differ}$$
$$d_4 \textit{ is the number of places where } B, C \textit{ differ}$$

Lemma 2: The vectors X and Y are unitary if and only if $d_1 = d_2 = d_3 = d_4 = \frac{N}{2}$.

Proof: It is easy to see that

$$\begin{aligned} X^{*}Y &= (A^T - jB^T)(C + jD) \\ &= (A^T C + B^T D) + j(A^T D - B^T C). \end{aligned}$$

Now using the same idea as in Lemma 1, we have that

$$X^{*}Y = [(N - 2d_1 + N - 2d_2)] + j[(N - 2d_3) - (N - 2d_4)]$$

Thus, we necessarily have that

$$X^{*}Y = [(2N - 2(d_1 + d_2)) + j\,2(d_4 - d_3)].$$

Thus, for X, Y to be unitary/orthogonal to one another, we must have that

$$N = (d_1 + d_2)\; and\; d_3 = d_4.$$

Note: One sufficient condition for unitarity/orthogonality of X, Y is that the vectors A, B, C, D lying on the real valued unit hypercube are such that $A^T C = 0,\ B^T D = 0,\ A^T D = 0,\ B^T C = 0$ i.e. Vector pairs (A, C), (B, D), (A, D), (B, C) are orthogonal. This can happen if $d_1 = d_2 = d_3 = d_4 = \frac{N}{2}$.

Now using the "identity swapping argument", it can be easily shown that the conditions $N = (d_1 + d_2)$ and $d_3 = d_4$ necessarily require that

$$d_1 = d_2 = d_3 = d_4 = \frac{N}{2}.$$

Hence this condition is necessary as well as sufficient. Q.E.D.

Suppose X, Y lie on the k-th order complex hypercube. Then, we have that the inner product of X, Y is expressed by the following formula.

$$X^{*}Y = 2K^2[(N - (d_1 + d_2) + j\,(d_4 - d_3)].$$

Note: It should be noted that the above inner product is a complex number with even integer real and complex parts. It is a real number if and only if $d_3 = d_4$.

- **Remark:** In the literature, there is an effort to study "complex Hadamard matrices" in which the elements are p^{th} roots of unity. But in view of the above discussion we are naturally led to the study of structured matrices whose elements are from the set

$$D = \{+1 + j1,\ +1 - j1,\ -1 + j1,\ -1 - j1\}.$$

Equivalently, motivated by the definition of complex hypercube, we are naturally led to the study of "novel complex Hadamard matrices" whose rows and columns are corners of the complex hypercube. Formally, we have the following definition.

Definition: A complex Hadamard matrix of order 'm', denoted by CH_m, is an m × m matrix whose elements belong to the set D such that

$$CH_m CH_m^{*} = 2mI_m$$

where I_m is the m x m identity matrix and CH_m^{*} is the conjugate transpose of CH_m. This definition is equivalent to saying that any two rows of H_m are unitary.

In view of the above definition and Lemma 4, we have the following interesting result

Lemma 3: The above novel complex Hadamard matrices of odd order do not exist.

Proof: Follows directly from Lemma 2 Q.E.D.

- **Motivation for complex Hadamard transform:** Consider a signal defined over finite support i.e. x[n], $0 \leq n \leq (N-1)$. It is well known that the Discrete Fourier Transform (DFT) of such a signal is given by

$$X(k) = \sum_{n=0}^{N-1} x[n] e^{\frac{-j2\pi nk}{N}} \; for \; 0 \leq k \leq (N-1).$$

In matrix-vector notation, the above equation can be rewritten as

$$\bar{X} = \bar{W}\bar{x}, \text{ where}$$
$$\bar{X} = [X(0)\, X(1) \ldots X(N-1)]^T$$
$$\bar{x} = [x(0)\, x(1) \ldots x(N-1)]^T \text{ and}$$

$\bar{W}$ is the DFT matrix whose elements are N^{th} roots of unity (i.e. complex numbers).
Thus, unlike Real Hadamard Transform, Discrete Fourier Transform (DFT) as a transform is defined not only for real valued signals but also for complex valued signals. Thus, we are motivated to define and study complex Hadamard transform in the following discussion.

- **Complex Hadamard transform:** From the discussion above and the real valued Hadamard transform definition, it is clear that complex Hadamard transform is a complex Hadamard matrix scaled by a normalization factor. i.e.

$$NCH_m = \frac{1}{\sqrt{2m}} CH_m.$$

Thus, it is clear that NCH_m is a unitary matrix. If such a matrix is constructed using the generalized Sylvester construction, it is also a Hermitian matrix. The order of such a matrix is a "power of 2".

Definition: The complex Hadamard transform of a complex valued vector $\bar{U}$ is obtained in the following manner:

$$\bar{V} = NCH_m \bar{U}.$$

Also, it is easy to see that (since NCH_m is unitary and Hermitian), the inverse complex Hadamard transform of $\bar{V}$ is given by $\bar{U} = NCH_m \bar{V}$.

- In view of the above discussion, it is easy to see that the complex Hadamard matrix NCH_m (i.e. also complex Hadamard transform) can be decomposed in the following manner

$$NCH_m = G + jH,$$

where G and H are real Hadamard matrices (i.e. real Hadamard transforms).

- Consider two real input signals i.e. two real valued vectors $\bar{A}, \bar{B}$. Let us form a complex signal $\bar{U}$ in the following manner i.e.

$$\bar{U} = \bar{A} + j\bar{B}$$

Now let us compute the complex Hadamard transform of the complex signal $\bar{U}$. It is easy to that

$$\bar{V} = NCH_m\bar{U} = (\bar{G}\bar{A} - \bar{H}\bar{B}) + j(\bar{G}\bar{B} + \bar{H}\bar{A}).$$

In a similar manner form another complex signal in the following manner:

$$W = \bar{B} + j\bar{A}.$$

The complex Hadamard transform of the complex signal $\bar{W}$ is given by

$$\bar{Z} = NCH_m\bar{W} = (\bar{G}\bar{B} - \bar{H}\bar{A}) + j(AGA + HB).$$

Thus, the complex valued vectors $\bar{V}, \bar{Z}$ can be decomposed in the following manner:

$$\bar{V} = Real\ Part\ of(\bar{V}) + j\ Imaginary\ Part\ of\ (\bar{V}).$$
$$\bar{Z} = Real\ Part\ of\ (\bar{Z}) + j\ ImaginaryPartof\ (\bar{Z}).$$

Using the above equations, we get the following results:

1. $\bar{G}\bar{A} = \frac{1}{2}(Real\ Part\ of\ (\bar{V}) + Imaginary\ Part\ of\ (\bar{Z})).$
2. $\bar{H}\bar{B} = \frac{1}{2}(Imaginary\ Part\ of\ (\bar{Z}) - Real\ Part\ of\ \bar{V}).$
3. $\bar{G}\bar{B} = \frac{1}{2}(Real\ Part\ of\ (\bar{Z}) + Imaginary\ Part\ of\ (\bar{V})).$
4. $\bar{H}\bar{A} = \frac{1}{2}(Imaginary\ Partof\ (\bar{V}) - Real\ Part\ of\ (\bar{Z})).$

- Thus, using the complex Hadamard transform of complex signals $\bar{U}, \bar{W}$ (formed from the real signals $\bar{A}, \bar{B}$), we are able to compute the real Hadamard transforms $\bar{G}\bar{A}$, $\bar{H}\bar{B}, \bar{G}\bar{B}\ and\ \bar{H}\bar{A}$.

The detailed properties and applications of complex Hadamard transform are discussed in [6]. It should be noted that real valued Hadamard transform finds many applications in research areas such as Digital Image Processing [4, 5].

- **Hadamard Transform: Threshold Decomposition:**

- It is well known that every finite dimensional linear operator has a matrix representation i.e. If "M" dimensional function of N variables $\bar{f}(X_1, X_2, \ldots X_N)$ satisfies the superposition property, then it can be represented in the following form:

$\bar{f}(X_1, X_2, \ldots X_N) = G\bar{X}$, where G is an M × N matrix and $\bar{X} = [X_1, X_2, \ldots X_N]^T$.

Using the principle of Threshold decomposition, the input $\bar{X}$ can be decomposed into binary signals. Suppose the components of vector $\bar{X}$ can assume say atmost L integer values. Then $\bar{X}$ can be decomposed in the following manner.

$\bar{X} = \overline{Y_1} + \overline{Y_2} + \cdots + \overline{Y_L}$, where $\overline{Y_i}$ are binary/Boolean vectors.

Thus $\bar{f}(X_1, X_2, \ldots X_N) = \bar{X} = G\overline{Y_1} + G\overline{Y_2} + \cdots + G\overline{Y_L}$.

Hence such linear functions can be computed using Threshold decomposition. If G is a Hadamard matrix, then the computation on the right hand side involves only additions and subtractions.

Remark: It should be noted that finite impulse response linear filters, involve filtering operation on signal values in a finite length window. Thus such linear filters can be implemented using Threshold decomposition, when the filter coefficients are integers/ rational numbers.

The above innovative idea of complex Hadamard Transform leads to the following novel research directions.

- **Complex Walsh, Haar and other functions:** The rows of complex Hadamard matrices (discussed above) are complex Walsh functions (could also be called complex Radamacher functions). Any pair of such functions f(t), g(t) are orthogonal on a finite domain, say [0,1] in the sense that

  $< \mathrm{f(t)}, \mathrm{g(t)} >= \int_0^1 f(t)\, g^*(t)\, dt = 0$, where $g^*(t)$ is the complex conjugate of g(t).

In a similar manner, complex Haar functions are defined and studied [RSM]. Also, one can consider complex valued functions of complex variable i.e. J(z), K(z) where "z" is a complex variable. The range of functions {J(.), K(.)} is the set $\{1 + \mathrm{j}1,\ 1 - \mathrm{j}1,\ -1 + \mathrm{j}1,\ -1 - \mathrm{j}1\}$. Specifically, one can consider the domain of functions J(.), K(.) to be the set E i.e. $\mathrm{E} = \{\mathrm{z}\colon \mathrm{z} = \mathrm{a} + \mathrm{j\,b},\ \text{where } a \in [0, 1]\ \ and\ b \in [0, 1]\}$ Orthogonality of such functions is defined in a similar manner. Other interesting complex valued orthogonal functions are discussed in [RSM].

It is well known that the concept of complex numbers is generalized to arrive at the concept of Quaternions and Octanions. Thus as in the case of "Complex hypercube", we introduce hypercube based on Quaternions whose components are constrained to be $\{+1 \text{ or } -1\}$. We study orthogonal Quaternions on such a hypercube. More generally we define and study hypercube based on Clifford algebra. The idea of Hadamard transform is thus generalized using such concepts.

Note: It is well known that every finite dimensional linear operator hesa matrix representation. Specifically, a structured matrix, such as "complex Hadamard matrix", corresponds to a linear operator. The columns of such a matrix are unitary and form a basis for the vector space defined over the field of complex numbers. Thus, the lemma 2 is an interesting contribution.

Traditionally, in the literature on Hadamard transform, concepts such as "sequency" are well defined. We attempt to introduce some coding theoretic concepts associated with a collection of (N of them) unitary basis vectors (lying on complex hypercube). It is possible to define the concepts of Hamming distance between two unitary vectors lying on the complex hypercube. The definition of distances $\{d_1, d_2, d_3, d_4\}$ in lemma 2 will provide insights into such effort. It is clear that the Hamming distance between orthonormal (real valued) basis vectors (column vectors of real Hadamard matrix) is always $\frac{N}{2}$.

4 Other Complex Transforms

In the spirit of the above discussion related to complex Hadamard transform, we can define and study various other complex transforms. For instance, let us consider the Haar transform [4, 5].

The basis functions of the Haar transform are the oldest and simplest known orthonormal wavelets. Haar transform can be expressed in the following matrix form:

$$T = HFH^T,$$

Where F, for instance, is an image matrix, H is an N × N Haar transformation matrix and T is the resulting N × N Haar transform of F. The transpose is required because H is not symmetric. The Haar transform, H contains the well known Haar basis functions, $h_k(z)$. They are defined over the continuous closed interval

$$z \in \left[0, 1\right] for\ k = 0, 1, 2, \ldots, N - 1,\ where\ N = 2^n\right].$$

The rows of Haar Matrix, H can be generated in a well known recursive manner. Our idea is to define and utilize complex Haar basis functions on the unit square [0,1] × [0,1]. The real and imaginary parts are the various possible real Haar basis functions. In a similar manner, complex Hough transform can be defined and studied [5]. Detailed efforts are documented in [6]. In reference [8], based on a different approach, a family of unified complex Hadamard transforms are discussed.

5 Conclusion

In this research paper, a novel complex Hadamard matrix is defined based on the concept of complex hypercube. Utilizing the complex Hadamard matrix, complex Hadamard transform is defined and studied. In a natural way, other complex transforms are proposed. It is expected that the results in this paper lead to novel theoretical research directions and practical applications.

References

1. Muezzinoglu, M.K., Guzelis, C., Zurada, J.M.: A new design method for the complex-valued multistate Hopfield associative memory. IEEE Trans. Neural Netw. **14**(4), 891–899 (2003)
2. Murthy, G.R., Praveen, D.: Complex-valued neural associative memory on the complex hypercube. In: IEEE Conference on Cybernetics and Intelligent Systems, Singapore, 1–3 December 2004
3. Murthy, G.R.: Multidimensional Neural Networks-Unified Theory. New Age International Publishers, New Delhi (2007)
4. Jain, A.K.: Fundamentals of Digital Image Processing. Prenctice Hall of India, New Delhi (2009)
5. Gonzalez, R.C., Woods, R.E.: Digital Image Processing, 3rd edn. Pearson Prentice Hall, Upper Saddle River (2008)

6. Murthy, G.R., Saramaki, T., Gabbouj, M.: Complex Hadamard/Haar transforms and applications. Manuscript in preparation
7. Gupta, R., Murthy, G.R.: Innovative structured matrices. Adv. Linear Algebra Matrix Theor. **3**(3), 17–21 (2013)
8. Rahardja, S., Falkowski, B.J.: Family of unified complex Hadamard transforms. IEEE Trans. Circ. Syst. II Analog Digital Sig. Process. **46**(8), 1094–1099 (1999)

A New Multivariate Time Series Transformation Technique Using Closed Interesting Subspaces

Sirisha G.N.V.G.[1(✉)] and Shashi M.[2]

[1] Department of CSE, S.R.K.R. Engineering College, Bhimavaram, India
sirishagadiraju@gmail.com
[2] Department of CS and SE, A.U. College of Engineering, Andhra University, Visakhapatnam, India
smogalla2000@yahoo.com

Abstract. Subspace clustering detects the clusters that are existing in the subspaces of the feature space. Density based subspace clustering defines clusters as regions of high density existing in subspaces of multidimensional datasets. This paper discusses the concept of closed interesting subspaces under density divergence context for multivariate datasets and proposes an algorithm to transform the multivariate time series to a symbol sequence using the closed interesting subspaces. The proposed transformation allows the applicability of any of the symbolic sequential mining algorithms to efficiently extract sequential patterns which capture the interdependencies and co-variations among groups of time series variables. The multivariate time series transformation technique is explained using a sample dataset. It is evaluated using a real world weather dataset obtained from Cambridge University. The representation power of the closed interesting subspaces and maximal interesting subspaces in transforming multivariate time series is compared.

Keywords: Closed interesting subspaces · Multivariate time series · Subspace clustering · Symbol sequence · Time series transformation

1 Introduction

Clustering aims at grouping objects such that all the objects in each group share similar characteristics. In traditional clustering all the variables/dimensions describing the objects are used in calculating the similarity between objects. In high dimensional spaces traditional clustering techniques fail to find meaningful clusters due to curse of dimensionality [1–3].

To overcome the curse of dimensionality subspace clustering techniques are proposed in the literature. Subspace clustering detects the clusters that are existing in subsets of dimensions of full dimensional space. Each subspace cluster S is defined as $S = O \times A$ where O is a subset of objects that are homogeneous in the subset of attributes A. Subspace clustering allows mining of overlapping clusters.

A time series is a sequence of real values measured over time. Multivariate time series is a group of time series. Multivariate time series exist in many domains like

R. Prasath et al. (Eds.): MIKE 2015, LNAI 9468, pp. 392–405, 2015.
DOI: 10.1007/978-3-319-26832-3_37

health informatics, economy, weather, hydrology, genetics, industrial process monitoring. Multivariate time series are of two types. The first is a group of homogeneous attributes like annual flow of water in different rivers. The second is a group of heterogeneous attributes like weather parameters including temperature, rainfall, etc. Studying multivariate time series helps us in understanding interdependencies and co-variations among groups of time series variables.

This paper discusses the concept of closed interesting subspaces in multidimensional spaces and a new multivariate time series transformation/representation technique that transforms a multivariate time series to a symbol sequence. This transformation extends the applicability of efficient sequential pattern mining algorithms to handle multivariate time series data to extract patterns that capture the inter relationships and co-variations among groups of time series variables. SCHISM [2] a grid and density based 2D subspace clustering algorithm which was developed for mining maximal interesting subspaces was adapted to mine closed interesting subspaces.

Algorithms like SAX [4] transform each time series to a separate symbol sequence. To our knowledge, ours is the first approach that uses subspace clustering to transform a multivariate time series to a single symbol sequence. The symbol sequence preserves the interdependencies and co-variations among groups of time series variables. This transformation allows us to apply algorithms from string processing, sequential mining and bioinformatics on the transformed multivariate time series for extracting different types of patterns. In the area of sequential mining there are many efficient algorithms that mine patterns like frequent episodes, frequent periodic patterns and frequent continuities from a single event (symbol) sequence. Very few algorithms exist which can mine these types of patterns from multi event sequences (multiple symbol sequences). The performances of the existing multi event sequence mining algorithms degrade enormously with the increase in number of sequences. This observation has motivated us to transform a multivariate time series to a single symbol sequence rather than multiple symbol sequences.

The proposed algorithm is explained using a sample dataset and its performance is tested using a real dataset. The real dataset is Cambridge daily weather collected from Cambridge University [5]. The paper also compares the representational power of maximal interesting subspaces with that of closed interesting subspaces in representing/transforming a multivariate time series.

2 Related Work

Many univariate time series representation/transformation techniques are proposed in the literature which includes Discrete Fourier transform (DFT), Discrete Wavelet Transform (DWT), Piecewise Linear Approximation (PLA), Piecewise Aggregate Approximation (PAA), Adaptive Piecewise Constant Approximation (APCA), and Singular Value Decomposition (SVD). These timeseries representation techniques help in finding the trends, shapes and patterns present in the univariate timeseries efficiently [6]. Though there are many univariate time series representation techniques, no single technique is suitable for all the data mining tasks.

Symbolic Aggregate Approximation (SAX) [4] converts a time series to a symbol sequence. SAX method first discretizes the time series using PAA. Symbol sequence is then obtained by mapping the PAA coefficients to symbols using predetermined breakpoints. Trend and Value based Approximation (TVA) is an extension to SAX representation which adds new symbols to represent the trend of the time series [7]. SAX and TVA are also univariate time series representation techniques. It is more complex to capture interdependencies and co-variations in multivariate time series using the above techniques.

Tanaka et al. used principal component analysis to transform a multivariate time series to univariate time series [8]. The univariate time series is then converted into a symbol sequence using SAX representation. The algorithm then mines the motif from the symbol sequence using minimum description length principle (MDL).

Zhuang et al. proposed an algorithm to find frequent temporal associations in multivariate time series [9]. In this approach first each time series is discretized using SAX representation to obtain a symbol sequence. Frequent closed patterns are obtained from each symbol sequence using suffix tree. All closed patterns that are covered by their closed super patterns are removed. Similar non-redundant closed patterns are grouped to form pattern clusters. Then the temporal association between these pattern clusters is found by expanding each pattern cluster into two level pattern cluster association, then three level pattern cluster association and so forth. A statistical significance measure is used to rank the discovered temporal associations as a post pruning step. The run time of this algorithm is exponential to the number of time series variables as well as maximum level of temporal association. The discovered temporal associations capture the local relations among multiple time series. The algorithm could detect temporal associations between four pattern clusters belonging to at most four time series variables beyond which the algorithm ran out of memory.

The approach proposed in this paper allows mining the relations between any numbers of time series variables. The proposed transformation allows application of efficient sequential pattern mining algorithms to extract patterns like frequent episodes, frequent continuities and periodic patterns from the symbol sequence. These patterns explain the relations, interdependencies and co-variations between groups of time series variables in a comprehensive way when compared to the local relations mined by Zhuang et al. approach.

3 Methodology

The representation/transformation technique proposed in this paper converts a multivariate time series to a symbolic sequence with each symbol representing a collection of co-occurring variable-value pairs. The transformation is done in two steps. The first step identifies the inter dependencies among the variables constituting the multivariate time series. This step involves the application of grid and density based subspace clustering algorithms. While most of the existing subspace clustering algorithms extract either all or maximal interesting subspaces, the proposed approach advocates the use of closed interesting subspaces for improved coverage and representation power. Each of the closed interesting subspaces mined in the first step is given a unique name. In the second step among all the potential closed interesting subspaces that can represent the set of variable-value pairs occurring at a given timestamp, the most representative closed subspace is chosen to represent the time stamp. Thus a multivariate time series is transformed into a single symbol sequence.

3.1 Closed Interesting Subspace Mining

Definition 1: (Subspace). A subspace is defined as an axis -aligned hyper-rectangle $[l_1, h_1] \times [l_2, h_2] \times \ldots\ldots\ldots \times [l_d, h_d]$ where $l_i = (aD_i)/\xi$, and $h_i = (bD_i)/\xi$, a,b are positive integers and $a < b < \xi$, ξ is the number of divisions of an axis, d is the number of dimensions and D_i is the range of *ith* dimension [2].

A d-dimensional subspace imposes d constraints on the data points it holds there by constraining the possible values of ith dimension of a data point within $[l_i, h_i]$.

If $h_i - l_i = D_i$, the subspace is unconstrained in dimension i whose range is given by D_i. A m-subspace is a subspace constrained in m dimensions, denoted as S_m [2]. The density of a subspace is the ratio of number of data points it holds to the total number of data points in the dataset.

In density based subspace clustering, an interesting subspace represents a combination of variable-value pairs that occur frequently together and whose density exceeds the given density threshold. Interesting subspace mining aims at finding all the interesting subspaces at all the dimensionalities. SCHISM [2] is a grid and density based maximal interesting subspace mining algorithm.

Density based subspace clustering algorithms which impose single density threshold for all dimensionalities fail to identify significant subspaces at all dimensionalities due to density divergence problem [2, 10]. So in order to find all interesting subspaces at all dimensionalities we have to use high density threshold for low dimensional subspaces and low density threshold for high dimensional subspaces. SCHISM algorithm overcomes the density divergence problem by setting different density thresholds at different subspace dimensionalities. These density thresholds are calculated using Chernoff-Hoeffding bounds.

Due to density divergence the apriori property is violated by those lower dimensional subspaces failing to satisfy their density thresholds even if they are part of a dense subspace of higher dimensionality. Hence unlike the traditional maximal frequent patterns representing all their sub-patterns, a dense maximal subspace may not imply all the subspaces that are subsets of it to be dense. Hence we proposed the concept of closed interesting subspaces in the context of density divergence. Density divergence is inevitable in subspace clustering.

The concept of closed pattern/subspace is framed by the authors as follows. As a natural consequence of density divergence, the density of a higher dimensional subspace is expected to reduce proportionately as a ratio of density thresholds of corresponding subspaces. A dense subspace of p-dimensions is closed if all of its extensions have their density reduced by more than the expectation in accordance with the ratio of density thresholds in corresponding subspaces.

Let S_p denote a dense subspace constrained in p dimensions.

S_p is closed if $\forall q > p, \frac{density(S_q)}{density(S_p)} < \frac{\text{density\_Threshold(q)}}{\text{density\_Threshold(p)}}$ where S_q is a dense subspace constrained in q dimensions and $S_p \subset S_q$ [11].

The authors have adapted SCHISM algorithm for extracting closed interesting subspaces under density divergence context. The density thresholds at different subspace dimensionalities are calculated in the same way as done in SCHISM algorithm. A closed interesting subspace represents a combination of variable-value pairs which gains significance due to their frequent co-occurrence and hence defines a pattern depicted by a unique symbol.

A d-dimensional multivariate time series contain a series of timestamps associated with d variable-value pairs. However, only a subset of these variable-value pairs that repeat at multiple timestamps have information content for pattern extraction. Hence the variable-value pairs occurring at a timestamp can be represented by the most representative pattern such that the loss of information is minimized.

According to Lin et al. a suitable representation/approximation greatly affects the ease and efficiency of time series data mining [12]. The utility of the data mining results obtained from the approximation depends on the quality of approximation/representation of the time series. The representation/approximation technique proposed in this paper approximates the variable-value pairs occurring at each timestamp with the most representative pattern such that the overall loss of information is minimized. A representative pattern is the closed interesting subspace that best represents the variable-value pairs occurring at the timestamp.

Accordingly the metrics for evaluating the algorithm used for transforming a multivariate time series into a symbol sequence are defined. They are Total Pattern Dissimilarity (TPD) which has to be minimized and Total Coverage (TC), Total Pattern Length (TPL) which are to be maximized.

3.2 Proposed Algorithm for Transforming a Multivariate Time Series to a Symbol Sequence

Algorithms 1 and 2 are explained using the sample dataset given in Table 1. Table 1 shows the discretized version of a multivariate time series containing nine time series. Discretization is done using equiwidth binning. On application of closed interesting subspace mining, the closed interesting subspaces found and their timestamp list are shown in Table 5.

Table 1. Discretized multivariate time series

Id	D1	D2	D3	D4	D5	D6	D7	D8	D9
1	6	11	21	31	41	51	68	71	81
2	6	12	21	31	41	52	68	72	82
3	6	11	21	31	40	55	61	73	83
4	7	11	21	31	41	55	65	74	84
5	5	11	21	31	41	55	65	75	85
6	8	12	22	31	41	55	65	76	86
7	6	13	21	32	41	55	65	70	80
8	5	16	26	36	41	55	68	77	87
9	6	17	20	37	41	56	65	78	88
10	9	19	29	39	49	55	65	79	89

Algorithm 1: Convert a multivariate time series to symbolic sequence

Input: D-dimensional Multivariate time series: with D variable-value pairs at each time stamp
Output: A symbolic representation of multivariate time series
Method:

1. Find closed interesting subspaces
   1.1. Discretize each variable using equiwidth binning and uniquely identify each interval with a symbol.
   1.2. The variable value pairs observed at a time stamp constitute D intervals/symbols and hence represent each time stamp as a data point in D dimensional space.
   1.3. Convert the data from horizontal to vertical format i.e. generate the list of timestamps for each interval
   1.4. Apply closed interesting subspace mining algorithm on the vertical database to get all the closed interesting subspaces. Each closed interesting subspace is represented by a symbol.
   1.5. For each closed interesting subspace find the list of timestamps where it occurs. This is a vertical database consisting of a set of <subspace_symbol, timestamp_list>
2. Find the best representative subspace at each time stamp of a multivariate time series
   2.1. Convert the vertical database obtained in step 1.5 to horizontal database. This gives us a multivariate symbolic representation of time series with each symbol representing a closed interesting subspace.
   2.2. At each time stamp, out of all the closed interesting subspaces that can represent the variable-value pairs at the given time stamp, choose the best representative closed subspace as described in algorithm 2 and substitute the corresponding symbol at the time stamp.

Algorithm 2: Find the Best Representative Subspace

Input: All the closed interesting subspaces found at a time stamp
Output: Representative subspace at the timestamp
Method:

1. For each closed interesting subspace that appear at the given timestamp find
   i) Number of mismatches between the intervals present at the time stamp and intervals defining the subspace
   ii) Distinctiveness of the subspace which is the ratio of dimensionality of the subspace to sum of participation scores of all the intervals defining the subspace. Participation score of an interval is defined as the number of closed interesting subspaces in which it occurs.
   iii) Density of the subspace
2. Find the best representative subspace
   The best subspace is defined as the one with
   Condition 1: Least number of mismatches with the intervals present at the time stamp
   Condition 2: If the least number of mismatches are equal for two or more closed subspaces, choose the one with maximum distinctiveness
   Condition 3: If the least number of mismatches and distinctiveness are same for two or more subspaces choose the one with maximum density

Original SCHISM algorithm and modified SCHISM algorithms are applied for generating maximal interesting subspaces. The subspaces mined by original SCHISM from the sample dataset are shown in Table 3 and the subspaces mined by modified SCHISM from the sample dataset are shown in Table 4. In the process of mining maximal interesting subspaces, Original SCHISM merges a newly mined subspace with an existing maximal interesting subspace if the similarity between the two exceeds a given merge threshold. The extensions of the newly mined subspace are not checked for interestingness. Some of the interesting subspaces will not be explored due to this merging. We have modified SCHISM algorithm to overcome this drawback. The modified algorithm is named as Modified SCHISM.

Table 2. Density thresholds at each subspace dimensionality for the sample dataset

Subspace Cardinality/Dimensionality	Density threshold
1	0.25
2	0.3493
3	0.3403
4	0.3394
5	0.339317
6	0.339308
7	0.339307
8	0.339307
9	0.339307

From Tables 3 and 4 we can notice that S6 a maximal interesting subspace found in Table 4 is missing in Table 3. This is due to the merging criterion followed by Original SCHISM. Original SCHISM, Modified SCHISM and Closed Interesting Subspace Miner use the same density thresholds. The density thresholds for the sample dataset are given in Table 2.

Table 3. Maximal Interesting Subspaces (MIS) Mined by Original SCHISM

Name of the subspace	Intervals defining the subspace	Density	Rid list
M1	11 21 31	0.4	1 3 4 5
M2	6 21	0.4	1 2 3 7
M3	6 41	0.4	1 2 7 9
M4	41 55 65	0.4	4 5 6 7
M5	31 55	0.4	3 4 5 6
M6	31 41	0.5	1 2 4 5 6
M7	21 55	0.4	3 4 5 7
M8	21 41	0.5	1 2 4 5 7

The best representative subspace at each time stamp is chosen using the algorithm "Find Best Representative Subspace". For example if we take the case of the closed interesting subspaces mined at each time stamp the best closed interesting subspace at each timestamp is chosen in the following way. At time stamp 1, C1,C2,C3,C4,C5, C11,C12,C13,C14,C18,C19,C20 are the closed subspaces that can represent timestamp 1(from Table 5), among them the subspaces C2 and C11 have minimum pattern dissimilarity of 6. Since they have the same pattern dissimilarity we compare their distinctiveness. C2 has distinctiveness score of 0.21428 and C11 has distinctiveness of 0.142. So by condition 2 of the algorithm, C2 is selected as best closed subspace at time stamp 1. Similarly among all the closed subspaces that can represent time stamp 2, C11 has least pattern dissimilarity of 6 so it is selected as best subspace. Similarly the representative subspaces at the rest of the timestamps are found.

Column 4 of Table 6 shows the best closed interesting subspace at each timestamp for the sample dataset. At each timestamp the best maximal interesting subspace among the maximal interesting subspaces mined by Original SCHISM algorithm and maximal interesting subspaces mined by the modified SCHISM algorithm are shown in columns 2 and 3 of Table 6 respectively.

Table 4. Maximal interesting subspaces (MIS) mined by modified SCHISM

Name of the subspace	Intervals defining the subspace	Density	Rid List
S1	11 21 31	0.4	1 3 4 5
S2	6 21	0.4	1 2 3 7
S3	6 41	0.4	1 2 7 9
S4	41 55 65	0.4	4 5 6 7
S5	31 55	0.4	3 4 5 6
S6	21 31 41	0.4	1 2 4 5
S7	21 55	0.4	3 4 5 7

Table 5. Closed interesting subspaces

Name of the subspace Id	Intervals defining the subspace Defining the Cluster	Density %	Rid List
C1	68	0.3	1 2 8
C2	11 21 31	0.4	1 3 4 5
C3	6 21	0.4	1 2 3 7
C4	6 41	0.4	1 2 7 9
C5	6	0.5	1 2 3 7 9
C6	41 55 65	0.4	4 5 6 7
C7	55 65	0.5	4 5 6 7 10
C8	41 65	0.5	4 5 6 7 9
C9	65	0.6	4 5 6 7 9 10
C10	31 55	0.4	3 4 5 6
C11	21 31 41	0.4	1 2 4 5
C12	21 31	0.5	1 2 3 4 5
C13	31 41	0.5	1 2 4 5 6
C14	31	0.6	1 2 3 4 5 6
C15	21 55	0.4	3 4 5 7
C16	41 55	0.5	4 5 6 7 8
C17	55	0.7	3 4 5 6 7 8 10
C18	21 41	0.5	1 2 4 5 7
C19	21	0.6	1 2 3 4 5 7
C20	41	0.8	1 2 4 5 6 7 8 9

Table 6. Best representative subspace at each time stamp

Time stamp	Best maximal subspace (Original SCHISM)	Best maximal subspace (Modified SCHISM)	Best closed subspace
1	M1	S1	C2
2	M2	S6	C11
3	M1	S1	C2
4	M1	S4	C2
5	M1	S4	C2
6	M4	S4	C6
7	M4	S4	C6
8			C16
9	M3	S3	C4
10			C7

4 Evaluation Metrics

The ability of a multivariate time series transformation technique in representing a multivariate time series is evaluated using the following metrics. The metrics used are Total pattern dissimilarity, Total coverage, Total pattern length.

Total Pattern Dissimilarity (TPD). Pattern Dissimilarity at a timestamp is defined as the ratio of the number of dimensions that are not represented by the most representative pattern at that timestamp to the dimensionality of the multivariate time series. It may be noted that due to missing values for certain attributes, a timestamp may be specified in less than d variable-value pairs and hence with reduced dimensionality. Total Pattern Dissimilarity is the sum of pattern dissimilarities at each timestamp constituting the time series. A representation technique with lesser TPD represents the information better.

Total Coverage (TC). A timestamp is covered by a pattern if variable-value pairs observed at the timestamp obey all the constraints imposed by the pattern. Total coverage is the ratio of the number of timestamps covered by one of the patterns discovered by the algorithm to the total number of timestamps constituting the multivariate time series. The representation technique with highest total coverage is most effective.

Total Pattern Length (TPL). The sum of lengths of the best representative patterns at all the timestamps of a multivariate time series is defined as the Total pattern length.

The Total pattern length specifies the number of intervals (aspects) of multivariate time series that are represented by the symbol sequence. The representation technique that has the maximum total pattern length has maximal representation power.

5 Performance Evaluation

The proposed algorithm advocates the use of closed interesting subspaces to transform/represent a multivariate time series. The performance of the algorithm is also tested when maximal interesting subspaces mined by original SCHISM (MIS by original SCHISM) and Modified SCHISM (MIS by Modified SCHISM) were used in transforming the multivariate time series. Table 7 gives the Total pattern dissimilarity, Total Coverage, Total Pattern Length, Total number of subspaces mined and Time in Seconds when all the three methods were applied on sample dataset given in Table 1.

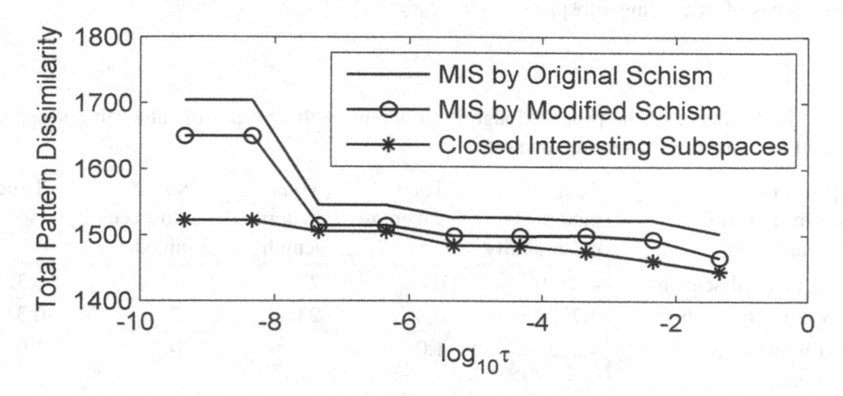

Fig. 1. Total pattern dissimilarity versus $\log_{10} \tau$ of the transformed multivariate time series generated using the three types of interesting subspaces

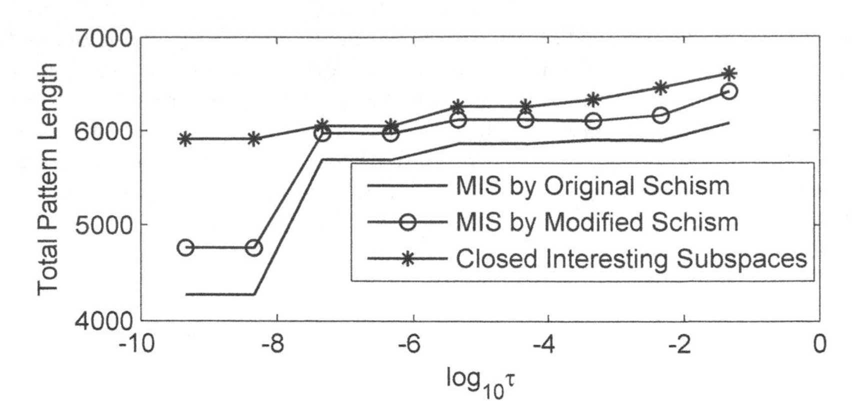

Fig. 2. Total pattern length versus $\log_{10} \tau$ of the transformed multivariate time series generated using the three types of interesting subspaces

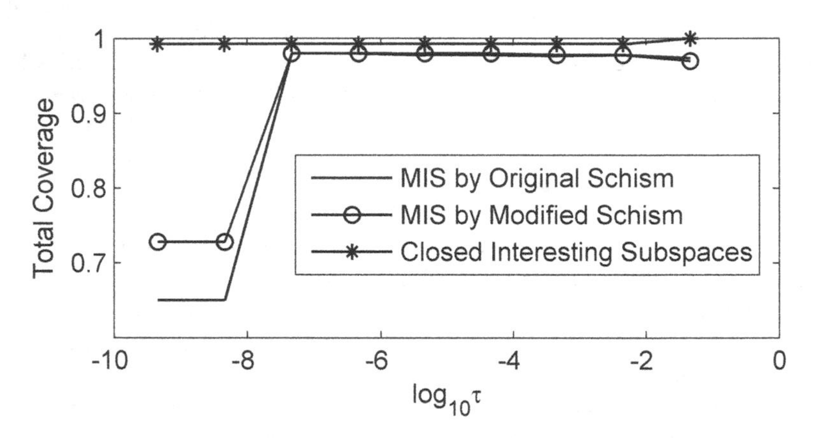

Fig. 3. Total coverage versus $\log_{10} \tau$ of the transformed multivariate time series generated using the three types of interesting subspaces

Table 7. Performance of the proposed algorithm when the three types of interesting subspaces were used to transform the sample dataset

Subspaces used to transform multivariate time series	Total pattern dissimilarity	Total coverage	Total pattern length	No. of subspaces mined	Time (sec)
MIS by original schism	4.8889	0.8	22	8	0.327
MIS by modified schism	4.7778	0.8	23	7	0.312
Closed interesting subspaces	4.222	1.0	27	20	1.014

The performance of the algorithm was tested on the weather dataset collected from University of Cambridge [5]. The dataset is described using 9 variables which are Temperature, Humidity, Dew point, Pressure, Average Wind Speed since previous recorded value, Most frequent Wind Direction since the previous recorded value, Sun shine in hours since midnight, Rain in mm since midnight, Maximum Windspeed since previous recorded value. Except wind direction all other variables are continuous valued variables. Wind direction is discrete valued variable having eight different distinct values which are N,NE,E,SE,S,SW,W,NW. The raw data consists of all the nine variable values collected at every half an hour of the day. Year 2012 weather data is collected which consisted of 9 time series each of length 17370. The data is pre-processed and summarized to four hour intervals. The length of the resultant multivariate time series is 2179.

Total Pattern Dissimilarity after transformation of multivariate time series to symbol sequence using closed interesting subspaces, maximal interesting subspaces mined by original SCHISM and maximal interesting subspaces mined by modified SCHISM is shown in Fig. 1. Figure 1 shows that when multivariate time series is transformed into symbol sequence using closed interesting subspaces the Total Pattern Dissimilarity is minimum at all values of τ. τ plays an important role in density thresholds calculation. Density thresholds are set using the formula proposed in SCHISM algorithm [2]. The performance of the proposed algorithm is tested for τ values ranging from $\frac{100}{n}, \frac{10}{n}, \frac{1}{n}, \frac{1}{10n}, \frac{1}{100n}$ and so on to $\frac{1}{1000000n}$ where n is the number of timestamps which is 2179 in the weather dataset.

Figure 2 shows Total Pattern Length of transformed multivariate time series obtained using all the three types of subspaces. The figure shows that the maximum Total Pattern Length is attained when multivariate time series is transformed using closed interesting subspaces. Figure 3 shows the total coverage for all the three transformations. The figure shows that the maximum total coverage is obtained when closed interesting subspaces are used to represent the multivariate time series.

Table 8. No. of Interesting subspaces mined and time taken in sec to mine the Subspaces at different values of τ where $\tau = \frac{X}{n}$, $n = 2179$

X	No. of interesting subspaces mined			Time in sec		
	CIS	MIS by original SCHISM	MIS by modified SCHISM	CIS	MIS by original SCHISM	MIS by modified SCHISM
100	242	116	123	15.3	8.9	6.3
10	216	123	128	14.7	8.9	6.2
1	215	123	132	14.6	9.1	6.2
0.1	214	125	135	14.4	9.1	6.0
0.01	214	125	135	14.4	9.1	6.0
0.001	211	128	141	14.2	9.2	6.5
0.0001	211	128	141	14.3	9.2	6.2
0.00001	210	35	54	14.2	5.1	3.9
0.000001	210	35	54	14.1	5.1	4.1

The transformation that maximizes the Total Pattern Length, Total Coverage and which minimizes the Total Pattern Dissimilarity will lead to less loss of information. Table 8 shows the Total number of subspaces mined and Total Time taken to mine the three types of interesting subspaces.

6 Conclusions

This research has proposed a new multivariate time series transformation technique that transforms a multivariate time series to a symbol sequence. Closed interesting subspaces are mined from the multivariate time series and are used in transforming the multivariate time series to symbol sequence. The symbol sequence preserves the inter dependencies and co-variations among groups of time series variables constituting the multivariate time series. The proposed transformation allows the applicability of symbolic sequential pattern mining algorithms and string processing algorithms on the transformed data. Patterns can be mined more efficiently by transforming multivariate time series to single symbol sequence rather than multiple symbol sequences as done in some of the state of art multivariate time series transformation techniques. Experimental results show that transforming multivariate time series using closed interesting subspaces leads to less loss of information when compared to transforming multivariate time series using maximal interesting subspaces.

References

1. Beyer, K., Goldstein, J., Ramakrishnan, R., Shaft, U.: When is nearest neighbor meaningful? In: Beeri, C., Bruneman, P. (eds.) ICDT 1999. LNCS, vol. 1540, pp. 217–235. Springer, Heidelberg (1998)
2. Sequeira, K., Zaki, M.: SCHISM: a new approach to interesting subspace mining. J. Bus. Intell. Data Min. **1**, 137–160 (2005)
3. Parson, L., Haque, E., Lui, H.: Subspace clustering for high dimensional data: a review. ACM SIGKDD Explor. Newslett. **6**, 90–105 (2004)
4. Lin, J., Keogh, E., Chiu, S.L.B.: A symbolic representation of time series, with implications for streaming algorithms. In: 8th ACM SIGMOD workshop on Research issues in data mining and knowledge discovery, pp. 2–11, ACM, NY, USA (2003)
5. Cambridge raw daily weather data. https://www.cl.cam.ac.uk/research/dtg/weather/index-daily-text.html
6. Hidaka, S., Yu, C.: Spatio-temporal symbolization of multidimensional time series. In: IEEE International Conference on Data Mining Workshops, IEEE Computer Society, Washington, DC, USA, pp. 249–256 (2010)
7. Esmael, B., Arnaout, A., Fruhwirth, R.K., Thonhauser, G.: Multivariate time series classification by combining trend-based and value-based approximations. In: Murgante, B., Gervasi, O., Misra, S., Nedjah, N., Rocha, A.M.A., Taniar, D., Apduhan, B.O. (eds.) ICCSA 2012, Part IV. LNCS, vol. 7336, pp. 392–403. Springer, Heidelberg (2012)
8. Tanaka, Y., Iwamoto, K., Uehara, K.: Discovery of time-series motif from multi-dimensional data based on MDL principle. J. Mach. Learn. **58**, 269–300 (2005)

9. Zhuang, D.E.H., Li, G.C.L., Wong, A.K.C.: Discovery of temporal associations in multivariate time series. IEEE Trans. Knowl. Data Eng. **26**, 2969–2982 (2014)
10. Chu, Y.-H., Huang, J.-W., Chuang, K.-T., Yang, D.-N., Chen, M.-S.: Density conscious subspace clustering for high-dimensional data. IEEE Trans. Knowl. Data Eng. **22**, 16–30 (2010)
11. Sirisha, G.N.V.G., Shashi, M.: Mining closed interesting subspaces to discover conducive living environment of migratory animals. In: Das, S., Pal, T., Kar, S., Satapathy, S.C., Mandal, J.K. (eds.). FICTA-2015. AISC, vol. 404, pp. 153–166. Springer, India (2015)
12. Ratanamahatana, C.A., Lin, J., Gunopulos, D., Keogh, E.J., Vlachos, M., Das, G.: Mining time series data. Data Mining and Knowledge Discovery Handbook, pp. 1049–1077. Springer, Berlin (2010)

S2S: A Novel Approach for Source to Sink Node Communication in Wireless Sensor Networks

Chhabi Rani Panigrahi[1](✉), Joy Lal Sarkar[1], Bibudhendu Pati[1], and Himansu Das[2]

[1] Department of Computer Science and Engineering, C.V. Raman College of Engineering, Bhubaneswar, India
{panigrahichhabi,patibibudhendu,das.himansu2007}@gmail.com, joy35032@rediffmail.com

[2] School of Computer Engineering, KIIT University, Bhubaneswar, India

Abstract. In Wireless Sensor Networks (WSNs) sensor nodes are deployed in various geographical areas. But, the main problem is the data collection from source nodes to the sink node in an energy efficient way and in data collection scenario it is also a challenging task to reduce inter-cluster communication cost which can balance the network traffic. To work with these challenges, in this work, we propose a source to sink node (S2S) communication algorithm where to reduce the communication overhead and to minimize the delay we used Cluster-Head(CH)-CH communication method where one CH forms coalition to another CH based on the distance of the sink node from each CH. The simulation results indicate the better performance of our approach as compared to the existing approaches and the results are validated through MATLAB.

Keywords: WSNs · Coalition formation · Energy consumption

1 Introduction

To send the information from source to sink node, WSNs need an important mechanism which can reduce the network traffic load as well as energy consumption by a node. There are various sensor nodes which are placed into several geographical locations [10]. From different nodes, one CH is selected to form cluster [12]. All time query may be generated in the network, so to reply each and every query nodes collect information from the environments. CH initiates to transmit information to the sink via another CH [5]. Data can be transmitted to the sink either in single hop or multi-hop [10] fashion. CH aggregates the information which are collected from the source node and is forwarded to the sink [5]. But, in that case one of the main problems is the limited energy of the sensor nodes [1]. Because sensor nodes are deployed in such areas where it is very difficult to change battery and once the battery of a sensor is over then this is called as *dead* node. In case of single-hop, the nodes which are the farthest from the CH always consume more energy but in case of multi-hop the nodes which are the closer to the CH consume more energy due to relaying [14].

R. Prasath et al. (Eds.): MIKE 2015, LNAI 9468, pp. 406–414, 2015.
DOI: 10.1007/978-3-319-26832-3_38

The multi-hop communication may require high packet transmission and large summary packets may be generated in WSNs. If CH to CH communication increases, it will result in increase in the inter-cluster communication cost [14] and also results in increased delay. For this, a mechanism is required to balance the network traffic as well as to reduce the delay. Therefore, we are interested to minimize the number of hops in between CH to CH communication. In this work, we use multi-hop technique for CHs communication where, aggregated information from the source node is transmitted to the sink via several CHs. For that, CHs form coalition and then send the information to the sink node.

The rest of the paper is organized as follows: Sect. 2 describes the related work and Sect. 3 presents the energy consumption model. Our proposed approach is described in Sect. 4. Section 5 presents the results obtained along with the analysis of results. Finally, we conclude the paper in Sect. 6.

2 Related Work

Several clustering approaches have been proposed for sending data to the sink node. In [7], authors proposed Low Energy Adaptive Clustering (LEACH) protocol where from different nodes one CH is selected for each cluster. All neighboring nodes send data to the CH. CH then aggregates the data and sends to the sink node. In [2], authors proposed a Stable Election Protocol (SEP) called as Deterministic-SEP (D-SEP) which is used to select CH in a distributed environment. In [8], authors proposed Balanced Energy-Efficient Grouping (BEEG) protocol where nodes are grouped based on the initial energy of nodes.

In [4], authors proposed a Markov model for a sensor network where to save energy a sensor maintains two operational modes: *sleep* and *active* mode. In case of sleep mode, sensor consumes less energy as compared to active mode. In [9], authors proposed Power-Efficient Gathering in Sensor Information Systems (PEGASIS) protocol which gives an useful method for data gathering in WSNs which used only one transmission to the base station. In [15], authors proposed Hybrid Energy-Efficient Distributed clustering (HEED) protocol which gives inter-cluster communication by multi-hop and HEED also maintains CH based on the residual energy labels of nodes.

In [11], authors proposed Distributed Energy-Efficient Clustering (DEEC) which calculates residual energy ratio and based on this CH are selected. In [3], authors proposed an Unequal Cluster-based Routing (UCR) protocol in WSNs which determines the process of cluster formation by assuming network-wide announcements.

3 Energy Consumption Model

In this section, we highlight the different factors which are the cause of energy consumption by sensors. The energy consumption model for a sensor node is described more briefly as follows:

- *Sensor sensing*
  To collect the information from the outside world, a sensor needs to be sensed first. To sense with the out side world, a sensor node n_i consumes it's energy denoted by $E_s(n_i)$ and is computed by using Eq. (1).

$$E_s(n_i) = p_{mesg} V_{supply} I_{sense} t_{sense} \tag{1}$$

  Where, t_{sense} is the time duration to for sensing including CH, I_{sense} is the total amount of current required for sensing activity for p_{mesg} bit packet sensing, and V_{supply} represents the supply voltage.
- *Sensor logging*
  A sensor node consumes it's energy for reading $p_m sg$ bit packet and also to write into the memory [6]. The total energy consumed due to sensor logging by node n_i denoted by $E_l(n_i)$ is computed by using Eq. (2).

$$E_l(n_i) = \frac{p_{msg} V_{supply}}{8} \left(I_{writing} t_{writing} + I_{reading} t_{reading} \right) \tag{2}$$

  Where, $I_{writing}$ and $I_{reading}$ denote the required current for writing and reading of one byte data at the time duration $t_{writing}$ and $t_{reading}$ respectively.
- *Transmitting and receiving data unit*
  For transmitting p_{msg} bits message, a sensor node n_i consumes it's energy denoted as $E_t(n_i)$ and is computed by using Eq. (3).

$$E_t(n_i) = \begin{cases} E_{el} . p_{msg} + l_{am} . p_{msg} . d^2 & ; When\ d \leq d_0 \\ E_{el} . p_{msg} + l_{fs} . p_{msg} . d^4 & ; When\ d > d_0 \end{cases} \tag{3}$$

  In Eq. (3), E_{el} is the energy consumption for transmitting and receiving data unit. In Eq. (3), l_{amp} and l_{fs} are the amplifier parameters of transmission corresponding to the multi-path fading model and free space model respectively. In Eq. (3), d represents the distance between any two nodes and d_0 is the threshold distance.

  Energy consumption for receiving p_{msg} bits message denoted as $E_r(n_i)$ and is computed by using Eq. (4).

$$E_r(n_i) = E_{el} . p_{msg} \tag{4}$$

- *Transition from sleep to active state*
  During transition from sleep to active state, a sensor node consumes energy. For n number of nodes the energy consumed due to transition from sleep to active state denoted as E_{sa} and is computed using Eq. (5).

$$E_{sa} = n e_{sa} \tag{5}$$

  For one node n_i, Eq. (5) can be rewritten as:

$$E_{sa}(n_i) = e_{sa} \tag{6}$$

Where, e_{sa} denotes the energy consumption by a sensor node from sleep to active state, while the cost of transition from active state to sleep state can be neglected [4].

Then the total energy consumed by a sensor node denoted as $E_{total}(n_i)$ and is given as in Eq. (7).

$$E_{total}(n_i) = E_s(n_i) + E_l(n_i) + E_t(n_i) + E_r(n_i) + E_{sa}(n_i) \tag{7}$$

If $E_{init}(n_i)$ be the node's initial energy then the residual energy denoted as $E_{residual}(n_i)$ and is computed by using Eq. (8).

$$E_{residual}(n_i) = E_{init}(n_i) - E_{total}(n_i) \tag{8}$$

4 Proposed Approach

In WSNs, every CH collects data from different sensor nodes in a cluster. The CH then aggregates the data and forwards to the sink via another CH. To reply the query when one CH wants to transmit information to the sink node, it generates a *Coalition Formation (CF)* message. This CH is named as *Source-CH*. Every other CHs can reply to this message if they heard a *CF* message and generate *Want to Join Coalition (WJC)* message. The *Source-CH* then selects it's coalition members based on the distance of the sink node from CHs and is called as a *Member-CH*. After getting the first coalition member, *Source-CH* sends packets to all it's members. The *Member-CH* initiates to send this packet to the sink node. If *Member-CH* does not find the sink node within it's radio range then *Member-CH* generates *CF* message and so on until sink node is discovered. If sink node is found then the packets are sent to the sink node.

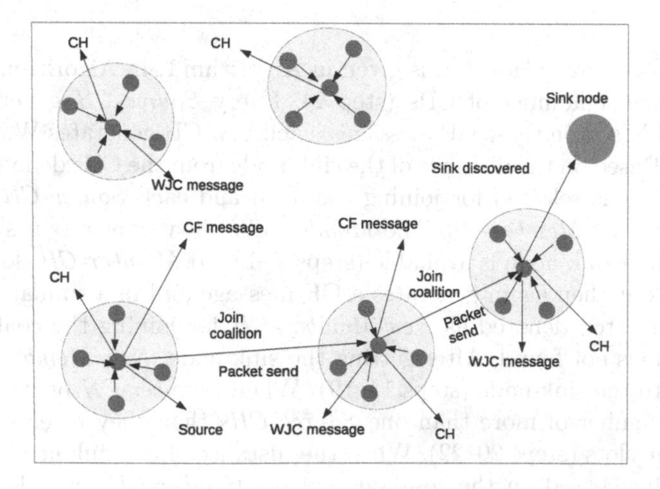

Fig. 1. Join coalition.

Figure 1 shows a scenario where CHs form coalition. From Fig. 1, it is clear that a *Source-CH* generates CF message and different CHs respond by WJC message based on the distance of the sink node from CHs. *Source-CH* selects it's *Member-CH* and sends packets. The *Member-CH* then generates *CF* message and it then selects another *Member-CH* and finally sends packets to the sink node.

The idea behind coalition formation of CHs is that it will reduce the network traffic as well as the energy consumption. To reply one query, there is no need to join coalition of all CHs. One CH can be a member of more than one *Source-CH.* In that scenario, *Member-CH* receives packets from different *Source-CHs* at different time slots. For example, in Fig. 2, one CH is the member of five *Source-CH*s. The *Member-CH* receives packets from *Source-CH*-1 at time slot (0, 1) and from *Source-CH*-2 at time slot (2, 3) and so on.

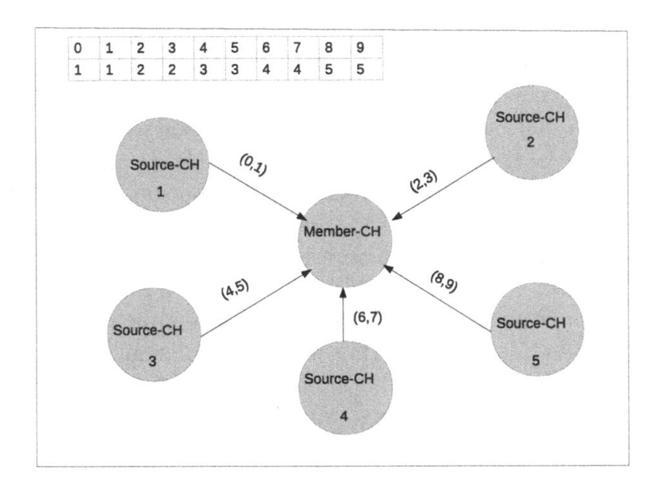

Fig. 2. One *Member-CH* from five *Source-CH*s.

The proposed S2S algorithm is given in Algorithm 1. In Algorithm 1, we consider there are n number of CHs (step 2). Every *Source-CH* generates a CF message and based on these CF messages, each new CH generates WJC message (steps 3–6). Based on the distance of the sink node from the CHs denoted as D^{CH} cluster member is selected for joining coalition and each *Source-CH* transmits it's packets to the *Member-CH.* The *Member-CH* then transmits it's packets to the sink node if sink node is available (steps 7–13). If *Member-CH* does not find the sink node it then again generates a CF message and in a similar way a new member is selected denoted as *New-Member-CH* for joining the coalition until the sink node is not found. After getting the sink node, *New-Member-CH* sends it's packets to the sink node (steps 14–19). When *Member-CH* or *New-Member-CH* is the member of more than one *Source-CHs* then they receive packets at different time slots (steps 20–22). When the distance of the sink node is different from CHs then based on the residual energy, *Member-CH* are selected (steps 23–25).

Algorithm 1. The S2S Communication Algorithm

```
1: procedure S2S(CH, sink, coalition)
2:   CH ← {1, 2, 3, ..., n}
3:   for each Source-CH do
4:     generates CF message
5:   for each new-CH do
6:     generates WJC message
7:   for ∀ new-CH which generates WJC message do
8:     if (D^CH_i > D^CH_j) then
9:       Member-CH ← D^CH_i
10:      Member_CH ← Source-CH_packets
11:      Last-Member-CH ← Member-CH
12:      if (Sink == true) then
13:        Sink ← Last-Member-CH_packets
14:        else
15:        for ∀ Last-Member-CH do
16:          Repeat steps 4 − 8
17:          New-Member-CH ← D^CH_i
18:          New-Member-CH ← Last-Member-CH_packets
19:        Sink ← New-Member-CH_packets
20:        for ∀ Last-Member-CH and ∀New-Member-CH do
21:          if Source-CH > 1 then
22:            Received packets at different time slots
23:    if (D^CH_i = D^CH_j) then ▷ When sink located at similar distance from CHs
24:      Calculate E_residual
25:      Member-CH ← E^max_residual
                                ▷ Sends packets at the same way by residual energy of CHs
```

5 Results and Analysis

We implemented our algorithm in MATLAB (version R2013a). To implement our algorithm, initially we used the model as described in Sect. 3 for selecting CH. After selection of CH, we then used a method as described in [13] to form clusters. To send information from source node to the sink node, all CHs are not necessary. We are interested to use those CHs which have the least distance from the sink node. If the distance of the sink node from two different CHs are same and both generate WJC message then based on the signal strength, a CH is selected. For that, we assume energy consumption of two different CHs are not same (Table 1).

A *Source-CH* generates CF message within a radius of $\sqrt{(A^2 + 4B^2)}$. Where, A is the width of the network area where sink node is located and B is the average region width of the sink node [14]. Each receiving CH generates WJC

Table 1. Parameters used in our proposed approach

Parameters	Values
Network Size	$200 \times 200\,\text{m}^2$
Sink	75×150
Number of nodes	1000
Cluster Radius	25 m
E_i	$3.8\,j$
E_{el}	$48\,mj/bit$
p_{msg}	$2000\,bit/sec$
Broadcast packet size	$17\,bytes$
E_{sa}	$0.31\,nj/bit$
l_{am}	$0.1\,nj/bit/m^2$
l_{fs}	$0.2\,nj/bit/m^4$
d	$\leq 10\,\text{m}$
e_{sa}	$0.31\,nj/bit$

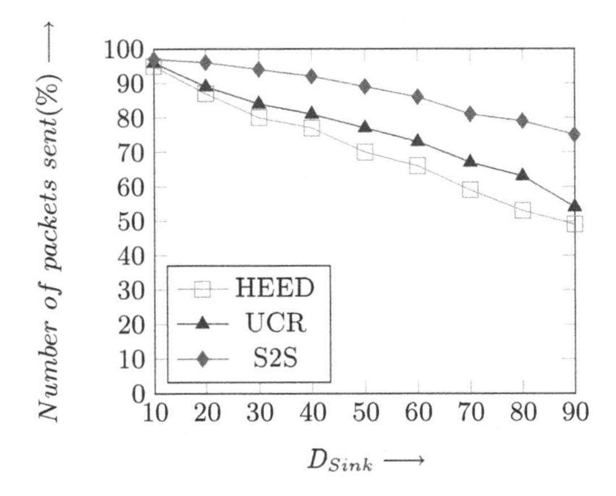

Fig. 3. Percentage of packets sent when the distance between *Source-CH* to the sink node is varied.

message. A *Source-CH* then selects its coalition member based on the distance of the sink node from that node. For computing the distance of the sink node from each cluster we used the concept as described in [14]. *Source-CH* may use multi-hop concepts for transmitting information to the sink node. To measure the effectiveness of our approach, we consider the parameters such as percentage of packets sent, number of hops used, and the average residual energy of nodes when the distance between *Source-CH* to the sink node is varied.

The graph as shown in Fig. 3 shows the percentage of packets sent when the distance between *Source-CH* to the sink node is varied. In the graph as shown in Fig. 3, X axis represents the distance between *Source-CH* to the sink node

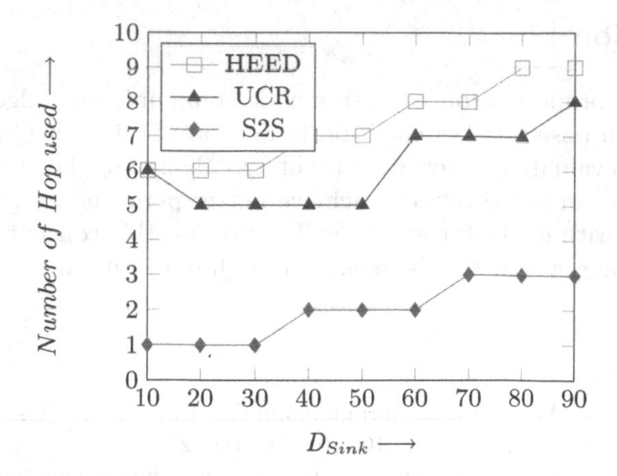

Fig. 4. Total number of hops used when the distance between *Source-CH* to the sink node is varied.

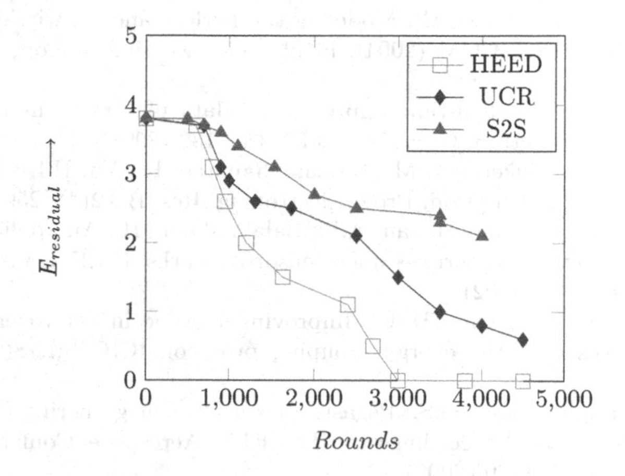

Fig. 5. Residual energy when distance between *Source-CH* to the sink node is varied.

denoted as D_{Sink} and Y axis represents the percentage of packets that are sent to the sink node. Figure 3 indicates that when D_{Sink} is 80, 21 % packets are lost in case of S2S algorithm whereas in case UCR and HEED, 37 % and 47 % of packets are lost respectively. The graph as shown in Fig. 3 indicates that S2S performs well as compared to existing approaches that is HEED and UCR. From the graph as shown in Fig. 4, it is also clear that S2S uses less number of *hops* as compared to HEED and UCR. This indicates that S2S results in reducing network traffic load and hence helps in reducing the energy consumption. We have also measured the average residual energy in all three approaches. The graph as shown in Fig. 5 indicates that the average residual energy is higher in case of S2S as compared to HEED and UCR.

6 Conclusion

In this work, we propose a source to sink node communication algorithm named as S2S algorithm based on the coalition formation of different CHs. Simulation is conducted to evaluate the performance of S2S algorithm. The results obtained indicate that the proposed S2S can achieve better performance as compared to the existing algorithms. In future, we will try to use this concept for communication in real-time, where WSNs work with high data rate applications.

References

1. Akyildiz, F., Su, W., Sankarasubramaniam, Y., Cayirci, E.: A survey on sensor networks. IEEE Commun. Mag. **40**, 102–114 (2002)
2. Bala, M., Awasthi, L.: Proficient d-sep protocol with heterogeneity for maximizing the lifetime of wireless sensor networks. I. J. Intell. Syst. Appl. **7**, 1–15 (2012)
3. Chen, G., Li, C., Ye, M., Wu, J.: An unequal cluster-based routing protocol in wireless sensor networks. Wirel. Netw. **15**, 193–207 (2007)
4. Chiasserini, C.-F., Garetto, M.: Modeling the performance of wireless sensor networks. In: IEEE INFOCOM (2004). http://www.ieee-infocom.org/2004/Papers/06_1.PDF
5. Gupta, S., Dave, M.: Real-time approach for data placement in wireless sensor networks. Int. J. Electron. Circ. Syst. **2**(3), 132–139 (2008)
6. Halgamuge, M.N., Zukerman, M., Ramamohanarao, K., Vu, H.L.: An estimation of sensor energy consumption. Prog. Electromag. Res. B **12**(4), 259–295 (2009)
7. Heinzelman, W.B., Chandrakasan, A.P., Balakrishnan, H.: An application-specific protocol architecture for wireless microsensor networks. IEEE Trans. Wirel. Commun. **1**(4), 660–670 (2002)
8. Liaw, J.J., Chang, L., Chu, H.-C.: Improving lifetime in heterogeneous wireless sensor networks with the energy grouping protocol. ICIC Int. **8**(9), 6037–6047 (2001)
9. Lindsey, S., Raghavendra, C.S.: Pegasis: power efficient gathering in sensor information systems. In: Proceedings of the IEEE Aerospace Conference, BigSky, Montana, pp. 1125–1130 (2002)
10. Diallo, O., Rodrigues, J., Sene, M.: Distributed data management techniques for wireless sensor networks. IEEE Trans. Parallel Distrib. Syst. **26**(2), 604–623 (2013)
11. Quing, L., Zhu, Q., Wang, M.: Design of a distributed energy- efficient clustering algorithm for heterogeneous wireless sensor networks. Comput. Commun. **29**, 2230–2237 (2006)
12. Sarkar, J.L., Panigrahi, C.R., Pati, B., Das, H.: A novel approach for real-time data management in wireless sensor networks. In: Proceedings of 3rd International Conference on Advanced Computing, Networking and Informatics, pp. 599–607. IEEE Computer Society Press (2015)
13. Vural, S., Ekici, E.: On multihop distances in wireless sensor networks with random node locations. IEEE Trans. Mob. Comput. **9**(4), 540–552 (2010)
14. Wei, D., Jin, Y., Vural, S., Moessner, K., Tafazolli, R.: An energy-efficient clustering solution for wireless sensor networks. IEEE Trans. Wirel. Commun. **10**(1), 3973–3983 (2011)
15. Younis, O., Fahmy, S.: A hybrid, energy-efficient, distributed clustering approach for ad hoc sensor networks. IEEE Trans. Mob. Comput. **3**(4), 366–379 (2004)

Establishing Equivalence of Expressions: An Automated Evaluator Designer's Perspective

K.K. Sharma[1](✉), Kunal Banerjee[2], and Chittaranjan Mandal[2]

[1] School of Information Technology, Indian Institute of Technology Kharagpur, Kharagpur, India
kks@sit.iitkgp.ernet.in

[2] Department of Computer Science and Engineering, Indian Institute of Technology Kharagpur, Kharagpur, India
kunalb@cse.iitkgp.ernet.in, chitta@iitkgp.ac.in

Abstract. Automated assessment of students' programs has become essential in the institutions where the intake of students is large to ensure fast and consistent evaluation. An automated evaluator compares a program written by a student with a model program supplied by the instructor and tries to evaluate the student's performance. In course of checking similarity between the two programs, the evaluator may sometimes have to determine whether some expression written in the student program assumes the same value as that of an equivalent expression in the model. Thus, determining equivalence between pairs of expressions is at the core of designing automated evaluators. This paper discusses different methods for determining equivalence between expressions involving various datatypes. Specifically, it proposes a novel technique to determine equivalence between expressions involving floating point and transcendental numbers, which have not been addressed in earlier literature to the best of our knowledge.

Keywords: Computer-assisted learning · Automated program evaluation · Equivalence checking · Floating point numbers · Transcendental numbers

1 Introduction

Many, if not all, engineering institutions provide mandatory courses on elementary programming for undergraduates. Consequently, the number of students enrolled in such subjects is high. Often, due to lack of sufficient and/or efficient staff, quick and consistent evaluation of students' programs may not be ensured in such places. As a result, automated evaluators which can speedily and consistently assess students' programs has become a necessity for sustenance of the modern academic needs. The problem of automated assessment of programs has been thoroughly studied as surveyed in [2,8].

The methods for automated assessment of students' programs can be broadly divided into two categorizes – dynamic and static. Dynamic analysis involves

R. Prasath et al. (Eds.): MIKE 2015, LNAI 9468, pp. 415–423, 2015.
DOI: 10.1007/978-3-319-26832-3_39

actual execution of a student's program over a supplied set of test cases and then checking the output for correctness. In this type of methods, a program is evaluated correct based on only the success or failure of the execution of the test cases because this method does not consider the way in which the program has been written to solve the problem. Neglecting how the problem has been attempted is the major drawback of dynamic analysis based approaches. Such mechanisms may ensue unfair grading because a program appearing to produce a right output may not meet the programming specification [22]. Moreover, novice programmers sometimes submit programs which do not generate any output or even worse could be the case where the submitted program never terminates (on entering into an infinite loop) when executed for dynamic testing. These situations make dynamic systems fail in awarding reasonable marks.

Static analysis based systems, on the contrary, do not execute a student program. In general, the control and data-flow graph (CDFG) of a student's program is compared with that of some standard program provided by the instructor and the assessment is done based on the extent of similarity between the two. For example, the work reported in [23] considers an intermediate representation form called *system dependence graph*, which captures a program's CDFG vividly, and tries to establish structural equivalence between the system dependence graphs obtained from the student and the instructor supplied programs. In the literature, there are also some static methods reported which do not do similarity checking of structures, they rather rely on logic based formulations, such as, the work presented in [19] which proposes a method to check the correctness of else-if constructs in programs by checking the ordering of the involved conditions in the constructs based on some ordering rules.

Recently, an automated evaluation scheme based on equivalence checking of the Finite State Machine with Datapath (FSMD) model [7] has been proposed in [20]. The FSMD model is suitable for statically checking equivalence of two programs whose datapaths contain only Boolean and integer datatypes. Since programs implementing algorithmic computations over integers involves the whole of integer arithmetic which is undecidable, majority of the FSMD equivalence checkers [3,4,10–12] employ a normalization technique [16] that tries to reduce two computationally equivalent expressions e_1 and e_2 to a syntactically identical form. This normalization technique, however, is not applicable to reason over finite precision datatypes (bit-vectors), user-defined datatypes, etc. In [5], the authors have proposed to augment the normalization module, wherever necessary, with an SMT solver to determine the validity of $e_1 = e_2$. They have demonstrated that the scope of equivalence checking can be extended to handle bit-vectors, user-defined datatypes and more sophisticated transformations by leveraging the capability of SMT solvers while keeping the basic equivalence checking framework intact. However, none of these techniques have been known to address the equivalence of floating point and transcendental numbers. Determining equivalence of expressions involving such values poses some interesting challenges, e.g., unlike real numbers (which can be thought of as floating point numbers with arbitrary precision), floating point numbers (of finite size) do not abide by the law of associativity and are not closed under addition. Moreover, no normal form is known to be proposed for transcendental numbers.

This paper introduces a method to determine equivalence of such expressions (with high assurance) which may be computationally equivalent but cannot be ascertained to be so by the known techniques. A preliminary version of the contents in this paper can be found in [17]. Let us consider two expressions $e_1(v_1, \ldots, v_n)$ and $e_2(v_1, \ldots, v_n)$, where $v_i, 1 \leq i \leq n$, are the same set of variables appearing in both the expressions; our method intelligently assigns values to these variables v_i and checks how close are the values computed by $e_1(v_1, \ldots, v_n)$ and $e_2(v_1, \ldots, v_n)$ while permitting a difference of ϵ; lesser the value of ϵ, higher is the assurance of the two expressions being equivalent. Note that floating point variables are one of the basic datatypes in all typed programming languages. Moreover, since students generally cover trigonometric and logarithmic functions in their twelfth standard, the instructors often tend to assign programming exercises involving such familiar functions, which essentially involve computations of transcendental numbers. Our work, therefore, proposes an augmentation of the FSMD based equivalence checking method presented in [20] by including floating point numbers and transcendental numbers in their datatpath for the purpose of assessing students' programming exercises.

The paper is organized as follows. Section 2 presents a synopsis of earlier techniques for deciding equivalence of expressions of various datatypes. Section 3 discusses our proposed evaluation scheme for floating point and transcendental numbers while underlining the challenges involved. A discussion on our preliminary results is given in Sect. 4. The paper is concluded in Sect. 5.

2 Related Work

Determining equivalence of two programs basically entails checking whether on giving the same inputs, the two programs produce the same outputs or not. Although this problem is undecidable in general, the problem can be proven to be sound [4,10,15] and even complete [14] for some restricted subsets. Since it has to be ascertained that the values output by both the programs are the same, checking equivalence of expressions is at the core of this problem. In this section, we outline the various procedures undertaken to determine equivalence of expressions involving different datatypes.

Boolean expressions are the easiest to check for equivalence. Any two Boolean expressions can be converted into either of the two canonical forms namely, conjunctive normal form and disjunctive normal form, and then they can be checked for equivalence. However, no such canonical form exists for expressions over integers and consequently, a normal form for such expressions has been proposed in [13] which reduces many of the computationally equivalent expressions over integers into identical syntactical form, e.g., $a \times a + 2 \times a \times b + b \times b$ is the normal form of $(a + b)^2$. This work has later been adopted in [16], some simplification rules for the normalization grammar has been proposed in [10]; recently, this grammar has been extended to include array references in [6].

The theory of real numbers and bit-vectors, on the other hand, is decidable. Effective solutions for expressions over such variables can be obtained from state-of-the-art SMT solvers. Accordingly, these SMT solvers have been adopted in [5] to handle such datatypes in the context of checking equivalence of programs. It is to be noted that user-defined datatypes actually encompass multiple variables of the same or different datatypes; consequently checking equivalence of two user-defined variables of the same sort requires application of separate rules for each of its constituent variables. SMT solvers, such as CVC4 [1], provide constructs which can readily capture user-defined datatypes as defined in high-level languages, such as C.

3 Proposed Evaluation Scheme

Apparently, floating point numbers can be treated in the same way as real numbers or bit-vectors; however, it is not so. In usual mathematics, addition and multiplication of real numbers (and bit-vectors) obeys law of associativity. By contrast, in computer science, the addition and multiplication of floating point numbers is not associative, as rounding errors are introduced when dissimilar-sized values are joined together. To illustrate this, consider a floating point representation with a 4-bit mantissa:

$$\begin{aligned}(1.000_2 \times 2^0 + 1.000_2 \times 2^0) + 1.000_2 \times 2^4 &= 1.000_2 \times 2^1 + 1.000_2 \times 2^4 \\ &= 1.001_2 \times 2^4\end{aligned} \quad \text{(i)}$$

$$\begin{aligned}1.000_2 \times 2^0 + (1.000_2 \times 2^0 + 1.000_2 \times 2^4) &= 1.000_2 \times 2^0 + 1.000_2 \times 2^4 \\ &= 1.000_2 \times 2^4\end{aligned} \quad \text{(ii)}$$

Clearly, the values computed by Eqs. (i) and (ii) are different even though they involve the same operands. It may be noted that techniques exist to minimize rounding errors, such as, Kahan summation algorithm [9].

Equivalence checking of expressions, both for floating point numbers as well as transcendental numbers, e.g., trigonometrical identities, is not trivial. Hence, there is a need for establishing their equivalence by computing their values at various points of interest and checking that the values of equivalent expressions are equal at those points. In case the values at all points could be checked to be equal, then it could be said that one expression overlays the other. Doing this, however, is not possible as this may require evaluation at infinitely many number of points. The equivalence, thus, can only be established in a probabilistic sense, which means that if evaluation of two expressions at a very large number of points results in equal values at all these points, then with a high confidence we can say that they are equivalent. We, thus, cannot claim equivalence with 100 % confidence, yet a high confidence value may be achieved by taking a very large number of sample points. We can, therefore, rely on Monte Carlo simulation to achieve such equivalence.

3.1 Example of Equivalence Checking

We consider the following well known identity by de Moivre:
$e^{i\theta} = cos\theta + isin\theta$

The left hand side (LHS) of the identity can be treated as one function and the right hand side (RHS) as the other. The range of θ for evaluation can be conveniently chosen from 0 to π. We can start with choosing 10 points randomly within the allowed range, generated using *rand* function, the values of the two functions can be computed at those points and their difference is also computed at each point. At all the points there will be no difference in the computed values of the expressions. This proves that at all these points the functions are equal.

3.2 Equivalence Checking with Monte Carlo Simulations with Some Known Properties

The range of equivalence of two univariate functions can be easier to establish if some of their properties are known, e.g., the roots, or the maxima and the minima. The range then could be between the two consecutive roots, or it could be the points between the minimum value to the maximum value. We illustrate this idea with the following examples.

Example 1. $x^2 + 3.5x + 3 = (x + 1.5)(x + 2)$

We treat the LHS as one function and the RHS as the other. As the roots of each function are x = -1.5 and x = -2, we can choose our range for equivalence checking from x = -2 to -1.5. A set of points can be chosen in the above range and the functions can be evaluated at each of them, and their difference can be shown to be zero at all the points chosen. □

Example 2. sin 2θ = 2 sinθ cosθ

As the minimum value and the maximum value of θ are 0 and $\pi/4$, so this can be chosen to be the range of θ. Proceeding as above, we can show that the LHS is equivalent to the RHS. □

3.3 Equivalence Checking with Fewer Samples

We may do better in some specific cases where the functions are or can be reduced in the form of polynomials. To illustrate this idea, we take the following expression.

Example 3. $x^2 + 4x + 3.75 = (x + 1.5)(x + 2.5)$

As we know the roots of the function on LHS and RHS are the same, viz., x= -1.5 and -2.5. Moreover, as the LHS is a second degree polynomial, so we need to test the values of the functions only for these two values. Hence, by just looking at the values at only these two points, viz., x= $-1.5, -2.5$, where both the functions have values 0, we can conclude their equivalence. □

To find the approximate location of a root, we can make use of the following theorem [21].

Matijasevic's Theorem: *Given a polynomial* $c_1x^n + c_2x^{n-1} + ... + c_nx + c_{n+1}$ *with a root at* $x = x_0$. *Let* c_{max} *be the largest absolute value of a* c_i, *then*

$|x_0| < (n+1)\frac{c_{max}}{|c_1|}$, *i.e., the roots of the polynomial must lie between the values* $\pm(n+1)\frac{c_{max}}{c_1}$.

However, calculating such bounds for multivariate polynomials is not possible.

3.4 Equivalence Checking with Approximate Valuations With fewer Samples

In case the LHS and RHS of a trigonometric identity can be reduced to approximate polynomials, then we can establish the equivalence by evaluating the polynomials at a number of points which is more than the maximum degree among the polynomials. We take an example below:

Example 4. $\frac{1-tan\theta}{1+tan\theta} = \frac{cos\theta - sin\theta}{cos\theta + sin\theta}$

The trigonometric ratios $sin\theta$, $cos\theta$ and $tan\theta$ can be represented by their infinite series expansions.

$\sin x = x - \frac{x^3}{3!} + \frac{x^5}{5!} - ...$ for every x

$\cos x = 1 - \frac{x^2}{2!} + \frac{x^4}{4!} - ...$ for every x

$\tan x = x + \frac{x^3}{3!} + \frac{2x^5}{15} + ...$ for $|x| < \pi/2$

We approximate the infinite series by retaining the terms upto the fifth degree in their respective polynomials. Thus we get the following from the above identity.

LHS $= \frac{(1-x-\frac{x^3}{3!}-\frac{2x^5}{15})}{(1+x+\frac{x^3}{3!}+\frac{2x^5}{15})}$ and

RHS $= \frac{(1-x-\frac{x^2}{2!}+\frac{x^3}{3!}+\frac{x^4}{4!}-\frac{x^5}{5!})}{(1+x-\frac{x^2}{2!}-\frac{x^3}{3!}+\frac{x^4}{4!}+\frac{x^5}{5!})}$

For x=0 both the LHS and RHS evaluate to 1.

When evaluated at various values of $x < \pi/2$, we will find that the values of LHS and RHS are close to each other. We can thus establish that they are equivalent within a permissible error ϵ, where ϵ is larger than the maximum difference between the computed values of the LHS and the RHS at any chosen point. □

We may thus arrive at a notion of the ϵ-cover of a given expression.

Definition 1. (ϵ-cover): *ϵ -cover of an expression may be defined as the function whose value when evaluated differs by at most ϵ, from the value of the given expression, at a large fraction of points chosen at random.*

Definition 2. (Probably Equivalent Expressions): *If two expressions lie within ϵ-cover, then they are said to be* probably equivalent expressions *within the permissible error of ϵ.*

The challenge here is that we want to be able to state the probable equivalence with high confidence value, and find out what should be the number of samples to be chosen for this. As it turns out, the relation of sample size with confidence interval may be given by Chernoff bound. In doing so we would also like to use as less sampling points as possible. Reducing the sample points may be possible in a case where the reference function is well behaved, e.g., piece-wise linear or piece-wise smooth between corner points. In such cases, checking equivalence only at corner points may suffice, thus reducing the number of sample points. In the absence of such well behaved functions, we may need to look at random sample points, such that when tested over those points, the equivalence can be established with some high confidence.

Reducing number of sample points may be possible if we can reduce the given functions to some polynomials with some finite terms and if we can find the roots of the polynomials, then the points for evaluating the polynomials could be just at the roots. So for a degree n polynomial, as it has n roots, checking at n points could be sufficient, all the points being the roots of the polynomial. We can take the following example to illustrate the above idea.

Example 5. $(x - 1.5)(x - 2)(x - 3.1) = (x^2 - 3.5x + 3)(x - 3.1)$

The roots of the function on the LHS are x= 1.5, 2, 3.1. When we evaluate the RHS at the values of x corresponding to the roots of LHS, we find that the RHS also evaluates to the same value at those points. Thus, we can establish equivalence only by evaluating at 3 points as the expressions on both the sides are polynomials of degree three. □

3.5 Exception Handling

Another import aspect of our scheme is to take precaution against exceptions, such as, division by zero and indeterminate forms (e.g., $0 \times \infty$), to avoid unexpected termination of our system. This can be done by executing the expressions within try-catch blocks designed for exception handling.

4 Evaluation of Proposed Scheme

We did some simulations for different expressions involving floating point and transcendental numbers. For brevity, here we elaborate our experiment on only the following logarithmic identity:

$ln((1 + x)^y) = y \times ln(1 + x)$.

The LHS of the identity was treated as one function and the RHS as the other. The natural logarithms are represented as polynomials and expanded till their seventh power. To start with 10 points were chosen randomly, generated using rand function, the values of the two functions were computed at those points and their difference was also computed at each point. At all the points the difference in the computed values of the expressions was very less – less than 0.001; consequently, these two expression are declared probably equivalent expressions within the permissible error of 0.001.

We have been able to establish equivalence/probably equivalence between pairs of expressions for all our test cases; note that for probably equivalent cases, the difference was always within a very small permissible error ϵ. Thus, the current method seems promising; integrating it within the FSMD based automated evaluator of [20] remains as our future goal to enhance its scope of application. Once integrated, the performance can be evaluated using the benchmark programs as suggested in [18].

5 Conclusion

Equivalence checking of expressions, both for floating point numbers and transcendental numbers, is not trivial as there does not exist any canonical form of representation for them. In this paper, we propose that equivalence (with high confidence) can be established in such cases by computing the expression values at various points of interest and checking that the values for a pair of expressions are equal at those points or within some small permissible range. Simulations can, however, be done only in the range of interest and that too at a finite number of points. Thus, although equivalence of two expressions may not be claimed with 100 % confidence, there exists a very high probability that they indeed are equivalent on being declared to be so by our method. We intend to do a more exhaustive experimentation with exponential and logarithmic functions as well as improve upon the notion of ϵ-cover in our future work. Integrating our technique within the framework of an automated evaluator remains as another future goal.

References

1. CVC4 - the smt solver. http://cvc4.cs.nyu.edu/web/
2. Ala-Mutka, K.M.: A survey of automated assessment approaches for programming assignments. Comput. Sci. Edu. **15**(2), 83–102 (2005). http://www.tandfonline.com/doi/abs/10.1080/08993400500150747
3. Banerjee, K., Karfa, C., Sarkar, D., Mandal, C.: A value propagation based equivalence checking method for verification of code motion techniques. In: ISED, pp. 67–71 (2012)
4. Banerjee, K., Karfa, C., Sarkar, D., Mandal, C.: Verification of code motion techniques using value propagation. IEEE Trans. CAD ICS **33**(8), 1180–1193 (2014)
5. Banerjee, K., Mandal, C., Sarkar, D.: Extending the scope of translation validation by augmenting path based equivalence checkers with SMT solvers. In: 18th International Symposium on VLSI Design and Test, pp. 1–6, July 2014
6. Banerjee, K., Sarkar, D., Mandal, C.: Extending the FSMD framework for validating code motions of array-handling programs. IEEE Trans. CAD ICS **33**(12), 2015–2019 (2014)
7. Gajski, D.D., Dutt, N.D., Wu, A.C., Lin, S.Y.: High-Level Synthesis: Introduction to Chip and System Design. Kluwer Academic, Boston (1992)
8. Ihantola, P., Ahoniemi, T., Karavirta, V., Seppälä, O.: Review of recent systems for automatic assessment of programming assignments. In: Koli Calling, pp. 86–93 (2010)
9. Kahan, W.: Pracniques: further remarks on reducing truncation errors. Commun. ACM **8**(1), 40 (1965)
10. Karfa, C., Sarkar, D., Mandal, C., Kumar, P.: An equivalence-checking method for scheduling verification in high-level synthesis. IEEE Trans. CAD ICS **27**, 556–569 (2008)
11. Karfa, C., Mandal, C., Sarkar, D.: Formal verification of code motion techniques using data-flow-driven equivalence checking. ACM Trans. Des. Autom. Electron. Syst. **17**(3), 30 (2012)

12. Karfa, C., Mandal, C., Sarkar, D., Pentakota, S.R., Reade, C.: A formal verification method of scheduling in high-level synthesis. In: ISQED, pp. 71–78 (2006)
13. King, J.C.: A program verifier. Ph.D. thesis, Pittsburgh, PA, USA (1970)
14. Lopes, N.P., Monteiro, J.: Automatic equivalence checking of UF+IA programs. In: Bartocci, E., Ramakrishnan, C.R. (eds.) SPIN 2013. LNCS, vol. 7976, pp. 282–300. Springer, Heidelberg (2013)
15. Manna, Z.: Mathematical Theory of Computation. McGraw-Hill Kogakusha, Tokyo (1974)
16. Sarkar, D., De Sarkar, S.: A theorem prover for verifying iterative programs over integers. IEEE Trans Softw. Eng. **15**(12), 1550–1566 (1989)
17. Sharma, K.K., Banerjee, K., Mandal, C.: Determining equivalence of expressions: an automated evaluator's perspective. In: 2015 IEEE International Conference on Technology for Education (T4E) (2015, accepted)
18. Sharma, K.K., Banerjee, K., Mandal, C., Vikas, I.: A benchmark programming assignment suite for quantitative analysis of student performance in early programming courses. In: 2015 IEEE International Conference on MOOC, Innovation and Technology in Education (MITE) (2015, accepted)
19. Sharma, K.K., Banerjee, K., Vikas, I., Mandal, C.: Automated checking of the violation of precedence of conditions in else-if constructs in students' programs. In: 2014 IEEE International Conference on MOOC, Innovation and Technology in Education (MITE), pp. 201–204 (2014)
20. Sharma, K.K., Banerjee, K., Mandal, C.: A scheme for automated evaluation of programming assignments using FSMD based equivalence checking. In: I-CARE, pp. 10:1–10:4 (2014)
21. Sipser, M.: Introduction to the Theory of Computation. PWS Publishing Company, Boston (1997)
22. Wang, T., Su, X., Ma, P., Wang, Y., Wang, K.: Ability-training-oriented automated assessment in introductory programming course. Comput. Edu. **56**(1), 220–226 (2011)
23. Wang, T., Su, X., Wang, Y., Ma, P.: Semantic similarity-based grading of student programs. Inf. Softw. Technol. **49**(2), 99–107 (2007). http://www.sciencedirect.com/science/article/pii/S0950584906000371

Data Driven Modelling for the Estimation of Probability of Loss of Control of Typical Fighter Aircraft

Antony Gratas Varuvel(✉)

Aeronautical Development Agency, Bangalore 560017, India
vaagratus@yahoo.co.in

Abstract. Loss of control of aircraft is one of the catastrophic safety critical events in the aerospace domain, which results usually into risks of loss of human lives and/or environmental hazards. Triggering of this undesired event could be at any level of hardware and/or software in the digital fly-by-wire fighter aircraft. The contributing factors for this undesirable event and the interrelationships among the basic events are to be carefully accounted, while estimating the loss of control of the fighter aircraft, probabilistically. Components which have the potential to cause failures are required to be treated carefully, by properly considering the failure modes of those components. This paper brings out, the data driven methodology to estimate the probability of control of the aircraft considering all the interdependent components along with the associated failure modes, which have the potential to trigger the occurrence of the undesired event-'Loss of Control of Aircraft'. The approach presented here would serve as a **guideline for estimating the PLOC of any types of aircraft**.

Keywords: Assessment · Estimation · Failure modes · Failure rate · Fighter aircraft · Fly-by-wire · PLOC · Safety critical

1 Introduction

Safety is a primary concern for the aerospace community and the certification agency, where human causalities are expected to result into fatal human and environmental consequences. Hence, the certification norms necessitate, safety as one of the design and demonstration criteria in the case of transportation and nuclear industries. It is very much essential to realize the system with adequate safety interlocks and features including fault tolerance with acceptable graceful degradation. For aircraft category, especially fighter aircraft, dependability is the most driving factor of design, without compromising on the manoeuvrability and performance. The loss of control of aircraft is considered, as one of the most safety critical events, which would otherwise result into loss of life and/or aircraft. In order to meet the certification norms, it is crucial to prove the probability of loss of control of the aircraft [PLOC], within acceptable limits. Refer [8]. Initially,

R. Prasath et al. (Eds.): MIKE 2015, LNAI 9468, pp. 424–436, 2015.
DOI: 10.1007/978-3-319-26832-3_40

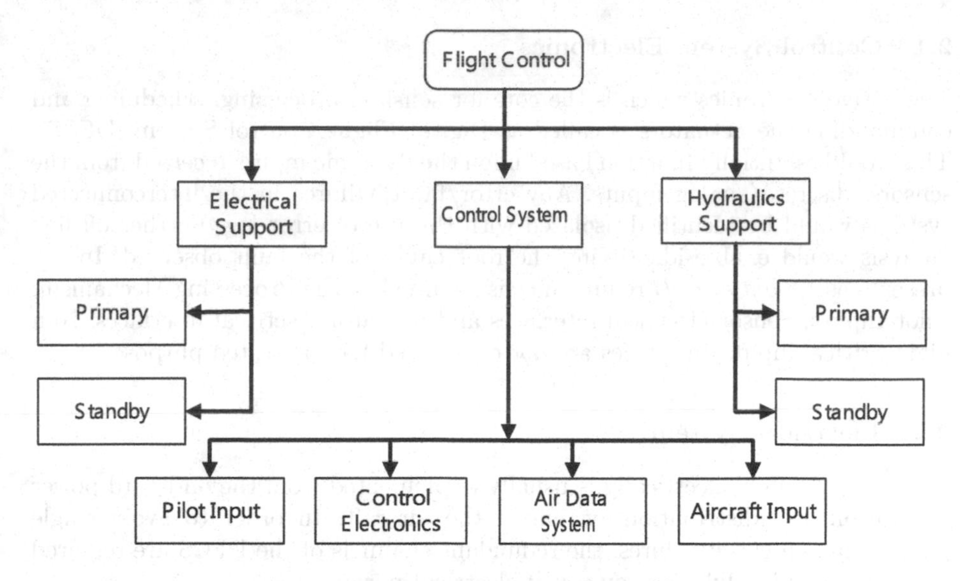

Fig. 1. FlightControl

PLOC could be estimated with the failure rate data of all the interrelated and interconnected systems/components followed by the assessment/demonstration of the same from the field data. The failure mode and failure mode ratio are to be appropriately chosen and logically related and modelled so as to establish the relationships of fault propagation and resultant failures. The main objective of this paper is to generalize the failure rate data driven methodology of estimating the PLOC and thereby providing the possible feedback to design team to further the design chosen or choosing other design alternatives to meet the safety goal mandated.

2 Flight Control-Typical Fighter Aircraft

A typical digital-fly-by-wire fighter class of aircraft, with redundancies built-in, sufficiently enough to mitigate/avoid any inadvertent catastrophic failure has been considered for this study and modelling. Flight control system when supported by hydraulics system and the electrical system of aircraft, would deliver the function of controlling the aircraft as intended and commanded by the Control System Electronics, which executes and commands the actuators, based on the dynamic external inputs and the attitude of the aircraft. Refer Fig. 1. Typical fighter aircraft would employ multiple levels of redundancy in the following:

- Sensing Mechanism.
- Processing Mechanism.
- Electrical Powering Scheme.
- Hydraulics Powering Scheme.
- Command & Control Scheme.

The major components of the Flight Control are summarized further.

2.1 Control System Electronics

The active electronics which is the core for sensing, processing, scheduling and commanding the actuators is called as Digital Flight Control System [DFCS]. This would essentially function based upon the dynamic inputs received from the sensors, discrete/analog inputs. Any error/fault/failures in the interconnected systems would be identified, isolated with the use of error log. Further off-line analysis would enable identifying the root cause of the fault observed. In the present case, *k-out of-n: G* redundancy is assumed on the Processing Mechanism, Pilot Inputs, Sensor electrical interfaces and actuator electrical interfaces. Non flight critical inputs/interfaces are not considered for the stated purpose.

2.2 Electrical System

Powering of DFCS electrically is usually implemented from the on-board power generation and distribution system of the aircraft. In order to avoid single point/common cause failures, the redundant channels of the DFCS are required to be powered by different sources of electrical power.

2.3 Hydraulics System

Actuation mechanism assumed for the fighter aircraft is hydraulics driven. Redundancy is also ensured in the hydraulics source and supply scheme, so as to avoid single point failure leading to loss of hydraulics to control the aircraft. As it would be unsafe, to land the aircraft, with the failure of total hydraulics system, electrically operated hydraulic pump is assumed as an option for the pilot to level the aircraft and eject safely, in case of failure of on board hydraulic system.

3 Probability of Loss Control [PLOC]

There are various events and/or combination of events which may eventually trigger the occurrence of the loss of control. Refer [8]. Considering criticality of Integrated Flight Control System, the above listed inter and intra-system interfaces, and the events which are having the potential to cause failure mode(s) in respect of the total loss of the aircraft are to be accounted while estimating the probability of occurrence. Following categorization of failures has been proposed.

- Common cause failures.
- Cascading failures.
- Dependent failures.
- Mutually exclusive failures.
- Mutually independent failures.
- Hardware Software Interaction failures.

4 Modelling Failure Types

In order to generalize all those failures, the categorization of failures have been proposed as in Sect. 3. Each and every failure types identified are to be expressed and to be modelled, mathematically. Brief explanation has been provided in the following subsections.

4.1 Common Cause Failures [CCF]

Occurrence of an event, if tend to trigger occurrence of more than one and all next higher level events, then the basic triggering event is termed as common cause event. In the present case, the electrical and hydraulic power failure to the interconnected components/actuators are treated as common cause failures. Mathematically, properties of these identified events are *same* and their occurrence is to be simulated simultaneously. Probability of common cause failure [CCF] is given by,

$$P(CCF) = \beta \times c_1 \times c_2 \times \cdots \times c_n \tag{1}$$

where,
β: Probability that a failure mode results in multiple failures
c_i: Failure rate of the component i.

4.2 Cascading Failures [CAF]

Cascading failures are those whose failure affect all the interconnected events. Their effects are unlimited, until reaches the undesired event. *Boolean OR Logic*, in simpler form, will ensure that the basic failure event propagates to the end event (making all other events as *don't cares*), by triggering all intermediate events.

$$P(CAF) = x(\vee y_i) \tag{2}$$

where,
x: Probability of cascading failure
y_i: Failure rate of the component i, which is linked with x.

4.3 Dependent Failures [DEF]

Failure of a component, if found to be interrelated with the failure of another component, then the basic triggering event is called as initiating event. The main difference between the common cause failure and the (inter)dependent failures is, in the former, the affected components are many, whereas in latter only one component which is in the next higher/immediate level is assumed to get affected. The dependency could be expressed in terms of *Binomial Process*. Standby, k-out of-n voter, load sharing are also treated as dependant failures. In general, the reliability of *k-out of-n voter* is expressed as below:

$$R(k,n) = \sum_{i=k}^{n} \binom{n}{i} p^i q^{n-i} \tag{3}$$

where,

R: Reliability of system with n components out of which k are to be successful

p: Probability of success of i.i.d. components

q: Probability of failure of i.i.d. components

n: Total number of components of the system

k: Minimum no. of components to be working, for a system to be in success state.

4.4 Mutually Exclusive Failures [MEF]

Mutually exclusive events are those whose failures, will nullify the occurrence of another mode of failure. Assuming that there are n failure modes of a component, the occurrence of first failure mode would rule out the remaining failure modes, which is described probabilistically, using the following equation,

$$P(X_i) = x_i . \prod_{j \neq i}^{n} (1 - x_j) \tag{4}$$

$$\sum_{i=i}^{n} x_i = 1 \tag{5}$$

where,

x_j: Probability of failure for the component x in mode j

i, j: Failure modes, $1 < i, j \leq n$.

4.5 Mutually Independent Failures [MIF]

Independent events, or failures while manifesting will be confined within the components and fault propagation will happen only to the next level, without affecting the associated circuitries or components. The effect of this failure is transmitted or transferred to the next higher level in the assembly/system. It would be easier to isolate and confine this types of failures by monitoring and control of the circuitry. These types of failures are very easily represented using the basic probability theory. In FTA [Fault Tree Analysis], *Boolean Logical* operation is performed to depict the failure mechanism and correlation. Refer [3–5].

$$P(MIF) = (x_i \vee y_i) \vee (x_i \wedge y_i). \tag{6}$$

4.6 Hardware Software Interaction Failures [HSF]

While it is easier to isolate and confine the faults which are purely manifestation of faults within the hardware, it would be extremely difficult to handle the hardware and software interaction failures. The erroneous execution of a flow path at a particular instant of time, given a set of input conditions, coupled with the faulty component in the hardware would produce adverse effects, which are practically difficult to control. Probabilistically, the failure distributions of both hardware and software are combined, by simulation. The software module RENO from Reliasoft Inc, has been utilized for modelling and simulation, to find the effective failure rate. [7] refers, for the aggregation of probability density functions with many *similar* distributions. Logarithmic opion pool is assumed for combining the probability distribution functions as in [7], which is reproduced below:

$$P(HSF) = p(\theta) \tag{7}$$

$$= k \times \prod_{i=1}^{n} p_i(\theta)^{w_i} \tag{8}$$

$$\sum_{i=i}^{n} w_i = 1 \tag{9}$$

k: Normalizing constant
w_i: Weights, satisfying $p(\theta)$, a probability distribution function.

4.7 Dormant Failures [DOF]

For the case of safety critical applications, it is imperative to design a mechanism, by which safety is ensured to an acceptable level. Dormant are those components/circuitries during the normal functioning and would be exercised, if and only if, all the redundant components have failed to function. The components are in non-operating state, in normal conditions and are expected to perform as intended, when warranted, with certainty. Standby failures are also could be treated as dormant failures. Due to the complexity of dormancy state of failure, it is analyzed and effective failure rate is estimated using Markov Process. Refer [6].

5 Failure Rate Data and Failure Modes

For the purposes of estimation of the probability, the basic failure rates of the triggering events are to be known a-priori. Estimation of probability would be more meaningful, if the failure modes of each and every component is known. Proportion of all the identified failure modes to the total failure rate of the component would be different, based on physical construction, mechanism of operation and interfaces of the component. Let us assume that, there are n failure modes of a component, having the total failure rate as λ. The failure mode ratio is defined as;

$$\sum_{i=i}^{n} \alpha_i = 1 \tag{10}$$

$$\sum_{i=i}^{n} \lambda_i = \lambda \tag{11}$$

$$\lambda_i = \lambda \times \alpha_i \tag{12}$$

α_i: Failure Mode Ratio of a component
n: Total number of failure modes, $1 < i \leq n$
λ_i: Failure rate of ith failure mode.

6 Estimation of PLOC Using FTA

In order to estimate the probability of loss of control of aircraft, the Fault Tree Analysis [FTA] approach would be well suited, owing to the nature of mathematical computation and the approach adopted in calculating the 'End event occurrence probability-The Loss of Control of Aircraft'. FTA is a top down approach, also termed as deductive reasoning analysis, by which the causes of failures/failure modes of intermediate events are developed logically. Loss of control of aircraft may result under;

- Failure of Electronic Controller.
- Failure of Air Data Sensors.
- Failure of Primary Actuators.
- Failure of Command Sensors.
- Failure of Motion Sensors.
- Loss of Hydraulics power.
- Loss of Electrical power.

Probability has been computed using the Binary Decision Diagram [BDD]. Refer [1,2]. FTA analysis software toolkit by Item Software, UK has been utilized for the modelling. Refer [3]. Crucial is the identification and logical modelling of failure modes and its relationships with the end event (Fig. 2).

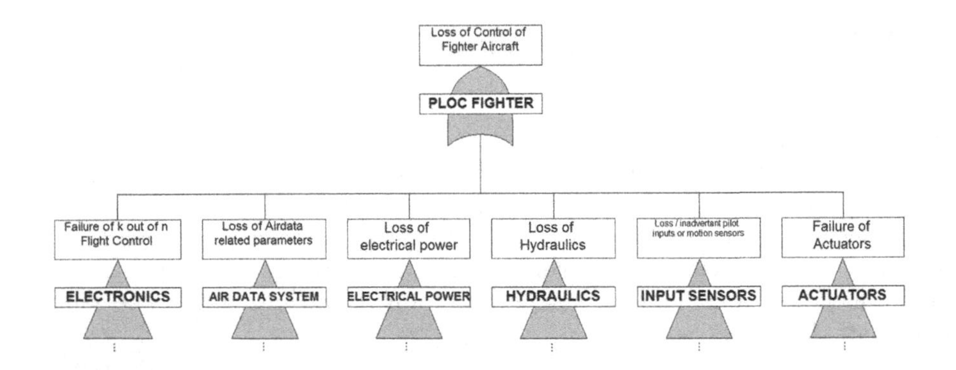

Fig. 2. FTA: Top Event

6.1 Modelling Electronic Controller Failure

The electronic computing system is assumed to be built with the k-out of-n: G level of redundancy, with every redundant system processing the same algorithm/program in a time synchronized manner. Typically, with this type redundancy, voting mechanism is provided within the software to monitor the health of the parameters and eliminate the failed source/input from further computation and command. In general, the redundancy of k-out of-n: G voting system is reproduced below,

$$R(k,n) = \sum_{i=k}^{n} \binom{n}{i} p^i q^{n-i} \tag{13}$$

The system will work, if at least 'k' components are working. Repair time is not considered and assumed to be zero. When the number of failures are more than '*n-k+1*', the system fails, gracefully.
Typically to assume a quadruplex redundant control electronics, the possible cases are reproduced in Table 1. [Op/Fail Op/Fail Op/Fail Safe/Failed].

Table 1. Success and Failure State of typical quadruplex system

No of failures	0 out of 4 : F	1 out of 4 : F	2 out of 4 : F	3 out of 4 : F	4 out of 4 : F
System State	No failure, Fully Operational	Failed, but Operational	Failed, Still Operational	Failed, in safe State	Aircraft Loss
No. of success	4 out of 4 : G	3 out of 4 : G	2 out of 4 : G	1 out of 4 : G	0 out of 4 : G

If the failure rate distribution is assumed to be exponential and if the components are i.i.d. [independent and identically distributed], then the Probability Density Function [PDF] and MTBF of *k out of n: G* configuration is given by,

$$\lambda(t) = \lambda.e^{\lambda(t)} \tag{14}$$

$$MTBF_{system} = \frac{1}{\lambda} \sum_{i=k}^{n} \frac{1}{i} \tag{15}$$

While modelling the Control Electronics, following assumptions were made:

- The active electronics is a *k-out-of n: G* system.
- Each channels and its components are i.i.d.
- Same software is executed in all the n channels in time synchronized manner.
- Voting mechanism is employed to contain and isolate faulty components/ channel.

Further to model and develop the FTA, any channel is assumed to contain 'x' modules. Of the 'x' modules, only '$(x$-$y)$' are assumed to be safety critical whose, failure would result into non-availability of the particular channel for computing and hence controlling. This channel would be eliminated by voting using the '*k-out-of n*' voter. The maximum combination of failure cases of a 'n' redundant system are computed using the possible combinations;

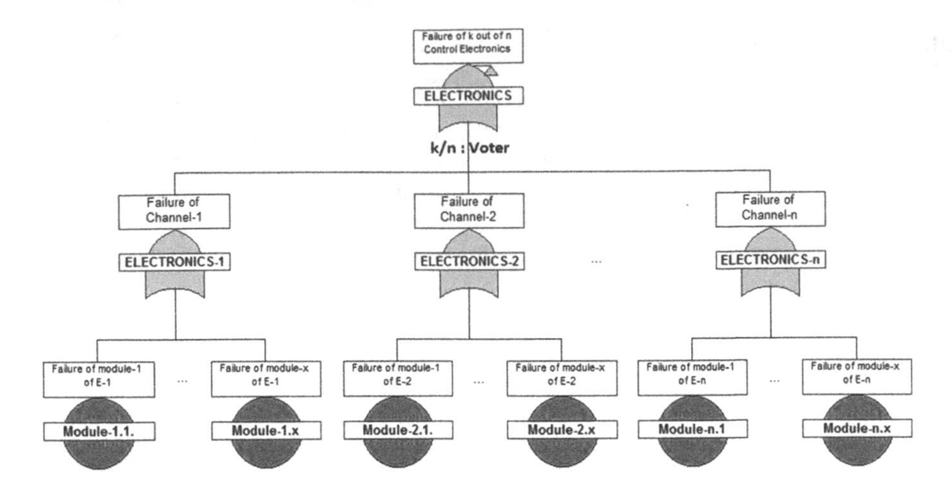

Fig. 3. Modelling Control System Electronics

$$C = n \times \binom{n}{x - y} \tag{16}$$

n: Total number of redundant channels

c: Total number of possible combinations of failures of all the n redundant channels

x: Total number of modules within a channel

y: Number of non critical modules within a channel

$$x - y > 2 \tag{17}$$

Assuming Power Supply Module and Digital Processing Module are highly critical in any active electronics functional item (Fig. 3).

6.2 Modelling Air Data Sensor Failures

Air data system, comprising group of sensors acquiring the outside free stream parameters, computes basic parameters to control the aircraft dynamically based on the altitude, attitude and drag. From the redundancy perspective of air data system, the main calculated parameters-Total Pressure, Static Pressure, Outside Air Temperature [OAT], & AoA [Angle of Attack] are modelled. These parameters are interdependent, meaning one parameter will form as input for the prediction of other parameter, and vice versa. Hence, Air Data System are modelled as given below in Fig. 4: The logical and functional portioning of air data sources are traced down to, physical/structural failures and electronics failures.

6.3 Modelling Primary Actuators Failures

Actuators are assumed to be hydraulic driven and direct drive valve type. Various modes of failures of actuators, which are not detected and could not be

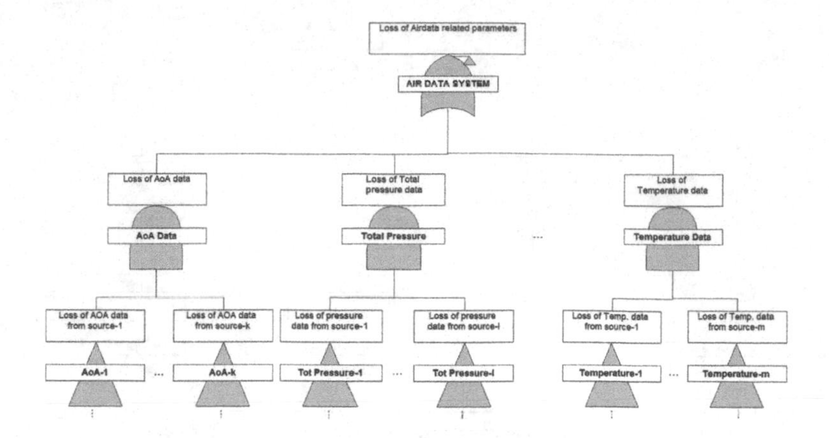

Fig. 4. Modelling air data system components in FTA

controlled are considered as catastrophic in nature and are only considered for the purposes of estimation of failure probability. The causes of major failure modes of actuators, viz., Hard Over failures and Bypass mode failures, are further developed to the lowest possible indenture level (Fig. 5).

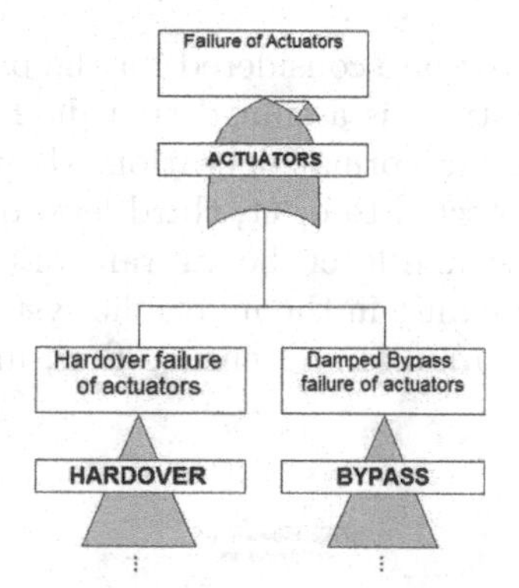

Fig. 5. Modelling Actuator failures in FTA

6.4 Modelling Input Sensors Failure

Various types of pilot inputs to control the aircraft, are Pitch, Roll & Yaw. As any inadvertent input in any of the pilot inputs in *latched* state would eventually lead the aircraft into uncontrollable state, the failure modes in each of the input sensors are treated as critical and developed to the lowest indenture level. While modelling the failure modes, mutually exclusiveness of the switches considering the applicable modes are modelled (Fig. 6).

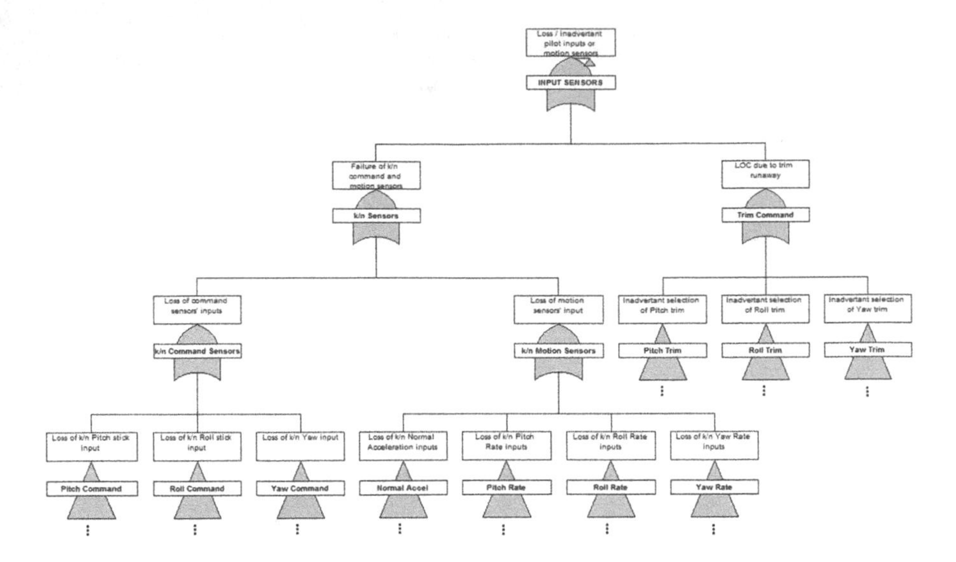

Fig. 6. Modelling Input Sensors failures in FTA

6.5 Modelling Hydraulics Failure in FTA

Hydraulically driven actuators are considered for the present case, where due to the complexity and criticality, it is assumed that dual redundancy is employed in the Hydraulics System, for normal operation. However, for the emergency state, where the pilot is expected to eject, third level of redundancy is required to be established to stabilize and level the aircraft, prior to ejection. In order to obviate the common cause failure in the hydraulic system, an electrical pump is expected to operate the hydraulics and ensure flow, under extreme emergency situations (Fig. 7).

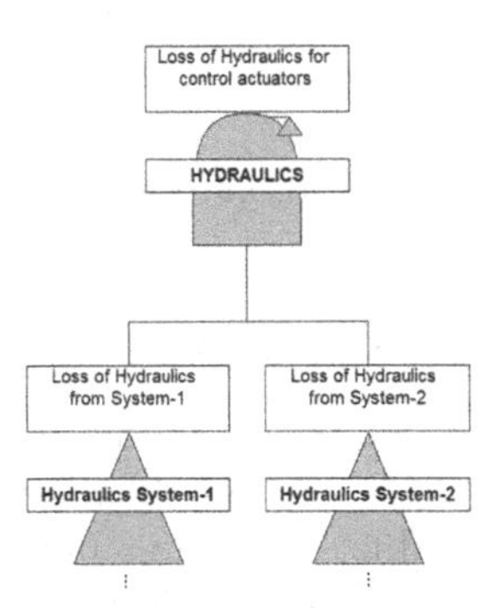

Fig. 7. Modelling Hydraulics failures in FTA

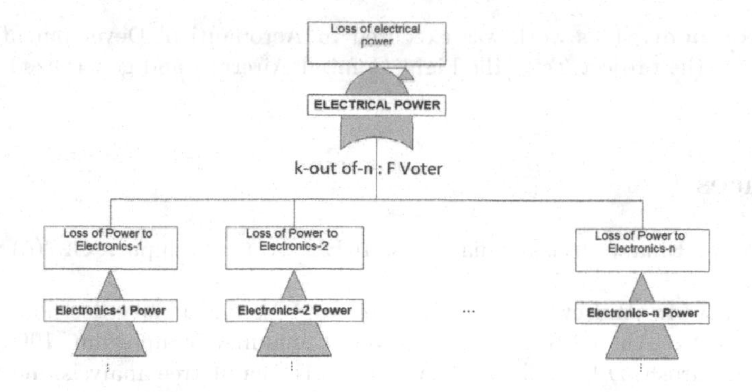

Fig. 8. Modelling Electrical Power failures in FTA

6.6 Modelling Electrical Failure in FTA

Electrical source is one of the common cause, which usually result into loss/ inadvertent operation of many associated electrical, electronic and electromechanical interfaces on aircraft. Hence common cause failure modes of the electrical components of the Power Generation and Distribution system are considered in the development of the FTA. Having considered a *k-out of-n* Control Electronics, the powering scheme is also expected to be *k-out of-n*, in order to isolate Active Electronics and its Power Source from the rest of the other redundant channels. This ensures that, though common cause is a major failure element, the failures arising out this kind, are confined within a single redundant channel (Fig. 8).

7 Conclusion

Complexities associated with, while modelling the hardware, software and hardware-software interaction failures, have been consolidated and enumerated and an analysis to estimate the Probability of Loss of Control of typical digital-fly-by-wire fighter aircraft has been carried out. The resultatnt PLOC is found to be better than 1×10^{-09}. Outcome of the analysis formed part of compliance to the design validation and certification, during the initial developmental stage of the aircraft. In the present work, the basic failure rate data were extracted from Handbook and approximated for the components, conservatively. However, as these data relies on generalization and approximation, the accuracy to the true value of the failure rate data may not be guaranteed. In order to obviate this flaw, it is proposed to assess the failure rate data of the components considered, using the real field data, using statistical means, systematically. This approach of assessment of failure rate from the field data would enable actual assessment of PLOC and also demonstration for compliance of the *estimated* PLOC value with the *assessed* PLOC.

Acknowledgment. This work was executed in Aeronautical Development Agency, Bangalore, for the project *Tejas*-the Light Combat Aircraft, and generalized for confidentiality.

References

1. Akers, S.B.: Binary decision diagrams. IEEE Trans. Comput. **C27**(6), 509–516 (1978)
2. Simalnoli, R.M., Andrews, J.D.: Fault tree analysis and binary decision diagrams. In: Proceedings Annual Reliability and Maintainability Symposium (1996)
3. Lee, W.S., Grosh, D.L., Tillman, F.A., Lie, C.H.: Fault tree analysis, methods and applications-a review. IEEE Trans. Reliab. **R–34**(3), 194–203 (1985)
4. Hessian, Jr., R.T., Salter, B.B., Goodwin, E.F.: Fault tree analysis for system design, development, modification and verification. IEEE Trans. Reliab. **39**(1), 87–91 (1990)
5. Geymayr, J.A.B., Ebecken, N.F.F.: Fault tree analysis-a knowledge engineering approach. IEEE Trans. Reliab. **44**(1), 37–45 (1995)
6. Hokstad, P., Frovig, A.T.: The modelling of degraded and critical and failures for components with dormant failures. Reliab. Eng. Syst. Saf. **51**, 189–199 (1996). Elsevier Science Limited
7. Clemen, R.T., Winkler, R.L.: Combining probability distributions from experts in risk analysis. Risk Anal. **19**(2), 187–203 (1999)
8. Belcastro, C.M.: Aircraft loss of control: analysis and requirements for future safety-critical systems and their validation. In: Proceedings of 8th Asian Control Conference (ASCC), May 2011

Ranking Business Scorecard Factor Using Intuitionistic Fuzzy Analytical Hierarchy Process with Fuzzy Delphi Method in Automobile Sector

S. Rajaprakash[1(✉)] and R. Ponnusamy[2]

[1] Department of Computer Science and Engineering, Aarupadai Veedu Institute of Technology, Vinayaka Mission University Chennai, Chennai, India
srajaprakash04@yahoo.com

[2] Department of Computer Science and Engineering, Rajiv Gandhi College of Engineering, Chennai, India
rponnusamy@acm.in

Abstract. Business scorecard is an integral part of human resource management in an industry or an organization and used to strengthen the functionality of the organization. It plays a vital role in promoting the business. Exploring the uncertainty creeping into various factors in business scorecard is an interesting challenge. In this work, we applied Intuitionistic Fuzzy Analytical Hierarchy Process (IFAHP) with Fuzzy Delphi method to analyse the uncertainty factors in business scorecard. Also we explore the importance of various factors by means of ranking using IFAHPwith Fuzzy Delphi method. The ranking scores are further used to strengthen the business scorecard.

Keywords: Intuitionistic fuzzy analytical hierarchy process · Fuzzy delphi method · Business scorecard · Human resource management

1 Introduction

In 1965 Fuzzy sets were introduced by Lotfi A. Zadeh. A fuzzy set is a class of objects defined by a membership function. Such a set is characterised by a membership (characteristic) function which assigns to each element a grade of membership in the interval [0, 1]. Fuzzy set introduces vagueness with the aim of reducing complexity by eliminating the sharp boundary dividing the members of the pair from non-members. This mapping associates each element in a set with a certain degree of membership. It can be expressed as a discrete value or as a continuous function. In fuzzy sets, each element is mapped by the membership function. The triangular and trapezoidal membership functions are commonly used for defining continuous membership functions [1]. The triangular fuzzy membership function is given by

$$\mu_A(x) = \begin{cases} \frac{(x-a_1)}{(a_m-a_2)} : a_1 \le x \le a_m \\ \frac{(x-a_2)}{(a_m-a_2)} : a_m \le x \le a_2 \end{cases} \quad (1)$$

R. Prasath et al. (Eds.): MIKE 2015, LNAI 9468, pp. 437–448, 2015.
DOI: 10.1007/978-3-319-26832-3_41

$$[\mu_A(x) = \begin{cases} \frac{(x-a_1)}{(a_1^{(1)}-a_1)} & : a_1 \leq x \leq a_1^{(1)} \\ 1 & : a_1^{(1)} \leq xa_2^{(1)} \,] \\ \frac{(x-a_2)}{a_2^{(1)}-a_2} & : a_2^{(1)} \leq x \leq a_2 \end{cases} \tag{2}$$

1.1 Fuzzy Analytic Hierarchy Process (FAHP)

In 1983, Laahoven proposed the Fuzzy Analytical Hierarchy Process (FAHP) [2]. It is a combination of fuzzy set theory and Analytic Hierarchy Process. In FAHP method, the ratio of the fuzzy comparison is able to better accommodate vagueness than AHP values.

1.2 Intuitionistic Fuzzy Set (IFS)

Intuitionistic fuzzy set was introduced by Atanassov [3]. The Intuitionistic fuzzy set theory is based on fuzzy set objects and their properties. $0 \leq \pi_A(x) \leq 1$ for each $x \in X$ $\mu_A(x) \in [0,1]$ is the membership function of the fuzzy set $A^1 : \mu_{A^1}(x) \in [0,1]$ is the membership of $x \in A^1$. The intuitionistic fuzzy set is defined by

$$A = \{\langle x, \mu_x, \nu_x \rangle | x \in X\}, 0 \leq \mu_x + \nu_x \leq 1 \tag{3}$$

where $\mu_A : X \rightarrow [0,1]$ and $\nu_A : X \rightarrow [0,1]$ s.t $\mu_A(x) \in [0,1]$ denotes the membership function and $\nu_A(x) \in [0,1]$ denotes the non-membership function. Obviously $A = \{\langle x, \mu_{A^1}(x), 1-\mu_{A^1}(x)\rangle | x \in X\}$ and $\pi_A(x) = 1-(\mu_x+\nu_x)$ is called the hesitation degree or degree of nondeterminacy of $x \in A$ or $x not \in A$. Szmidt and kacprzyk [4] point out that when calculating the distance between two IFSs, we cannot omit $\pi_A(x)$. We consider that $\alpha = (\mu_\alpha, \nu_\alpha, \pi_\alpha)$ is an intuitionistic fuzzy value where $\mu_\alpha \in [0,1]$ and $\nu_\alpha \in [0,1], \mu_\alpha + \nu_\alpha \leq 1$. According to the szmidt and kacprzyk [4] put forth a function in mathematical form

$$\rho(\alpha) = 0.5(1 + \pi_\alpha)(1 + \mu_\alpha) \tag{4}$$

The α means its contain all positive information included. Therefore intuitionistic fuzzy set mainly based on membership function and non membership function and hesitation degree.

1.3 Intuitionistic Relation

Let R be the relation in the intuitionistic values on the set $X = \{x_1, x_2 ... x_n\}$ and represented by matrix $R = (M_i^k)_{n\times n}$, where $M_{ik} = \langle (x_i, x_k), \mu(x_i, x_k), \nu(x_i, x_k)\rangle i$, $k = 1, 2, 3, \cdots, n$. Let Assume that $M_{ik} = (\mu_{ik}, \nu_{ik})$ and $\pi(x_i, x_k) = 1-\mu(x_i, x_k)-\nu(x_i, x_k)$ is interpreted as an indeterminacy degree. The notion of intuitionistic fuzzy $t - norm$ and $t - conorm$ is as found in Deschrijver *et* al. [5] The intuitionistic fuzzy triangular norms was studied by Xu [2]. He introduced the following operations:

Table 1. Comparison scale [2]

Linguistic value	Scale	Linguistic scale
9	0.9	Extreme Important
7	0.8	Very Strong Important
5	0.7	Strong Important
3	0.6	Moderately Important
1	0.5	Equal Preference
1/3	0.4	Moderately not Important
1/5	0.3	Strong not Important
1/7	0.2	Very strong not Important
1/9	0.1	Extreme not Important

1. $M_{ik} \bigoplus M_{lm} = (\mu_{ik} + \mu_{lm} - \mu_{ik}\mu_{lm}, \nu_{ik}\nu_{lm})$
2. $M_{ik} \bigotimes M_{lm} = (\mu_{ik}\mu_{lm}, \mu_{ik} + \mu_{lm} - \nu_{ik}\nu_{lm})$

In our work, we applied the Intuitionistic fuzzy AHP with Delphi method, over the business scorecard in the Auto mobile sector of India. Based on the scale given in Table 1, we are going to apply DIFAHP in the business scorecard and finally rank the factors that influences the business scorecard.

1.4 Fuzzy Delphi Method

Fuzzy Delphi method has been studies well in the literature by Kaufman and Gupta [6]. The generalization of fuzzy Delphi method is as follows:

1. Identify experts based on the domain and make the expert panel members
2. Using experts' opinion, categorize the attributes. Using the attributes, prepare the questionnaires.
3. Using the questionnaire, get the first set of the suggested attributes.
4. From the attributes, compute the mean [7]. Then deviation is calculated between mean and each expert's opinion [it is also a fuzzy number]. The deviation is sent to be each expert for re-evaluation.
5. In the second round, a new fuzzy number is received from the experts. Next, the same procedure is repeated (step-2) until two successive means become very close; else the Delphi expert will take the final decision.

2 Past Work

Satty [8] introduced the AHP approach for decision making. Atanassov [3] proposed the intuitionistic fuzzy sets and its applications. The heat produced by fans in the system can be controlled by the intuitionistic fuzzy logic approach. In this work, the heat of the fan is calculated with the help of intuitionistic fuzzy rules applied in an inference engine using defuzzification method by Mahman akkram *et* al. [9]. The Intuitionistic fuzzy sets are used in some medical application by Eulalia Szmidt *et* al. [10]. As a generalisation of fuzzy sets, a new definition of distance between two intuitionistic fuzzy sets has been given by Atanassov by *et* al. [4]. Using the intuitionistic fuzzy analytic hierarchy process the environmental decision in the best drilling fluid(mud) for drilling operation has been by Rehan Sadiq *et* al. [11]. Rajaprakash *et* al. [12] studied the customer satisfaction in the automobile sector using the intuitionistic fuzzy analytic hierarchy process. Yen cheng chen *et* al. [13] studied the hotel and atmosphere usage using Delphi fuzzy Analytical Hierarchy Process in two phases: the first one by the Delphi method and the second one by AHP. The selection of best DBMS among several candidates in the Turikish National Identity Card Management project was done using the Fuzzy AHP by F.Ozgur Catak *et* al. [14]. The Fuzzy AHP evaluation of the E-commerce in order manage and determine the drawbacks and opportunities was studied by Feng Kong *et* al. [15] Mohammad Izadikhah [16] studied the supplier selection problem under incomplete and uncertain information environment using TOPSIS Method. The prediction of highest and lowest temperature by back propagation neural networks training for abnormal weather alerts has been studied by a fuzzy AHP and rough set. In this work, we compared the fuzzy AHP and rough set as guided by Dan Wang *et* al. [7] using the FAHP students expectation in the present education system in Tamilnadu, India. Fuzzy Delphi Method and Fuzzy Analytic Hierarchy process is applied to determine the critical factors of the regenerative technologies and to find the degree of each important criterion as the measurable indices of the regenerative technologies proposed by Yu-Lung Hsu *et* al. [17] Lazim Abdullah *et* al. [18] have been studied the human capital indicator and ranking by using IFAHP to evaluate the four main indicators of human capital. Diagnosis progress in bacillus colonies identification in the medical domain using the intuitionistic fuzzy set theory has been studied by Hoda Davarzani *et* al. [18] The preference of Customer requirement factors in automobile sector using IFAHP with delphi method was studied by s.Rajaprakash *et* al. [19] Intuitionistic Fuzzy Delphi Method is used as forecasting tool based on experts suggestion. Tapan Kumar *et* al. [20] have used triangular fuzzy numbers and aggregated based on the opinion of the experts.

3 Methodology

1. Based on the requirements, expert panel was formed. In our work, we are used 10 experts for Delphi Method.
2. Based on the suggestions of the experts, the values are converted into intuitionistic value (comparison scale Table 4 and then the construction of the comparison matrix is carried out.
3. According to Xu *et* al. [21], check the consistency of the matrix intuitionistic preference relations as given below:

$R = (M_{ik})_{n\times n}$ with $M_{ik} = (\mu_{ik}, \nu ik)$ is multiplicative consistent if

$$\mu_{ik} = \begin{cases} 0 & if(\mu_{it}, \mu_{tk}) \in \{(0,1),(1,0)\} \\ \frac{\mu_{it}\mu_{tk}}{\mu_{it}+\mu_{tk}+(1-\mu_{it})(1-\mu_{tk})} & otherwise \end{cases} \tag{5}$$

$$\nu_{ik} = \begin{cases} 0 & if(\nu_{it}, \nu_{tk}) \in \{(0,1),(1,0)\} \\ \frac{\nu_{it}\nu_{tk}}{\nu_{it}+\nu_{tk}+(1-\nu_{it})(1-\nu_{tk})} & otherwise \end{cases} \tag{6}$$

In the fuzzy preference relation, the following statements are equivalent: [21]

$$b_{ik} = \frac{b_{ik}b_{tk}}{b_{ik}b_{tk} + (1-b_{ik})(1-b_{lk})} \quad i,t,k = 1,2,3... \tag{7}$$

$$b_{ik} = \frac{\sqrt[n]{\prod_{s=1}^{n} b_{ik}b_{tk}}}{\sqrt[n]{\prod_{s=1}^{n} b_{is}b_{sk}} + \sqrt[n]{\prod_{s=1}^{n} b_{is}b_{sk}}} \quad i,k = 1,2,...n \tag{8}$$

$$\bar{\mu}_{ik} = \frac{\sqrt[k-i-1]{\prod_{t=i+1}^{k-1} \mu_{it}\mu_{tk}}}{\sqrt[k-i-1]{\prod_{t=i+1}^{k-1} \mu_{it}\mu_{tk}}} \quad k > i+1 \tag{9}$$

$$\bar{\nu}_{ik} = \frac{\sqrt[k-i-1]{\prod_{t=i+1}^{k-1} \nu_{it}\nu_{tk}}}{\sqrt[k-i-1]{\prod_{t=i+1}^{k-1} \nu_{it}\nu_{tk}}} \quad k > i+1 \tag{10}$$

4. The distance between intuitionistic relations [4] is calculated using

$$d(M, \bar{M}) = \frac{1}{2(n-1)(n-2)} \sum_{t=1}^{n}\sum_{k=1}^{n} (|\bar{\mu}_{ik} - \mu_{ik}| + |\bar{\nu}_{ik} - \nu_{ik}| + |\bar{\pi}_{ik} - \pi_{ik}|) \tag{11}$$

5. The priority of the intitionistic preference relation is calculated by the following method suggested by Zeshuri Xu [21]:

$$W_i = \frac{\sum_{k=1}^{n} M_{ik}^{1}}{\sum_{i=1}^{n}\sum_{k=1}^{n} M_{ik}^{1}}$$

$$W_i = \left[\frac{\sum_{k=1}^{n} \mu_{ik}}{\sum_{i=1}^{n}\sum_{k=1}^{n} [1-\nu_{ik}]}, 1 - \frac{\sum_{k=1}^{n} [1-\nu_{ik}]}{\sum_{i=1}^{n}\sum_{k=1}^{n} \mu_{ik}} \right] \tag{12}$$

6. After finding weights of all levels, perform ranking of the weights by using the formula (4); then find preference ranking.

4 Illustrative Work

The above work illustrated in the areas of business scorecard of Human Resource Management department in the automobile sector at Chennai. Here, we are ranking the factors of the business scorecard using the above method. In this work, the data were collected from the car manufacturing company.

4.1 Observation from the Experts

Based on the observation from the experts, the business score card is used to promote the business and its future requirements via formulating the critical needs of the business with meticulously framed metrics to cater their clients in the most efficient way as well as coping up with the latest technologies. The important factors are Profit, Growth, People, and Reputation. This process is adopted half yearly once for effective monitoring and implementation. The hierarchy is based on the experts' suggestion as shown in the Fig. 1.

4.1.1 Profit: The employees should adjust their expenses and head count according to the sales and profits they generated. Here we are classified this into three factors: Increase or maintain shareholder return, maintain projected revenue growth and collections, maintain projected efficiency in the use of fixed and variables.

1. Main shareholder return: Actual requirements is identified and the Key Research Area (KRA) has been fixed to develop the same, when this has been successfully implemented. Then the shareholders will invest further into the business for multiplying their investments.
2. Maintain projected revenue growth: The business score cards help to attain the projected growth in terms of revenue at the start up of the project with their will defined KRA's.
3. Maintain projected efficiency in use of fixed and variable: The fixed and variable assets have to be efficiently utilised to achieve the projected target.

4.1.2 Growth: The organization should work for the business growth by understanding the business goals and requirement, fitting into their Key Resource Areas and identify the Key Result Areas as well. The three important factors to grow up are: improve material yield, improve Value Added Per Employee (VAPE), improve Value Added Per Employee Cost (VAPEC).

1. Improve VAPE: The merits and calculations help to identify per employee value addition to the business. It is a detailed study of the contributions by every employee.
2. Improve VAPEC: It is a calculation which helps to the organization to identify the return earned through each individual employee. It is based on the actual factors like sales and plan.

4.1.3 People: The HR Department should work for the development of their people and welfare through conducting various Employee Engagement Activities. Here we have three important attributes: Welfare, Employee Engagement and Salary

1. Welfare: It is the process of providing the employees all facilities and amenities in order to work in the organisation with an effective output.
2. Employee Engagement: It is the process of bring up all employees together and involving them in one agenda there by creating an atmosphere of one culture with one goal. It helps retention of employee in an organisation.
3. Salary: It is the main key of motivation for an employee. It is fixed based on the merits of the employee in terms of his education experience and skill level as will in accord with ovaries trend.

4.1.4 Reputation: The organization should identify the potential of the business and manage the latest technology and competition thereby, bringing in competitive products, people and price as well contributing to the brand building of the organization.

1. Vision and Mission: The organization will have their own preformed principles vision as mission in which the company wants to excel and reinforce that all their employees should follow. This principal will make the organisation to grow up further.
2. Culture and Values: The climate of the organisation has to pleasantly maintained, which is possible only when the organization sets its own professional environment with value, ethics and cultures.
3. Organisation Transformation: This is a process of adapting and adjusting according to the changing environment, technologies by bring in a lot of new innovative and creative ideas to develop the organisation. It is only a source to withstand the demand and expectations of the clients in the present scenario.

4.2 Business Scorecard in Level-1

In order to find the business scorecard in level-1, four attributes are available. Based on the experts' opinion, the first initial Table 2 has been formed.

The mean values are calculated. The deviations of experts' opinion from the calculated mean values are given below Table 3.

Here the Delphi experts are not satisfied with the deviation Table 3. Therefore, opinion is sent back to the experts for one more opinion.

The deviation from the mean is calculated as given in Table 5.

Now the Delphi expert is satisfied with the above deviation Table 5. Based on the expert suggestion, the first intuitionistic preference relation matrix for the business scorecard is formed as shown below:

$$M = \begin{pmatrix} (0.5,0.5) & (0.5,0.4) & (0.4,0.6) & (0.5,0.6) \\ (0.7,0.3) & (0.5,0.5) & (0.5,0.4) & (0.5,0.5) \\ (0.5,0.4) & (0.5,0.4) & (0.5,0.5) & (0.4,0.5) \\ (0.6,0.4) & (0.6,0.4) & (0.5,0.4) & (0.5,0.5) \end{pmatrix}$$

check the consistence preference relations using the above formula (9) and (10), we can get the multiplicative fuzzy relation Matrix($\bar{M}$).

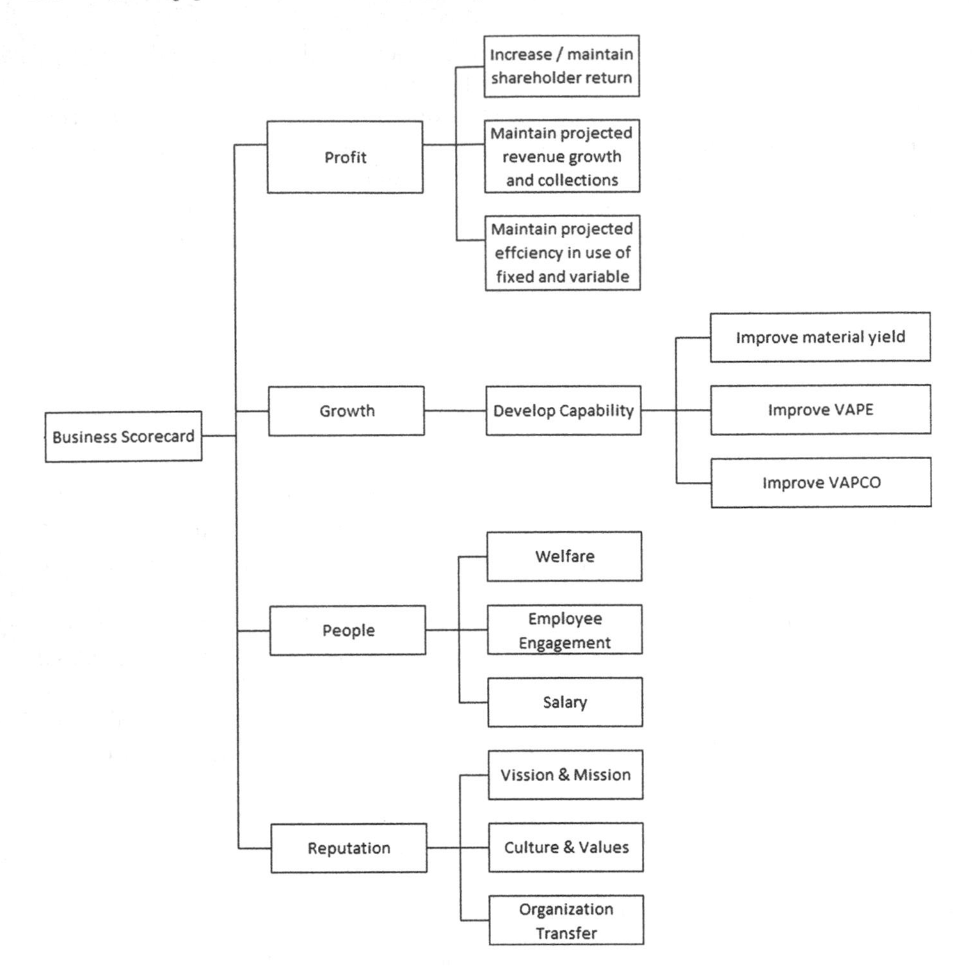

Fig. 1. Business scorecard - hierarchy

$$\bar{M} = \begin{pmatrix} (0.5, 0.5) & (0.5, 0.4) & (0.4494, 0.5) & (0.4, 0.51) \\ (0.6, 0.5) & (0.5, 0.5) & (0.5, 0.4) & (0.5025, 0.4449) \\ (0.5, 0.449) & (0.4, 0.5) & (0.5, 0.5) & (0.4, 0.5) \\ (0.5, 0.4) & (0.449, 0.50254) & (0.5, 0.4) & (0.5, 0.5) \end{pmatrix}$$

Then using the Eq. (11), the distance between intuitionistic relation is calculated as $d(\bar{M}, M) = 0.09578$ which is less than τ. Here, we will fix the threshold value as $\tau = 0.1$. Therefore, the above matrix is consistent. The next step is calculating the weight of all attributes using the Eq. (12). It is given in Table 6 and using the Eq. (4), we will get the preference(P) of all attributes. It is given in Table 7. We can see that the first preference is quality, second is delivery, third is cost, and last one is service. Similarly we calculated the weights for other levels.

Table 2. Delphi 1

Experts	BS1 to BS2		BS1 to BS3		BS1 to BS4		BS2 to BS3		BS2 to BS4		BS3 to BS4	
1	0.5	0.4	0.4	0.2	0.6	0.3	0.4	0.3	0.5	0.3	0.6	0.4
2	0.6	0.3	0.5	0.3	0.5	0.4	0.3	0.4	0.7	0.3	0.6	0.4
3	0.7	0.3	0.3	0.2	0.5	0.3	0.5	0.3	0.5	0.5	0.5	0.6
4	0.3	0.6	0.6	0.3	0.6	0.3	0.4	0.4	0.5	0.4	0.7	0.4
5	0.4	0.5	0.2	0.3	0.3	0.6	0.5	0.5	0.6	0.4	0.5	0.4
6	0.5	0.4	0.3	0.4	0.5	0.5	0.6	0.4	0.3	0.5	0.6	0.5
7	0.6	0.3	0.4	0.2	0.5	0.4	0.7	0.3	0.5	0.4	0.5	0.6
8	0.7	0.2	0.3	0.3	0.6	0.5	0.3	0.4	0.6	0.4	0.6	0.4
9	0.5	0.4	0.2	0.5	0.4	0.4	0.5	0.5	0.7	0.4	0.4	0.5
10	0.4	0.5	0.7	0.3	0.3	0.6	0.5	0.2	0.6	0.4	0.5	0.6

Table 3. Delphi2

Experts	BS1 to BS2		BS1 to BS3		BS1 to BS4		BS2 to BS3		BS2 to BS4		BS3 to BS4	
1	0.02	−0.01	−0.01	0.1	−0.12	0.13	0.07	0.07	0.05	0.1	−0.05	0.08
2	−0.08	0.09	−0.11	0	−0.02	0.03	0.17	−0.03	−0.15	0.1	−0.05	0.08
3	−0.18	0.09	0.09	0.1	−0.02	0.13	−0.03	0.07	0.05	−0.1	0.05	−0.12
4	0.22	−0.21	−0.21	0	−0.12	0.13	0.07	−0.03	0.05	0	-0.15	0.08
5	0.12	−0.11	0.19	0	0.18	−0.17	−0.03	−0.13	−0.05	0	0.05	0.08
6	0.02	−0.01	0.09	−0.1	−0.02	−0.07	−0.13	−0.03	0.25	−0.1	−0.05	−0.02
7	−0.08	0.09	−0.01	0.1	−0.02	0.03	−0.23	0.07	0.05	0	0.05	−0.12
8	−0.18	0.19	0.09	0	−0.12	−0.07	0.17	−0.03	−0.05	0	−0.05	0.08
9	0.12	−0.11	−0.31	0	0.18	−0.17	−0.03	0.17	−0.05	0	0.05	−0.12
10	0.12	−0.11	−0.31	0	0.18	−0.17	−0.03	0.17	−0.05	0	0.05	−0.12

5 Empirical Result

Based on the suggestions given by the experts of the business scorecard hierarchy is formed in Level-1. Four attributes (factors) are ranked (Table 7) as follows: Profit, Growth, People and, Reputation of the organization. In the Level-2, the key attribute of the industry is Profit which is based on three factors and its ranked as follows: Increase or Maintain Shareholder return,Maintain Projected efficiency is use of fixed and variable and Maintain Projected revenue growth and collection. The next attribute is growth of the industry. It is mainly classified in to three factors and its ranked as follows: improve material yield, improve VAPCO and Improve VAPE. The next attribute is People. It have three factors and its ranked as follows: welfare, salary, employment engagement. The last Attribute is reputation, it has three main attributes and its raked as follows: Vision and Mission, Organisation Transfer and Culture and values. Therefore,

Table 4. Delphi3

Experts	BS1 to BS2		BS1 to BS3		BS1 to BS4		BS2 to BS3		BS2 to BS4		BS3 to BS4	
1	0.6	0.4	0.4	0.4	0.6	0.3	0.4	0.4	0.52	0.5	0.3	0.5
2	0.6	0.4	0.3	0.5	0.6	0.4	0.4	0.4	0.6	0.6	0.4	0.4
3	0.5	0.4	0.6	0.2	0.6	0.2	0.5	0.4	0.4	0.4	0.4	0.6
4	0.6	0.6	0.4	0.4	0.6	0.3	0.4	0.4	0.6	0.4	0.4	0.6
5	0.6	0.4	0.5	0.4	0.5	0.6	0.5	0.5	0.5	0.5	0.3	0.6
6	0.4	0.2	0.4	0.4	0.5	0.5	0.6	0.4	0.6	0.5	0.5	0.4
7	0.6	0.4	0.4	0.4	0.5	0.4	0.7	0.3	0.5	0.5	0.4	0.3
8	0.4	0.4	0.5	0.5	0.4	0.4	0.4	0.4	0.5	0.5	0.5	0.5
9	0.4	0.4	0.3	0.4	0.4	0.4	0.6	0.4	0.4	0.5	0.4	0.6
10	0.5	0.5	0.4	0.3	0.4	0.5	0.5	0.4	0.4	0.6	0.5	0.4

Table 5. Delphi4

Experts	BS1 to BS2		BS1 to BS3		BS1 to BS4		BS2 to BS3		BS2 to BS4		BS3 to BS4	
1	−0.08	0.01	0.02	−0.01	−0.09	0.1	0.1	0	−0.01	0 0.11	−0.01	0
2	−0.08	0.01	0.12	−0.11	−0.09	0	0.1	0	−0.098	−0.1	0.01	0.09
3	0.02	0.01	−0.18	0.19	−0.09	0.2	0	0	0.102	0.1	0.01	−0.11
4	-0.08	−0.19	0.02	−0.01	−0.09	0.1	0.1	0	−0.098	0.1	0.01	−0.11
5	−0.08	0.01	−0.08	−0.01	0.01	−0.2	0	−0.1	0.002	0	0.11	−0.11
6	0.12	0.21	0.02	−0.01	0.01	−0.1	−0.1	0	−0.098	0	−0.09	0.09
7	−0.08	0.01	0.02	−0.01	0.01	0	−0.2	0.1	0.002	0	0.01	0.19
80	.12	0.01	−0.08	−0.11	0.11	0	0.1	0	0.002	0	−0.09	−0.01
90	.12	0.01	0.12	−0.01	0.11	0	−0.1	0	0.102	0	0.01	−0.11
10	0.02	−0.09	0.02	0.09	0.11	−0.1	0	0	0.102	−0.1	−0.09	0.09

Table 6. Weight

weight	μ	ν
W(BS1)	0.21763	0.72733
W(BS2)	0.2474	0.702077
W(BS3)	0.2118	0.73375
W(BS4)	0.22939	0.71467

Table 7. Rank

Attribute	$\rho(\alpha)$	P
Profit(PS1)	0.88958	1
Growth(PS2)	0.85299	2
People(PS3)	0.852624	3
Reputation(PS4)	0.840044	4

from the value $\rho(\alpha)$ and Tables and Diagrams, we can get the preference ranking of the attribute of the automobile sector using the IFAHP Fuzzy Delphi method.

6 Conclusion

In this work, we combine Intiutionistic Fuzzy Analytical Hierarchy Process and Fuzzy Delphi Method to analyse the business scorecard in the automobile sector in India. The Major part of IFAHP With Fuzzy Delphi Method include the following. In Delphi Method, questionnaires were framed based on the suggestions and opinions obtained from the experts in the automobile sector. In this work, we are categories the business scorecard. At each and every level, we rank the (preference) factors of business scorecard. The major disadvantage of our work is in identifying the experts and getting opinions from them will take a huge amount of time. The outcome is useful for the automobile sector and it may improve the industrial standard and economy of the company.

References

1. Klir, G.J.: Fuzzy Set and Fuzzy Logic Theory and Application. PTR Publisher, New York (1995)
2. Xu, Z., Liao, H.: Intuitionistic fuzzy analytic hierarchy process. IEEE Trans. Fuzzy Syst. **22**(4), 749–761 (2014)
3. Atanassov, K.T.: Intuitionistic fuzzy sets. Fuzzy Sets Syst. **20**(1), 87–96 (1986)
4. Szmidt, E., Kacprzyk, J.: Distances between intuitionistic fuzzy sets. Fuzzy Sets Syst. **114**(3), 505–518 (2000)
5. Deschrijver, G., Cornelis, C., Kerre, E.: On the representation of intuitionistic fuzzy t-norms and t-conorms. Notes Intuit. Fuzzy Sets **8**(3), 1–10 (2002)
6. Kaufmann, A., Gupta, M.M.: Fuzzy Mathematical Models in Engineering and Management Science. Elsevier Science Inc., New York (1988)
7. Carlsson, C., Fullr, R.: On possibilistic mean value and variance of fuzzy numbers. Fuzzy Sets Syst. **122**(2), 315–326 (2001)
8. Saaty, T.: The Analytic Hierarchy Process, Planning, Priority Setting, Resource Allocation. McGraw-Hill, New york (1980)
9. Akram, M., Shahzad, S., Butt, A., Khaliq, A.: Intuitionistic fuzzy logic control for heater fans. Math. Comput. Sci. **7**(3), 367–378 (2013)
10. Szmidt, E., Kacprzyk, J.: Intuitionistic fuzzy sets in some medical applications. In: Reusch, B. (ed.) Fuzzy Days 2001. LNCS, vol. 2206, pp. 148–151. Springer, Heidelberg (2001)
11. Sadiq, R., Tesfamariam, S.: Environmental decision-making under uncertainty using intuitionistic fuzzy analytic hierarchy process (IF-AHP). Stoch. Env. Res. Risk Assess. **23**, 75–91 (2009)
12. Rajaprakash, S., Ponnusamy, R., Pandurangan, J.: Determining the customer satisfaction in automobile sector using the intuitionistic fuzzy analytical hierarchy process. In: Prasath, R., O'Reilly, P., Kathirvalavakumar, T. (eds.) MIKE 2014. LNCS, vol. 8891, pp. 239–255. Springer, Heidelberg (2014)
13. Chen, Y.C., Yu, T.H., Tsui, P.L., Lee, C.S.: A fuzzy AHP approach to construct international hotel spa atmosphere evaluation model. Qual. Quant. **48**(2), 645–657 (2014)
14. Catak, F.O., Karabas, S., Yildirim, S.: Fuzzy analytic hierarchy based DBMS selection in turkish national identity card management project. Int. J. Inf. Sci. Tech. (IJIST) **2**(4), 29–38 (2012)

15. Kong, F., Liu, H.: Applying fuzzy analytic hierarchy process to evaluate success factors of e-commerce. Int. J. Inf. Syst. Sci. **1**(3), 406–412 (2005)
16. Izadikhah, M.: Group decision making process for supplier selection with topsis method under interval-valued intuitionistic fuzzy numbers. Adv. Fuzzy Sys. **2012**, 1 (2012)
17. Hsu, Y.L., Lee, C.H., Kreng, V.B.: The application of fuzzy delphi method and fuzzy ahp in lubricant regenerative technology selection. Expert Syst. Appl. **37**(1), 419–425 (2010)
18. Abdullah, L., Jaafar, S., Taib, I.: Intuitionistic fuzzy analytic hierarchy process approach in ranking of human capital indicators. J. Appl. Sci. **13**(3), 423–429 (2013)
19. S.Rajaprakash, R.ponnusamy, J.: Intuitionistic fuzzy analytical hierarchy process with fuzzy delphi method. Global journal of pure and applied mathematics (3) (2015) 1677–1697
20. Roy, T.K., Garai, A.: Intuitionistic fuzzy delphi method: More realistic and interactive forecasting tool. Notes Intuit. Fuzzy Sets **18**(50), 37–50 (2012)
21. Xu, Z.: Intuitionistic preference relations and their application in group decision making. Inf. Sci. **177**(11), 2363–2379 (2007)

Text and Citations Based Cluster Analysis of Legal Judgments

K. Raghav[1(✉)], Pailla Balakrishna Reddy[1], V. Balakista Reddy[2], and Polepalli Krishna Reddy[1]

[1] IIIT-Hyderabad, Hyderabad, Telangana State, India
raghav.k@research.iiit.ac.in, pkreddy@iiit.ac.in
[2] NALSAR University of Law, Hyderabad, Telangana State, India
{balakrishnar,balakista}@gmail.com

Abstract. Developing efficient approaches to extract relevant information from a collection of legal judgments is a research issue. Legal judgments contain citations in addition to text. It can be noted that the link information has been exploited to build efficient search systems in web domain. Similarly, the citation information in legal judgments could be utilized for efficient search. In this paper, we have proposed an approach to find similar judgments by exploiting citations in legal judgments through cluster analysis. As several judgments have few citations, a notion of paragraph link is employed to increase the number of citations in the judgment. User evaluation study on the judgment dataset of Supreme Court of India shows that the proposed clustering approach is able to find similar judgments by exploiting citations and paragraph links. Overall, the results show that citation information in judgments can be exploited to establish similarity between judgments.

Keywords: Legal judgments · Citation · Link based analysis · Clustering

1 Introduction

The amount of available text-data in legal domain is vast and continuously growing which makes it challenging to deal with. Apart from the size of data, the inherent complexity of legal domain demands better and more sophisticated methods to process legal documents to satisfy information need of legal practitioners. Building efficient search approaches in legal domain is an active research area.

In the literature, efforts have been made to address the challenges of information overload in the area of web search. A web page has links (references) to other web pages often called hyperlinks. It is considered that the existence of a link between two web pages indicates a relationship between the two pages [10]. In the area of web search and retrieval, these links provide important information and efficient search systems have been developed by exploiting these links [4,10] available in the web pages. Also, efficient methods have been developed

R. Prasath et al. (Eds.): MIKE 2015, LNAI 9468, pp. 449–459, 2015.
DOI: 10.1007/978-3-319-26832-3_42

for effective organization and retrieval of web pages by extending clustering [9] and classification [5] approaches. Several efforts have been made to build efficient search systems based on communities formed by links. In addition, efforts have also been made in information retrieval to link one topic with other topics by forming hyperlinks among the topics by carrying out text-based comparison [11,20].

Legal systems are generally based on one of the two basic systems of law, viz., civil law and common law. In civil law, core principles are codified into a referable system which serves as the primary source of law. As opposed to civil law, the common law is a law developed by judges through previous judgments, i.e., courts have interpreted the law in individual cases alongside using a referable system of rules as a source of law. A legal judgment is a closed case. It is a text document, which explains the formal decision made by a court following a lawsuit. Similar to web domain, links can be observed in legal judgments in the form of a citation network in which one judgment is said to be connected to another judgment when it cites the prior judgment. This citation information in legal judgments could be utilized for efficient search.

In this paper, we have made an effort to find the related judgments through cluster analysis by considering the judgments in a common law system. In particular, we consider citation information and propose an approach to find similar legal judgments. By analyzing judgments delivered by the Supreme Court of India it has been found that a considerable number of legal judgments have only a few citations. Similar to the notion of links among topics [11,20], we employed the notion of paragraph link to group similar judgments at a paragraph level. We have applied clustering approach on the judgments dataset by considering the citations, paragraph links and by combining both citations and paragraph links. We show that it is possible to establish similarity between judgments by exploiting citations.

In addition, we propose a clustering approach based on citations. Document clustering by using vector space model is a well studied approach. In general, it is accomplished by representing each data object as n-dimensional feature vector with each coordinate of the vector being a term in vocabulary [18]. In case of judgments with only citation information, application of typical clustering algorithms like *K-means* is difficult, as computing the mean or central node of the cluster is difficult. We propose clustering approach by employing the notion of multiple central judgments to represent the cluster. By conducting experiments on real world dataset, it has been shown that the proposed clustering approach can be useful in establishing similarity between judgments by utilizing citation information.

The rest of the paper is organized as follows. In the next section, we discuss the related work. In Sect. 3, we explain the proposed approach. In Sect. 4, we discuss the experimental results. The last section contains conclusion and future work.

2 Related Work

Extensive work has been done in the area of web search and information retrieval by exploiting the text based content in the web pages. Traditional methods to compare two documents treat documents as bag-of-words where each term is weighted according to TF-IDF score [20]. The vector space model [18] is a popular approach to model the documents and then *cosine* similarity method is employed to compare two documents. In the survey [24], a taxonomy of clustering techniques, like agglomerative and partitional techniques have been provided. They identify recent advancements in this domain and present various applications of clustering algorithms in the area of information retrieval.

In web domain, efficient search systems have been developed by exploiting links [4,10]. The link based approaches have been explored to extract communities [12]. An effort [7] has been made to identify related web pages by using the connectivity information in the web.

In [19], an approach has been proposed by considering the document as a collection of segments of themes or topics. In that approach, to improve the search performance, similar paragraphs in multiple documents are linked in order to provide attention to concepts captured at the paragraph level.

In legal domain, several efforts are being made to build better information extraction approaches. A machine learning based approach for retrieval of prior cases has been studied in [2]. A navigation model [25] has been proposed to browse through legal issues by exploiting the legal citation network in the form of a semantic network. A probabilistic graphical model [21] for automatic text summarization has been proposed in legal domain. An approach [23] has been proposed to perform automatic categorization of case laws into high level categories. Karypis et al. [6] conducted clustering experiments (hard clustering, soft clustering and hierarchical clustering) using several kinds of law firm data for building decision support system to help legal experts. In [16], classification based recursive soft clustering algorithm with built in topic segmentation was proposed by employing metadata such as topical classification, document citations and click stream data from user behavior databases. The importance of exploiting link information in legal judgments has been demonstrated by analyzing sample pairs of judgments [13–15].

In this paper, we have made an effort to establish similarity among legal judgments through cluster analysis by exploiting citation information and paragraph links.

3 Proposed Approach

In this section, we present the basic idea and the approach for clustering judgments using citations.

3.1 Basic Idea

In the web domain, links (or URLs) have been exploited for efficient search. Existence of links is considered as an important feature which is widely exploited for search. Suppose a web page X has a link to web page Y, we say X has an out-link to Y and Y has an in-link from X. As per cocitation-based similarity, two web pages are considered similar if they have common in-links above a certain threshold. As per bibliographic coupling, two web pages are similar if they have common out-links above certain threshold. Link based approaches have been proposed to extract authoritative resources, assign ranks to pages, extract cohesive communities and crawling. There are also efforts to develop improved information retrieval approaches by exploiting thematic similarity between similar paragraphs and similar sentences [19].

Similar to the web page, a legal judgment also has citations. As judgments are very credible documents, citations indicate significant association between two judgments.

Based on the analysis of real world data, it has been observed that considerable number of judgments have few citations. In order to increase the number of citations of a judgment, we employ the notion of paragraph links (PLs) between judgments and perform clustering using induced citations between judgments. The PLs captures the intricate legal concepts discussed at a minute level in the paragraphs of a legal judgment.

It is possible to cluster documents by applying similarity measures such as *cosine* similarity between document vectors and apply *K-means* or other agglomerative clustering algorithms. But, the center of a cluster is difficult to define in the case of citation-based similarity measures. In the next sub-section, we propose a clustering approach by considering multiple central documents as representative documents of the cluster.

We have carried out the clustering analysis by considering the following methods.

- **Text based clustering:** In this approach, the terms present in the text judgments are used for performing similarity analysis among judgments using clustering. We use standard *K-means* clustering algorithm for judgments. As a part of preprocessing, we removed stop words from the document corpus and applied porter stemming algorithm [17] for suffix-stripping of words in the corpus. After assigning weights to terms with TF-IDF scores, we use the iterative partition based *K-means* clustering algorithm [8]. We determine the value of k by using Bayesian information criterion method [22] which is a statistical approach for finding natural model selection for the dataset. It provides information regarding the natural number of clusters suitable for the dataset.
- **Citations based clustering:** From each judgment, we remove all text and only keep citation information. By applying the proposed clustering approach (proposed in the next sub-section), we obtain the clusters.
- **Paragraph links (PLs) based clustering:** We divide the judgment into a set of paragraphs. For each judgment, we compute the similarity of each paragraph with each paragraph of other judgments. A paragraph link (PL) is established between two judgments if they have more than a threshold number of similar

paragraphs. By considering only PLs for each judgment, clusters are obtained by applying the proposed clustering approach.
- **Combination of citations and PLs based clustering:** We cluster the judgments by considering both citations and PLs information.

3.2 Clustering Judgments Using Citations

We propose a clustering method by considering the citation information in judgments. Clustering the judgments having only citations is a difficult task because of two reasons. One is the notion of center cannot be clearly defined in the case of clustering using citations. It is also observed that there are some important judgments to which many judgments have common citations with them. We develop a clustering approach by considering several representative judgments as the center of the cluster. The judgment which establishes high connectivity with other members in the cluster is chosen as cluster representative. The proposed methodology consists of three steps, which are explained below.

Step 1: Finding Initial Clusters: We form the first cluster with the first judgment. For each other judgment J, we compute *Jaccard coefficient* similarity of $J's$ citations and citations of each existing cluster. The judgment J is inserted into the cluster with which it establishes the maximum similarity. If the judgment is not similar to any of the existing clusters, a new cluster is created with the judgment as its only element.

Step 2: Refinement of Clusters: The refinement consists of two steps.

- **Finding representative judgments:** For each cluster generated in the first step, we find the k representative judgments in the cluster based on number of judgments it is connected within the cluster. We say two judgments are connected if they a have common citation.
- **Assigning judgments to clusters:** We compute the *Jaccard coefficient* similarity for every judgment with each of the citation sets of representative nodes of each cluster. We assign the judgment to the cluster with maximum similarity.

Step 3: Termination Condition: After the completion of the second step, we compute the number of judgments that changed the cluster during this refinement iteration and the number of representative judgments which changed clusters. If the number of judgments changing clusters are less than a threshold value or the number of representative judgments changing clusters are less than a threshold value, then we stop the refinement step and return the current clusters as the final set of clusters. Otherwise, Step 2 is followed.

About Convergence: The clustering algorithm we employed is similar to partitional clustering algorithms like *K-means* and *K-medoids*. So, the algorithm converges similar to *K-means*. The convergence properties of partitioning approaches like *K-means* [3] have been explored in literature.

4 Experiments

In this section, we explain the dataset, preprocessing, evaluation metrics and results.

4.1 Dataset and Preprocessing

We have conducted experiments using the dataset consisting of judgments delivered by the Supreme Court of India [1] from 1970 to 1993. The dataset consists of 3, 738 judgments.

Structure of a Legal Judgment: A legal judgment is a text document. The following are the important components of a legal judgment: Petitioner, Respondent, Names of Judges, Date of Judgment, Citation, Act and Headnote. Here, *Petitioner* is the one who presents a petition to the court. *Respondent* is the entity against which/whom an appeal has been made. *Act* indicates the brief category of the judgment. The *headnote* contains a brief summary of the judgment. Supreme Court Reports (SCR) is an official reporter for the judgments delivered by the Supreme Court. Supreme Court Cases (SCC) and All India Reporter (AIR) are some prominent private reporters. There are two types of citations for a judgment, namely, *out-citations* and *in-citations*. The *out-citations* of a judgment are the external references made by the current judgment. The *in-citations* are the references made to the current judgment by other judgments. For example, if a judgment X refers to judgment Y to provide the decision, then we say judgment X has an out-citation to Y and Y has an in-citation from X. Out-citations and in-citations of a judgment together are referred as citations of a judgment.

Extracting Citations: We extract all the citations from the headnote section of the judgments using regular expressions for the format used by reporters.

Extracting Paragraph Links (PLs): We consider a paragraph as a text between two consecutive (<p>) html tags. We extract all the paragraphs in the headnote section of the judgment. Then we remove stop words and perform stemming [17] on all the words in the corpus. We choose only those paragraphs which have between 20 and 60 words. After extracting the paragraphs for all the judgments, we find TF-IDF based *cosine* similarity of every paragraph with every other paragraph in the dataset. We call two paragraphs as similar if they have a *cosine* similarity $\geq$0.5. We establish a PL for each pair of judgments, if they have at least three similar paragraphs, which has been studied as a good estimate for similarity between text documents in the approach [19].

4.2 Evaluation Metrics

We evaluate the clustering process by utilizing expert scores for 47 random pairs of judgments provided to legal experts. The domain experts were asked to provide a similarity score between 0 to 10 for each pair, based on the utility in making legal decisions. A similarity score of 0 indicates that there is no similarity between the two judgments and no utility to the legal practitioner in making decisions. A similarity score of 10 indicates that they are similar to each other and is of good utility to legal practitioner.

For each clustering method, a judgment pair is classified as true-positive (TP), if the expert rating ≥ 5 and the clustering method assigned the judgments to the same cluster and as true-negative (TN), if the expert rating ≤ 3 and the clustering method assigned the judgments to different clusters. It is classified as false-positive (FP), if the expert rating ≤ 3 and the clustering method assigned the judgments to the same cluster and as false-negative (FN), if the expert rating ≥ 5 and the clustering method assigned the judgments to the different clusters. We report the effectiveness of clustering using binary classification measures of Precision as $\frac{TP}{TP+FP}$, Recall as $\frac{TP}{TP+FN}$ and F1 score as $\frac{2\times TP}{(2\times TP)+FP+FN}$.

4.3 Results of Text Based Clustering

We consider all the judgments as text documents and cluster them using *K-means* clustering algorithm. Two documents are compared by applying *cosine* similarity to the corresponding vectors of terms with TF-IDF weights. The Bayesian information criterion (BIC) [22] approach has been used to determine the natural number of clusters. The variation of BIC with number of clusters, k is shown in Fig. 1. The number of clusters is set as 13 based on the elbow point.

The summary of the results obtained by text based clustering for $k = 13$, and $k = 60$ are provided in the second and third column of Table 1 respectively. We present the average results of 10 runs of the algorithm. All the 47 pairs (100 %) participate in the evaluation. The average values of Precision, Recall, and F1 score for $k = 13$ are 0.84, 0.68 and 0.75 respectively. The average values of Precision, Recall, and F1 score for $k = 60$ are 0.81, 0.64 and 0.71 respectively.

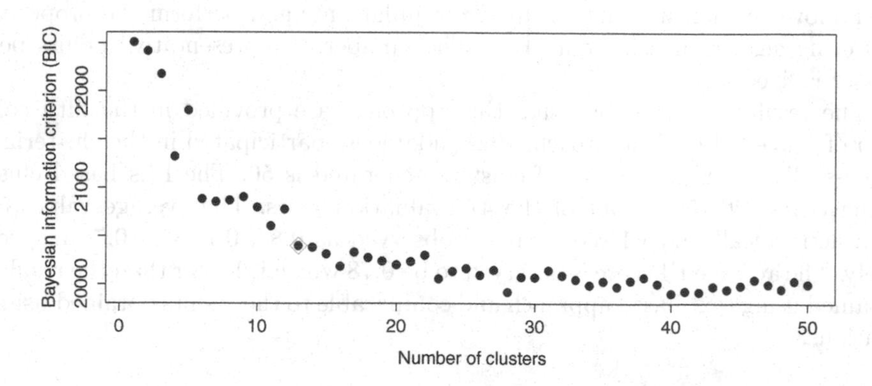

Fig. 1. Number of clusters Vs Bayesian information criterion (BIC)

Table 1. Comparison of clustering results

Parameter	Clustering based on				
	Text ($k = 13$)	Text ($k = 60$)	Citations	PLs	Citations + PLs
Number of judgments covered	3738	3738	1508	1928	2704
Average number of clusters	13	60	140	56	62
Total number of evaluation pairs	47	47	47	47	47
Pairs participated in evaluation	47(100 %)	47(100 %)	38(80 %)	26(56 %)	45(96 %)
Average Precision	0.84	0.81	0.86	0.84	0.81
Average Recall	0.68	0.64	0.73	0.73	0.75
Average F1 score	0.75	0.71	0.79	0.78	0.80

4.4 Results of Clustering Using Citations

For each judgment, we keep only the citations (both in-citations and out-citations) of the judgment and provide them as input to the proposed clustering algorithm. The judgments with at least 3 citations are used. We apply the proposed clustering approach and termination condition is set when we have less than 5 % of the judgments changing clusters. It has been observed that the algorithm converges in less than 15 iterations. We vary the number of representative points from 1 to 6 and for each value of the representative points we run 10 trials of the algorithm. The number of clusters generated and quality measures are recorded each time. The number of representative points per cluster is fixed as 4 based on the best F1 score observed.

The fourth column of Table 1 summarizes the results of this approach. In this approach, 1508 judgments participated in the clustering process. The number of clusters generated is 140. Citations based clustering covers 38 (80 %) pairs out of the 47 evaluation pairs. The average values of Precision, Recall and F1 score are 0.86, 0.73 and 0.79 respectively.

4.5 Results of Clustering Using Paragraph Links (PLs)

We remove citation information from the judgments and perform the proposed clustering algorithm using only PLs. The number of representative points per cluster is fixed as 5.

The results obtained by using this approach are provided in the fifth column of Table 1. In this approach, 1928 judgments participated in the clustering process. The average number of clusters generated is 56. The PLs based clustering covers 26 (56 %) out of the 47 evaluation pairs. The average values of Precision, Recall and F1 score can be observed as 0.84, 0.73 and 0.78 respectively. The average F1 score is observed to be 0.78 which is better than the results obtained using text based approach and comparable to the results obtained using citations.

4.6 Results of Clustering by Combining Citations and PLs

We use both the citations and PLs as input to the clustering approach and the results obtained are shown in last column of Table 1. The number of representative points is set as 6 based on the best F1 score.

By combining citations and PLs we cover 2704 judgments. The average number of clusters generated is 62. Clustering using PLs covers 45 (96 %) out of the 47 evaluation pairs. The average values of Precision, Recall and F1 score can be observed as 0.81, 0.75 and 0.80 respectively. The average F1 score is observed to be 0.80 which is better than the results obtained using text based clustering and comparable to the results obtained using only citations or only PLs.

4.7 Summary of the Results

From Fig. 2, it can be observed that a significant number of judgments have only a few natural citations. Also, there are significant number of judgments with a few PLs. By combining the citations and PLs, the total number of citations is increased in significant number of judgments. As shown in third column of Table 1, only, 1, 508 judgments participate in the clustering using citations among the 3, 738 judgments in the dataset. By introducing PLs among the judgments, we are able to include 1, 982 judgments in clustering process. By combining the citations and PLs we are able to achieve more coverage by performing clustering for 2, 700 judgments.

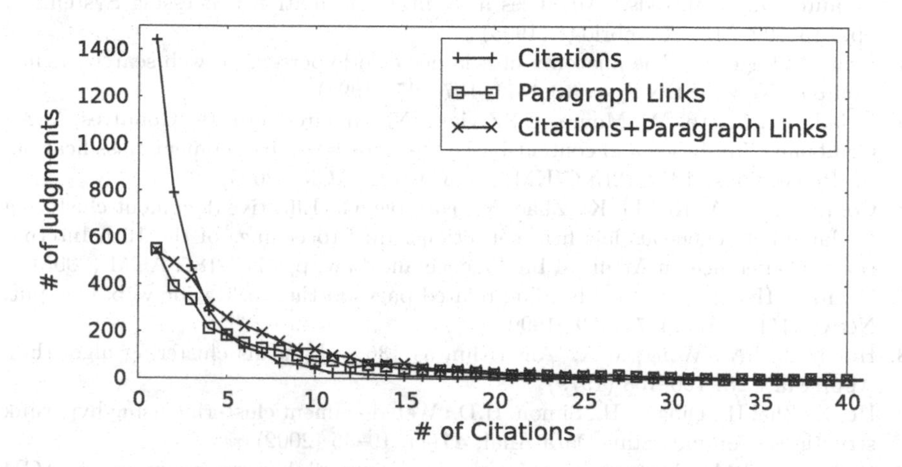

Fig. 2. Number of citations Vs Number of judgments

Overall, the experimental results show that the citation information is a key feature which can be exploited to find similar judgments. The results also indicate that the PLs can be exploited to establish similarity between judgments. It has been observed that citation based methods can be used for establishing similarity among legal judgments by clustering.

5 Conclusion and Future Work

In this paper, we have proposed an approach to cluster the judgments by exploiting citation information to compute the similarity between the judgments.

For clustering judgments using only citation information, we have employed the notion of multiple representative judgments for each cluster. To increase the number citations of a judgment, we have used a notion of paragraph link. The user evaluation study results, show that the citation-based approach is effective in establishing similarity between legal judgments. As a part of future work, we are planning to conduct a detailed user evaluation study. In addition, we would like to exploit *Acts* information in judgments for better clustering. We would also like to investigate the building of better search system by exploiting clusters to provide better utility for the legal practitioners.

References

1. The Supreme Court of India Judgments. http://www.liiofindia.org/in/cases/cen/INSC/
2. Al-Kofahi, K., Tyrrell, A., Vachher, A., Jackson, P.: A machine learning approach to prior case retrieval. In: Proceedings of the 8th International Conference on Artificial Intelligence and Law, pp. 88–93. ACM (2001)
3. Bottou, L., Bengio, Y.: Convergence properties of the k-means algorithms. In: Tesauro, G., et al. (eds.) Advances in Neural Information Processing Systems 7, pp. 585–592. MIT, Cambridge (1995)
4. Brin, S., Page, L.: The anatomy of a large-scale hypertextual web search engine. Comput. Netw. ISDN Syst. **30**(1–7), 107–117 (1998)
5. Calado, P., Cristo, M., Moura, E., Ziviani, N., Ribeiro-Neto, B., Gonalves, M.A.: Combining link-based and content-based methods for web document classification. In: Proceedings of the 12th CIKM, pp. 394–401. ACM (2003)
6. Conrad, J.G., Al-Kofahi, K., Zhao, Y., Karypis, G.: Effective document clustering for large heterogeneous law firm collections. In: Proceedings of the 10th International Conference on Artificial Intelligence and Law, pp. 177–187. ACM (2005)
7. Dean, J., Henzinger, M.R.: Finding related pages in the world wide web. Comput. Netw. **31**(11–16), 1467–1479 (1999)
8. Hartigan, J.A., Wong, M.A.: Algorithm as 136: a k-means clustering algorithm. Appl. Stat. **28**, 100–108 (1979)
9. He, X., Zha, H., Ding, C.H., Simon, H.D.: Web document clustering using hyperlink structures. Comput. Stat. Data Anal. **41**(1), 19–45 (2002)
10. Kleinberg, J.M.: Authoritative sources in a hyperlinked environment. J. ACM **46**(5), 604–632 (1999)
11. Knoth, P., Novotny, J., Zdrahal, Z.: Automatic generation of inter-passage links based on semantic similarity. In: Proceedings of the 23rd International Conference on Computational Linguistics, pp. 590–598. Association for Computational Linguistics (2010)
12. Kumar, R., Raghavan, P., Rajagopalan, S., Tomkins, A.: Trawling the web for emerging cyber-communities. Comput. Netw. **31**(11–16), 1481–1493 (1999)
13. Kumar, S.: Similarity Analysis of Legal Judgments and applying Paragraph-link to Find Similar Legal Judgments. Master's thesis, International Institute of Information Technology Hyderabad (2014)
14. Kumar, S., Reddy, P.K., Reddy, V.B., Singh, A.: Similarity analysis of legal judgments. In: Proceedings of 4th Annual ACM COMPUTE 2011, pp. 17:1–17:4. ACM (2011)

15. Kumar, S., Reddy, P.K., Reddy, V.B., Suri, M.: Finding similar legal judgements under common law system. In: Madaan, A., Kikuchi, S., Bhalla, S. (eds.) DNIS 2013. LNCS, vol. 7813, pp. 103–116. Springer, Heidelberg (2013)
16. Lu, Q., Conrad, J.G., Al-Kofahi, K., Keenan, W.: Legal document clustering with built-in topic segmentation. In: Proceedings of the 20th CIKM, pp. 383–392. ACM (2011)
17. Porter, M.: An algorithm for suffix stripping. Program Electron. Libr. Inf. Syst. **14**(3), 130–137 (1980)
18. Salton, G., Wong, A., Yang, C.S.: A vector space model for automatic indexing. Commun. ACM **18**(11), 613–620 (1975)
19. Salton, G., Allan, J., Buckley, C., Singhal, A.: Automatic analysis, theme generation, and summarization of machine-readable texts. In: Card, S.K., Mackinlay, J.D., Shneiderman, B. (eds.) Readings in Information Visualization, pp. 413–418. Morgan Kaufmann Publishers Inc., San Francisco (1999)
20. Salton, G., Buckley, C.: Term-weighting approaches in automatic text retrieval. Inf. Process. Manag. **24**(5), 513–523 (1988)
21. Saravanan, M., Ravindran, B., Raman, S.: Improving legal document summarization using graphical models. In: Proceedings of the JURIX 2006, pp. 51–60. IOS (2006)
22. Schwarz, G.: Estimating the dimension of a model. Ann. Stat. **6**(2), 461–464 (1978)
23. Thompson, P.: Automatic categorization of case law. In: Proceedings of the 8th International Conference on Artificial Intelligence and Law, pp. 70–77. ACM (2001)
24. Xu, R., Wunsch II, D.: Survey of clustering algorithms. Trans. Neur. Netw. **16**(3), 645–678 (2005)
25. Zhang, P., Koppaka, L.: Semantics-based legal citation network. In: Proceedings of the 11th International Conference on Artificial Intelligence and Law, pp. 123–130. ACM (2007)

Vision-Based Human Action Recognition in Surveillance Videos Using Motion Projection Profile Features

J. Arunnehru(✉) and M. Kalaiselvi Geetha

Speech and Vision Lab, Department of Computer Science and Engineering, Annamalai University, Annamalainagar, Tamilnadu, India
{arunnehru.aucse,geesiv}@gmail.com

Abstract. Human Action Recognition (HAR) is a dynamic research area in pattern recognition and artificial Intelligence. The area of human action recognition consistently focuses on changes in the scene of a subject with reference to time, since motion information can prudently depict the action. This paper depicts a novel framework for action recognition based on Motion Projection Profile (MPP) features of the difference image, representing various levels of a person's posture. The motion projection profile features consist of the measure of moving pixel of each row, column and diagonal (left and right) of the difference image and gives adequate motion information to recognize the instantaneous posture of the person. The experiments are carried out using WEIZMANN and AUCSE datasets and the extracted features are modeled by the GMM classifier for recognizing human actions. In the experimental results, GMM exhibit effectiveness of the proposed method with an overall accuracy rate of 94.30 % for WEIZMANN dataset and 92.49 % for AUCSE dataset.

Keywords: Video surveillance · Human action recognition · Frame difference · Feature extraction · Gaussian mixture models

1 Introduction

In two decades, the smart visual surveillance system is a significant technology to make sure the security and safety in both public and private regions like airports, railway stations, petrol/gas stations, bus stands, banks, and commercial buildings due to terrorist activities and other societal problems. The analysis of human action recognition has been generally utilized in the field of pattern recognition and computer vision [1], whose point is to consequently segment, keep and distinguish human action progressively.

The recognition of human actions opens a number of interesting application areas such as automatic visual surveillance, health care activities, content based retrieval, human-computer-interaction (HCI), robotics and gesture recognition with conceptual information from videos [2,3]. The actions performed by human

R. Prasath et al. (Eds.): MIKE 2015, LNAI 9468, pp. 460–471, 2015.
DOI: 10.1007/978-3-319-26832-3_43

are reliant on many factors, such as the view invariant postures, outfits, lighting changes, occlusion, shadows, changing backgrounds and camera movements. Any human action can be represented by short video sequences showing a distinct set of the corresponding motion of the body. Recognizing and classifying the strange and suspicious actions from natural actions will be helpful in envisioning the threats and terrorist attacks [4].

1.1 Outline of the Work

This paper deals with human action recognition, which aims to identify action from the video sequences. The proposed approach is evaluated using WEIZMANN and AUCSE datasets. Difference image is produced by subtracting the successive frames in order to obtain the motion information. Thus, the motion projection profile (MPP) features are extracted from the difference image as a feature set. The extracted features are modeled by GMM classifiers for training and testing using 10-fold cross validation technique. The rest of the paper is structured as follows. Section 2 reviews related work. Section 3 provides an overview of the proposed feature extraction method. Section 4 explains the properties of GMM classifier and the experimental results on WEIZMANN and AUCSE datasets are presented in Sect. 5. Finally, Sect. 6 concludes the paper.

2 Related Work

The vision based action recognition becomes the significant goal to discriminate the actions automatically. The purpose is to build a general framework for the representation and recognition of dissimilar actions. Despite the information that recognition of single human actions have been compensated by a mere amount of attention, there has been extensive prior work and survey on human action recognition [1–4]. The following section discusses the previous work on human action recognition. Computer vision and pattern recognition methods [5], consist of object detection, feature extraction, feature selection, classification and it is being successfully used in many activity recognition systems. Image-processing techniques [6] such as analysis, detection of motion, shape, texture, color, optical flow and interest points [7], have also been found to be proficient. Bobick and Davis [8] proposed a technique of motion energy images and motion-history images based on Hu moments to construct action templates interpreted by the mean and covariance matrix of the moments. The Mahalanobis distance measure is used to calculate the distance between the moment descriptions of the input action for recognition. Wang et al. [9] presented a novel technique based on motion boundary histograms, that have shown better performance by classifying different actions. Moreover, Chundi Mu et al. [10] presented an enhanced technique for extracting histogram of motion vectors from high definition video sequences. Iglesias-Ham, et al. [11] proposed a technique based on 2D binary projections of the shape over the time and human actions is recognized by a rapid matching algorithm which considers the spatial distribution of the 2D projection of the shape. Vezzani et al. [12] proposed a technique for action recognition based on projection histograms of the foreground mask that gives adequate information to understand the dynamic posture of the person and then the HMM framework has been used for action recognition. In [13] presents a novel approach for automatic human activity recognition based on maximum motion patterns extracted from the temporal difference image by selecting the Region of Interest (ROI) and the experiments were carried out on the publicly available datasets like KTH and WEIZMANN.

3 Proposed Approach

The workflow of the proposed approach is shown in Fig. 1. The videos are smoothed by the Gaussian filter with a kernel of size 3×3 and variance $\sigma = 0.5$. It is essential to pre-process all video sequences eliminate noise for fine feature extraction and classification. The motion information is identified by subtracting the successive frames. The Motion Projection Profile (MPP) features are extracted from the difference image (temporal difference), representing various levels of an individuals posture. The motion projection profile method finds out the measure of moving pixel of each row, column and diagonal (left and right) pixels of the difference frame and they give sufficient motion information to recognize the instantaneous posture of the person as discussed in Sect. 3.1. Then extracted motion projection profiles of row, column, left diagonal and right diagonal features are resampled to a fixed number of bins, where bin value is empirically fixed to 8 and the values are concatenated into a distinctive vector, in order to obtain a 32-dimensional feature vector for each frame in an action video sequence. The extracted features are modeled by GMM for action recognition. In this work, WEIZMANN and AUCSE datasets are used in order to evaluate the effectiveness of these datasets on GMM classifier.

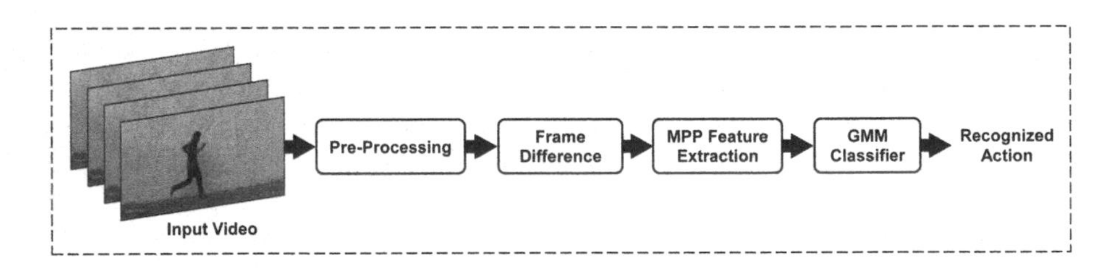

Fig. 1. Overview of the proposed approach

3.1 Feature Extraction

The extraction of discriminative feature is the most vital problem in human action recognition, that represents the significant information that is essential for further study. The following sections present the representation of the feature extraction method used in this work.

Frame Difference. To identify the person motion across a sequence of images, the current image is subtracted either by the previous frame or successive frame of the image sequences called as temporal difference. The difference images are obtained by applying thresholds to eliminate pixel changes due to camera noise, changes in lighting conditions, etc. This method is extremely adaptive to detect the motion region corresponding to moving objects in dynamic scenes and superior for extracting momentous feature pixels. The temporal difference image obtained by simply subtracting the previous frame t with current frame is a time $t+1$ on a pixel by pixel basis. The extracted motion pattern information is considered as the Region of Interest (ROI). Figure 2(a) and (b) illustrate the consecutive frames of the AUCSE dataset. The resulting difference image is shown in Fig. 2(c). $D_t(x, y)$ is the difference image, $I_t(x, y)$ is the pixel intensity of (x, y) in the tth frame, h and w are the height and width of the image correspondingly. The sample difference image for various activities is shown in Fig. 3. Motion information D_t or difference image is considered using

$$D_t(x, y) = |I_t(x, y) - I_{t+1}(x, y)| \quad (1)$$
$$1 \leq x \leq w, 1 \leq y \leq h$$

$$T_k(x, y) = \begin{cases} 1, & \text{if } D_k(x, y) > t; \\ 0, & \text{Otherwise}; \end{cases} \quad (2)$$

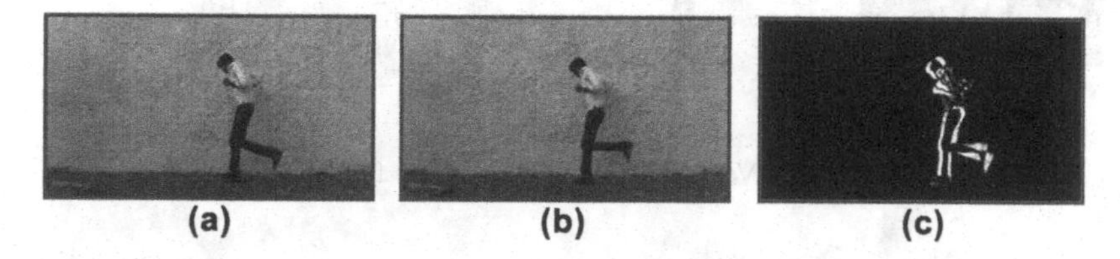

Fig. 2. (a), (b) Two consecutive frames from running action (c) Difference image of (a) and (b)

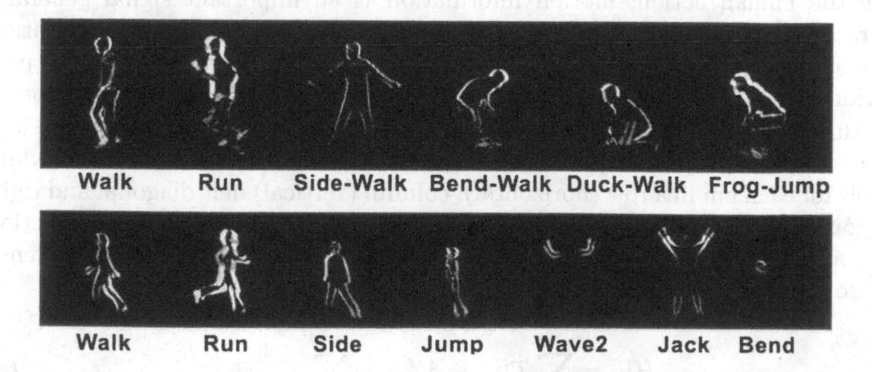

Fig. 3. Sample motion frames from AUCSE dataset (top row) and WEIZMANN dataset (bottom row)

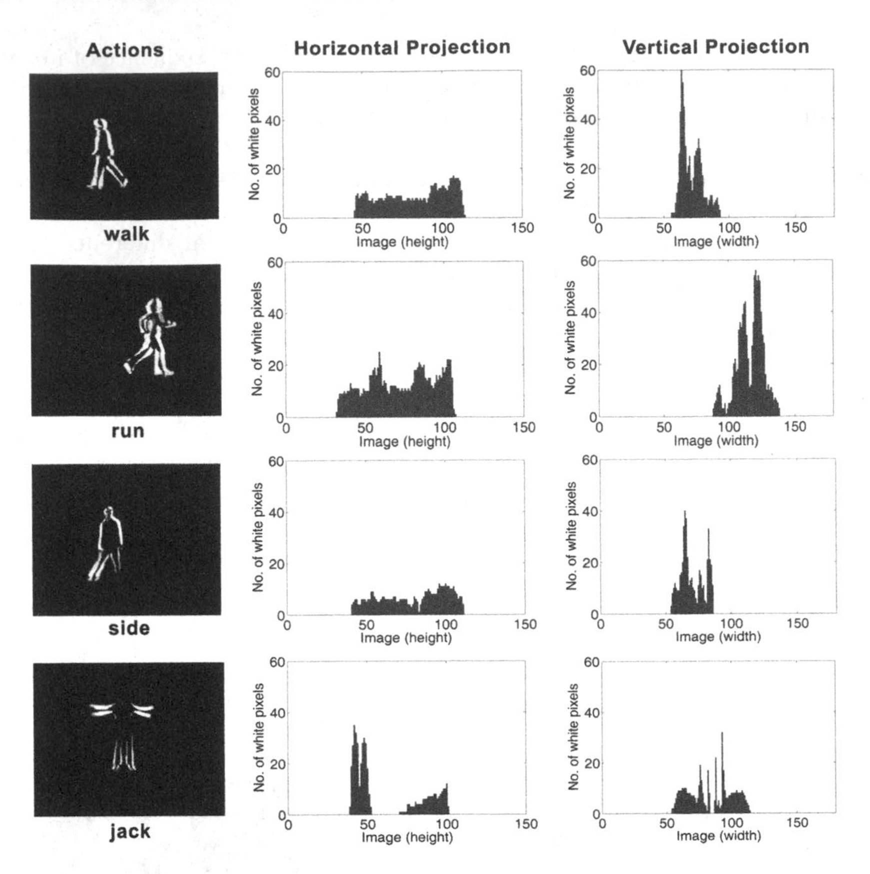

Fig. 4. Horizontal and Vertical projection profile for the various actions

Motion Projection Profile (MPP) Features for Action Recognition. The procedure for extracting the features is explained in this section. To recognize the human action, motion information is an important signal generally extracted from video sequences. Projection profiles are compact representation of images, since much valuable information is retained in the projection. The projection profiles are extracted from the difference image that consists of motion information only [14]. The row (horizontal), column (vertical), left diagonal and right diagonal projection profiles are obtained by finding the number of white pixels for each bin in a row (horizontal), column (vertical), left diagonal and right diagonal four directions respectively, as shown in Figs. 4 and 5. The projection $H[i]$ along the rows and the projection $V[j]$ along the columns of a difference image are mathematically defined by

$$H[i] = \sum_{j=0}^{n-1} T[i,j] \; ; \; V[j] = \sum_{i=0}^{m-1} T[i,j] \tag{3}$$

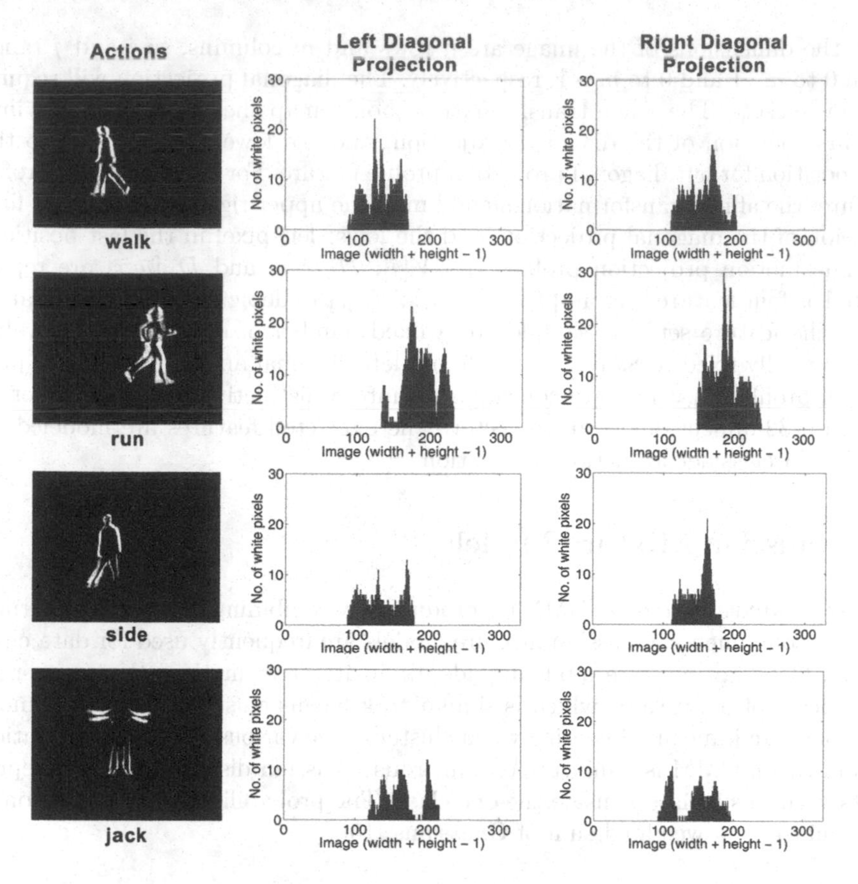

Fig. 5. Left diagonal projection profile and Right diagonal projection profile for the various actions

Where T is the difference image and n and m are the height and width of the difference image respectively.

For diagonal projections of an object requires only the area moments and the position thus computed from the horizontal and vertical projections are defined by

$$A_m = \sum_{j=0}^{m-1} V[j] \; ; \; A_n = \sum_{i=0}^{n-1} H[i] \tag{4}$$

$$\overline{y} = \frac{\sum_{i=0}^{n-1} iH[i]}{A_n} \; ; \; \overline{x} = \frac{\sum_{j=0}^{m-1} jV[j]}{A_m} \tag{5}$$

Where A_m and A_n are an area of the difference image. The diagonal projection profile is to compute the index for the histogram bucket for the current row and column. Let the row and column be denoted by i and j, respectively. Suppose that the dimensions of the image are n rows and m columns, so i and j range from 0 to $n-1$ and 0 to $m-1$, respectively. The diagonal projection will require $n+m$ buckets. The affine transformation should map the upper left pixel into the first position of the diagonal projection, and the lower right pixel into the last position for left diagonal projection profile feature. For right diagonal profile feature, the affine transformation should map the upper right pixel into the first position of the diagonal projection, and the lower left pixel in the last position.

The motion projection profiles $H[i]$,$V[j]$, $D_L[\overline{x},\overline{y}]$ and $D_R[\overline{y},\overline{x}]$ are represented as four feature vectors $\{f_H,\ f_V, f_{D_L},\ f_{D_R}\}$ to depict the difference frame. Then the feature set is re-sampled to a fixed number of bins, where bin value is empirically fixed to 8. The row, column, left diagonal and right diagonal projection profile bin values are concatenated into a distinctive vector, in order to obtain a 32-dimensional feature vector. The extracted features are modeled by the GMM classifier for action recognition.

4 Gaussian Mixture Models

Gaussian Mixture Models (GMM) is produced by combining multivariate normal density components. Gaussian mixture models are frequently used for data clustering. Gaussian mixture modeling adopts an iterative method that converges to a local optimum value which is similar to k-means clustering. GMM is more suitable than k-means clustering when clusters have various sizes and correlation within them. GMM is a mixture of numerous Gaussian distributions and represents various subclasses inside the one class. The probability density function is determined as a weighted sum of Gaussians [15].

$$p(x|\Theta) = \sum_{k=1}^{K} \alpha_k p_k(x|\theta_k) \tag{6}$$

where k is the number of mixtures and $\sum_{k=1}^{K} \alpha_k = 1$ are the mixture weights and where $\Theta = \{\alpha_1, ..., \alpha_k, \theta_1, ..., \theta_k\}$ where individual component is a multivariate Gaussian density.

5 Experimental Results

In this section, the proposed method is evaluated using WEIZMANN and AUCSE datasets. The experiments are carried out in MATLAB 2013a in Windows 7 Operating System on a computer with Intel Core i7 Processor 3.40 GHz supported by 8 GB RAM. The obtained MPP features are fed to GMM to build a model for each action and these models are used to test the performances of the classifier.

5.1 WEIZMANN Dataset

The Weizmann Institute of Science [16] provides this dataset in 2005 that consist of 10 types of human actions namely bend, jumping-jack (or shortly jack), jump-forward-on-two-legs (or jump), jump-in-place-on-two-legs (or pjump), run, gallop sideways (or side), skip, walk, wave one- hand (or wave1) and wave-two-hands (or wave2) performed by nine individuals in an outdoor environment. The video sequences are recorded in homogeneous background with a static camera, deinterlaced 50 fps. In total, this dataset contains 90 low-resolution video clips of 180×144 pixels and have a duration of 1 second in average. The sample frames of WEIZMANN action sequence are illustrated in Fig. 6.

Fig. 6. Example frames of the WEIZMANN dataset

5.2 AUCSE Dataset

The Faculty of Engineering and Technology (Computer Science and Engineering), Annamalai University created AUCSE action dataset in 2014. It contains 6 action categories: walking, running, side-walk, bend-walk, duck-walk and frog-jump performed by five different individuals in an outdoor environment. The video sequences are recorded under homogeneous background with a static camera with 30 fps frame rate. The action sequences were downsampled to the standard resolution of 960×540 pixels and have a duration of 3 seconds in average. In total, the dataset consists of 60 video samples. This dataset is very challenging due to small variation in action pose interpretation and the sample frames of the AUCSE action sequence are illustrated in Fig. 7.

5.3 Performance Evaluation

As explained in Sect. 3.1, the 32-dimensional MPP features are extracted. The performance of the proposed feature method on GMM classifier is evaluated using 10-fold cross-validation approach. The performance evaluation is done with

Fig. 7. Example frames of the AUCSE dataset

statistical metrics like Precision, Recall (Sensitivity), F-Score and Specificity, where, tp and fp are the number of true positive and false positive predictions of the class and tn and fn are the number of true negative and false negative predictions. Accuracy (A) shows the overall correctness of the activity recognition. Precision (P) is a measure of exactness. Recall (R) or Sensitivity gives how good an action is identified correctly. F-Score is the harmonic mean of Precision and Recall. Finally, Specificity (S) gives a measure of how good a method is identifying negative action correctly. The statistical measures of Accuracy, Precision, Recall (Sensitivity), F-Score and Specificity is defined as follows

$$Accuracy = \frac{tp + tn}{tn + fp + tp + fn} \tag{7}$$

$$Precision = \frac{tp}{tp + fp} \tag{8}$$

$$Recall(Sensitivity) = \frac{tp}{tp + fn} \tag{9}$$

$$Specificity = \frac{tn}{tn + fp} \tag{10}$$

$$F - Score = 2\,\frac{Precision \times Recall}{Precision + Recall} \tag{11}$$

5.4 Evaluation on WEIZMANN Dataset

The average recognition accuracy is 94.30 % on the WEIZMANN dataset and the corresponding confusion matrix is shown in Fig. 8. The diagonal of the confusion matrix represents the percentage of instances that were classified correctly. The each action class instance are represented by the rows and the action class predicted by the classifier is represented by the columns. The actions like walk, run, side, skip, jump, jack, bend and wave2 are classified well with accuracy greater than 93 %. From this, pjump and wave1 actions are confused as bend, where these two actions intuitively seem hard to differentiate and it needs further attention.

The performance evaluation results are tabulated in Table 1, which shows that the proposed MPP approach has a good precision, recall (sensitivity), F-score and Specificity for the GMM classifier on WEIZMANN dataset.

	WK	RN	SD	SK	JP	PJ	JK	BD	W1	W2
WK	96.4	1	0.8	0.5	1.1	0	0.2	0	0	0
RN	0	96.6	1.4	1.1	0.9	0	0	0	0	0
SD	2.3	0.7	95.2	1.6	0	0	0.2	0	0	0
SK	2.3	0	1.4	95.6	0.7	0	0	0	0	0
JP	0.7	0.2	0	1.1	97.8	0.2	0	0	0	0
PJ	0	0	0	0	0	90.2	0.6	6.6	1	1.6
JK	0	0	0	0	0	1.8	94.9	1.2	0.4	1.7
BD	0	0	0	0	0	1.3	0	94.2	2.6	1.9
W1	0	0	0	0	0	0	0	9.6	89.7	0.7
W2	0	0	0	0	0	1.8	0.5	3.3	0.7	93.7

Fig. 8. Confusion matrix of recognition accuracy (%) for the WEIZMANN dataset using GMM classifier, where WK=Walk, RN=Run, SD=Side, SK=Skip, JP=Jump, PJ=Pjump, JK=Jack, BD=Bend, W1=Wave1, W2=Wave2.

Table 1. Performance measure (%) of the WEIZMANN dataset on GMM classifier

Actions	Precision	Recall	F-Score	Specificity
Walk	96.29	96.45	96.37	99.49
Run	97.13	96.58	96.86	99.79
Side	96.28	95.17	95.72	99.66
Skip	95.60	95.60	95.60	99.59
Jump	97.12	97.77	97.45	99.72
Pjump	93.62	90.16	91.86	99.35
Jack	98.84	94.86	96.81	99.82
Bend	79.17	94.18	86.02	97.52
Wave1	95.53	89.69	92.52	99.47
Wave2	94.19	93.7	93.94	99.27
Average	**94.38**	**94.42**	**94.31**	**99.37**

5.5 Evaluation on AUCSE Dataset

The average recognition accuracy is 92.49 % on the AUCSE dataset and the corresponding confusion matrix is shown in Fig. 9. The diagonal of the confusion matrix represents the percentage of instances that were classified correctly. The each action class instance are represented by the rows. The action class predicted by the classifier is represented by the columns. The actions like run, side, bend-walk and frog-jump are almost classified well with accuracy greater than 91 %. From this, walk and duck-walk actions are confused as side and bend-walk respectively and it requires further attention.

The performance evaluation results are tabulated in Table 2, which shows that the proposed MPP approach has a very good precision, recall (sensitivity),

	WK	RN	SD	BW	DW	FJ
WK	89.9	1.3	5.7	0.8	0.6	1.7
RN	0.5	98.1	0.9	0.5	0	0
SD	1.3	0.2	94	1.8	1.7	1
BW	0.4	0.2	3.2	91	3.6	1.6
DW	0.2	0.3	3.2	2.9	90.6	2.8
FJ	0.6	0.6	1.5	1.5	3.1	92.7

Fig. 9. Confusion matrix of recognition accuracy (%) for the AUCSE dataset using GMM classifier, where WK=Walk, RN=Run, SD=Side, BW=Bend-Walk, DW=Duck-Walk, FJ=Frog-Jump

Table 2. Performance measure (%) of the AUCSE dataset on GMM classifier

Actions	Precision	Recall	F-Score	Specificity
Walk	94.25	89.87	92.01	99.28
Run	94.33	98.09	96.18	99.55
Side	93.2	93.96	93.58	97.05
Bend-Walk	91.23	90.96	91.09	98.27
Duck-Walk	89.89	90.6	90.24	97.91
Frog-Jump	93.09	92.71	92.9	98.51
Average	**92.67**	**92.7**	**92.67**	**98.43**

F-score and Specificity for the GMM classifier on AUCSE dataset. It is clear from the Tables 1 and 2 that the proposed method performs well on both WEIZMANN and AUCSE datasets.

6 Conclusion

An efficient framework for action recognition task on surveillance videos using the Motion Projection Profile (MPP) as a feature is proposed in this paper. The MPP feature was obtained from temporal difference image and it represents the measure of moving pixel of each row, column and diagonal (left and right) of the difference image and they give adequate motion information to recognize the instantaneous posture of the person. Experiments are conducted on WEIZMANN and AUCSE datasets and the performance of MPP feature is evaluated using GMM classifier. From the experimental results, it is observed that GMM shows a recognition accuracy of 94.30 % for WEIZMANN dataset and 92.49 % for AUCSE dataset. The experimental results have demonstrated that the MPP method has a promising performance on two datasets: WEIZMANN and AUCSE.

References

1. Vishwakarma, S., Agrawal, A.: A survey on activity recognition and behavior understanding in video surveillance. Vis. Comput. **29**(10), 983–1009 (2013)
2. Weinland, D., Ronfard, R., Boyer, E.: A survey of vision-based methods for action representation, segmentation and recognition. Comput. Vis. Image Underst. **115**(2), 224–241 (2011)
3. Hassan, M., Ahmad, T., Liaqat, N., Farooq, A., Ali, S.A., et al.: A review on human actions recognition using vision based techniques. J. Image Graph. **2**(1), 28–32 (2014)
4. Poppe, R.: A survey on vision-based human action recognition. Image Vis. Comput. **28**(6), 976–990 (2010)
5. Duda, R.O., Hart, P.E., et al.: Pattern Classification and Scene Analysis, vol. 3. Wiley, New York (1973)
6. Gonzalez, R.C.: Digital Image Processing. Pearson Education, India (2009)
7. Laptev, I.: On space-time interest points. Int. J. Comput. Vis. **64**(2–3), 107–123 (2005)
8. Bobick, A.F., Davis, J.W.: The recognition of human movement using temporal templates. IEEE Trans. Pattern Anal. Mach. Intell. **23**(3), 257–267 (2001)
9. Wang, H., Kläser, A., Schmid, C., Liu, C.-L.: Action recognition by dense trajectories. In: 2011 IEEE Conference on Computer Vision and Pattern Recognition (CVPR), pp. 3169–3176. IEEE (2011)
10. Mu, C., Xie, J., Yan, W., Liu, T., Li, P.: A fast recognition algorithm for suspicious behavior in high definition videos. Multimedia Syst., 1–11 (2015)
11. Iglesias-Ham, M., García-Reyes, E.B., Kropatsch, W.G., Artner, N.M.: Convex deficiencies for human action recognition. J. Intell. Robot. Syst. **64**(3–4), 353–364 (2011)
12. Vezzani, R., Baltieri, D., Cucchiara, R.: HMM based action recognition with projection histogram features. In: Ünay, D., Çataltepe, Z., Aksoy, S. (eds.) ICPR 2010. LNCS, vol. 6388, pp. 286–293. Springer, Heidelberg (2010)
13. Arunnehru, J., Geetha, M.K.: Automatic activity recognition for video surveillance. Int. J. Comput. Appl. **75**(9), 1–6 (2013)
14. Arunnehru, J., Geetha, M.K.: Human activity recognition based on projected histogram features in surveillance videos using tree based classifiers. Int. J. Appl. Eng. Res. **9**(21), 4950–4954 (2014)
15. McLachlan, G., Peel, D.: Finite Mixture Models. Wiley, New York (2004)
16. Blank, M., Gorelick, L., Shechtman, E., Irani, M., Basri, R.: Actions as space-time shapes. In: Tenth IEEE International Conference on Computer Vision, 2005, ICCV 2005, vol. 2, pp. 1395–1402. IEEE (2005)

A Web-Based Intelligent Spybot

Pruthvi Raj[1], N. Rajasree[1], T. Jayasri[1], Yash Mittal[2], and V.K. Mittal[1](✉)

[1] Indian Institute of Information Technology Chittoor, Sri City, India
{pruthvi.m14,rajasree.n14,jayasri.t14,vkmittal}@iiits.in
[2] Jaypee University of Information Technology, Solan, India
yashmittal@hotmail.co.in

Abstract. Robots have been making inroads to human life in almost all spheres. Spybots can be immensely useful for unmanned surveillance and covert spying operations. If online streaming of the spied data can be made feasible, that would be an added advantage. In this paper, we propose an unmanned Spybot that can be controlled remotely from web-page based commands, using a WiFi network. It can also stream back the spied data, that could be video or images, over the WiFi network. A prototype Spybot is developed. Users can give the control instructions from a web-page. The WiFi module on board the Spybot receives these commands from the web-page and passes those to the microcontroller. The microcontroller interprets the control commands, and generates control signals to operate the Spybot as per the user's commands. The captured images and video data is sent back to a smartphone over WiFi network. Performance evaluation is carried out, to measure various limits of operation of the Spybot. The results are encouraging. The proposed Spybot can have potential applications for military, security forces and surveillance purposes.

Keywords: Intelligent spybot · WiFi module · Web-page based remote control operation · IP camera · Smartphone

1 Introduction

Robots are powerful machines that are immensely useful in different walks of human life. They do the tasks which may be potentially risky to human life. Robots enhance the quality of our lives by performing tedious works and providing assistance to people with disabilities. Some of the first robots in the 1940's were used to handle the radioactive materials. In the early 1960's the robots were used to pick up and place objects in new locations. Many robots simulate several human functions. Spybots can also be used to gain access to places stealthily which may be difficult for human operators or may pose a risk to human life.

Robots can be operated in two ways: *wired* and *wireless*. In *wired* robots there is a direct connection between the robot and the computer, that limits the range of operation of the robot. With the advancement in technology, *wireless* controls have gained attention. Wireless technology can be used for control operations in

R. Prasath et al. (Eds.): MIKE 2015, LNAI 9468, pp. 472–481, 2015.
DOI: 10.1007/978-3-319-26832-3_44

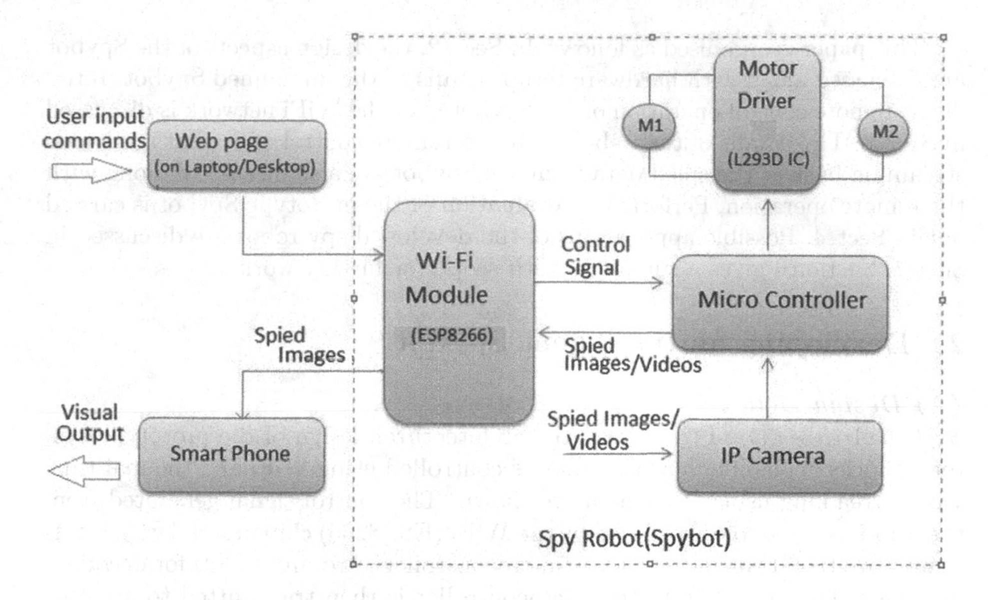

Fig. 1. Architecture of the *'Intelligent Spy Robot'* [M1: Motor 1 and M2: Motor 2]

many ways, such as IR, RF communications, Bluetooth, WiFi etc. [2]. Amongst these, WiFi has the most wide range of applications, compared to other wireless communication methods. WiFi is a wireless LAN technology for computer networking that facilitates users to transfer data over the internet, without using any wired medium [4].

Using WiFi, it is possible to control the Spybot through the internet. In this type of communication, a wireless network adapter in a computer converts the digital data into radio signals, and a WiFi unit on board the Spybot converts these signals into digital form. Institute of Electrical and Electronics Engineers (IEEE) has come up with a set of standards for these wireless networks and WiFi communication technology, generally referred as IEEE 802.11 [6]. WiFi modules offer greater data transfer speeds and wider operating range. Hence, one can communicate more data at faster rates and to longer distances, for operation of the Spybot [5].

In this paper, we develop a Spybot using web-based controls that are transmitted to the Spybot over a WiFi network. A WiFi module (ESP8266) on board the Spybot is used. This chip is a serial module with a built-in TCP/IP stack. The Spybot can be controlled by the user through multiple control options given through a web-page based control console. These control commands are received by the WiFi module on board the Spybot, and are further transferred to the microcontroller module. The path of the robot can be monitored by an IP camera using video-streaming, so that the Spybot can enter spaces that are difficult to access otherwise. The paper also discusses the steps required for hardware implementation and software setup of the Spybot. The remote control operation for navigating the Spybot to locations which are not in the line of sight, is also discussed.

This paper is organised as follows. In Sect. 2, the design aspects of the Spybot are discussed along with hardware setup details of the unmanned Spybot. Intelligent remote control operation of the Spybot over the WiFi network is discussed in Sect. 3. The details of the web-page based remote control of Spybot and video streaming back of the spied data from the Spybot are also discussed, along with the remote operation. Performance evaluation of the prototype Spybot is carried out in Sect. 4. Possible applications of the developed spy robot are discussed in Sect. 5. Section 6 gives a summary, with scope for further work.

2 Developing an Intelligent Spybot

(A) Design Details

System Architecture: Fig. 1 shows the architectural design of the prototype system. Blocks explaining how the robot is controlled using WiFi and the real-time video streaming using IP camera are shown. The control signal generated from the web-based console is recieved by the WiFi (ESP8266) chip in serial way. Commands are then transmitted to the microcontroller (Arduino UNO) for decoding purposes. The output from the microcontroller is then transmitted to a motor driver chip (L293D, shown in Fig. 3), which amplifies the current signal and gives the minimum voltage required to drive the motors. Video streaming of the spied images or video data is carried out using the IP camera to a WiFi connected device such as a desktop, laptop or any other mobile device.

Table 1. Key components

Sl	*Components*
1	Robot Chassis with 2 wheels and Castor wheel
2	Microcontroller (Arduino Uno)
3	WiFi Module (ESP8266)
4	Voltage Regulator (IC LD1117v33)
5	Motor Driver IC (L293D)
6	Two DC Motors
7	Smartphone
8	IP Camera

Key Components: The key components used are listed in Table 1. The Spybot uses Arduino UNO, an 8-bit microcontroller, that operates at a clock speed of 16 MHz. Arduino possesses real-time and analog capabilities, in addition to simpler interfacing of sensors, modules and devices attached to it. For WiFi connectivity, the ESP8266 chip is the best low cost module, that has the ability to act as a host as well as a client.

(a) Microcontroller (Arduino UNO) Board: The Arduino Uno board, a microcontroller board based on the ATMega328, is designed and developed specifically

for the purpose of learning and experimentation [3]. It has total 28 pins - 14 digital input/output (of which 6 can be used as PWM outputs), 6 analog inputs, a 16 MHz ceramic resonator, a USB connection, a power jack, an ICSP header and a Reset button. The board can be powered by a USB port or by an AC-to-DC adapter or a battery source. Various shields like prototype shield, WiFi shield, Bluetooth shield etc. can be added to the Arduino Uno board. Apart from the clock speed of 16 MHz, the microcontroller has 32 KB of Flash memory, 2 KB of SRAM and 1 KB of EEPROM.

(b) WiFi Module (ESP8266): For transmission of the control signals and spied data between the user and the Spybot, ESP8266 WiFi module is used. This is a complete System-on-Chip (SoC) with integrated TCP/IP protocol stack that can give any microcontroller access to a WiFi network [7]. Each ESP8266 module comes pre-programmed with a firmware set of AT(*ATtention*) commands, which are used for controlling the wired modems, GSM/GPRS modules and mobile phones. The ESP8266 is a cost effective module and can be integrated with different sensors. The firmware implements TCP/IP, the full 802.11 b/g/n WLAN MAC protocol, and WiFi Direct (P2P) specification. It also has an Integrated low power 32-bit CPU, used as application processor. This ESP8266 WiFi module is low-cost (approx INR 500) which makes it usable for appliances related to spying, at low cost. It can also be used as an expendable module.

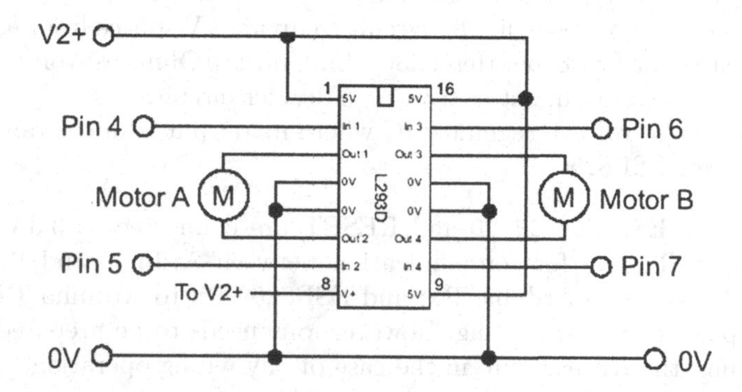

Fig. 2. Circuit Diagram for the connections of *'Motor Driver'* (L293D)

(c) L293D Motor Driver IC: A dual H-bridge motor driver integrated circuit, L293D acts as a current amplifier. It takes a low-current control signal and provides a higher-current signal which is used to drive the motors. In its common mode of operation, two DC motors can be driven simultaneously by it, both in forward and reverse direction. The motor operations of two motors can be controlled by input logic at pins 2 & 7 and 10 & 15 as shown in Fig. 2. Input logic 00 or 11 will stop the corresponding motor, while logic 01 and 10 will rotate it in clockwise and anti clockwise directions, respectively. Pins 1 & 9 (corresponding to the two motors) must be HIGH for motors to start operating.

Table 2. Movements of the spybot

(a) Control button	(b) Left motor	(c) Right motor	(d) Spybot movement
Front	Forward	Forward	Forward
Back	Backward	Backward	Backward
Left	Halt	Forward	Leftward
Right	Forward	Halt	Rightward

(d) DC Motors: The DC motors are fixed on the chassis. Two ends of each DC motor are connected to the output pins 3 & 6 of the L293D IC for left motor, and to 14 & 11 for the right motor (Fig. 2). Four input pins of the IC L293D, 2 & 7 on the left and 15 & 10 on the right regulate the rotation of the motor, connected across both sides. The motors rotate on the basis of the inputs provided across the input pins as logic 0 or 1, which is given to the microcontroller (Arduino UNO). Further information on the buttons available, configuration of the DC motors and the resulting motion of the Spybot are listed in Table 2.
(e) ESP8266 Logic Level Shifter: Since Arduino Uno microcontroller works at 5 V and ESP8266 WiFi module works at 3.3 V, there are 3 ways to connect these:

(i) *Option A:* Directly plug in the Tx-Rx pairs between Arduino and ESP8266, but there is a potential risk of damage.
(ii) *Option B:* Use a voltage divider circuit to divide 5 V source from Arduino to 3.3 V source for ESP8266. Here, 330 Ohm and 180 Ohm resistors, connected in parallel connection, act as a voltage divider circuit.
(iii) *Option C:* Use voltage regulator IC where its output is in the range that is required for ESP8266.

Wiring Scheme: ESP8266 `CH_PD` and RESET are connected to 3.3 V for logic HIGH. The Tx-Rx pair for communication between Arduino and WiFi module are: ESP8266 Tx to Arduino Rx; and ESP8266 Rx to Arduino Tx (Fig. 3). Remaining pins should be floating. However, one needs to be prepared to occasionally ground the RESET pin, in the case of any wrong operation.
(B) Operation Details
Necessary details required for the implementation and functioning of the proposed Spybot are shown using the detailed circuit diagram given in Fig. 3. The configuration code for initialising and operating the WiFi module is given in Fig. 4. Further operation details are discussed in the following section.

3 Intelligent Control Operation

(A) Web-based Control Operation
Main advantage of the web-based control is that the user can give instructions from the web, since it doesn't require any serial communications using USB. These commands are further passed on to the microcontroller, in the form of an HTTP request given to ESP8266. Next, the Arduino board is programmed

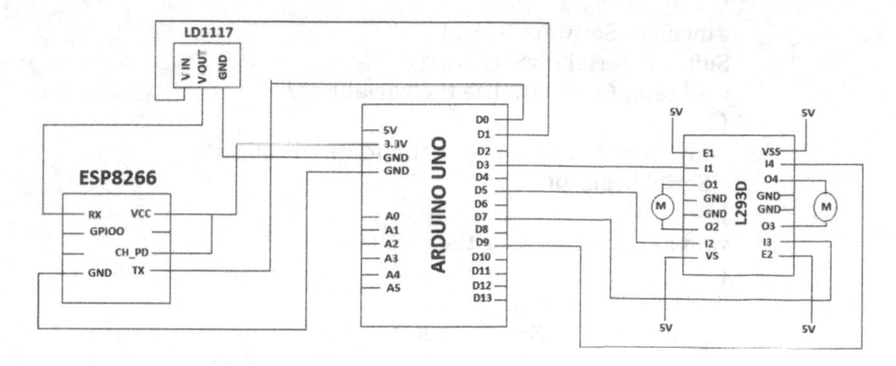

Fig. 3. Circuit diagram for connecting the Microcontroller (Arduino Uno), WiFi module (ESP8266) and Motor Driver (L293D)

(using embedded C) for movements of the robot in forward, backward, left or right directions.

Depending upon the button we click on the web-page, the Spybot can be manoeuvred in the user's desired direction. Once the Arduino is programmed, video streaming to the browser by the Spybot can be initialized at the click of a button.

(B) WiFi Control Operation

A remote control vehicle is controlled remotely such that does its motion w.r.t an origin external to device, is not restricted. In this paper, remote operation of the Spybot is performed using WiFi communication that involves transfer of information from the microcontroller to a robot wirelessly. After connecting the ESP8266 to the microcontroller (as shown in Fig. 3), one needs to configure the WiFi Module using the AT commands for wireless network [1]. The configuration code shown in Fig. 4, can be used for this purpose.

Configuration of ESP8266: The code shown in Fig. 4, needs to be transferred to the microcontroller (Arduino UNO), to configure the WiFi module. In this code, the Tx line from ESP8266 is connected to Arduino's pin 2 and Rx line from the ESP8266 is connected to Arduino's pin 3. The baud rate is fixed to 9600 (which can be changed, if desired) while ensuring that both NL and CR are selected in the line ending pop-up menu, at the bottom of the serial monitor. The details of AT commands and steps to be followed are:

Step 1: In the terminal, type AT and press enter (AT is not case sensitive but other commands are). If the response is not "OK", change the baud rate and try again.

Step 2: Ensure the module is in a known state by issuing a reset command (AT+RST). The response should be ready on the last line.

Step 3: Check the firmware version, by using the Command (AT+GMR). The response should be the firmware version of the module.

Step 4: AT+CWMODE? returns the mode of operation of the module. CWMODE returns an integer designating the mode of operation either 1, 2

```
#include<SoftwareSerial.h>
SoftwareSerial mySerial(2,3);
void setup() /*Initialize the variables*/
{
  mySerial.begin(9600); /*baud rate is 9600*/
  Serial.begin(9600);
}
void loop() /*Look for the input*/
{
 if(mySerial.available())
  Serial.write(mySerial.read());
 if(Serial.available())
 mySerial.write(Serial.read());
}
```

Fig. 4. Configuration code for the WiFi module

or 3. Here, 1. station mode (client), 2. AP Mode (Host) and 3. AP with station mode (ESP8266 has a dual mode)

Step 5: Enable the module to act as both a "Station" and "Access Point" by giving the command AT+CWMODE=3. Another reset may be needed here.

Step 6: Check the available WiFi networks by giving the command (AT+CWLAP). All the available networks are displayed in response.

+CWLAP: (0,"", 0)
+CWLAP: (3,"HOMEWiFi",-80)
+CWLAP: (3,"PublicWiFi",-51)
OK

Step 7: Connect to a suitable WiFi access point by giving the following command which consist of access point name and password.

AT+CWJAP="access-point-name", "Password"
The response should be OK.

Step 8: Check the IP address of the module by the command (AT+CIFSR).
It returns the IP address as the response.

(C) Streaming back the spied Images/Video over WiFi network

Video streaming is necessary to drive the robot without line of sight. There are various methods to achieve this. In this paper, live video streaming is achieved using an Internet Protocol (IP) Camera, to obtain true real-time streaming. It requires an IP Camera compatible with WiFi. The IP cameras are designed to send the spied video or image data to the microcontroller board, and over WiFi network to a smartphone or another computer, for displaying the spied data.

In order to achieve this, an app needs to be installed (IP Cam Controller) on the smart phone or laptop. After installing the app, the camera in use needs to be configured. Launching the app for the first time on a laptop will automatically navigate to the new camera screen. Below steps can be followed,

Table 3. Control range of spybot

(a) Connection type	(b) Antenna type	(c) ESP8266 range
Connecting to TP-Link Router	External WiFi Antenna	479 m
	PCB-mounted Antenna	366 m
Connecting to Dish Antenna	External WiFi Antenna	4.2 km
	PCB-mounted Antenna	3.7 km

Step 1: Choose the camera type closest to the IP Camera in use.

Step 2: Give a name to the camera.

Step 3: If the camera is a DVR, a camera number may need to be input, which normally corresponds to the stream number, channel number, etc.

Step 4: Enter the hostname or IP address of the camera. One needs to ensure that in order to access the camera over the internet, the internet IP address is used and not the internal IP address.

Step 5: Enter the port number on which the camera is hosted. If the camera is behind a WiFi router, port forwarding may have to be configured on the router.

Step 6: Give the user name and password of the camera (not the router) and choose the image resolution for the video feed.

After setting up the camera, the video or images data captured by the camera will be available for display on the Android device.

4 Performance Evaluation

Performance evaluation of the Spybot is carried out using multiple parameters, for example, by measuring the range up to which the Spybot maintains a steady signal for transmission. Other parameters on which the performance is evaluated here are: the connection type, the antenna type and the range of the WiFi module (ESP8266).

In Table 3, the first column (a) shows the type of connection either to *router* or to *dish antenna*. For each type of connection, two types of antennae (in column (b)) can be connected: *external WiFi antenna* and *PCB-mounted antenna*. In column (c), the average range of the WiFi module (ESP8266) is estimated. Connecting to the WiFi module through a router (TP Link), the user is able to ping the module at 479 m with a huge antenna soldered on the module, and at 366 m with the PCB-mounted antenna. In order to test out the maximum range of the WiFi module, it is linked with the dish antenna. Range of operation was evaluated by measuring it for good reception for 5 times, and average values are computed. The user is able to move an average approximately 4.2 km away from the module with an external WiFi antenna and can still ping it. The range for the normal PCB-mounted antenna on average is approximately 3.7 km.

5 Possible Applications and Future Scope

There are many applications of the intelligent Spybot whose prototype is developed. These kind of vehicles may move in places where humans may not be able to reach. Few real world applications using the key concept of the intelligent Spybot are discussed in this section.

(a) *Military Applications:* The first military applications of unmanned vehicles appeared in 19th century to transport equipment, carry weapons, etc. Intelligent Spybots can be immensely useful in covertly gathering intelligence inputs about the enemy locations.
(b) *Spy Bots:* The camera in the Spybots can detect the objects in the surrounded areas, and can take images of the objects in places where human beings can't reach. The night vision mode is used to capture the video or images even in the night or smoke.
(c) *Search and Rescue Operations:* There are some situations where the environment is hazardous or not suitable for human intervention, for example in search and rescue operations, sending an unmanned vehicle or an intelligent Spybot is the best option.
(d) *Surveillance:* Such intelligent Spybots can also be used in civil operations such as monitoring and surveillance. By doing so, the cost involved in the case of human workforce can be drastically reduced.

The prototype Spybot developed in this paper can further be improved and enhanced to meet suitable needs according to its application. One can integrate GPS module with the robot which delivers the data such as latitude, longitude coordinates, etc. It can also be used to locate the position of the spy robot. Further, the path traversed by the spy robot can be displayed on the Google maps. One can also make this spy robot as an unmanned aerial vehicle for a wide range of applications, in aerial photography, hurricane hunting, 3-D mapping, protecting wild life, surveillance, security and inspections etc. It is also possible to control this robot using voice commands. These Spybot technologies may also be used to implement electronic wheel chairs for physically disabled people.

6 Summary and Conclusion

An intelligent spy robot is developed in this paper, which can be controlled by giving the control instructions from a web-page based console. The robot can be guided with a certain freedom and autonomy without using wires and can be used to enter confined spaces when needed. The WiFi module (ESP8266) is used as an easy-to-use programmable device, and is of low cost. The Arduino Uno microcontroller used, which is essentially a tiny computer, is compatible with a large number of modules. By mounting the ESP8266 WiFi module on it, the microcontroller can communicate with any device having a WiFi receiver.

One major limitation in the case of WiFi networks is that a network needs to be set up first, either using a router or a dish. Other issues arising with

WiFi usage are the limited range and potential security threats due to multiple clients being connected to the network (one-to-many). These limitations can be overcome with the use of a GSM module which can help to establish connection between the microcontroller and the GSM-GPRS system. A SIM (Subscriber Identity Module) may be required to communicate over the network, making the communication (one-to-one) and more secure as compared to WiFi. No network needs to be set up here, since there is direct satellite communication. This may also solve the problem of limited range, since one GSM module can be securely connected to another anywhere in the world. The proposed solution would make use of this module in the next updated version of the Spybot, for which the work is in progress.

This paper also presents a comprehensive view of the WiFi network technologies in the Spybot. Few other technologies are also available that may be added at a later date. This paper can be very helpful to fellow researchers who aim to conduct further tests and experiments on remote controlled Spybots.

References

1. AllAboutEE: ESP8266 Arduino LED Control (2015). http://allaboutee.com/2015/01/02/esp8266-arduino-led-control-from-webpage/. Accessed on 19 Jul 2015
2. Yeole, A.R., Bramhankar, S.M., Wani, M.D., Mahajan, M.P.: Smart phone controlled robot using ATMEGA328 microcontroller. Int. J. Innovative Res. Comput. Commun. Eng. **3**, 191–197 (2015)
3. Arduino: Arduino - Arduino Board Uno (2015). https://www.arduino.cc/en/Main/arduinoBoardUno. Accessed on 17 Jul 2015
4. Hernández, C., Poot, R., Narváez, L., Llanes, E., Chi, V.: Design and implementation of a system for wireless control of a robot. Int. J. Comput. Sci. Issues **7**, 191–197 (2010)
5. Dixit, S.K., Dhayagonde, S.B.: Design and implementation of e-Surveillance robot for video monitoring and living body detection. Int. J. Sci. Res. Publ. **4**(4), 1–3 (2014)
6. IEEE: IEEE Standard for Information technology-Telecommunications and information exchange between systems Local and metropolitan area networks-Specific requirements Part 11: Wireless LAN Medium Access Control (MAC) and Physical Layer (PHY) Specifications. IEEE Std 802.11-2012 (Revision of IEEE Std 802.11-2007), pp. 1–2793, Mar 2012
7. Sparkfun: WiFi Module - ESP8266 (2015). https://www.sparkfun.com/products/13678. Accessed on 20 Jul 2015

Evidential Link Prediction Based on Group Information

Sabrine Mallek[1(✉)], Imen Boukhris[1], Zied Elouedi[1], and Eric Lefevre[2]

[1] LARODEC, Institut Supérieur de Gestion de Tunis,
Université de Tunis, Tunis, Tunisia
sabrinemallek@yahoo.fr, imen.boukhris@hotmail.com, zied.elouedi@gmx.fr
[2] University Lille Nord de France, Uartois EA 3926 LGI2A, Lille, France
eric.lefevre@univ-artois.fr

Abstract. Link prediction has become a common way to infer new associations among actors in social networks. Most existing methods focus on the local and global information neglecting the implication of the actors in social groups. Further, the prediction process is characterized by a high complexity and uncertainty. In order to address these problems, we firstly introduce a new evidential weighted version of the social networks graph-based model that encapsulates the uncertainty at the edges level using the belief function framework. Secondly, we use this graph-based model to provide a novel approach for link prediction that takes into consideration both groups information and uncertainty in social networks. The performance of the method is experimented on a real world social network with group information and shows interesting results.

Keywords: Social network analysis · Link prediction · Uncertain social network · Group information · Belief function theory

1 Introduction

Social networks are usually conceptualized as a graph representation that provides a mapping of the ties relating the social structures. They are very dynamic and alter quickly over specific time intervals. New connections are established continuously between the network nodes. One of the most popular researches in social network analysis that studies social networks evolving is link prediction. It addresses the problem of predicting the existence of new or missing connections in social networks.

Yet, existing methods for link prediction are devoted to social networks under a certain framework. In fact, most methods assume the links to have binary values, either 1 (exist) or 0 ($\neg$ exist). Still, the structure of social network critically depends on the accurate structure of the data. That is, sparse distortions affect considerably the analysis results. As pointed out in [2,7], social networks data are frequently noisy and missing, they are also prone to errors of observation (e.g., missing information about the nodes and/or edges from the data).

R. Prasath et al. (Eds.): MIKE 2015, LNAI 9468, pp. 482–492, 2015.
DOI: 10.1007/978-3-319-26832-3_45

Hence, one would have to deal with two possible problems: take all the nodes and edges into account risking the possibility of considering erroneously false ones into the network or remove all the uncertain nodes and/or edges risking the issue of missing nodes and edges [7]. Furthermore, the unreliability of the tools used for the construction of the social network can lead to distortions [6]. On that point, we propose to incorporate uncertainty into the graph structure of social networks.

Generally, most existing studies consider weighted networks with integer values. Yet, one way for representing an uncertain network is to weight the edges with values in [0, 1] to depict the degrees of uncertainty regarding the links' existence [6]. In fact, several real world social networks are characterized by shifting degrees of uncertainty, more particularly the large scale ones [2]. For this reason, we embrace the theory of belief functions [4,11] as a general framework for reasoning under uncertainty. We use its assets for the handling of imprecision in data and the modeling of partial and total ignorance to quantify the degrees of uncertainty into the edges of the social network.

Furthermore, we develop a new approach for inferring new links in a social network characterized by uncertain edges based on group information and structural neighborhood measures of nodes. In fact, most of the existing methods are based on the local node neighborhood and global paths measures. Yet, these latter do not take into account a very important aspect in social networks which is its community structure. The participation of actors in social groups can bring important information concerning their characteristics and thus, may enhance the prediction task. To this end, we propose a method that performs exclusively with the belief function tools. The degrees of uncertainty of the similar nodes from the common shared groups are considered. They are revised, transferred and combined as independent sources of information and are afterward employed to get an outlook on the existence of a new link.

The rest of the paper is organized as follows: Sect. 2 gives a brief survey of the link prediction problem and the existing approaches. Section 3 provides some basic knowledge of the belief function theory notations and definitions. In Sect. 4, we present our evidential link-based graph model for a social network under an uncertain framework. Section 5 reveals the proposed approach for link prediction under the belief function framework. Section 6 illustrates the proposed method and Sect. 7 gives the experimental results. Finally, Sect. 8 concludes the paper.

2 Link Prediction

Due to its great applicability, link prediction constitutes a rich research area and has attracted many researchers from various fields. Namely, in social networks, link prediction is a basic task in social relationships formation. It can be applied to infer the new relations to be formed in the future, expose links which already exist but are not apparent, or even assist users to make new connections.

In most common formulation, the link prediction problem can be defined as follows [8]: given a current state of the social network graph in time t, the aim is to accurately infer the potential edges to be added to the unlinked pairs of nodes given a snapshot of the social network during the time interval $[t, t']$. It may also be considered as the problem of deriving the missing links of the network. In fact, one may construct a social network from a given observable data and try to derive the invisible links that are likely to exist. Most of the state-of-the-art link prediction methods have focused on two groups of network information that can be categorized into local (node neighborhood) and global (path) information. Local information-based approaches use the local similarities of the nodes characteristics in the network. These latter may be the essential attributes, i.e., gender, age, interests, or structural indices which are based solely on the network structure, i.e., common neighbors that two nodes share. Yet, nodes' attributes are generally not available or hidden, thus the majority of local approaches use metrics based on the structural similarities. The global approaches use the proximity of the nodes in the network, they employ metrics based on the ensemble of paths to determine the closest nodes in the network. The intuition is that the more close two nodes are in the network, the more they tend to be linked or to influence each other in the future.

The main advantage of these measures is that they are generic, they can be applied to networks from several fields. While the global methods perform better than the local ones, some path based metrics are time consuming as they inquire for the topological information of the whole network which is in many cases not available. Besides, a relevant aspect characterizing social networks is not considered which is the participation of the actors in social groups (clusters, communities). In fact, in several social networks, users are involved in many social groups at the same time. Thus, hybrid methods that use local and cluster information have been proposed [16,18,19]. That is, our proposed method is based on local and group similarity measures. Thus, we recall in this section some state-of-the-art structural measures based on local and group information.

2.1 Local Information Based Measures

Some of the base measures from the literature are "Common Neighbors" [10], "Jaccard's Coefficient" [5] and "Adamic/Adar" [1]. Let $\tau(v_i)$ denote the set of neighbors of the node v_i in the social network. The common neighbors measure denoted by $CN(v_i, v_j)$ characterizes the number of common neighbors between two nodes v_i and v_j. It is defined as:

$$CN(v_i, v_j) = |\tau(v_i) \cap \tau(v_j)| \tag{1}$$

On the other hand, the Jaccard's Coefficient considers all the neighbors of the pair (v_i, v_j). It is defined as follows:

$$JC(v_i, v_j) = \frac{|\tau(v_i) \cap \tau(v_j)|}{|\tau(v_i) \cup \tau(v_j)|} \tag{2}$$

The Adamic/Adar measure denoted by $AA(v_i, v_j)$ weights the contribution of each common neighbor v_k by the inverse of the logarithm of its degree, it is defined as:

$$AA(v_i, v_j) = \sum_{v_k \in (\tau(v_i) \cap \tau(v_j))} \frac{1}{log|\tau(v_k)|} \quad (3)$$

2.2 Group Information Based Measures

Structural similarity measures based on group information use both local structure of the nodes and group information, they include the Common Neighbors of Groups (CNG) and Common Neighbors Within and Outside of Common Groups (WOCG) [18,19]. Let $\Lambda^G_{v_i v_j}$ denote the set of common neighbors of the pair (v_i, v_j) belonging to the group G. The CNG depicts the size of the set of common neighbors of (v_i, v_j) that belong to at least one group G to which v_i or v_j is part of. It is defined as:

$$S^{CNG}_{v_i v_j} = |\Lambda^G_{v_i v_j}| \quad (4)$$

Let $\Lambda_{v_i v_j} = \Lambda^{WCG}_{v_i v_j} \cap \Lambda^{OCG}_{v_i v_j}$ be the set of common neighbors of (v_i, v_j) such that $\Lambda^{WCG}_{v_i v_j}$ is the set of common neighbors within common groups (WCG) and $\Lambda^{OCG}_{v_i v_j}$ is the set of common neighbors outside the common groups (OCG). The WOCG measure is defined as:

$$s^{WOCG}_{v_i v_j} = \frac{|\Lambda^{WCG}_{v_i v_j}|}{|\Lambda^{OCG}_{v_i v_j}|} \quad (5)$$

3 Belief Function Framework

The belief function theory [4,11] is a general framework for the representation and management of uncertain evidence. Let Θ be the frame of discernment, an exhaustive and finite set of mutually exclusive events associated to a given problem. 2^Θ is the power set of Θ, it includes all the possible subsets and formed unions of events, and the empty set which matches the conflict. A basic belief assignment (*bba*), denoted by m, is the mass assigned to an event given a piece of evidence. It is defined as:

$$m : 2^\Theta \rightarrow [0,1]$$
$$\sum_{A \subseteq \Theta} m(A) = 1 \quad (6)$$

A *bba* with at most one focal element A different from Θ is called a simple support function (ssf). It is defined as [13]:

$$\begin{cases} m(A) & = 1 - \omega, \forall A \subset \Theta \\ m(\Theta) & = \omega, \omega \in [0,1] \end{cases} \quad (7)$$

Beliefs can be fused using combination rules. In particular, the conjunctive rule of combination permits to combine evidence given by two reliable and distinct sources of information characterised by two *bba*'s m_1 and m_2. It is denoted by $\textcircled{\cap}$ and is defined by [14]:

$$m_1 \textcircled{\cap} m_2(A) = \sum_{B,C \subseteq \Theta : B \cap C = A} m_1(B) \cdot m_2(C) \tag{8}$$

While combining evidence on Θ, it is important to take into consideration the reliability of the evidence. For that, a so-called discounting mechanism can be performed [11]:

$$\begin{aligned} {}^{\alpha}m(A) &= (1-\alpha) \cdot m(A), \text{ for } A \subset \Theta \\ {}^{\alpha}m(\Theta) &= \alpha + (1-\alpha) \cdot m(\Theta) \end{aligned} \tag{9}$$

where $\alpha \in [0, 1]$ represents the discount rate (coefficient).

Let Θ and Ω be two disjoint frames of discernment. In order to establish the relation between them, one may use a multi-valued mapping [4]. In fact, a multi-valued mapping function denoted by τ, permits to bring together to different frames of discernment the subsets $B \subseteq \Omega$ that can possibly match under τ to a subset $A \subseteq \Theta$:

$$m_\tau(A) = \sum_{\tau(B)=A} m(B) \tag{10}$$

The Transferable Belief Model (TBM), proposed by Smets [15], is one of the well-known interpretations of the belief function theory. In the TBM, decision making is performed at the pignistic level where beliefs are transformed into probabilities using pignistic measures denoted by $BetP$ [12]:

$$BetP(A) = \sum_{B \subseteq \Theta} \frac{|A \cap B|}{|B|} \frac{m(B)}{(1 - m(\emptyset))}, \text{for all } A \in \Theta \tag{11}$$

4 Evidential Link-Based Social Network

Most social networks graphs include nodes and edges that are assumed to be certain 1 (exist) 0 ($\neg$ exist). The authors in [6] highlighted the importance of incorporating uncertainty when dealing with social networks and proposed to weight the strengths of the links by probabilities. In [3], the authors proposed a belief social network where the nodes, edges and messages are weighted by *bba*'s. The aim is to detect the nature of a message that flows through the network. Yet, the purpose of this work is to treat the uncertainty upon the links. Thus, we introduce our evidential link-based social network graph model where uncertainty is encrypted using the belief function theory. Each edge v_iv_j has assigned a *bba* defined on $\Theta^{v_iv_j} = \{E_{v_iv_j}, \neg E_{v_iv_j}\}$ denoted by $m^{v_iv_j}$, i.e., $E_{v_iv_j}$ means that v_iv_j exists and $\neg E_{v_iv_j}$ means it is absent. That is, an evidential link-based social network graph is defined as $G(V, E)$ where: $V = \{v_1, \ldots, v_{|V|}\}$ is the set of nodes, and E is the set of edges: An edge $v_iv_j \in E$ has assigned a *bba* $m^{v_iv_j}$ that depicts the degree of uncertainty regarding its existence.

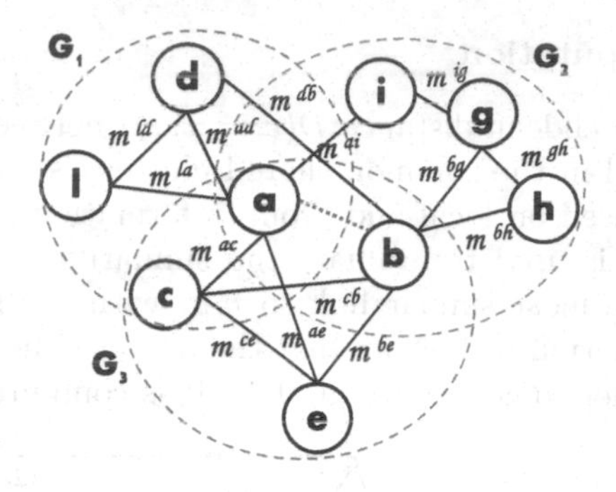

Fig. 1. A social network graph with *bba*'s weighted edges and group belonging nodes

Figure 1 illustrates an example of such a *bba*'s edge weighted graph structure. In fact, instead of weighting the links by either 1 or 0 to demonstrate whether or not a link is existent, we ascribe a *bba* with values in $[0, 1]$ to quantify the degree of uncertainty about the link existence. Note that a link v_iv_j is represented if the pignistic probability $BetP^{v_iv_j}(E_{v_iv_j}) > 0.5$ which means that its likelihood to exist is greater than 50 %.

5 Evidential Link Prediction Based on Group Information

Our proposed method for link prediction uses node neighborhood and group information given a snapshot of a graph. In fact, an earlier phase for the partitioning of the network into groups needs to be applied, most works apply algorithms for communities detection with low computational cost. Yet, this makes the prediction quality dependent to the community detection algorithm performance. The authors in [18,19] proposed to eliminate this dependency by using the natural information of groups, i.e., the information from groups of interests to which users participate to. Thus, each edge v_iv_j has assigned a feature vector that corresponds to the structural similarity measures based on local and group information as explained in Sect. 2. CN (Eq. 1), JC (Eq. 2), AA (Eq. 3), CNG (Eq. 4) and WOCG (Eq. 5) are employed as similarity measures since they are simple and have proved their efficiency in several social networks domains [10,18,19]. The feature vector is used to compute the similarity between the link to be predicted and its neighbors belonging to the shared groups. The intuition is that in many real world social networks, users with similar experiences or interests are more likely to share a relationship than those that do not share common characteristics. The most similar link is subsequently considered as a source of information. Our formulation of the link prediction problem is as follows:

Given a current state of the graph $G(V, E)$ at time t, predict the existence of an edge v_iv_j between two the unlinked nodes (v_i, v_j) at $t + 1$ by considering the relationships shared in their common groups. To this end, we propose a method for the prediction of the existence of a link between (v_i, v_j) based on the steps outlined below.

5.1 Distance Computation

At a first place, the Euclidean distance $D(v_iv_j, v_kv_l)$ between the link v_iv_j and each link v_kv_l included in the common shared clusters is computed. Structural similarity measures based on local and group information are used as features. That is, $D(v_iv_j, v_kv_l)$ is used to evaluate the similarity between v_iv_j and the neighboring links. The most similar link to v_iv_j with the smallest distance is considered in the prediction task. Note that the distance metric is divided by its maximum value in order to get values in [0, 1]. It is computed as follows:

$$D(v_iv_j, v_kv_l) = \frac{\sqrt{\sum_{s=1}^{n}(x_{v_iv_j}^s - y_{v_kv_l}^s)^2}}{D_{max}} \quad (12)$$

where s is the index of a structural similarity metric, $x_{v_iv_j}$ and $y_{v_kv_l}$ are respectively its values for v_iv_j and v_kv_l and D_{max} is the maximum value of the Euclidean distance.

5.2 Reliability Computation

In order to quantify the degree of reliability of the most similar link, a discounting operation (Eq. 9) is applied using the value given by the distance measure as a discount coefficient denoted by $\alpha = D(v_iv_j, v_kv_l)$. In fact, the more similar the two links are, the more reliable the similar link is, i.e., if the two links are totally similar $D(v_iv_j, v_kv_l) = 0$ then v_kv_l is considered as a totally reliable source of evidence i.e., $\alpha = 0$. Thus, $m^{v_kv_l}$ is discounted as follows:

$$\begin{cases} {}^{\alpha}m^{v_kv_l}(\{E_{v_kv_l}\}) = (1-\alpha)\cdot m^{v_kv_l}(\{E_{v_kv_l}\}) \\ {}^{\alpha}m^{v_kv_l}(\{\neg E_{v_kv_l}\}) = (1-\alpha)\cdot m^{v_kv_l}(\{\neg E_{v_kv_l}\}) \\ {}^{\alpha}m^{v_kv_l}(\Theta^{v_kv_l}) = \alpha + (1-\alpha)\cdot m^{v_kv_l}(\Theta^{v_kv_l}) \end{cases} \quad (13)$$

Note that when there is more than one most similar link, i.e., two links with smallest equal distances, the link with the highest mass on the event "exist" is chosen since the degree of certainty of its existence would be higher.

5.3 Information Transfer and Fusion

To transfer the discounted *bba* of the most similar link v_kv_l to the frame $\Theta^{v_iv_j}$, a multi-valued operation denoted by τ: $\Theta^{v_kv_l} \rightarrow 2^{\Theta^{v_iv_j}}$ is applied to bring together the elements as follows:

- The discounted mass ${}^{\alpha}m^{v_kv_l}(\{E_{v_kv_l}\})$ is transferred to $m_{v_kv_l}^{v_iv_j}(\{E_{v_iv_j}\})$;
- The discounted mass ${}^{\alpha}m^{v_kv_l}(\{\neg E_{v_kv_l}\})$ is transferred to $m_{v_kv_l}^{v_iv_j}(\{\neg E_{v_iv_j}\})$;
- The discounted mass ${}^{\alpha}m^{\Theta^{v_kv_l}}(\Theta^{v_kv_l})$ is transferred to $m_{v_kv_l}^{v_iv_j}(\Theta^{v_iv_j})$.

Where $\alpha = D(v_iv_j, v_kv_l)$ and $m_{v_kv_l}^{v_iv_j}$ denotes the *bba* of v_iv_j on $\Theta^{v_iv_j}$ given the most similar link, here v_kv_l.

Upon transferring ${}^{\alpha}m^{v_kv_l}$ to $2^{\Theta^{v_iv_j}}$, the *bba* of v_iv_j is updated given the new evidence obtained from the most similar link. To accomplish this, the initial *bba* $m^{v_iv_j}$ and $m^{v_iv_j}_{v_kv_l}$ are combined using the conjunctive rule of combination (Eq. 8). This step is essential, as it permits to fuse the information provided by the most similar link and treat it as an independent source of evidence.

5.4 Decision Making

At last, the pignistic probability $BetP^{v_iv_j}(E_{v_iv_j})$ is computed (Eq. 11) to make a decision about the existence of the link v_iv_j on the graph. As a matter of fact, when $BetP^{v_iv_j}(E_{v_iv_j}) > 0.5$ it means that the likelihood that a link exist between v_i and v_j at $t+1$ has probability $> 50\,\%$, it would be absent otherwise.

6 Illustration

To illustrate our link prediction approach, we try to predict the existence of a new link between the pair of nodes (a, b) presented in Fig. 1. To do so, the edge *ab* is assumed to be present in the graph in order to be able to compare its structural attributes and those of the other links belonging to the shared groups G_2 and G_3. Thus, the neighboring links in the common shared groups are: $ai, ac, ae, bg, bh, hg, be, bc, ce, gi$. We apply the steps presented in Sect. 5.

Step 1: At first, we compute the Euclidean distance between *ab* and each neighboring link included in the common groups G_1 and G_2 shared between a and b using Eq. 12. The results are reported in Table 1.

Table 1. Distance between *ab* and the links in the common shared groups of Fig. 1

Distance	*ac*	*ae*	*bc*	*be*	*bg*	*bh*	*ce*	*hg*	*ai*	*gi*
ab	0.282	0.282	0.283	0.283	0.551	0.599	0.357	0.638	1	1

Hence, the most similar links to *ab* are *ac* and *ae*. That is, we have to use one of them to update m^{ab}. Suppose we have *bba*'s allocated as follows:

$$\begin{cases} m^{ab}(\{E_{ab}\}) = 0.35 \\ m^{ab}(\{\neg E_{ab}\}) = 0.42 \\ m^{ab}(\Theta^{ab}) = 0.23 \end{cases}, \begin{cases} m^{ac}(\{E_{ac}\}) = 0.65 \\ m^{ac}(\{\neg E_{ac}\}) = 0.2 \\ m^{ac}(\Theta^{ac}) = 0.15 \end{cases} \text{and} \begin{cases} m^{ae}(\{E_{ae}\}) = 0.55 \\ m^{ae}(\{\neg E_{ae}\}) = 0.25 \\ m^{ae}(\Theta^{ae}) = 0.2 \end{cases}$$

Thus, *ac* is chosen as a source of information since $m^{ac}(\{E_{ac}\}) > m^{ae}(\{E_{ae}\})$.

Step 2: The next step is to discount the *bba* m^{ac} using $D(ab, ac)$ to quantify its degree of reliability. We denote $\alpha = D(ab, ac)$ the discount rate. Thus, ${}^{\alpha}m^{ac}$ after the discounting operation is: ${}^{\alpha}m^{ac}(\{E_{ac}\}) = (1 - 0.282) \cdot 0.65 = 0.4667$, ${}^{\alpha}m^{ac}(\{\neg E_{ac}\}) = (1 - 0.282) \cdot 0.2 = 0.1436$ and ${}^{\alpha}m^{ac}(\Theta^{ac}) = 0.282 + (1 - 0.282) \cdot 0.15 = 0.3897$.

Step 3: When the discounted mass of the most similar link is transferred using the τ function (Eq. 10), the mass of ab given ac is: $m_{ac}^{ab}(\{E_{ab}\}) = 0.4667$, $m_{ac}^{ab}(\{\neg E_{ab}\}) = 0.1436$ and $m_{ac}^{ab}(\Theta^{ab}) = 0.3897$. To update the *bba* of the link ab, m^{ab} and m_{ac}^{ab} are fused by applying the conjunctive rule of combination (Eq. 8). Thus, we get: $m^{ab} \textcircled{\cap} m_{ac}^{ab}(\{E_{ab}\}) = 0.407$, $m^{ab} \textcircled{\cap} m_{ac}^{ab}(\{\neg E_{ab}\}) = 0.257$, $m^{ab} \textcircled{\cap} m_{ac}^{ab}(\Theta^{ab}) = 0.09$ and $m^{ab} \textcircled{\cap} m_{ac}^{ab}(\emptyset) = 0.246$.

Step 4: Finally, the pignistic probability $BetP^{ab}$ (Eq. 11) is computed to make a decision on the link existence between the nodes a and b. Thus, $BetP(E_{ab}) = 0.575$ and $BetP(\neg E_{ab}) = 0.425$. Hence, there is 57 % chance that a link may exist between a and b. That is, a link would be schematized in the graph representation.

7 Experiments

In order to test our approach for link prediction, it is necessary to consider an uncertain social network. Yet, uncertain social network data are not available. Thus, we preprocessed a real world social network of 4K nodes and 88K edges of Facebook friendships obtained from [9] in order to transform it into an uncertain social network.

7.1 Network Pre-processing

To transform the social network into a belief-link based social network, we follow two major steps: (1) we generate three snapshots of the network from the data (2) then we simulate mass functions on the basis of the three first graphs to get a belief link-based version of the social network.

Graphs Generation. At first, we create three graphs from the data by removing randomly a portion of the edges. That is, we get three graphs that we call $G(t-2)$, $G(t-1)$ and $G(t)$. Indeed, this technique is widely used in the link prediction literature. In several works, a number of edges is pruned from the graph so that they will be considered in the prediction process [17,20].

Mass Functions Simulation. In order to generate the belief link-based version of the social network, we weight each link v_iv_j by a simulated *bba* regarding its links existence on the basis of $G(t-2)$, $G(t-1)$ and $G(t)$ as follows:

- If v_iv_j exists in the three graphs $G(t-2)$, $G(t-1)$ and $G(t)$ then a ssf $m^{v_iv_j}$ is assigned such that $m^{v_iv_j}(\{E_{v_iv_j}\}) \in [2/3, 1]$;
- If v_iv_j exists in $G(t-2)$ and $G(t)$ or $G(t-1)$ and $G(t)$ then a mass $m^{v_iv_j}$ is generated such that $m^{v_iv_j}(\{E_{v_iv_j}\}) \in [1/3, 2/3[$, $m^{v_iv_j}(\{\neg E_{v_iv_j}\}) \in]0, 1/3]$;
- If v_iv_j exists only in $G(t)$ then a mass function $m^{v_iv_j}$ is assigned such that $m^{v_iv_j}(\{E_{v_iv_j}\}) \in]0, 1/3]$, $m^{v_iv_j}(\{\neg E_{v_iv_j}\}) \in [1/3, 2/3]$;

Table 2. The prediction results measured by the precision

	G_1	G_2	G_3
Precision	0.54	0.56	0.57

- If v_iv_j does not exist in $G(t)$ and exists in $G(t-2)$ and $G(t-1)$ then a ssf $m^{v_iv_j}$ is assigned such that $m^{v_iv_j}(\{\neg E_{v_iv_j}\}) \in\ ]1/3, 2/3]$;
- If v_iv_j exists only in $G(t-2)$ or in $G(t-1)$ then a ssf $m^{v_iv_j}$ is assigned such that $m^{v_iv_j}(\{\neg E_{v_iv_j}\}) \in\ ]0, 1/3]$.

7.2 Results

To test our proposed link prediction method, we produce three different belief link-based versions of the social network that we call G_1,G_2 and G_3. We evaluate the accuracy of our link prediction algorithm using the precision measure. It expresses the number of correctly predicted existent links n_c with respect to the set of analyzed links n. It is defined as follows:

$$precision = \frac{n_c}{n} \tag{14}$$

Table 2 gives the obtained precision values for the three experiments. As illustrated, the prediction quality measured by the precision gives values higher than 50 % reaching a maximum performance of 57 % for G_3. Hence, validity and performance of the new approach is empirically confirmed.

8 Conclusion

In this paper, we have provided an uncertain graph-based model for social networks whose edges are valued with mass functions given by the belief function theory. Furthermore, we have proposed a novel link prediction approach that takes into consideration both uncertainties in data and group information in social networks. Our method is exclusively based on the belief function framework tools, evidence from the neighbors of the common groups is revised, transferred and combined to successfully predict new connections. As part of future work, extension to the case of both uncertain nodes and edges would be considered. Also, comparison with existing methods is left open for future work.

References

1. Adamic, L.A., Adar, E.: Friends and neighbors on the web. Soc. Netw. **25**(3), 211–230 (2003)
2. Adar, E., Ré, C.: Managing uncertainty in social networks. Data Eng. Bull. **30**(2), 23–31 (2007)

3. Ben Dhaou, S., Kharoune, M., Martin, A., Ben Yaghlane, B.: Belief Approach for Social Networks. In: Cuzzolin, F. (ed.) BELIEF 2014. LNCS, vol. 8764, pp. 115–123. Springer, Heidelberg (2014)
4. Dempster, A.P.: Upper and lower probabilities induced by a multivalued mapping. Ann. Math. Stat. **38**, 325–339 (1967)
5. Jaccard, P.: Étude comparative de la distribution florale dans une portion des Alpes et des Jura. Bull. Soc. Vaudoise Sci. Nat. **37**, 547–579 (1901)
6. Johansson, F., Svenson, P.: Constructing and analyzing uncertain social networks from unstructured textual data. In: Özyer, T., Erdem, Z., Rokne, J., Khoury, S. (eds.) Mining Social Networks and Security Informatics. Lecture Notes in Social Networks, pp. 41–61. Springer, Netherlands (2014)
7. Kossinets, G.: Effects of missing data in social networks. Soc. Netw. **28**, 247–268 (2003)
8. Liben-Nowell, D., Kleinberg, J.: The link-prediction problem for social networks. J. Am. Soc. Inf. Sci. Technol. **58**(7), 1019–1031 (2007)
9. McAuley, J.J., Leskovec, J.: Learning to discover social circles in ego networks. In: NIPS, pp. 548–556 (2012)
10. Newman, M.E.J.: Clustering and preferential attachment in growing networks. Phys. Rev. E **65**, 025102 (2001)
11. Shafer, G.: A Mathematical Theory of Evidence. Princeton University Press, Princeton (1976)
12. Smets, P.: The transferable belief model for quantified belief representation. In: Smets, P. (ed.) Handbook of Defeasible Reasoning and Uncertainty Management Systems., pp. 267–301. Springer, Netherlands (1988)
13. Smets, P.: The canonical decomposition of a weighted belief. In: Proceedings of the Fourteenth International Joint Conference on Artificial Intelligence, IJCAI 1995, vol. 14, pp. 1896–1901 (1995)
14. Smets, P.: Application of the transferable belief model to diagnostic problems. Int. J. Intell. Syst. **13**(2–3), 127–157 (1998)
15. Smets, P., Kennes, R.: The transferable belief model. Artif. Intell. **66**(2), 191–234 (1994)
16. Soundarajan, S., Hopcroft, J.: Using community information to improve the precision of link prediction methods. In: Proceedings of the 21st International Conference Companion on World Wide Web, pp. 607–608. ACM (2012)
17. Valverde-Rebaza, J., de Andrade Lopes, A.: Exploiting behaviors of communities of twitter users for link prediction. Soc. Netw. Anal. Min. **3**(4), 1063–1074 (2013)
18. Valverde-Rebaza, J.C., de Andrade Lopes, A.: Link prediction in complex networks based on cluster information. In: Barros, L.N., Finger, M., Pozo, A.T., Gimenénez-Lugo, G.A., Castilho, M. (eds.) SBIA 2012. LNCS, vol. 7589, pp. 92–101. Springer, Heidelberg (2012)
19. Valverde-Rebaza, J.C., de Andrade Lopes, A.: Link Prediction in Online Social Networks Using Group Information. In: Murgante, B., Misra, S., Rocha, A.M.A.C., Torre, C., Rocha, J.G., Falcão, M.I., Taniar, D., Apduhan, B.O., Gervasi, O. (eds.) ICCSA 2014, Part VI. LNCS, vol. 8584, pp. 31–45. Springer, Heidelberg (2014)
20. Zhang, Q.M., Lü, L., Wang, W.Q., Zhu, Y.X., Zhou, T.: Potential theory for directed networks. PLoS ONE **8**(2), e55437 (2013)

Survey of Social Commerce Research

Anuhya Vajapeyajula[1(✉)], Priya Radhakrishnan[2], and Vasudeva Varma[2]

[1] MIT, Cambridge, MA 02139, USA
anuhyav@mit.edu
[2] IIIT, Hyderabad, India
priya.r@research.iiit.ac.in, vv@iiit.ac.in

Abstract. Social commerce is a field that is growing rapidly with the rise of Web 2.0 technologies. This paper presents a review of existing research on this topic to ensure a comprehensive understanding of social commerce. First, we explore the evolution of social commerce from its marketing origins. Next, we examine various definitions of social commerce and the motivations behind it. We also investigate its advantages and disadvantages for both businesses and customers. Then, we explore two major tools for important for social commerce: Sentiment Analysis, and Social Network Analysis. By delving into well-known research papers in Information Retrieval and Complex Networks, we seek to present a survey of current research in multifarious aspects of social commerce to the scientific research community.

1 Introduction

$14 billion dollars, or 5 % of all online revenue, are expected to come from social commerce in[1] 2015. Social commerce, a rapidly growing branch of commerce, originates from social media marketing and hence marketing.

Marketing is an integral part of commerce that has evolved greatly throughout the years. Before the 1400's, marketing was mainly word-of-mouth. Mid-1400's, print advertising become popular. Advertisements were published in newspapers, magazines, billboards, and posters. However, with the invention of the telephone and the radio in late 1800's, marketing soon shifted to these devices as it's main form of communication. Radio advertisements and telemarketing were the most popular methods of marketing in addition to print advertising. Soon, television became popular and marketing spread to this platform too. By the early 2000's, the computer started prevailing and marketing online became highly profitable. Spam emails and online ads spread. Then came Web 2.0 where user generated content started increasing as blogs and other forms of social media became popular. Soon, marketers started realizing consumers trusted other consumers' inputs more than they trusted advertising campaigns run by marketers. Thus, social media marketing began, eventually giving rise to social commerce.

A. Vajapeyajula—worked done during internship at IIIT-Hyderabad.

[1] http://blog.hubspot.com/marketing/social-ecommerce-revenue-infographic

R. Prasath et al. (Eds.): MIKE 2015, LNAI 9468, pp. 493–503, 2015.
DOI: 10.1007/978-3-319-26832-3_46

In this paper, we try to present a survey of social media commerce where we do not assume the reader has any previous background in social commerce. We seek to clarify what social commerce is and what are the motivations behind it before analyzing the technology driving it.

2 Social Commerce

2.1 What is Social Commerce?

Origins Yahoo is credited with coining the term social commerce in 2005. It introduced the term as it released its "Shoposphere"[2] which allowed users to add items that they wanted to buy to pick lists. Other users were then able to comment on and rate the pick lists. Users could also share their lists with family and friends. From then, social commerce has been researched by many and defined in multiple ways.

Definition. Researchers haven't agreed upon one standard definition of social commerce but social media and commercial activities are at the core of definition [12]. Social commerce involves activities and transactions via social media environment. It supports the use of social interactions and user content contributions [11]. Simply put, social commerce is where the sellers are individuals, not firms [18]. Firms don't market directly to customers but rather take advantage of customers' willingness to share their experiences online to market their products. Social commerce connects to sellers as well as users [5]. [17] defines social commerce as a type of e-commerce that uses online media that supports social interaction to support online buying and selling of products and services. This aspect of social media is what differentiates social commerce from e-commerce.

Domain of S-commerce. Social commerce is considered by most researchers to be a subset of e-commerce which includes a social component [3,12,17,20,23]. Furthermore, while e-commerce is popular among males, social commerce is more female-oriented [3]. [23] expand the domain of social commerce to include not just transactions but exchange-related activities that occur before, during, and after a focal transaction. Both consumer-side and firm-side activities fall under the domain of social commerce. Firm-related activities include participating in public social networks by advertising, market researching, creating brand and product awareness campaigns, etc. Firms have also been creating their own social networks such as Oracle's Connect and IBM's Beehive [20]. Consumer-side activities include sharing and liking a business' post, participating in social contests, making purchases using s-commerce, etc.

2.2 What Drives Social Commerce?

Social Support. The rise of social commerce depends primarily on social support offered in a Computer Mediated Social Environment(CMSE). Social support is

[2] http://www.ysearchblog.com/2005/11/14/social-commerce-via-the-shoposphere-pick-lists/

defined by [12] as an individual's experiences of being cared for and helped by people in the social group. The greater the social support on a social network, the more likely the user is to participate in social media commerce. [12] found that social support plays a greater role in continuance intention of social commerce than the website quality of the social media site. They concluded that frequent sharing of helpful and supportive information can strengthen relationship quality between users. Closely related to social support is subjective norm: an individual's perception of whether or not people important to them think a specific behavior should be performed [17]. Subjective Norm also positively affects a user's social commerce intention.

Social Commerce Strategies. There are a variety of social commerce strategies in place. Firstly, companies use social media, email, web sites, print, mobile sites/applications, SMS text message among other vehicles for marketing their services to customers [8]. In a survey of 500 companies, it was found that some of the most effective incentives for increasing social media interaction are discount codes for "Liking" or "Following" a brand page, In-store discounts, and social contests. Social contests are successful because they involve the public, which helps humanize a brand [14]. Companies also reported using daily deals, ratings, reviews, and product recommendations, wish lists, curation, and user-generated photos to further encourage social commerce [8]. Warby Parker, a business that sells glasses, allows home try-ons of their products where customers can order several pairs of glasses, try them on, and post them to the Warby Parker Facebook page asking other users on social media to help them decide which pair to choose [15]. This strategy is very common where users are encouraged to upload pictures of them using a product because images are retweeted much more than just text posts, and users are more likely to purchase a product if they see their friends using it [6]. Viral marketing is another popular strategy employed where even if the topic of the advertisement, most often a video, isn't related to the product or service directly, a logo is added at the end which helps to spread the name of business and garner some attention from the media [23]. Viral marketing is effective when influential users engage in a campaign and pass along information to their followers or friends. Social media marketing is more trusted by users because it propagates from user to user [6].

2.3 Advantages and Disadvantages of Social Commerce

Advantages of Social Commerce. There are certain advantages to using social commerce. The primary advantage is users find it pleasurable [5]. A secondary advantage is that users of s-commerce do not need to physically go anywhere. Everything can be bought online and even shopping with friends can be done online. Consumers can consider their friends' inputs and opinions before purchasing a product or service. Another positive factor is the discounts and deals offered when using social commerce [5].[3] Groupon, a popular e-commerce marketplace, employs this strategy to sell products where if a certain number of people buy

[3] https://www.groupon.com/

a coupon, they receive a "groupon" with a greater value than the amount they spent on it. For example, a customer can buy a restaurant coupon for $15 but spend up to $30 on food at the restaurant. Other benefits of using social commerce include faster vendors' responses to complaints, customers assisting other customers, and engaging directly with vendors among others [12].

Social Advantages. However, discounts are not always the first and foremost factor for using group social shopping sites such as Groupon, LivingSocial, Plum District and Half Off Depot. [9] found that group shopping on these sites was mainly focused around social activities such as event planning, building relationships, and identity construction. Furthermore, social media is also used for soliciting advice before buying a product as observed in the Facebook page of Warby Parker [15]. [12] categorize social media marketing into several activities such as social shopping, ratings and reviews, recommendation and referrals, forums and communities, and social advertising campaigns.

Vendors' Benefits. There are also benefits to vendors such as saving money on customer service, testing new products/ideas easily, learning about customers, easily comparing to competitors, users marketing, improved service/product design, etc. [12]. Using a group shopping site (e.g. Groupon) can be expensive for businesses which face a loss when they offer more expensive services at cheap prices. However, the exposure the businesses gain when many customers sign up for a "groupon" helps them spread their customer base and get more recognition. Similary, social media marketing really proves to be successful with the buzz mechanism when users pass information to other users making a campaign go viral [2].

Disadvantages. However, there are a few disadvantages to businesses and customers. Companies may not be able to justify spending on s-commerce and abandon it all together like Wal-Mart because a lot of the benefits are intangible [20]. Invasion of privacy is a major issue where social networks such as Facebook will sell members' information to advertisers. For example, Facebook's attempt at tracking users' web history resulted in a class-action lawsuit [22]. Insufficient security is another issue that harms both businesses and customers. Fraud, a problem in e-commerce, also affects s-commerce where fake accounts are used to dupe people and steal identities. Companies may also face problems over violating intellectual property laws [20]. Lastly, as in e-commerce, keeping reviews and recommendations honest is another problem in s-commerce [10].

3 Tools for Social Commerce

With the rise of social commerce, there is a need for research in tools for understanding social media such as sentiment analysis and social network analysis. In this section of the paper, we explore important technologies that are crucial to using social media effectively.

3.1 Sentiment Analysis

Opinion mining and sentiment analysis are used interchangeably to describe the process of extracting posts from Web 2.0 and automatically analyzing them to determine the sentiment of users towards a certain product, services, and/or companies. Opinion mining is a broad term which refers to extracting data from Web 2.0 and analyzing it for various applications such as identifying trends. Sentiment analysis refers to taking the same data but analyzing it further for emotions on a certain topic.

The first step to any opinion mining system is data extraction. Twitter offers a public API where tweets pertaining to a specific search query can be retrieved. The second step is data pre-processing and normalization. Often times, users' posts on social media contain non standard language such as slang words, grammatical errors, and spelling errors. Most researchers clean up the data by eliminating repetitive words, excessive punctuation, correcting spelling errors, expanding abbreviations, and fixing other grammatical errors [21].

After preprocessing, data is then analyzed using a variety of techniques. In text mining, a document (such as a blog) is often split into smaller segments using either classification or clustering techniques. For microblogging, clustering isn't necessary and a part of speech tagger, the next step, can be immediately implemented after preprocessing. This linguistics approach is useful for creating a matrix for keywords and features. Adverbs and adjectives are most indicative of sentiment. There are many exisitng sentiment lexicons for these words. SentiWordNet[4] is popular resource for identifying the polarity of an adjective. However, negations can reverse the sentiment of text. Thus, negations ('not' good), and intensifiers ('*very*' good) are also tracked and sentiment scores are adjusted accordingly. Since sentiment of words depends on the context they are used in, topics of the data must also be considered [13]. For example, "unpredictable" may be negative when used in the context of automobiles such as "unpredictable steering," yet the same term can be positive when used in context of movie reviews to describe an "unpredictable plot" [21]. Most researchers have used domain-specific sentiment lexicons to simply analyze if the sentiment is positive or negative but some researchers have also classified the sentiment in various bipolar or unipolar emotions.

The final step is presenting the analyzed data. For trend analysis, polarity isnt a required measurement and instead, after clustering/classification, various statistics are used on the topics to figure out the most trending topics on a social media site. A tree map which indicates volume and sentiment (if analyzed) is presented with the data. Another visual tool is a word cloud in which the largest term indicates the highest popularity.

The results of opinion mining and sentiment analysis are used in a variety of applications in social commerce. Companies can "listen in" on social media to determine people's opinions on their products and adjust their product assortment accordingly. For example, Converseon[5] is a company that offers data consisting of

[4] http://sentiwordnet.isti.cnr.it/

[5] http://converseon.com/

comments around the web on their client's brands and services so that the client can understand public sentiment on various aspects of their products and services [16]. For example, in 2003, Kraft monitored public sentiment over trans fat and decided to cut down trans fat from their products [19]. Perhaps an even more lucrative application of social media analysis is using it to predict stock price movements [20,22]. Finally, trend analysis is often used in the design process and for tailoring marketing strategies. Marketers can design their advertisements based on opinions extracted from social media in order to stay updated and connect to customers better.

3.2 Social Networks Model

Another useful analysis for social commerce is social network analysis. In order to analyze networks, [18] first defined four characteristics of social commerce marketplaces: (1) Sellers are individuals instead of firms. (2) Sellers create product assortments as personalized online shops. (3) Sellers can create links between personalized shops. (4) Sellers' incentives are based on making commissions from sales by their shops. In social commerce, sellers often don't own merchandise but simply manage it. Uber[6] is an example of this model where the company doesn't own any taxis but offers a taxi service to customers. A network structure that is most successful for social commerce in one that is expansive and has many hyperlinks. This ensures few dead ends so that customers can easily find appealing shops and products before leaving the marketplace. Their research further showed that shops with more links going into them and less links going out of them generate more revenue. However, if links going out of a shop are reciprocated by the shops they are going to, then the outgoing links aren't necessarily harmful to the shop. Yet incoming links from shops that are highly interconnected don't generate much traffic into the shop. Overall, the presence of network adds economic value to online marketplaces [18].

A social network is treated as a graph where users are the nodes and the relations between them are the edges. Also called complex networks, in this analysis the document is first preprocessed, clustered, and classified. Preprocessing consists of removing duplicate nodes (a user has multiple profiles), inactive nodes (user has an account but stopped using it), and artificial nodes (spam accounts with malicious intentions) [1].

The next step is to create mapping rules and transform the data into a graph. A node can be a hub or an authority. Hubs are users who follow others whereas authorities are users who are followed. While nodes are almost always users, edges can be classified by many different types of relations. Researchers note that while explicit relations are most often used, implicit relations can be helpful in predictive modeling. Explicit relations are those in which users knowingly interact with another user by adding them as a friend or following them on social media sites, messaging them, posting on their wall or tweeting at them, etc. Implicit relations is still a new area of research but consists of tracking shared interests and activities of those users who dont have an explicitly defined

[6] https://www.uber.com/about

relation [1,2,20,24]. For example, two users on Facebook may like a lot of the same pages or belong to the same groups but not be friends. Implicit relations are of interest because social friendships often form between similar people with shared interests and can be used in predictive modeling.

After establishing the mapping rules, data is analyzed in a variety of ways. Component analysis usually labels components as weakly connected or strongly connected. Weakly connected components are important because they are important for understanding the network structure on massive graphs. Strongly connected components are useful in identifying if there is a strong core to the network where users are well connected and interact frequently. Various other measurements and statistics aid in characterizing a core. Network size, network density, average degree, average path length, diameter, modularity, and average clustering coefficient are frequently used in analysis. For example, a large network is often thought to lack trust and strong relations among users. Another example is viral marketing which is most effective when the core is penetrated along with the networks influential users. Influential users are often found using algorithms such as PageRank. The more mentions a user has, the more influential he or she is assumed to be.

As mentioned previously, viral marketing and other social marketing strategies greatly utilize social network analysis. It can also be used to avoid churn, loss of customers. Social network analysis aids in determining if there is an unexpected amount of churn so that businesses can rethink their advertising campaigns and execute customer management measures [1]. It can also be used in predictive modeling where businesses can predict which other users will leave based on their relations to the lost customer. If an influential user leaves, businesses can assume several other users will also leave. Networks can also be used to route customer services differently through social media. [1] suggest assigning trust and reputation scores to users where businesses can easily differentiate spam comments from comments that require immediate action based the customers previous actions. Social search is another useful application where search results could be based on what other members of a users network search for.

Network analysis of Flickr, LiveJournal, YouTube, and Orkut found that all the networks have a large, densely connected core. The path lengths from node to node are also fairly short (2–6 hops). This information suggests that if marketers or business want to reach a wide audience quickly, they have to affect users in the core of the network. Also, viral campaigns will be relatively easy to achieve since path lengths are short. Thus, users can easily pass along information through their networks. For this reason, companies have also been creating their own internal social networks to generate ideas and receive feedback from their employees [20]. Analysis of the world's largest e-commerce marketplace in China, Taobao, by [7] showed that there is an increasing positive relationship between trade volume and message volume. Taobao has an integrated messaging system where buyers can message sellers and other members on the site. Furthermore, the number of messages sent increased logarithmically as price increased. Similarly, a direct relationship exists between the seller's rating and the prices at which products are sold [7].

4 Suggestions for Future Research

While social commerce is becoming more prominent, more research is required in various aspects of it to fully understand the field and ensure the success of a company's social commerce strategy. One area that many businesses haven't considered much is using input from customers to design and improve products. Companies have used social networks for this purpose internally with their employees but not frequently with the common public. Companies could initiate an open design concept where they could initiate polls and surveys on features they would like to include in a product before designing the product. This way both users and businesses benefit where users can specify what they would like to see in a product and businesses can receive help and suggestions in the design processes.

Another area of lesser research is social media marketing. Combined with social network analysis, businesses could learn to identify and advertise to various cluster centers for easy propagation of information among a social network. Since people trust other users more than marketers for input on products and services, strategies could be revised to incorporate comments from social media sites into the advertising [20]. Customer service could also be greatly improved through the use of social media. Since other users can see complaints on a business' Facebook and how the business responds, more research is recommended on how network analysis can used to make the whole process more efficient. Perhaps a social media page can be customized to user based on the activities of other members of the user's network. Product recommendations could be greatly improved with the use of social media analysis. An interesting application of trend analysis is businesses creating deals on sites such as Groupon based on the products or topics that are trending. This could allow businesses to profit by offering deals in trending products or topics.

Furthermore, research on what makes a social network suitable of social commerce is an interesting topic to explore. Facebook accounts for a majority of the purchases on social media but research on why isn't extensive. Twitter was found to be easier to use by businesses to implement social commerce strategies but Facebook drives more purchases. Thus, there exists a discrepancy between businessess' social media platform preferences and the effectiveness of them [4]. A social commerce strategy successful on one social media platform may not be effective on another platform. Hence, research on tailoring strategies to different platforms is recommended to ensure the success of social commerce implementation. Privacy concerns must also be addressed since network analysis often reveals a lot of information about a user.

In the field of sentiment analysis, irony and emoticons are some aspects that have not been fully explored yet. Sarcastic posts are often eliminated in data analysis because of the difficulty in detecting sarcasm. Emoticons and excessive punctuation are also removed and cleaned up but these features can provide valuable data to sentiment analysis. Especially, when users post comments that don't express sentiment in the text but express it through an emoticon or punctuation. For example, a user could post the phrase "iPhone6 :D" or "You are in Paris?!" and no sentiment would be derived through traditional sentiment analysis but analysis of the emoticon would reveal happiness and analysis of the punctuation could suggest wonder and excitement. Though emoticons are gaining momentum as research topic of interest, punctuation has still not been explored as a sentiment-expressing feature. More research in these topics along with improving sentiment lexicons and fully automating the analyzing process would make social commerce an accessible tool for more companies and businesses. See [20] for more suggestions in social commerce research.

5 Conclusion

In this paper, we presented a review of various aspects of social commerce from the technology perspective and suggested areas of future research. We presented a review of well-known papers in Sentiment Analysis, and Complex Networks, along with Social Media Commerce in order to enable researchers to obtain an adequate picture of social commerce in order to contribute to the field. However, we did not fully explore the advertising perspective and investigate customer psychology. A vast field of research exists on improving marketing strategies, and ensuring the success of social media commerce from a business' perspective.

Social media is a great medium for businesses to market and sell their products since many people spend a significant amount of their time on various social media sites. A third of adult Internet users under 30 are getting their information from what their friends post on social media sites instead of business homepages[7]. With this in mind, we sought to establish a basic understanding of social commerce that explores both the technology and intentions behind it. Social media is used for a variety of reasons and social commerce strategies can be greatly improved with this understanding.

References

1. Bonchi, F., et al.: Social network analysis and mining for business applications. ACM Trans. Intell. Syst. Technol. **2**(3), 22:1–22:37 (2011). doi:10.1145/1961189.1961194. http://doi.acm.org/10.1145/1961189.1961194. ISSN: 2157–6904
2. eca@cs.stir.ac.uk. Erik, C., et al.: Sentic Computing for social media marketing. Multimed. Tools Appl. **59**(2), 557–577 (2012). http://libproxy.mit.edu/login?url=http://search.ebscohost.com/login.aspx?direct=true&AuthType=cookie,sso,ip,uid&db=aci&AN=75163262&site=eds-live. ISSN: 13807501
3. Chingning, W., Ping, Z.: The evolution of social commerce: the people, management, technology, and information dimensions. Commun. Assoc. Inf. Syst. 31, 105–127 (2012). ISSN: 15293181. http://libproxy.mit.edu/login?&http://search.ebscohost.com/login.aspx?direct=true&AuthType=cookie,sso,ip,uid&db=bth&AN=86652306&site=eds-live
4. Collier, M.: Social Media Commerce For Dummies. City: For Dum- mies, (2012). ISBN: 9781283803984

[7] http://www.gigya.com/blog/10-stats-driving-the-future-of-social-commerce/

5. eunhee0103@empas.com. Kim, E.H. and ysyeob@hanmail.net Yu, S.Y.: In uential Factors on Consumers' Purchase via Social Commerce: Use Motives, Benefits and Cost. Adv. Inf. Sci. Serv. Sci. 5(15), 2013, pp. 170–178. ISSN: 19763700. http://libproxy.mit.edu/login?url=http://search.ebscohost.com/login.aspx?direct=true&AuthType=cookie,sso,ip,uid&db=aci&AN=97949515&site=eds-live
6. Alabhya, F., et al.: Analyzing trends in social media marketing. Int. J. Comput. Sci. Manag. Stud. 14(11), 1–5 (2014). ISSN: 22315268. http://libproxy.mit.edu/login?=http://search.ebscohost.com/login.aspx?direct=true&AuthType=cookie,sso,ip,uid&db=aci&AN=100673754&site=eds-live
7. Guo, S., Wang, M., Leskovec, J.: The role of social networks in online shopping: information passing, price of trust, and consumer choice. In: Proceedings of the 12th ACM Conference on Electronic Commerce. EC 2011. San Jose, California, USA: ACM, 2011, pp. 157–166. ISBN: 978-1-4503-0261-6.10.1145/1993574.1993598. http://doi.acm.org/10.1145/1993574.1993598
8. Hauss, D.: Strides in social commerce. In: Retail Touchpoints, April 2014
9. Hillman, S., et al.: Shared joy is double joy: the social practices of user networks within group shopping sites. In: Proceedings of the SIGCHI Conference on Human Factors in Computing Systems. CHI 2013. Paris, France: ACM, 2013, pp. 2417–2426. ISBN: 978-1-4503-1899-0.10.1145/2470654.2481335. http://doi.acm.org/10.1145/2470654.2481335
10. Kugler, L.: Keeping online reviews honest. Commun. ACM57. 11, 20–23 (2014). ISSN 0001-0782.10.1145/2667111. URL: http://doi.acm.org/10.1145/2667111
11. Liang, T.-P., Turban, E.: Introduction to the special issue social commerce: a research framework for social commerce. Int. J. Electron. Commerce 16.2, 5–14, January 2011.10.2753/JEC1086-4415160201. URL: http://dx.doi.org/10.2753/JEC1086-4415160201. ISSN 1086–4415
12. Liang, T.-P., et al.: What drives social commerce: the role of social support and relationship quality. Int. J. Electron. Commerce 16.2, 69–90 (2011). ISSN 1086–4415.10.2753/JEC1086-4415160204. URL: http://dx.doi.org/10.2753/JEC1086-4415160204
13. Lin, C., He, Y.: Joint sentiment/topic model for sentiment analysis. In: Proceedings of the 18th ACM Conference on Information and Knowledge Management. CIKM 2009. Hong Kong, China: ACM, 2009, pp. 375–384. ISBN: 978-1-60558-512-3.10.1145/1645953.1646003. URL: http://doi.acm.org/10.1145/1645953.1646003
14. Mancuso, J., Stuth, K.: Social media refresh. Mark. Insights 26(4) 1–5, 2014. ISSN: 10408460. URL: http://libproxy.mit.edu/login?url=http://search.ebscohost.com/login.aspx?direct=true&AuthType=cookie,sso,ip,uid&db=bth&AN=102625545&site=eds-live
15. Said, K., et al.: Framing the conversation: the role of facebook conversations in shopping for eyeglasses. In: Proceedings of the 17th ACM Conference on Computer Supported Cooperative Work & #38; Social Com- puting. CSCW 2014. Baltimore, Maryland, USA: ACM, pp. 652–661 (2014). ISBN: 978-1-4503-2540-0.10.1145/2531602.2531683. URL: http://doi.acm.org/10.1145/2531602.2531683
16. Schweidel, D.A., Moe, W.W.: Listening in on social media: a joint model of sentiment and venue format choice. J. Mark. Res. (JMR) 51(4), 387–402 (2014), ISSN: 00222437. URL: http://libproxy.mit.edu/login?url=http://search.ebscohost.com/login.aspx?direct=true&AuthType=cookie,sso,ip,uid&db=bth&AN=98572966&site=eds-live

17. dshin@skku.edu Dong-Hee, S.: User experience in social commerce: in friends we trust. Behav. Inf. Technol. 32(1), 52–67 (2013). ISSN: 0144929X. URL: http://libproxy.mit.edu/login?= http://search.ebscohost.com/login.aspx?direct=true&AuthType=cookie,sso,ip,uid&db=aci&AN=85041467&site=eds-live
18. Stephen, A.T., Toubia, O.: Deriving Value from Social Commerce Networks. J. Mark. Res. (JMR) 47(2), 215–228 (2010). ISSN: 00222437. URL: http://libproxy.mit.edu/login?
19. Terdiman, D.: Why companies monitor blogs. In: CNET (2006). URL: http://www.cnet.com/news/why-companies-monitor-blogs/
20. Turban, E., Bolloju, N., Liang, T.-P.: Social commerce: an e-commerce perspective. In: Proceedings of the 12th International Conference on Electronic Commerce: Roadmap for the Future of Electronic Business. ICEC 2010, pp. 33–42. ACM, Honolulu, Hawaii, USA (2010). ISBN: 978-1-4503-1427-5.10.1145/2389376.2389382. URL: http://doi.acm.org/10.1145/2389376.2389382
21. Turney, P.D.: Thumbs up or thumbs down?: semantic orientation applied to unsupervised classification of reviews. In: Proceedings of the 40th Annual Meeting on Association for Computational Linguistics. ACL 2002. Philadelphia, Pennsylvania: Association for Computational Linguistics, pp. 417–424 (2002).10.3115/1073083.1073153. URL: http://dx.doi.org/10.3115/1073083.1073153
22. Vascellaro, J.E.: Facebook settles class-action suit over beacon service. Wall Street J. (2009). URL: http://www.wsj.com/articles/SB125332446004624573
23. Yadav, M.S., et al.: Social commerce: a contingencyframework for assessing marketing potential. J. OfInteractive Mark. 27.Soc. Media Mark., 311–323. (2013). ISSN: 1094–9968. URL: http://libproxy.mit.edu/login?url=http://search.ebscohost.com/login.aspx?direct=true&AuthType=cookie,sso,ip,uid&db=edselp&AN=S1094996813000364&site=eds-live
24. Yang, C.C., et al.: Identifying implicit relationships between social media users to support social commerce. In: Proceedings of the 14th Annual International Conference on Electronic Commerce. ICEC 2012. Singapore, Singapore: ACM, 2012, pp. 41–47. ISBN: 978-1-4503-1197-7.10.1145/2346536.2346544. URL: http://doi.acm.org/10.1145/2346536.2346544

Refine Social Relations and Differentiate the Same Friends' Influence in Recommender System

Haitao Zhai(✉) and Jing Li

University of Science and Technology of China, Hefei, China
haitaoz@mail.ustc.edu.cn, lj@ustc.edu.cn

Abstract. Social relations has been widely used in recommender system to improve accuracy of recommendations. Most works consider influence from overall friends simultaneously when recommending, and to each item the same friend always has equal influence. However, existing models fail to be consistent with real life recommendations, because in real life only a part of friends can affect our decisions, and we couldn't be influenced by the same friends on everything. So in this paper, we use machine learning way to infer truly influential friends in a mixed friends circle. And to different items we use relevance to differentiate the same friend's influence. A model, Topic-based Friends Refining Probabilistic Matrix Factorization (TFR-PMF), is proposed to check the performance of our theory. Through experiments on public data set, we domenstrate that our method can increase the accuracy of recommendation by 6.5 %, comparing with models that do not filter unrelated friends' influence.

Keywords: Refine friends · Social rating network · PMF · Recommender system

1 Introduction

As an effect tool to solve problems of information overload, recommender system (RS) has been popularly used in E-commerce websites such as Amazon, Epinions and Douban. However, the problem of cold start (new user or item in RS with little historical behavior) and sparsity of data set (low percent of rated user-item pairs over the whole user-item pairs of RS) have deeply limited the performance of RS [1].

Luckily, with raid development of social network, the utilization of social links among users [2], as an extra data source, brings opportunities of improving accuracy of RS. Recently, some social-relation based models, such as Bayesian inference-base model [3], Social Regularization [4], Weighted Trust [5] and PWS [6], have been proposed. These models consider both users' history behaviors and social relations when making recommendations, and they have a common assumption that a user's taste is similar to his friends and influenced by them.

R. Prasath et al. (Eds.): MIKE 2015, LNAI 9468, pp. 504–514, 2015.
DOI: 10.1007/978-3-319-26832-3_47

However, the accuracy of recommendations is still not high enough. We think reason lies in that existing models always take overall friends into consideration when calculating influence from social network. While in real life we may follow different friends' advices in different tastes. For example, Alice, Bob and John are friends, Alice likes A, B and C, Bob likes A and B, John likes C, when recommending things similar to C to Alice, we may follow John's advice in real life while existing models will accept Bob's. It's irrational that recommender systems are inconsistent with real life recommendations.

In this paper, in order to increase the accuracy of recommendations, we refine truly influential friends on different tastes by machine learning way first. And then, in each taste we use SocialMF [4] to calculate influence from those truly influential friends and get predicted ratings. At last, to distinguish same friends' influence on different items, we weight predicted ratings to get the final recommendations. As a friend may decide user's choice on one thing, but may have no effect on another, it's hard to find correct friends who really dominate users' decisions. For this, we try to use reasons why users build social links to infer the truly influential friends by PLSA model.

Main contributions of this paper include:

1. We find some inconsistent traits of existing recommender systems, in which RS considers too much effect from unrelated friends when considering influence from social relations.
2. We propose TFR-PMF model for personalized recommendations, which can solve the problem of finding truly influential friends on different tastes. And it can be more consistent with real life recommendations.
3. We do experiments on public data set with TFR-PMF model to check our thought, and the results show that our method can improve the accuracy of recommendation by 6.5 %, comparing with models that do not filter unrelated friends' influence.

The remainder of this paper is organized as follows: In Sect. 2, we review some related work about recommender approaches. In Sect. 3, we will describe how to infer truly influential friends and our model in details. Experiments are showed in Sect. 4, and then we make a conclusion and introduce some future work in Sect. 5.

2 Related Work

Collaborative Filtering is one of the most popular recommendation methods, which focus on user-item rating matrix [7]. It can be divided into two approaches: memory-based algorithms and model-based. Memory-based approaches try to use different similarity strategies to search similar users or items to get predictions [8,9]. While the model-based approaches [10,11] employ statistical and machine learning techniques to train models from the known data.

Among the model-based approaches, PMF [2] presumes that ratings of users and items are related to a small number of factors respectively, and this model factorizes the original rating matrix into two low-rank latent factors to get the final predictions. Many evidences show that PMF model outperforms other CF approaches. However, assuming that all the users are independent and identically distributed, the PMF and traditional CF approaches ignore the connection among users, and are inconsistent with the real word recommendations that people are affected by each other.

Recently, several approaches considering social relations among users have been proposed [5,6,12]. All these methods have a common assumption that users are likely to have similar taste with their friends in social network. Previous works show that social relations are benefit to improve the accuracy of recommendations. However, these works consider influence from overall friends simultaneously and ignore the fact that to different tastes we may trust different friends. In [13] Yang et al. presents circle-based recommender system to find influential friends related to domains, but it's only fit for categories that have been obviously divided, and its method is built on simple map relations, which could not reflect the internal nature of why those friends can affect the users. TopRec [14] tries to use experts-guided ways to filter out items and users, but it focuses on top-N recommendations, and its result is human-incomprehensible and computing costly.

3 Influential Friends Refining and TFR-PMF Model

3.1 Problem Definition

Notations used in this paper are shown in Table 1. We use $\mathrm{U} = \{\mathrm{u}_1, \mathrm{u}_2, ..., \mathrm{u}_M\}$ as the set of users and $\mathrm{V} = \{\mathrm{v}_1, \mathrm{v}_2, ..., \mathrm{v}_N\}$ as the set of items, the user-item rating matrix is $\mathbf{R}_{M \times N}$. Items in RS are things that users are going to buy or watch, like movies in a movie site or goods in a shopping site. If u_i gives a rating to v_j, R_{ij} means the rating score, while R_{ij} is 0 if u_i doesn't give a rating to v_j. $\mathbf{S}_{M \times M}$ implies users' social relations where S_{ij}=1 if u_i and u_j are friends while 0 otherwise. To items, every item has some tags, for example, a item with tags like "commodity41 :: women's clothing Jewelry" means that it's a commodity about women's clothing. In this paper, we use $\mathrm{T} = \{\mathrm{t}_1, \mathrm{t}_2, ...\mathrm{t}_L\}$ to donate the set of tags.

Definition 1: Topic. A topic in our paper is a subset of tags which can present their meaning. Suppose the number of topics is K, then a topic Z_k is composed of tag subset T_k, where k=1, 2, ..., K. So we have $\mathrm{L}_k = \mid \mathrm{T}_k \mid$ tags in each topic and the set of topics is represented as $\mathrm{Z} = \{\mathrm{Z}_1, \mathrm{Z}_2, ..., \mathrm{Z}_K\}$.

Definition 2: Problem Definition. In this paper, we try to find influential friends of each user with the change of items, that means given rating matrix R, social relationships S and item tags T, our goal is: to each user u_i, when the target item v_j changes, to find those friends $\mathrm{u}_c \in \mathrm{S}_i$ who will deeply affect predicted rating R_{ij}.

Table 1. Symbols used in this paper and their descriptions

Symbol	Description	Symbol	Description
U	A set of users	M	The number of users
V	A set of items	N	The number of items
$R_{M\times N}$	Rating matrix expressed by users on items	L	The number of items' tags
$S_{M\times M}$	Matrix of social relations	K	The number of topics
T	A set of items tags	T_e	Test data set
Z	A set of topics	α,β	The tradeoff parameters in the objective function
$\hat{R}_{M\times N}$	The predicted rating matrix	I	The indicator function
d	The dimension of latent feature	X^T	The transposition of matrix X
X^t	The variable corresponding to X in topic t	ψ	The threshold to filter topics
ρ	The topic numbers		

3.2 Inferring Truly Influential Social Relations

In this section, we try to use why users build social links to infer those truly influential friends who can directly affect users' choice. We think that if a user builds social links with others for some kind of things, those friends may affect his decision on similar things. And we deem that the common tastes between users can reflect the reasons why users build social links, because if users have no common tastes, they can hardly become friends. So our goal is to infer those common tastes. Here our tastes are interpretable, which is formally defined as topics in Definition 1, while traditional interests in RS are latent factors.

We use probabilistic latent semantic analysis (PLSA) model to detect users' tastes from their history behaviors, and tag users with these tastes, then we use common tastes among friends to infer the reasons why users build social links.

3.2.1 PLSA

Probabilistic latent semantic analysis [13] has been widely used in text mining tasks, which treats documents as a set of topics, and each topic is a set of words. Here in our model, we treat user as a set of topics and a topic is a set of tags, which are defined in Definition 1. And the Bayesian network of our method is shown as Fig. 1.

P(Z|U) represents users' probabilistic distribution on topics and P(T|Z) represents topics' probabilistic distribution on tags. They are the results of PLSA, which should be learned. EM algorithm is employed to train our model by get the max value of (1) and (2) is used to calculate the posterior probability of implicit variable in the current parameter:

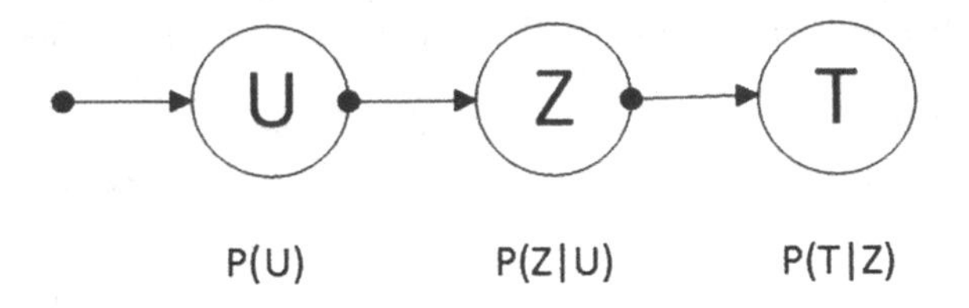

Fig. 1. U are users, T are tags and Z are latent topics which give a distribution over tags.

$$L = \sum\nolimits_{i=1}^{N} \sum\nolimits_{j=1}^{M} n(u_i, t_j) \bullet \log[\sum\nolimits_{k=1}^{K} p(t_j|z_k)p(z_k|u_i)] \tag{1}$$

$$P(z_k|u_i, t_j) = \frac{p(t_j|z_k)p(z_k|u_i)}{\sum_{l=1}^{K} p(t_j|z_l)p(z_k|u_i)} \tag{2}$$

In (1), $n(u_i,t_j)$ is the frequency of tag t_j appearing in topics of u_i, which is similar to word frequency in documents clustering. By training of EM algorithm, we will get the probabilities P(Z|U) and P(T|Z). And then, P(Z|U) is used to cluster users and their friends, which is a probabilistic matrix $P_{N\times K}$, and each row of $P_{N\times K}$ means the probabilistic distribution on every topic of user u_i. We choose those topics whose $P_{ij} > \psi$ to tag users, ψ is the threshold. Then, we can classify users into different topics like [13]. The difference is that our method utilizes users' tags by learning to classify users. As the same to clustering users, we use similar way to part item set.

In our method, using tags which can reflect items' content and characteristic, the produced topics are human comprehensible. Through the meaning of tags, we can easily understand the senses of topics.

3.2.2 Choosing of $n(u_i,t_j)$

$n(u_i,t_j)$ is the frequency of items' tags which appear in topics that users are interested in. It reflects the degree of attention that user u_i pays to items tagged with t_j. The choosing of $n(u_i,t_j)$ deeply affects the preference of trained topics and then decides the results of predicted ratings. So a good measure to choose $n(u_i,t_j)$ is very important. In our method, we define Score Of Attention to calculate $n(u_i,t_j)$, which is shown as (3):

$$SOA(u_i, t_j) = \frac{\sum_{t=1}^{N_t} R_{it}}{\sum_{k=1}^{L} |N_k|} \tag{3}$$

R_{it} is the score of u_i rating on items tagged with t_j, N_j is the item set tagged with t_j, L is the number of tags and $|N_k|$ represents the number of items tagged with t_k. We think that if a user gives a high score to items with t_j, this user may have a higher attention to this kind of items. Besides, if a user has a high proportion of ratings to items with t_j, he may care more about that kind of items. So we use (3) to combine these two elements.

3.3 Model Training

By training of PLSA based on items tags, we can get the topics of users who are interested in. And then we can divide users and items into relevant topics. In this section, we will train our model in each topic and get the predicted ratings. In each topic, we use socialMF [4] model to learn users' and items' feature by minimizing the square error in (4), and then use these feature to get the predicted ratings.

$$\begin{aligned} &L(R,U,V,S) = \\ &\frac{1}{2}\sum_{i=1}^{M_t}\sum_{j=1}^{N_t} I_{ij}(R_{ij}^t - \hat{R}_{ij}^t)^2 \\ &+\frac{\beta}{2}\sum_{i=1}^{M_i}((U_i^t - \sum_{k=1}^{M_t}\bar{I}_{i,k}^t S_{i,k}^t U_k^t)(U_i^t - \sum_{k=1}^{M_t}\bar{I}_{i,k}^t S_{i,k}^t U_k^t)^T) \\ &+\frac{\lambda}{2}(||U^t||_F^2 + ||V^t||_F^2) \end{aligned} \quad (4)$$

In (4), M_t is users in topic t, N_t is items in topic t, I_{ij} is the indictor whether user u_i rated on item v_j and $\bar{I}_{ik}^t$ represents whether u_i and u_k are friends, $S_{i,k}^t$ is the similarity of user u_i and u_k in topic t, $||U^t||_F^2$ and$||V^t||_F^2$ are Frobenius norm avoiding overfitting. $\hat{R}_{ij}^t$ is the predicted ratings of user u_i to item v_j, which is calculated by the way of (5).

$$\hat{R}_{ij}^t = r^t + U_i^t V_j^{tT} \quad (5)$$

In (5), r^t is set as the average ratings of users on topic t, U_i^t and V_i^t are the latent feature of users and items.

3.4 Final Predicted Ratings

After training of above model, in each topic we can get a predicted ratings $\hat{R}_{ij}^1$, $\hat{R}_{ij}^2$, ..., $\hat{R}_{ij}^K$. In this part, we will solve the problem of how to get the final predicted ratings based on those got ratings. As each user and item belong to many topics, we use (6) to calculate the final ratings. It is based on the relevance between topics and items. The higher the relevance is, the bigger the weight will get.

$$R_{ij} = \sum_{m=1}^{K} \tilde{I}_{jk} p(t_j|z_k)^* \hat{R}_{ij}^t \quad (6)$$

$\tilde{I}_{jk}$ is the indicator whether item v_j belongs to topic Z_k, $p(t_j|z_k)^*$ is probability distribution of item's tag t_j on topic z_k after normalizing, and it reflects the relevance between items and topics. The process of normalization is showed as (7).

$$p(t_j|z_k)^* = \frac{p(t_j|z_k)}{\sum_{m=1}^{K} p(t_j|z_m)} \quad (7)$$

As (6) shows, the final predicted ratings are related to (user, item) pairs. To different users, we will get different $\hat{R}_{ij}^t$, which can reflect the influence of different friends. And to different items, the same user, whose influence is considered in $\hat{R}_{ij}^t$, will get different weights by $p(t_j|z_k)^*$. And in this way the same friend will have different degrees of impact on the same user,according to different items.

4 Experiments

In this section, we will investigate the result of TRF-PMF model, and compare it with other famous models on real-world data set. In our experiments, we will check those questions:

(1) Whether refining influential friends and treating the same friend differently are good for recommendations?
(2) Can our automatically learning methods infer the truly influential friends effectively, and get better performance than existing models which find influential friends by categories?
(3) How many topics should we choose that our model can get the best performance, and what kind of users will get biggest benefit from our model?

4.1 Datasets

The data set we use is Yelp. Yelp is one of the biggest consumer review websites in America, it allows people to comment reviews and provides social networking functionality to connect users. People in Yelp can share their insight and suggestions on any products, here we call items, and assign them numeric ratings in the range of 1 to 5. In this way, users can contribute their knowledge and experience to other people, especially their friends.

We use the Yelp data set in [1], which has eight subsets and is publicly available from [15]. Some statistics of these data sets are exhibited in Table 2. And in our experiments, we randomly select 80 % of the review data to train models and compare their performances using the rest 20 % of the ratings.

4.2 Performance Measures

We adopt the common metric in our experiments, Root Mean Square Error (RMSE), to evaluate quality of our proposed approach and other popular methods. The metrics RMSE is defined as:

$$RMSE = \sqrt{\frac{1}{|T_e|}\sum_{i,j}(R_{ij} - \hat{R}_{ij})^2} \tag{8}$$

Table 2. Statistics of the Datasets

Category	Active life	Beautysvc	Home service	Hotels travel	Nightlife	Pets	Restaurants	Shopping
Users	5327	5466	2500	4712	4000	1624	2000	3000
Items	7495	8495	3213	5883	21377	1672	32725	16154
Ratings	24395	21345	5180	21658	99878	3093	91946	33352
Relations	372571	323412	124289	330788	110406	53441	24757	92456
Rating sparsity	99.94 %	99.95 %	99.94 %	99.92 %	99.88 %	99.89 %	99.86 %	99.93 %
Relation sparsity	98.69 %	98.92 %	98.01 %	98.51 %	99.31 %	97.97 %	99.38 %	98.97 %

Table 3. RMSE on yelp

Category	Active life	Beautysvc	Home services	Hotels travel	Nightlife	Pets	Restaurants	Shopping
BaseMF	1.0578	1.1770	1.3784	1.1238	1.0602	1.2305	1.0555	1.1375
SocialMF	1.0469	1.1662	1.3659	1.0935	1.0559	1.2250	1.0434	1.1145
CircleCon1	1.0437	1.1692	1.3621	1.0904	1.0568	1.2253	1.0417	1.1141
CircleCon2a	1.0242	1.1597	1.3454	1.0832	1.0506	1.2241	1.0401	1.1093
CircelCon2b	1.0219	**1.1527**	1.3493	1.0857	1.0497	1.2161	1.0399	1.1086
CircleCon3	**1.0207**	1.1550	**1.3408**	**1.0797**	**1.0485**	**1.2132**	**1.0382**	**1.1078**
TFR-PMF	**1.0182**	**1.1514**	**1.3238**	**1.0636**	**1.0205**	**1.1503**	**1.0127**	**1.0828**

From the definitions, we can find that a smaller RMSE value means a better performance.

4.3 Evaluation

4.3.1 Comparative Algorithms

BaseMF: This model is the basic matrix factorization approach proposed in [2], which doesn't consider social relationships.

SocialMF: SocialMF considers both social relations between users and personal bias when making recommendations. But it always uses all social links available simultaneously in the dataset.

CircleCon: This method is proposed in [13] to infer domain-obvious trust circle of social network, and includes four different metrics to measure trust: CircleCon1, CircleCon2a, CircleCon2b, CircleCon3.

In all our experiments, we set the dimensionality of latent space to be d $=10$ and normalized parameter $\alpha=0.1$ as [13]. The weight of social relationships in the second part of (4) is set $\beta=15$.

4.3.2 Experimental Result

The results of RMSE on Yelp are shown in Table 3. By comparing socialMF and our models, we can find that refining influential friends and treating the same friends differently are necessary for recommendations. And by comparing CircleCons and TFR-PMF, we believe that the automatically learned topics can reflect the nature of why those friends can affect users' decisions, and it's good for more accurate recommendations than dividing friends circle by simple rating maps. From the result, we find that in category pet, the RMSE of our method decreases 6.52 % comparing to BaseMF, 6.10 % comparing to SocialMF, 5.41 % comparing to the best performance of CircleCon. But in data set beautysvc, our method may not get a big improvement. That's maybe users' common tastes of

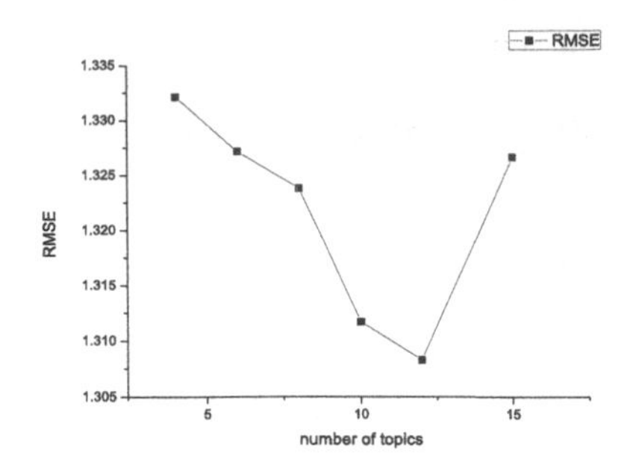

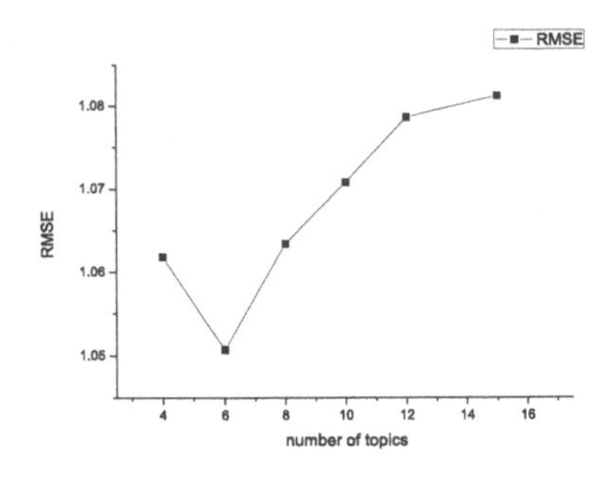

Fig. 2. Performance for different topic numbers on home services(left) and hotels travel(right)

beautysvc are centralized on one topic and our method could not filter influential friends effectively, while in other data sets users' common tastes are more decentralized and our method can work well.

4.4 Discussion and Analysis

4.4.1 Choice of Topics Number

The number of topics ρ decides the species of tags to describe users, and it influences the result of refining influential friends which then affects the final predicted ratings. So the choice of suitable number of topics is very important (in our above experiments, we set the number of topics equal to 8). Figure 2 shows the RMSE of choosing different number of topics on data sets home services and hotels travel, which we choose as examples, and in other Yelp data sets there are similar conclusions. From Fig. 2, we find that a too small or too large ρ is not good for a high accuracy. The reason may be that if topics number is small, users' common tastes are more centralized, and we can't distinguish truly influential friends effectively, so the result may be unsatisfying. While the number of topics is too large, topics to reflect users' taste get more, but information to reflect each topic may be less, and it will negatively affect the final results.

4.4.2 Users Getting Benefits in Our Model

To find which kind of users can get the biggest benefit from our model, we make a statistic on the distribution of users in Yelp data set, based on the amount of users' friends, we also choose home services and hotels travel as examples. Figure 3 is the distributions of users' number of friends on home services and hotels travel, while Fig. 4 is their RMSE. In Fig. 3, home services has 750 users who have no friends and hotels travel has 1002. And In Fig. 4, we choose CircleCon3 as represents of CircleCon, which has the best performance. We calculate RMSE of different kinds of users in test dataset, and from the result we can conclude that our method is more suitable for those users who have many friends, while other models may get a poor performance. And it's the main contributor

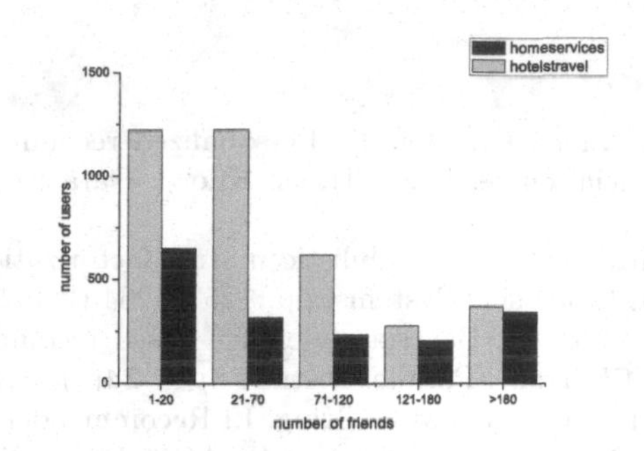

Fig. 3. Distribution of friends on home services and hotels travel

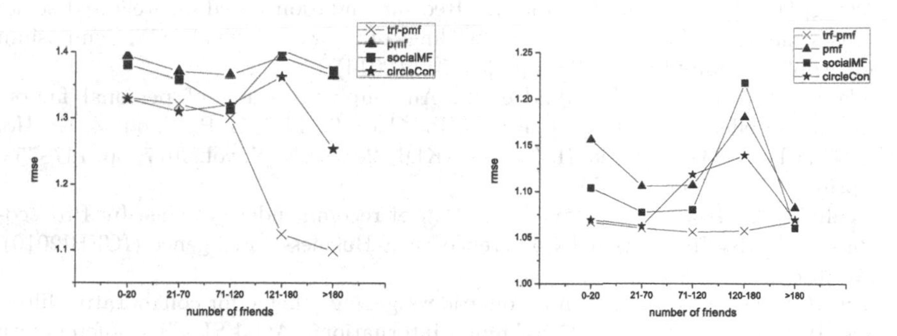

Fig. 4. RMSE distribution on different number of friends in home services(left) and hotels travel(right)

to get a lower RMSE for our model. Besides, we are surprised to find that even to those users who have few friends, our model can get a lower RMSE. That means even to those users, they may also not consider all friends' advices. Finally, our model may not work well to those users whose friends are too many, such as more than 180 in Fig. 4(right), that's maybe this kind of users is too small and noisy, a causal factor may make a big impact.

5 Conclusion and Future Work

In this paper, we do some work on improving recommendation accuracy in social rating network, by refining truly influential friends of each user and changing the influence from the same friend. Our idea is that, to different items we use reasons why users build social links to infer different most influential friends of a user, which is tailored for (user,item) pairs. Compared with existing models, our inferred influential friends are domain-unrelated and can be learned. We check our thought by TFR-PMF model, and the results show that our method can improve the accuracy of recommendation over existing approaches, which use mixed social relations or other methods. In the future, we will do some works on scalability of our model and more accurate methods to find truly influential friends in one taste.

References

1. Qian, X., Feng, H., Zhao, G., Mei, T.: Personalized recommendation combining user interest and social circle. IEEE Trans. Knowl. Data Eng. **26**(7), 1763–1777 (2014)
2. Mnih, A., Salakhutdinov, R.: Probabilistic matrix factorization. In: Advances in Neural Information Processing Systems, pp. 1257–1264 (2007)
3. Yang, X., Guo, Y., Liu, Y.: Bayesian-inference-based recommendation in online social networks. IEEE Trans. Parallel Distrib. Syst. **24**(4), 642–651 (2013)
4. Ma, H., Zhou, D., Liu, C., Lyu, M.R., King, I.: Recommender systems with social regularization. In: Proceedings of the Fourth ACM International Conference on Web Search and Data Mining. ACM, pp. 287–296 (2011)
5. Wang, D., Ma, J., Lian, T., Guo, L.: Recommendation based on weighted social trusts and item relationships. In: Proceedings of the 29th Annual ACM Symposium on Applied Computing. ACM, pp. 254–259 (2014)
6. Wang, Z., Yang, Y., Hu, Q., He, L.: An empirical study of personal factors and social effects on rating prediction. In: Cao, T., Lim, E.-P., Zhou, Z.-H., Ho, T.-B., Cheung, D., Motoda, H. (eds.) PAKDD 2015. LNCS, vol. 9077, pp. 747–758. Springer, Heidelberg (2015)
7. Tuzhilin, A.: Towards the next generation of recommender systems. In: Proceedings of the 1st International Conference on E-Business Intelligence (ICEBI2010), Atlantis Press (2010)
8. Jin, R., Chai, J.Y., Si, L.: An automatic weighting scheme for collaborative filtering. In: Proceedings of the 27th Annual International ACM SIGIR Conference on Research and Development in Information Retrieval. ACM, pp. 337–344 (2004)
9. Linden, G., Smith, B., York, J.: Amazon. com recommendations: item-to-item collaborative filtering. Internet Comput. **7**(1), 76–80 (2003)
10. OConnor, M., Herlocker, J.: Clustering items for collaborative filtering. In: Proceedings of the ACM SIGIR Workshop on Recommender Systems, vol. 128, Citeseer (1999)
11. Hofmann, T.: Latent semantic models for collaborative filtering. ACM Trans. Inf. Syst. (TOIS) **22**(1), 89–115 (2004)
12. Jamali, M., Ester, M.: A matrix factorization technique with trust propagation for recommendation in social networks. In: Proceedings of the Fourth ACM Conference on Recommender Systems. ACM, pp. 135–142 (2010)
13. Yang, X., Steck, H., Liu, Y.: Circle-based recommendation in online social networks. In: Proceedings of the 18th ACM SIGKDD International Conference On Knowledge Discovery And Data Mining. ACM, pp. 1267–1275 (2012)
14. Zhang, X., Cheng, J., Yuan, T., Niu, B., Lu, H.: Toprec: domain-specific recommendation through community topic mining in social network. In: Proceedings of the 22nd International Conference on World Wide Web, International World Wide Web Conferences Steering Committee, pp. 1501–1510 (2013)
15. http://smiles.xjtu.edu.cn

User Similarity Adjustment for Improved Recommendations

R. Latha(✉) and R. Nadarajan

Department of Applied Mathematics and Computational Sciences,
PSG College of Technology, Coimbatore, India
{lathapsg,nadarajan_psg}@yahoo.co.in

Abstract. Recommender systems are becoming more and more attractive in both research and commercial communities due to Information overload problem and the popularity of the Internet applications. Collaborative Filtering, a popular branch of recommendation approaches, makes predictions based on historical data available in the system. In particular, user based Collaborative Filtering largely depends on how users rate various items of the database and the success of such a system largely relies on pair wise similarity between users. However popular items may give a negative effect on choosing similar users of the target user. The proposed work namely User Similarity Adjustment based on Item Diversity (USA_ID) is designed to achieve personalized recommendations by modifying user similarity scores, for the purpose of reducing the negative effects of popular items in user based Collaborative Filtering framework. A Recommender system is focusing exclusively on achieving accurate recommendations i.e., providing the most relevant items for the needs of a user. From user's perspective, they would not be interested when they are facing monotonous recommendations even if they are accurate. Whilst much research effort is spent on improving accuracy of recommendations, less effort is taken on analyzing usefulness of recommendations. Novelty and Diversity have been identified as key dimensions of recommendation utility. It has been made clear that greater accuracy leads to lower diversity which results in accuracy-diversity trade off in personalized recommender systems. The proposed work provides an approach to increase the utility of a Recommender system by improving accuracy as well as diversity. Experiments are conducted on the bench mark data set MovieLens and the results show efficiency of the proposed approach in improving quality of predictions.

Keywords: Recommendation · Collaborative filtering · Novelty · Diversity · Utility

1 Introduction

Recommender Systems (RS) are helping people to identify their preferences from large collection of candidate objects. They are used in variety of applications such

R. Prasath et al. (Eds.): MIKE 2015, LNAI 9468, pp. 515–525, 2015.
DOI: 10.1007/978-3-319-26832-3_48

as online recommendation of books [5], CDs [18], movies [10], news [12] and many others. RSs are now popular both commercially and in the research community [21]. Many commercial applications like Amazon.com^{TM} (www.amazon.com), Netflix (www.netflix.com), etc., make use of recommendations in order to attain business profits. RSs can be viewed as personal information retrieval in which there is no explicit query to express user's wish rather implicit information about user's interest. RSs are getting more and more attraction from electronic commerce domains as they have potential value in business. Research communities from Machine Learning, Data Mining, Information Retrieval and Statistics are working on RS domain.

RSs are based on one of three strategies [3]. They are Collaborative Filtering (CF) [20] and Content Based Filtering [19](CB) and a hybrid of both the approaches [8]. CB creates a profile for each user or product to characterize its nature. The profiles are used to associate users with matching products [19]. An alternative to CB is CF which relies only on past user behavior in the form of previous transactions or product ratings [13]. Collaborative Filtering analyzes relationships between users and inter-dependencies among products to identify new user-item associations.

CF approach can be designed based on either user similarities or item similarities that are derived from historical data [22]. In CF, the prediction is based on a database of past purchases, or ratings made by the system users [19]. The ratings given by users for various items are available in the system. Each rating shows how much an item is liked by a particular user. The task of a CF based recommender system is to predict how much a user likes an item which is currently unrated.

The most common form of CF is the neighborhood-based approach (also known as k Nearest Neighbors) [15]. The neighborhood CF techniques can be user based or item based [22]. These KNN techniques identify items that are likely to be rated similarly or like-minded people with similar history of rating or purchasing, in order to predict unknown relationships between users and items. Merits of the neighborhood-based approach are intuitiveness, sparing the need to train and tune many parameters, and the ability to easily explain the reasoning behind a recommendation [4]. Item-based approach looks into set of items similar to the target item i and selects k most similar items i_1,i_2,...,i_k. Once the most similar items are found, the prediction is computed by taking a weighted average of the target user's ratings for those similar items [20].

User-based Collaborative Filtering approach assumes that users who agreed on preferred objects in the past will tend to agree in the future. User-based approach looks into the set of users who share similar preferences with the target user u and selects k most similar users u_1, u_2,... u_k and makes prediction for the target item i based on the preferences given by them for the item i.

The ultimate goal of any RS is to satisfy user's requirement [11]. Improving accuracy of predictions has been the motive for RS research community for few decades. At the same time there is a growing demand in the user population to receive interesting rather than accurate recommendations. Recommending same kind of items ever to a user results in monotonic behavior of Recommender Systems [17]. Always recommending highly similar items to the items already rated by a user brings no novelty to the user [3]. In particular novelty and diversity in recommendations are identified as key dimensions in user satisfaction in many application domains [23]. For E-Commerce websites, recommending variety of items has the potential to make more profits by increasing sales diversity [6].

In user based Collaborative Filtering, similarity between users is the core part of recommendation. If two users correlate in more number of items then their similarity score is high. The similarity score will be less if they co-rate less number of items. Popular items would have been rated by many users. So even if there is no actual correlation between two users, since they have co-rated popular items, their similarity score is influenced by the popular items. There fore the proposed work reduces the negative impact of popular items in user similarity computations.

The objective of the proposed technique is to overcome the limitations of accurate recommendations by giving a proper trade off between accuracy and diversity of recommendations. The proposal

- calculates global popularity score of each item based on ratings available in the system, from which each item's global diversity score is calculated
- modifies pair-wise user similarity which is calculated from the historical data using Pearson Correlation and Cosine similarity with the help of global diversity score of items to reduce the adverse effect of popular items in user based collaborative filtering framework
- empirically shows the usefulness of the proposed user similarity modification approach on the benchmark data set,namely MovieLens

The rest of the paper is organized as follows. Section 2 describes the state of the art user based Collaborative Filtering framework, Sect. 3 discusses about related work of diversity enhancement available in the literature, Sect. 4 describes the proposed method of diversity enhancement, Sect. 5 discusses about experimental evaluations and Sect. 6 gives future development of the work and conclusion.

2 Existing User-Based Collaborative Filtering Framework

This section describes user based Collaborative Filtering framework [16] which is used as base line technique for making predictions.

In user based CF, given a target user u, users who share similar rating pattern with u are considered as neighbors and their ratings are used to predict the unrated items of u. The effectiveness of user based CF methods depends on pairwise similarity scores between users. Each user profile (row vector) is sorted by its dis-similarity towards the target user's profile. Ratings by similar users contribute to predict the target item rating. Most commonly used metrics for calculating similarity between items are Cosine similarity [22] and Pearson Correlation coefficient [3].

Pearson Correlation coefficient to compute similarity between each pair of users is formulated as

$$S_{u,v} = \frac{\sum_{i \in I}(r_{u,i} - \bar{r}_u) \times (r_{v,i} - \bar{r}_v)}{\sqrt{\sum_{i \in I}(r_{u,i} - \bar{r}_u)^2} \times \sqrt{\sum_{i \in I}(r_{v,i} - \bar{r}_v)^2}} \tag{1}$$

where I is the set of items rated by both users u and v. $r_{u,i}$ is the rating provided by user u for item i. $\bar{r}_u$ is the average rating of user u. $S_{u,v}$ can be between -1 and $+1$.

By treating each user profile as a vector in a high dimensional space, Cosine similarity calculates the similarity score between two users as the cosine of the angle between the two corresponding user profile vectors. Cosine similarity between users u and v is calculated as

$$S_{u,v} = \frac{\sum_{i \in I} r_{u,i} \times r_{v,i}}{\sqrt{\sum_{i \in I} r_{u,i}^2} \times \sqrt{\sum_{i \in I} r_{v,i}^2}} \tag{2}$$

The most important step in a Collaborative Filtering system is to generate the output interface in terms of predictions [20]. Once the set of most similar users of the target user u is identified the next step is to calculate target user $u's$ rating for an item i with the help of the ratings provided by those neighbours for the item i. Predictions can be made based on the weighted average of known ratings as given below

$$P_{a,i} = \bar{r}_a + \frac{\sum_{u \in K} S_{u,a} \times (r_{u,i} - \bar{r}_u)}{\sum_{u \in K} S_{u,a}} \tag{3}$$

where $P_{a,i}$ is the predicted value of target user a for item i and K is the set of Top k similar users of the target user a.

3 Related Work

Making only accurate recommendations is not always useful to users. For example, recommending only popular items (e.g., blockbuster movies that many users tend to like) could obtain high accuracy, but also can lead to a decline of other aspects of recommendations, including recommendation diversity [2]. Recommending long-tail items to individual users can intensify this effect. Thus, more consumers would be attracted to the companies that carry a large selection of long tail items and have long tail strategies, such as providing more diverse recommendations [7].

Diversification is defined as the process of maximizing the variety of items in recommendation lists [1]. In [24], Ziegler et al. did a large scale online study, and their experimental results show that users' overall satisfaction with recommendation lists not only depends on accuracy, but also on the range of interests covered. They also found that human perception can only capture a certain level of diversification inherent to a list.

Temporal diversity is an important facet of recommender systems [17]. The authors showed that how CF data changes over time by performing a user survey and they evaluated three CF algorithms from the point of view of diversity in the sequence of recommendation lists produced over time.

In [23], the author classifies diversity into two types, namely, aggregate diversity and individual diversity. The first case accounts for how different are items in a recommendation list for a user, which is normally the notion of diversity employed in most works. Nevertheless, aggregate diversity is understood as the total amount of different items, a recommendation algorithm can provide to the community of users.

Brynjolfsson et al. [6] demonstrated that recommendations would increase sales of the items in the long tail, resulting in the improvement of aggregate diversity in contrast to individual diversity. Herlocker [16] proposed aggregate diversity measure to be the percentage of items that the recommender system is able to make recommendations for (often known as coverage). Gediminas Adomavicius et al. [2], talked about the importance of aggregate diversity in recommendation. They proposed diversity-in-Top-N metric which can serve as an indicator of the level of personalizations provided by a recommender system.

A common approach to diversified ranking is based on the notion of maximal marginal relevance (MMR) [9]. Marginal relevance is defined as a weighted combination of the two metrics namely, accuracy and diversity in order to account for the trade-off between them. A method called PLUS (Power Law adjustments of User Similarities) is proposed in [14] to achieve personalized recommendations. PLUS makes use of power function to adjust user similarity scores for the purpose of reducing adverse effects of popular objects in the user based Collaborative Filtering framework. The proposed work (USA_ID) aims to reduce the negative effect of popular items in order to improve the quality of recommendations.

4 Proposed Technique

The objective of a personalized recommender system is to rank a set of items for a given user so that highly ranked items are more preferred by the user [14]. In order to achieve this, the proposed technique called User Similarity Adjustment based on Item diversity (USA_ID) modifies the pair-wise similarity between users. The modification is done to reduce the negative impact of popular items which is expected to be rated by most of the users.

4.1 User Similarity Computation

If there are m users who have given ratings for n items, then the ratings data can be represented as an mXn matrix with rows representing users and columns representing items. The matrix is called user-item rating matrix R. Each element $r_{u,i}$ is an ordinal value ranging from R_{min} to R_{max}. Unrated values are considered to be zero.

For the given mXn user item rating matrix R, user-user similarity matrix can be represented as an mXm symmetric matrix S. The matrix rows and columns represent users and each $S_{u,v}$ represents the similarity between user u and user v. More specifically given the profiles of users u and v, the similarity between them is given by

$$S = (s_{u,v})_{mXm} \tag{4}$$

where $s_{u,v}$ can be Pearson Correlation Coefficient or Cosine similarity as defined in Sect. 2.

4.2 Computing Popularity Score of Items

Global popularity score of an item i is defined as the ratio of number of users who rated for the item to the total number of users in the system. Popularity score of an item will be more for items which have been rated by many users. Popularity score of item i, pop_i is defined as

$$pop_i = \frac{|U_i|}{m}, r_{u,i} > 0, u \in u_1, u_2, ...u_m \tag{5}$$

where U_i is the set of users who rated for item i and $r_{u,i}$ is the rating assigned by the user u for the item i and m is the total number of users. From popularity score of item i, one can compute global diversity score of item i as

$$div_i = 1 - pop_i \tag{6}$$

4.3 User Similarity Modification

The basic assumption in user based Collaborative Filtering is that users with similar preferences will have similar preferences in future. Therefore predictions are made based on the preferences given by close neighbours of the target user.

Similarity between two users is based on how they agree while giving preferences for various items of the domain. In an extreme case, popular items should have been rated by many users, and thus the chance of any two users to correlate on those items is high [14]. As a consequence of this, less similarity score is assigned to two users when they correlate only on popular items. Even though such users are less preferred in prediction process, still they have impact on the quality of predictions. So the proposed technique gives a discount to the pair wise similarity between two users who have correlated only in popular items. Thus the pairwise similarity of such users is multiplied by the diversity score of the items in which they correlate. The proposed technique adjusts user similarity values which can be calculated using Pearson Correlation or Cosine similarity.

In order to do so, the similarity between each pair of users is modified as

$$T_{u,v} = S_{u,v} \times \frac{1}{t} \sum_{i \in C} div_i, \quad r_{u,i} > 0, r_{v,i} > 0, \tag{7}$$

where C is the set of items co-rated by users u and v and t is the cardinality of the set C. $T_{u,v}$ is the modified user similarity between users u and v. The modification is done for every element of the similarity matrix S. Further the prediction computations are done based on the modified user similarity values which is given as

$$P_{a,i} = \bar{r}_a + \frac{\sum_{u \in K} T_{u,a} \times (r_{u,i} - \bar{r}_u)}{\sum_{u \in K} T_{u,a}} \tag{8}$$

where $P_{a,i}$ is the predicted value of active user a and item i and K is the set of Top k similar users of the active user a.

5 Experimental Results

This section discusses about the data set used, evaluation metrics and the effectiveness of the proposed approach, USA_ID.

5.1 Data Set Used

The experiments are conducted on Movielens 100k (www.Movielens.com), which is a standard data set for discussing the efficiency of Collaborative Filtering techniques.The data set contains ratings given by 943 users for 1683 items. The ratings are in the range 1 to 5. Total number of ratings available is 100000. We split the data set into two sets namely train with 80 % of the ratings of the original rating matrix and test with the remaining 20 % of the ratings. Five cross validation is done on the data set to report the results.

5.2 Evaluation Metrics Used

In order to measure accuracy of the predictions, two categories of techniques namely accuracy and diversity metrics are adopted. In order to prove accuracy of the proposed approach two classification accuracy measures namely *Precision* and *Recall* [16] are considered. Precision is defined as the ratio of relevant items selected to number of items selected. Precision represents the probability that a selected item is relevant.

$$Precision = \frac{N_{rs}}{N_s} \tag{9}$$

Recall is defined as the ratio of relevant items selected to total number of relevant items available. *Recall* represents the probability that a relevant item will be selected.

$$Recall = \frac{N_{rs}}{N_r} \tag{10}$$

where N_{rs} is the number of relevant items retrieved, N_s is the number of items retrieved and N_r is the total number of relevant items in the data set.

In order to prove diversity of the proposed approach two metrics namely, ILD (Intra List Diversity) and MN (Mean Novelty) are used. Ziegler, et al. [24]

Table 1. Improvements of the proposed approach for Top10 recommendations

Criteria	Similarity	UserSim	PLUS	USA_ID
Precision	Cosine	0.0054	0.0058	0.0068
	Pearson	0.0042	0.0048	0.0052
Recall	Cosine	0.0490	0.0495	0.0536
	Pearson	0.0399	0.0410	0.0416
ILD	Cosine	0.8056	0.8089	0.8467
	Pearson	0.7912	0.7966	0.8378
MN	Cosine	0.1934	0.1956	0.4367
	Pearson	0.1855	0.1899	0.4302

introduced the ILD to assess the topical diversity of recommendation lists, which is computed in terms of decreasing ILS (Intra List Similarity). The authors suggested that ILS is an efficient measure that complements existing accuracy measures to capture user satisfaction. ILS is calculated as

$$ILS_u = \frac{\sum_i \sum_j Sim_{i,j}}{n \times (n-1)}, \quad i, j \in RL, i \neq j \tag{11}$$

where n is the total number of items recommended, RL is the Top k recommended list to the user u and $Sim_{i,j}$ is the similarity score between item i and item j. The similarity score used here is Cosine similarity as discussed in [20]. Higher score denotes lower diversity. If Sim used is normalized to the range 0 to 1, then ILD can be computed as

$$ILD_u = \frac{\sum_i \sum_j (1 - Sim_{i,j})}{n \times (n-1)}, \quad i, j \in RL, i \neq j \tag{12}$$

Next metric for evaluating the diversity of the system is Mean Novelty (MN) [14]. For each item, we calculate the fraction of users that have rated the item and obtain the information content of the item as the negative logarithm of the fraction. Let $D_u(k)$ is the Top k ranking subset of items of the user u. Given topk items recommended to a user, we average the information content of all items to obtain novelty of the system to the user. The mean novelty of the system is calculated as the average novelty over all users as

$$MN(k) = -\frac{1}{|U|} \sum_{u \in U} \frac{1}{k} \sum_{i \in D_u(k)} log\, f_i \tag{13}$$

where f_i is the fraction of users that have collected the ith item and U is the set of users.

Table 2. Improvements of the proposed approach for Top20 recommendations

Criteria	Similarity	UserSim	PLUS	USA_ID
Precision	Cosine	0.0098	0.0996	0.0104
	Pearson	0.0060	0.0061	0.0085
Recall	Cosine	0.0987	0.1109	0.1187
	Pearson	0.0923	0.0956	0.1201
ILD	Cosine	0.8142	0.8198	0.8387
	Pearson	0.7988	0.8022	0.8324
MN	Cosine	0.1853	0.1876	0.4328
	Pearson	0.1688	0.1698	0.4277

5.3 Improvement of Recommendation Accuracy and Utility

This section compares the efficiency of the proposed method USA_ID with actual user similarity computed from the historical data as given in (2) and (3) which is referenced as $UserSim$, and $PLUS$, discussed in [14].

For each user TopK recommendations are considered for the discussion about the efficiency of the proposed approach. Experiments are done for three values of K namely 10, 20, and 50. The comprehensive comparison between the approaches is shown in Tables 1, 2 and 3 for Top10, Top20 and Top50 recommendations respectively. In Table 1, USA_ID with Pearson Correlation yields an improvement of 21 % and 15 % over UserSim and PLUS respectively on Precision measure. USA_ID with Cosine gives an improvement of 18 % and 8 % over UserSim and PLUS respectively on Precision measure.

USA_ID with Pearson Correlation provides an improvement of 8 % and 7 % over UserSim and PLUS respectively with respect to Recall measure. USA_ID with Cosine provides an improvement of 4 % and 2 % over UserSim and PLUS respectively with respect to Recall measure.

USA_ID with Pearson Correlation offers an improvement of 4.8 % and 4.4 % over UserSim and PLUS respectively with respect to ILD measure. USA_ID with Cosine offers an improvement of 5.5 % and 5 % respectively over UserSim and PLUS with respect to ILD measure.

USA_ID with Pearson Correlation yields an improvement of 55 % and 54 % over UserSim and PLUS respectively with respect to MN. USA_ID with cosine yields an improvement of 57 % and 56 % over UserSim and PLUS respectively with respect to MN. The similar improvements are reported in Tables 2 and 3 for Top20 and Top50 recommendations. We observe from the tables that USA_ID significantly improves the performance of the recommender systems in terms of accuracy measured by *Precision*, *Recall* as well as diversity measured by ILD, MN.

Table 3. Improvements of the proposed approach, for Top50 recommendations

Criteria	Similarity	UserSim	PLUS	USA_ID
Precision	Cosine	0.0230	0.0299	0.0321
	Pearson	0.0243	0.0298	0.0378
Recall	Cosine	0.3910	0.4100	0.4167
	Pearson	0.4001	0.4121	0.4129
ILD	Cosine	0.8121	0.8145	0.8521
	Pearson	0.8021	0.8093	0.8402
MN	Cosine	0.1858	0.1898	0.4450
	Pearson	0.1798	0.1802	0.4310

6 Conclusion

Novelty and Diversity in recommendations are considered as significant dimensions to attract users. This work presents an approach called USA_ID which modifies user similarity to reduce the negative impact of popular items of the domain. Standard user based Collaborative Filtering frame work is used to execute the prediction computations. Experiments are conducted on the standard data set Movielens. Results show that USA_ID is effective in improving accuracy and diversity of recommendations. The proposed method can further be investigated to check its applicability in Item based Collaborative Filtering framework. The modified similarity score of the users can be modeled as a graph, and the graph properties can be analyzed to improve the quality of predictions further.

References

1. Adamopoulos, P., Tuzhilin, A.: On unexpectedness in recommender systems: or how to expect the unexpected. In: Workshop on Novelty and Diversity in Recommender Systems (DiveRS 2011), at the 5th ACM International Conference on Recommender Systems (RecSys 2011), pp. 11–18, Chicago, Illinois, USA. ACM (2011)
2. Adomavicius, G., Kwon, Y.: Maximizing aggregate recommendation diversity: A graph-theoretic approach. In: Proceedings of the 1st International Workshop on Novelty and Diversity in Recommender Systems (DiveRS 2011), pp. 3–10 (2011)
3. Adomavicius, G., Tuzhilin, A.: Toward the next generation of recommender systems: a survey of the state-of-the-art and possible extensions. IEEE Trans. Knowl. Data Eng. **17**(6), 734–749 (2005)
4. Bell, R.M., Koren, Y.: Improved neighborhood-based collaborative filtering. In: KDD Cup and Workshop at the 13th ACM SIGKDD International Conference on Knowledge Discovery and Data Mining (2007)
5. Bogers, T., Van Den Bosch, A.: Fusing recommendations for social bookmarking web sites. Int. J. Electron. Commer. **15**(3), 31–72 (2011)
6. Brynjolfsson, E., Hu, Y., Simester, D.: Goodbye pareto principle, hello long tail: the effect of search costs on the concentration of product sales. Manage. Sci. **57**(8), 1373–1386 (2011)

7. Brynjolfsson, E., Hu, Y., Smith, M.D.: Research commentary-long tails vs. superstars: the effect of information technology on product variety and sales concentration patterns. Inf. Syst. Res. **21**(4), 736–747 (2010)
8. Burke, R.: Hybrid recommender systems: survey and experiments. User Mod. User-Adap. Interact. **12**(4), 331–370 (2002)
9. Carbonell, J., Goldstein, J.: The use of mmr, diversity-based reranking for reordering documents and producing summaries. In: Proceedings of the 21st Annual International ACM SIGIR Conference on Research and Development in Information Retrieval, pp. 335–336. ACM (1998)
10. Carrer-Neto, W., Hernández-Alcaraz, M.L., Valencia-García, R., García-Sánchez, F.: Social knowledge-based recommender system. Application to the movies domain. Expert Syst. Appl. **39**(12), 10990–11000 (2012)
11. Castells, P., Vargas, S., Wang, J.: Novelty and diversity metrics for recommender systems: choice, discovery and relevance. In: International Workshop on Diversity in Document Retrieval (DDR 2011) at the 33rd European Conference on Information Retrieval (ECIR 2011), pp. 29–36. Citeseer (2011)
12. Chuanmin, M., Xiaofei, S., Jing, M., Xin, Z.: Collaborative filtering algorithm based on random walk with choice. sekeie-14, pp. 192–196 (2014)
13. Ekstrand, M.D., Riedl, J.T., Konstan, J.A.: Collaborative filtering recommender systems. Found. Trends Hum.-Comput. Interact. **4**(2), 81–173 (2011)
14. Gan, M., Jiang, R.: Improving accuracy and diversity of personalized recommendation through power law adjustments of user similarities. Decis. Support Syst. **55**(3), 811–821 (2013)
15. Herlocker, J.L., Konstan, J.A., Borchers, A., Riedl, J.: An algorithmic framework for performing collaborative filtering. In: Proceedings of the 22nd Annual International ACM SIGIR Conference on Research and Development in Information Retrieval, pp. 230–237. ACM (1999)
16. Herlocker, J.L., Konstan, J.A., Terveen, L.G., Riedl, J.T.: Evaluating collaborative filtering recommender systems. ACM Trans. Inf. Syst. (TOIS) **22**(1), 5–53 (2004)
17. Lathia, N., Hailes, S., Capra, L., Amatriain, X.: Temporal diversity in recommender systems. In: Proceedings of the 33rd international ACM SIGIR Conference on Research and Development in Information Retrieval, pp. 210–217. ACM (2010)
18. Linden, G., Smith, B., York, J.: Amazon. com recommendations: Item-to-item collaborative filtering. Internet Comput. **7**(1), 76–80 (2003)
19. Pazzani, M.J.: A framework for collaborative, content-based and demographic filtering. Artif. Intell. Rev. **13**(5–6), 393–408 (1999)
20. Sarwar, B., Karypis, G., Konstan, J., Riedl, J.: Item-based collaborative filtering recommendation algorithms. In: Proceedings of the 10th international conference on World Wide Web, pp. 285–295. ACM (2001)
21. Shani, G., Gunawardana, A.: Evaluating recommendation systems. In: Ricci, F., Rokach, L., Shapira, B., Kantor, P.B. (eds.) Recommender Systems Handbook, pp. 257–297. Springer, New York (2011)
22. Su, X., Khoshgoftaar, T.M.: A survey of collaborative filtering techniques. Adv. Artif. Intell. **2009**, Article No. 4 (2009)
23. Vargas, S., Castells, P.: Rank and relevance in novelty and diversity metrics for recommender systems. In: Proceedings of the Fifth ACM Conference on Recommender Systems, pp. 109–116. ACM (2011)
24. Ziegler, C.N., McNee, S.M., Konstan, J.A., Lausen, G.: Improving recommendation lists through topic diversification. In: Proceedings of the 14th International Conference on World Wide Web, pp. 22–32. ACM (2005)

Enhancing Recommendation Quality of a Multi Criterion Recommender System Using Genetic Algorithm

Rubina Parveen[1], Vibhor Kant[2(✉)], Pragya Dwivedi[3], and Anant K. Jaiswal[4]

[1] Krishna Engineering College, Ghaziabad, India
ruby.pasha@gmail.com
[2] The LNM Institute of Information Technology, Jaipur, India
vibhor.kant@gmail.com
[3] Motilal Nehru National Institute of Technology, Allahabad, India
pragya.dwijnu@gmail.com
[4] Amity University, Noida, India
akjayswal@amity.edu

Abstract. Recommender system (RS) the most successful application of Web personalization helps in alleviating the information overload available on large information spaces. It attempts to identify the most relevant items for users based on their preferences. Generally, users are allowed to provide overall ratings on experienced items but many online systems allow users to provide their ratings on different criteria. Several attempts have been made in the past to design a RS focusing on the ratings of a single criterion. However, investigation of the utility of multi criterion recommender systems in online environment is still in its infancy. We propose a multi criterion RS based on leveraging information derived from multi-criterion ratings through genetic algorithm. Experimental results are presented to demonstrate the effectiveness of the proposed recommendation strategy using a well-known Yahoo! Movies dataset.

Keywords: Multi-criterion decision making · Machine learning · Genetic algorithm · Recommender system · Multi-criterion ratings

1 Introduction

With the advent of numerous products in the market especially online, the recommender systems are no further an unfamiliar area for researchers as well as for the customers. The diversity in the features of same kind of products and the choices available in the marketplace or the e-commerce sites makes the selection procedure for the customers even more challenging. Recommender systems have contributed to a great deal to resolve this issue. Capturing some specific knowledge about the product features and knowledge about the customer's preferences (ratings given to experienced products, user contextual information or the demographic information) can show the way to make better choices. The prerequisite of RS is to process this available information for providing appropriate suggestions to customers. The idea

R. Prasath et al. (Eds.): MIKE 2015, LNAI 9468, pp. 526–535, 2015.
DOI: 10.1007/978-3-319-26832-3_49

is to make the system proficient enough to recommend the most appropriate items or information amongst available quantity.

Since the last decade several techniques have been deduced to make recommender system increasingly efficient. The most widely used techniques in the area of RS are collaborative filtering (CF); content based filtering (CBF) and hybrid filtering [1]. Recommendation through collaborative filtering technique is based on the interests of other similar users while in content based approach recommendations are generated on the basis of the user preferences in the past and content information. Collaborative and content based filtering are the most popular filtering techniques in the field of RSs. Both have individual weaknesses and advantages and are complementary to each other. As a result, various hybrid RSs based on collaborative and content based techniques are developed to overcome their individual weaknesses. A rich survey on hybrid RS is presented by Robin Burke [2] where several hybridization techniques are proposed. Amongst the above mentioned techniques, collaborative filtering (CF) is the most widely used technique and has gained a significant attention of the researchers due to the ease in implementation and prominent outcomes.

Though a lot of research has been done in the area of recommender systems, there is a continuous evolution going on to make these systems more efficient to their utmost level [3–5]. Multi criterion recommender system (MCRS) is another prominent step towards this aim [6]. So far, the majority of existing recommender systems obtains an overall numerical rating as input information for the recommendation algorithm. This overall rating depends only on one single criterion that usually represents the overall preference of user on item [7]. In contrast to this, in multi-criteria recommender system, the input is taken as the ratings provided by the users based on several aspects of the item. We can also call this as multi component ratings as referred in [8]. The idea of MCRS is to make a move towards understanding the user's interest in a more efficient and fine grained manner. The most typical example can be ratings given to a movie by the viewers based on several aspects namely story, visual, acting and direction etc. Similarly restaurants rated upon their cleanliness, service, cuisines and vicinity etc. These examples also show the importance of various criteria ratings for generating more efficient suggestions to users.

Adomavicious and Kwon [6] provided the very initial proposal related to this area. The idea was to design a system based on the ratings on more than one dimensions and not just a single dimensional rating. The focus was on the better learning of the system from the given set of data. Their research offered two approaches and the experiment was conducted on movies rated on different dimensions. Our work extends these approaches with the help of genetic algorithm aiming to obtain more precise and accurate results.

The rest of the paper is structured as follows: Sect. 2 provides the back-ground related to our work. In Sect. 3, the proposed scheme is presented. Computational experiments and results are presented in Sect. 4. Finally, in the last Section, we conclude our work with some future research directions.

2 Background

In this section, we describe some background on multi-criterion recommender systems using different recommendation techniques.

2.1 Recommender Systems

Recommender systems are part of many e-commerce sites in the current internet scenario. Availability of numerous products on these sites results in puzzled customers hence leads to selection of wrong alternatives sometimes. This problem actuates the requirement of such recommender systems.

Unlike MCRS, Recommender system is much widely explored area. Here, the items for a particular user are recommended on the basis of a single attribute of the item for example its overall rating or ranking. According to Adomavicius and Tuzhilin [1] and Nikos and Costopoulou [9], let C be the set of all users and S the set of all possible items that can be recommended. They defined a utility function $U^c(s)$ *as* $U^c(s){:}C \times S \rightarrow R^+$. This function measures the appropriateness of recommending an item s to user c. It is assumed that this function is not known for the whole $C \times S$ space but only on some subset of it. Therefore, in the context of recommendation, we want for each user $c \in C$ to be able to:

- Estimate (or approach) the utility function $U^c(s)$ for an item s of the space S for which $U^c(s)$ is not yet known; or
- Choose a set of items $S' \subset S$ that will maximize $U^c(s)$:

$$c \in C, s = \max_{s \in S'} U^c(s) \quad (1)$$

Therefore by maximizing the above function $U^c(s)$ for the user-item pair, we can get the topmost recommendable item/items for the user.

2.2 Multi-criterion Recommender System

Single criterion based recommender systems proved to be a landmark in the field of web personalization. Further, by embedding some machine learning techniques like genetic algorithms, fuzzy logics etc. into the traditional recommendation algorithms like collaborative filtering and content based filtering, many researchers presented remarkable outcomes in this area.

In addition to the overall (single) rating, there is a need of some additional information for better understanding of the user preferences. Integrating this information into the existing recommender systems leverages the systems with more accuracy and efficiency. Manouselis and Costopoulou [9] elaborated on the analysis of MCRS and their classification. Utility of an item for a particular user can also be computed on the basis of more than one attribute of an item in the area of MCRS. The recommendation problems therefore become multi-attribute or multi-criterion problems. Hence these recommendation problems can be solved by formulating them as multi-criterion decision making problems (MCDM) [10]. There are three steps in utilizing any decision-making technique involving numerical analysis of alternatives:

- Determining the relevant criteria and alternatives.
- Attaching numerical measures to the relative importance of the criteria and to the impacts of the alternatives on these criteria.
- Processing the numerical values to determine a ranking of each alternative.

Our proposed system employs multi-criterion decision making in a three step process. Moreover, we apply genetic algorithm approach aiming to extract optimal priorities of a user on various criteria for enhancing the recommendations accuracy. The next section in the paper is focused on the proposed scheme.

3 Proposed Recommendation Framework

The fact worth bearing in mind in multi-dimensional ratings is that the user rates an item giving his main concern to some specific dimensions than others. This can vary from user to user according to one's personal interest and choice. For example, in a movie recommender system, if a user wants to rate a movie on dimensions likes story, acting, direction and visuals etc. then he may give his priorities in decreasing order on these dimensions as follows: 4 for story, 3 for acting, 2 for direction and 1 for visuals. He may be more particular about some dimensions and less about others. Thus it is required to confer the overall rating for a movie as some conclusion based on the criteria ratings. Our proposed scheme is based on the assumption that the overall rating of an item is not just another rating but it is some "aggregation" function *f* of the multi criteria ratings of that item [6] i.e.

$$r_o = f(r_1 \ldots, r_k) \tag{2}$$

where r_o is the overall rating giving to an item and $r_1, \ldots, r_k$ are the individual k criteria ratings. The task is to find a suitable relationship between these individual criteria ratings and the overall rating because each user has different priorities on various criterions.

3.1 Proposed System

We partition our proposed system framework into three following steps:

1. Step *1 (Prediction of N multi-criterion ratings):* We have decomposed the k-dimensional multi-criterion ratings into individual ratings converting the problem into k single criteria based recommendation system problem i.e. $R: User \ x \ Item \rightarrow R_o x \ R_1 x \ldots x R_k$ is converted into $R: User \ x \ Item \rightarrow R_i$ *(where* $i = 1 \ldots k$*)* [6]. Any single criterion recommendation technique (collaborative, content-based, hybrid etc.) can be applied to predict these ratings individually. Collaborative filtering is the most popular filtering technique in the field of single criterion recommender system. After decomposing multi criteria ratings into k individual criteria ratings, we have

employed collaborative filtering for the prediction of these individual unknown ratings in our approach. After finding individual predicted criteria ratings, the aggregation of these individual ratings to get overall rating is an important task in multi criteria recommender system because users have different priorities on various criteria. We have used genetic algorithm for extracting priorities of users on these criteria. The necessary components of genetic algorithm are described in the next step (Fig. 1).

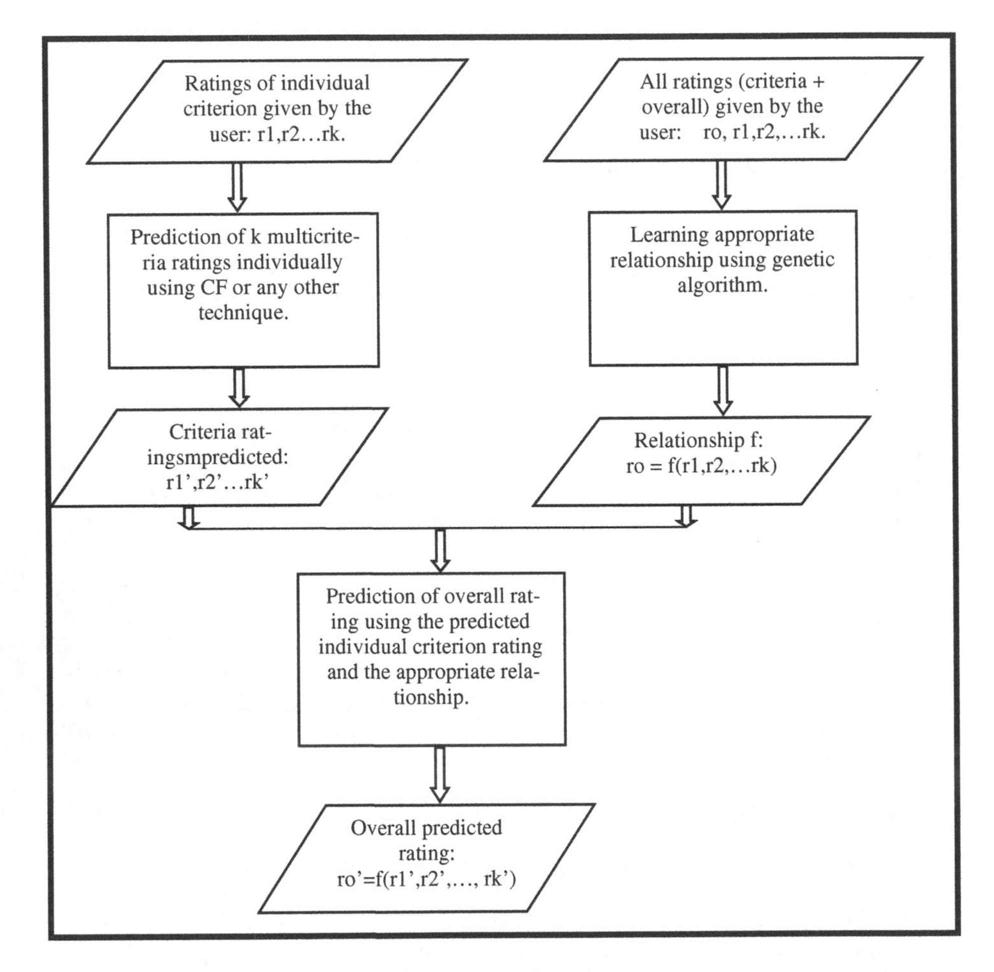

Fig. 1. Proposed MCRS framework

2. *Step 2 (Learning the function)*: We obtain optimized weights for individual criterion for each user through genetic algorithm. As mentioned above, we have used genetic

algorithm approach to find weights on various criterion of every user. Genetic algorithm is an effective search method based on the technique of natural selection and genetics. It is an adaptive, efficient search method and suitable for solving optimization tasks, especially for the large search space. The feature weight of user u_a is represented as a set of weights, weight $((u_a) = [w_i]$ where $i = 1, \ldots, k$ where k is the number of available dimensions. When the weight for any feature is zero, that feature is not considered for further calculation. This enables feature selection to be adaptive to each user's preference. The main purpose of employing GA here is to search best fitted chromosomes that can represent the weights given by the user to each dimension. The main steps of the proposed GA are as follows:

- *Chromosome representation:* Chromosome representation is the mapping of transferring a candidate solution into a chromosome consisting of several genes of binary bits, integers or real numbers etc. depending on optimization problem. In our scheme the genotype of w_i is a string of double vector values. This means that we represent each chromosome by a double vector.
- *Initial Population:* The two popular aspects of population used in GA are *initial population* and *population size*. For each and every problem the population size depends upon the complexity of the problem. Ideally the size of the problem should be as large as possible in order to explore maximum possible search space. Here we chose the initial population randomly and the size of the population is 20
- *Genetic Operators:* The basic operators used in GA are crossover, mutation, encoding and selection etc. Amongst them crossover and mutation are the two most popular operators. Crossover allows formation of two fresh offspring by allowing two parent chromosomes interchanging important meaningful genetic materials and mutation is used to maintain the inherited diversity of the population by introducing a completely new member into the latest population.

  Crossover: Crossover is the process of taking two parent solutions and producing from them a child. There are several methods for crossover for example, single point, two-point, uniform, three-parent crossover etc. In this work, we have chosen traditional single-point crossover. Here, a cross site or a crossover point is selected randomly along the length of the mated strings and bits next to the cross site are exchanged.

  Mutation: After crossover the strings are subjected to mutation. Mutation plays the role of recovering the lost genetic materials as well as for randomly distributing genetic information. It randomly alters one or more genes of a selected chromosome. Mutation also has several methods amongst which we have chosen adaptive feasible. Adaptive feasible randomly generates directions that are adaptive with respect to the last successful or unsuccessful generation.

  Selection: Selection is the process of choosing two parents from the population for crossing. Roulette Wheel Selection is one of the traditional GA selection technique that is being used in our proposed scheme.
- *Fitness Function*: Fitness function measures the excellence of each chromosome after the initialization of population. It is the most significant part of GA. The choice of a fitness function is a challenging job for GA algorithm. Fitness

function is a factor which drives the GA process towards the convergence of the optimal solution. Chromosomes having more fitness values are allowed to breed and mix their datasets with any of several techniques, producing a new generation that will be even better. In our approach fitness function is evaluated by using statistical techniques to minimize the difference between the overall rating and the aggregation of the individual ratings. We formulated the fitness function as follows

$$fitness = \sum_{j=1}^{Nc} \left| \left(r_{o(overall)} - \frac{w_1 r_{i1} + \cdots + w_k r_{ik}}{(w_1 + \cdots + w_k)} \right) \right| \quad (3)$$

Where N_c is the total number of the items rated by the user u_c in the training set and r_o is the overall rating given to an item. Our aim is to minimize the difference between the overall rating and the aggregation function using genetic algorithm to obtain the most optimized weights.

- *Stopping Criteria:* In our scheme, we have pre-assigned number of generations as stopping criteria. After 50 generations, the chromosomes with the minimum fitness value are taken as final output. These values are assigned to each dimension by the user. In the present approach there are four values (four criteria) to be taken as final output.

3. *Step 3 (Predicting the overall ratings)*: Finally as mentioned earlier the overall ratings r'_o in the testing data are calculated by using the aggregation function

$$r'_{o(overall)} = \frac{w_1 r_{i1} + \cdots + w_k r_{ik}}{(w_1 + \cdots + w_k)} \quad (4)$$

4 Experimental Setup and Result Analysis

4.1 Design of Experiments

From the Yahoo! Movies dataset [7], we extracted only those users who have rated at least 20 movies and only those movies that are rated by at least 20 users (YH-20-20). Hence we ended up on 186 users, 181 movies and 6398 ratings out of 62156 total ratings. To remove the bias of proposed scheme, we split YH-20-20 dataset into different samples namely Sample 1, Sample 2 and Sample 3. Each sample has 50 users and their ratings on items. The sample data is further divided into two separate datasets randomly 80 and 20 %. The larger one (Training Dataset) is used for learning the system and the smaller (Testing Dataset) is used for calculating the ratings. We have also used the full dataset YH-20-20 for computing the absolute difference of predicted ratings and overall ratings to make a user level comparison. These calculated ratings are matched with the actual ratings given by the user for analyzing the performance of the proposed system.

4.2 Performance Evaluation

In order to test the performance of our multi-criterion recommendation strategy, we measure system's accuracy using three evaluation metrics namely, mean absolute error, precision and recall. We also calculated the overall rating with the traditional Collaborative filtering (CF) approach to compare the proposed multi-criteria recommendation system scheme results (MC-RS).

- *Mean Absolute Error (MAE):* Mean absolute error for each user is calculated as:

$$MAE\left(u_c\right) = \sum_{j=1}^{Nc} \left| \frac{\left(r_{o(overall)} - r'_{o(overall)}\right)}{Nc} \right| \tag{5}$$

where u_c user in consideration and N_c is set of movies rated by the user u_c. We have also calculated the overall MAE of the system.

$$MAE = \frac{1}{N} \sum_{j=1}^{N} \left| r_{o(overall)} - r'_{o(overall)} \right| \tag{6}$$

Where N is the total number of overall ratings

- *Precision*: Precision, measuring correctness of recommendation, is defined as the ratio of the number of selected items to the number of recommended items

$$Precision = \frac{Number\ of\ relevant\ items\ recommended}{Total\ Number\ of\ Items\ Recommended} \tag{7}$$

- *Recall:* Recall can be used as a measure of the ability of the system to all relevant resources.

$$Recall = \frac{Number\ of\ relevant\ items\ recommended}{Total\ Number\ of\ relevant\ items} \tag{8}$$

4.3 Experiments

To demonstrate the feasibility and effectiveness of the proposed scheme, we computed the overall rating with the traditional collaborative filtering approach (CF-RS) to compare the proposed multi-criteria recommendation system scheme (MC-RS) in terms of various performance measures. Table 1 shows the comparisons of MAE and Table 2 shows the comparisons of precision and recall on different Samples. A lower value of MAE and higher values of precision & recall show the effectiveness of any scheme. From Table 1, it is clear that our proposed scheme MC-RS performs better in comparison to CF-RS in terms of MAE because lower value of MAE. In the similar manner, Table 2 demonstrates the effectiveness of our proposed scheme MC-RS in terms of precision & recall.

To demonstrate effectiveness of our MC-RS scheme on each user level, we have depicted the absolute difference between user rating and predicted rating in Fig. 2 where series 1 and series 2 show total absolute differences between the overall ratings given by each user with MC-RS ratings and with CF-RS respectively. Figure 2 clearly shows the higher absolute difference in series 2 in comparison to series 1 reflecting the effectiveness of proposed scheme MC-RS on YH-20-20 dataset.

Table 1. Comparison of MAE of CF-RS and MC-RS

Datasets	Schemes	
	CF-RS (MAE)	MC-RS (MAE)
Sample 1	2.077	**2.003**
Sample 3	2.110	**2.037**
Sample 3	2.053	**1.981**

Table 2. Comparison of precision and recall of CF-RS and MC-RS

Datasets	Measures	Scheme	
		CF-RS	MC-RS
Sample 1	**Precision**	0.7589	0. 7854
	Recall	0.5462	0.5937
Sample 2	**Precision**	0.7614	0.7976
	Recall	0.5534	0.5889
Sample 3	**Precision**	0.7631	0.8002
	Recall	0.5594	0.5890

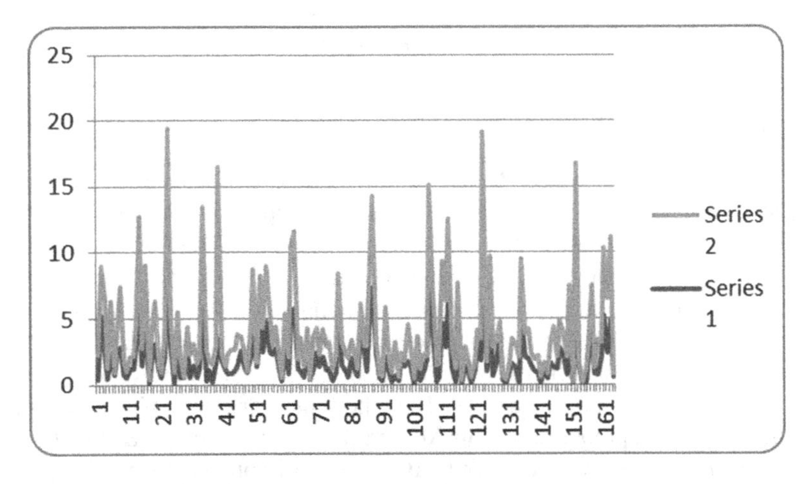

Fig. 2. Absolute differences between the ratings

5 Conclusions

Our work in this paper is an attempt towards finding an optimal relation between the overall ratings and individual criterion ratings in multi-dimensional datasets. We introduce a naive approach in multi-criterion recommendation systems by employing genetic algorithm in addition to the traditional CF. Our approach bridges the gap between the ratings of each dimension and the overall rating. The experimental study is conducted on the real world data of Yahoo! Movies to make our work pragmatic. The outcomes clearly indicate that proposed scheme outperforms the CF-RS. In our future work we plan to extend the current approach by employing fuzzy logic and statistical techniques to learn appropriate aggregation function for combining individual criterion ratings [11].

References

1. Adomavicius, G., Tuzhilin, A.: Toward the next generation of recommender systems: a survey of the state-of-the-art and possible extensions. IEEE Trans. Knowl. Data Eng. **17**(6), 734–749 (2005)
2. Burke, R.: Hybrid recommendation systems: survey and experiments. User Model. User-Adap. Inter. **12**(4), 331–337 (2002)
3. Kant, V., Bharadwaj, K.K.: A user-oriented content based recommender system based on reclusive methods and interactive genetic algorithm. In: Proceedings of the Seventh International Conference on Bio-Inspired Computing: Theories and Applications (BIC-TA 2012) Advances in Intelligent Systems and Computing, vol. 201, pp. 543–554. Springer, India (2013)
4. Kant, V., Bharadwaj, K.K.: Enhancing recommendation quality of content-based filtering through collaborative predictions and fuzzy similarity measures. In: Proceeding of the International Conference on Modeling, Optimization and Computing (ICMOC 2012), Procedia Engineering, vol. 38, 939–944 (2012)
5. Kant, V., Bharadwaj, K.K.: Fuzzy computational models of trust and distrust for enhanced recommendations. Int. J. Intell. Syst. **28**(4), 332–365 (2013)
6. Adomavicius, G., Kwon, Y.O.: New recommendation techniques for multicriteria rating systems. IEEE Intell. Syst. **22**(3), 48–55 (2007)
7. Lakiotaki, K., Matsatsinis, N.F., Tsoukias, A.: Multi-criteria user modeling in recommender systems. IEEE Intell. Syst. **26**(2), 64–76 (2011)
8. Sahoo, N., Krishnan, R., Dunkan, G., Callan, J.P.: Collaborative filtering with multi component rating for recommender systems. In Proceedings of the Sixteenth Annual Workshop on Information Technologies and Systems (WITS 2006) (2006)
9. Manouselis, N., Costopoulou, C.: Analysis and classification of multi-criteria recommender systems. World Wide Web **10**(4), 415–441 (2007)
10. Triantaphyllou, E., Shu, B., Sanchez, S., Ray, T.: Multi-criteria decision making: an operations research approach. In: Webster, J.G. (ed.) Encyclopedia of Electrical and Electronics Engineering, vol. 15, pp. 175–186. Wiley, New York (1998)
11. Kant, V., Bharadwaj, K.K.: Incorporating fuzzy trust in collaborative filtering based recommender systems. In: Panigrahi, B.K., Suganthan, P.N., Das, S., Satapathy, S.C. (eds.) SEMCCO 2011, Part I. LNCS, vol. 7076, pp. 433–440. Springer, Heidelberg (2011)

Adapting *PageRank* to Position Events in Time

Abhijit Sahoo(✉), Swapnil Hingmire, and Sutanu Chakraborti

Department of Computer Science and Engineering,
Indian Institute of Technology, Chennai, India
{asahoo,swapnilh,sutanuc}@cse.iitm.ac.in

Abstract. In this paper, we order events in time by using evidence present in their partial orders. We propose an algorithm named *TimeRank*, a variant of *PageRank*, for this task. *PageRank* operates on the hyperlink graph and orders the web pages according to their importance. We identify limitations of *PageRank* in the context of temporally ordering the nodes. We draw an analogy between the notion of importance in *PageRank* to the notion of *recency* in *TimeRank*. We evaluate *TimeRank* using the *Citation Graph* of scientific publications of physics and propose a baseline method to compare *TimeRank* and *PageRank*. The baseline method ranks the nodes according to their number of immediate predecessors without considering the higher order transitive relations among the events. Evaluation results suggest that *TimeRank* outperforms both the baseline method and *PageRank* in this task.

1 Introduction and Related Works

In the real world, we often encounter situations where we need to order things temporally but we only have partial information about them. For example, to order past events of our life in time, we often relate them with other pivot events for which the date is known, using AFTER or BEFORE relations. We try to combine the evidence from the partial orders to arrive at an ordering of all the events. We devise an algorithm named as *TimeRank*, a variant of *PageRank* [6], that does the same by assigning a *TimeRank* score to each event. *TimeRank* can be operated on an *Event Graph* where the events are used as nodes and the temporal ordering between the events are represented as the directed edges.

TimeRank can have interesting applications in estimating the occurrence time of events in history, automatic biography compilation and creating a timeline of events from history documents. In history, there are uncertainties associated with the occurrence of events. For example, the birth year of *Gautama Buddha*, the founder of Buddhism, is uncertain[1]. The occurrence time of an event is estimated based on evidence of its associated events that are known in the history. Thus, the new evidence can potentially reorder certain events in time and the ordering can be more accurate as more evidence is used. *TimeRank* can model this progressive reduction of uncertainty. It can combine the evidence from higher order associations of events to arrive at an ordering of events. *TimeRank*

[1] https://en.wikipedia.org/wiki/Gautama_Buddha.

R. Prasath et al. (Eds.): MIKE 2015, LNAI 9468, pp. 536–542, 2015.
DOI: 10.1007/978-3-319-26832-3_50

can be applied to order biographical events extracted from the digital trails of a person like emails, search logs and his information from social medias. Similarly, we can extract events from the text of a history document and can represent them as the nodes of an *Event Graph.* The history document should present the events in a chronological order which can be exploited to obtain the partial temporal orders among the events. As the events are presented using plain text in the document, there is a high chance of presenting some events in out of chronological order. Earlier events can be referred at a later part of the document which introduces uncertainty in the partial orders of the events and results in circularity among the events in the *Event Graph. TimeRank* can potentially resolve the temporal order using the available evidence in the event's higher order associations.

TimeRank is built upon the *PageRank* algorithm. To best of our knowledge, no extension of *PageRank* has been proposed so far to realize temporally ordering of events. Some related works are mentioned below. Mani et al. [4] anchors the clauses from text and orders them in a sentence level. Vrotsou [7] mines the sequence of daily events to observe the evolution and trends of the events. The notion of events in [4,7] are entirely different from our notion of notable events. O'Madadhain et al. [5] ranks individuals on a social network using the event sequences of interactions among users in the network, Berberich et al. in [1] and Jiang et al. in [2] try to improve *PageRank* using time as a component, unlike our algorithm which orders the events in time.

2 Overview

An *Event Graph*, $G = (V, E)$ where v_i is an event and $e_{ij} = 1$, if v_i has a directed edge to v_j. An *Event Graph* is similar to the hyperlink graph of the web where the web pages and their hyperlinks are analogous to the events and their order of occurrence respectively. *PageRank* [6] orders the web, according to the importance of web pages using the hyperlink graph. In the context of *PageRank*, a web page is important if it is pointed to by other important pages. Analogously, an event in an *Event Graph* is recent if it is pointed to by several recent events. In this setting, can we deploy *PageRank* on G to order the events in a chronology using the evidence present in the directed edges of G? Although an *Event Graph* is similar to a hyperlink graph, there is a significant difference. Unlike a hyperlink graph, it is important to respect transitivity in an *Event Graph.* For example, in a hyperlink graph, if a node A points to node B and B points to node C then A does not necessarily intend to point to C. This is not true in an *Event Graph.* In an *Event Graph* if an event A occurs before an event B and B occurs before an event C then one can infer that A must occur before C. The idea of *TimeRank* is motivated by this observation. The approach is presented in Sect. 2.1. We evaluate *TimeRank* using *Citation Graphs*, the details are in Sect. 2.2.

Table 1. Ranking of nodes using *TimeRank* and *PageRank* on the graph shown in Fig. 1(a)

Time marker →	Least recent	– – – – – – →						Most recent
Expected rank order (sorted in ascending order of recency)	1	2	3	4	5	7	8	6
PageRank rank order	1	2	5	7	8	3	6	4
TimeRank rank order	1	2	3	4	5	7	8	6

2.1 Approach

As discussed above, *PageRank* has limitations when applied to the task of ordering nodes of an *Event Graph* in time due to the implicit transitive relations among events. To address this problem, *TimeRank* explicitly adds virtual edges to the graph. The virtual edges capture transitivity of events. With the addition of virtual edges, all the indirect predecessors of any node become its direct predecessors and can pass their *recency* scores directly to it. We analyze the situation in detail using the toy example as shown in Fig. 1(a). To apply *TimeRank* on the graph in Fig. 1(a), we augment the graph with second order virtual edges (as shown in 1(b)) and apply *PageRank* on the augmented graph. Table 1 shows that the rank order obtained by *PageRank* for the graph as shown in Fig. 1(a), varies from the expected rank order whereas *TimeRank* gives the correct order.

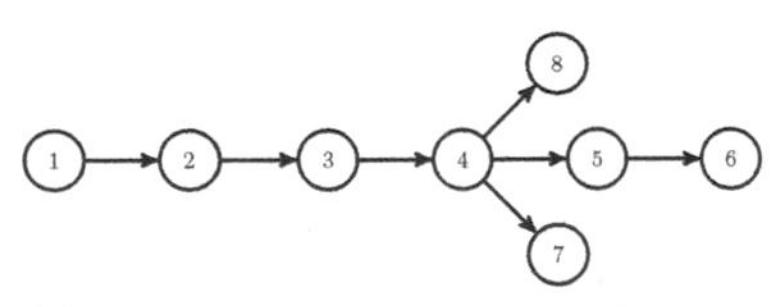

(a) *Event Graph* before adding virtual edges

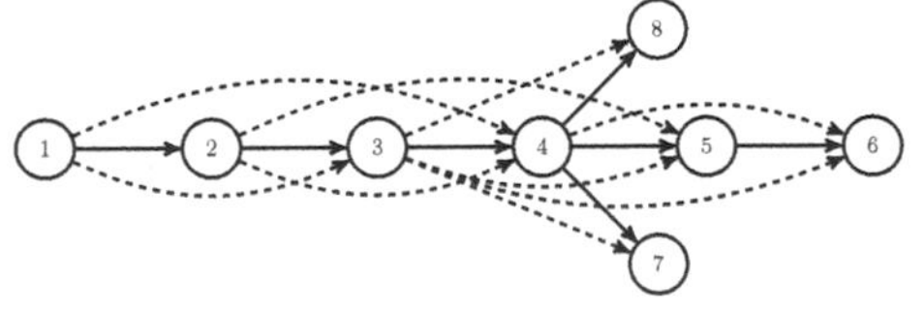

(b) *Event Graph* after adding second order virtual edges

Fig. 1. *Event Graphs* with and without virtual edges

2.2 Evaluation

We empirically evaluate *TimeRank* on a *Citation Graph. Citation Graph*s are constructed using research articles as nodes and their citations as directed edges. The reason for choosing *Citation Graph*s for evaluating *TimeRank* is that they contain the required temporal properties as in the *Event Graph* i.e., the later articles cite older articles. The temporal ordering information is present in the directed edges of the *Citation Graph* as partial orders. *TimeRank* combines the evidence present in the directed edges and orders all the nodes in time. To perform the evaluation, we pretend that actual years of publications are not available, only the citation information is available regarding the *Citation Graph.* The order is obtained by *TimeRank.* The time of appearance of each node in the network is known which we use as the ground truth to compare and evaluate the rank order output of *TimeRank*.

The *Citation Graphs* are directed acyclic graphs, unlike real-world *Event Graphs* that can contain cycles. The noise and uncertainty present in the order of occurrence of the events that are extracted from plain text result in those cycles to appear, as illustrated in Sect. 1. To simulate the properties of real-world *Event Graph*, we introduce random noise in the *Citation Graph*. The random noise maps to the arbitrary edges which introduce uncertainty and circularity into the *Event Graph*. With adding different percentages of such noise, we operate and evaluate *TimeRank* on the *Citation Graphs*. We compare the output of *TimeRank* with a basic inlink counting (BIC) method as described below in Sect. 3. The BIC method orders the nodes only by the number of immediate predecessors and does not take the higher order associations into consideration. We use *PageRank* to considers the higher order associations, but it does not consider the information from the implicit transitive relations. *TimeRank* considers both higher order associations and transitive relations. The empirical results suggest that *TimeRank* outperforms both BIC and *PageRank*, in the context of ordering events in time.

3 Experiments

Data. We use the *Citation Graph* of *High-energy physics theory citation network* from *Stanford Network Analysis Project*[2]. The complete *Citation Graph* has 27770 nodes, 352807 edges, with a diameter of 13 and 90-percentile effective diameter of 5.3. We took a subgraph of the *Citation Graph* with 11444 nodes and 81088 edges for our experiments.

Baseline Method. We use basic inlink counting (BIC) as our baseline method. In this method, the nodes are ranked according to their numbers of inlinks. In our *Citation Graph*, a directed edge from node A to node B exists if B cites A. Here, the assumption is that the recent nodes will have a higher number of inlinks than older nodes. In the context of *TimeRank* the statement of circularity is: a node is recent if it is pointed to by several recent nodes. The obtained rank orders using *TimeRank*, *PageRank* and BIC are compared against the actual rank order that is obtained using the actual publication dates of articles in the *Citation Graph*. It can be seen that *TimeRank* and *PageRank* capture higher order temporal relations that BIC fails to capture, and *TimeRank* can capture transitive relations that *PageRank* fails to capture.

Measure. We use *Kendall rank correlation coefficient* [3] to compare two rank orders. *Kendall rank correlation* measures the similarity between two rank orders as follows. In a rank order $(x_1, y_1), (x_2, y_2)....(x_n, y_n)$, any two pairs of observation (x_i, y_i) and (x_j, y_j) are said to be *concordant* if $x_i > x_j$ and $y_i > y_j$ or if

[2] https://snap.stanford.edu/.

$x_i < x_j$ and $y_i < y_j$. Any two pairs of observations (x_i, y_i) and (x_j, y_j) are said to be discordant if $x_i > x_j$ and $y_i < y_j$ or if $x_i < x_j$ and $y_i > y_j$. The *Kendall tau rank correlation coefficient* is defined as the difference between the number of concordant and discordant pairs divided by the total number of possible pairs. The range of the correlation coefficient is from −1 to +1. In our experiment, x_i is the obtained rank and y_i is the correct rank of an event. The range of the correlation coefficient is from −1 to +1.

4 Results

TimeRank(TR) and *basic inlink count (BIC)* method are operated on the *Citation Graph* with the addition of different order of transitive virtual edges. TR^k and BIC^k refer to *TimeRank* and BIC of kth order respectively. TR^0 and BIC^0 do not add virtual edges to the graph and TR^0 is same as *PageRank*. BIC^1 and TR^1 add first order transitive virtual edges, and BIC^2 and TR^2 add second order transitive virtual edges to the graph. All the methods are operated on the *Citation Graph* with the addition of different percentages of random noise to it.

Kendall's tau coefficients between rank orders of different methods are compared. The results as shown in Table 2 is divided into three parts. The first part shows the comparison between BIC^0 and *PageRank* or TR^0, the second part shows the comparison between BIC^1 and TR^1 and the third part shows the comparison between BIC^2 and TR^2. We can see in all the 6 cases the performance of $TR^1 > PR$, $TR^2 > TR^1$, $TR^1 > BIC^1$ and $TR^2 > BIC^2$. Interestingly, the BIC^0 performs better than *PageRank* in 2 out of 6 cases. Some selected results of BIC^0, TR^0 and BIC^2, TR^2 are presented using the histogram as shown in Fig. 2. Note that *PageRank*(PR) is same as TR^0.

Table 2. Presenting the comparison between *TimeRank* (TR^k), *PageRank* (PR) and Basic Inlink Count (BIC^k) method.

Percentage	Methods					
of noise	BIC^0	TR^0/PR	BIC^1	TR^1	BIC^2	TR^2
0	0.3385	0.4294	0.3745	0.4941	0.4403	0.5514
3.0	0.3345	0.3938	0.3613	0.4489	0.4032	0.4612
5.0	0.3314	0.3714	0.3526	0.4209	0.3799	0.421
10.0	0.3279	0.3361	0.3418	0.3738	0.3575	0.3745
20.0	0.3183	0.2879	0.323	0.3307	0.3283	0.3393
30.0	0.3094	0.265	0.3086	0.3092	0.3085	0.3135

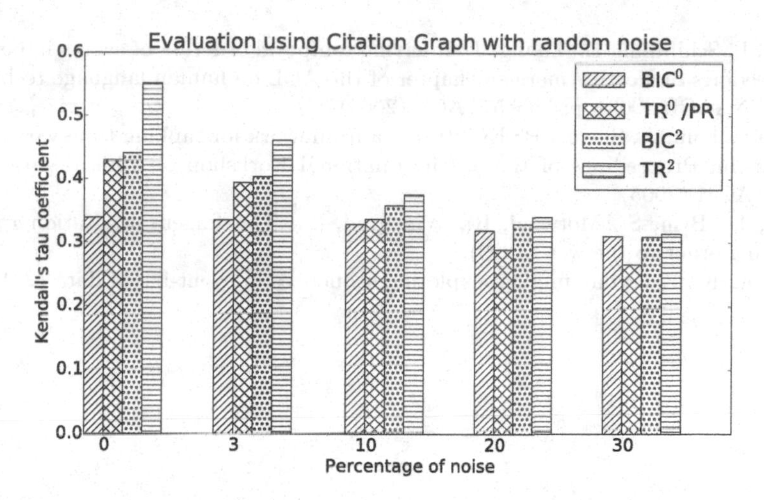

Fig. 2. Histograms showing the comparison between *TimeRank* (TR^k), *PageRank* (PR) and basic inlinks count (BIC^k) method.

5 Conclusion and Discussion

We propose an algorithm named *TimeRank* that can temporally order the nodes of an *Event Graph*. We use *PageRank* for the task, but it does not consider the information present in the implicit transitive relations among the events. To make use of this information, *TimeRank* adds virtual edges to the graph. We evaluated *TimeRank* on *Citation Graph* and compared it against *PageRank* and our proposed baseline method named as BIC. BIC does not take the higher order associations into consideration. *TimeRank* outperforms both *PageRank* and BIC.

TimeRank can be applied on text data of a history book to order the historical events in time. The index of the book can be used to construct an *Event Graph* as the events of a history book are mentioned on the index page along with their page number of occurrence. The events must be presented in a chronological order in the book from whose index the partial orders of the events are extracted. *TimeRank* can be used to order the events in the *Event Graph*. We can evaluate the ordering of *TimeRank* using the actual chronology of events.

References

1. Berberich, K., Vazirgiannis, M., Weikum, G.: T-Rank: time-aware authority ranking. In: Leonardi, S. (ed.) WAW 2004. LNCS, vol. 3243, pp. 131–142. Springer, Heidelberg (2004)
2. Jiang, H., Ge, Y.-X., Zuo, D., Han, B.: Timerank: a method of improving ranking scores by visited time. In: 2008 International Conference on Machine Learning and Cybernetics, vol. 3, pp. 1654–1657 (2008)
3. Kendall, M.G.: A new measure of rank correlation. Biometrika **30**(1–2), 81–93 (1938)

4. Mani, I., Schiffman, B., Zhang, J.: Inferring temporal ordering of events in news. In: Proceedings of North American chapter of the ACL on human language technology (HLT-NAACL 2003), pp. 55–57. ACL (2003)
5. O'Madadhain, J., Smyth, P.: Eventrank: a framework for ranking time-varying networks. In: Proceedings of the 3rd international workshop on Link discovery, pp. 9–16. ACM (2005)
6. Page, L., Brin, S., Motwani, R., Winograd, T.: The Pagerank citation ranking: bringing order to the web (1999)
7. Vrotsou, K.: Everyday mining: exploring sequences in event-based data (2010)

After You, Who? Data Mining for Predicting Replacements

Girish Keshav Palshikar(✉), Kuleshwar Sahu, and Rajiv Srivastava

TCS Innovation Labs - TRDDC, Tata Consultancy Services Limited,
54B Hadapsar Industrial Estate, Pune 411013, India
{gk.palshikar,kuleshwar.sahu,rajiv.srivastava}@tcs.com

Abstract. This paper proposes a new class of data mining problems in which agents replace their current object (*predecessor*) by another object (*replacement* or *successor*); the problem is to discover the knowledge used by the agents in identifying suitable successors. While such replacement data is available in many practical applications, in this paper we explore a problem in *HR analytics*, viz., replacing person in a key position in a project by another most suitable person from other employees. We propose unsupervised (distance-based) algorithms for finding suitable replacements. We also apply several standard classification techniques. This paper is the first in applying *metric learning* algorithms to a problem in HR analytics. We compare the approaches using a real-life replacement dataset from a multinational IT company. Results show that metric learning is a promising approach that captures the implicit knowledge for replacement identification.

1 Introduction

In many practical situations, each *agent*, posseses an *object* (*predecessor*); then some agents replace their object by another object (*replacement* or *successor*). Agents use their own knowledge in identifying the "right" replacement. This knowledge is agent-specific and not externally visible; we assume that many *knowledge elements* are common across many agents. Both the predecesor and replacement objects are described in terms of the same *features* or *attributes*. Given a dataset of such *replacements*, it is interesting to find the frequently occurring patterns in the replacements, and thereby gain some understanding of the knowledge used by the agents for finding the right replacements. Objects replaced may be houses, cars, mobile service providers, insurance policies or even people (in a team). In this paper, we consider one particular application, with a focus on the workforce in IT industry. Often, a person working in a key position or role in a project has to be released due to reasons such as: resignation, retirement, transfer, long leave, re-allocation etc. It then becomes necessary to identify the "most suitable" replacement for this person, from among the pool of available employees. Identifying the "most suitable" replacement is important for a smooth taking over of responsibilities, transfering knowledge, ensuring continuity and minimizing impact on the project and customer satisfaction. In modern

R. Prasath et al. (Eds.): MIKE 2015, LNAI 9468, pp. 543–552, 2015.
DOI: 10.1007/978-3-319-26832-3_51

organizations, *replacement identification* is an important HR process. Replacements are often sub-optimal due to urgency and manual search over a limited pool of reachable employees. The HR process is manual, subjective, there is little use of objective data and no learning from the past replacements.

In this paper, we propose a data-driven analytics-based approach to solving the problem of replacement identification. A goal is to minimize the search for the "best" successor (i.e., minimize the pool of plausible candidate employees examined) and maximize the chances of finding the "best" replacement. Section 2 formalizes the problem of replacement identification. Section 3 introduces the real-life case-study dataset used. Section 4 presents various approaches for replacement identification: unsupervised, classification-based and a novel metric-learning based) and reports their results on the case-study dataset. Section 5 presents related work and Sect. 6 discusses conclusions and future work.

2 Problem Formulation

We are given a database $E = \{\mathbf{e}_1, \mathbf{e}_2, \ldots\}$ of employees and a particular employee $\mathbf{e} \in E$ to be replaced (predecessor). Each employee $\mathbf{e}_i$ is described in terms of the same set of n variables $V = \{V_1, \ldots, V_n\}$ i.e., each $\mathbf{e}_i = (x_1, \ldots, x_n)$, where x_j is a value for the variable V_j. Some of the n variables may be categorical. The goal is to identify a subset $H(\mathbf{e}) \subseteq E$ containing K "most suitable" replacements for $\mathbf{e}$ (K is given). $H_{\mathbf{e}}$ can be thought of as the short-listed set of K possible replacements. For better control of the human factors involved in people management, the final replacement will still be manually identified from $H(\mathbf{e})$.

Suitability of any employee in E as a possible replacement for $\mathbf{e}$ is measured by a score function (unknown, in general) $f : E \times E \rightarrow [0, 1]$, where $f(\mathbf{e}, \mathbf{e}')$ denotes how suitable employee $\mathbf{e}'$ is as a possible replacement for predecessor $\mathbf{e}$. We need to estimate this score function. We assume that f is a similarity function i.e., the more similar an employee is to a given predecessor, the higher are its chances of being chosen as the replacement. We often have available a dataset of actual replacements $D = \{(\mathbf{x}_1, \mathbf{y}_1), \ldots, (\mathbf{x}_N, \mathbf{y}_N), \}$ of N (predecessor, replacement) pairs, where each $\mathbf{x}_i, \mathbf{y}_i \in E$. Note that D includes neither the set of plausible candidates nor the set of short-listed candidates for any of the replacements. We can now have the following approaches for replacement identification. (1) Ignoring D, use a geometric distance measure (such as Euclidean or Mahalanobis) to identify K employees "nearest" to $\mathbf{e}$. (2) Learn a classifier from D (after adding some negative examples of replacements to D), apply it to each employee in E, and choose a subset of K employees with predicted label +1. (3) Use metric learning to learn the score function using D, then use it to identify K employees "nearest" to $\mathbf{e}$.

Note that D consists of only positive examples of replacement. Considering *all* other employees as negative examples for every replacement is wasteful. So we propose a multi-step approach with the following steps ($\mathbf{e}$ is the predecessor employee to be replaced). We focus on developing an analytics-based approach for automating the first two steps. The third step is manual for better control and to allow taking human factors into account.

1. In the first step (*candidate identification*) only a subset $SIM(\mathbf{e})$ of the remaining employees are identified, where each employee in $SIM(\mathbf{e})$ is considered as "plausible" replacement for e and the actual replacement is chosen only from this subset. The employees in $SIM(\mathbf{e})$ are candidate replacements for **e**; each employee in $SIM(\mathbf{e})$ is supposed to be "highly similar" to **e** in terms of some well-defined criteria.
2. In the second step, called *replacement short-listing*, once $SIM(\mathbf{e})$ is identified, a small K-size subset $H(e)$ of $SIM(e)$ is identified as the "most suitable" replacement short-list for **e** (usually, $K = 5$, 10, or 15). Often, these top K most suitable replacements in $H(\mathbf{e})$ are ranked in terms of their suitability as a replacement for **e**.
3. In the third step, called the *final selection*, an employee **r** in $H(\mathbf{e})$ is identified as the *actual replacement* for **e**. This step is often manual, subjective and consists of human-centric HR processes (such as interviews and feedback).

3 Case-Study Dataset

We have data from a particular business unit of a large multi-national IT services organization for the two-year period 01-Apr-2011 to 31-Mar-2013. First part D_1 of the dataset consists of 546 pairs of employees (predecessor, replacement), where the predecessor is always in the PROJECT_LEADER role. The reason for this focus is because PROJECT_LEADER is the most crucial leadership role in a project and carries the overall responsibility of delivering the project within the specified time, efforts and budget, and also meeting stringent quality criteria. D_1 does not include any data about the candidate set $SIM(\mathbf{e})$ nor about the successor short-list set $H(\mathbf{e})$, for any predecessor **e** (we only know the final (actual) successor **r** for each **e**). Each pair of employees in D_1 can be considered as a *positive* example of replacement. The second part D_2 of the dataset consists of employees (from this business unit only), who were neither predecessors nor replacements in the first dataset i.e., they are the remaining employees. Each employee, including predecessors and replacements, is described using the data variables (i.e., features); some are shown below:

V5	Text	Project type: most experienced in
V6	Text	Project type: second most experienced in
V7	Text	Project nature: most experienced in
V8	Text	Project nature: second most experienced in
V9	Text	Project application area: most experienced in
V10	Text	Project application area: second most experienced in
V11	Text	Project technology platform: most experienced in
V12	Text	Project technology platform: second most experienced in

Each employee has demographic attributes such as gender, age, or *grade*. Work in each account (i.e., customer) is organized as a set of software development *projects* for that customer. Each project has a start and end date and a designated team, mostly consisting of software engineers. Each team member in a project has a specific *role* (e.g., PROJECT_LEADER, REQ_ANALYST,

DB_DESIGNER, DEVELOPER, TESTER) and carries out tasks in accordance with the assigned role. A subset of roles is designated as *leadership roles* and experience of an employee in them is counted as *leadership experience*. A total of 4528 projects were active in this period, on which a total of 1597918 person-days efforts were spent by 5132 distinct employees. The actual number of employees varies from month-to-month; hence if an employee **e** leaves in a particular month, only employees present in that month need to be considered for replacement for **e**. Over time, each employee works in many different projects. Each project has several attributes such as *project type* (e.g., `Production Support, Maintenance, Development, Reengineering, Migration` etc.), *project nature* (e.g., `T&M, Turnkey, Consulting`), *technology platforms* (e.g., `Unix, AS 400, Mainframe, MS-Windows`), *application area* (e.g., `Groupware, E-Commerce, DW, Systems Management`). Values in the record for each employee vary with time; e.g., age or experience i.e., there is actually a sequence of records for each employee, one record for each unit of time (e.g., quarter). When a successor is chosen for a predecessor at a particular point in time, the records of the corresponding quarter are used for identifying candidate employees and for preparing the successor short-list. Salient patterns in dataset D_1 are:

1. 166 (30.4 %) replacements had more experience than predecessor; 231 (42.3 %) replacements had less experience than predecessor; and 149 (27.3 %) replacements had "similar" experience (±1 year) as predecessor.
2. 165 (30.2 %) replacements were older than predecessor; 223 (40.8 %) replacements were younger than predecessor; and 158 (28.9 %) replacements had "similar" age (±1 year) as predecessor.
3. 47 (8.6 %) replacements had more leadership experience than predecessor; 216 (39.6 %) replacements had less leadership experience than predecessor; and 283 (51.8 %) actual replacements had "similar" leadership experience (±10 %) as predecessor.
4. 98.7 % replacements were from the same business unit as predecessor.
5. 98.2 % replacements have the same project nature as the predecessor; i.e., `e.V7 = r.V7` or `e.V7 = r.V8` or `e.V8 = r.V7` or `e.V8 = r.V8`.
6. 93.0 % replacements have same technology platform as predecessor; i.e., scriptsize `e.V11 = r.V11` or `e.V11 = r.V12` or `e.V12 = r.V11` or `e.V12 = r.V12`.
7. 93.0 % replacements have same application area as predecessor; i.e., `e.V9 = r.V9` or `e.V9 = r.V10` or `e.V10 = r.V9` or `e.V10 = r.V10`.
8. 91.9 % replacements have same project type as predecessor; i.e., `e.V5 = r.V5` or `e.V5 = r.V6` or `e.V6 = r.V5` or `e.V6 = r.V6`.
9. 90.5 % replacements are within the same account as the predecessor.
10. 76.0 % replacements are either from the same grade as the predecessor or one grade above or below that of the predecessor.
11. 72.7 % replacements are from the same location as the predecessor
12. 52.9 % replacements had the same role as the predecessor.
13. 50.2 % replacements are within the same project as the predecessor.
14. 42.7 % replacements had similar age as the predecessor (±2 years)

4 Solution Approaches

4.1 Identification of Plausible Candidate Replacements

Let E denote the set of employees available at the time when **e** needs to be replaced. The first need is to identify a subset $SIM(\mathbf{e}) \subseteq E$ of plausible candidate replacements for **e**. We use the patterns found in actual replacements as a guide to define some *domain rules* to identify the candidate subset $SIM(\mathbf{e})$ for **e**; Fig. 1. The sub-condition `Same project_type` stands for the pattern (8) given above;

```
1. Same project_type AND |grade difference| ≤ 1 AND same account (64.1)
2. Same project_type AND |grade difference| ≤ 1 AND different account (5.3)
3. Same project_type AND |total_exp difference| ≤ 4 AND same account (65.6)
4. Same project_type AND |total_exp difference| ≤ 4 AND different account (5.5)
5. Same project_nature AND |grade difference| ≤ 1 AND same account (67.6)
6. Same project_nature AND |grade difference| ≤ 1 AND different account (6.0)
7. Same project_nature AND |total_exp difference| ≤ 4 AND same account (68.7)
8. Same project_nature AND |total_exp difference| ≤ 4 AND different account (6.0)
9. Same application_area AND |grade difference| ≤ 1 AND same account (64.5)
10. Same application_area AND |grade difference| ≤ 1 AND different account (5.9)
11. Same application_area AND |total_exp difference| ≤ 4 AND same account (65.2)
12. Same application_area AND |total_exp difference| ≤ 4 AND different account (5.7)
13. Same technology_platform AND |grade difference| ≤ 1 AND same account (64.8)
14. Same technology_platform AND |grade difference| ≤ 1 AND different account (5.7)
15. Same technology_platform AND |total_exp difference| ≤ 4 AND same account (65.9)
16. Same technology_platform AND |total_exp difference| ≤ 4 AND different account (5.7)
```

Fig. 1. Rules for identifying candidates for a given predecessor (%recall in brackets).

similarly for other sub-conditions (even numbered rules cover rare replacement patterns). With these domain rules, we can design several different strategies for creating $SIM(e)$ for any given e. For a rule R_i, let $S_i(\mathbf{e}) \subseteq E$ denote those employees which satisfy rule R_i. Strategies **Singular** and **Exhaustive** yield large sets of plausible candidates employees for any given predecessor; whereas **Random-All** gives at most $m \times M$ employees in the set $SIM(\mathbf{e})$, where m is the number of rules used.

1. **Singular**: Use only one particular rule (say R_i): then $SIM(\mathbf{e}) = S_i(\mathbf{e})$.
2. **Random-All**: Select a subset $A_i(\mathbf{e})$ of M employees from the set $S_i(\mathbf{e})$. Take a union of these subsets: $SIM(\mathbf{e}) = A_1(\mathbf{e}) \cup A_2(\mathbf{e}) \ldots$.
3. **Exhaustive**: $SIM(\mathbf{e}) = S_1(\mathbf{e}) \cup S_2(\mathbf{e}) \ldots$.

4.2 Distance-Based Unsupervised Approach

algorithm distance_based_replacement_identification
input K, M, employee e (predecessor)
input $D_2 = \{\mathbf{x}_1, \ldots, \ldots, \mathbf{x}_m\}$ set of all employees
output $H \subseteq D_2$ of K feasible replacements (initially empty)
Identify $SIM(\mathbf{e})$ using a suitable strategy // **Singular**, **Random-All** or **Exhaustive**
Rank employees in $SIM(\mathbf{e})$ in terms of their "similarity" with e
$H :=$ the top K employees from this ordered set
return H

A distance (or similarity) based approach for replacement identification is shown above. We use any one of the many possible ways to compute the similarity of each employee in this set $SIM(\mathbf{e})$ with the given predecessor e. Using only the numeric variables, we can use Euclidean or Mahalanobis distance. If we are using only categorical variables, then we can use an appropriate distance measure. If we are using both, then we use a weighted average of the two distances (weights are specified by the user). Currently, we use Hamming distance to compute the distance between two employees in terms of only the categorical variables.

To determine the accuracy of the algorithm over the training dataset D_1, we modify the algorithm to add e to the set A and then check if e is present in the set S; if yes, we count it as a success and otherwise, as a failure. The % of predecessors in D_1 for which the algorithm succeeds is then the accuracy of the algorithm (i.e., this is the recall$K$). Input M controls how many negative examples are picked using each business rule (note that employees that satisfy a business rule are already "similar" to the predecessor, since each business rule is derived from a pattern observed in actual replacements. Input K controls how many feasible candidates are reported for the final (manual) step (3): final selection. Table 1 shows the results obtained using the unsupervised replacement identification algorithm, under various parameter settings. We use the following ways to compute "similarity" between a predecessor and any other employee:

1. **Hamming**: use only textual variables and use Hamming distance
2. **Euclidean**: use only numeric variables and use Euclidean distance
3. **Mahalanobis**: Use only numeric variables and use Mahalanobis distance
4. **Weighted-H-E**: Use all variables and use a weighted sum of Hamming and Euclidean distances between textual and numeric variables respectively
5. **Weighted-H-M**: Same as above, except use Mahalanobis distance instead of Euclidean

For the strategy **Random-All**, as K increases, the %recall improves. As M increases, the %recall drops because adding more "similar" examples confuses the distance computations. Numeric variables seem much less important than the categorical attributes, which seem to matter the most in replacement selection. We experimented with various weightages for Hamming distance and numeric distance; Table 1 shows the results for the weight vector $(0.8, 0.2)$. For brevity, results for only 4 domain rules are shown for **Singular**. The **Exhaustive** uses only the odd-numbered rules.

4.3 Classification

It is not clear how to adapt the usual classification framework for replacement prediction. Firstly, each training instance in the replacement dataset D_1 consists of a *pair* of objects, rather than a single object with a class label. Secondly, there are no *negative* training examples. Obviously, for any predecessor $\mathbf{e}$ in D_1, the pairs $(\mathbf{e}, \mathbf{e}')$ for all possible employees $\mathbf{e}'$ in the dataset D_2 can be considered as a negative example. However, this yields extremely large number of negative examples for every positive replacement pair and an extreme class imbalance. One way to solve the first problem is to translate each pair of employees into a single object consisting of the differences between the corresponding variables. For each pair $(\mathbf{x}, \mathbf{y})$, where $\mathbf{x} = (x_1, \ldots, x_n)$ and $\mathbf{y} = (y_1, \ldots, y_n)$, create a single object $(x_1 - y_1, \ldots, x_n - y_n)$. For categorical variables, the difference is 1 if both values match exactly; otherwise it is 0. If $\mathbf{y}$ is the actual replacement for $\mathbf{x}$, then this object has the class label $+1$; if $\mathbf{y}$ was some employee other than the actual replacement for $\mathbf{x}$, then this object has the class label -1. We adapt the standard cross-validation procedure to our task. We randomly divide the dataset D_1 of 546 replacements into 2 parts: 80 % (437) for training and 20 % (109) for testing. We randomly selected 437 other employees as negative training examples. Thus we have 437 vectors labeled +1, where each vector consists of the difference between the vectors for a predecessor and his/her actual successor. We also have 437 vectors labeled −1, where each vector consists of the difference between the vectors for a predecessor and a randomly selected employee. We train a classifier on this dataset. Accuracy on the training data itself is good. We used the classifiers in the WEKA tool [2] for experiments.

For testing, we created new datasets using the various candidate identification strategies discussed earlier. For example, using **Random-All** with $M = 5$, we created $16 \times 5 = 80$ records for each predecessor $\mathbf{e}$, and also added the actual successor to this set. A record in the testing dataset then consists of the difference between $\mathbf{e}$ and one of the 80 randomly selected employees that meet the particular domain rule. Thus, for a particular predecessor $\mathbf{e}$ (among 109), there are 81 records in the testing dataset (note the class imbalance). For a particular predecessor $\mathbf{e}$ (among 103 in the testing dataset), we use a probabilistic classifier (trained on the training dataset) to predict the probability that a particular record in the training dataset belongs to class +1, sort these 81 records in terms of this probability and check if the actual successor is among the top K. If yes, then we count this as a success for that particular predecessor; otherwise a failure. We repeat this procedure for 5 times (5-fold cross validation) and report the average accuracy of prediction on the training dataset. Table 1 shows the results; for brevity we have omitted results for **Exhaustive**. For **Random-All**, the nearest-neighbour classifier (we used the number of neighbours = 20) shows the best prediction accuracy. As expected, generally the accuracy numbers are higher for supervised classification than unsupervised prediction.

4.4 Metric Learning

Dataset of replacements gives hints about the "similarity" function used by HR executives in identifying replacements. *Metric learning* is about automatically learning a task-specific distance function $\hat{d}(\mathbf{x}, \mathbf{y})$ in a supervised manner. We use one approach for metric learning [12]. Suppose we are given a dataset S of similar pairs objects and a dataset D of dissimilar pairs of objects. *Mahalanobis distance* denotes any distance function of the form $d_A(\mathbf{x}, \mathbf{y}) = \sqrt{(\mathbf{x} - \mathbf{y})^T A (\mathbf{x} - \mathbf{y})}$, where A is some positive semi-definite matrix. The idea is to identify matrix A such that the sum of distances between similar objects is minimized and the sum of distances between dissimilar objects is maximized. In [12], this is formalized and solved as a convex optimization problem in two separate cases: A is assumed to be a diagonal matrix (solved using Newton-Raphson) or A is assumed to be a full matrix (solved using a gradient-based iterative algorithm). We treat pairs (predecessor, actual replacement) as similar and (predecessor, randomly-selected-employee) as disimilar. The testing is same as for unsupervised distance, where we use the matrix A learnt in the training phase to compute the distance d_A between pairs of employees. We use only the numerical attributes to learn the 13×13 matrix A. Table 1 shows the results for the

Table 1. %Recall of various replacement identification methods.

Method	Candidate strategy	$K = 5$	$K = 10$	$K = 15$
Hamming (unsupervised)	**Random-All** $M = 5$	**56.78**	**71.43**	**80.04**
	Random-All $M = 10$	47.80	61.54	67.77
	Singular R_1	25.09	36.26	43.04
	Singular R_3	25.65	35.24	42.07
	Singular R_5	24.18	35.53	42.12
	Singular R_7	24.68	34.07	41.44
	Exhaustive	22.89	32.60	39.93
Euclidean (unsupervised)	**Random-All** $M = 5$	24.91	38.64	46.89
	Random-All $M = 10$	17.77	26.01	33.15
	Singular R_1	15.02	21.98	26.56
	Singular R_3	14.58	19.93	25.09
	Singular R_5	13.74	19.78	23.26
	Singular R_7	12.89	18.42	21.55
	Exhaustive	11.54	17.22	20.15
Mahalanobis (unsupervised)	**Random-All** $M = 5$	18.68	30.59	39.38
	Random-All $M = 10$	11.54	19.78	25.09
	Singular R_1	11.36	17.22	19.05
	Singular R_3	12.36	16.24	18.27
	Singular R_5	9.52	14.65	17.03
	Singular R_7	10.50	15.10	16.76
	Exhaustive	9.89	14.29	15.93
Weighted-H-E; (0.8, 0.2) weightage (unsupervised)	**Random-All** $M = 5$	42.12	56.04	63.55
	Random-All $M = 10$	52.2	66.12	72.89
	Singular R_1	24.36	30.95	37.73
	Singular R_3	23.62	32.29	38.19
	Singular R_5	23.26	30.04	36.45
	Singular R_7	22.84	30.76	36.83
	Exhaustive	22.16	28.57	34.62
Weighted-H-M; (0.8, 0.2) weightage (unsupervised)	**Random-All** $M = 5$	53.30	70.33	76.92
	Random-All $M = 10$	43.22	57.51	65.93
	Singular R_1	20.70	29.85	37.36
	Singular R_3	21.96	29.89	36.72
	Singular R_5	19.96	28.21	35.90
	Singular R_7	21.55	29.10	35.54
	Exhaustive	20.15	27.11	34.62
Naive Bayes (supervised)	**Random-All** $M = 5$	41.76	53.66	60.99
	Random-All $M = 10$	47.80	56.41	61.36
	Singular R_1	35.17	42.50	49.27
	Singular R_3	45.46	56.18	61.26
	Singular R_5	47.44	56.05	60.63
	Singular R_7	46.41	56.18	59.49
Nearest Neighbour (supervised)	**Random-All** $M = 5$	62.83	76.01	81.14
	Random-All $M = 10$	45.07	51.30	56.80
	Singular R_1	52.94	63.56	70.89
	Singular R_3	44.51	52.98	59.01
	Singular R_5	43.97	50.57	55.33
	Singular R_7	44.02	49.54	55.08
Random Forest (supervised)	**Random-All** $M = 5$	42.87	51.85	58.44
	Random-All $M = 10$	42.31	50.92	55.14
	Singular R_1	34.26	42.87	49.10
	Singular R_3	39.84	49.80	55.45
	Singular R_5	41.57	50.00	54.22
	Singular R_7	41.79	49.53	53.59
Metric-learning	**Random-All** $M = 5$	26.37	40.10	48.16
	Random-All $M = 10$	19.78	28.57	34.06
	Singular R_1	14.83	20.51	26.19
	Singular R_3	14.20	18.45	24.35
	Singular R_5	13.36	17.94	23.80
	Singular R_7	12.70	16.94	21.54

diagonal case. Diagonal entries of the matrix A learned by the algorithm are: $1.07, 1.02, 0.91, 0.97, 1.00, 0.03, 1.26, 0.95, 0.97, 1.01, 0.99, 1.00, 0.99$; e.g., $A_{3,3} = 0.91$ means the weight of the third numeric variable V_{15} in computing the distance. Compared to the unsupervised replacement identification 1, metric learning shows better performance than Mahalanobis distance and comparable to Euclidean distance. Results of learning full matrix A are similar, which means the distance used by humans in replacement identification (in this case-study) is close to Euclidean (for numeric attributes).

5 Related Work

The problem of mining replacement datasets to build predictive models is new. A related problem is that of *churn* or *attrition* where an agent drops his/her association with an entity; e.g., customer churn, employee attrition. The focus there is to build predictive models to understand the reasons of this "churn". There is no information about what the entity was replaced with. Thus churn problems are different from the replacement problem proposed in this paper. [10] presents a case study where several predictive models for employee churn (attrition) are compared. Due to its impact on costs, revenue, growth and brand-value, customer churn is a serious problem particularly in volatile products and services markets, such as mobile phones [1,3,11], insurance [8], subscription services [4] and banking [6]. [9] use logit regression, decision trees, neural networks, and boosting to build predictive models of customer churn in telecom and to identify incentives for subscribers to improve retention and maximize profitability. [8] presents a case-study of customer churn in insurance and a technique to discover features suitable for building customer churn models. [4] builds a predictive model of newspaper subscription churn using SVM. [7] applies survival analysis techniques to build a predictive model for customer churn in telcom. [6] uses random regression forests to build predictive models of customer churn in a Belgian financial institution. [5] provides a survey of metric learning. We used the metric learning algorithm in [12] for learning a Mahalanobis-style distance function.

6 Conclusions and Further Work

This paper proposed a new class of data mining problems in which agents replace their current object by another object; a goal is to discover the knowledge used by the agents in identifying the suitable successors. While such replacement data is available in many practical applications, in this paper we explored a problem in *HR analytics*: replacing person in a key role in a project by another most suitable person from a pool of employees. We compared the approaches using a real-life people replacement dataset from a multinational IT company. Although we have only reported the results for employees in PROJECT_LEADER role, we have experimented with other roles; the results are similar. The implementation of these algorithms is undergoing active real-life trials. As expected, the supervised approaches perform better than the unsupervised approaches. Metric learning shows up as a promising approach to capture the implicit knowledge used for replacement identification. We will explore other supervised approaches. We are experimenting with other distance measures e.g., Kolmogorov complexity-based information distance and association-based distance for categorical variables. We are experimenting with other metric learning approaches and are exploring how to adapt metric learning to work with categorical data. We want to extend these ideas to *succession planning* in human resources management and to identify replacements to improve the overall "goodness" of the team composition.

Acknowledgments. We thank Dr. Ritu Anand, VP (HR), Preeti Gulati and other HR executives in TCS for enthusiastically supporting this work. Thanks to our team members for discussions and other help.

References

1. Archaux, C., Martin, A., Khenchaf, A.: An SVM based churn detector in prepaid mobile telephony. In: Proceedings of International Conference on Information and Communication Technologies: From Theory to Applications (2004)
2. Hall, M., Frank, E., Holmes, G., Pfahringer, B., Reutemann, P., Witten, I.H.: The WEKA data mining software: an update. SIGKDD Explor. **11**(1), 10–18 (2009)
3. Hung, S.-Y., Yen, D.C., Wang, H.-Y.: Applying data mining to telecom churn management. Expert Syst. Appl. **31**, 515–524 (2006)
4. Van den Poel, D., Coussement, K.: Churn prediction in subscription services: an application of support vector machines while comparing two parameter-selection techniques. Expert Syst. Appl. **34**(9), 313–327 (2008)
5. Kulis, B.: Metric learning: a survey. Found. Trends Mach. Learn. **5**(4), 287–364 (2012)
6. Lariviere, B., den Poel, D.V.: Predicting customer retention and profitability by using random forests and regression forests techniques. Expert Syst. Appl. **29**, 472–484 (2005)
7. Junxiang, L.: Predicting customer churn in the telecommunications industry - an application of survival analysis modeling using sas. In: Proceedings of SAS User Group International, pp. 114–127 (2002)
8. Morik, K., Köpcke, H.: Analysing customer churn in insurance data – a case study. In: Boulicaut, J.-F., Esposito, F., Giannotti, F., Pedreschi, D. (eds.) PKDD 2004. LNCS (LNAI), vol. 3202, pp. 325–336. Springer, Heidelberg (2004)
9. Mozer, M.C., Wolniewicz, R., Grimes, D.B., Johnson, E., Kaushansky, H.: Predicting subscriber dissatisfaction and improving retention in the wireless telecommunications industry. IEEE Trans. Neural Netw. **11**(3), 690–696 (2000)
10. Saradhi, V.V., Palshikar, G.K.: Employee churn prediction. Expert Syst. Appl. **38**(3), 1999–2006 (2011)
11. Wei, C.-P., Chiu, I.-T.: Predicting customer retention and profitability by using random forests and regression forests techniques. Expert Syst. Appl. **23**, 103–112 (2002)
12. Xing, E.P., Ng, A.Y., Jordan, M.I., Russell, S.: Distance metric learning, with application to clustering with side-information. In: Proceedings of Advances in Neural Information Processing Systems, vol. 15, pp. 505–512 (2003)

Tri-Axial Vibration Analysis Using Data Mining for Multi Class Fault Diagnosis in Induction Motor

Pratyay Konar, Parth Sarathi Panigrahy, and Paramita Chattopadhyay(✉)

Department of Electrical Engineering, Indian Institute of Engineering Science and Technology, Shibpur, India
{pratyaymaithon, parth.panigrahy}@gmail.com, paramita_chattopadhyay@yahoo.com

Abstract. Induction motor frame vibration is believed to contain certain crucial information which not only helps detecting faults but also capable of diagnosing different types of faults that occur. The vibration data can be in radial, axial and tangential directions. The frequency content of the three different directions are compared and analyzed using data mining techniques to find the most informative vibration data and to extract the vital information that can be effectively used to diagnose multiple induction motor faults. The vibration data is decomposed using powerful signal processing tools like Continuous Wavelet Transform (CWT) and Hilbert Transform (HT). Statistical features are extracted from the decomposition coefficients obtained. Finally, data mining is applied to extract knowledge. Three types of data mining tools are deployed: sequential greedy search (GS), heuristic genetic algorithm (GA) and deterministic rough set theory (RST). The classification accuracy is judged by five types of classifiers: k-Nearest Neighbors (k-NN), Multilayer Perceptron (MLP), Radial Basis Function (RBF) and Support Vector Machine (SVM), and Simple logistic. The benefits of using all the tri-axial data combined for vibration monitoring and diagnostics is also explored. The results indicate that tri-axial vibration combined provides the most informative knowledge for multi-class fault diagnosis in induction motor. However, it was also found that multi-class fault diagnosis can also be done quite effectively using only the tangential vibration signal with the help of data mining knowledge discovery.

Keywords: Data mining · Tri-axial vibration · Fault diagnosis · GS · GA · RST

1 Introduction

Ever-increasing demand in power, performance, and safety, along with frequent failures resulting in financial losses have made fault detection and diagnostics in rotating machinery a challenging task [1]. Recent researches focuses on implementing on-board condition monitoring systems, which has driven the researchers to design hardware friendly algorithms with minimum use of resources [2]. To ensure trouble free operations in power industry, accurate detection and diagnosis of different induction motor

R. Prasath et al. (Eds.): MIKE 2015, LNAI 9468, pp. 553–562, 2015.
DOI: 10.1007/978-3-319-26832-3_52

faults is very important. Vibration emanating from motor frame has become a prime candidate for developing efficient monitoring algorithms, as most of the failures that occur in induction motor (stator, rotor and bearings) show their symptoms as changes in the frequencies and amplitudes of vibration signatures. The vibration signals have been widely investigated in detail to understand the baseline frequencies to detect various types of faults [3]. Vibration monitoring has already contributed tremendously in the field of condition monitoring and provides reliable detection of electrical problems in motors and no other technique can, as effectively detect mechanical problems in all types of rotating machines [4].

The motor condition monitoring science is moving towards an automated computerized scheme, trying to remove human experts from the condition monitoring process [4]. The development of automatic diagnostics methods for electrical machine condition monitoring is still in its infancy and despite the considerable amount of work in this field, there is still much scope of work especially in the area of feature extraction using powerful signal processing tools.

The vibration data can be in radial, axial or in tangential directions. Each vibration data from three directions contain unique features about various types of faults that occur [5]. However, the incipient fault features are often weak and buried in the signal, so it is difficult to detect them using the traditional methods. Moreover, today's motors are associated with variable frequency drives (VFDs), making traditional analytical methods inappropriate and obsolete [5]. In addition, vibration in induction motors is a very complex subject and often contaminated with noise, making the extraction of useful information or features a challenging task [4]. In spite of these various challenges, the fault frequency information/knowledge can be extracted from the vibration signal using data mining and knowledge discovery.

Data mining has emerged as a very promising tool in handling real world data which are often vague and redundant. Technically, data mining is the process of finding correlation or patterns in large relational database and then making crucial decisions depending upon the information acquired [6]. Different computational tools such as Rough-set theory (RST), Fuzzy-set theory, Fuzzy-rough set and genetic algorithm (GA), are utilized to develop the data mining algorithms.

In the present work an attempt is made, to extract knowledge from the tri-axial vibration data using three different data mining techniques like sequential search based Greedy search (GS), heuristic search based GA and deterministic search based RST. Powerful signal processing tools like CWT and HT are used to extract the fault frequency information hidden in the vibration signals. Five types of classifier are used to validate the robustness of the knowledge acquired.

2 Tri-axial Frame Vibration Analysis

Tri-axial frame vibration in radial, axial, and tangential directions are analyzed and compared to study the information content in the signals that can be used to develop a more efficient fault diagnostic algorithm for detecting multi-class induction motor fault. The information/features obtained from tri-axial vibration are also combined to investigate and compare the diagnostic capability when only one vibration signal is taken.

2.1 Experimental Setup and Data Acquisition

Spectra Quest's Machinery Fault Simulator (MFS) [7] is used here as the experimental set-up to extract the vibration signatures of common motor faults. The Machinery Fault Simulator (MFS) has seven three phase induction motors; one is healthy and rests are of same specification but with various faults. Rating of the each motor is 1/3 hp, 190 V, 50 Hz, 2980 rpm. Seven different motor conditions were considered viz. healthy, broken rotor bar, rotor unbalance, bowed rotor, bearing fault, stator fault and voltage unbalance.

The set up has the provision for collecting tri-axial frame vibration signal of the induction motor using factory calibrated [7] piezoelectric type accelerometer (100 mV/g) probes through data acquisition system. The vibration data were collected at a sampling frequency of 5120 Hz under no-load (0.05649 Nm), 1-unit load (0.22596 N-m), 2-unit load (0.45193 N-m), 3-unit load (0.67790 N-m) and 4-unit load (0.90387 N-m). The load was given with the help of brake-clutch mechanism and belt pulley system [7]. The supply frequency was set to 50 Hz. During experimentation, multi-sample data were taken to ensure that the training and validation data are statistically independent. The schematic representation of the proposed scheme in shown in Fig. 1.

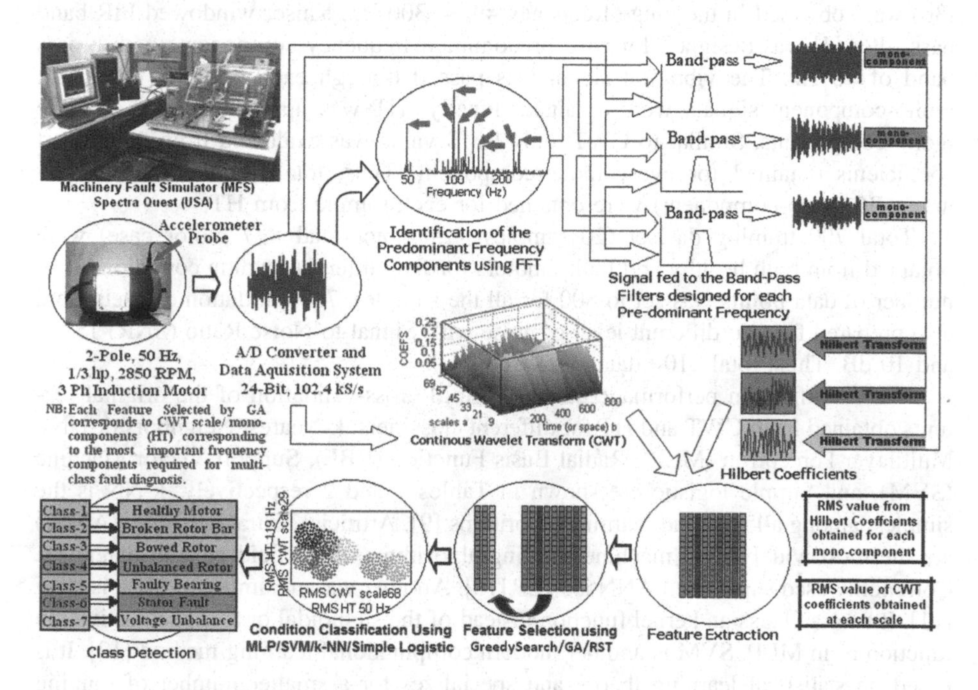

Fig. 1. The schematic representation of the proposed scheme

2.2 Feature Extraction Using CWT and HT

To extract the subtle information hidden in vibration signal contaminated by noise, two powerful signal-processing tools: CWT and HT are used. CWT breaks the signal into its 'wavelets', scaled and shifted versions of the '*mother wavelet*' and can be defined as the sum over all time of the signal multiplied by scaled, shifted versions of the wavelet function Ψ [4]. The Hilbert transform is used to create an analytic signal from a real signal. The instantaneous amplitude and instantaneous frequency can be easily computed from the analytic signal [4]. The complete information about the signal can be extracted by acquiring mono-component signals. Detailed theoretical background is beyond the scope of this paper and can be found in literature [4].

For CWT scales 12 $\sim$ 68 were selected to extract information from the signals in the frequency range 49 $\sim$ 300 Hz. 'db8' was used as mother wavelet for CWT to obtain improved frequency resolution since the frequency can be determined via more number of vanishing moments [8]. Only Root Mean Square (RMS) values were evaluated from the coefficients obtained for every sample and for all loading conditions and treated as attributes/features. One statistical feature was selected to keep the number of features equal to the number of scales selected for CWT. Thus, total 57 attributes (1 feature $\times$ 57 scales) were obtained for each sample using CWT.

For successful application of HT, it is necessary to have mono-component signal. To obtain mono-component signals the predominant frequency components were identified using FFT as shown in Fig. 1. 36 predominant frequencies components (f1 to f36) were obtained in the range frequency 49 $\sim$ 300 Hz. Kaiser windowed FIR band pass filter [4] was designed for each predominant frequency components with a pass band of $\pm$5 Hz. The vibration signal was passed through each of these filters and mono-component signals were obtained. Finally, HT was applied to each of these mono-components. Similar to CWT, only RMS value was evaluated from the Hilbert coefficients obtained for every mono-component. Thus, total 36 attributes (1 feature $\times$ 36 mono-component) were obtained for each sample from HT.

Total 700 training dataset (20 samples $\times$ 5 motor load $\times$ 7 motor case) were obtained from both healthy and faulty motors for five different loading conditions. The number of data-points was set to 800 for all the samples. 700 validation datasets were also prepared for two different level of noise with Signal-to-Noise Ratio (SNR) 15 dB and 10 dB. Thus, total 2100 datasets were obtained.

The classification performance under 10-fold cross-validation of the original features obtained from CWT and HT for different classifiers: k-Nearest Neighbors (k-NN), Multilayer Perceptron (MLP), Radial Basis Function (RBF), Support Vector Machine (SVM), and Simple logistic are shown in Tables 1 and 2 respectively. k-NN is the simplest among all machine-learning algorithms [9]. Artificial neural networks (ANNs) are the most widely used machine learning algorithms in the field of fault diagnosis. Commonly used supervised ANN is MLP [10]. Another variant similar to MLP is RBF [11]. It uses a Gaussian kernel function instead of the sigmoidal or S-shaped activation function as in MLP. SVM is another modern computational learning method [12]. It is based on statistical learning theory and specializes for a smaller number of training samples. It has become very popular since it does have the problem of over fitting and suitable for data with high-dimensional feature spaces. Decision tree based algorithm is also quite popular. One such method is Simple Logistic classifier, a Logistic Model Tree (LMT) based on the idea of combining two popular techniques tree induction methods and linear models and then using logistic regression instead of linear regression. It forms 'model trees' that contain linear regression functions at the leaves [13].

Table 1. Classification performance of subsets obtained from tri-Axial using CWT

Feature set	Radial Frame Vibration	Axial Frame Vibration	Tangential Frame Vibration	Tri-Axial Frame Vibration
Attribute extracted using CWT	57	57	57	171
k-NN (%)	94.52	94.38	99.33	100
MLP (%)	93.61	93.52	97.42	99.90
RBF (%)	81.95	74.42	82.76	98.52
SVM (%)	97.47	94.90	99.04	99.76
Simple logistic (%)	91.52	90.19	97.28	100

Table 2. Classification performance of subsets obtained from tri-axial using HT

Feature set	Radial Frame Vibration	Axial Frame Vibration	Tangential Frame Vibration	Tri-Axial Frame Vibration
Attribute extracted using HT	36	38	45	119
k-NN (%)	99.76	99.67	99.95	100
MLP (%)	98.85	98.80	99.42	100
RBF (%)	99.47	99.47	99.90	100
SVM (%)	99.04	99.04	99.76	100
Simple logistic (%)	98.90	98.85	99.42	99.95

Interestingly it is observed from the Tables 1 and 2 that tangential frame vibration provides better classification accuracy as compared to others. In spite of having one third feature sets, its performance is comparable with all tri-axial vibrations taken together.

2.3 Tri-axial Frame Vibration Analysis Using Data Mining

Real world data are often vague and redundant, creating lot of problems for powerful machine learning technique to take decision accurately [1]. In the results presented in Tables 1 and 2, RBF NN has faced problem in classifying the data. The features extracted using CWT contains highly redundant information. For HT, although the features extracted are non-redundant, it contains a lot of irrelevant information. It should be noted, that some features from a subset with high relevancy might be redundant. On the other hand, some features from the subset with low relevancy value may be quite important for effective fault diagnosis [3]. An intelligent feature selection or data-mining algorithm is essential to find the best feature subset that will reduce the dimensionality of the original feature space preserving same knowledge that can be effectively used to diagnose multiple induction motor faults. In the present work three different approaches based on sequential search (Greedy search), heuristic search (Genetic Algorithm) and deterministic search (Rough Set Theory) are used.

Greedy search (GS) is a form of hill climbing search where we select the local move that leads to the largest improvement of the objective function. At each step, an attribute is added or deleted from the attribute set and stops when the addition or removal of any remaining attributes results in decrease in evaluation [14]. Genetic Algorithms (GA) is an optimization technique that mimic the process of natural evolution. The simplest GA, originally proposed by Goldberg [15] is used in this work. Rough Set theory was introduced by Z. Pawlak as a mathematical approach to handle vagueness [16]. Unlike other intelligent methods, rough set analysis requires no external parameters and uses only the information presented in the given data. Every rough set is associated with two crisp sets, called lower and upper approximation, which represents black and white conceptual thinking and thus simplifies the logic used with this kind of thinking process. The algorithm for using RST as a feature selection tool can be found in [5].

The feature selection capability of Greedy Search (GS), Genetic Algorithm (GA) and Rough Set Theory (RST) are used to extract the fault related features from two different types of data sets obtained using CWT and HT. The feature subset obtained can be used to select the relevant CWT scales and thus also the fault frequencies which can efficiently detect different induction motor faults. Consequently, the redundant information and computation time can be considerably reduced along with the burden on the classifiers.

The fault detection accuracy of different classifier for the subsets obtained from CWT feature pool using GS, GA and RST under 10-fold cross-validation is given in Tables 3, 4 and 5 respectively.

Table 3. Classification performance of subsets obtained from CWT using GS

Feature set	Radial Vibration Frame		Frame Vibration Axial		Tangential Frame Vibration	
Attribute selected using greedy search	57	16	57	16	57	16
k-NN (%)	94.52	93.14	94.38	93.67	99.33	93.67
MLP (%)	93.61	92.57	93.52	87.23	97.42	95.71
RBF (%)	81.95	87.33	74.42	74.61	82.76	87.42
SVM (%)	97.47	97.57	94.90	92.33	99.04	97.57
Simple logistic (%)	91.52	92.04	90.19	88.00	97.28	96.42

Table 4. Classification performance of subsets obtained from CWT using GA

Feature set	Radial Frame Vibration		Axial Frame Vibration		Tangential Frame Vibration	
Attribute selected using GA	57	16	57	16	57	16
k-NN (%)	94.52	93.71	94.38	96.71	99.33	94.28
MLP (%)	93.61	93.85	93.52	93.52	97.42	96.28
RBF (%)	81.95	87.14	74.42	79.19	82.76	91.85
SVM (%)	97.47	96.00	94.90	93.47	99.04	96.80
Simple logistic (%)	91.52	92.28	90.19	90.14	97.28	96.71

Table 5. Classification performance of subsets obtained from CWT using RST

Feature set	Radial Frame Vibration		Axial Frame Vibration		Tangential Frame Vibration	
Attribute selected using RST	57	16	57	16	57	16
k-NN (%)	94.52	85.95	94.38	80.42	99.33	98.47
MLP (%)	93.61	91.80	93.52	86.61	97.42	94.28
RBF (%)	81.95	80.52	74.42	66.14	82.76	85.23
SVM (%)	97.47	93.19	94.90	83.47	99.04	97.38
Simple logistic (%)	91.52	88.95	90.19	84.38	97.28	92.52

The classification performances of different classifier for the subsets obtained from HT using GS, GA and RST under 10-fold cross-validation are shown in Tables 6, 7 and 8 respectively.

Table 6. Classification performance of subsets obtained from HT using GS

Feature set	Radial Frame Vibration		Axial Frame Vibration		Tangential Frame Vibration	
Attribute selected using Greedy Search	36	17	38	17	45	17
k-NN (%)	99.76	99.38	99.67	99.00	99.95	99.52
MLP (%)	98.85	98.33	98.80	97.90	99.42	97.95
RBF (%)	99.47	99.28	99.47	98.90	99.90	99.52
SVM (%)	99.04	99.76	99.04	99.61	99.76	99.80
Simple logistic (%)	98.90	98.09	98.85	98.00	99.42	98.23

Table 7. Classification performance of subsets obtained from HT using GA

Feature set	Radial Frame Vibration		Axial Frame Vibration		Tangential Frame Vibration	
Attribute selected using GA	36	17	38	17	45	17
k-NN (%)	99.76	99.14	99.67	98.90	99.95	99.52
MLP (%)	98.85	97.09	98.80	94.90	99.42	97.67
RBF (%)	99.47	98.57	99.47	98.28	99.90	99.23
SVM (%)	99.04	99.47	99.04	99.47	99.76	99.71
Simple logistic (%)	98.90	97.52	98.85	96.04	99.42	97.61

The information/features obtained from tri-axial vibration are also combined to investigate and compare the diagnostic capability when only one vibration signal is taken. Here also GS, GA and RST are used to extract useful knowledge from the combined tri-axial feature subset obtained from both CWT and HT. The classification performance of different classifiers for the subsets obtained from CWT and HT using the three feature selection/reduction tools under 10-fold cross-validation are shown in Tables 9, 10 and 11.

Table 8. Classification performance of subsets obtained from HT using RST

Feature set	Radial Frame Vibration		Axial Frame Vibration		Tangential Frame Vibration	
Attribute selected using RST	36	18	38	18	45	17
k-NN (%)	99.76	98.67	99.67	95.42	99.95	99.14
MLP (%)	98.85	94.61	98.80	89.95	99.42	96.19
RBF (%)	99.47	97.00	99.47	93.14	99.90	98.00
SVM (%)	99.04	99.09	99.04	98.14	99.76	99.52
Simple logistic (%)	98.90	93.71	98.85	88.61	99.42	96.00

Table 9. Classification performance of subsets obtained from tri-axial using GS

Feature set	CWT		HT	
Attribute selected using greedy search	171	21	119	27
k-NN (%)	100	99.90	100	100
MLP (%)	99.90	99.90	100	99.85
RBF (%)	98.52	99.19	100	99.95
SVM (%)	99.76	100	100	99.90
Simple logistic (%)	100	99.90	99.95	99.90

Table 10. Classification performance of subsets obtained from tri-axial using GA

Feature set	CWT		HT	
Attribute selected using GA	171	21	119	27
k-NN (%)	100	99.67	100	99.95
MLP (%)	99.90	99.85	100	99.85
RBF (%)	98.52	97.52	100	99.85
SVM (%)	99.76	99.80	100	99.90
Simple logistic (%)	100	99.95	99.95	99.76

Table 11. Classification performance of subsets obtained from tri-axial using RST

Feature set	CWT		HT	
Attribute selected using RST	171	21	119	27
k-NN (%)	100	96.14	100	99.85
MLP (%)	99.90	97.14	100	99.67
RBF (%)	98.52	90.95	100	99.76
SVM (%)	99.76	97.80	100	99.90
Simple logistic (%)	100	98.00	99.95	99.52

The number of features selected from each tri-axial direction when three axis vibration signals are combined is shown in Table 12.

Table 12. Number of attributes selected from each tri-axial direction

CWT			HT		
Radial Frame Vibration	Axial Frame Vibration	Tangential Frame Vibration	Radial Frame Vibration	Axial Frame Vibration	Tangential Frame Vibration
6 Attributes	**6 Attributes**	**9 Attributes**	**7 Attributes**	**7 Attributes**	**13 Attributes**
28.5 %	**28.5 %**	**42.85 %**	**25.9 %**	**25.9 %**	**48.15 %**

2.4 Conclusions

The results obtained clearly indicates that data mining provides ample scope of extracting vital information buried in motor frame vibration signals along with application of powerful signal processing tools like CWT and HT. The useful knowledge extracted not only helped to reduce the number of features drastically but also improved the classification performance to a considerable extent. The knowledge has also helped in understanding the information content in each vibration signals analysed and identify the most relevant CWT scales and mono-components for HT thereby reducing the computational cost incurred in extracting fault related features using CWT/HT, making them suitable for real time applications. Experimental results show that the tangential frame vibration is the most informative, whereas the axial is the least. Statistically speaking, for the subset selected from combined tri-axial signals using data mining techniques, the features from tangential direction accounted for 42 % of all the features in case of CWT and 48 % for HT selected features. It has also been found that the information content in tangential vibration signal is alone enough to give the same classification accuracy as all tri-axial frame vibration signals taken together. Among the different data mining techniques used although GS and GA provides promising results, both approaches have a tendency to get trapped at local peaks. The RST based deterministic approach shows poor performance due to loss of information caused by discretization, but does not suffer from the problem of being trapped into local optima. It has a promising future as a data mining tool.

Acknowledgments. The authors are thankful to Council of Scientific and Industrial Research (CSIR) for their support for continuation of this project. The authors are also thankful to AICTE and TEQIP-I (BESU, Shibpur unit), Govt. of India for their financial support toward the project.

References

1. Tsypkin, M.: Induction motor condition monitoring: vibration analysis tech-nique – a practical implementation. In: IEEE International Electric Machines and Drives Conference (IEMDC), pp. 406–411 (2011)
2. Kar, C., Mohanty, A.R.: 'Vibration and current transient monitoring for gearbox fault detection using multiresolution Fourier transform. J. Sound Vibr. **2**, 109–132 (2008) (Elsevier)

3. Singh, G.K., Ahmed, S.A.K.S.: Vibration signal analysis using wavelet transform for isolation and identification of electrical faults in induction machine. Electr. Power Syst. Res. **68**(1), 119–136 (2004)
4. Konar, P., Chattopadhyay, P.: Multi-class fault diagnosis of induction motor using Hilbert and Wavelet Transform. Appl. Soft Comput. **30**, 341–352 (2015)
5. Seshadrinath, J., Singh, B., Panigrahi, B.K.: Investigation of vibration signatures for multiple fault diagnosis in variable frequency drives using complex wavelets. Electr. Power Syst. Res. IEEE Trans. Power Electron. **29**(2), 936–945 (2013)
6. Konar, P., Sil, J., Chattopadhyay, P.: Knowledge extraction using data mining for multi-class fault diagnosis of induction motor. Neurocomputing **166**, 14–25 (2015)
7. Spectra Quest, Inc., USA, Machinery Fault Simulator (MFS)—User Manual (2005)
8. Konar, P., Chattopadhyay, P.: Feature extraction using wavelet transform for multi-class fault detection of induction motor. J. Inst. Eng. (India): Ser. B **95**(1), 73–81 (2014)
9. Aha, D.W., Kibler, D., Albert, M.K.: Instance-based learning algorithms. Mach. Learn. **6**, 37–66 (1991)
10. Hagan, M.T., Demuth, H.B., Beale, M.: Neural Network Design. PWS Publishing Company, Boston (2002)
11. Sundararajan, N., Saratchandran, P., Ying Wei, L.: Radial Basis Function Neural Networks with Sequential Learning MRAN and Its Applications. Progress in Neural Processing. World Scientific Publishing Co. Pte. Ltd., Singapore (1999)
12. Konar, P., Chattopadhyay, P.: Bearing fault detection of induction motor using wavelet and Support Vector Machines (SVMs). Appl. Soft Comput. **11**(6), 4203–4211 (2011)
13. Landwehr, N., Hall, M., Frank, E.: Logistic model trees. Mach. Learn. **59**, 161–205 (2005)
14. Caruana, R.A., Freitag,D.: Greedy attribute selection. In: Proceedings 11th International Conference on Machine Learning, pp. 28–36 (1994)
15. Goldberg, D.E.: Genetic Algorithms in Search, Optimization and Machine Learning. Addison-Wesley, Boston (1989)
16. Konar, P., Saha, M., Sil, J., Chattopadhyay, P.: Fault diagnosis of induction motor using CWT and rough-set theory. In: IEEE Symposium on Computational Intelligence in Control and Automation (CICA), pp. 17–23 (2013)

An Efficient Text Compression Algorithm - Data Mining Perspective

C. Oswald[1(✉)], Anirban I. Ghosh[2], and B. Sivaselvan[1]

[1] Department of Computer Engineering, Indian Institute of Information Technology, Design and Manufacturing Kancheepuram, Chennai, India
{oswald.mecse,anir1ghosh}@gmail.com, sivaselvanb@iiitdm.ac.in

[2] Department of Information Technology, IIIT Allahabad, Allahabad, India

Abstract. The paper explores a novel compression perspective of Data Mining. Frequent Pattern Mining, an important phase of Association Rule Mining is employed in the process of Huffman Encoding for Lossless Text Compression. Conventional Apriori algorithm has been refined to employ efficient pruning strategies to optimize the number of pattern(s) employed in encoding. Detailed simulations of the proposed algorithms in relation to Conventional Huffman Encoding has been done over benchmark datasets and results indicate significant gains in compression ratio.

Keywords: Apriori · Compression ratio · Frequent pattern mining · Huffman encoding · Lossless compression

1 Introduction

Technological advances have resulted in enormous amount of data like text, audio, video, etc. being transferred through Internet medium. The challenge lies in efficiently storing these massive data in reduced size. This leads to compression of large files according to the type of data and their quality. Compression is a technique used for reducing the number of bits needed to store or transmit data. Lossy compression methods allow inexact approximation even if the original data like image, audio, and video does not have any redundancy [1]. It achieves better compression ratio than lossless techniques. Few of them include JPEG, MPEG, MP3, PGF, codec2, etc [1]. Lossless compression technique allows reconstruction of original data from compressed data. Typical examples include text documents, executable programs and source code. Some of the available techniques are discussed in Sect. 2.

This work focuses on one of the seminal algorithm namely Huffman Encoding [2]. Huffman Encoding is used entirely or as an intermediary phase by some commercial solutions. Gilbert et al. proved that Huffman code has the shortest average code length resulting in minimum length codes [3]. Several models based on Statistical, Dictionary and Sliding Window for lossless compression have been proposed [1]. These methods do not consider symbol frequencies for encoding.

R. Prasath et al. (Eds.): MIKE 2015, LNAI 9468, pp. 563–575, 2015.
DOI: 10.1007/978-3-319-26832-3_53

The major drawback of these models is the huge memory space to store dictionary data structure, where a large static collection of words is involved. Most of these above said lossless algorithms employ character/word based encoding.

The paper explores the scope of Data Mining in the domain of data compression. Data Mining is the non-trivial extraction of implicit, hidden and potentially useful information from large volumes of data [4]. Naren Ramakrishnan et al. explored various recurrent perspectives of Data Mining like Compression, Induction, Querying, Approximation and Search [5]. Data Mining can be viewed as a compression technique wherein the mined knowledge set(pattern) can be viewed as a condensed representation of the original data. Types of Data Mining techniques are Association Rule Mining, Classification, Clustering, Outlier Analysis, etc [4]. They are widely applied in Medical Diagnosis, Publication databases, Personal Recommendation Systems like Amazon, Ebay, Priority Inbox(Gmail), etc.

Frequent Pattern Mining(FPM) is an important and non-trivial phase in Association Rule Mining followed by Rule Generation [4]. FPM mines frequent itemsets occurring together in transactional databases. Let $I = \{i_1, i_2, i_3, \dots, i_n\}$ be a set of items, and a *transaction database* $D = < T_1, T_2, T_3, \dots, T_m >$, where $T_i (i \in [1...m])$ is a transaction containing a set of items in I. The *support* of a *pattern* X, where X is a set of items, is the number of transactions containing X in D. An itemset(X) is frequent if its support is not less than a user-defined minimum support($min\_supp = \alpha$). Apriori, a seminal algorithm in FPM uses prior knowledge which is, "*All nonempty subsets of a frequent itemset must also be frequent*" [6]. FPM algorithms concentrate on mining all possible frequent patterns in the Transactional Database.

According to the general law of compression, frequently occurring patterns should get shorter codes than the non frequent patterns. This work contributes a novel approach to text compression namely Frequent Pattern based Huffman Encoding(FPH), wherein Conventional Huffman is modified to employ FPM process in the code assignment/generation process. The proposed work employs efficient strategies to mine frequently occurring phrases in the text T, which can be used to design an efficient Huffman Encoding. The paper is organized as follows. Related studies are presented in Sect. 2. Section 3 formally presents the problem. Section 4 presents the proposed algorithms. Details of the datasets, results and performance discussion are given in Sect. 5. Section 6 concludes with further research directions.

2 Related Work

Foundations by Claude Shannon and Robert Fano regarding systematic assignment of codewords resulted in Entropy encoding. Codes are assigned on the basis of probabilities of a character's occurrence with shortest codes for frequent characters. Shannon-Fano coding is a seminal algorithm that focuses on constructing a prefix code that is based on a set of symbols and their probabilities [7]. However, it is suboptimal as it doesn't achieve optimal codeword length. It ensures that the codeword lengths are within one bit of the value which is theoretically ideal, i.e., $-log(P(x))$, where $P(x)$ is the probability of occurrence of the symbol x. Similar work by Huffman in 1952, was Huffman Encoding which is a slight modification of the previous method [2]. It is a prefix free coding which gives variable length codes to source symbols by assigning short codes to more frequent symbols. It is simple, efficient and assigns optimal codes for individual symbols.

Other Variable Length codes to name a few are Run Length Encoding(RLE), Arithmetic Encoding, Adaptive/Dynamic Huffman Encoding, Prediction by Partial Matching(PPM), Self-Deleting Codes, Start/Stop code, Elias codes, Location Based Encoding, Move to Front Encoding, Burrow Wheeler Transform, Fibonacci Code, etc [1,8–11]. These methods are either single character/word based encoding schemes. Also, the quality of compression depends on how the statistical model is built.

The next major class of compression is sliding window algorithms. The seminal work on this happened with Lempel-Ziv-77(LZ77) which is from the family of Lempel-Ziv algorithms [12]. A good review of these algorithms and its analysis can be found in [1,13]. In these algorithms, a dictionary is employed to encode each string as a token by selecting strings from input data. The dictionary can be static/dynamic. A sliding window is the basis for this method. The window to the input file is maintained by the entropy encoder and as strings of symbols are seen, it shifts the input in that window from right to left. Variants of LZ77 include LZR, LZHuffman, LZSS, LZB, SLH, ROLZ, LZPP, QIC-122, LZRW4, LZMA, LZJB, LZX, LZO, LZH, Statistical Lempel-Ziv, etc [1,13]. The variants of LZ77 work on the assumption that patterns in the input data occur close together which may not be true in all the cases.

Next variation is the LZ78 class of algorithms [14]. In these algorithms, instead of buffers and window, a dictionary is maintained which stores the previously encountered strings. To maintain the dictionary, a tree is used which limits the size of available memory and its decoding is more complex than LZ77. Variations of the LZ78 family and other algorithms include LZFG, LZRW1, LZW, GIF, ZIP, LZAP, LZMW, LZJ, LZY, LZP1, LZP2, LZWL, Repetition Finder, DMC, Context Tree Weighting, Switching method, bzip2, PAQ, RAR, Win RAR, GZIP, EXE Compression, DCA, CRC, V.42*bis*, UNIX Compress, LZC, LZT, etc [1,13]. Most of these above mentioned methods rely only on word based encoding and suffer from the limitations of poor compression, huge memory requirement and time inefficiency.

A short discussion on FPM algorithms follows. Apriori aims at generating all itemsets which occur frequently. Apriori first generates the set of all frequent (1-length) itemsets L_1 by scanning the transaction database. It then iteratively generates candidate itemsets C_k from L_{k-1}. Next, it prunes the itemsets whose subsets are infrequent. This is iterated until no more candidate itemsets can be generated. The result is the set of all frequent itemsets present in the Transactional Database. Other algorithms in FPM are FP-growth [15], Equivalence Class Transformation(ECLAT)[16], Dynamic Itemset Counting(DIC) [17], DHP [18], Diffsets [19], Counting Inference [20], H-Mine [21], RELIM [22], Opportunistic

Projection [21], Partitioning [23], Sampling [21], etc. A detailed survey on various FPM algorithms is in [21,24].

3 Problem Definition

Let us look at some notations introduced in this paper. Absolute frequency(f_{abs}) of a pattern p in T, is defined as $|\{p|p \sqsubseteq T\}|$, which is the number of occurrences of p as a sequence in T. P denotes the set of all patterns with their respective absolute frequency f_{abs}. The modified frequency(f_{mod}) of a pattern p' is defined as, $|\{p'|p' \sqsubseteq T\}|$, which is the number of occurrences of p' in a non-overlapping manner in the text T(T is modelled as an array A) where T contains individual arrays and size of every array is at least 1. The patterns with modified frequency constitute set P' and $|P'| \leq |P|$. f_{mod} maintains the modified frequency of pattern(s) considering the issue of overlapping characters between patterns when they occur as sequence in T.

For instance, given T = *abracadabra* and $\alpha = 2$, the pattern "*a*" from P has $f_{abs} = 5$, but its $f_{mod} = 1$. The f_{mod} of a pattern depends on the f_{mod} of it superpatterns in the set P'. The f_{mod} of pattern "*a*" is calculated after considering the f_{mod} of pattern "*abra*" and then deleting "*abra*" from the array A. If the frequent pattern "*abra*" in P'(also a superpattern of "*a*") with $f_{mod} = 2$ is given priority for encoding, *a*'s count has already been considered four times in the pattern "*abra*", which leads to f_{mod} of pattern "*a*" being 1.

The problem of Frequent Pattern based Huffman Encoding(FPH) for a Text is formally defined as, "Given an input file T of size z, we need to generate the set(P) of all patterns(frequent and candidate) with its respective $f_{abs}(\geq \alpha)$ in T. Classical Huffman Encoding is applied over P', which is constructed from P, to generate the file T' of size z' where $z' << z$."

4 Proposed Frequent Pattern Based Huffman Encoding(FPH) Algorithm

To encode every pattern in text T, the algorithm has to mine the set of all patterns from T, which forms the essential phase of the encoding process. Every pattern in the code table should be used in the conventional Huffman Encoding process. Our algorithm eliminates patterns which will not be used in the encoding process thereby reducing the code table size where the patterns are stored. Apriori algorithm best suites our approach in mining frequent patterns from T. Apriori pruning strategy is extended where the redundant patterns at level k are pruned at level $k+1$. The redundant patterns refer to those phrases(sequence of characters) generated at level k, where its superpattern at level $k+1$ has the same f_{abs}. This enables us to give preference to lengthier patterns for encoding thus reducing code table size.

Algorithms like FP growth and its successors involve expensive operations for checking redundant patterns because we need to wait till the complete construction of the data structure where the entire set of frequent patterns are

Algorithm 1 (Frequent Pattern Generation). *Efficient Pruning based Apriori Mining of Frequent Patterns.*

Input: An Input Text T and α.

Output: The FPH Tree consisting of patterns generated, including frequent patterns and the corresponding codeword for every patterns in the code table.

Method:

Scan T to find C_1 and include them to a set P
k=2, $L_2 \leftarrow (e \in C_2) \wedge (count.e \geq \alpha)$
while ($L_{k-1} \neq \emptyset$) **do**
  for each ($s_1 \in L_{k-1}$) **do**
    for each ($s_2 \in L_{k-1}$) **do**
      if (($k-2$) length prefix of s_2 = ($k-2$) length suffix of s_1) **then**
        $C_k = C_k \cup (e$ =(first character of s_1 + s_2)) //Add pattern e to the Candidate pattern list
      end if
    end for
  end for
  for each ($e \in C_k$) **do**
    if (e is a pattern in T) **then**
      $e.f_{abs}++$;
      add e to P if $e.f_{abs} \geq \alpha$ // P - Set containing patterns with its corresponding f_{abs}
    end if
  end for
  Increment k
end while
for all ($e \in L_k$) **do**
  for each ($e' \in L_{k-1}$) **do**
    if (e's ($k-1$) length prefix and suffix has the same $e.f_{abs}$) **then**
      remove the pattern from P
    end if
  end for
end for
find_f_{mod}(P);
Construct a priority Queue(Q) to accommodate all patterns generated
for every pattern (e in P) **do**
  push e to Q
end for
while (Q.*empty*()! = $TRUE$) **do**
  Extract two minimum frequencies from Q and construct a parent node which has the summed up values of the frequencies and push the new node's value into Q
end while
Return the root of the FPH Tree

Algorithm 2. *Allocation of codes to patterns for encoding.*

Input: Root of the FPH Tree
Output: Encoded/Compressed Text T'.
Method:
Step 1. While $(P' \neq \emptyset)$ and for every $(p' \in P')$, store p' with its respective codeword in descending order of its pattern length and then by ascending order of its *ASCII* values into an array E.

Step 2. For every p' in the array E, if there is a match in T, encode p' and return T'.

Procedure *find_f_{mod}(P): Find the modified frequency f_{mod} of each patterns in the set P*

Input: $P = \{p_1, \ldots, p_x\}$
Output: $P' = \{p'_1, \ldots, p'_y\}$; $|P'| \leq |P|$

Method:

List L;
L.push(A) // T is modelled as an array A
Sort every pattern p in P in the descending order of pattern length and in ascending order of ASCII values
while (L.empty()!= TRUE) **do**
  for (each p in P) **do**
    for (j in 1 to sizeof(L)) **do**
      if ($p \sqsubseteq L[j]$) **then** //$\sqsubseteq$ denotes sequence
        $p'.\,f_{mod}$ ++ //p is assigned to p' while its f_{mod} is found.
      end if
      $A \leftarrow L$.top(); L.pop()
      Delete h characters from A where $h \leftarrow$ sizeof(p)
      Split A into $A_1 = (c_1, c_2, \ldots, c_{j-h})$ and $A_2 = (c_j, \ldots, c_n)$
      Add A_1 and A_2 to L
    end for
    if ($p'.f_{mod} \neq 0$) **then**
      if ($d == 0$) *FPH1(p')*
        else if ($d == 1$) *FPH2(p')*
      end if
    end if
  end for
end while

generated. But in Apriori, as and when the levels are constructed, the frequent patterns are generated. File T undergoes m scans to generate patterns of length 1 to maximum of $\frac{n}{min\_supp}$, where n is the number of characters in the text and m is the maximum length of the frequent patterns generated. Two of our FPH strategies are explained in Algorithm 1 and 2. Table 1. illustrates our proposed FPH Encoding. For a pattern p in P, the split-up of arrays in its f_{mod} calculation process can be seen from Figs. 1, 2 and 3.

Procedure *FPH1(p′) To select patterns satisfying min_supp*

for (each pattern p' in P') **do**
  if $(p'.f_{mod} \geq \alpha) \wedge (p' \in L_{k\geq 2}) \mid\mid (p' \in C_1)$ **then**
    $P' = P' \cup p'$ //Add p' to the set P'
  end if
end for
return P'

Procedure *FPH2(p′) To select patterns satisfying the condition* $f_{mod} \neq 0$

for (each pattern p' in P') **do**
  if $p' \in C_{k\geq 1}$ **then**
    $P' = P' \cup p'$ //Add p' to the set P'
  end if
end for
return P'

Table 1. Frequent patterns in T

Input Text(T): *eat to live, not live to eat;* $\alpha = 2$.	
Candidate Patterns(C_k) with its f_{abs}	**Frequent Patterns(L_k) with its f_{abs}**
C_1: (1-length characters) = {e-4, a-2, t-5, ′ ′-6, o-3, l-2, i-2, v-2, ,-2, n-1, .-1}	No pruning in L_1
C_2 = {ea-2, at-2,. . . ,t.-1}	L_2={ea-2, at-2, t′ ′-2, ′ ′t-2, to-2, o′ ′-2, ′ ′l-2, li-2 iv-2, ve-2}
C_3={eat-2,. . . ,at.-1}	L_3 = {eat-2, ′ ′to-2, to′ ′-2, ′ ′li-2, liv-2, ive-2}
$L_2 = L_2$ - {ea-2, at-2, ′ ′t-2, to-2, o′ ′-2, ′ ′l-2, li-2, iv-2, ve-2} ⇒ L_2={t′ ′-2}	
C_4={eat′ ′-1,. . . ,eat.-1}	L_4= {′ ′to′ ′-2, ′ ′liv-2, live-2}
$L_3 = L_3$ - { ′ ′to-2, to′ ′-2, ′ ′liv-2, liv-2, ive-2} ⇒ L_3={eat-2}	
C_5 = {eat′ ′t-1,. . . ,′ ′eat.-1}	L_5 = {′ ′live-2}
$L_4 = L_4$ - {′ ′liv-2, live-2} ⇒ L_4= {′ ′to′ ′-2}	
C_6 = {eat′ ′to-1,. . . ,o′ ′eat.-1}	$L_6 = \emptyset$
P(arranged in descending order of length of p and ascending order of ASCII values = {′ ′live-2, ′ ′to′ ′-2, eat-2, t′ ′-2, ′ ′-6, ,-1, .-1, a-2, e-4, i-2, l-2, n-1, o-3, t-5, v-2 }	

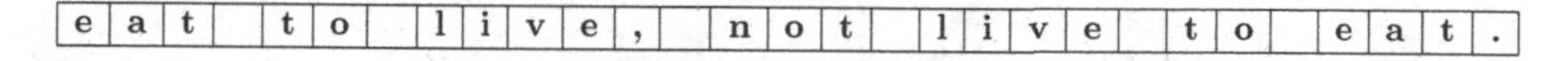

Fig. 1. Text T as an array A

e	a	t		t	o

(a) Array A_1

,		n	o	t

(b) Array A_2

	t	o		e	a	t	.

(c) Array A_3

Fig. 2. Array split-up after finding f_{mod} of pattern ′ ′*live* in A in Fig. 1.

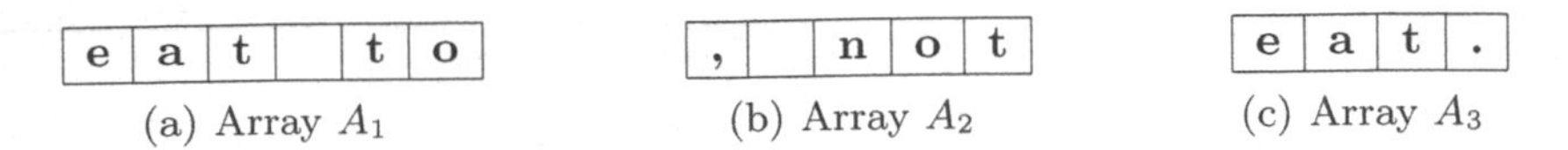

(a) Array A_1 (b) Array A_2 (c) Array A_3

Fig. 3. Array split-up after finding f_{mod} of pattern $'\ 'to'\ '$ in Fig. 2.

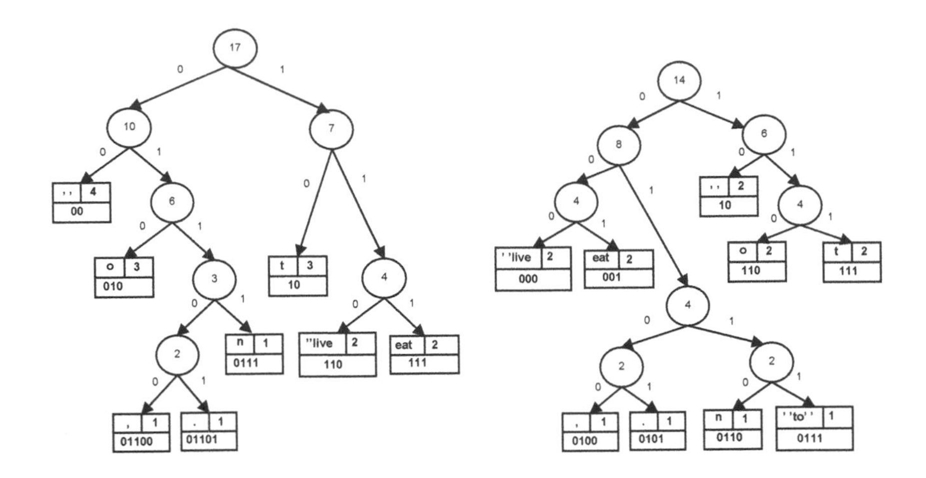

Fig. 4. FPH1 Tree **Fig. 5.** FPH2 Tree

The FPH tree is constructed by taking patterns from P' as shown in Figs. 4 and 5. Every pattern is used to create a minimum priority queue(Q). The tree is constructed in the same strategy as conventional Huffman Encoding. The patterns and its associated codes are stored in a code table in the descending order of pattern length and in ascending order of ASCII values. The ASCII ordering is done because T may contain alpha-numerals and special characters. The decoding happens in the same way as Conventional Huffman Decoding. We focus on encoding phrases(sequence of characters) occurring frequently based on a given min_supp so that our proposed algorithms achieves better compression ratio than Huffman Encoding.

5 Results and Discussions

Simulation is performed on an Intel Xeon processor E5-2600 v3 family with 128GB Main Memory and 1TB Hard disk on CentOS Platform. FPH Encoding and Conventional Huffman Encoding are implemented using C++ 11 standard. The algorithms have been tested over Calgary, Canterbury, Silesia benchmark Corpus as given in Table 2 [25,26]. Compression Ratio for various benchmark datasets that has been tested, is given in Table 3.

Table 2. Description of the various benchmark datasets

File	Size [Bytes]	Description	Type of data
bib	111,261	Bibliographic files	Text data
book1	768,771	Text of the book *Far from the madding crowd*	English text
book2	610,856	Text of the book *Principles of computer speech*	English text
news	377,109	A Usenet news batch file	English text
plrabn12	481,861	*Paradise Lost* by John Milton	English text
ptt5	513,216	Picture number 5 from the CCITT Facsimile	Image
e.coli	4,638,690	Complete genome of the *E.Coli* bacterium	binary data
bible	4,047,392	The King James version of the bible	English text
world192	2,473,400	The CIA world fact book	English text
vldb	1,455,480	Bibliographic files of conference on VLDB	Text data
dickens	10,192,446	Collected works of Charles Dickens	English text

Table 3. Compression ratios for various benchmark datasets

File Name	*min_supp*(%) for max C_r	# Patterns in P' for FPH1	Conventional Huffman C_r	FPH1 C_r	FPH2 C_r	Time taken for FPH1(sec)	C_r Efficiency (%) of FPH1
bib	0.01	1189	1.5236	**2.3713**	**2.0671**	25.53	55.63
book1	0.01	1386	1.61613	**2.3202**	**2.1241**	250.13	43.54
book2	0.01	1351	1.6570	**2.3713**	**2.1593**	189.30	43.10
news	0.01	1366	1.5285	**1.9753**	**1.8269** at 0.02 %	83.58	29.23
plrabn12	0.01	1322	1.7349	**2.3706**	**2.1375** (at 0.03 %)	102.9	36.63
ptt5	1	159	4.7520	**4.7878**	**4.7837** (at 0.5 %)	52.92	0.75
bible	0.01	1188	1.8203	**2.8304**	**2.7824**	24272	55.48
E.Coli	30	4	3.8167	**3.9974**	**3.9907**	71565.1	4.73
world192	0.01	1281	1.5750	**2.3784**	**2.3840**	16509.5	51.008
vldb	0.01	615	1.5097	**3.6883**	**3.6556**	2424.3	144.30
Dickens	0.01	1333	1.7487	**2.4205**	**2.3860**	49668.8	38.53

Figure 6. highlights the efficiency in terms of Compression Ratio(C_r) of the proposed technique in relation to Conventional Huffman Encoding at varying *min_supp* values for bible dataset. The Compression ratio C_r is defined as :

$$C_r = \frac{\text{Uncompressed size of Text}}{\text{Compressed size of Text}}$$

Compressed size denotes the total size of the code table and the encoded text size post FPH compression. As can be seen from Figs. 6, 7, 8 and 9, increased *min_supp* values result in degraded compression (as a result of reduced number of frequent patterns at higher support values). It can be seen from Fig. 6. for bible corpus, the maximum compression ratio for FPH1 and FPH2 approach is 2.8304 and 2.7824 at a *min_supp* of 0.01 % as opposed to Conventional Huffman ratio of 1.8203. The proposed algorithms FPH1 and FPH2 achieves C_r efficiency of 55.48 % and 52.85 % respectively in relation to Conventional Huffman(CH). The C_r efficiency is defined as,

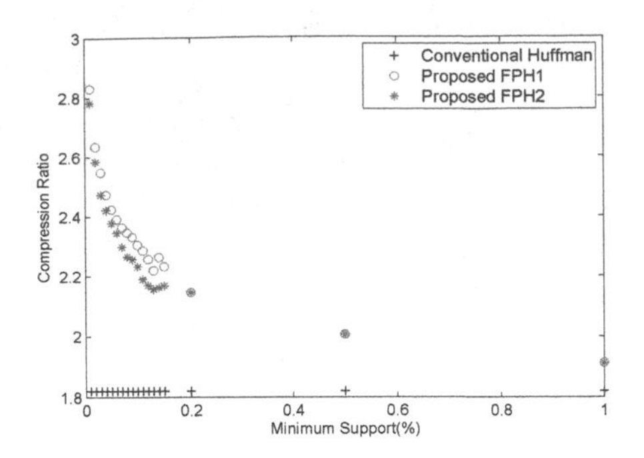

Fig. 6. *min_supp* vs Compression Ratio for Bible

Fig. 7. *min_supp* vs Time for Bible

$$C_r \text{ Efficiency}(\%) = \frac{C_r(FPH) - C_r(CH)}{C_r(CH)} \times 100$$

It is observed that FPH1 and FPH2 achieve better compression ratio than Conventional Huffman. The number of patterns generated reduces on increasing the *min_supp*. The code table comprising the frequent patterns is stored along with their binary codes from which the Huffman tree is generated for decompression. Thus on increasing the *min_supp*, the size of the code table decreases as the number of frequent pattern decreases. However, the benefits of reduced code table size are nullified by the increase in encoded size of input text data.

In bible, FPH1 approach performs better than FPH2, because it has only frequent patterns of length ≥ 1 which satisfies *min_supp* and candidate patterns(C_1) of length 1 in set P'. This reduces the number of patterns thereby reducing code table size in FPH1 than in FPH2. In FPH2, the set P' contains all patterns of length ≥ 1 with $f_{mod} > 0$ which also includes frequent patterns satisfying *min_supp*. This drastically increases the number of patterns thereby increasing code table size to a large extent in FPH2, even though there is a slight decrease in encoded size compared to FPH1. Similar trend can be observed for other datasets as well. The maximum C_r efficiency is achieved in vldb dataset which is 144.30 % because of the dense occurrence of frequent patterns. Efficient compression ratio for all the tested datasets for both FPH1 and FPH2 is achieved for *min_supp* values in the range of 0.01 % to 1 %.

Figure 7. highlights the relation between C_r and the time taken to compress T. In bible corpus, in FPH1 in the support range of 0.01 % to 1 %, the maximum C_r is attained at 0.01 % with a time of 24272 sec. as opposed to the Conventional Huffman time of 40916.9 sec. For few *min_supp* values, our algorithm takes lesser time than Conventional Huffman Encoding as can be seen from Fig. 7. Even though the patterns generated are more than the Conventional Huffman in this range of support values, the encoding time reduces because of the ability to capture lengthier patterns and so the time for traversing T is less.

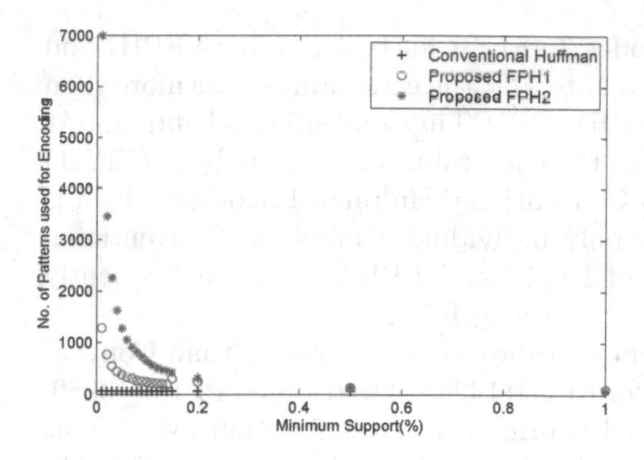

Fig. 8. *min_supp* vs #Patterns for Bible

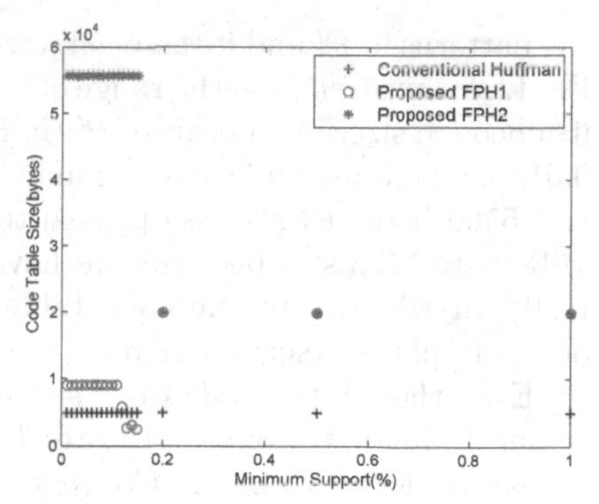

Fig. 9. *min_supp* vs Code table for Bible

In Conventional Huffman Encoding, the time to encode the text goes up, since it needs to encode individual symbols which will make the algorithm to traverse the text more number of times. In FPH2 approach for the bible dataset, time taken is always lesser than the Conventional Huffman strategy for which the same reasoning discussed above holds true. For few other *min_supp* in FPH1, time taken by our approach is greater than Conventional Huffman, since the number of patterns are huge. This is because, the time taken to generate the patterns in P and in P' adds up the cost because of m scans of the database where m is the maximum length of the frequent patterns generated. The time complexity for the Apriori process to construct P for text T is $O(n^3)$ which contributes to the time taken. The same trend is observed in book1 and other tested corpus for both approaches in *min_supp* range of 0.01 to 1 %.

Figure 8 shows the relation between *min_supp* vs number of patterns in the set P' for bible dataset. For both FPH1 and FPH2, for *min_supp* values in the range of 0.01 to 1 %, $|P'| < |P|$. This is because of f_{mod}, which gives the exact count of patterns needed for encoding without overlapping. This helps us not to store patterns that go without encoding. In P, patterns overlap with one another as we have f_{abs} which leads to exhaustive/huge number of patterns. In FPH1, for the maximum C_r attained at *min_supp* 0.01 %, the patterns are reduced to 1188 from 8216 which is 85.5 % reduction in the count. More number of patterns are retained in P' in FPH2 than in FPH1, because we consider all patterns that satisfy $f_{mod} > 0$. Similar results can be seen in other datasets as well. In all the tested datasets, maximum reduction in $|P'|$ is achieved at the *min_supp* where maximum C_r is attained. For *min_supp* from 0.2 to 1 % for a few datasets, $|P| = |P'|$. This is because when the *min_supp* is increased, the number of patterns in both the sets converges to the Conventional Huffman strategy as mostly there will be 1-length characters.

Figure 9. highlights the analysis between *min_supp* and code table size in bytes for bible corpus. Code table refers to the size of the memory occupied by the patterns in P' and its associated codes. In Fig. 9, for bible, for both FPH1 and FPH2, for *min_supp* in the range of 0.01 % to 1 %, the code table size is more than its encoded size. It is because $|P'|$(in FPH) >> $|P'|$ in Conventional Huffman. At 0.01 %, where we achieve maximum C_r, the code table size is 9021 bytes(FPH1) and 5569 bytes(FPH2) as opposed to Conventional Huffman Encoding which is 4918 bytes. This is because, we have only individual symbols in Conventional Huffman whereas in the code table of FPH1 and FPH2, we have frequently occuring phrases(sequences of characters) of length ≥ 1.

Even though the code table size of our approaches are greater than Conventional Huffman, the encoded size of T for FPH1(1420943 bytes) and FPH2(139891 bytes) are lesser than the Conventional Huffman which is 2218450 bytes. This contributes to a greater C_r in our approaches for *min_supp* in the range of 0.01 % to 0.1 %. The same trend is seen in other datasets as well. Since $|P'|$ is more in FPH2 than FPH1 for all datasets, code table size of FPH2 increases leading to a reduction in C_r than FPH1. This makes FPH1 approach to perform slightly better than FPH2 but in turn, both are efficient than Conventional Huffman technique for all corpus.

6 Conclusion

The paper has presented a novel approach to lossless text compression, employing FPM in Huffman Encoding. Efficient pruning strategies to handle the pattern base has been employed. Results indicate the significant gains in Compression Ratio. Future work shall concentrate on efficient data structures for finding support count and time efficiency. Scope of the proposed algorithm in the context of lossy compression techniques shall also be explored.

References

1. David, S.: Data Compression: The Complete Reference, 2nd edn. Springer, New York (2004)
2. Huffman, D.A.: A method for the construction of minimum redundancy codes. Proc. IRE **40**(9), 1098–1101 (1952)
3. Gilbert, E.N., Moore, E.F.: Variable-length binary encodings. Bell Syst. Tech. J. **38**(4), 933–967 (1959)
4. Han, J., Kamber, M.: Data Mining: Concepts and Techniques. Morgan Kaufmann, San Francisco (2000)
5. Ramakrishnan, N., Grama, A.: Data mining: from serendipity to science - guest editors' introduction. IEEE Comput. **32**(8), 34–37 (1999)
6. Agarwal, R., Srikant, R.: Fast algorithms for mining association rules in large databases. In: Bocca, J.B., Jarke, M., Zaniolo, C. (eds.) VLDB 1994, Proceedings of 20th International Conference on Very Large Data Bases, pp. 487–499. Santiago de Chile, Chile, Morgan Kaufmann (1994)
7. Shannon, C.E.: A mathematical theory of communication. ACM SIGMOBILE Mob. Comput. Commun. Rev. **5**(1), 3–55 (2001)
8. Pountain, D.: Run-length encoding. Byte **12**(6), 317–319 (1987)

9. Witten, I.H., Neal, R.M., Cleary, J.G.: Arithmetic coding for data compression. Commun. ACM **30**(6), 520–540 (1987)
10. Vitter, J.S.: Design and analysis of dynamic huffman codes. J. ACM (JACM) **34**(4), 825–845 (1987)
11. Moffat, A.: Implementing the PPM data compression scheme. IEEE Trans. Commun. **38**(11), 1917–1921 (1990)
12. Ziv, J., Lempel, A.: A universal algorithm for sequential data compression. IEEE Trans. Inf. Theor. **23**(3), 337–343 (1977)
13. Deorowicz, S.: Universal lossless data compression algorithms. Philosophy Dissertation Thesis, Gliwice (2003)
14. Ziv, J., Lempel, A.: Compression of individual sequences via variable-rate coding. IEEE Trans. Inf. Theor. **24**(5), 530–536 (1978)
15. Han, J., Pei, J., Yin, Y., Mao, R.: Mining frequent patterns without candidate generation: a frequent-pattern tree approach. Data Min. Knowl. Discov. **8**(1), 53–87 (2004)
16. Goethals, B.: Survey on frequent pattern mining. manuscript (2003)
17. Brin, S., Motwani, R., Ullman, J.D., Tsur, S.: Dynamic itemset counting and implication rules for market basket data. In: ACM SIGMOD Record, vol. 26, pp. 255–264. ACM (1997)
18. Park, J.S., Chen, M.S., Yu, P.S.: An effective hash-based algorithm for mining association rules. ACM SIGMOD Rec. **24**, 175–186 (1995)
19. Zaki, M.J., Gouda, K.: Fast vertical mining using diffsets. In: Proceedings of the Ninth ACM SIGKDD International Conference on Knowledge Discovery and Data Mining, pp. 326–335. ACM (2003)
20. Bastide, Y., Taouil, R., Pasquier, N., Stumme, G., Lakhal, L.: Mining frequent patterns with counting inference. ACM SIGKDD Explor. Newsl. **2**(2), 66–75 (2000)
21. Han, J., Cheng, H., Xin, D., Yan, X.: Frequent pattern mining: current status and future directions. Data Min. Knowl. Discov. **15**(1), 55–86 (2007)
22. Borgelt, C.: Keeping things simple: finding frequent item sets by recursive elimination. In: Proceedings of the 1st International Workshop on Open Source Data Mining: Frequent Pattern Mining Implementations, pp. 66–70. ACM (2005)
23. Savasere, A., Omicinski, E.R., Navathe, S.B.: An efficient algorithm for mining association rules in large databases. In: VLDB (1995)
24. Borgelt, C.: Frequent item set mining. Wiley Interdisc. Rev.: Data Min. Knowl. Discov. **2**(6), 437–456 (2012)
25. Calgary compression corpus datasets. `corpus.canterbury.ac.nz/descriptions/` Accessed: 23 July 2015
26. Silesia dataset. http://sun.aei.polsl.pl/sdeor/index.php?page=silesia Accessed: 23 July 2015

Identifying Semantic Events in Unstructured Text

Diana Trandabăţ(✉)

Faculty of Computer Science, University "Al. I. Cuza" of Iasi, Iaşi, Romania
dtrandabat@info.uaic.ro

Abstract. Semantics has always been considered the hidden treasure of texts, accessible only to humans. Artificial intelligence struggles to enrich machines with human features, therefore accessing this treasure and sharing it with computers is one of the main challenges that the natural language domain faces nowadays. This paper represents a further step in this direction, by proposing an automatic approach to extract information about events from unstructured texts by using semantic role labeling.

Keywords: Artificial intelligence · Natural language parsing · Semantic roles · Machine learning for semantic labeling

1 Introduction

Attracted by the potential applications, more and more researchers from the artificial intelligence field submerged into the natural language processing domain. Thus, one further step in the human-computer interaction is the use of human languages instead of some pre-defined expressions. In order to teach a computer to understand a human speech, language models need to be specified and created from human knowledge. While still far from completely decoding hidden messages in political speeches, computer scientists, electrical engineers and linguists have all joined efforts in making the language easier to be learned by machines.

Our approach uses semantic role analysis in order to establish the roles that entities have in different contexts, and what are the temporal, modal or local constraints that determine or restrict an event to take place. A semantic role represents the relation-ship between a predicate and an argument. Semantic parsing, by identifying and classifying the semantic entities in context and the relations between them, has great potential on its downstream applications, such as text summarization or machine translation.

In this paper we propose a system which, starting from an input entity, extracts web pages found on a Google search for a particular entity, selects the snippets that contain the input entity, and then performs semantic role labeling to extract the relations between the entity and its context. Thus, our system creates a contextual map by identifying the role an entity plays in different contexts, as well as the roles played by words frequently co-occurring with the input entity.

The motivation behind the work presented in this paper is the need to create a map of structured context related to a specific entity (e.g. a company or product name, an event, etc.). Through this map, the concepts that are usually in relation to the searched

R. Prasath et al. (Eds.): MIKE 2015, LNAI 9468, pp. 576–585, 2015.
DOI: 10.1007/978-3-319-26832-3_54

input entity are highlighted, together with their specific role (which can be of type Cause, Effect, Location, Time, etc.), thus providing a good material for social analyses, market research or other marketing purposes.

The paper is structured in 5 sections. After an introduction in the field of semantic role analysis, we briefly present the current work in Sect. 2. Section 3 introduces the overall application, describing the intermediary steps, while Sect. 4 presents our approach for a Semantic Role Labeling system and evaluates it. The final section draws the conclusions of this paper and discusses further envisaged developments.

2 Semantic Role Analysis

Natural language processing is a key component of artificial intelligence. All content elements of a language are seen as predicates, i.e. expressions which designate events, properties of, or relations between, entities. The predication represents the mechanism that allows entities to instantiate properties, actions, attributes and states. More precisely, the linking between a phenomenon and individuals is known as predication. Predicates are not treated as isolated elements, but as structures, named predicate frames or semantic frames. Fillmore in [8] defined six semantic roles: Agent, Instrument, Dative, Factive, Object and Location, also called deep cases. His later work on lexical semantics led to the conviction that a small fixed set of deep case roles was not sufficient to characterize the complementation properties of lexical items, therefore he added Experiencer, Comitative, Location, Path, Source, Goal and Temporal, and then other cases. This ultimately led to the theory of Frame Semantics [7], which later evolved into the FrameNet project [1].

The semantic relations can be exemplified within the Commercial Transaction scenario, whose actors include a buyer, a seller, goods, and money. Among the large set of semantically related predicates, linked to this frame, we can mention *buy*, *sell*, *pay*, *spend*, *cost*, and *charge*, each indexing or evoking different aspects of the frame.

In the last decades, hand-tagged corpora that encode such information for the English language were developed (VerbNet [12], FrameNet [1] and PropBank [18]). For other languages, such as German, Spanish, and Japanese, semantic roles resources are being developed.

For role semantics to become relevant for language technology, robust and accurate methods for automatic semantic role assignment are needed. Automatic Labeling of Semantic Roles is defined as identifying frame elements within a sentence and tag them with appropriate semantic roles given a sentence, a target word and its frame [13]. Most general formulation of the Semantic Role Labeling (SRL) problem supposed determining a labeling on (usually but not always contiguous) substrings (phrases) of the sentence s, given a predicate *p*.

In recent years, a number of studies, such as [3, 4, 9], have investigated this task. Role assignment has generally been modeled as a classification task: A statistical model is trained on manually annotated data and later assigns a role label out of a fixed set to every constituent in new, unlabelled sentences. The work on SRL has included a broad spectrum of probabilistic and machine-learning approaches to the task, from probability estimation [9], through decision trees [22] and support vector machines [10, 19], to memory-based learning [15]. While using different statistical frameworks, most studies have largely converged on a common set of features to base their decisions on, namely syntactic information (path from predicate to constituent, phrasal type of constituent) and lexical information (head word of the constituent, predicate).

As for extracting semantic events and relations, worth noticing is the work done by the Watson group at IBM on relationship extraction and snippets evaluation with applications to question answering [20], the work in [5] for sport texts, but also the work in [11], where frame-semantic parsing and event extraction are considered structurally identical tasks and a frame-semantic parsing system is used to predict event triggers and arguments.

3 Towards Semantically Interpreting Texts

The goal of the application we propose is to extract the context in which an entity occurs in web documents, together with the relations that the searched entity establishes with frequently co-occurring words. The basic architecture of our application contains a series of specialized modules, as follow:

3.1 User Input Acquiring Module

This module prompt the user for an input entity, usually representing a company or product name, an event name, a person, etc. A drop down list of the most frequent searched entities is offered as suggestion.

3.2 Web Page Retrieval Module

This module extract from the web the first n (in our tests n = 200) web pages found on a Google search for the input entity. Google search options restricting the web search can be applied, such as selecting articles from newspapers, blogs or in a specific language.

3.3 Snippet Extraction Module

Using the Google snippet suggestion and some simple heuristics, the paragraphs containing the input entity are selected. A simple anaphora resolution method, based on a set of reference rules, is applied to the web pages, in order to link all entities to their referees.

The anaphoric system we used is a basic rule-based one, focusing on named entity anaphoric relations. Thus, we developed a rule-based system that performs the following actions:

- identifies a subset of a named entity with the full named entity, if it appears as such in the same text. For instance, Caesar is identified with Julius Caesar if both entities appear in the same text. Similarly, the President of Romania and the President are considered anaphoric relations of the same entity, if they appear in a narrow word window in the text.

- solves acronyms using a gazetteer we have initially built over the Internet, and which is continuously growing in size. For instance, *United States of America* and *USA* are co-references.
- searches for different addressing modalities and matches the ones that are similar. For instance, *John Smith* is co-referenced with *Mr. Smith*, and *Mary and John Smith* is co-referenced with *The Smiths*, or *The Smith Family*.
- solve pronominal anaphora in a simplistic way. Thus, if a pronoun (i.e. *she*, *he*, *him*, *his* etc.) is found in the text, and in the preceding sentence an entity is found, then we create an anaphoric link between the pronoun and its antecedent. A similar rule exists for companies, where the pronoun *it* may be linked to *the Insurance Company*, for instance.

3.4 Snippet Cleaning Module

After relevant paragraphs (containing the input word) are extracted, functional words such as (and, so, a, etc.) are eliminated from these paragraphs, since, being statistically too frequent for any kind of texts, they do not convey any useful information for semantic role labeling. A list of these functional words, created by using word frequencies in large corpora, is used.

3.5 Semantic Role Labeling Module

This module performs semantic role labeling on the obtained paragraphs, in order to identify the role the entity in question and the related entities plays. It will be described in details in the following section and evaluated in Sect. 6.

3.6 Module for the Creation of a Map of Concepts

This modules works in two steps: first, it extracts from the semantic role analysis the relations between the searched entity and the neighbor entities, creating a list of relations. Secondly, it generalizes the concepts that are found to be in relation with the searched entity across all extracted paragraphs, using the WordNet [6] hierarchy. Semantic distances using Wordnet have been previously investigated [2].

The next section presents the core module of this architecture, the semantic role labeling system, developed by training a set of supervised machine learning algorithms for several languages.

4 Our Semantic Role Labeling Approach

In order to detach semantic information from texts, we built a supervised SRL system, which we named PASRL – Platform for Adjustable Semantic Role Labeling. Several machine learning techniques and feature sets have been tested using the algorithms implemented in the Weka toolkit [25]. We used 12 of the most common machine learning algorithm that are available in Weka, such as decision trees, SVMs, memory-based learners, etc. A full description of the machine learning algorithm from Weka used for PASRL can be found in [24]. Each learning algorithm is trained with a set of features, the performances of the different obtained models are compared, and the best one is selected. The output of PASRL is a Semantic Role Labeling System which can be used to annotate new texts.

Using training data[1] preprocessed with syntactic and dependency information, PASRL is composed of two main sub-systems: a Predicate Prediction module and an Argument Prediction module.

4.1 Predicate Prediction Module

The first module of the semantic role labeling system is the predicate prediction module. The Predicate Prediction module system is composed out of three sub-modules:

- *Predicate Identification* this module takes the syntactic analyzed sentence and decides which of its verbs and nouns are predicational (can be predicates), thus for which ones semantic roles need to be identified;
- *Predicate Sense Identification* – once the predicates for a sentence are marked, each predicate need to be disambiguated since, for a given predicate, different sense may demand different types of semantic roles;
- *Joint Predicate and Predicate Sense Identification* – jointly identifies the predicates and their senses (the two above sub-modules).

Predicate Identification Task. Our semantic role labeling system uses the PropBank annotation of semantic roles. Since predicational words are not just verbs, beside PropBank [18] for the verbal frames, NomBank [14] is also used for nouns. Using syntactic annotation, consisting of marked dependency relations, and also the resources PropBank and NomBank, the system first tries to identify the words in the sentence that can behave as semantic predicates, and for which semantic roles need to be found and annotated (the Predicate Identification module). This module relies mainly on the external resources, thus the verbs that are in PropBank (have semantic frame annotation) are likely to be semantic predicates, those which aren't, are not predicational verbs, thus cannot have semantic arguments. For example, the verb to be has no annotation in PropBank, since it is a state and not an action, predicational, verb. Similarly, the NomBank is used to sort nouns that can behave as predicates from those that cannot have semantic arguments.

The predicate identification program transforms the annotated input into training instances for the ML algorithms in order to identify which nouns or verbs from the input are predicates. For each noun or verb in the sentence, an instance is created with a set of features from a syntax-based vector space, inspired from the features usually used for Semantic Role Labeling [17], and a binary class label (the candidate word is or not a predicate for the considered sentence). The features used for the Predicate Prediction

[1] The training data consists of manually annotated data from Penn Treebank.

are detailed in [23]. The output of this module is the input file, where each verb or noun that behaves as a predicate is annotated.

After the predicates from the input sentence are identified, the next module is successively applied for all the predicates found in the sentence, in order to identify for all of them all and only their arguments. For example, for the sentence:

```
The assignment of semantic roles depends on the number of
predicates.
```

two predicates are identified by the Predicate Identification module (assignment and depend). The next module will be applied two times, once for the identification of the sense and semantic arguments of the assignment predicate, and once for the depend predicate.

The output will therefore provide the two annotations:

[The assignment of semantic roles]$_{ARG0}$ [depends]$_{TARGET}$ [on the number of predicates]$_{ARG1}$.
[The assignment]$_{TARGET}$ [of semantic roles]$_{ARG1}$ depends on the number of predicates.

Predicate Sense Identification Module. Since this module considers that the predicational words are already identified, a binary feature is predicate is extracted from the training file and added to the feature set created for the Predicate Identification task, and Weka classifiers are again ran to classify each predicate with its PropBank/NomBank role set. This module is needed in order to select the types of semantic roles specific to the sense the predicate has. The sense annotation in PropBank and NomBank is similar to some extent to the sense annotation from WordNet, with the observation that the classification in sense classes (role sets in PropBank's terminology) is centered less on the different meanings of the predicational word, and more on the difference between the sets of semantic roles that two senses may have. The senses and role sets in PropBank for a particular predicate are usually subsumed by WordNet senses, since the latter has a finer sense distinction.

Joint Predicate and Predicate Sense Identification. Instead of running the Predicate Identification and the Predicate Sense Identification processes successively, we tested running them simultaneously, using the same features presented above and as class the predicate role set similar to the one used for Predicate Sense Identification.

4.2 Argument Prediction Module

After running the first module of the semantic role labeling system (either pipelined or joint), the Argument Identification task is called in order to assign to each syntactic constituent of a verb/noun its corresponding semantic role. The instances in this case are not only the nouns and verbs, but every word in the sentence.

The Argument Prediction module performs argument prediction, based on the dependency relations previously annotated with the MaltParser [16] and the Predicate

Prediction output. The input of this module contains syntactic information (part of speech and syntactic dependencies), predicate and predicate role set, and in the output each syntactic dependent of the verb is labeled with its corresponding role. This module uses as external resources PropBank and NomBank frame files and a list of frequencies of the assignment of different semantic roles in the training corpus. The features used are described in [23].

5 Evaluating PASRL

An evaluation metric for semantic roles have been proposed within CoNLL shared task on semantic role labeling [21]. The semantic frames are evaluated by reducing them to semantic dependencies from the predicate to all its individual arguments. These dependencies are labeled with the labels of the corresponding arguments. Additionally, a semantic dependency from each predicate to a virtual ROOT node is created. The latter dependencies are labeled with the predicate senses.

The evaluation of the PASRL performance was computed using 10-fold cross-validation on the training set. For each task, PASRL evaluates all the machine learning algorithms used against the gold-annotated corpus, and the best performing algorithm is saved in a configuration file.

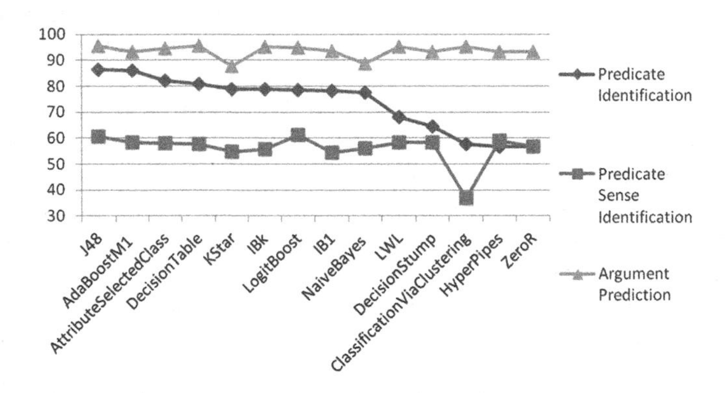

Fig. 1. Performance of the analyzed machine learning algorithms for the SRL tasks performed by PASRL

The evaluation was performed considering the number of correctly classified labels and correctly identified predicates.

Figure 1 presents an overview of the performance of different machine learning algorithms for each of the SRL tasks for English: Predicate Predication (using the Predicate Identification and the Predicate Sense Identification modules, not the joint learning module) and Argument Prediction. One can clearly see that in the Argument Prediction task the algorithms give best results, and the Predicate Sense Identification task is the most difficult task to model. Detailed results for the Predicate Prediction task and its sub-tasks are presented below.

For the Predicate Identification task, running the classifiers with the default weights of Weka for the English dataset, their results ranged from 86 % correctly identified predicates to 56 %. The algorithm that performs best is the J48 classifier (decision tree) and the one that performs worst is the simple ZeroR classifier (see Table 1).

Table 1. Top 5 ML algorithms evaluated with 10-fold cross-validation for English, for the Predicate Identification task

ML algorithm	10-fold cross validation
J48	86.323
AdaBoostM1	86.016
AttributeSelectedClass	82.169
DecisionTable	80.921
KStar	78.97

Using boosting techniques with J48 as base classifier could improve further the module's performance. Changing the default weights of the classifiers can modify their performances, but we believe that the hierarchy will not change substantially. However, this remains a direction to address in a further work. The best performing model (J48 in this case, which is a decision tree learning method) is saved and will be used when the Predicate Identification module will be called from the configuration file for annotating an unlabeled text.

The results for the Predicate Sense Identification Task are considerably worse than the ones for Predicate Identification task. However, we notice that J48 is still among the best algorithms, and memory based algorithms generally perform badly on this subtask.

Instead of running the Predicate Identification and the Predicate Sense Identification processes successively, we tested running them simultaneously, using the same features presented above. Figure 2 shows the difference in the performance obtained by using the pipelined Predicate Identification and Predicate Sense Identification module, as compared to the module that jointly learn Predicate and Predicate Sense Identification. One can notice that, although the results are significantly worse for the joint learning task, the algorithms that perform best for the pipelined task have still good performance in the joint task.

When considering the 5 best classifiers for all the sub-tasks trained on the whole training data, evaluated using 10-fold cross-validation, we can observe that J48 and Decision Tables are among the 5 best algorithms, which suggests that some algorithms may perform better than others for Semantic Role Labeling.

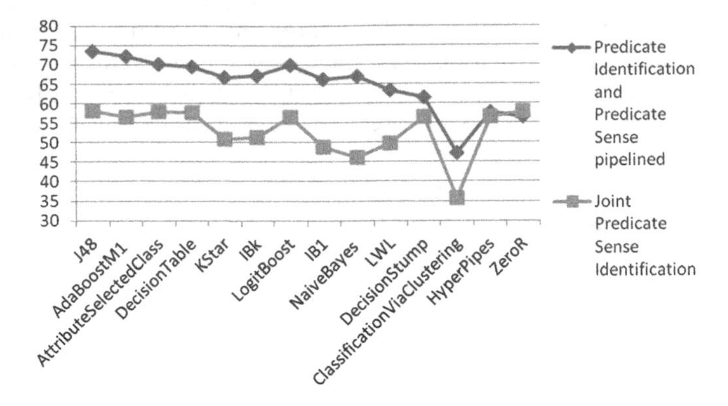

Fig. 2. Performance of the predicate prediction system using (1) the pipelined predicate identification and predicate sense identification module tans (2) the joint predicate and predicate sense identification

6 Conclusions

This paper presented a semantic role labeling system developed using supervised machine learning algorithms from the Weka framework. This system is used in an application that monitors the contexts in which a specific entity appears in web texts and the relations it has with other co-occurring concepts. The developed SRL platform can be used for different languages, provided that a training corpus annotated with semantic roles is available. After testing several classifiers on different sub-problems of the SRL task (Predicate Identification, Predicate Sense Identification, Predicate and Sense Identification, Argument Prediction), the proposed system chooses the algorithm with the greatest performance and returns a Semantic Role Labeling System (a sequence of trained models to run on new data).

The envisaged application of our system is in the marketing field, where the related concepts our system offers can be used to monitor the reaction of the consumer to different changes in a company's marketing policies.

References

1. Baker, C.F., Fillmore, C.J., Lowe, J.B.: The Berkeley FrameNet project. In: Proceedings of COLING-ACL, Montreal, Canada (1998)
2. Budan, I.A., Graeme, H.: Evaluating wordnet-based measures of semantic distance. Comput. Linguist. **32**(1), 1347 (2006)
3. Chen, J., Rambow, O.: Use of deep linguistic features for the recognition and labeling of semantic arguments. In: Proceedings of EMNLP 2003 (2003)
4. Chu-Carroll, J., Fan, J., Schlaefer, N., Zadrozny, W.: Textual resource acquisition and engineering. IBM J. Res. Dev. **56**(3/4), 4:1 (2012)
5. Chen, C.-M., Chen, L.-H.: A novel approach for semantic event extraction from sports webcast text. Multimedia Tools Appl. **71**(3), 1937–1952 (2014)
6. Fellbaum, C. (ed.): WordNet: an Electronic Lexical Database. MIT Press, Cambridge (1998)

7. Fillmore, C.J.: Frame semantics. In: Linguistics in the Morning Calm, pp. 111–137. Hanshin Publishing, Seoul (1982)
8. Fillmore, C.J.: The case for case. In: Bach, E., Harms, R. (eds.) Universals in Linguistic Theory, pp. 1–88. Holt, Rinehart, and Winston, New York (1968)
9. Gildea, D., Jurafsky, D.: Automatic labeling of semantic roles. Comput. Linguist. **28**(3), 245–288 (2002)
10. Hacioglu, K., Wayne, W.: Target word detection and semantic role chunking using support vector machines. In: Proceedings of HLT/NAACL-03 (2003)
11. Surdeanu, M., Johansson, R., Meyers, A., Marquez, L., Nivre, J.: CoNLL-2008 shared task on joint parsing of syntactic and semantic dependencies. In: Proceedings of CoNLL-2008 (2008)
12. Judea, A., Strube, M.: Event extraction as frame-semantic parsing. In: Proceedings of *SEM 2015, Denver, Colorado (US), June 2015
13. Levin, B., Hovav, M.R.: Argument Realization. Research Surveys in Linguistics Series. Cambridge University Press, Cambridge (2005)
14. Marquez, L., Carreras, X., Litkowski, K.C., Stevenson, S.: Semantic role labeling: an introduction to the special issue. Comput. Linguist. **34**(2), 145–159 (2008)
15. Morante, R., Daelemans, W., Van Asch, V.: A combined memory-based semantic role labeler of English. In: Proceedings of CoNLL, pp. 208–212 (2008)
16. Nivre, J.: An efficient algorithm for projective dependency parsing. In: Proceedings of the 8th International Workshop on Parsing Technologies (IWPT 2003), pp. 149–160 (2003)
17. Pado, S., Lapata, M.: Dependency-based construction of semantic space models. Comput. Linguist. **33**(2), 161–199 (2007)
18. Palmer, M., Gildea, D., Kingsbury, P.: The proposition bank: an annotated corpus of semantic roles. Comput. Linguist. **31**(1), 71–106 (2005)
19. Meyers, A., Reeves, R., Macleod, C., Szekely, R., Zielinska, V., Young, B., Grishman, R.: The NomBank Project: an Interim Report, HLT-NAACL 2004 Workshop: Frontiers in Corpus Annotation (2004)
20. Pradhan, S., Hacioglu, K., Krugler, V., Ward, W., Martin, J.H., Jurafsky, D.: Support vector learning for semantic argument classification. Mach. Learn. J. **60**(13), 11–39 (2005)
21. Schlaefer, N., Chu-Carroll, J., Nyberg, E., Fan, J., Zadrozny, W., Ferrucci, D.: Statistical source expansion for question answering. In: Proceedings of CIKM 2011 (2011)
22. Surdeanu, M., Harabagiu, S., Williams, J., Aarseth, P.: Using predicate-argument structures for information extraction. In: Proceedings of ACL2003, pp 8–15, Tokyo (2003)
23. Trandabăţ, D.: Mining Romanian texts for semantic knowledge. In: Proceedings of Intelligent Systems and Design Application Conference, ISDA2011, pp. 1062–1066, Cordoba, Spain (2011). ISSN: 2164-7143, ISBN: 978-1-4577-1676-8
24. Trandabăţ D.: Natural language processing using semantic frames. PhD thesis (2010). http://students.info.uaic.ro/~dtrandabat/thesis.pdf
25. Witten, I.H., Frank, E.: Data Mining: Practical Machine Learning Tools and Techniques, 2nd edn. Morgan Kaufmann, Burlington (2005)

Predicting Treatment Relations with Semantic Patterns over Biomedical Knowledge Graphs

Gokhan Bakal[2] and Ramakanth Kavuluru[1,2](✉)

[1] Division of Biomedical Informatics, Department of Biostatistics, University of Kentucky, Lexington, KY, USA
[2] Department of Computer Science, University of Kentucky, Lexington, KY, USA
{mgokhanbakal,ramakanth.kavuluru}@uky.edu

Abstract. Identifying new potential treatment options (say, medications and procedures) for known medical conditions that cause human disease burden is a central task of biomedical research. Since all candidate drugs cannot be tested with animal and clinical trials, *in vitro* approaches are first attempted to identify promising candidates. Even before this step, due to recent advances, *in silico* or computational approaches are also being employed to identify viable treatment options. Generally, natural language processing (NLP) and machine learning are used to predict specific relations between any given pair of entities using the distant supervision approach. In this paper, we report preliminary results on predicting treatment relations between biomedical entities purely based on semantic patterns over biomedical knowledge graphs. As such, we refrain from explicitly using NLP, although the knowledge graphs themselves may be built from NLP extractions. Our intuition is fairly straightforward – entities that participate in a treatment relation may be connected using similar path patterns in biomedical knowledge graphs extracted from scientific literature. Using a dataset of treatment relation instances derived from the well known Unified Medical Language System (UMLS), we verify our intuition by employing graph path patterns from a well known knowledge graph as features in machine learned models. We achieve a high recall (92 %) but precision, however, decreases from 95 % to an acceptable 71 % as we go from uniform class distribution to a ten fold increase in negative instances. We also demonstrate models trained with patterns of length ≤ 3 result in statistically significant gains in F-score over those trained with patterns of length ≤ 2. Our results show the potential of exploiting knowledge graphs for relation extraction and we believe this is the first effort to employ graph patterns as features for identifying biomedical relations.

1 Introduction

Biomedical processes are inherently composed of interactions between various types of entities involved. Typically, these interactions are captured, for computational convenience, as binary relations connecting a subject entity to an object entity through a predicate (or relation type). For example, the relation

R. Prasath et al. (Eds.): MIKE 2015, LNAI 9468, pp. 586–596, 2015.
DOI: 10.1007/978-3-319-26832-3_55

("Tamoxifen", *treats*, "Breast Cancer") indicates that the subject entity Tamoxifen is related to the object entity breast cancer via the relation type or predicate *treats*. Besides *treats*, relations with other types of associative predicates such as *causes, prevents,* and *inhibits* are also interesting for biomedical research. Often different relations are put together to derive new relations, also termed knowledge discovery. Given that we have established that relations are central to biomedical research, a natural question that arises is how we obtain these relations. Relations that are already discovered and considered common knowledge in the clinical and biomedical communities are typically manually recorded and distributed in public knowledge bases like the Unified Medical Language System (UMLS) Metathesaurus [9]. However, relations that are not well known and accepted by the scientific community but are being discovered by particular individuals are often reported in research articles that are subject to peer review. Given the exponential growth [3] of scientific literature, it is unrealistic to manually review all articles published on a given topic. Hence, natural language processing (NLP) techniques have been increasingly used to *extract* biomedical relations from free text documents. For instance, the treatment relation example discussed earlier in this paragraph may be extracted from the sentence – "We conclude that Tamoxifen therapy is more effective for early stage breast cancer patients." However, NLP extractions are essentially based on evidence present in particular sentences and are prone to two types of errors. First, the NLP techniques themselves might not be foolproof and second the evidence found in a particular sentence might be circumstantial and not something that is universally accepted. However, extraction of the same relation from multiple sentences might be indicative of the strength of the relation if it is being reported by multiple research teams.

In this paper, we take a completely different approach to predict relations between arbitrary pairs of biomedical entities input to our predicate specific models. We refrain from the NLP approaches that simply look at individual sentences to extract a potential relation. Instead, we build a large graph of relations (given each relation translates to a labeled edge) extracted using NLP approaches from scientific literature and use semantic path patterns over this graph to build models for specific predicates. That is, instead of looking at what a particular sentence conveys, we model the prediction problem at a global level and build models that output probability estimates of whether a pair of entities participate in a particular relation. Our models are trained with graph pattern features over a well known knowledge graph extracted from scientific literature.

In our approach, a different model needs to be trained for each predicate. The ability to identify potentially viable drugs, procedures, and other therapeutic agents for treating different conditions that cause disease burden among humans is at the heart of biomedical research. So in this paper, we focus on the *treats* predicate and build models that achieve a recall of over 92 % with a precision of 71 % even with ten times as many negative examples in the test set. Our method generalizes to other predicates and can also complement other lexical and syntactic pattern based distant supervision [4] approaches for relation extraction.

In the rest of the paper, we first discuss the primary motivation for our efforts and some related work in Sect. 2. We provide the details of the knowledge graph used in our experiments and specify the graph pattern features used in our models in Sect. 3. Next, we present the details of our experiments including a discussion of our results in Sect. 4 and conclude with some remarks on limitations and future work in Sect. 5.

2 Related Work

As we discussed in Sect. 1, NLP approaches can be used to extract relations from particular sentences using the linguistic structure of a sentence (syntactic/dependency parse) especially involving the spans of named entities that occur in it [2]. However, in our current approach, we take a global approach to predicting treatment relations between any two entities without looking at particular sentences that contain them.

The primary motivation for our effort is the distant supervision [4] paradigm (also called weak supervision) where any sentence containing a pair of entities known to participate in a relation (based on an external knowledge base) is assumed to manifest that relation in natural language. Hence an external knowledge base of relations is used to search for corresponding entity pairs in all sentences from a corpus. Many lexical and syntactic features surrounding both entity spans in those sentences form features for a multiclass classifier trained on sentences that contain pairs from the external knowledge base. At test time, each pair of entities that co-occurs in at least one sentence, but is not already known to be related according to the external knowledge base, is a candidate input pair to the classifier that predicts the best predicate for that pair using features extracted from all sentences that contain that pair. This is different from conventional approaches that look at every sentence that contains an entity pair and determine whether a particular relation is expressed in it. The distant supervision approach rather looks at all sentences containing the pair and makes a decision based on evidence gleaned from all of them.

Distant supervision offers a great alternative in cases where the number of predicates is large and when hand labeling thousands of sentences with relations is impractical. However, it suffers from two important issues.

1. Not all sentences that contain a pair of entities express the particular relation recorded in the knowledge base. That is, just because a sentence has a pair of entities, it does not automatically mean that the sentence is discussing the particular relation between them that we have in the knowledge base.
2. Negative examples for training, which are typically derived from entity pairs absent in the knowledge base, may not necessarily be true negative examples given the knowledge base could simply be incomplete.

Although later approaches [11,13–15] addressed these issues to some extent, they still persist given they are inherent to the distant supervision paradigm. Because our effort relies on graph patterns over knowledge graphs extracted from text, we obviate the issue of whether a particular sentence containing a training entity pair expresses our relation. Although we rely on NLP based extractions for building the knowledge graph over which the patterns are obtained, the graph pattern abstraction disassociates direct dependence on relations in particular sentences. Our method also largely avoids deriving negative examples based on potentially missing relations from the knowledge base by leveraging NLP extractions from over 20 million biomedical research article abstracts. Thus our approach, although inspired by distant supervision, overcomes its main issues.

3 Semantic Patterns over Knowledge Graphs

In this section, we describe our main approach to relation extraction using semantic graph patterns. Our basic intuition is simple: different entity pairs participating in a particular relation type (that is, linked via a specific predicate) are potentially connected in "similar" ways to each other where the connections are paths between them in knowledge graphs extracted from scientific literature. This is analagous to the NLP variant where a particular type of relation manifests with certain lexico-syntactic patterns surrounding the entity pair mentions in free text, which is the central idea exploited in distant supervision. Coming to our approach, we need two important components:

1. a broad scoped and large knowledge graph over which paths connecting candidate entity pairs can be obtained and
2. an approach to identify similar paths connecting entities so we can abstract or "lift" specific paths to high level semantic graph patterns to be subsequently used as features in a supervised classifier.

3.1 The SemMedDB Knowledge Graph

For this effort, we build a large knowledge graph of relations obtained from SemMedDB [1,7], a large database of over 70 million binary relationships extracted from over 20 million biomedical citations (titles & abstracts). SemMedDB is a public resource made available by the US National Library of Medicine (NLM), which uses a rule based NLP program SemRep [8,12] to extract "semantic predications" from biomedical text. SemMedDB is produced by running the SemRep program on all biomedical citations made available thorough the PubMed search system. The relations recorded in this database are called semantic predications given SemRep normalizes textual mentions of entities to unique UMLS Metathesaurus concepts (that is, performs named entity recognition) and the predicates are also based upon those available in the UMLS semantic network [5]. Each of the UMLS concepts also has at least one semantic type [6], which is essentially a classification system to categorize different biomedical entities. As such, the 70 million relations in SemMedDB represent a semantic summary of over 20 million biomedical citations. Our knowledge graph is essentially a directed graph with labeled edges formed from the relations in

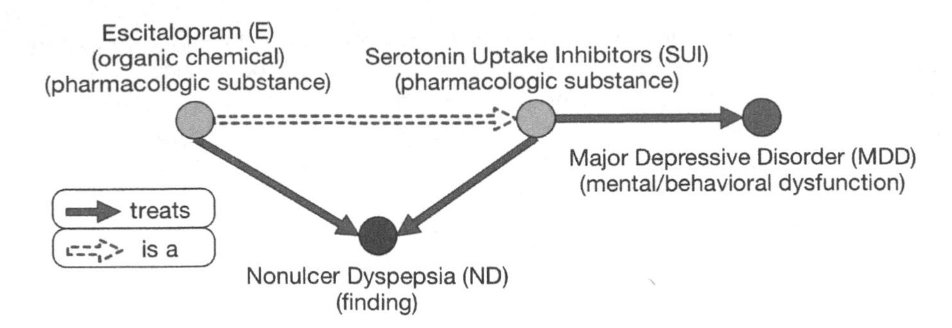

Fig. 1. A small subgraph of the SemMedDB knowledge graph

SemMedDB. The scope of this graph is very broad in a thematic sense given its edges are not limited to a particular biomedical subject. It is also large in that it has around 13 million unique edges[1] connecting over 3 million nodes.

3.2 Specific Paths to Semantic Graph Patterns

To abstract specific paths between entities over the SemMedDB graph to semantic patterns, we use an intuitive heuristic – simply replace the concepts along the path with their corresponding semantic type sets (given a concept can have more than one type) and retain the directions of the edges and edge labels as they are. For example, consider a segment of the SemMedDB graph showing a couple of paths between the drug `Escitalopram (E)` and the condition `major depressive disorder (MDD)` in Fig. 1. We only employ simple paths (that is, without cycles) and ignore directionality when computing paths (but retaining it after paths are identified). Thus we have the following two paths between `E` and `MDD`: (E, is_a, SUI, treats, MDD) and (E, treats, ND, treats^{-1}, SUI, treats, MDD), where SUI stands for `Serotonin Uptake Inhibitors` and ND denotes `Nonulcer Dyspepsia`. For notational convenience we encode the reverse direction with a superscript of -1 on the predicate. To obtain the patterns, we replace the specific entities with their semantic type sets. Thus the corresponding two patterns are:

$$(\{oc, ps\}, is\_a, \{ps\}, treats, \{md\})$$

$$(\{oc, ps\}, treats, \{f\}, treats^{-1}, \{ps\}, treats, \{md\}),$$

where *oc*, *ps*, *f*, and *md* are abbreviations of the semantic types organic chemical, pharmacologic substance, finding, and mental/behavioral dysfunction respectively. By replacing specific entities with their semantic types we aim to capture high level patterns that connect candidate entity pairs. Although we just showed two paths, there are usually many others with a variety of edge types

[1] Although SemMedDB has 70 million relations, there are many duplicates given a relation can be extracted from multiple sentences due to the semantic mapping to UMLS concepts and semantic network predicates.

(over 50 different predicates) connecting related entities. We reiterate that our main hypothesis is that these patterns will act as highly discriminative features in identifying entity pairs that participate in a particular type of relationship.

In this paper, we are exclusively interested in predicting treatment relationships and hence we chose this particular example from Fig. 1. The two example patterns we show here have a nice high level meaning. In the first pattern, we see that a pharmacologic substance (SUI) is a hypernym for another (E) and is known to treat a dysfunction (MDD). In the second pattern, two pharmacologic substances (SUI and E) both treat a common second condition (ND) while one of them (SUI) treats the target condition (MDD). However, in general, the patterns themselves do not need to have interesting or meaningful interpretations, but when considered together they should be reasonably predictive of the particular predicate that is of interest to us. In this specific example, it turns out that the treatment relationship also holds for our candidate pair (E, MDD). Essentially, we expect to leverage machine learned models to automatically weight different patterns based on their predictive power rather than human experts having to manually identify interesting patterns, a highly impractical task with the explosion of biomedical knowledge.

Before we move on to our experiments, we should mention that although we refer to the SemMedDB graph as a knowledge graph, the precision of NLM's SemRep tool used to build SemMedDB is known to be around 75 % [1]. However, the advantage of our approach is that our prediction is not directly dependent on the correctness of each and every relation in the knowledge graph, rather on the general patterns found within it. Hence any knowledge graph with reasonable quality will suffice although high quality graphs should yield better results.

4 Prediction with Graph Pattern Features

We describe experiments conducted and results obtained using semantic graph patterns connecting candidate pairs as discussed in Sect. 3. We derive our dataset from the UMLS Metathesaurus's MRREL table [9, Chapter 2] that has over 11 million *manually curated* relations that are sourced from different biomedical terminologies. Among these we also have several TREATS relations which are used for our experiments. We needed an external human vetted resource like the relations in UMLS given our knowledge graph is derived from a computationally curated relation database. We curated a set of around **7000 unique treatment relations** (entity pairs connected through the *treats* predicate) connecting UMLS concepts from the MRREL table. We divided this positive example dataset into 80 % (5600) training set and 20 % (1400) test set splits.

4.1 Selection of Negative Examples

Selecting negative examples for distant supervision is a challenge as discussed in Sect. 2 given the incomplete nature of manually curated knowledge bases – that is, just because a pair is not participating in a treatment relation in such a knowledge base, does not automatically mean it is a negative example. A second problem is that, in general, any two biomedical entities are not going to have a treatment relation between them. That is, it is not even worth exploring arbitrary pairs of entities to find potential treatment associations and researchers are not usually interested in such pairs. So we carefully choose negative training examples using the following two steps.

1. Every predicate in the UMLS semantic network, including *treats*, has a set of domain/range semantic type constraints. That is, NLM based on expert consultation prescribes which types of entities can take the role of the subject and object in treatment relations. All such possible and allowable subject-object semantic entity type combinations for each predicate are available in three tables with the SRSTR prefix [9, Chapter 5] in the UMLS. We first randomly select a pair of entities (from over 3 million unique UMLS concepts) that satisfies these domain/range constraints.
2. For each pair selected in step 1, we check if there is a *treats* relation between its entities either in the UMLS MRREL table or in the SemMedDB relation database. If it does not hold in our knowledge bases, we include it as a negative example in our dataset.

This two step process selects fairly hard-to-classify negative examples since they satisfy the domain/range constraints but don't participate in a treatment relationship. Checking for membership in both the UMLS and SemMedDB resources minimizes concerns surrounding incomplete knowledge bases. Since we want to predict treatment relations based on graph patterns, if the knowledge graph already has a *treats* edge between our candidate pairs, the prediction could become trivial and the whole process would be self-deceiving. So we deleted any existing *treats* edges between entities in all training/test positive pairs from the knowledge graph (note that negative example selection already ensures this) to guarantee a fair analysis of the predictive ability of graph patterns.

4.2 Experiments and Results

For the experiments, we selected the same number of positive and negative examples (80 % of 7000 = 5600) for training. It is straightforward to see that the ground truth or the true treatment relation space is very sparse and most candidate pairs (even if they satisfy domain/range constraints) will belong to the negative class. However, if this sparsity is incorporated as is in the training data, the classifiers might not have enough to learn about predictive patterns for the positive class, which is of most interest to us. Thus we chose equal number of pairs in the training dataset. However, for testing the resultant models, we gradually increased the number of negative examples in the test set to as many as ten times that of the positive class size. We also conducted experiments with patterns of lengths one, two, and three to see the potential of patterns of varying lengths. From our literature review, there are no efficient implementations for computing *all* simple paths of an arbitrary length between two given nodes in large graphs, although many well known algorithms (e.g., modified breadth first search) exist for identifying shortest paths. In general, finding all simple paths becomes extremely expensive with lengths greater than three simply because the number of such paths could increase drastically in dense graphs. Our implementation for lengths ≤ 3 is based on straightforward heuristics that maintain precomputed lists of neighbors for each node in the knowledge graph. Specifically, to determine length two paths between nodes $e1$ and $e2$, we simply look at nodes in $\mathcal{N}(e1) \cap \mathcal{N}(e2)$ where $\mathcal{N}(e)$ denotes neighbors of node e. To identify length three paths, we look for edge membership for pairs in $\mathcal{N}(e1) \times \mathcal{N}(e2)$ in our knowledge graph.

Table 1. Test set scores with 43,246 patterns of length ≤ 2

Imbalance	Precision	Recall	F-score
$\|N\| = \|P\|$	0.970	0.857	0.910
$\|N\| = 2 \cdot \|P\|$	0.900	0.857	0.878
$\|N\| = 5 \cdot \|P\|$	0.816	0.857	0.836
$\|N\| = 10 \cdot \|P\|$	0.716	0.857	0.780

Table 2. Test set scores with patterns of length ≤ 3 with feature minimum frequency thresholds of 1000 (97,864 patterns) and 500 (384,417 patterns)

Imbalance	Min. frequency: 1000			Min. frequency: 500		
	Precision	Recall	F-score	Precision	Recall	F-score
$\|N\| = \|P\|$	0.951	0.919	0.935	0.955	0.922	0.939
$\|N\| = 2 \cdot \|P\|$	0.900	0.919	0.909	0.911	0.922	0.916
$\|N\| = 5 \cdot \|P\|$	0.809	0.919	0.861	0.825	0.922	0.871
$\|N\| = 10 \cdot \|P\|$	0.695	0.919	0.792	0.708	0.922	**0.801**

We trained our models with the semantic pattern features using the well known logistic regression classifier available through the Python Scikit-Learn [10] machine learning library. When we conducted experiments with paths of only length one, we found a total of 25 unique patterns used as features. Regardless of the number of negative examples, the precision was over 99 % with a recall of 23 %. It is clear that length one pattern based models yield very low recall. Next we show the results on the test sets with patterns of lengths ≤ 2 in Table 1 and lengths ≤ 3 in Table 2. In both tables we keep the positive example set P constant and add many examples to increase the size of the negative example set N. From Table 1, we notice there are a total of 43,246 unique patterns for lengths one or two and the precision drops to 71.6 % while the recall stays at 85.7 % with the ten fold increase in $|N|$. The number of patterns increased drastically when we also included those with length three. Hence we needed to apply a minimum feature frequency threshold which parameterizes the minimum number of training pairs that should be connected with a particular pattern for it to be included in the final feature space. This is akin to how a minimum frequency is imposed for n-gram features for text classification. We experimented with thresholds of 1000, 500, and 250 that result in feature sets with sizes 97,864, 384,417, and 1,101,106 patterns, respectively. We only show the results for the first two thresholds in Table 2, which indicates a recall improvement of nearly 7 % compared with simply using length ≤ 2 patterns from Table 1. This recall gain was obtained at the expense of only 1–2 % loss in precision. Between the two thresholds, we notice that the smaller 500 threshold gives a precision gain of 1.3 % and a recall gain of 0.3 % compared with the larger threshold from the last row of the table.

We did not show the results of experiments we did with a threshold of 250, which led to over a million features, because the performance slightly dipped compared with the larger thresholds shown in Table 2. In both tables, we see that the recall stays constant even as precision goes down when more negative examples are added to the test set. That is, the number of false negatives stayed constant as the imbalance in test set increased. This is not surprising given the model was not changed during these experiments and only new negative examples were added to increase class imbalance in the test set.

We conducted an additional set of experiments where we obtained repeated measurements for the last row cases in both tables (min. frequency 500 case for Table 2) with train-test splits from hundred distinct shuffles of our full dataset. Based on these experiments the 95 % confidence intervals for precision are 71.11 ± 0.002 (length ≤ 2) and 68.30 ± 0.005 (length ≤ 3) and for recall are 86.14 ± 0.002 (length ≤ 2) and 92.80 ± 0.001 (length ≤ 3). These results establish the generalizability of our results and show that adding patterns of length 3 results in more than 6 % improvement in recall with a less than 3 % loss in precision, which is a reasonable compromise given the recall increase is double that of the loss in precision. The confidence intervals for the F-score are 77.90 ± 0.001 (length ≤ 2) and 78.66 ± 0.003 (length ≤ 3), which clearly indicate that the improvement in F-score is also statistically significant given the intervals do not overlap.

Although we cannot predict or quantify the true distribution of treatment relations, we think the imbalance will be more extreme than the 1:10 ratio we tried in our experiments. While additional experiments are justified, even with what we were able to show, we conclude that semantic graph patterns are definitely useful in predicting treatment relations and have the potential to complement NLP based approaches and also aid in relation extraction for other predicates.

5 Concluding Remarks

In this paper, we employed semantic graph patterns connecting pairs of candidate entities as the sole set of features to predict treatment relations between them. We exploited a well known biomedical relation database to build a knowledge graph with over 13 million edges. We then used the knowledge graph to derive features and also select good negative training instances for experiments. Evidenced by the results presented in Sect. 4, we have successfully verified our hypothesis that semantic patterns over knowledge graphs can be powerful predictors of treatment relations. Our central idea is straightforward, intuitive, and naturally generalizes to other predicates besides *treats*, and also to other domains of interest where knowledge graphs of reasonable quality are available. Next, we identify some limitations of our effort and discuss future research directions to address them.

- When a node has multiple semantic types (like `Escitalopram` in Fig. 1), instead of considering them as a compound set type in the semantic abstraction process (see Sect. 3.2), we can relax them into multiple semantic patterns each with the constituent primitive types. This can drastically prune the feature space and might result in semantically higher level patterns than we are able to obtain now.
- From Tables 1 and 2, it is clear that with increasing class imbalance, the precision significantly goes down – with a ten fold increase in negative test examples, precision dropped by 25 %. This presents both an opportunity and a challenge: some of the false positives (FPs) leading to low precision might actually be indicating potential new treatments while others might be simply wrong. The challenge is to separate these two types of FPs; solution might involve semi-automatic approaches using classifier scores and biomedical domain experts. Features that are weighted heavily by the classifier might also give us insights into ranking FPs in descending order of their potential for representing an actual new treatment relation.
- A thorough manual analysis of FPs and false negatives (FNs) can also help us identify broad classes of such errors that can be targeted using specific tailored heuristics including some form of pre-processing of the knowledge graph and post-processing of the results. A more thorough analysis of top features and errors for different classes of treatment relations depending on each allowed semantic type combination can also help us further fine-tune these heuristics based on the semantics types of input pair.
- We strongly believe our work complements NLP approaches in that the graph patterns can be used as additional features along with the lexico-syntactic features typically employed for weakly supervised relation extraction. Besides combining our methods with NLP approaches, employing more sophisticated machine learning techniques especially ensemble approaches [16] might help in improving the overall results.

Acknowledgments. Thanks to anonymous reviewers for their helpful comments that helped improve the paper. The project described in this paper was supported by the National Center for Advancing Translational Sciences (UL1TR000117). The content is solely the responsibility of the authors and does not necessarily represent the official views of the NIH.

References

1. Kilicoglu, H., Shin, D., Fiszman, M., Rosemblat, G., Rindflesch, T.C.: SemMedDB: a pubmed-scale repository of biomedical semantic predications. Bioinformatics **28**(23), 3158–3160 (2012)
2. Kim, S., Liu, H., Yeganova, L., Wilbur, W.J.: Extracting drug-drug interactions from literature using a rich feature-based linear kernel approach. J. Biomed. Inform. **55**, 23–30 (2015)
3. Lu, Z.: PubMed and beyond: a survey of web tools for searching biomedical literature. Database J. Biol. Databases Curation (2011)
4. Mintz, M., Bills, S., Snow, R., Jurafsky, D.: Distant supervision for relation extraction without labeled data. In: Proceedings of the Joint Conference of the 47th Annual Meeting of the ACL and the 4th International Joint Conference on Natural Language Processing of the AFNLP, pp. 1003–1011. Association for Computational Linguistics (2009)
5. National Library of Medicine. Current Hierarchy of UMLS Predicates. http://www.nlm.nih.gov/research/umls/META3_current_relations.html
6. National Library of Medicine. Current Hierarchy of UMLS Semantic Types. http://www.nlm.nih.gov/research/umls/META3_current_semantic_types.html
7. National Library of Medicine. Semantic MEDLINE Database.http://skr3.nlm.nih.gov/SemMedDB/
8. National Library of Medicine. SemRep - NLM's Semantic Predication Extraction Program. http://semrep.nlm.nih.gov
9. National Library of Medicine. Unified Medical Language System Reference Manual. http://www.ncbi.nlm.nih.gov/books/NBK9676/
10. Pedregosa, F., Varoquaux, G., Gramfort, A., Michel, V., Thirion, B., Grisel, O., Blondel, M., Prettenhofer, P., Weiss, R., Dubourg, V., Vanderplas, J., Passos, A., Cournapeau, D., Brucher, M., Perrot, M., Duchesnay, E.: Scikit-learn: machine learning in python. J. Mach. Learn. Res. **12**, 2825–2830 (2011)
11. Riedel, S., Yao, L., McCallum, A.: Modeling relations and their mentions without labeled text. In: Balcázar, J.L., Bonchi, F., Gionis, A., Sebag, M. (eds.) ECML PKDD 2010, Part III. LNCS, vol. 6323, pp. 148–163. Springer, Heidelberg (2010)
12. Rindflesch, T.C., Fiszman, M.: The interaction of domain knowledge and linguistic structure in natural language processing: interpreting hypernymic propositions in biomedical text. J. Biomed. Inform. **36**(6), 462–477 (2003)
13. Ritter, A., Zettlemoyer, L., Etzioni, O., et al.: Modeling missing data in distant supervision for information extraction. Trans. Assoc. Comput. Linguist. **1**, 367–378 (2013)
14. Surdeanu, M., Tibshirani, J., Nallapati, R., Manning, C.D.: Multi-instance multi-label learning for relation extraction. In: Proceedings of the 2012 Conference on Empirical Methods in Natural Language Processing, pp. 455–465. Association for Computational Linguistics (2012)
15. Xu, W., Hoffmann, R., Zhao, L., Grishman, R.: Filling knowledge base gaps for distant supervision of relation extraction. In: Proceedings of the 51st Annual Meeting of the Association for Computational Linguistics, pp. 665–670. Association for Computational Linguistics (2013)
16. Zhou, Z.-H.: Ensemble Methods: Foundations and Algorithms. CRC Press, Boca Raton (2012)

A Supervised Framework for Classifying Dependency Relations from Bengali Shallow Parsed Sentences

Anupam Mondal and Dipankar Das(✉)

Computer Science and Engineering, Jadavpur University, Kolkata, India
link.anupam@gmail.com, ddas@cse.jdvu.ac.in

Abstract. Natural Language Processing, one of the contemporary research area has adopted parsing technologies for various languages across the world for different objectives. In the present task, a new approach has been introduced for classifying the dependency parsed relations for a morphologically rich and free-phrase-ordered Indian language like Bengali. The pair of dependency parsed relations (also referred as *kaarakas* 'cases') are classified based on different features like *vibhaktis* (inflections), Part-of-Speech (POS), punctuation, gender, number and post-position. It is observed that the consecutive and non-consecutive occurrences of such relations play a vital role in the classification. We employed three different machine-learning classifiers, namely NaiveBayes, Sequential Minimal Optimization (SMO) and Conditional Random Field (CRF) which obtained the average F-Scores of 0.895, 0.869 and 0.697, respectively for classifying relation pairs of three primary *kaarakas* and one primary *vibhakti* relation. We have also conducted the error analysis for such primary relations using confusion matrices.

Keywords: Dependency relations · *Kaaraka* · *Vibhakti* · Machine-learning classifiers

1 Introduction

Dependency Parsing, a challenging task for processing any natural language seems an obvious milestone while dealing with morphologically rich and free-phrase ordered languages, especially Indian Languages. Bengali, the seventh popular language[1] in the World, second in India and the national language of Bangladesh is morphologically rich and resource constrained. Thus, to the best of our knowledge, at present, there is no such full-fledged parser available in Indian languages and especially for Bengali. However, Bengali is one of the important Indo-Iranian languages spoken by a population that now exceeds 211 million or 3.11 % of the world population. Geographically, Bengali-speaking population percentages[2] are as follows: Bangladesh (over 95 %), Indian States of Andaman and Nicobar Islands (26 %), Assam (28 %), Tripura (67 %), and West Bengal (85 %). The development of parsers for Indian languages in general

[1] http://listverse.com/2008/06/26/top-10-most-spoken-languages-in-the-world/.
[2] https://en.wikipedia.org/wiki/List_of_languages_by_number_of_native_speakers.

R. Prasath et al. (Eds.): MIKE 2015, LNAI 9468, pp. 597–606, 2015.
DOI: 10.1007/978-3-319-26832-3_56

and Bengali in particular is difficult and challenging as the language is (1) inflectional language providing the richest and most challenging sets of linguistic and statistical features resulting in long and complex word forms, and (2) relatively free phrase order and less computerized compared to English [2].

Till date, due to the scarcity of reliable annotated data, it is observed that several attempts that were used to develop the parsers for Indian languages mainly depend on linguistic rules [2, 13, 14, 17, 18]. A hybrid dependency parser that proposed two stage parsing system [1] and a data driven parser that identifies the dependency relations between chunks in a sentence using Treebank are found in the literature. The respective researchers conducted their experiments to improve the mistakes of the data driven parser based on the effects of case frames. However, none of the approaches has considered the classification of the relation pairs using machine learning approaches. Therefore, the present task aims to identify the chunks and phrases and their intra relationships from sentences using a data-driven approach. In addition, we have also classified the dependency relations that are considered as the prerequisites towards developing a full-fledged parser. It is observed that different relations like *kaaraka* and *vibhakti* use to play an important role for constructing the sentences. In Bengali grammar, *kaaraka* is the relationship between verb and noun or verb and pronoun in a sentence. There are seven different *kaaraka* relations such as *kartaa, karma, karana, sampradana, apadana, nimito, adhikarana* that are represented in the paper as K1, K2, K3, K4, K5, K6 and K7, respectively. Here, we have dealt with only three *kaaraka* relations e.g., *kartaa* (K1), *karma* (K2) and *adhikarana* (K7) as per the frequency. The examples of the *kaaraka* have been illustrated in Fig. 1.

Kaaraka:

Kartaa Kaaraka:-

Chatok kator swarey daake. চাতক কাতর স্বরে ডাকে।

(*The bird is singing very sadly.*)

Illustration: The bird "chatok" (চাতক) is active.

Karma Kaaraka:-

Ami tomakei khunjchilam. আমি তোমাকেই খুঁজেছিলাম।

(*I was searching you only.*)

Illustration: It emphasizes "you" (তোমাকেই) because it makes a sense that, the person I was searching for is 'you'.

Adhikarana Kaaraka:-

Takata ghore pore thakte dekhechi. টাকাটা ঘরে পড়ে থাকতে দেখেছি।

(*I have seen the money in the room.*)

Illustration: Where I have seen the money? The answer is room (ঘরে).

Fig. 1. *Kaaraka* examples with Illustration

In case of making the Bengali sentences, *vibhakti* plays an important role. There are presently ten symbols that are considered for indicating the *vibhaktis* in Bengali viz. *ay, ke, re, te, sunyo* etc. and are represented as R1, R2, R3, …, R10, respectively in the

Vibhakti:

<u>য় *(ay) Vibhakti:-*</u>

<u>Takai ki na hoy!</u> টাকায় কি না হয়! (*What can't be bought by money?*)

Illustration: টাকায় " টাকা (money) + য় (ay) = টাকায় (by money) ", য় (ay) *Vibhakti* is used.

<u>কে (ke) *Vibhakti:-*</u>

<u>Maake aar desher matike valobaste sekho.</u> মাকে আর দেশের মাটিকে ভালবাসতে শেখ। (*Learn how to love mother and motherland.*)

Illustration: মাকে " মা (mother) + কে (ke) = মাকে (to mother) ", (কে) ke *Vibhakti* is used and in মাটিকে " মাটি (motherland) + কে (ke) = মাটিকে (to motherland) ", (কে) ke *Vibhakti* is used.

<u>তে (te) *Vibhakti:-*</u>

<u>Somitite kichu dite hbe.</u> সমিতিতে কিছু দিতে হবে। (*Some money has to be given to the Committee.*)

Illustration: সমিতিতে " সমিতি (Committee) + তে (te) = সমিতিতে (Somitite) (to the Committee) ", তে (te) *Vibhakti* is used.

Fig. 2. *Vibhakti* examples with Illustration

annotated corpus [12]. It also provides the information related to the respective *kaarakas* as shown in Fig. 2. Therefore, in order to deal with such relations using a machine learning framework, we always need to extract the linguistic features at different levels of granularities (word, chunk and or sentence).

The first obvious question was how to select the important *kaarakas* in order to identify the dependency parsed relations. To start with the top four frequent dependency relations (K1, K2, K7 and R6), we have initiated the inclusion of associated features viz. Part-of-Speech (POS), punctuations, number, gender and post- position for implementing the machine learning framework. An exhaustive error analysis with respect to different classifiers and mode of operations were performed to achieve the maximum F-Scores of 52 %, 42 %, 45 % and 69 % for K1, K2, K7 and R6 dependency relations, respectively.

The rest of the Sections are as follows. In Sect. 2, we have discussed the related attempts made in developing the parsing technologies for Bengali and other Indian languages. Preprocessing of the corpus and selection of top-frequent relations are discussed under Sect. 3. In the next Section, we have mentioned the methodologies to extract the related features which were supplied with the relations. The system framework for classifying the dependency relations is discussed in Sect. 5 while in Sect. 6 we have analyzed the errors in terms of the confusion matrices. Finally, Sect. 7 concludes the task and mentioned the possibilities of future scopes.

2 Related Work

In literature survey, we have found the development of a predictive parser in an efficient way for morphological rich and free word order languages viz. Bengali [3, 15]. The identification of structured Bengali sentences purposes the symbols (constituents) based on Context Free Grammar (CFG) rules. The recognition of Bengali grammar from the sentences was a contributory attempt of this task for the reason of availability of different grammars. In contrast, the grammar driven parser was developed for Bengali language which achieved a score near to 90 % [4] in a Shared Task[3]. A group of researchers were trying to generate a new dataset for reducing the gap between structured and unstructured form of data with the help of Treebank's which contains approximately 1500 sentences. They had not used the developed dataset for extracting linguistics rules for the task. A comparative analysis had done between grammar driven and data driven approaches of Bengali language for developing a dependency parser of Bengali language [10, 11].

Lexical Functional Grammar (LFG) [6] based linguistic phenomena has been applied in a wide range for Bengali language. The Constituent phrase Structure (C-structure) and Functional structure (F-structure) is considered as primary features for LFG technique.

In concern, the Paninian model and dependency based framework were introduced as an effective technique for parsing the Bengali sentences [5]. The researchers were taking help of demand-source concept under Paninian grammar with six different types of *kaaraka* and verb. The *kaaraka* and verb group of words are treated as source and demand groups, respectively for the task where the dependency tree root indicates as verb along with appropriate *kaaraka* labels. Several researchers had analyzed the dependency parsers [7, 18] for Indian languages and remarked that the development of dependency parsers can be carried out either using grammar driven approaches or data driven approaches [16]. In case of morphologically rich and free word order languages, the grammar driven approach is difficult than the data driven approach. In several cases, Malt parser[4] has been used as transition based approach for dependency parsing and it mainly consists of the transition and classifier based on prediction approaches.

In this report, we have introduced the dependency relations (*kaaraka, Vibhakti*) based classification approaches with several features viz. POS, punctuation, number etc. for developing a Bengali dependency parser.

3 Resource Preparation

3.1 Corpus

In order to develop a dependency parser for any language, we need to identify the linguistic rules that guide us how to relate different chunks of a sentence using the grammar of that language. The effect of morphological richness and free phrase ordered

[3] http://ltrc.iiit.ac.in/mtpil2012/.

[4] www.maltparser.org.

<af=e, drel=nmod: NP2/name=NP>
<af=keu, drel=k1: VGF/name=NP2>
<af=biRayZa, drel=k7: VGF/name=NP3>
<af=Agrahl, drel=k1s: VGF/name=JJP>

Fig. 3. A sample of dependency relations of words based on Shakti Standard Format (SSF)

structure make the rule identification difficult. Therefore, in the present task, we have mainly tried to design the language dependent rules without considering the structure of the sentences in the input corpus. We have observed that in Bengali language, the words of a sentence appear in the form of any of the seven *kaarakas* and ten *vibhaktis* as per the guidelines [8]. The dependency relation (*drel*) tag of a sentence has shown in Fig. 3. We have adopted two different techniques based on consecutive and non-consecutive occurrences of such primary dependency relations, as described in the next section.

3.2 Selection of Consecutive and Non-consecutive Occurrences

In order to derive the classification features, we have evaluated the occurrence probabilities of the dependency relations from the corpus in terms of consecutive and non-consecutive appearances. In case of consecutive dependency relation, we have considered the dependency relations of neighboring words where as two or three word gap was considered for identifying dependency relations in case of non-consecutive occurrences. We have observed that, in case of morphologically rich and free phrase ordered language, the occurrence probability of the dependency relation (*kaaraka*) is high in case of consecutive words. Similarly, the non-consecutive presence of the dependency relations also plays a crucial role. The non-consecutive appearances help to identify the implicit co-reference exists among the long-distanced words in order to develop a full-fledged dependency parser. Table 1 has illustrated the dependency relations of consecutive and non-consecutive appearances of the words whereas Fig. 4 shows the steps to identify consecutive dependency relations and similar steps we have considered for identifying the non-consecutive relations.

In order to implement any data-driven model, we need to analyze the data based on different statistics prior to start applying the supervised algorithms. In the present report, the whole corpus was collected from the articles published in newspapers, text books by a group of members of IIIT-H and annotated with different relations based on *kaarakas* (e.g. K1, K2, K7) and *vibhaktis* (e.g. R1, R2, R6) [9]. The corpus was provided by IIIT-H in a shared task challenge[5] in order to build the shallow parser for Bengali. We split the corpus in three different sets, namely training, development and test randomly with a distribution of 50 %, 20 % and 30 %, respectively. The important distributions of the sentential relations, POS tags and their combinations in these three sets are mentioned in Table 2 with their corresponding distributions. Therefore, we attempted to identify other features that are available from the annotated corpus.

[5] http://shiva.iiit.ac.in/SPSAL2007/.

Table 1. Important dependency relation combinations for consecutive (non-consecutive) words

Important relation combinations	Training	Development	Test
K1-K2	139 (376)	20 (53)	44 (106)
K2-k2	126 (274)	7 (22)	83 (165)
R6-K1	121 (234)	11 (14)	77 (182)

Step1: *Take a list of dependency relations and termed it as a DRL.*
DRL= { K1, K2, K3, K4, K5, K6, K7, R1, R2, R3, R4, R5, R6, R7, R8, R9, R10, vmod and ccof }
Step2: *Extract the dependency relation from each of the words of a sentence and store in a list called L.*
Step3: *Pick up two consecutive relations,* R_i *and* R_{i+1} *from the list L.*
Step3.1: *If both* R_i *and* R_{i+1} *belong to DRL, take the relation-pair* R_i *and* R_{i+1} *as our candidate pair.*
Step3.2: *Else move to next relation pair,* R_{i+1} *and* R_{i+2}.
Step4: *Repeat until the list L is exhausted.*

Fig. 4. Steps for identifying the consecutive dependency relations

Table 2. Important primary relations, their POS tags and combinations in Training, Development and Test Data Sets

	Training	Development	Test
Words	2329	660	1854
Sentences	700	150	280
Top 4 relations			
K1	734	174	252
K2	756	101	386
K7	395	91	273
R6	446	58	297
POS tags			
Noun	795	165	569
Pronoun	384	61	77
Unk	709	76	332
POS tag combinations			
Adverb-Noun	869	175	623
Noun-Pronoun	1179	226	673

4 Feature Extraction

While analyzing the training data with 700 sentences, we have found that a total 2329 instances are present in the top-4 relations (K1, K2, K7 and R6). These relations are appeared with an average of 3.3 relational instances per sentence and therefore considered as our key instances. The distributions of such top-4 relations are mentioned in

Table 3. Important Feature analysis for (Training/Development/Test) Datasets

Relations on resource ($T/$D/$Te)	POS (f1)			Punc (f2)	Gender (f3)		Number (f4)			Post position (f5)	
	Adj	Adv	Noun	unk	Sg	Pl	4	5	a	D	O
K1 (734/174/252)	47/44/7	22/1/5	211/61/128	**265/34/79**	323/77/140	**37/10/15**	6/0/3	0/0/0	0/0/0	359/87/154	4/0/1
K2 (756/101/386)	**80/5/93**	15/4/13	**295/41/179**	206/18/73	**346/44/177**	14/4/10	7/1/3	21/6/6	3/2/0	348/46/183	12/2/4
K7 (395/91/273)	36/6/20	**25/5/3**	157/39/153	81/8/58	148/39/152	0/0/3	0/1/0	1/2/1	4/0/3	195/50/164	0/1/2
R6 (443/58/297)	9/2/2	12/0/6	141/24/136	**157/16/121**	221/31/141	32/6/14	0/0/0	0/0/0	6/0/11	3/0/1	250/37/153
Total (2329/660/1854)	172/57/122	74/10/27	795/165/596	709/76/331	1038/191/610	83/20/42	13/2/6	22/8/7	13/4/14	905/183/502	266/40/160

$T → Training $D → Development $Te → Test Punc (f2) → Punctuation
Adj → Adjective Sg → Singular Adv → Adverb Pl → Plural a → any number

Table 3. After an initial investigation on the training, test and development data sets, we extracted five features (POS, punctuation, Gender, Number and post-position) that play important roles in distinguishably identifying the top-4 primary relations. The POS tag feature produces remarkable output for identifying K1, K2 and K7 relations. Mainly, the adjective, adverb, noun, verb and WQ tags are notable for identifying the relation pairs of K1-K2 and K2-K2 pairs. In case of gender feature, K2 mainly appears as singular whereas K1 represents the plural. The above derived observations played the vital roles for designing the dependency parser for Bengali language.

5 System Framework

We have used the Weka[6] tool and employed two different classifiers viz. NaiveBayes and SMO for classifying the relations. Along with the extracted features described in the previous section, we also included the consecutive and non-consecutive occurrences of the relation pairs and their POS tag combinations as features for developing the classification framework. It is observed that the inclusion of the features related to the consecutiveness improves the accuracy of the system as illustrated in Table 4. We have adopted four different modes of operations namely Use training set, Supplied test set, Cross validation Folds-10 and Percentage split 66 % on each of the classifiers in Weka toolkit. NaiveBayes classifier produced the remarkable accuracy (70 %), average precision, recall and F-measure with top 6 features and all features with respect to all types of operations. Similarly, in case of SMO classifier, the accuracy (75 %), average precision, recall and F-measure are notable with top 5, top 6 and all features set with respect to all types of operations.

In addition to different classifiers in Weka, we also used Conditional Random Field (CRF[7]) for classifying the primary dependency relations. The precision and F-measure with respect to top 4 features are high for K1 whereas recall is low with top 6 features for relation K2. In case of identifying the R6 relation using CRF, the precision, recall and F-score are notable with top 5, top 6 and all features. The detail observations of precision, recall and F-Score for all primary relations (K1, K2, K7 and R6) along with secondary relations (SR) is mentioned in Table 5.

[6] www.cs.waikato.ac.nz/ml/weka.

[7] nlp.stanford.edu/software/CRF-NER.shtml.

Table 4. System generated results with important mode of operation for different classifiers

A	B	C	D	E	F
NaiveBayes classifier					
Cross-validation Folds-10 [661]	9#	498	75.34	0.80	0.753
	8$	496	75.04	0.79	0.750
	7@	439	66.41	0.64	0.664
	6**	412	62.33	0.60	0.623
	5***	399	60.36	0.57	0.604
SMO classifier					
Use training set [661]	9#	577	87.29	0.89	0.873
	8$	575	86.99	0.88	0.870
	7@	558	84.42	0.86	0.844
	6**	524	79.27	0.80	0.793
	5***	524	79.27	0.80	0.793

all features $ top 6-features @ top 5-features
** top 4-features *** top 3-features
A → Important Mode of Operation [No. of Instances]
B → No. of Attributes (No. of features)
C → No. of Correctly Classified Instances
D → Avg. Precision E → Avg. Recall F → Avg. F-Measure

Table 5. System generated important results based on CRF tool

Dependency relation with no. of occurrence		Precision	Recall	F-Score
K1 (734)	All 7	0.431	0.6	0.5
	Top 5	0.434	0.7	0.5
	Top 6	0.478	0.7	0.6
K2 (756)	All 7	0.518	0.3	0.4
	Top 5	0.522	0.3	0.4
	Top 6	0.444	0.4	0.4
K7 (396)	All 7	0.570	0.3	0.4
	Top 5	0.588	0.4	0.4
	Top 6	0.696	0.3	0.4
R6 (443)	All 7	0.713	0.7	0.7
	Top 5	0.714	0.7	0.7
	Top 6	0.673	0.6	0.6
SR (1986)	All 7	0.976	1	0.9
	Top 5	0.978	1	0.9
	Top 6	0.975	1	0.9

6 Error Analysis

We have also conducted an error analysis based on the confusion matrices for the classified dependency relations in the form of graphical representation. Figure 5 shows the occurrences of the relations (K1, K2, K7, R6 and SR) in the confusion matrices for different classifiers, NaiveBayes, SMO and CRF respectively with respect to all features. However, we have observed that the occurrences are high when K1 relation appears as K2, K2 appears as K1 or K7, K7 appears as K1 and R6 as SR (Secondary relations) relation for their important modes of operation for all classifiers.

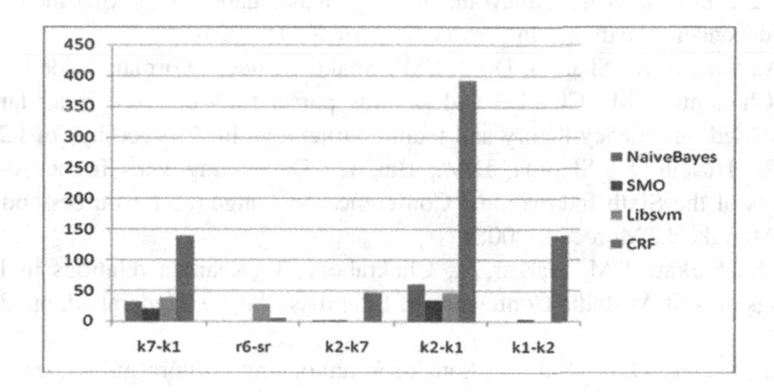

Fig. 5. Confusion matrix for different classifiers w.r.t all features of important modes

7 Conclusion and Future Work

In this paper, we have introduced the approaches for classifying the dependency parsed relations based on *kaarakas* and *vibhaktis* for the morphologically rich and free-word order language, Bengali. The consecutive and non-consecutive techniques have been used for identifying the important dependency relations from the sentences. The dependency relations based on chunks or phrases also gives satisfactory output. Finally, we prepared a machine-learning framework for classifying the dependency relations followed by an exhaustive error analysis that shows crucial insights towards developing a full-fledged parser.

In future, we will include the semantic relationships for extracting the suitable chunks from the sentences which can guide to develop the full-fledged dependency parser in efficient manner.

References

1. Dhar, A., Chatterji, S., Sarkar, S., Basu, S.: A hybrid dependency parser for Bangla. In: Proceedings of the 10th Workshop on Asian Language Resources, COLING Mumbai, pp. 55–64, India (2012)

2. Ghosh, A., Bhaskar, P., Das, A., Bandyopadhyay, S.: Dependency parser for Bengali. In: JU System at ICON (2009)
3. Chatterji, S., Sonare, P., Sarkar, S., Roy, D.: Grammar driven rules for hybrid Bengali dependency parsing. In: Proceedings of ICON 2009 NLP Tools Contest: Indian Language Dependency Parsing, Hyderabad, India (2009)
4. Das, A., Shee, A., Garain, U.: Evaluation of two Bengali dependency parsers. In: Proceedings of the Workshop on Machine Translation and Parsing in Indian Languages (MTPIL), COLING, pp. 133–142 (2012)
5. Garain, U., De. S.: Dependency Parsing in Bangla. IGI Global (2013)
6. Haque, M.N., Khan, M.: Parsing Bangla using LFG. In: Proceedings of Association for Computational Linguistic (1997)
7. Kosaraju, P., Kesidi, S.R., Ainavolu, V.B.R., Kukkadapu, P.: Experiments on Indian language dependency parsing. In: Proceedings of ICON (2010)
8. Bharati, A., Sangal, R., Sharma, D.M.: SSF: Shakti Standard Format Guide (2007)
9. Das, D., Choudhury, M.: Chunker and shallow parser for free word order languages: an approach based on valency theory and feature structures. In: Proceedings of ICON (2004)
10. Begum, R., Husain, S., Sharma, D.M., Bai, L.: Developing verb frames in Hindi. In: Proceedings of the Sixth International Conference on Language Resources and Evaluation (LREC), Marrakech, Morocco (2008)
11. Chatterji, S., Sarkar, T.M., Sarkar, S., Chakrabory, J.: Kaaraka relations in Bengali. In: Proceedings of 31st All-India Conference of Linguists (AICL), Hyderabad, pp. 33–36, India (2009)
12. Bharati, R., Sangal, D.M., Bai, L.: AnnCorra: annotating corpora guidelines for POS and chunk annotation for Indian languages. Technical report (TR-LTRC-31), LTRC, IIIT Hyderabad, India (2006)
13. Ghosh, A., Das, A., Bhaskar, P., Bandyopadhyay, S.: Bengali parsing system. In: ICON NLP Tool Contest (2010)
14. Rao, P.R.K., Vijay, S.R.R., Vijaykrishna, R., Sobha, L.: A text chunker and hybrid POS tagger for Indian languages. In: Proceedings of IJCAI Workshop on Shallow Parsing for South Asian Languages (2007)
15. De, S., Dhar, A., Garain, U.: Structure simplification and demand satisfaction approach to dependency parsing in Bangla. In: Proceedings of ICON 2009 NLP Tools Contest: Indian Language Dependency Parsing, Hyderabad, India (2009)
16. Bandyopadhyay, S., Ekbal, A., Halder, D.: HMM based POS tagger and rule-based chunker for Bengali. In: Proceedings of NLPAI Machine Learning Workshop on Part of Speech and Chunking for Indian Languages (2006)
17. Das, D., Ekbal, A., Bandyopadhyay, S.: Acquiring verb subcategorization frames in Bengali from corpora. In: Li, W., Mollá-Aliod, D. (eds.) ICCPOL 2009. LNCS, vol. 5459, pp. 386–393. Springer, Heidelberg (2009)
18. Begum, R., Husain, S., Dhwaj, A., Sharma, D.M., Bai, L., Sangal, R.: Dependency annotation scheme for Indian Languages. In: Proceedings of the Third International Joint Conference on Natural Language Processing (IJCNLP), Hyderabad, India (2008)

Learning Clusters of Bilingual Suffixes Using Bilingual Translation Lexicon

K.M. Kavitha[1,3(✉)], Luís Gomes[1,2], and José Gabriel P. Lopes[1,2]

[1] NOVA Laboratory for Computer Science and Informatics (NOVA LINCS), Faculdade de Ciências e Tecnologia, Universidade Nova de Lisboa, 2829-516 Caparica, Portugal
k.mahesh@campus.fct.unl.pt, luismsgomes@gmail.com, gpl@fct.unl.pt
[2] ISTRION BOX-Translation & Revision, Lda., Parkurbis, 6200-865 Covilhã, Portugal
[3] Department of Computer Applications, St. Joseph Engineering College, Vamanjoor, Mangaluru 575 028, India
kavitham@sjec.ac.in

Abstract. By learning bilingual suffixation operations from translations using an existing bilingual lexicon with near translation forms we can improve its coverage and hence deal with the OOV entries. From this perspective, we identify bilingual stems, their bilingual morphological extensions (bilingual suffixes) and subsequently clusters of bilingual suffixes using known translation forms seen in an existing bilingual translation lexicon. We rely on clustering to enable safer translation generalisations. The degree of co-occurrence between two bilingual morphological extensions with reference to common bilingual stems determines if each of them should fall in the same cluster. Results are discussed for language pairs English-Portuguese (EN-PT) and English-Hindi (EN-HI).

1 Introduction

Large coverage translation lexicons are crucial for natural language processing systems such as machine translation, tagging and cross-language information retrieval. However, in spite of the availability and accessibility to raw texts, linguistic resources such as high-coverage bilingual lexicons are limited for less-studied language pairs such as English (EN) - Hindi (HI). On the other hand, the problem of OOV bilingual entries still prevail in well-researched languages. This is because not all possible forms of a word might be seen in the corpora used for translation extraction. Also, any particular technique for translation extraction does not guarantee to extract most of the possible translation pairs not found in the (parallel) corpora used for their acquisition. Furthermore, they are not able to extract everything. Source-target asymmetry with respect to language pairs further adds to the problem. Above all, new words are often encountered as language vocabulary tend to grow. Automatically learning and generalising word and multi-word structures is hence important.

Human ability to comprehend new words is attributed to his/her ability to analyse them based on previously seen known forms. Similar observation

R. Prasath et al. (Eds.): MIKE 2015, LNAI 9468, pp. 607–615, 2015.
DOI: 10.1007/978-3-319-26832-3_57

extends to bilingual word pairs or translations. Studies from the bilingual perspective show that, by identifying frequent forms or morphological similarities occurring in seen translation examples, learning bilingual suffixation operations and productively combining the bilingual evidences thus identified, new variants for observed translation forms might be suggested [1]. Furthermore, by learning the bilingual inflection classes [2], precise translation generalisations can be achieved. Thus, the existing lexicon can be automatically augmented by adding unseen translation variants suggested based on previously seen forms, thereby reducing the need for larger training data or corpora.

From this perspective, in this paper, we discuss an approach for learning clusters of bilingual suffixes so as to enable precise translation generalisation and safer generation of new bilingual pairs that are different from but similar to known forms. The rules for formation of new bilingual pairs are identified with an attempt to deal with variants of the known bilingual pairs that are orthographically similar. Productive bilingual segments consisting of bilingual stems and their bilingual morphological extensions (bilingual suffixes) are identified and clusters are formed such that all bilingual stems sharing same bilingual suffixes are grouped together. By simple concatenation of stem-pairs and suffix-pairs in a chosen cluster, we then generate out-of-vocabulary translations that are identical to, but different from the previously existing translations, thereby completing the existing lexicon.

2 Background

In our previous work, we proposed the bilingual approach to learning morph-like units for suggesting new translations [1], that relies on morphological information and pairing of bilingual evidence. Specifically, the fundamental idea is to identify and extract orthographically and semantically similar bilingual segments, as for instance, *'good'* ⇔ *'acCh'*, seen in known translation examples, such as, *'good'* ⇔ *'acChA'*, *'good'* ⇔ *'acChe'* and *'good'* ⇔ *'acChI'*, together with their bilingual extensions constituting dissimilar bilingual segments (bilingual suffixes), *''* ⇔ *'A'* | *'e'* | *'I'*[1]. The common part of translations that conflates all its bilingual variants[2] represents a bilingual stem (*'good'* ⇔ *'acCh'*). The dissimilar parts of the translations contributing to various surface forms represent bilingual suffixes (*''* ⇔ *'A'* | *'e'* | *'I'*). A pair of such bilingual extensions represent bilingual suffix replacement rules. Further, set of bilingual suffixes representing bilingual extensions for a set of bilingual stems together form bilingual suffix clusters[3], hence allowing safer translation generalisation.

In the afore-mentioned approach, the bilingual stems characterised by suffix pairs (features) are clustered using the clustering tool, CLUTO[4]. The clustering

[1] Note the null suffix in EN corresponding to gender and number suffixes in HI.

[2] Translations that are lexically similar.

[3] A *suffix cluster* may or may not correspond to Part-of-Speech such as noun or adjective but there are cases where the same suffix cluster aggregates nouns, adjectives and adverbs.

[4] http://glaros.dtc.umn.edu/gkhome/views/cluto

is based on partition approach and requires that the number of partitions are explicitly specified before clustering [1]. Frequent associations between word suffixes have been observed to be crucial in inducing correct morphological paradigms [3]. Extending this observation with respect to bilingual morphological extensions, we use the co-occurrence scores between bilingual morphological extensions to determine if they should belong to the same cluster.

3 Proposed Approach

We focus on clustering word-to-word translations, by treating a translation lexicon itself as a parallel corpus. The degree of co-occurrence between two bilingual morphological extensions (bilingual suffixes) with reference to common bilingual stems is used in deciding if each of them should fall in the same cluster. As the suffix-pair based co-occurrence statistics is used in clustering the bilingual translations, number of partitions need not be anticipated in advance.

3.1 Learning Bilingual Segments

Learning bilingual segments using near translation forms (bilingual variants) closely follows the bilingual learning approach and is based on learning bilingual suffixes and suffixation operations [1]. Further, over-segmentation and redundant bilingual suffixes are tackled through an additional loop that performs the suffix containment check by examining if one candidate bilingual suffix is enclosed within another. A true compound bilingual suffix (a combination of multiple candidate bilingual suffixes) is retained based on the observation that the strength of a compound bilingual suffix is less than the strengths of the bilingual suffixes composing it [4].

By applying the bilingual approach for learning bilingual stems and suffixes [1] on a bilingual lexicon containing word-to-word translations, we obtain the following resources as output:

List of Bilingual Stems and Bilingual Suffixes: These represent the list of bilingual stems (columns 3, 4 in Tables 3 and 4) and bilingual suffixes (Table 1) with their observed frequencies in the training dataset. Sample bilingual stems include *'plant'* ⇔ *'paudh'*, *'boy'* ⇔ *'laDak'*. Sample bilingual suffixes include *('', 'I')*, *('', 'A')*, *('s', 'oM')* and so forth and are attached to 10,743, 29,529 and 226 different bilingual pairs respectively. These lists aid in identifying bilingual stems and bilingual suffixes when a new translation is given.

Bilingual Suffixes Grouped by Bilingual Stems: This represents which set of bilingual suffixes attach to which bilingual stem. In Table 3[5], *('', 'A')*, *('s', 'oM')* are bilingual suffixes which attach to the same bilingual stem *'plant'* ⇔ *'paudh'* contributing to the surface forms *'plant'* ⇔ *'paudhA'* and *'plants'* ⇔ *'paudhoM'*. Each such grouping indicates the bilingual suffix replacement rules

[5] 2nd line in each row shows the transliterations for HI terms.

Table 1. Bilingual Suffixes undergoing frequent replacements in EN-HI

Bilingual Suffixes	Bilingual Suffixes (Hindi Suffixes transliterated)	Frequency
('', 'ी')	('', 'I')	10743
('', 'ा')	('', 'A')	29529
('ion', 'ा')	('ion', 'A')	457
('er', 'ा')	('er', 'A')	428
('ity', 'ा')	('ity', 'A')	286
('s', 'ों')	('s', 'oM')	226
('ity', 'ता')	('ity', 'tA')	223

that enable one translation form to be obtained using the other. For instance, from the above grouping, it follows that replacing the null suffix *''* with *'s'* and the suffix *'A'* with *'oM'* in the bilingual pair *'plant'* ⇔ *'paudhA'*, the bilingual pair *'plants'* ⇔ *'paudhoM'* can be obtained. In other words, in the bilingual pair *'plant'* ⇔ *'paudhA'*, appending *'s'* at the end of the EN word *'plant'* and replacing the suffix *'A'* with *'oM'* in its translated form (in HI) *'paudhA'*, yields the bilingual pair *'plants'* ⇔ *'paudhoM'*. Table 2 provides a clear instantiation of such groupings with more examples.

Table 2. Bilingual suffixes grouped by bilingual stems for EN-HI

Bilingual Stems		Bilingual Suffixes			
('nation', 'राष्ट्र')	:	('al', 'ीय'),	('alism', 'ीयता'),	('ality', 'ीयता'),	('alist', 'ीयतावादी')
('nation', 'rAShTr')	:	('al', 'Iya'),	('alism', 'IyatA'),	('ality', 'IyatA'),	('alist', 'IyatAvAdI')
('test', 'परीक्ष')	:	('', 'ा'),	('er', 'क'),	('ers', 'कों')	
('test', 'parIksh')	:	('', 'A'),	('er', 'k'),	('ers', 'koM')	

3.2 Clusters of Bilingual Suffixes

Our intention behind clustering is to generalise the bilingual suffix replacement rules, by looking for other stem pairs that go through the same transformation. A cluster is typically made of bilingual suffixes that attach to a list of bilingual stems in the input lexicon. Below we discuss the use of Bilingual Suffix Co-occurrence score in learning bilingual suffix clusters.

Preliminary Clusters. After all the bilingual suffixes that attach to a bilingual stem have been grouped as mentioned in the previous section, all bilingual stem pairs sharing same set of bilingual suffixes (and hence undergoing similar transformations) are further grouped forming a cluster. Preliminary clusters are thus obtained by initially grouping bilingual stems sharing identical bilingual suffix replacement rules. For instance, in the Table 3, we may see that both the bilingual stems *'plant'* ⇔ *'paudh'* and *'boy'* ⇔ *'laDak'* share the bilingual suffix replacement rule *('','A')* and *('s', 'oM')* and hence belong to the same cluster.

Bilingual Suffix Co-occurence Score. Bilingual suffix co-occurrence score represents the number of times a bilingual suffix $(s_{i_{L1}}, s_{i_{L2}})$ has co-occurred with another bilingual suffix $(s_{j_{L1}}, s_{j_{L2}})$ in the bilingual lexicon. Two bilingual suffixes are said to co-occur if they attach to a common bilingual stem. The co-occurrence scores between different bilingual suffixes for the language pair EN-HI are shown in the Table 3; the co-occurrence score between the bilingual suffixes *('', 'A')* and *('s', 'oM')* is 27, implying that they co-occur with 27 distinct bilingual stems. Similarly, in the Table 4, for the language pair EN-PT, we may see that the co-occurrence score between *('ence', 'ência')* and *('ences', 'ências')* is 65.

Table 3. Translation patterns representing bilingual suffix replacement rules with the bilingual suffix co-occurrence scores for EN-HI

Bilingual Suffixes	Bilingual Suffix Co-occurrence Score	Bilingual Stems	
('', '◌ा'), ('s', '◌ों') ('', 'A'), ('s', 'oM')	27	('plant', 'पौध') ('plant', 'paudh')	('boy', 'लड़क') ('boy', 'laDak')
('', '◌ी'), ('s', '◌ों') ('', 'I'), ('s', 'oM')	27	('job', 'नौकर') ('job', 'naukar')	('archer', 'धनुषधार') ('archer', 'dhanuShadhaar')
('', '◌ा'), ('er', 'क') ('', 'A'), ('er', 'k')	32	('test', 'परीक्ष') ('test', 'parIksh')	('print', 'मुद्र') ('print', 'mudr')
('', '◌ा'), ('s', '◌े') ('', 'A'), ('s', 'e')	10	('month', 'महीन') ('month', 'mahIn')	('curtain', 'पर्द') ('curtain', 'pard')

Table 4. Translation patterns with the bilingual suffix co-occurrence scores for EN-PT

Bilingual Suffixes	Bilingual Suffix Co-occurrence Score	Bilingual Stem Instances	
(ence, ência), (ences, ências)	65	(prefer, prefer)	(recurr, ocorr)
(al, ais), (al, al)	568	(compartment, compartiment)	(department, departament)
(e, er), ('', ir)	0	-	-
(ed, ida), (ed, idas)	75	(acclaim, aplaud)	(dismiss, demit)
(ed, ada), (ed, adas)	318	(affirm, confirm)	(adjust, ajust)

Bilingual Suffix Clusters. Based on the notion that the bilingual suffixes that co-occur more frequently are likely to be good candidates for a cluster, the candidate bilingual suffixes in the preliminary clusters are retained if the bilingual suffix co-occurrence score between two bilingual suffixes in the cluster is above the set threshold. For EN-HI, we set this threshold to 3 and for EN-PT it is set to 5. Algorithm 1 shows the steps involved in clustering.

Definitions. Let L be a Bilingual Lexicon.
Let L1, L2 be languages with alphabet set Σ_1, Σ_2.
Let $S_{StemPair}$ represent the set of bilingual stems and $S_{SuffixPair}$ be the set of bilingual suffixes.

Algorithm 1. Learning Bilingual Suffix Clusters

1: **procedure** LEARN–BILINGUALSUFFIXCLUSTER
2: **for** each input bilingual pair $(a_{L1}, a_{L2}) \in L$, where,
3: $(a_{L1}= {p1}_{L1}+{s1}_{L1})$, $(a_{L2}= {p1}_{L2}+{s1}_{L2})$, and
4: $({p1}_{L1}{p1}_{L2}) \in S_{StemPair}$ and $({s1}_{L1}, {s1}_{L2}) \in S_{SuffixPair}$ **do**
5: Set Suffix-Class-String= $({s1}_{L1}, {s1}_{L2})$
6:
7: **for** every bilingual pair, $(b_{L1}, b_{L2}) \in L$, such that,
8: $b_{L1}= {p1}_{L1}+{s2}_{L1}$ and $b_{L2}= {p1}_{L2}+{s2}_{L2}$, where,
9: $({p1}_{L1}{p1}_{L2}) \in S_{StemPair}$ and $({s2}_{L1}, {s2}_{L2}) \in S_{SuffixPair}$ **do**
10:
11: **if** Co-occurence-Score(($({s1}_{L1}, {s1}_{L2})$,$({s2}_{L1}, {s2}_{L2})$) $\geq$ threshold **then**
12: Set Suffix-Class-String=Suffix-Class-String.$({s2}_{L1}, {s2}_{L2})$
13: Add Suffix-Class-String to $S_{Suffix-Class}$, bilingual suffix set
14: **end if**
15: **end for**
16: **end for**
17: **for** each Suffix-Class-String $S_1 \in S_{Suffix-Class}$, where,
18: $S_1=(({s1}_{L1}, {s1}_{L2}), ({s2}_{L1}, {s2}_{L2}), ..., ({sn}_{L1}, {sn}_{L2})).({p1}_{L1}, {p1}_{L2})$, and
19: $(({s1}_{L1}, {s1}_{L2}), ({s2}_{L1}, {s2}_{L2}), ..., ({sn}_{L1}, {sn}_{L2})) \in S_{SuffixPair}$,
20: $({p1}_{L1}{p1}_{L2}) \in S_{StemPair}$ **do**
21: Set Merged-Suffix-Class-String = $(({s1}_{L1}, {s1}_{L2}), ({s2}_{L1}, {s2}_{L2}), ..., ({sn}_{L1}, {sn}_{L2})).({p1}_{L1}.{p1}_{L2})$
22:
23: **if** $\exists$ suffix class string $S_2 \in S_{Suffix-Class}$, such that,
24: $S_2=(({s1}_{L1}, {s1}_{L2}), ({s2}_{L1}, {s2}_{L2}), ..., ({sn}_{L1}, {sn}_{L2})).({p2}_{L1}{p2}_{L2})$,
25: and $({p2}_{L1}{p2}_{L2}) \in S_{StemPair}$ **then**
26: Merged-Suffix-Class-String=Merged-Suffix-Class-String.$({p2}_{L1}{p2}_{L2})$
27: Add Merged-Suffix-Class-String to $S_{Cluster}$,
28: the set of Bilingual Suffix clusters.
29: **end if**
30: **end for**
31: **end procedure**

In the Algorithm 1, by 'Suffix-Class-String' we mean a string consisting of all bilingual suffixes $((s_{i_{L1}}, s_{i_{L2}})$, $1 \leq i \leq n$, separated by commas preceding the dot in Step 18 of Algorithm 1) along with the bilingual stem $((p_{i_{L1}} p_{i_{L2}})$ following the dot in Step 18 Algorithm 1) to which those suffixes attach. Each row in Table 2 may thus be interpreted in the above specified form as follows:

('al', 'Iya'), ('alism', 'IyatA'), ('ality', 'IyatA'), ('alist','IyatAvAdI'). ('nation', 'rAShTr').

('', 'A'), ('er', 'k'), ('ers', 'koM'). ('test', 'parIksh').

Then, $S_{Suffix-Class}$ represents the set of such strings.

'Merged-Suffix-Class-String' represents a 'Suffix-Class-String' consisting of all bilingual suffixes along with all bilingual stems sharing those suffixes. An example is ('', 'A'), ('er', 'k'), ('ers', 'koM'). ('test', 'parIksh'). ('print', 'mudr'). Here, ('print', 'mudr') is another bilingual stem that shares the same transformations ('', 'A'), ('er', 'k') and ('ers', 'koM') as the bilingual stem ('test', 'parIksh').

4 Experiments

4.1 Data Set

EN-HI. For the experiments with EN-HI, the bilingual pairs representing word-to-word translations taken from an EN-HI bilingual lexicon was used. Approximately 90 % of the entries in the lexicon were acquired from the dictionary[6]. The remaining (10 %) entries were partly compiled manually and partially using the Symmetric Conditional Probability (SCP) based statistical measure [5] from the aligned parallel corpora[7]. We rely only on the bilingual lexicon without the use of any other larger resources for bilingual learning.

EN-PT. For EN-PT, the lexicon of bilingual entries (word-to-word translations) was extracted from the aligned parallel corpora[8] using various extraction techniques [6–9].

In our experiments, as training data, we used 52 K and 210 K bilingual lexicon entries for EN-HI and EN-PT language pairs, respectively.

4.2 Results and Discussion

We obtained a total of 143 clusters for EN-HI and 63 clusters for EN-PT. For both EN-PT and EN-HI, the smallest cluster consisted of only one bilingual suffix replacement rule, i.e., a pair of bilingual suffixes. For EN-PT, the largest cluster representing the *('', 'er')* group consisted of 15 different bilingual suffixes that are

[6] http://sanskritdocuments.org/hindi/dict/eng-hin$_$unic.html/ www.dicts.info www.hindilearner.com
[7] EMILLE Corpus - http://www.emille.lancs.ac.uk/
[8] DGT-TM - https://open-data.europa.eu/en/data/dataset/dgt-translation-memory
Europarl - http://www.statmt.org/europarl/
OPUS (EUconst, EMEA) - http://opus.lingfil.uu.se/

Table 5. Clustering statistics

Language Pairs	Clustering Approach	Number of Clusters	Generation Precision
EN-HI	Partition Approach (Proposed in Kavitha et al, 2014)	224	0.81
	Proposed here	143	0.84
EN-PT	Partition Approach (Proposed in Kavitha et al, 2014)	50	0.90
	Proposed here	63	0.88

shared by 717 different bilingual stems. For EN-HI, the largest cluster comprised of 5 different bilingual suffixes. Tables 3 and 4 respectively show sample clusters (with partial entries of bilingual suffixes for EN-PT[9]) for each of the language pairs EN-HI and EN-PT.

The clustering results are evaluated by examining the applicability of induced segments and clusters in suggesting new translations accurately. We first complete the translation lexicon with missing bilingual pairs using bilingual stems and bilingual suffixes induced from known bilingual pairs. Generation of missing translations is purely concatenative - involving simple concatenation of bilingual stems and suffixes and is achieved using the bilingual segments learnt from the training data and the associated bilingual suffix clusters [1]. The newly suggested translations are then evaluated for correctness. Table 5 shows the results obtained in completing the lexicon for missing forms. The precision for translation generation is calculated as the fraction of correctly generated bilingual pairs to the total number of bilingual pairs generated. In completing the translation lexicon for missing forms, where both bilingual stems and bilingual suffixes are known, the precision achieved for translation generation reaches 84.02% when compared to the precision of 81.31% obtained using the Kavitha's *et al.* approach [1] for EN-HI and 88 % for EN-PT, which is 2 % below the precision obtained using the partition approach [1].

Given an unseen translation, we may use this procedure to generate all possible translation forms by first predicting the bilingual stems and bilingual suffixes followed by classification of the given translation into one of the induced classes [10]. A simple technique for such classification could be based on the longest matching suffix approach [2].

5 Conclusion

In this paper we have discussed the use of co-occurrence score between two bilingual suffixes to determine if two bilingual suffixes should fall in the same cluster.

[9] In the Table 4, only two bilingual suffixes are shown per cluster although the original clusters contains varying number of bilingual suffixes ranging from 2 to 15 for EN-PT and from 2 to 5 for EN-HI.

In clustering the bilingual translations, this enables the bilingual suffixes to be grouped without having to suppose the number of clusters prior to clustering. Further, evaluation based on the correctness of newly suggested translations shows precision closer to that achieved using the partition based approach [1].

Acknowledgements. K.M. Kavitha and Luís Gomes acknowledge the Research Fellowship by FCT/MCTES with Ref. nos., SFRH/BD/64371/2009 and SFRH/BD/65059/2009, respectively, and the funded research project ISTRION (Ref. PTDC/EIA-EIA/114521/2009) that provided other means for the research carried out. The authors thank NOVA LINCS, FCT/UNL for the support, SJEC for providing the financial assistance to participate in MIKE 2015, and ISTRION BOX - Translation &Revision, Lda., for providing the data and valuable consultation.

References

1. Karimbi Mahesh, K., Gomes, L., Lopes, J.G.P.: Identification of bilingual segments for translation generation. In: Blockeel, H., van Leeuwen, M., Vinciotti, V. (eds.) IDA 2014. LNCS, vol. 8819, pp. 167–178. Springer, Heidelberg (2014)
2. Lindén, K.: Assigning an inflectional paradigm using the longest matching affix. In: Mitään ongelmia, E., Wiberg, M., Koura, A. (eds.) Juhlakirja Juhani Reimanille 50-vuotispäiväksi 23.1.2008. Turku 2008 (2008)
3. Desai, S., Pawar, J., Bhattacharyya, P.: A framework for learning morphology using suffix association matrix. In: WSSANLP-2014, pp. 28–36 (2014)
4. Dasgupta, S., Ng, V.: Unsupervised word segmentation for bangla. In: Proceedings of ICON, pp. 15–24 (2007)
5. Da Silva, J.F., Lopes, G.P.: Extracting multiword terms from document collections. In: Proceedings of the VExTAL: Venezia per il Trattamento Automatico delle Lingue, pp. 22–24 (1999)
6. Brown, P.F., Pietra, V.J.D., Pietra, S.A.D., Mercer, R.L.: The mathematics of statistical machine translation: Parameter estimation. Computat. Linguist. **19**(2), 263–311 (1993)
7. Lardilleux, A., Lepage, Y.: Sampling-based multilingual alignment. Proc. RANLP **2009**, 214–218 (2009)
8. Aires, J., Lopes, G.P., Gomes, L.: Phrase translation extraction from aligned parallel corpora using suffix arrays and related structures. In: Lopes, L.S., Lau, N., Mariano, P., Rocha, L.M. (eds.) EPIA 2009. LNCS, vol. 5816, pp. 587–597. Springer, Heidelberg (2009)
9. Gomes, L., Pereira Lopes, J.G.: Measuring spelling similarity for cognate identification. In: Antunes, L., Pinto, H.S. (eds.) EPIA 2011. LNCS, vol. 7026, pp. 624–633. Springer, Heidelberg (2011)
10. Kavitha, K.M., Gomes, L., Lopes, J.G.P.: Bilingually motivated segmentation and generation of word translations using relatively small translation data sets. In: Proceedings of the PACLIC29 (Accepted) (2015)

Automatic Construction of Tamil UNL Dictionary

Ganesh J.[1(✉)], Ranjani Parthasarathi[1(✉)], and Geetha T.V. [2(✉)]

[1] Department of Information Science and Technology, College of Engineering, Anna University, Chennai 600025, Tamilnadu, India
{ganesh13689,ranjani.parthasarathi}@gmail.com
[2] Department of Computer Science and Engineering, College of Engineering, Anna University, Chennai 600025, Tamilnadu, India
tv_g@hotmail.com

Abstract. In this paper, we propose an automatic tool for creating dictionary entries of Tamil words for the Universal Networking Language (UNL). Dictionary plays a crucial role in many NLP applications especially in machine translation (MT) systems. However, creating dictionary entries manually is a time consuming process. Moreover the UNL dictionary consists of additional features such as semantic constraints and attributes. To address this complex task, we propose a domain specific approach where the dictionary entries are created automatically using other word-based resources such as WordNet, bilingual dictionaries, and the UNL ontology. For the source of domain specific words, we use domain specific documents from the web. The resources used for extracting meaningful words from the documents are: Morphological analyzer, to extract the grammatical information of a given word, WordNet, to identify the semantics of the given word and UNL KB (Knowledge Base) to obtain the semantic constraints of a given word. Semantic constraints help to know the tense mood and aspect of the given word. Sometimes these semantic constraints may not be determined correctly by the automatic process. In such cases, a semantic similarity based filtering method based on UNL ontology is used to remove the incorrect dictionary entries. Thus, this automatic dictionary tool handles words semantically and also improves the correctness of the dictionary.

Keywords: Agaraadhi (online tamil dictionary) · Tamil wordnet · English wordnet · Morphological analyzer · UNL ontology

1 Introduction

For many natural languages, generic dictionaries giving lexical information pertaining to linguistic characteristics of the lexical units such as pronunciation, definition, etymology, grammatical category, etc., are created by linguistic experts manually. Computationally, such dictionaries are of great value, for many natural language processing (NLP) tasks. Recently, many artificial languages to represent and process knowledge have been designed for aiding NLP tasks. The Universal Networking Language (UNL) is one such artificial computer language designed by the UNDL foundation [15], to process information and knowledge. It is designed to semantically represent knowledge in a language-independent

R. Prasath et al. (Eds.): MIKE 2015, LNAI 9468, pp. 616–630, 2015.
DOI: 10.1007/978-3-319-26832-3_58

manner, in the form of concepts, semantic constraints, and relations. The UNL is made up of Universal Words (UWs), which constitute its vocabulary. UNL has a language independent component called the UNL Knowledge Base or the UNL ontology, and language dependent Universal Word (UW) based dictionaries. The UW dictionaries give the UWs and the corresponding natural language words. They can be used by people and computers as they give synonyms across many languages, linked through UWs. The idea behind the UNL system is that it can act as an interlingua to represent knowledge in a language-independent fashion, and facilitate moving from any language to any other language in an easy manner.

However, this requires one huge task of creating the UW dictionaries for each and every natural language. Creating these manually is definitely a time consuming process. Hence, we need a fully or semi-automatic process to be designed to address this task. The good news in this direction, is that there are many linguistic resources that have been created for other purposes. Thus, if we can come up with a process that leverages these resources, we can provide a viable solution to this humongous task. This is the approach we take in this paper, and develop a UW dictionary for Tamil.

We use domain-specific web documents as the source of the words for which the UW dictionary is to be created. We utilize various external Tamil knowledge sources as Tamil dictionary (Agaraadhi [13]) and Tamil WordNet [14] to obtain the synonyms. Agaraadhi is a "Tamil-Tamil-English" dictionary, in the sense that it gives the Tamil synonyms of a given word, and the corresponding English meaning. We also use a Morphological analyzer to obtain the root word, part-of-speech and certain attributes [12]. The UNL semantic constraints are obtained from the UNLKB. Redundant and incorrect entries are eliminated by a similarity checking mechanism using the UNL ontology. It is to be noted that this method can be adopted for other languages as well. We have focused on building the dictionary for specific domains, but the methodology is generic enough to work for all domains.

The remainder of this paper is organized as follows. In Sect. 2, we give some more details of the UNL system. In Sect. 3, we describe the existing approaches on UW dictionary creation. In Sect. 4, we describe our proposed process for Automatic Generation of Tamil-UNL Dictionary. In Sect. 5, we present an analysis of our proposed approach.

2 Universal Networking Language (UNL)

UNL is an artificial language that converts any natural language text into a deep semantic representation. UNL is a directed acyclic graph representation and consists of nodes and edges in which node represents concepts and edges represent relation between the concepts respectively. UNL consists of three main components Universal words, UNL relations and attributes. UNL consists of 46 semantic relations which are used to describe the relationship between the concepts in a sentence. UNL attributes are used to represent the mood, tense, aspect etc. The process of converting a natural language sentence to its equivalent UNL graph representation is called enconversion and the process to convert from UNL to natural language is called deconversion.

The UNL dictionary has a standard format proposed by UNDL foundation and the format is given below.

[NLW]{ID}"UW"(ATTR….) < FLG,FRE,PRI > Where,

NLW is the headword of the natural language lexical item.

ID is the unique identifier of the entry.

UW is the Universal word (English word from UNL knowledge base).

ATTR is the list of linguistic features of the NLW; these are set according to the UNL tagset. It also includes the affixation, linear and sub-categorization rules in the generation dictionary.

FLG is the three-character language code according to ISO 639-33.

FRE is the frequency of NLW in natural texts. It is used in natural language analysis (NL-UNL).

PRI is the priority of the NLW. It is used in natural language generation (UNL-NL).

COMMENT is any comment necessary to clarify the mapping between NL and UNL entries. For Example:

[இல்லம்]{1}"home/icl>place"(POS=Noun,LEX=Noun)<ta_14849,3,2>

3 Related Work

Since our focus is on building the UNL dictionary for different domains, this section discusses similar existing work carried out for different languages and for various purposes. Most of the earlier works focus mainly on semi-automatic creation of UNL dictionary entries using WordNet to provide the semantic intelligence and Morphological analyzer to extract the POS tags and root form of the given word.

One of the semi-automatic approaches developed for the Portuguese language by Ribeiro et al. [1] uses WordNet P.T lexical database and encoding of morphological rules. It ports information from the Portuguese WordNet database to the Portuguese UNL Dictionary, in a semi-automatic way. Similarly for the semantic UNL based text summarization proposed by Mangairkarasi and Gunasundari [2], the UNL dictionary is generated semi-automatically with human judgment. A general approach for creating Automatic Generation of Bilingual Dictionaries using intermediary languages and comparable corpora has been proposed by Gamallo Otero and Pichel Campose [3]. This work uses an unsupervised method to derive a new bilingual lexicon by using existing resources and it is validated with the comparable corpora.

An automatic method has been proposed by Verma and Bhattacharyya [4]. This aims at generating document specific UNL dictionary for Hindi, Marathi and English using language specific WordNet and a set of morphological rules. Verma and Bhattacharyya [5] have also proposed another approach for automatic lexicon generation through WordNet. Here, the dictionary entries are generated in the standard UNL format and associated with the syntactic and semantic properties. This system uses Word Sense Disambiguation (WSD), inferencer and the knowledge base of the UNL in addition to WordNet. In the work to generate semantic net like expressions from text documents for Bangla language proposed by Ali et al. [6] the UNL dictionary is the main resource, which is used to convert Bangla natural language text into UNL using a method called

Predicate preserving parser. In another approach, Bangla UNL dictionary structure has been generated by Mridha et al. [7] using a Bangla words morphological analyzer with a set of rules. The UNL structure is created for Bangla roots, Krit prottoy (primary suffix) and Kria Bivokti (verbal suffix). Mridha et al. [8] has proposed an approach to generate grammatical attributes for Bangla words using UNL. This is used to develop a Bangla word dictionary for Enconversion and Deconversion.

Similarly Alansary et al., [16] have proposed a UNL approach for translating the metadata of books in a library information system. Here, the UW dictionary is generated by transliteration and human based translation methods.

For the Tamil language, a UNL dictionary for the tourism domain has been created manually for the purpose of enconversion and deconversion (Balaji J et al. [9]).

To our knowledge, there is no existing method for automatic creation of UW dictionary for Tamil. Hence this paper proposes such an idea.

4 Automatic Generation of Tamil – UNL Dictionary

The key factor that we exploit is that Tamil is a morphologically rich language in which the morphological suffixes convey most of the lexical, syntactic and semantic information. By using this information, we can assign the semantic constraints easily and accurately with the help of the UNLKB.

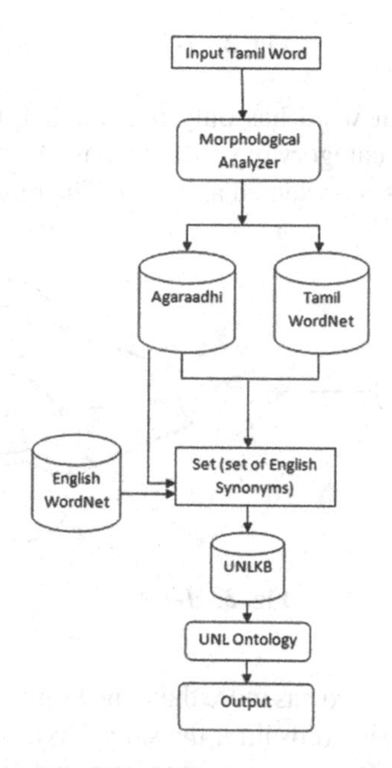

Fig. 1. Automatic dictionary flow diagram

Thus, if we take a document consisting of inflected words, and run it through a morphological analyser, we can get the root words, along with their POS, case ending, and certain other attributes. The root words are then checked in a Tamil-Tamil-English dictionary, and a Tamil WordNet, to obtain their synonyms. These words are then matched with an English WordNet to obtain equivalent English words. These words are then checked for in the UNLKB/UNL ontology to obtain the corresponding UWs and the corresponding UNL dictionary entries are created. This process is shown in Fig. 1, and described in Sect. 4.2. The next subsection gives the different cases that need to be handled in this process.

4.1 Categories of Dictionary Entries

In this process there are seven different cases that occur out of which five give us direct unique entries without any ambiguity. But in two cases there are ambiguities which need to be resolved. We make use of the UNL ontology to resolve the ambiguity.

Category 1: The Tamil input word has only one English meaning in Agaraadhi and has only one semantic constraint in UNLKB. It creates single entry for the Tamil word in this category. This is simply represented as *1-1-1* and the diagrammatic representation is given below (Fig. 2).

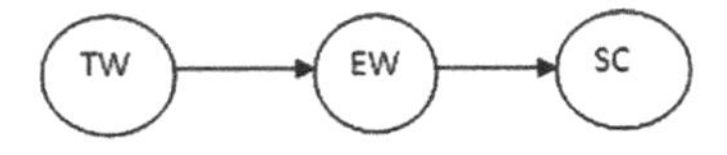

Fig. 2. *1-1-1*

Category 2: The Tamil input word has only one English meaning and has multiple semantic constraints. For this category of word, it creates as many entries as the number of semantic constraints and is represented as *1-1-N.* The representation is Fig. 3.

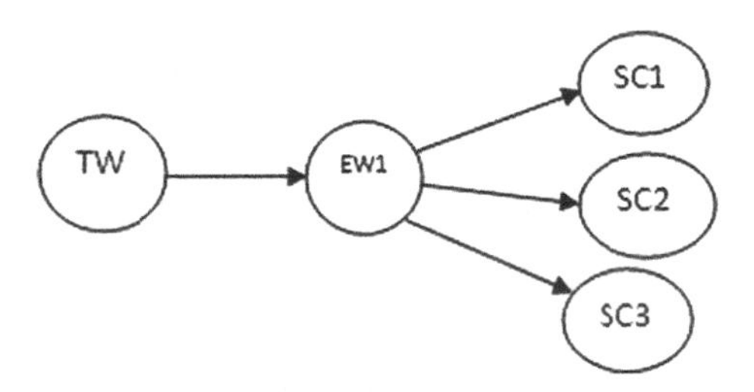

Fig. 3. *1-1-N*

Category 3: The Tamil input word has more than one English meaning, say for example two meanings and both English words infer the same UNL semantic constraint. Words' falling under this category creates dictionary entries as many as the English words. This is represented as *1-N-1* and the diagrammatic representation is shown in Fig. 4.

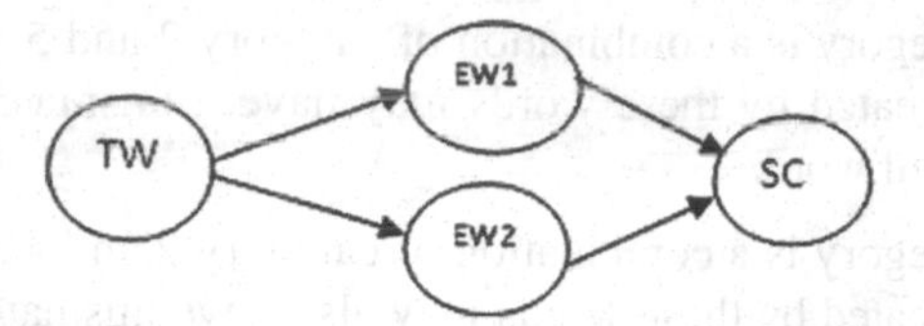

Fig. 4. *1-N-1*

Category 4: This category of words has more than one English meaning and each English word has only one UNL semantic constraint for each English word. In this case dictionary entries are created for each of the English words, and it can be represented as *1-N-N* and the diagrammatic representation is Fig. 5.

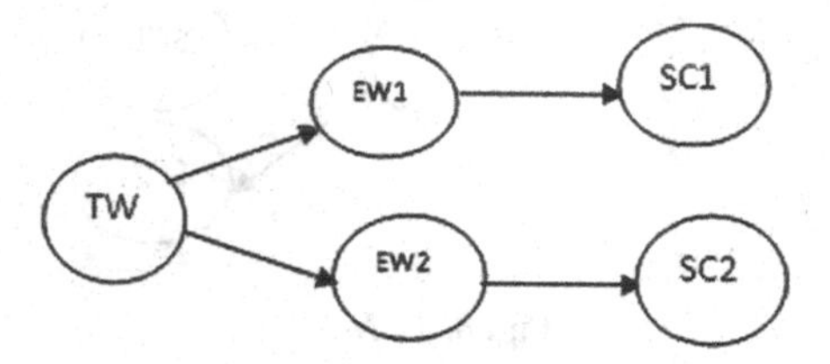

Fig. 5. *1-N-N*

Category 5: This category is a special case of category 1. Here we check whether the English word has more Tamil synonyms other than the given Tamil word. This category of words creates as many entries as the number of Tamil words. This can be represented as *N-1-1* and the diagram representation is Figs. 6 and 7.

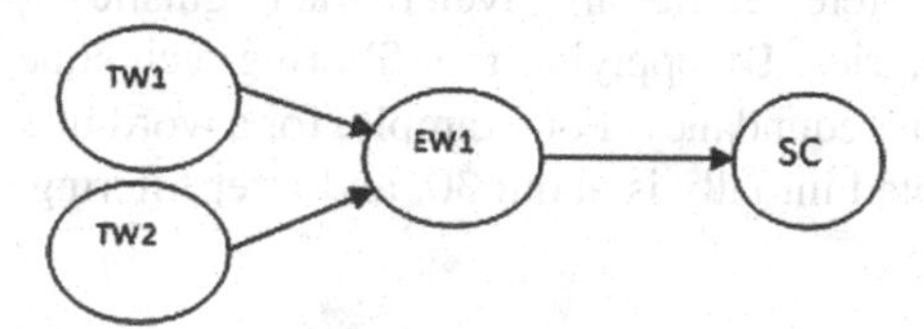

Fig. 6. *N-1-1*

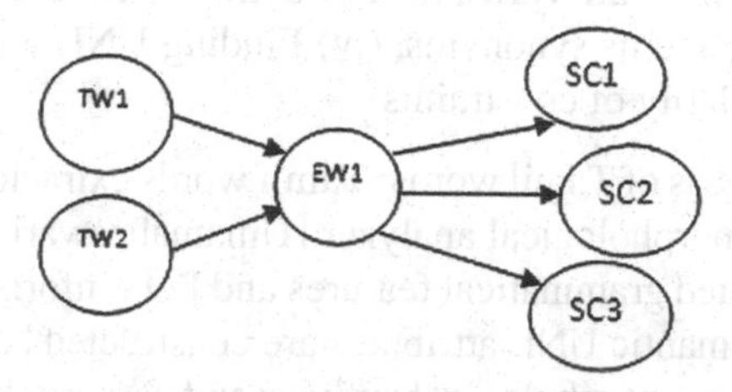

Fig. 7. *N-1-N*

Category 6: This category is a combination of category 2 and 5 and is represented as *N-1-N*. The entries created by these words may have a mismatch with the semantic constraints of the Tamil word.

Category 7: This category is a combination of category 2 and 4 and is represented as *1-N-N*. The entries created by these words may also have mismatch with the semantic constraints of the Tamil word (Fig. 8).

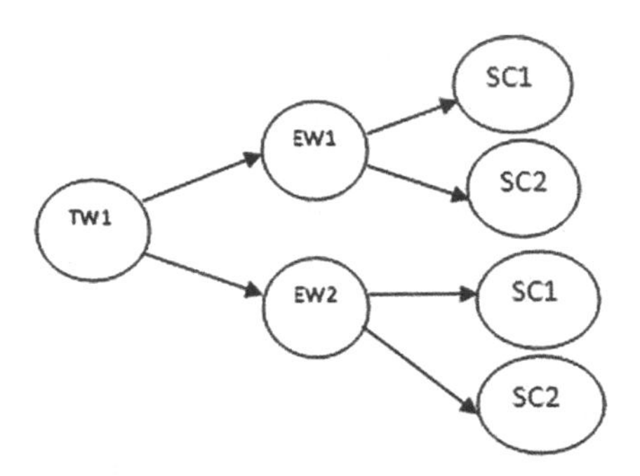

Fig. 8. *1-N-N*

Since the categories 6 and 7 have mismatch of semantic constraints with the Tamil word, we need to evaluate their correctness manually by linguistic experts. This is a laborious task as there could be many incorrect entries. To reduce the number of incorrect entries, we use the UNLKB ontology and filter the entries. We first identify the parent of constraints for each entry using the UNLKB ontology and based on the parent frequency, we filter the entries of those whose parents are less frequent. The results of the UNLKB ontology filtered entries are given to the linguistic experts for verifying the correctness of those entries. By applying this filtering technique we can increase the accuracy and reduce the redundancy. For example, for a word like "adai (அடை)", the number of entries created initially is about 30, and after filtering with the UNL KB, it is reduced to 12.

4.2 Dictionary Generation Process

The method of constructing the UNL dictionary consists of the following steps: (i) Morphological analysis of Tamil words, (ii) Extraction of Tamil Synonyms, (iii) Extraction of English meaning and its synonyms, (iv) Finding UNL semantic constraints, (v) Checking semantic similarity of constraints.

(i) Morphological analysis of Tamil words: Tamil words extracted from the documents are given to Tamil morphological analyzer [Umamaheswari et al.] to extract a root word and its associated grammatical features and POS information. Moreover most of the lexical and semantic UNL attributes are constructed based on morphological analyzer's output (i.e.) morphological suffixes and case markers are extracted from the morphological analyzer of a given input word. There are three cases that could occur here. Case a) the analyzer produces correct root word as output; this is a straight forward approach. Case b) the word is not handled by Morphological Analyzer, i.e. unknown words (for example named entities). In this case, we search for that word in the dictionary (Agaraadhi) directly and if it is found we proceed with the next step. Case c) the Morphological Analyzer produces incorrect output, where the root word may or may not be valid. In either case, we take the root word (since we are interested in the root word, and not the inflected word), and check in the dictionary. A meaningful word would find a match in the dictionary, while an erroneous root word will not and will be discarded.

For example:

(a) illathil → illam + Noun
<in the house> → house + Noun
(b) ottapanthayam → unknown
<running race> → unknown
(c) vibaritham → vibara + Adjective
<accidental> → details + Adjective

(ii) Extraction of Tamil Synonyms: The morphologically analyzed Tamil root word is given to the Agaraadhi dictionary [Elanchezhiyan et al.] and Tamil WordNet [Rajendran] to extract the synonyms of the corresponding Tamil root word. Two sets of synonyms words A and W, are obtained from Agaraadhi and Tamil WordNet, respectively. The two sets A & W are intersected to form a new set which contains the words common to both A & W. This step helps to identify words that are close in meaning and sense.

A → refers to the synonyms which is collected from Agaraadhi.
W → refers to the synonym which is collected from WordNet.
AWi → refer to the collection of common words from set A and W
For example:

Agaradhi → illam, thainadu, uttkalam, vittil, mugappu, agam, viidu, oyevidam, iyalidam, manai
WordNet → illam, agam, viidu, manai, kappagam
AWi → (common words) illam, agam, viidu, mania

(iii) Extraction of English meaning and its Synonyms: The English words and its synonyms are extracted for the corresponding words in the set AW_i. Each word in the set AW_i is given as input to the Agaraadhi dictionary to obtain the English meanings and given these English meanings we find the synonyms of those words using English WordNet. Now each Tamil word contains a set of English words which in turn can have another set of synonyms associated with that collection. Among these sets, the common terms are extracted by intersecting all the sets.

For example let each AW_i be a set consisting of four Tamil words: $AW_I = \{TW1, TW2, TW3, TW4\}$.

Each word TWi is given to Agaraadhi to get all the English meanings. Then for each TWi we will get a set consisting of English words.

(i.e.) S(TW1) = {EM1, EM2, EM3}
S(TW2) = {EM1, EM2, EM3, EM4, EM5 } S(TW3) = {EM1, EM2, EM3, EM4}
S(TW4) = {EM1, EM2, EM3, EM4, EM5 }

Now the sets are intersected [S(TW1) ∩ S(TW2) ∩ S(TW3) ∩ S(TW4)] to obtain a single set of common English words for the corresponding Tamil words in the set $S(EM_i)$.

$S(EM_i)$ = {EM1, EM2}

For Example:

illam → Home, House, Hostel
agam → Unit, Home, Subject, House, Heart, chest, inner
viidu → home, nest, house, building, Bower, heaven
manai → house, habitat, place, Home, aerie
$S(EM_i)$ = {Home, House}

(iv) Finding UNL Semantic Constraints: The set of English words obtained in S (EM_i) is then passed into the UNL Knowledge Base [11] to obtain the possible set of UNL semantic constraints. It is to be noted that the UNL semantic constraints are classified based on four different concepts. They are (i) Nominal concepts, (ii) Attributive concepts, (iii) Adjectival concepts and (iv) Predicate concepts. The predicate concepts are then sub classified into be_verbs, do_verbs, and occur_verbs. This classification is to narrow down the searching of semantic constraints of a word efficiently. Hence we follow two different strategies to extract the constraints (a) General search – search is done in the whole UNLKB, and (b) POS based search – search is done based on the POS of the word in the respective concepts. Thus UNL semantic constraints are obtained for the set of English words in $S(EM_i)$.

POS based search example:

Home → icl>person, icl>place, icl>country
House → icl>building, icl>person, icl>company

General search example:

Home → icl>person, icl>place, icl>country
House → icl>building, icl>person, icl>company, agt>person, agt>person, obj>thing, agt>thing, obj>person.

(v) Semantic Similarity of Constraints: In cases where a single Tamil word maps to more than one constraint (cases 6 and 7 discussed above), it needs to be disambiguated. This is achieved by finding the semantic similarity of the constraints using the UNL Ontology. The UNL ontology is depicted as a tree structure giving parent-child relationship. For each constraint, we check its parent, and count the number of common constraints. This count is called the similarity score. We choose the constraints which have the highest score, and filter out the others as non-relevant. This is then verified manually by linguists.

Similarity Score = Sum (Max. no. of constraints occurring in the parents)
For Example:

Home → icl > place
House → icl > building

(vi) UNL Attributes: We classify the UNDL tags with respect to POS information obtained from morphological analyzer. Certain UNDL tags are common to all types of POS tags such as Noun, Verb, Adjective, Adverb, Postposition and Pronoun. List of UNL tags common to all POS types are POS, LEX. The tag values are assigned automatically for each entry.

5 Results and Evaluation

We have tested our system by using general crawled documents from various domains like tourism, arts, sports, entertainment, news etc. The crawled corpus contains 2300 documents with an overall count of 6,00,000 words. From this we have extracted 85,771 unique words. This input was given to the morphological analyzer, and it extracted 8,818 unique root words. With this set of unique root words, the number of entries created for the first five cases – without ambiguity – is given in Table 1. A total of 15180 and 11945 entries have been created in the POS and general approach.

Table 1. Results of first five cases using POS and General approach

S.no	Case	UNLKB constraints	
		Unique POS based entries	Unique general entries
1.	*I-I-I*	124	196
2.	*I-I-N*	1116	2692
3.	*I-N-I*	1030	817
4.	*I-N-N*	1587	1941
5.	*N-I-I*	12994	7865
6.	*Overall Total*	15180	11945

The POS wise breakup of the unique entries observed in each case on applying POS based approach and the general approach, are shown in Figs. 9, 10, 11, 12, 13, 14. The Fig. 15 gives the results for the last two cases. A total of 5254 and 7641 entries are created in the POS and general approach. On filtering the entries using UNL Ontology, the number of entries is reduced to 4361 and 5267, respectively. Of these entries 2351 and 1245 are found to be correct by manual verification.

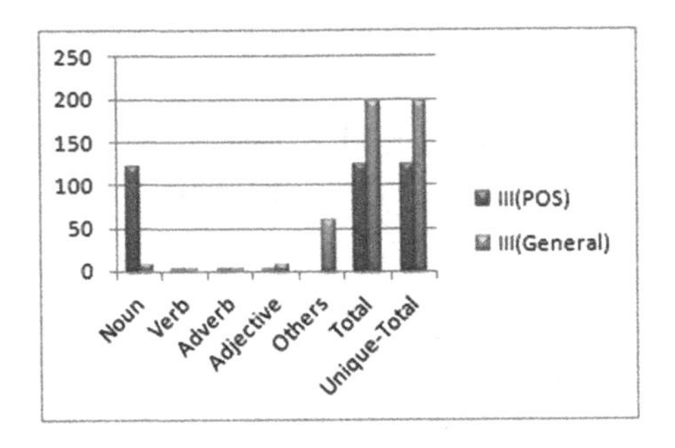

Fig. 9. *I-I-I* – case unique entries

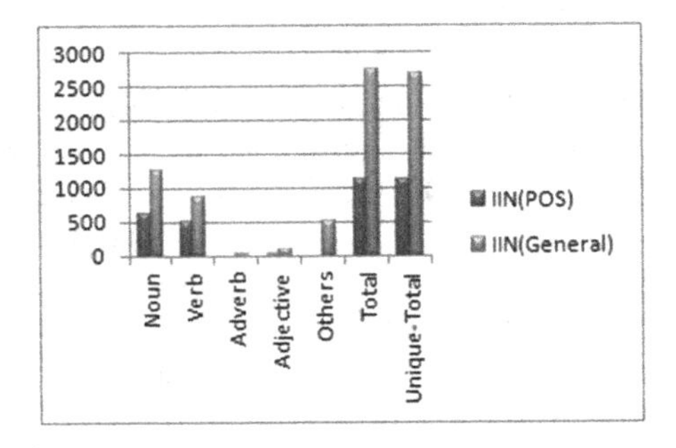

Fig. 10. *I-I-N* - case unique entries

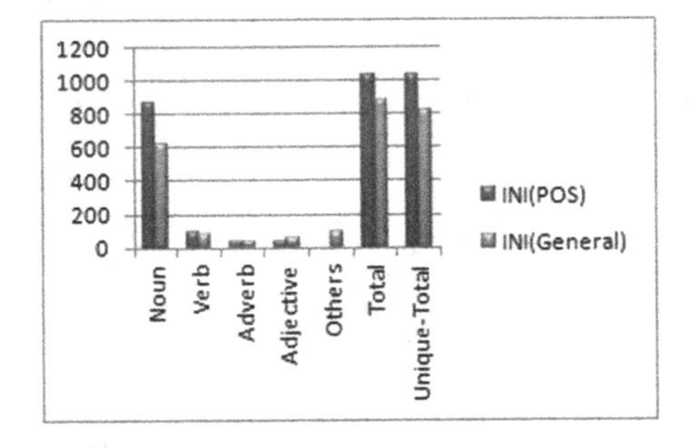

Fig. 11. *I-N-I* – case unique entries

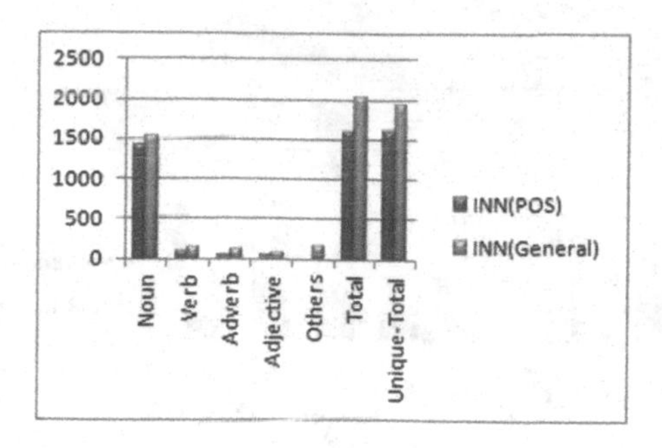

Fig. 12. *I-N-N* - case unique entries

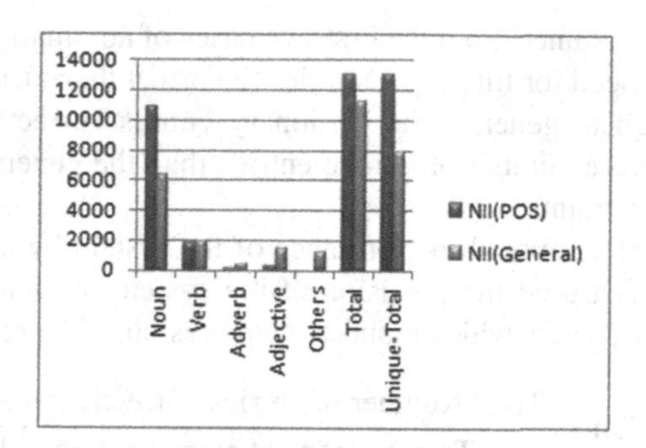

Fig. 13. *N-I-I* – case unique entries

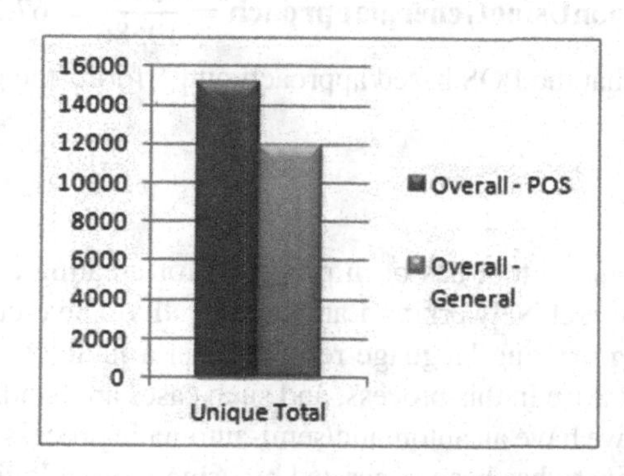

Fig. 14. Overall – total of first five cases

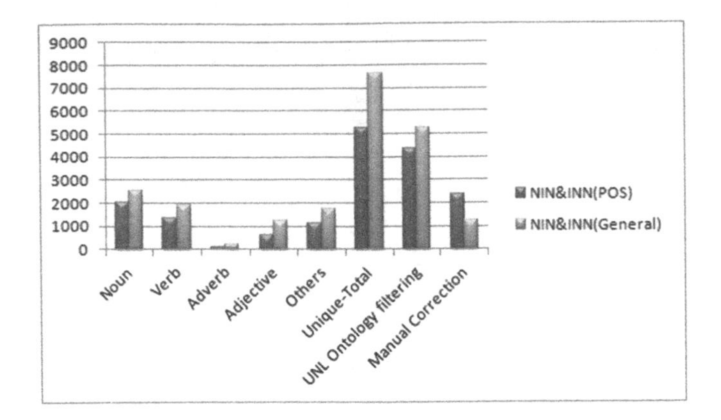

Fig. 15. *N-I-N and I-N-N cases*

The output that is obtained from the first five cases of automatic approach is highly accurate. There is no need for filtering and rules to correct the entries as the conditions are more than enough to generate the dictionary entries correctly. The POS based approach generates more number of unique entries than the general approach since it narrows down the constraints.

We have measured the overall performance of the system by manual checking for correct entries. We measured the precision of the system by manually checking the entries. It is the last two cases which reduces the precision. The precision is defined as:

$$\textbf{Precision} = \frac{\text{Total Number of entries correctly generated}}{\text{Total number of entries generated}}$$

$$\textbf{PrecisionUsingPOSapproach} = \frac{17531}{20434} = 85.79\%$$

$$\textbf{PrecisionUsingGeneralapproach} = \frac{13190}{19586} = 67.34\%$$

It can be seemed that the POS based approach outperforms the general approach.

6 Conclusion

In this paper, an automatic tool has been proposed for creating a dictionary entry of Tamil words for Universal Networking Language (UNL). The dictionary entries have been created by using existing language resources, in a judicious manner. There are some ambiguities that arise in this process, and such cases are handled with the manual help of experts. Thus we have an automatic/semi-automatic process for UNL dictionary generation. The dictionary has been generated by using morphological analyzer, Tamil and English WordNet for extracting their corresponding synonyms and Tamil-English dictionary, UNLKB, and rule based filtering techniques for filter out the incorrect entries. The ambiguities have been tackled by measuring the semantic similarity using UNL

Ontology so as to obtain the appropriate semantic constraints for a given input word. The results obtained on a limited set of documents are encouraging. This is being investigated on a larger scale. We plan to make the tools and results available after verification, as an open resource.

This method is suitable for other morphologically rich languages, as long as other language-specific resources such as WordNet are available. With a large effort in creating WordNets for various languages world-wide [17, 18], we believe that it is possible to use this approach for a majority of languages.

References

1. Ribeiro, C., Santos, R., Chaves, R.P., Marrafa, P.: "Semi-Automatic UNL dictionary generation using WordNet.PT. In: Universidade de Lisboa, CLUL CLG – Computation of Lexical and Grammatical Knowledge Research Group
2. Mangairkarasi, S., Gunasundari, S.: Semantic based text summarization using universal networking language. Int. J. Appl. Inf. Syst. **3**(8), 18–23 (2012)
3. Gamallo Otero, P., Pichel Campose, J.R.: Automatic generation of bilingual dictionaries using intermediary languages and comparable corpora. In: Gelbukh, A. (ed.) CICLing 2010. LNCS, vol. 6008, pp. 473–483. Springer, Heidelberg (2010)
4. Verma, N., Bhattacharyya, P.: Automatic generation of multilingual lexicon by using WordNet. In: The Proceedings of Convergences 2003, International Conference on the Convergence of Knowledge, Culture, Language and Information Technologies (2003)
5. Verma, N., Bhattacharyya, P.: Automatic lexicon generation through WordNet. In: Global WordNet Conference (2004)
6. Ali, M.N.Y., Ripon, S., Allayear, S.M.: "UNL based Bangla natural text conversion – predicate preserving parser approach. Int. J. Comput. Sci. Issues **9**, 259–265 (2012)
7. Mridha, M.F., Nur, K.M., Banik, M., Huda, M.N.: Structure of dictionary entries of Bangla morphemes for universal networking language (UNL). Int. J. Comput. Inf. Syst. Ind. Manag. Appl. 746–754 (2011)
8. Mridha, M.F., Nur, K.M., Banik, M., Huda, M.N.: Generation of attributes for Bangla words for universal networking language (UNL). Int. J. Adv. Comput. Sci. Appl. **2**, 1–7 (2011)
9. Balaji, J., Geetha, T.V., Parthasarathi, R., Karky, M.: Article: morpho-semantic features for rule-based Tamil enconversion. Int. J. Comput. Appl. **26**(6), 11–18 (2011)
10. Dhanabalan, T., Geetha, T.V.: UNL deconverter for Tamil. In: The International Conference on the Convergence of Knowledge, Culture, Language and Information Technologies (2003)
11. UNDL. 2011. Universal networking digital language. http://www.undl.org/. Accessed 28 September 2011
12. Umamaheswari, E., Ranganathan, K., Geetha T.V., Parthasarathi, R., Karky, M.: Enhancement of morphological analyzer with compound, numeral and colloquial word handler. Tamil Computing Lab (TaCoLa), College of Engineering Guindy, Anna University, Chennai
13. Elanchezhiyan, K., Karthikeyan, S, Geetha, T.V., Parthasarathi, R., Karky, M.: Agaraadhi: a novel online dictionary framework. In: 10th International Tamil Internet Conference of International Forum for Information Technology in Tamil
14. Rajendran,S.: Tamil WordNet, Department of Linguistics Tamil University, Thanjavur
15. UNL Ontology 2011. http://www.undl.org/unlsys/uw/UNLOntology.html

16. Alansary, S., Nagi, M., Adly, N.: A library information system (LIS) based on UNL knowledge infrastructure. In: Proceedings of the Universal Networking Language Workshop in conjunction with 7th International Conference on "Computer Science and Information Technology (2009)
17. Pushpak Bhattacharyya IndoWordNet, Lexical Resources Engineering Conference 2010 (LREC 2010), May 2010
18. Vossen, P.: EuroWordNet: a Multilingual Database with Lexical Semantic Networks. Spriger, Berlin (1998)

A New Approach to Syllabification of Words in Gujarati

Harsh Trivedi(✉), Aanal Patel, and Prasenjit Majumder

Dhirubhai Ambani Institute of Information and Communication Technology, Gandhinagar, India
{harshjtrivedi94,ptl.aanal,prasenjit.majumder}@gmail.com
http://daiict.ac.in

Abstract. This paper presents a statistical approach for automatic syllabification of words in Gujarati. Gujarati is a resource poor language and hardly any work for its syllabification has been reported, to the best our knowledge. Specifically, lack of enough training data makes this task difficult to perform. A training corpus of 14 thousand Gujarati words is built and a new approach to syllabification in Gujarati is tested on it. The maximum word and syllable level accuracies achieved are 91.89 % and 98.02 % respectively.

Keyword: Syllabification Gujarati CRF

1 Introduction

This paper explains in detail a supervised system developed to split words written in Gujarati into its constituent syllables. This process of word syllabification has important applications in real world text processing. It plays an important role in speech synthesis and recognition [6] and is required for effective text-to-speech (TTS) systems. It is useful in calculating readability indices like Flesh Kincaid, Gunning Fox and SMOG, which require to count the number of syllables in the word. Besides, because of the dynamic nature of any language, a dictionary look-up for syllabification can never suffice. Even more, there is hardly any digital resource for word syllabification in Gujarati. And that is why, to address above mentioned problem an automated system to syllabify Gujarati words can be of immense use.

Many efforts for syllabification in various languages have already been made. Generic principles of the syllabification include Maximum Onset Principle [5], Legality Principle [3] and Sonority Principle [11]. A 99 % efficient statistical approach syllabification was proposed by Mayer [8], which involved counting of the syllables in order know about the best split possible. Hammond [4] showed how to use Optimality Theory for effective syllabification. Another approach demonstrated a discriminative approach that uses Support Vector Machine and Hidden Markov Model together for syllabification with 99.9 % and 99.4 % accuracies in English and German respectively [1]. Conditional Random Fields have also

R. Prasath et al. (Eds.): MIKE 2015, LNAI 9468, pp. 631–639, 2015.
DOI: 10.1007/978-3-319-26832-3_59

been used for syllabification [10]. All these approaches have been predominantly demonstrated in English and European languages.

To the best of our knowledge, no text based data-driven approach has been done on Gujarati and hence this paper attempts to make effort in this direction.

The paper is organized as follows. Section 2 describes the process of data collection and talks about CRF approach adapted for Gujarati, which is used for bootstrapping the training data. Section 3 details on our probabilistic approach for syllabification of Gujarati words. Section 4 elaborates on evaluation details. Conclusion and future scopes are described in Sect. 5.

The system input and output examples are shown in Fig. 1 with symbol ("hyphen") denoting the syllabic break. Transliteration to English is shown for the purpose of readability. They are not part of the input or output of the system.

Input	**Output**
ખજૂર */khajur/*	ખ-જૂર */kha-jur/*
હોળી */holi/*	હો-ળી */ho-li/*

Fig. 1. Input/Output examples

2 Data Collection

Six thousand words and their syllabification were collected from "Babu Suthar's Gujarati dictionary" [12]. For expanding the corpus to more words, an iterative bootstrapping model was used. Two iterations were conducted and suggestions of 4 thousand words in each iteration were generated using CRF based trained model. These suggestions were rectified by a Gujarati linguist and were added back to training data. The new 4 thousand words selected in each iteration were taken as the most frequently occurring words from a Gujarati newspaper corpus [9]. For rectification of faulty suggestions, an online interface was made and was provided to a language expert to make this process faster. As a result, a corpus of about 14 thousand word syllabifications had been generated.

2.1 Using Conditional Random Fields to Bootstrap Data

Modeling the problem of automatic syllabification as sequential tagging problem and usage of Conditional Random Fields [7] for the same has been done before. These approaches have been well demonstrated in languages following Roman script [2,10,14]. The method involves learning to predict the sequence of output labels (syllabic break or not prior/post character) by taking as input sequence of characters of word and their respective feature set.

In our implementation, each unicode character in the word is labeled 'S' if it marks the beginning of the syllable and 'F' otherwise. For example, for the syllabification 'શુ-ભેચ્-છા', the tags are: શ(S), ુ(F), ભ(S), ે(F), ચ(F), ્(F), છ(S), ા(F).

The software that we use as an implementation of Conditional Random Fields is CRF++ [13].

Unlike Roman script, Gujarati alphabet can be categorized in vowels, consonants and *matras*. *Matras* sound like vowel, but they do not exist isolated. They represent vowel-like sounds that are preceded by a consonant. As observed, these set of characters play an important role in deciding position of the syllabic breaks and hence are included in feature vector for CRF.
For feature vector for each character, categorization was done as follows:

- **Vowels:** { અ, આ, ઇ, ઈ, ઉ, ઊ, એ, ઍ, ઐ, ઓ, ઑ, ઔ, અં }
- **Consonants:** { ક, ખ, ગ, ઘ, ચ, છ, જ, ઝ, ટ, ઠ, ડ, ઢ, ત, થ, દ, ધ, ન, પ, ફ, બ, ભ, મ, ય, ર, લ, વ, શ, સ, ષ, હ, ળ, ક્ષ, જ્ઞ }
- **Matras:** { ◌ા, ◌િ, ◌ી, ◌ુ, ◌ૂ, ◌ે, ◌ૅ, ◌ૈ, ◌ો, ◌ૉ, ◌ૌ, ◌ં, ◌્, ◌ૃ, ◌ૄ, ◌ઁ, ◌ં, ◌ઃ , ◌઼, ૨ }

Table 1. Combinations of context features tried

Context window	Character n-grams	Total context features
-1 to 1	1 to 3	6
-2 to 2	1 to 5	15
-3 to 3	1 to 7	28
-4 to 4	1 to 9	45
-5 to 5	1 to 11	66

For example, for the context window -1 to 1 : { w[-1], w[0], w[1], w[-1]/w[0], w[0]/w[1], w[-1]/w[0]/w[1] } are taken as context features. Context window of -3 to +3 with character 1-gram to 7-grams used as context features turn out to give best results. No significant gain was achieved there-after by increasing context features.

3 Our Approach

The Subsect. 3.1 describes a probabilistic model for syllabification, while the Subsect. 3.2 describes an add-on method that when applied prior to the former method, contributes to improve the overall performance.

3.1 Predicting Maximum Probable Syllabification

This approach focuses on statistically predicting the most probable syllabification from all the possible syllabifications of a word. It attempts to calculate the probability of each of the possible syllabification being correct and chooses the one with maximum probability. A word with n characters, can theoretically have 2^{n-1} possibilities of syllabification owing to n-1 positions where a '-' can be kept. However, increasing word size increases search space exponentially. Hence, an assumption is taken considering the language constructs to reduce the search space.

It was observed[1] that the Gujarati speakers always pronounce the *matra* along with the preceding consonant or vowel. Hence, an **assumption** was made, that a vowel or a consonant along with all its subsequently occurring *matra's* would always fall into the same syllable, eliminating the possibility of break between them. Using this assumption, a word can be broken down into **units**, each unit being unbreakable any further. For example, શુભેચ્છા /*shubhechcha*/, the units of this words are:

- શુ (શ+ુ) /*shu*/
- ભે (ભ+ે) /*bhe*/
- ચ્ (ચ+્) /*ch*/
- છા (છ+ા) /*chha*/

and the possible syllabifications (2^{4-1}) are:

- શુ-ભે-ચ્-છા
- શુભે-ચ્-છા
- શુ-ભેચ્-છા
- શુભેચ્-છા
- શુ-ભે-ચ્છા
- શુભે-ચ્છા
- શુ-ભેચ્છા
- શુભેચ્છા

Computing Scores: Let word(W) be composed of n units, $W = u_1.u_2...u_n$ and let S_i be the candidate syllabification to calculate the score for. S_i would be composed of same units as word(W) but with '-' (syllabic break) present between 2 consecutive units. Hence for test syllabification S_i, we can define a set:

$$Pairs(S_i) = \{X_i | X_i = u_i u_{i+1} \ or \ u_i \text{-} u_{i+1}\} \tag{1}$$

where,
$u_i u_{i+1}$ indicates absence of syllabic break between units u_i and u_{i+1}
$u_i \text{-} u_{i+1}$ indicates presence of syllabic break between units u_i and u_{i+1}
and using it, probability of S_i being the correct syllabification is approximated as the product of probabilities of having a break or no break at each possible place:

$$P(S_i) \approx \prod_{X_i \in Pairs(S_i)} P(X_i) \tag{2}$$

where,

$$P(X_i = \text{"}ab\text{"}) = \begin{cases} \frac{N_s(\text{"}ab\text{"})}{N_w(\text{"}ab\text{"})} & \text{if } N_w(\text{"}ab\text{"}) \neq 0 \\ 0.5 & \text{otherwise} \end{cases} \tag{3}$$

$$P(X_i = \text{"a-b"}) = \begin{cases} \frac{N_s(\text{"a-b"})}{N_w(\text{"}ab\text{"})} & \text{if } N_w(\text{"}ab\text{"}) \neq 0 \\ 0.5 & \text{otherwise} \end{cases} \tag{4}$$

[1] Assumption verified and corrected by a Gujarati linguist.

$N_s(X)$ = Number of syllabifications containing expression X from training corpus
$N_w(X)$ = Number of words containing expression X from training corpus
For example,

$$\text{P(શુભે-ચ્છા)} \approx \text{P(શુભે)} \times \text{P(ભે-ચ્)} \times \text{P(ચ્છા)}$$

Finally out of all the possible syllabifications which does not contain any illegal syllable[2], one with the maximum score is chosen.

3.2 Predicting First and Last Syllable

Some character sequences when occur at beginning or ending of a word are always spoken separately from the word and hence form first or last syllables of that word respectively. The idea behind this can be understood more in English context. For example, 'non', 'un', 'ex' are common prefixes[3] and 'ism', 'ist', 'less', 'ness' are common suffixes in English which when occur at beginning / ending of a word, always form first and last syllables of word respectively. A similar pattern is followed in Gujarati also. 'શ્રી', 'બિન' and 'લીન', 'રણ' are the examples of common prefixes and suffixes respectively in Gujarati.

Instead of feeding system with static prefixes and suffixes, an attempt was made to make supervised model that learns about the important prefix and suffix from the given training data, which when appear at beginning / ending of the word, must be the first and last syllable of the word respectively. The remaining portion can be syllabified using the previous approach to find maximum probable syllabification. To define how important a prefix or suffix is, 2 entities are defined.

Prefix Score (S_p): It is the probability of the sequence of letters being first syllable, given the word is starting with that particular sequence. For example for sequence $(u_1u_2..u_k)$:

$$S_p(u_1u_2...u_k) = \begin{cases} \frac{N_{fs}(u_1u_2...u_k)}{N_{ws}(u_1u_2...u_k)} & \text{if } N_{ws}(u_1u_2...u_k) \neq 0 \\ \frac{1}{k+1} & \text{otherwise} \end{cases} \quad (5)$$

$N_{fs}(x)$ is number of words with first syllable 'x' in training data
$N_{ws}(x)$ is number of words that start with expression 'x' in training data

Suffix Score (S_s): It is the probability of the sequence of letters being last syllable, given the word is ending with that particular sequence. For example for sequence $(u_1u_2..u_k)$:

---

[2] Illegal syllables are the character sequences which do not occur as a syllable in the training data.

[3] Any reference to prefix and suffix in this paper henceforth would refer to first and last syllable of the word respectively.

$$S_p(u_1u_2...u_k) = \begin{cases} \frac{N_{ls}(u_1u_2...u_k)}{N_{we}(u_1u_2...u_k)} & \text{if } N_{we}(u_1u_2...u_k) \neq 0 \\ \frac{1}{k+1} & \text{otherwise} \end{cases} \quad (6)$$

$N_{ls}(x)$ is number of words with last syllable 'x' in training data
$N_{we}(x)$ is number of words that end with expression 'x' in training data

First and last syllables of all words from training data were extracted and prefix and suffix score for them were precomputed respectively. For a sequence of characters if the prefix/suffix score is above a specified threshold, then it is called **confident prefix/suffix** under that threshold respectively.

Application of Precomputed Scores on the Word: For a given word it is checked whether character sequence of any length starting from first character serve as a confident-prefix or not. Similarly, an attempt is also made to find the confident suffix. When confident prefix/suffix is found in the word, it is taken as first or last syllable respectively and rest of the word is syllabified with previous approach. If multiple confident prefix / suffix are found, one with maximum score should be selected and if none are found, nothing is to be done. Also, if confident prefix marks first split such that the split occurs between consonant and *matra*, then such confident prefix is not chosen because it would contradict our original assumption. The same is followed for confident suffix. Figure 2 shows process of finding confident prefix in word 'બિ ન ઉ પ યો ગી'.

Prefix Score	score > threshold ?
P(બિ)	No
P(બિ ન)	**Yes**
P(બિ ન ઉ)	No
P(બિ ન ઉ પ)	No
P(બિ ન ઉ પ યો)	No
P(બિ ન ઉ પ યો ગી)	No

Fig. 2. Example of finding confident prefix(es) in a word

4 Evaluation

To define syllabic break, each character would be tagged as 'S' if it marks beginning of the syllable and 'F' otherwise. For example, for syllabification ('abc-d-efg'), sequential tags would be a(S), b(F), c(F), d(S), e(S), f(F), s(F). Evaluation of results has been done in 2 ways. Percentage of words syllabified entirely correctly **(W)**, and percentage of the sequential tags **(S)** detected correctly.

For each of the experiments, 10 fold cross-validation has been done using corpus of about 14 thousand words in Gujarati.

Table 2. Evaluation examples

Actual Tags	Predicted Tags	Compare	Word Accuracy (W)	Syllabic Accuracy (S)
ab-cd-ef (SFSFSF)	abc-d-ef (SFFSSF)	(SFSFSF) (SFFSSF) (✓✓✗✗✓✓)	0/1	4/6
ab-cd-ef (SFSFSF)	ab-cd-ef (SFSFSF)	(SFSFSF) (SFSFSF) (✓✓✓✓✓✓)	1/1	6/6

4.1 Results and Error Analysis

Once the training data was built, 10 fold cross validation on 14 thousand words was done using CRF based approach and the approach[4] mentioned in Sect. 3. For prefix-suffix approach maximum result was obtained at threshold (th = 0.95)

Table 3. Results

	CRF based appraoch	Maximum Probable Approach	Maximum Probable + Prefix/Suffix Approach (th = 0.95)
Word Accuracy **(W)**	89.56 %	88.98 %	91.89 %
Syllabic Accuracy **(S)**	97.58 %	97.36 %	98.02 %

A sample of 10 thousand words was taken to analyse prefix-suffix approach. When operated at threshold 0.95, in total of 1963 words confident prefix was detected out of which 1877 (95.6 %) were correct. Similarly, in total of 8602 words confident suffix was detected out of which 8043 (93.5 %) were correct. These values show the accuracies of prefix-suffix algorithm to detect first and last syllables correctly.

On 10 thousand sampled words, when prefix-suffix approach was applied as an add-on to maximum probable approach, some words turned from wrongly tagged to correctly tagged and vice-versa. Table 4 summarizes the results.

Credibility of Prefix-Suffix approach: To support the claim of prefix-suffix model improving the overall performance, a paired T-test was done. 15 groups, each of 20 random sample words were taken and the number of words tagged

[4] Maximum Probable approach refers to method described in Subsect. 3.1 and Prefix/Suffix approach refers to add-on method described in Subsect. 3.2.

Table 4. Change of correctness of predicted syllabification on application of Prefix-Suffix approach on Maximum probable approach (10 thousand words)

	Prefix detected	Suffix detected
wrong to right	86	342
right to wrong	8	7

correctly prior and later to the application of prefix-suffix model were noted. When this data was passed to one-tailed paired T-test, T-value of 1.709 and P-value of 0.054 were obtained. This shows with almost 95 % confidence that that this improvement is not by a mere chance or randomness.

Exceptions: The proposed system will fail to work when the training data is not enough to disambiguate between situations of keeping or not keeping a syllabic break at some position. This can be due to lack of enough training data or certain exceptions intrinsic to the language itself. For certain words, the assumption made in Sect. 3 does not hold. For example, syllabifications 'મોર-િ-ચો', 'મોર-િ-યો', 'ફ-ટક-ી-યુ', 'ફસ-િયો' include a syllabic break between consonant and following *matra* which is very unnatural in training data. Such words invariably fail to get syllabified correctly. However, the existence of such words are negligible and does not affect the overall performance adversely.

5 Conclusion and Future Scopes

Gujarati syllabification data has been bootstrapped from a smaller data-set using CRF model. The resulting data was verified and corrected by a linguist. This demonstrates the use of CRF for Gujarati syllabification. A new approach for syllabification is then tested on this data and compared with the CRF results. The proposed model works quite good at word and syllable level accuracies 91.89 % and 98.02 % respectively. These results are very much comparable with CRF results and hence is offered as its alternative approach for Gujarati syllabification which works on simple statistical calculations.

The assumption underlying this approach to syllabification is followed roughly by 99.34 % of 14 thousand words which shows its soundness. This assumption can also be extended to other Indian languages like Hindi, Bengali, Marathi etc. An active work for building of Hindi and Bengali corpus of syllabified words is going on and as a future work the same procedures can be tested and compared on these languages.

Acknowledgements. We are thankful to Dr. Nilotpala Gandhi (HOD in Department of Linguistics, Gujarat University) for her invaluable help in bootstrapping the training data by scrutinizing and rectifying the mistakes predicted by our model. We also acknowledge the occasional but really helpful discussions with Parth Mehta. Finally, the project could not have been possible without the contribution and initial thrust provided by Shubham Patel.

References

1. Bartlett, S., Kondrak, G., Cherry, C.: On the syllabification of phonemes. In: Proceedings of Human Language Technologies: The 2009 Annual Conference of the North American Chapter of the Association for Computational Linguistics, pp. 308–316. Association for Computational Linguistics (2009)
2. Dinu, L.P., Niculae, V., Sulea, O.-M.: Romanian syllabication using machine learning. In: Habernal, I., Matoušek, V. (eds.) TSD 2013. LNCS, vol. 8082, pp. 450–456. Springer, Heidelberg (2013)
3. Goslin, J., Frauenfelder, U.H.: A comparison of theoretical and human syllabification. Lang. Speech **44**(4), 409–436 (2001)
4. Hammond, M.: Parsing syllables: Modeling ot computationally. arXiv preprint cmp-lg/9710004 (1997)
5. Kahn, D.: Syllable-based generalizations in English phonology, vol. 156. Indiana University Linguistics Club Bloomington (1976)
6. Kiraz, G.A., Möbius, B.: Multilingual syllabification using weighted finite-state transducers. In: The Third ESCA/COCOSDA Workshop (ETRW) on Speech Synthesis (1998)
7. Lafferty, J., McCallum, A., Pereira, F.C.: Conditional random fields: Probabilistic models for segmenting and labeling sequence data (2001)
8. Mayer, T.: Toward a totally unsupervised, language-independent method for the syllabification of written texts. In: Proceedings of the 11th Meeting of the ACL Special Interest Group on Computational Morphology and Phonology, pp. 63–71. Association for Computational Linguistics (2010)
9. Palchowdhury, S., Majumder, P., Pal, D., Bandyopadhyay, A., Mitra, M.: Overview of FIRE 2011. In: Majumder, P., Mitra, M., Bhattacharyya, P., Subramaniam, L.V., Contractor, D., Rosso, P. (eds.) FIRE 2010 and 2011. LNCS, vol. 7536, pp. 1–12. Springer, Heidelberg (2013)
10. Rogova, K., Demuynck, K., Van Compernolle, D.: Automatic syllabification using segmental conditional random fields. Comput. Linguist. Neth. J. **3**, 34–48 (2013). http://www.clinjournal.org/node/37
11. Selkirk, E.O.: On the major class features and syllable theory (1984)
12. Suthar, B.: Gujarati- english learner's dictionary, 10 August 2015. http://ccat.sas.upenn.edu/plc/gujarati/guj-engdictionary.pdf
13. Kudo, T.: Crf++: Yet another crf toolkit. https://taku910.github.io/crfpp/
14. Trogkanis, N., Elkan, C.: Conditional random fields for word hyphenation. In: Proceedings of the 48th Annual Meeting of the Association for Computational Linguistics, pp. 366–374. ACL 2010, Association for Computational Linguistics, Stroudsburg, PA, USA (2010). http://dl.acm.org/citation.cfm?id=1858681.1858719

A Support Vector Machine Based System for Technical Question Classification

Shlok Kumar Mishra(✉), Pranav Kumar, and Sujan Kumar Saha

Department of Computer Science and Engineering,
Birla Institute of Technology Mesra, Ranchi 835215, India
shlokkumarmishra@gmail.com

Abstract. This paper presents our attempt on developing a question classification system for technical domain. Question classification system classifies a question into the type of answer it requires and therefore plays an important role in question answering. Although the task is quite popular in general domain, we were unable to find any question classification system that classifies the questions of a technical subject. We defined a technical domain question taxonomy containing six classes. We manually created a dataset containing 1086 questions. Then we identified a set of features suitable for the technical domain. We observed that the parse structure similarity plays an important role in this classification. To capture the parse tree similarity we employed the tree kernel and we proposed a level-wise matching approach. We have used these features and dataset in a support vector machine classifier to achieve 93.22 % accuracy.

Keywords: Question classification · Technical question classification · Support vector machine · Question answering

1 Introduction

Question answering systems are often treated as the next step of information retrieval. Question answering systems allow users to pose questions in natural language and aim to cater the information need. To serve the succinct answer, the question answering system should understand the expectation of the question. Question classification may play an important role here. The purpose of question classification is to assign a question to an appropriate category from the set of predefined categories which constitute the Question Taxonomy. Therefore question classification helps to reduce the search space by filtering out a wide range of candidate answers.

Due to immense importance of the problem, several research groups and researchers have attempted to solve this problem. Most of the previous work has been done in the field of generic domain (Breck et al. 1999; Ittycheriah et al. 2000; Hovy et al. 2002; Li and Roth 2002). Several taxonomies have been proposed for general domain. Among these the most widely

R. Prasath et al. (Eds.): MIKE 2015, LNAI 9468, pp. 640–649, 2015.
DOI: 10.1007/978-3-319-26832-3_60

accepted taxonomy has been given by Li and Roth (2002) who proposed a two layered taxonomy with 6 coarse classes and 50 fine classes.

However when we looked into the technical domain we were unable to find sufficient work. But due to recent popularity of technical question answering web forums, it has become exceedingly important to work in technical domain question classification.

In this paper we present our work on development of a question classification system in technical domain. As we were unable to find any taxonomy for technical questions in the literature, we first attempt to define a taxonomy. After studying various types of questions on different websites and question papers, we proposed a taxonomy applicable for this domain. The proposed taxonomy contains 6 classes, namely: define, describe, difference, enumerate, advantages and reason. Then we manually annotated a training data containing 1086 questions. We then found set of suitable features and used the support vector machine (SVM) classifier. During the SVM training we found that the parse structure similarity between two target questions plays an important role in classification. We tried two possible strategies for parse tree similarity. First strategy is based on the popular tree kernel. As the second approach we propose a level-wise similarity based approach. So, apart from the system development, the additional contribution of the paper is the comparison of both the approaches. In our experiments we observed that the proposed level-wise comparison based approach was superior to the tree kernel in the task.

2 Related Work

Here we present a brief overview on question classification tasks in the literature. Question classification task is quite popular and substantial effort has been devoted to develop such systems. Defining the question taxonomy is the basic step of question classification and several classification taxonomies have been defined in the literature. The question taxonomies vary in number of classes. For example, the taxonomy proposed by Breck et al. (1999) contains 17 classes, Ittycheriah et al. (2000) deiňĄned 31 types taxonomy in two level hierarchical structure, and the taxonomy of Hovy et al. (2002) contains 196 classes. The most widely accepted taxonomy is proposed by Li and Roth (2002), which is a two layered taxonomy with 6 coarse classes and 50 fine classes. The coarse classes are: abbreviation, entity, description, human, location and numeric. Lally et al. (2005) classified question in 12 categories in the Jeopardy! Domain. Hyo-jung-oh et al. (2005) presented a taxonomy with 10 classes. These 10 classes were mainly focused on descriptive type questions.

Once the taxonomy is defined, the next step is to build the classifier. The classification techniques can be broadly classified into two categories: rule based and machine learning based. The former one (e.g., Singhal et al. 1999; Hermjakob 2001) uses various types of rules using the words occurring in the question, parse information, semantic information etc. The machine leaning based approach became widely popular after the release of the UCIC dataset by

Li and Roth (2002) which contains 5,952 manually labeled questions using the aforementioned 56 class taxonomy. They have also developed the first machine learning system using the dataset which uses the SNoW learning architecture. Later, several machine learning algorithms have been used for the development of classification systems using the dataset. For example, Zhang and Lee (2003) used linear support vector machines (SVM) with all question n-grams as feature. Hacioglu and Ward (2003) used linear SVM with question bigrams and error-correcting codes. Suzuki et al. (2003) proposed the HDAG kernel for question classification using SVM. Huang et al. (2008) used combination of SVM and maximum entropy (MaxEnt) with question head words and their hypernyms as features to obtain 89.2 % of accuracy on fine classes of UCIC dataset. Hyo-jung-oh et al. (2005) used pattern extraction from pre-tagged corpus and extraction of Descriptive Indexing Unix using pattern matching techniques. They achieved F score of about 52 % on top1 and 68 % on top5.

The system of IBM Watson (Lally et al. 2005) detected question classes using a variety of techniques. The classes they considered were, definition, categorization, fill-in-the-blank, abbreviation, puzzle (answer require derivation, synthesis etc.), etymology, verb (question ask a verb), translation, number, bond (common between multiple entries), multiple-choice, and date. The recognizers are independent of one another; hence, more than one class can be potentially associated with any particular question. There are some pairwise incompatibilities, which were enforced by a consolidator process that runs after recognition and removes the less preferred of any mutually inconsistent class pair. They detected some classes such as Puzzle, Bond, Fill-in-the-blank with the help of regular expressions. Other classes such as Etymology, Translation etc. were identified by syntactic rules that match predicate-argument structure. Some other classes such as Number and Date were identified by lexical answer type (LAT). They tested their system on about 3500 questions and achieved f-score of 63.7. The accuracy of the definition class, which is quite frequent in the dataset, is poor - achieving a f-score of 49.7.

3 Technical Question Taxonomy

We have already discussed that several question taxonomies have been defined for the task of question classification in general domain but we were unable to find any taxonomy defined for technical questions. Hence we defined our own taxonomy for technical questions. For this task, we studied and analyzed a huge number of question papers belonging to various subjects of computer science. We tried to figure out what is expected as an answer to a given question. Based on that expectation, we classified the questions into six classes. These classes are defined below.

3.1 Question Taxonomy

A. Define: Consists of questions that ask for a definition or a brief explanation of an exact entity.

Example:- *What is meant by Dynamic Hashing?*
Explain the concept of transaction atomicity.
B. Enumerate/List: This class contains questions that expect several key-points about the subject matter or a list of entities.
Example:- *What are different types of end users?*
List the applications of wrapper class in Java.
C. Description: Comprises of those questions which need descriptive answers, explanations or discussions regarding a topic.
Example:- *Discuss how generic units are implemented.*
Explain the layered network architecture.
D. Difference: Comparison between two entities possibly in a tabulated manner.
Example:- *Explain the difference between Logical and physical address.*
Differentiate between a hub and a switch.
E. Reason/Cause/Manner: Answers contain the cause for a phenomenon or the way a particular phenomenon occurs. Mostly these are how and why type questions.
Example:- *Give two reasons why caches are useful.*
Why are page sizes always powers of 2?
F. Advantage/Disadvantage: This class has questions which ask for advantage/disadvantage or both related to an entity. Also advantages of an entity over the other or disadvantages of one against some other entity.
Example:- *Justify the advantages that image processing boards have.*
What are the advantages of binary search.

All possible technical domain questions may not be covered by this taxonomy. A few more classes are required to be incorporated in the taxonomy to cover all possible questions. Here we would like to mention that our focus remains on the development of the classifier instead of proposing a complete standard taxonomy for technical questions. Therefore, the classes, for which sufficient data is not available, are not considered in the taxonomy.

3.2 Corpus Creation

As already mentioned that openly available training dataset is not available for technical domain. Therefore we had to prepare a question classification dataset. To do this we considered various subjects of computer science and engineering and collected questions from several universities. We found that several questions are ambiguous and may belong to multiple classes. But our classifier forcibly assigns a single class-label to each question. Therefore annotator agreement is essential during the data preparation. To achieve this we employed 3 human annotators to label the questions into their respective class. We accepted only those questions which were labelled into same class by all the annotators. All other questions were rejected. The final dataset contains 1086 questions. In Table 1 we present the class-wise distribution of the dataset. Here we would like to mention that in this task only the single line questions have been considered. Multi-line questions, numerical questions, questions containing graph or figure have not been selected in the training data.

Table 1. Classification dataset

Class	Total	Train	Test
Define	178	135	43
Enumerate	192	152	40
Description	233	180	53
Difference	203	163	40
Reason	185	145	40
Advantage	95	75	20
Total	1086	850	236

4 Our Approach for Classification

We have used support vector machine (Cortes and Vapnik 1995) to build the classifier. SVM has proven very successful in text classification. Also in the question classification task SVM has been found to be quite popular. Several high performance systems on UCIC data use SVM.

4.1 Support Vector Machine

SVM is a supervised learning classifier. SVM aims to find a hyperplane (or, decision surface) that separates the positive and negative examples by maximizing the margin. The margin is defined as the distance between the hyperplane and nearest positive and negative samples. These samples are termed as the support vectors. By default SVM is a binary classifier. But it can be utilized for multi-class classification by using one-vs-all like methods. For the task we use openly available LIBSVM toolkit (Chang and Lin 2011).

4.2 Feature Set

Performance of SVM is largely dependent on the quality of the features chosen for learning. In the literature, a number of features can be found that have been used commonly for the question classification task. Initially we also used such common features that are mentioned below.

Unigram: Words occurring in the questions play a major role in classification. Most of the works in general domain question classification use all the words (bag-of-words) in the dataset as feature. Similarly, we also used the unique words in the dataset as feature.

Word n-gram: In addition to the unigram features, we have also used word bigrams and trigrams as features. In this technical domain ngrams may provide valuable information as many terms and concepts are represented as ngram but the individual words in the n-grams are not having much importance. We use all n-grams and/or frequent n-grams as feature.

Wh-words and Question Words: The wh word in the question provides an important clue. For example, 'where' deals with location class, 'who' implies the person class etc. But when we focused on the technical domain, we observed that here the role of the wh-words is restricted. Use of wh-words is less in questions compared to general domain and highly frequent who and where (in general domain) are very rare in technical domain. Rather, another set of question words are dominant. For example: define, explain, discuss, write, differentiate etc. We make a list of such question terms and use the list as feature.

Term Excluded Unigram: We observed that the unigram feature set was very high dimensional as the dataset contained a large number of unique words. And use of a large feature set in a small dataset often creates overfitting and accuracy degrades. In order to reduce the dimensionality we planed to identify the technical terms and remove the terms from the feature set. For identifying the technical terms we use the method discussed in the Sect. 4.3.

Parse Structure: We observed that a high amount of structural similarity exists among the questions belonging to a particular class in the question taxonomy. That is, a particular class possesses a set of common parse trees that cover majority of the questions. Therefore parse tree information may play an important role in the classification. To capture the parse tree similarity we have used two approaches: one is the popular tree kernel based approach and other one is the proposed level wise matching based approach. These approaches are discussed in the following subsections.

4.3 Term Identification and Question Normalization

We observed that the number of unique feature values became too large because of the presence of wide variety of technical terms in the questions. Almost every question in the dataset contained one or more technical term and majority of these terms contained multiple tokens. These term tokens do not have considerable contribution in classification but these lead to increase the dimensionality of the feature space in a high ratio. As the size of the training data is limited, a very high-dimensional feature space fails to achieve expected performance. Therefore we had to normalize the questions by identifying the technical terms and replacing these by a term variable.

Identification of technical terms from the questions is not easy due to the complex and ambiguous nature of terms. Term identification itself is a research task. As in the article we focus on question classification, we do not want to devote much time in term identification. Simple approach of term identification can be a list based approach, but creation of a list that contains all possible technical terms is very difficult. And the machine learning based approach requires training data, creation of which is costly and time-consuming.

To identify the technical terms from the questions we defined a set of parse-tree based rules. We used the Stanford Parser (Socher et al. 2013) to generate the parse tree of the questions and by analyzing the parse trees we manually identified the rules. For example, *the noun phrase having depth greater than half of the parse-tree depth is a term.* Another rule is, *if the head-noun of the noun phrase is plural then we use a special plural indicator for the term.*

4.4 Parse Structure Matching with Tree Kernel

Tree Kernel measures the similarity of two trees by counting the number of their common tree fragments. Tree fragment is defined as any subtree that includes more than one node, with the restriction of entire rule production must be included. It could be defined as

$$k(T1, T2) = \sum_{n_1 \in N_1} \sum_{n_2 \in N_2} C(n_1, n_2)$$

where N1 and N2 are sets of nodes in two syntactic trees T1 and T2, and C(n_1,n_2) equals to the number of common fragments rooted in nodes n1 and n2. However, to enumerate all possible tree fragments is an intractable problem. The tree fragments are thus implicitly represented, and with dynamic programming, the value of C(n_1,n_2) can be efficiently computed as follows:

$$C(n_1, n_2) = \begin{cases} 0 \text{ if } n_1 \neq n_2 \\ 1 \text{ if } n_1 = n_2 \text{ on terminal nodes} \\ \lambda, \text{if } n_1 \neq n_2 \text{ on pretermial nodes} \\ \prod_{j=1}^{nc(n_1)} [1 + C(ch(n_1, j), ch(n_2, j))] \end{cases}$$

One of the variation of tree kernel is the Syntactic Tree Kernel (STK). STK satisfies the constraint remove 0 or all children at a time. Another type of tree kernel is Partial Tree Kernel (PTK). PTK is combination of STK and string kernel. PTK uses Syntactic Tree Kernel and String Kernel with weighted gaps on nodes children. Smoothed Partial Tree Kernels uses same idea as the Syntactic Semantic Tree Kernel but the similarity is extended to any node of the tree. The tree fragments are those generated by PTK. Basically it extends PTK with similarities.

4.5 Syntactic Structure with Level-Wise Matching Approach

By analyzing the similarity scores, we realised that the tree kernel may not be a good tool to measure the similarity we are trying to capture. Because, the tree kernel tries to match similarity between all the possible subtrees, and lower level subtrees often do not play any major role in classification. Hence we propose a new approach for capturing Syntactic Tree Structure.

In this approach we first prune the leaf nodes. Now the parse tree contains only the internal tags, that is the parts-of-speech labels and the phrase labels. Further the two target trees are compared level-wise. We count the number of matches between two parse trees in a particular level and sum up all the match-values and finally normalize the values.

Next we observed that the parse structure sometimes changes due to the occurrence of the terms. A term can be a single word or can be long containing several tokens. Long terms sometimes lead to make the parse tree erroneous. To handle this, first we replaced the technical terms by a single token term variable "TERM" and then generate the parse tree. Then the parse trees are matched using the aforementioned approach.

Performance of the feature is largely dependent on the accuracy of the parser. Stanford parser is very popular but surprisingly we observed that the parser is not much accurate to parse these technical domain questions. It often fails to generate accurate parse tree for a large number of questions. For example, "State the 8-queens problem", "Explain Cook's theorem" etc. resulted erroneous parse trees. By analyzing these erroneous parse trees, we observed that the major reason behind the error is due to the question terms. So in order to generate correct parse tree, we replaced such question terms by another question term (that generates correct parse tree) from same category. For example, 'state' is replaced by 'define'.

5 Result and Discussion

In order to find the best feature set for the task we ran a set of experiments using the candidate features (presented in Sect. 4) individually or in combination. Table 2 summarizes the results.

Table 2. Accuracy in different features

Feature set	Accuracy (%)
F1: Unigram	83.05
F2: Unigram + n-gram (all)	82.62
F2 with selected n-gram, question term	85.17
F4: F3 with technical term and question term replacement	88.56
F5: Parse Tree (of exact questions)	80.51
F6: F3 + Parse Tree	77.97
F7: Parse Tree with normalization and question term replacement	86.44
F8: F4 + F7	93.22
F9: Tree Kernel	83.41

From the table we can observe that the highest accuracy we achieve in the question classification system is 93.22 % (F8). This accuracy is achieved when we have used bag-of-word, selected n-grams, question terms, technical terms are identified and replaced by term variable, and parse trees of the normalized questions. Parse tree based features have shown significant effectiveness in this domain. Only parse tree features (F5) are able to achieve an accuracy of 80.51 %.

The leaf level of the parse trees are not included in the features, that means individual words are not included in the feature set. When we used the parse trees of the normalized questions, the accuracy has reached to 86.44 %. But surprisingly on using all unigrams and n-grams and the original parse trees of the questions, the accuracy has considerably degraded (F6: 77.97 %). This might be due to the fact that the size of the training data is not sufficient and we have used a large number of features. Furthermore, some of the classes are ambiguous, larger training data is required to capture them properly. For example, the description class itself achieves high accuracy but has a tendency to attract other classes, like enumerate and difference, to fall in it. After testing our system using various features we tried to find accuracy using tree kernel. However the accuracy achieved by the tree kernel was significantly less as compared to our system. One of the possible reasons for lower accuracy of tree kernel is maybe due the fact that tree kernel tries to match similarity between all the possible subtrees, while in our task similarity between all the parse trees is not needed, only similarity between full parse tree is needed.

6 Conclusion

In this paper we have presented our attempt on developing a question classification system in technical domain. For the task we defined a question taxonomy, created a dataset and identified a set of suitable features. In our experiments we have achieved significant accuracy. However, the system opens several directions to work on in future, to make the system more accurate and robust. In this system we have considered only single line questions. But in technical domain most of the questions are multi-line, contain numerical data and figures. Also a better taxonomy can be defined. The dataset is also required to be enlarged to make it robust and more accurate.

References

Breck, E., Burger, J., Ferro, L., House, D., Light, M., Mani, I.: Publications manual. In: Proceedings of The Eighth Text Retrieval Conference (TREC8) (1999)

Chang, C.C., Lin, C.J.: LIBSVM 2011. ACM Trans. Intell. Syst. Technol. **2**(27), 1–27 (2011)

Cortes, C., Vapnik, V.N.: Support vector networks. Mach. Learn. **20**, 273–297 (1995). 1981

Hovy, E., Hermjakob, U., Ravichandran, D.: A question/answer typology with surface text patterns. In: Proceedings of the Human Language Technology (HLT) Conference (2002)

Huang, Z., Thint, M., Qin, Z.: Question classification using head words and their hypernyms. In: Proceedings of the Conference on EMNLP, pp. 927–936 (2008)

Hacioglu, K., Ward, W.: Question classification with support vector machines and error correcting codes. In: Proceedings of HLT-NACCL 2003 (2003)

Hermjakob, U.: Parsing and question classification for question answering. In: ACL 2001 Workshop on Open-Domain Question Answering (2001)

Suzuki, J., Taira, H., Sasaki, Y., Maeda, E.: Question classification using HDAG kernel. In: ACL 2003 Workshop on Multilingual Summarization and Question Answering (2003)

Ittycheriah, A., Franz, M., Zhu, W.-J., Ratnaparkhi, A.: IBM's statistical question answering system. In: The Ninth Text REtrieval Conference (TREC9) (2000)

Li, X., Roth, D.: Question classification using head words and their hypernyms. In: Proceedings of the Conference on EMNLP, pp. 927–936 (2002)

Socher, R., Bauer, J., Manning, C.D., Ng, A.Y.: Parsing with compositional vector grammars. In: Proceedings of ACL 2013 (2013)

Singhal, A., Abney, S., Bacchiani, M., Collins, M., Hindle, D., Pereira, F.: QAT & T at TREC-8. In: The Eighth Text REtrieval Conference (TREC8) (1999)

Zhang, D., Lee, W.S.: Question classification using support vector machines. In: Proceedings of the ACM SIGIR Conference on Research and Development in Information Retrieval, pp. 26–32 (2003). 1999

Oh, H.J., Lee, C.H., Kim, H.J., Jang, M.G.: 2005 Descriptive question answering in encyclopedia. In: Association for Computational Machinery (2005)

Lally, A., Prager, J.M., McCord, M.C., Boguraev, B.K., Patwardhan, S., Fan, J., Fodor, P., Chu-Carroll, J.: Question analysis: how watson reads a clue (2005)

Moschitti, A.: State-of-the-art tree kernels in natural language processing. In: ACL (2012)

Shared Task on Sentiment Analysis in Indian Languages (SAIL) Tweets - An Overview

Braja Gopal Patra[1], Dipankar Das[1], Amitava Das[2], and Rajendra Prasath[3(✉)]

[1] Department of Computer Science and Engineering,
Jadavpur University, Kolkata India
{brajagopal.cse,dipankar.dipnil2005}@gmail.com
[2] Department of Computer Science and Engineering,
IIIT, Sri City, Chittoor India
amitava.santu@gmail.com
[3] Department of Computer and Information Science,
Norwegian University of Science and Technology, NO-7491 Trondheim, Norway
drrprasath@gmail.com

Abstract. Sentiment Analysis in Twitter has been considered as a vital task for a decade from various academic and commercial perspectives. Several works have been performed on Twitter sentiment analysis or opinion mining for English in contrast to the Indian languages. Here, we summarize the objectives and evaluation of the sentiment analysis task in tweets for three Indian languages namely Bengali, Hindi and Tamil. This is the first attempt to sentiment analysis task in the context of Indian language tweets. The main objective of this task was to classify the tweets into positive, negative, and neutral polarity. For training and testing purpose, the tweets from each language were provided. Each of the participating teams was asked to submit two systems, constrained and unconstrained systems for each of the languages. We ranked the systems based on the accuracy of the systems. Total of six teams submitted the results and the maximum accuracy achieved for Bengali, Hindi, and Tamil are 43.2 %, 55.67 %, and 39.28 % respectively.

Keywords: Sentiment analysis · Tweets · Bengali · Hindi · Tamil

1 Introduction

Sentiment Analysis or Opinion Mining from electronic texts is a hard semantic disambiguation problem [1]. Sentiment analysis refers to the process of identifying the subjective responses or opinions about a specific topic. It is observed that sentiment analysis has become a main stream research during the last two decades with an immense possibility from the perspectives of both industry and academia [5]. Sentiment analysis task has been performed on English [6–8] as well as on Indian languages [3, 4] for the plain texts.

A part of this work was done at University College Cork, National University of Ireland, Cork, Ireland.

R. Prasath et al. (Eds.): MIKE 2015, LNAI 9468, pp. 650–655, 2015.
DOI: 10.1007/978-3-319-26832-3_61

On the other hand, the evolution of social media texts such as blogs, micro-blogs (e.g., Twitter), and chats (e.g., Facebook messages) has created not only many new opportunities for information access and language technology, but also many new challenges, making it one of the prime present-day research areas[1]. In case of social texts, the presence of misspellings, poor grammatical structure, emoticons, acronyms, and slang are very common and thus, making the task of sentiment analysis from these texts more difficult. Sentiment analysis becomes more challenging in case of the Twitter text when people try to project their sentiment using only 140 characters. Indeed sentiment analysis on social media text is a hot research discipline in present days, but most of the efforts so far have been made on English. Tasks like sentiment analysis in tweets [8], classifying figurative tweets [6], strength of the sentiment in figurative tweets [7] have been performed in English.

The shared task: *Sentiment Analysis in Indian Language tweets (SAIL-2015)* patronizes the Indian researchers to work on automatic sentiment analysis for their own languages by providing them relevant data. Prime motivation of the *SAIL-2015* is to gather researchers, experts and practitioners together to discuss, collaborate and instigate the sentiment analysis research particularly for Indian languages, which involves resource creation, sharing and future collaboration. In this task, we, the organizers provide tweets for three Indian languages namely Bengali, Hindi, and Tamil annotated with positive, negative, and neural polarity. The main objective of this task is to classify the tweets into positive, negative, and neutral categories.

In the remainder of this paper, we described the sentiment analysis task and the process of creating training and test data for three languages in Sect. 2. In Sect. 3, we presented the results of the participating systems and also analyzed their contributing features and results. Finally, we concluded our study with future steps and work in Sect. 4.

2 Task Description and Data Preparation

2.1 Task Description

Here, we describe the shared task of *SAIL-2015*. Given a tweet, the participants are asked to determine whether it expresses a positive or a negative or a neutral sentiment. If any tweet expresses both positive and negative sentiment, then the stronger one should be chosen. We asked the participants to submit two systems for each of the languages namely constrained and unconstrained systems. In case of the constrained system, the participants are only allowed to use corpus supplied by the organizers and at most the *SentiWordNet* for Indian languages by Das and Bandyopadhyay [1]. No external resource is allowed to develop or be used for the constrained systems. In contrast, the unconstrained system, the participants were allowed to use any external resource (POS tagger, NER, Parser, and additional data) to train their system and they have to mention those resources explicitly in their task reports, accordingly.

[1] http://amitavadas.com/SAIL/index.html.

2.2 Dataset

We collected the training and test datasets from Twitter over a period of three months. It is difficult to search tweets, specifically in Bengali or Hindi or Tamil. Therefore, we followed an interesting approach to collect the training and test data. First, we collected the monolingual corpus for each of the languages manually on different topics. Then, we removed the stop words and prepared a word frequency list. We searched each of the words exists in the frequency list in Twitter and collected the maximum of 2000 tweets for each of the words. We used the TWITTER4J[2], a Java implementation of Twitter API to download the tweets. The duplicate tweets were removed and the statistics of the training and test dataset are given in Table 1. We also counted the number of smiles or emoticons in each of the classes after normalizing them. For example, we normalized the happy smiley having multiple brackets, i.e. we converted ':)))))))))' to :). The statistics of the smiley for each of the classes separately is given in Table 2. We can observe that the usage of smiley is more in case of Tamil compared to Bengali and Hindi.

The undergraduate students annotated these data voluntarily. The annotators are the native speakers of the above mentioned languages. We employed 12, 4 and 2 annotators for annotating the Bengali, Hindi, and Tamil language tweets, respectively. Examples from each of the languages are given in Fig. 1.

Table 1. Data statistics

		Positive	**Negative**	**Neutral**	**Total**
Bengali	Training	277	354	368	999
	Test	213	151	135	499
Hindi	Training	168	559	494	1221
	Test	166	251	50	467
Tamil	Training	387	316	400	1103
	Test	208	158	194	560

Table 2. Smiley count for each class

	Training			**Test**		
	Positive	**Negative**	**Neutral**	**Positive**	**Negative**	**Neutral**
	+ve/− ve	**+ve/− ve**	**+ve/− ve**	**+ve/− ve**	**+ve/− ve**	**+ve/− ve**
Bengali	21/0	10/15	14/4	18/9	10/15	10/10
Hindi	16/2	12/4	18/2	15/0	5/15	0/0
Tamil	24/3	10/7	28/6	22/8	15/9	12/0

[2] http://twitter4j.org/en/index.html.

3 Results and Discussion

3.1 Results

Initially, 21 teams from different institutes all over India have registered for the task and finally, six teams succeeded to submit the results. We calculated the accuracy of the positive, negative, neutral tweets individually as well as total with respect to different teams for all the submitted systems. The team IDs, individual results obtained for each polarity and the total accuracy are shown in Table 3.

Table 3. Results of each team as per evaluation criteria (B: Bengali, H: Hindi, T: Tamil, C: *Constrained*, U: *Unconstrained*)

Team names	Positive	Negative	Neutral	Total accuracy
JUTeam_KS (B_C)	23.94	60.26	47.41	41.2
JUTeam_KS (B_U)	21.13	63.58	45.18	40.40
JUTeam_KS (H_C)	2.41	88.45	22.0	50.75
JUTeam_KS (H_U)	3.61	82.87	28.00	48.82
IIT-TUDA (B_C)	23.47	59.60	56.30	**43.2**
IIT-TUDA (B_U)	24.88	54.30	55.56	42.0
IITTUDA (H_C)	9.04	73.70	64.0	49.68
IITTUDA (H_U)	4.22	69.72	68.0	46.25
ISMD (H_C)	4.22	58.17	72.0	40.47
ISMD (H_U)	1.81	42.63	72.0	31.26
AmritaCENNLP (B_C)	27.23	65.56	0.0	31.4
AmritaCENNLP (H_C)	36.14	64.94	2.0	47.96
AmritaCENNLP (T_C)	57.77	40.51	0.52	32.32
AMRITA-CEN (B_C)	29.58	34.44	39.26	33.6
AMRITA-CEN (H_C)	45.79	57.37	80.0	**55.67**
AMRITA-CEN (T_C)	29.81	26.58	59.79	**39.28**
AMRITA (H_C)	17.47	54.59	68.0	42.83

Bengali: দেশে এখনো আড়াই কোটি লোক নিরক্ষর: প্রাথমিক ও গণশিক্ষামন্ত্রী (*Negative*)
Transliteration: Deshe ekhono adai koti lok nirakhor: prathamik o ganasikhamontri
Translation: Till date, two and half crores of people are illiterate in the Nation : Primary and Mass Education Minister

Hindi: भारत के पूर्व राष्ट्रपति एपीजे अब्दुल कलाम को चीन की पेकिंग यूनिवर्सिटी में पढ़ाने का न्यौता मिला है। (*Positive*)
Transliteration: Bharat ke purb rastrapati APJ Abdul Kalam ko China ki Peking University main padane ka niyota mila hai |
Translation: The Former President of India APJ Abdul Kalam got an invitation for teaching at Peking University, China.

Tamil: கடவுள் என்று தனியாக ஒருவர் இல்லை...!! உன் கைகாசில் ஒரு குழந்தைக்குசாக்லேட் வாங்கிக்குடுத்துவிட்டு அந்த குழந்தையின் முகத்தைபார் !! (*Positive*)
Transliteration: kadavul endru thaniyaga oruvar illai ... !! un kaikasil oru kizhandaikku chaaklet vanghikkuduththuvittu andha kuzhandaiyin mugaththaippaar!!
Translation: There is no one like GOD ... !! Offer a chocolate to a kid by spending your own money and then look at the face of the kid !!

Fig. 1. Examples of Indian language tweets

3.2 Discussion

Total of four teams have submitted the results for Bengali. It is observed that the maximum accuracy achieved for the Bengali language is 43.2 % by the team *IIT-TUDA*. For Hindi, total six teams have submitted the results and among them *AMRITA-CEN* achieved the maximum accuracy of 55.67 %. Only two teams have submitted the results for Tamil language and *AMRITA-CEN* is achieved the maximum accuracy of 39.28 %. Most of the teams used the *SentiWordNet* that was developed in [1] for the constrained system. Teams have used the features like *hash tags*, *re-tweet*, *TF-IDF scores of n-grams*, *links*, *question marks*, *exclamatory marks*, *smiley lists* and *SentiWordNet* for the sentiment analysis task.

The teams have used several well-known supervised classification algorithms like Decision Tree, Naïve Bayes, Multinomial Naïve Bayes and Support Vector Machines (SVMs). We observed that, the accuracies of the unconstrained systems are less compared to the constrained systems. The main reason may be the unavailability of basic NLP tools like POS taggers and NER for Indian language tweets, specifically.

The accuracies of the systems for the Indian language tweets are less as compared to the systems for English tweets as mentioned in [8]. The reason may be the scarcity of the sentiment lexicons for Indian languages tweets. However, a good number of sentiment lexicons available for the Hindi, Bengali and Tamil, but these are collected from the plain texts and not specialized for tweets. Because, in case of tweets, there are many spelling variations, acronyms and emoticons that make the sentiment analysis task more difficult and challenging as compared to other tasks on traditional sentiment analysis. It is also difficult to collect the monolingual Indian language tweets as most of the cases the tweets are written using English alphabets and sometime tweets are code mixed. The annotation of such monolingual tweets based on sentiment expressions requires the involvement of manpower and time.

4 Conclusion and Future Work

We described the first shared task organized for the sentiment analysis of Indian languages for Twitter data. This time, 21 teams have registered for the task and six teams successfully submitted their results using different features and machine learning algorithms. In future, we will try to draw the attentions from more number of teams to participate in this task. We hope this shared task will facilitate more research on the sentiment analysis for Indian language tweets by focusing different research challenges associated with it.

We are planning for a new edition of sentiment analysis task in Indian Languages tweets in the coming year, where focus will be on the Code-Mixed tweets. We are also planning to prepare data for classifying the figurative tweets for Indian Languages.

References

1. Das, A., Bandyopadhyay, S.: SentiWordNet for Indian languages. In: Proceedings of the 8th Workshop on Asian Language Resources (COLING 2010), Beijing, China, pp. 56–63 (2010)
2. Das, A., Bandyopadhyay, S.: Dr. sentiment knows everything! In: Proceedings of the 49th Annual Meeting of the Association for Computational Linguistics: Human Language Technologies (ACL/HLT 2011 Demo Session), Portland, Oregon, USA, pp. 50–55 (2011)
3. Das, D., Bandyopadhyay, S.: Word to Sentence Level Emotion Tagging for Bengali Blogs. In: Proceedings of the ACL IJCNLP-2009, Suntec, Singapore, pp. 149–152 (2009)
4. Das, D., Bandyopadhyay, S.: Labeling emotion in Bengali blog corpus – a fine grained tagging at sentence level. In: Proceedings of the 8th Workshop on Asian Language Resources (COLING 2010), Beijing, China, pp. 47–55 (2010)
5. Patra, B.G., Takamura, H., Das, D., Okumura, M., Bandyopadhyay, S.: Construction of emotional lexicon using potts model. In: Proceedings of the International Joint Conference on Natural Language Processing (IJCNLP 2013), Japan, pp. 674–679 (2013)
6. Reyes, A., Rosso, P.: On the difficulty of automatically detecting irony: beyond a simple case of negation. Knowl. Inf. Syst. **40**(3), 595–614 (2014)
7. Ghosh, A., Li, G., Veale, T., Rosso, P., Shutova, E., Barnden, J., Reyes, A.: Semeval-2015 task 11: sentiment analysis of figurative language in Twitter. In: Proceedings of the 9th International Workshop on Semantic Evaluation (SemEval 2015), Co-located with NAACL, Denver, Colorado, pp. 470–478 (2015)
8. Rosenthal, S., Nakov, P., Kiritchenko, S., Mohammad, S., Ritter, A., Stoyanov, V.: SemEval-2015 task 10: sentiment analysis in Twitter. In: Proceedings of the 9th International Workshop on Semantic Evaluation (SemEval 2015), Co-located with NAACL, Denver, Colorado, pp. 451–463 (2015)

Sentiment Classification: An Approach for Indian Language Tweets Using Decision Tree

Sudha Shanker Prasad, Jitendra Kumar, Dinesh Kumar Prabhakar(✉), and Sukomal Pal

Department of Computer Science and Engineering, Indian School of Mines, Dhanbad, India
{sudha.shanker,jitendrakumar}@cse.ism.ac.in, {dinesh.nitr,sukomalpal}@gmail.com
http://www.ismdhanbad.ac.in/computer-science-engineering

Abstract. This paper describes the system we used for Shared Task on Sentiment Analysis in Indian Languages (SAIL) Tweets, at MIKE-2015. Twitter is one of the most popular platform which allows users to share their opinion in the form of tweets. Since it restricts the users with 140 characters, the tweets are actually very short to carry opinions and sentiments to analyze. We take the help of a twitter training dataset in Indian Language (Hindi) and apply data mining approaches for analyzing the sentiments. We used a state-of-the-art Data Mining tool Weka to automatically classify the sentiment of Hindi tweets into positive, negative or neutral.

Keywords: Sentiment analysis · Polarity identification · NLP · Machine learning

1 Introduction

Sentiment Analysis is one of the major areas of focus in Natural Language Processing (NLP). It deals with opinion mining to determine the sentiment regarding the various topics/subjects in discussion. The basic task is classification of piece of text stating an opinion on an issue with one of the two opposing sentiments (Thumbs up/positive or Thumbs down/negative).

Earlier sentiment detection was adopted to analyze the sentiments of long textual contents such as letters, emails, reviews etc. With the proliferation of Internet and Web 2.0, application such as micro-blogging websites (Facebook, Twitter, Whatsapp etc.), forums and social networking, research in this field has gained momentum. Now-a-days people frequently use social media to engage, discuss, criticize and to express their views on a variety of subjects. This results in large amount of user generated data which carries a lot of information on different topics/subjects.

R. Prasath et al. (Eds.): MIKE 2015, LNAI 9468, pp. 656–663, 2015.
DOI: 10.1007/978-3-319-26832-3_62

Das and Chen [5] and Tong [15], referred the term "sentiment" as automatic analysis of evaluative text and tracking of predictive judgments. Work in this area originated early from its applications in blogs [1] and product/movie reviews [16]. As Internet is penetrating into different areas of life, traditional ways are changing the paradigm to online versions of the same. Microblogging is such a field where people actively share their sentiments and views over a variety of topics. Usually they have there own method of expressing their sentiments over a particular topic.

Sentiment detection plays an important role in extracting the sentiments hidden inside the text. Sentiment classification aims to determine and categorize the view of a writer into one of pre-defined classes of sentiments. It is useful for the customers who are trying to know about a product or service they are planning to buy, or marketers researching on public opinion of a product or the company etc.

In this paper we describe our participation to the Shared Task on Sentiment Analysis in Indian Languages (SAIL) Tweets, at MIKE-2015. The task concerned with detection of sentiments in a set of Hindi tweets and then classifying them into negative, positive or neutral tweets based on their intrinsic sentiments. In order to accomplish the task, we used C4.5 Decision Tree algorithm. J48 is an open source Java implementation of the C4.5 algorithm in Weka data mining tool.

Our paper is organized as follows. Section 2 focuses some of the works already done in this area. In Sect. 3, we discuss our approach to the problem. Section 4 describes the system description and implementation. Section 5 reports our classification results. Finally, Sects. 6 and 7 deal with discussion and conclusion respectively.

2 Related Work

Plenty of work has been done in the area of sentiment analysis for tweets and blogs that includes early work in this area by Turney [16] and Pang [9] for detecting the polarity of product reviews. Classification of document can be done in a multi-way scale based on polarity, which was attempted by Pang [8] and Synder [14] and focuses on classifying a movie review either as positive or negative review based on star rating, while Synder analysis on restaurant reviews on various aspects of a restaurant like food and ambiance.

Aspect/Feature based sentiment analysis model tries to determine the sentiments based on features and aspect of entities. Several subtask e.g. identification of entities, extraction of feature and opinion about the feature either as positive, negative or neutral. More discussion about can be found in Liu's NLP Handbook chapter "Sentiment Analysis and Subjectivity" [7].

Since the growing popularity of Twitter, researchers have focused their work in this area within last few years. They have applied the existing methodologies of sentiment analysis to study the sentiments of tweets. The above task is greatly supported by machine learning methods in natural language processing and information retrieval. Lin and Kolcz [13] used logistic regression classifiers learned from hashed byte 4-grams as features. The feature extractor considers the tweet as a raw byte array. Rodrgues et al. (2013) and Clark et al. [13] proposed the use of classifiers ensembles at the expression-level, which is related to Contextual Polarity Disambiguation.

WordNet based, dictionary based, corpus based or generative approaches for developing SentiWordNet(s) for Indian languages: Bengali, Hindi and Telugu is done by Das and Bandyopadhyay [2]. Emotion analysis on blog texts has been carried out in Bengali language for word to sentence level emotion tagging by Das and Bandyopadhyay [3,4].

In this perspective, the sentiment label (positive, negative, or neutral) is applied to a specific phrase or word within the tweet and does not necessarily match the sentiment of the entire tweet.

3 Our Approach

We have considered polarity identification as classification problem and applied Decision Tree (DT) which is a predictive model. It maps the observations of an item to its conclusion about the item target value.

A Decision tree summarizes the information from the training set and predicts correct classification on test data.

3.1 C4.5 Decision Tree Algorithm

The C4.5 is one of the DT algorithm developed by Quinlan [11]. It is an extension of ID3 algorithm and uses the Shannon Entropy. Here, the Shannon Entropy has been used to find information gain in C4.5 algorithm. The calculated Information Gain ratio contained by the data, helps to make Decision Tree and to predict the targets.

C4.5 is a classification algorithm based on decision tree approach uses the information gain ratio that is evaluated by entropy [12]. The test feature i.e. the feature selection at each node in the tree is selected using information gain ratio. The attribute with the highest information gain ratio is chosen as the test feature for the current node. Let D be a set consisting of $(D_1, D_2, ...)$ data instances. Suppose the class label attribute has m distinct values defining m distinct classes, C_i (for i= 1,..., m). Let $|D_j|$ be the number of sample of D in class C_i. The expected information needed to classify a given sample is:

$$Splitinfo_A(D) = -\sum\left(\left(\frac{|D_j|}{|D|}\right) * \log\left(\frac{|D_j|}{|D|}\right)\right) \tag{1}$$

$$Gain\ ratio(A) = \frac{Gain(A)}{Splitinfo_A(D)} \tag{2}$$

where,

$$Gain(A) = Info(D) - Info_A(D) \tag{3}$$

$$Info(D) = -\sum P_i * log_2(P_i) \tag{4}$$

and

$$Info_A(D) = -\sum(\frac{|D_j|}{|D|}) * Info(D_j) \quad (5)$$

where,

P_i = probability of distinct class C_i, D =data Set, A=sub-attribute from attribute, $(\frac{|D_j|}{|D|})$=act as weight of jth partition. In other words, Gain(A) is the expected reduction in entropy caused by knowing the value of feature A.

4 System Description

4.1 Dataset and Tool Used

The corpus available for training contains 1162 tweets with different annotations namely Positive, Negative and Neutral and the Test Data contains 467 tweets [10]. The whole training and test corpus is a collection of Hindi tweets provided by SAIL-2015. We have used Weka [17] tool in our system for classification purpose.

4.2 Proposed Methodology

The following steps will elucidate the process of proposed system which is discussed in this paper.

I Pre-processing of Tweets
II Model Creation and Training
III Sentiment Classification

Above steps are explained in details.

I. **Pre-processing of Tweets.** Usually, when large quantity of information is to be included within a limited space, there is some compromise. The tweets, therefore, often have no grammatical structure, improper way of using punctuation marks, contains intentional misspellings (also unintentional) and abbreviations. So, pre-processing is one of the important steps to be carried on raw twitter data. This step deals with cleaning of raw twitter data by removing URL links, user names, punctuation marks. Hashtags are maintained in the tweets(i.e. #flood) after removing the hash(#) symbol.

II. **Model Creation and Training.** The annotated and filtered training data is used to create a decision tree using C4.5 decision tree algorithm. For this we select the training file and the algorithm (C4.5) to be used to train the model (Fig. 1).

The classification is a 3-step process as given below.

(a) Model Construction (Learning): Build a model from the training set.
(b) Evaluation of Model (Accuracy): Accuracy of the model is verified on the basis of classifying training data.

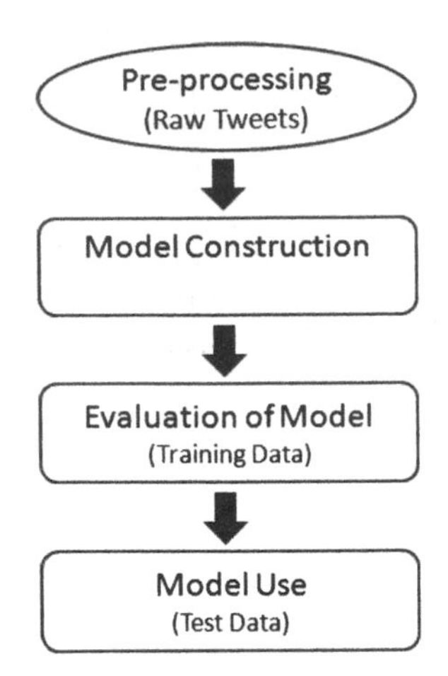

Fig. 1. Steps involved in classification

(c) Model Usage (Classification): Model classify the test data into various classes.

Learning is the first step done on training data and is used to create a model. Accuracy of model is evaluated on training data also known as testing phase. Each Model is used to classify the test data set.

During the Model construction phase, we revisited some of the parameters of the C4.5 decision tree algorithm to improve overall performance of the system.

III. **Sentiment Classification.** Classification of test data is being done with the help of trained model. Each tweet from the test data is classified based on sentiments.

5 Results

We submitted two runs: *Constrained* and *Unconstrained.* Here we show the results of both the runs. As we considered given task as a classification problem, initially we tried with two different classifiers namely Naive Bayes and Decision Tree model (C4.5). Experiments were done on the training dataset w. r. t. with equal number of tweets from each categories (i.e. positive, negative and neutral). The average F-score on entire training dataset were 0.527 and 0.804 for Naive Bayes and C4.5 respectively and We observed that C4.5 performed better on the development dataset as shown in Fig. 2. Hence we applied C4.5 classifier for further experiment.

5.1 Constrained

The constrained run, uses only the corpus provided by SAIL-2015 task coordinators and no other resources. The performance (precision, recall and F_1-score) of our system is shown in Table 1. The Accuracy for our run is 40.47 %, for Positive 4.22 %, for Negative 58.17 % and for Neutral: 72.0 % Table 2.

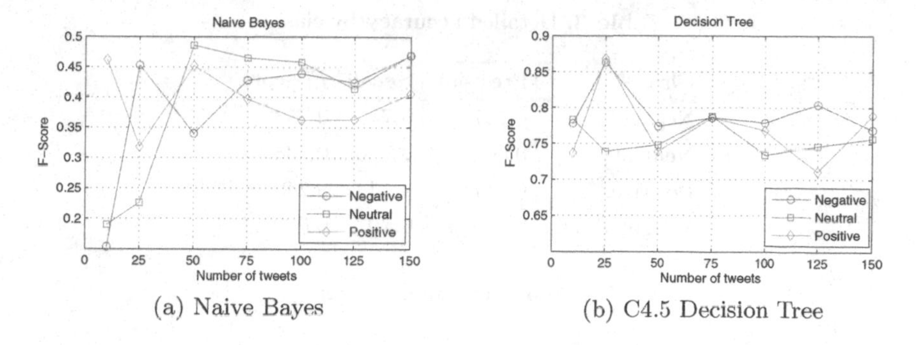

(a) Naive Bayes

(b) C4.5 Decision Tree

Fig. 2. Performance comparison of naive bayes and C4.5 decision tree

Table 1. Detailed accuracy by class

Class	Precision	Recall	F_1-score
Negative	0.835	0.896	0.852
Neutral	0.785	0.892	0.835
Positive	0.894	0.386	0.539
Weighted avg.	0.822	0.815	0.804

Table 2. Confusion matrix

a	**b**	**c**	<–*classified as*
457	65	4	**a = Negative**
49	431	3	**b = Neutral**
41	53	22	**c = Positive**

5.2 Unconstrained

In unconstrained runs we used Hindi stopwords along with the provided corpus. The performance measure (precision, recall and F_1-score) of the system in this approach is shown in Table 3. The Accuracy is 31.26 % Positive 1.80 %, Negative 42.63 % and Neutral 72.0 % (Table 4).

6 Discussion

As shown above, we achieved better results in J48 (C4.5 algorithm) for Hindi tweets than other classifiers such as (Naive Bayes, ZeroR etc.) within Weka. For both the submissions, therefore, we used the same classifier with and without Hindi stopwords. But with the removal of stopwords, the performance of the classifier degraded. The performance degraded possibly because stopwords also add some meaning to the context. Removal of these stopwords may alter the sentiment of the tweets. In addition to this, as tweets are small messages,

Table 3. Detailed accuracy by class

Class	Precision	Recall	F_1-score
Negative	0.747	0.759	0.753
Neutral	0.664	0.830	0.738
Positive	0.917	0.144	0.249
Weighted avg.	0.735	0.707	0.680

Table 4. Confusion matrix

a	b	c	<–*classified as*
399	126	1	**a = Negative**
81	401	1	**b = Neutral**
54	77	22	**c = Positive**

so for sentiment analysis task it is not appropriate. The accuracy of classification of test data greatly depends on context and size of training data in each classes. Overall, the accuracy of our classifier is good in classifying neutral and negative tweets but the accuracy is poor in case of positive tweets. One possible reason for this is imbalance in the number of tweets in each class in the training set. There are only 153 positive tweets as compared to 526 negative and 483 neutral tweets. So the classifier is biased towards neutral and negative sentiments.

In addition to this, sarcasm is one of the important reason that affects the classification of negative tweets as sarcastic tweets express a positive sentiment towards a negative situation [4,6]. Also, tweets with sad contents usually lead to negative polarity and tweets with romantic contents usually lead to positive polarity often producing wrong classifications.

Limitations. We filtered out some special characters, symbols and words from the corpus and trained model on cleaned tweets. Hence, system may not classify correctly when the test tweets are unprocessed. Emoticons and Hashtags carry important knowledge features in dataset and they are not considered properly in our study. Also, our system is not well adapted for handling negation terms.

7 Conclusion

We describe our participation to the Shared Task on Sentiment Analysis in Indian Languages (SAIL) Tweets, at MIKE-2015. Sentiment Classification of twitter dataset especially in Indian language is a challenging problem due to limited language resources available. Nevertheless, it has been studied for past few years with different feature set and classifiers. We used a machine learning techniques performing reasonably well for sentiment classification in Hindi tweets. A distinctive feature of our approach is that we used Decision Tree algorithm for the classification task. However, there are still much scope left for improvement. Possibly, performance can be improved if we use list of SentiWords whose polarity is defined which we can be attempted in future.

References

1. Bautin, M., Vijayarenu, L., Skiena, S.: International sentiment analysis for news and blogs. In: ICWSM (2008)
2. Das, A., Bandyopadhyay, S.: Sentiwordnet for indian languages, pp. 56–63. Asian Federation for Natural Language Processing, China (2010)
3. Das, D., Bandyopadhyay, S.: Word to sentence level emotion tagging for bengali blogs. In: Proceedings of the ACL-IJCNLP 2009 Conference Short Papers, pp. 149–152. Association for Computational Linguistics (2009)
4. Das, D., Bandyopadhyay, S.: Labeling emotion in bengali blog corpus-a fine grained tagging at sentence level. In: Proceedings of the 8th Workshop on Asian Language Resources, p. 47 (2010)
5. Das, S., Chen, M.: Yahoo! for amazon: extracting market sentiment from stock message boards. In: Proceedings of the Asia Pacific Finance Association Annual Conference (APFA), vol. 35, p. 43, Bangkok, Thailand (2001)
6. Ghosh, A., Li, G., Veale, T., Rosso, P., Shutova, E., Reyes, A., Barnden, J.: Semeval-2015 task 11: sentiment analysis of figurative language in twitter. In: International Workshop on Semantic Evaluation (SemEval-2015) (2015)
7. Liu, B.: Sentiment analysis and subjectivity. In: Handbook of Natural Language Processing, vol. 2, pp. 627–666 (2010)
8. Pang, B., Lee, L.: Seeing stars: exploiting class relationships for sentiment categorization with respect to rating scales. In: Proceedings of the 43rd Annual Meeting on Association for Computational Linguistics, pp. 115–124. Association for Computational Linguistics (2005)
9. Pang, B., Lee, L., Vaithyanathan, S.: Thumbs up?: sentiment classification using machine learning techniques. In: Proceedings of the ACL-2002 Conference on Empirical Methods in Natural Language Processing, vo. 10, pp. 79–86. Association for Computational Linguistics (2002)
10. Patra, B.G., Das, D., Das, A., Prasath, R.: Shared task on sentiment analysis in indian languages (sail) tweets - an overview. In: Proceeding of the Mining Intelligence and Knowledge Exploration (MIKE-2015) (2015)
11. Quinlan, J.R.: C4. 5 Programs for Machine Learning. Elsevier, New York (2014)
12. Sharma, S., Agrawal, J., Sharma, S.: Classification through machine learning technique: C4. 5 algorithm based on various entropies. Int. J. Comput. Appl. **82**(16), 28–32 (2013)
13. Silva, N.F., Hruschka, E.R., Hruschka Jr., E.R.: Biocom usp: tweet sentiment analysis with adaptive boosting ensemble. In: SemEval 2014, p. 123 (2014)
14. Snyder, B., Barzilay, R.: Multiple aspect ranking using the good grief algorithm. In: HLT-NAACL, pp. 300–307 (2007)
15. Tong, R.M.: An operational system for detecting and tracking opinions in on-line discussion. In: Working Notes of the ACM SIGIR 2001 Workshop on Operational Text Classification, vol. 1, p. 6 (2001)
16. Turney, P.D.: Thumbs up or thumbs down?: semantic orientation applied to unsupervised classification of reviews. In: Proceedings of the 40th Annual Meeting on Association For Computational Linguistics, pp. 417–424. Association for Computational Linguistics (2002)
17. Witten, I.H., Frank, E.: Data Mining: Practical Machine Learning Tools and Techniques. Morgan Kaufmann, San Francisco (2005)

Sentiment Classification for Hindi Tweets in a Constrained Environment Augmented Using Tweet Specific Features

Manju Venugopalan[1(✉)] and Deepa Gupta[2]

[1] Department of Computer Science, Amrita School of Engineering, Amrita Vishwa Vidyapeetham, Bengaluru Campus, Bengaluru, India
manjusreekumar2007@gmail.com

[2] Department of Mathematics, Amrita School of Engineering, Amrita Vishwa Vidyapeetham, Bengaluru Campus, Bengaluru, India
g_deepa@blr.amrita.edu

Abstract. India being a diverse country rich in spoken languages with around 23 official languages has always left open a wide arena for NLP researchers. The increase in the availability of voluminous data in Indian languages in the recent years has prompted researchers to explore the challenges in the Indian language domain. The proposed work explores Sentiment Analysis on Hindi tweets in a constrained environment and hence proposes a model for dealing with the challenges in extracting sentiment from Hindi tweets. The model has exhibited an average performance with cross validation accuracy for training data around 56 % and a test accuracy of 43 %.

Keywords: Sentiment analysis · Machine learning · Twitter · Feature extraction

1 Introduction

The current era being termed as internet era has led to the exponential growth of online data available. The technological advancements help us be connected anywhere anytime which opens the scope of retrieving information or expressing yourself in seconds. This immense amount of online data is always a huge corpus for researchers who are interested in data analytics. Natural Language Processing (NLP) is one such domain used to train machines to attain the capability of manipulating natural languages at an expert level. The sub-domain of NLP which focuses on extracting the sentiments from raw-data is referred to as opinion mining or sentiment analysis and it owes a lot to the human instinct to find what others feel. Opinion-rich resources such as social networking sites, online review sites, personal blogs etc. have led to an immense amount of research activity in the field of opinion mining and summarization, which deals with the computational treatment of opinion, sentiment, and subjectivity in text and hence automatically mines and organizes opinion from heterogeneous information sources. The wide range of applications ranging from a tourist persuading a travelers opinion on a tourist destination to companies exploring the immense corpora of product reviews which in turn serves for product design or improvement have led to growing popularity of Sentiment Analysis.

R. Prasath et al. (Eds.): MIKE 2015, LNAI 9468, pp. 664–670, 2015.
DOI: 10.1007/978-3-319-26832-3_63

Most of the research in the area has been in English and a few other foreign languages like Chinese, Arabic etc. Being a culturally diverse country, with around 23 official languages and sources of data extraction being available, there is a huge scope for exploration of opinion summarization using Indian languages. Opinion summarization on English corpora alone in our country wouldn't reflect the majority view point. This emphasizes the need and importance of research in this area. As far as Indian languages are concerned minimal work has happened as of now mainly owing to lack of opinionated text data available. Social networking sites like Twitter are a huge source of opinionated data. Twitter supports six Indian languages and Hindi tweets have been available for more than a year. Hindi being the national language of India and the most widely spoken language in the country definitely demands more research exploration.

The proposed approach has attempted to explore a constrained sentiment analysis model which doesn't utilize any other resources other than the train data. Weighted unigrams, tweet specific features, punctuations captured from the train data and machine learning techniques are employed to the train the model.

2 Related Works

Attempts have been made to extract the sentiment from Indian languages, but to a very small extent. Initial approaches utilized MT tools to translate the text documents in the source language to English and hence utilize the classifiers or lexicons created in English to extract sentiment. Poor translation and unavailability of translators for many language pairs constrained further research in this area. Another approach focused on mapping WordNets [1] of a language pair and hence use the classifier of the mapped language. The success of this method to a great extent is dependent on the similarity of the concepts in both the languages. Training classifiers using parallel corpus has been attempted for a foreign language pair which turned out to be a good strategy but applicable only when parallel corpus is available. Phrase level polarity identification techniques using machine learning classifiers like SVM [2] have exhibited an average performance. The lack of corpora was projected to be the reason for low accuracy using machine learning classifiers and hence lexicon approaches gained popularity. SentiWordNet equivalents [3] have been built and are being modified for Hindi, Telugu and Bengali languages even though a thorough evaluation of these resources is yet to be done. A group headed by Prof Pushpak Bhattacharya has designed a Hindi WordNet available online bringing together different lexical and semantic relations between the Hindi words. The group has also contributed to building Hindi polarity labeled corpora in movie and tourism domain but comparatively restricted in size. Das and Bandyopadhyay proposed [4] a computational technique for developing SentiWordNet for Bengali using a English-Bengali dictionary and Sentiment Lexicons available in English. The authors in [5] proposed four approaches for deriving word polarities. The first approach attempts an interactive game strategy, the second a bi-lingual dictionary based approach developed for English and Indian Languages. WordNet expansion using antonym and synonym relations is attempted in the third approach and the fourth approach uses a pre-annotated corpus for learning. Das and Bandyopadhyay developed [6] a method for labeling emotional expressions in Bengali blog corpus. Classification of words is performed into six emotion classes (happy, sad, surprise, fear, disgust, anger) according to three categories of intensities (low, general and high). The authors have also attempted opinion summarization from Bengali news corpus implementing page rank and k-means algorithms [7]. Hindi Subjective Lexicon and Hindi WordNet have been utilized for identifying semantic orientation of adjectives and adverbs in [8]. Different approaches to perform Sentiment Analysis in Hindi have been suggested in [9] which include classifier based, translation to English and using English lexicon resources, building a Hindi-SentiWordNet etc. The Conditional Random Field (CRF) based Subjectivity Detection approach [10] was meant to be applied for any language with minimum linguistic knowledge but still has scope of improvement. Sharma et al. [11] have attempted sentiment classification on Hindi movie reviews using a dictionary based approach which classify reviews as positive, negative or neutral.

The availability of opinionated corpora in Indian languages to a great extent has limited the research in this area. Most of the works have been based upon machine translation techniques or mapping resources available in English to other languages and hence creating SentiWordNet equivalents in other languages. The applications of Sentiment Analysis being wide spread, the availability of structured opinionated corpora would surely open doors to research works in other directions. The proposed work attempts designing a model which extracts sentiment to the possible extent in an environment constrained to use only the training corpora. The work has drawn its insights from [12] partially which has attempted a sentiment classification model on tweets in English.

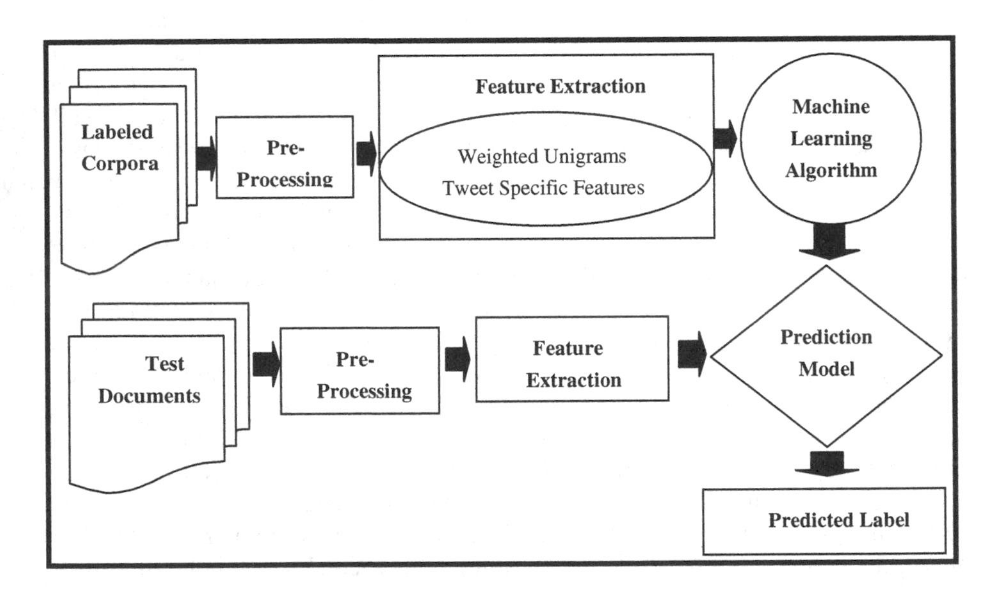

Fig. 1. Schematic diagram of proposed approach

3 Proposed Approach

The proposed approach for sentiment classification is depicted in Fig. 1. The Pre-Processing, Feature Extraction and Sentiment Classification phases are explained in detail in the following subsections.

3.1 Pre-processing Module

The streamed tweets are initially subjected to pre-processing which takes them through a series of stages.

Hyper Link Removal: URL links that are often mentioned as part of the tweet do not contribute to the sentiment expressed and hence need to be removed.

Username Removal: The username which begins with the @ symbol are a part of the tweet which needs to be removed as they do not contribute to the sentiment.

3.2 Feature Extraction

This module emphasizes on extracting the differentiating features that are identified to train the classifier so as to classify tweets into positive, negative and neutral tweets.

TFIDF Score of Unigrams: The TFIDF score is a statistical measure to identify the words inclined to any one of the classes under consideration. This feature is generally found to outperform the basic unigram feature. The TF-IDF score of a unigram in a tweet is calculated as per Eq. 1 where *tf*, the term frequency is the count of occurrence of w_i in the tweet, N the total number of tweets which is the total count of positive, negative and neutral tweets, and *df* denotes document frequency, the number of tweets containing w_i

$$tf - idf(w_i) = tf(w_i) \times \log_2(\frac{N}{df_{w_i}}) \tag{1}$$

Features Exclusive to Tweets: The restricted length of tweets confined to 140 characters poses a greater challenge for sentiment extraction when compared to formal text reviews. This emphasizes the need for exploiting each and every character in the tweet. The tweet specific features which seem to play an important role in expressing sentiment have been identified and exploited.

Emoticons: The practice of using emoticons to express sentiment in tweets is tapped here. Faces like :), :(, >:(which express happy, sad and angry sentiment respectively are a few in the list. These act as carriers of sentiment and are a considered to be a valid feature. The presence of these emoticons adds a weight to the feature which characterizes a positive, negative or neutral sentiment.

Hashtags: These are an important feature of twitter which increases the visibility of your tweets and enables grouping of tweets on related topic. The presence of a hash tag for e.g. #KashmirFloods can express an orientation towards a positive, negative or neutral sentiment and hence the feature is exploited. Hashtags are ranked based on their frequency of occurrence in the train data and the high ranked hashtags are learnt for their inclination towards any class. The presence of these hashtags adds a weight to the feature, proportional to its probability of occurrence in each class, which learns its affinity towards one of the classes.

Punctuations: Exclamation marks, inverted quotes to emphasize a phrase, question marks, combination of question mark and exclamations to express ambiguity etc. are a few among the punctuations in a tweet which significantly express emotions. This feature represents a weight corresponding to the contribution of the presence of all considered punctuations in a tweet.

3.3 Sentiment Classification

Thus unigram TF-IDF scores and tweet specific features are combined to train a classifier for a three class sentiment classification of tweets. The classifier thus trained based upon a machine learning algorithm is capable of predicting the classes of tweets given for testing.

4 Results and Analysis

The train data consisted of 168 positive, 559 negative and 495 neutral Hindi tweets from generic domains. The positive train data was comparatively smaller in volume. The test data consisted of 467 tweets. The tweets are characterized by an average word length of 14 to 15 words mostly confined to one sentence.

The SVM and the decision tree based J48 classifier implemented in Weka, the popular data mining tool has been trained using the proposed model. The system exhibited a tenfold cross-validation accuracy of 55.9 % by the SVM classifier and 51.9 % by J48 classifier on the training data. The test results of the SVM classifier averaged at 42.83 % with an accuracy of 17.47 % for the positive class, 54.59 % for the negative class and 68 % for the neutral class. The results are tabulated in Table 1.

The relative dip in the positive accuracies owes a large extent to the limited positive tweets in the train data. The proposed approach being subjected to a constrained environment has been able to perform to a considerable extent. Pure lexicon based approaches which have proved to be challenging for tweets when supported by this methodology would elevate the results to a much better scenario. Most of the techniques for Sentiment Classification have focused on binary classification and the features considered are more oriented towards differentiating positive and negative classes. Hence features which uniquely identify the neutral class need to be figured out.

Table 1. Accuracies on train and test data

Data	Classifier model	Positive class accuracy %	Negative class accuracy %	Neutral class accuracy %	Average accuracy %
Training	SVM	34.5	62.07	56.36	55.9
Training	Decision tree	25	59.57	52.5	51.9
Testing	SVM	17.47	54.59	68	42.83

5 Conclusion and Future Work

The proposed approach attempted a sentiment classifier model constrained to use only the train data.TF-IDF scores of unigrams and tweet specific features were exploited to the maximum extent to extract sentiment from the tweets. The tweet specific features which include emoticons, punctuations, hash tags etc. have proved to act as complementary sources to extract twitter sentiment. The model would have performed better on a larger train corpus. The hashtag feature captured from tweets would contribute significantly in a domain oriented approach. The features which differentiate neutral tweets need to be explored. The work can be extended by developing a hybrid model inculcating POS (parts of speech) and lexicon features to the proposed approach. The POS features which contribute largely to the three way class differentiation need to be identified through feature selection. The polarity values of the sentiment bearing words extracted from Hindi SentiWordNets available would certainly contribute as enhancing resources.

References

1. Balamurali, A.R., Joshi, A., Bhattacharya, P.: Cross-lingual sentiment analysis for indian languages using linked WordNets. In: Proceedings of COLING, pp 73–81 (2012)
2. Das, A., Bandyopadhyay, S.: Phrase-level polarity identification for Bangla. Int. J. Comput. Linguist. Appl. (IJCLA) **1**(1–2), 169–182 (2010)
3. Das, A., Bandyopadhyay, S.: SentiWordNet for Indian languages. In: 8th Workshop on Asian Language Resources (ALR), Beijing, China, pp. 56–63, 21–22 August 2010
4. Das, A., Bandyopadhyay, S.: SentiWordNet for Bangla. In: Knowledge Sharing Event-4: Task 2 (2010)
5. Das, A., Bandyopadhay, S.: Dr sentiment creates SentiWordNet(s) for Indian languages involving internet population. In: Proceedings of Indo-wordnet Workshop (2010)
6. Das, A., Bandyopadhay, S.: Labeling emotion in Bengali blog corpus - a fine grained tagging at sentence level. In: Proceedings of 8th Workshop on Asian Language Resources, Beijing, pp. 47–55 (2010)
7. Das, A., Bandyopadhay, S.: Opinion summarization in Bengali: a theme network model. In: Second International Conference on Social Computing, pp. 675–682. IEEE (2010)
8. Narayan, D., Chakrabarti, D., Pande, P., Bhattacharyya, P.: An experience in building the indowordnet – a wordnet for Hindi. In: Proceeding of First International Conference on Global WordNet (2002)

9. Joshi, A., Balamurali, A.R., Bhattacharyya, P.: A Fallback strategy for sentiment analysis in Hindi: a case study. In: Proceedings of the 8th ICON (2010)
10. Das, A., Bandyopadhyay, S.: Subjectivity detection in English and Bengali: a CRF based approach. In: Proceedings of 7th International Conference on NLP, (ICON) (2009)
11. Sharma, R., Nigam, S., Jain, R.: Polarity detection of movie reviews in Hindi Language. Int. J. Comput. Sci. Appl. **4**(4), 49–57 (2014)
12. Venugopalan, M., Gupta, D.: Exploring sentiment analysis on Twitter data. In: Eighth International Conference on Contemporary Computing (IC3), JIIT, Noida, 20–22 August 2015. (Accepted)

AMRITA_CEN-NLP@SAIL2015: Sentiment Analysis in Indian Language Using Regularized Least Square Approach with Randomized Feature Learning

S. Sachin Kumar(✉), B. Premjith, M. Anand Kumar, and K.P. Soman

Centre for Excellence in Computational Engineering and Networking, Amrita Vishwa Vidyapeetham, Coimbatore, India
sachinnme@gmail.com

Abstract. The present work is done as part of shared task in Sentiment Analysis in Indian Languages (SAIL 2015), under constrained category. The task is to classify the twitter data into three polarity categories such as positive, negative and neutral. For training, twitter dataset under three languages were provided Hindi, Bengali and Tamil. In this shared task, ours is the only team who participated in all the three languages. Each dataset contained three separate categories of twitter data namely positive, negative and neutral. The proposed method used binary features, statistical features generated from SentiWordNet, and word presence (binary feature). Due to the sparse nature of the generated features, the input features were mapped to a random Fourier feature space to get a separation and performed a linear classification using regularized least square method. The proposed method identified more negative tweets in the test data provided Hindi and Bengali language. In test tweet for Tamil language, positive tweets were identified more than other two polarity categories. Due to the lack of language specific features and sentiment oriented features, the tweets under neutral were less identified and also caused misclassifications in all the three polarity categories. This motivates to take forward our research in this area with the proposed method.

1 Introduction

Sentiment Analysis is a hot main-stream research in Natural Language Processing (NLP). The sentiment carries several information like emotions - love, happy, surprise, sadness, anger etc., opinions about social issues, major events, products, brands, services etc., human moods/attitude identified using polarity of a text comment - positive or negative etc. People express their sentiments in English or in their own mother tongue through social media sites like facebook, blogs, twitter etc. Everyday massive volume of text content is generated through social media. These text contents are favorites to researchers, governments and companies as the contents provide recommendations and suggestions regarding

R. Prasath et al. (Eds.): MIKE 2015, LNAI 9468, pp. 671–683, 2015.
DOI: 10.1007/978-3-319-26832-3_64

a product or service from a company, carries a positive or negative opinion about a movie or these text content gives a direct feedback. Forums are the discussion place where different people shares their opinion and this raises its importance from the company point-of-view. The sentiments provides an option for the people to take a better decision or to improve their views. The explosive growth of information happened with the advancement in technology and it gave several option to the people to share their sentiments and opinions by exchanging and interacting on their views. The social networking sites like facebook, twitter, google+ etc. acts as the medium or platform for this type of content generation. It facilitates real time sharing of opinions about different topics and issues of interest. These sites will have petabytes of data stored and analyzing these data creates enormous job and business opportunities [1–3]. Several work has been carried out in extracting the sentiments, opinion mining, classification of sentiments etc. The importance of research work in this field lies in the fact that these methods helps to understand the sentiments or opinion about a topic in few seconds by analyzing large volume of content [4–6]. Since large volume of data is generated every day, the analysis of these text content is performed using various computational approaches. Considering the importance of sentiment analysis, several thousands of articles are been published stating various aspects on dealing the analysis task.

The present work is done as part of shared task in Sentiment Analysis in Indian Language (SAIL) 2015, Constrain category. The task is to identify the polarity of the given tweet of a given language. The main objective of the share task is to motivate researchers to do sentiment analysis in their own native language. The task contained three classes (positive, negative, neutral) twitter data in three languages - Hindi, Bengali and Tamil. In Sect. 2, provides a literature review about sentiment analysis in Indian scenario. Section 3 provides a view about the proposed method and Sect. 4 discusses about result.

2 Related Work

Several approaches had been tried in the past related to Sentiment analysis. The graph based approach that uses graphs to represent text and perform tweet collections using specific keywords [7]. Some works has focused to bring lexicon based senti wordnets and corpora in emotions in languages such as English, Chinese, Hindi, Bengali [5–10]. Several other work focused to bring polarity based sentiment lexicon word list such as SentiWordNet, Subjectivity Word list, WordNet-Affect list, Taboada's adjective list for languages like English, Japanese, Hindi, Bengali, Marathi, Tamil, Telugu etc. [5,11–14]. In [15–17] proposes various techniques to develop dictionaries relating sentiment words. The research work tries to explore the use of standard machine learning methods such as Maximum Entropy, Naive Bayes, Support Vector Machines. The experiment tried several feature combinations like presence of unigram, unigram frequency, bigrams, combination of unigram and bigram, adjectives, unigram with part-of-speech tags, positions, frequent unigrams. The paper compares the machine learning approach with baseline system generated by human.

Even though majority of research work in sentiment analysis is happening in English, Chinese, German, Japanese, Spanish etc., tremendous research is going on for sentimental analysis in Indian languages like Hindi, Bengali, Marathi, Punjabi, Manipuri, Tamil, Malayalam etc. Due to complexity in Hindi text processing, only few system exists which docs sentimental analysis. The research in this area requires good lexicon dictionaries, taggers, parsers, annotated corpora's, complete HindiSentiWordNet etc. Several contexts exist where words used are same. However the meaning of the word changes based on the context. The problem of sentiment analysis thus need to be semantically disambiguated [18]. In [18], authors proposes a framework for sentiment analysis on Hindi language using HindiSentiWordNet and also improves the SentiWordnet by adding the missing senti words in Hindi. In their approach, the missing senti words are translated to English and searched in English SentiWordNet and its polarity is retrieved. In [19], an annotated corpus for Hindi movie reviews was developed. It discusses on improving the HindiSentiWordNet by adding more opinion words into it and proposed new rules to handle discourse relations and negation. A fall-back strategy for sentiment analysis was proposed in [20]. It proposes three different approach; one method uses a machine learning based model developed using annotated corpora of Hindi movie review. In the second approach the Hindi text content was translated to English to perform sentiment classification. The third approach uses a score based strategy to classify the sentiments using Hindi Senti WordNet. In [21], proposed several approaches to generate SentiWordNets for Indian Languages using WordNet, dictionary, corpus to generate. For validating the developed Senti WordNet, an internet based interactive game was created in order to attract internet users. The paper shows the statistics of text content (news and blogs) in Bengali language used for experiment, the subjectivity classification accuracy based on senti wordnet for English and Bengali. In [22–24], provides an extensive survey on opinion mining and sentiment analysis in Hindi language.

In [25], it discusses naive bayes based polarity identification on movie review in Punjabi language. It utilizes unigram, bigram and combination of both as feature representation and does sentiment classification at sentence level. The sentiment analysis is also used to understand the human interaction in meetings or conferences as their can occur several comments, positive and negative opinion [26]. The human interactions are modeled using tree structure and analyzed with tree pattern matching algorithm. The method was experimented on Tamil documents and it instead considering the complete document, specific subjects are classified as positive or negative opinions. In [27], proposes sentiment analysis of movie review data in Tamil language by considering frequency count as feature. The collected data was hand tagged as positive and negative. The paper provides a comparison of machine learning algorithms such as Multinomial Naive Bayes, Bernolli Naive Bayes, Logistic Regression, Random Kitchen Sink and Support Vector Machines, in classifying Tamil movie reviews. The paper [28] is an initial work towards sentiment analysis in Malayalam language at sentence level. The paper proposes method to extract the emotional response (positive and negative response) or mood from the text occurring at different situations. A word set which shows the positive or negative responses were created and used an unsupervised method to classify the text. The corpus developed was hand tagged and semantic association was calculated using Semantic Orientation from Point-wise Mutual Information (PMI) method. In [29], proposes a rule based approach for performing sentiment analysis on Malayalam movie review data. They developed the corpus from websites containing Malayalam movie review content. The various stages of the implementation comprises of corpus collection, tokenization with sandhi splitting, negation rules and sentiment generation. Another research work [30] proposes sentiment extraction from documents related to novels, product reviews and movie reviews. The features were generated by considering the semantic orientation of the content and based on rules the polarity of the sentence is identified which is used to determine the polarity of the document. The paper [31] presents an identification of sentiments (positive, negative or neutral) of lexicon verbs in Manipuri language. The corpus contains the letters send to editors of daily newspapers. A Conditional Random Field (CRF) based part-of-speech tagger was used to tag the text content in a sentence and was used to disambiguate the verb to find the correct sentiments. The paper [32] presents the initial research work in the sentiment analysis in Indian Language. They proposed a technique that uses WordNet synonym graph to incorporate linguistic knowledge in identifying the polarity of document. The classification task was performed using support vector machines. The paper [33] presents a cross-lingual based sentiment classification for Indian Languages using features generated from WordNet synsets. Support vector machine method was used to learn the model from these features. It considered text content in Hindi and Marathi language for performing experiments. The cross-lingual way of sentiment analysis avoids the issues in machine translation based sentiment analysis as this approach will get effected by the translation accuracy.

3 Proposed Method

This paper presents a regularized least square based sentiment analysis over SAIL 2015 twitter dataset in Hindi, Bengali and Tamil language [34]. The various stages of the proposed method, shown in Fig. 1, are preprocessing of raw twitter data, feature generation, finding weight matrix using regularized least square method and projection to find the polarity.

3.1 Preprocessing

The training data contains symbols as shown in Fig. 2. These symbols are removed in the preprocessing stage. In this stage, along with the removal of the symbols, all the URLs, hash tags, numbers, dates, tweet names with @, English words are replaced with <URL>, <# tag>, <@name>, <num>, <date> and <eng>.

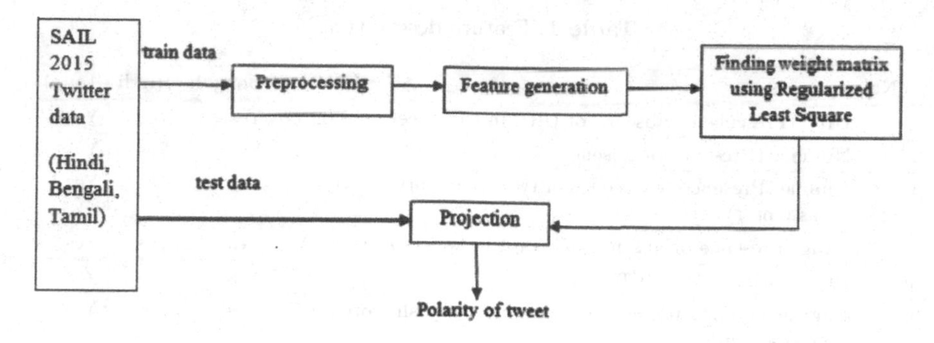

Fig. 1. Overall stages of proposed method

Fig. 2. Symbols found in twitter data provided shared task SAIL 2015

3.2 Feature Generation

From each tweets, 24 different feature category, tabulated as in Table 1, are generated. Two types of features are generated - binary features and statistical features which is derived from the SentiWordNet. The feature categories from 1 to 18 and 24 denotes the binary features (presence or absence of the feature category as 0 or 1). The 24th feature category denotes the presence or absence of words in the tweet as 0 or 1. To generate this feature, corresponding to each tweet language (Hindi, Bengali and Tamil), a word list without duplicate (or unique) is prepared. Based on the words in the tweet, a binary '1' is marked at the index location of that word in the word list. Each language will have different lengths of word list as shown in Table 2. The size of the 24th feature category thereby depends on the size of unique word list for each language. Therefore the total features for Hindi, Bengali and Tamil will be 6905, 3423, and 8592. After the feature is generated, due to the sparse nature of the data, a nonlinear random Fourier mapping is performed. This creates compact feature space where a linear classification can be performed.

The statistical features, 19–23 as in Table 1, are generated with the help of SentiWordNet. if the SentiWordNet does not contain any words, the corresponding feild will be marked 0. In this way, for all the words in a single tweet, a polarity score will get associated and this can be used to find a total positive polarity or total negative polarity score corresponding to each tweet which can be considered as a feature. Z-score is the measure of the how many standard deviations an observation is away from the mean. A positive Z-score indicates, the number of standard deviations above the mean and a negative Z-score is the number of standard deviations below the mean. If Z-score is 0, it means observation is same as mean. Z-score is associated with Normal distribution. Z-score is used to standardize the data in normal distribution to a standard normal

Table 1. Feature description

Sl.No	Features	Bengali	Hindi	Tamil
1	URL (Presence or absence of URL in the tweet as 0 or 1)	Yes	Yes	Yes
2	Numbers(Presence or absence of numbers as 0 or 1)	Yes	Yes	Yes
3	@name (Presence or absence of tweet ids starting with @ as 0 or 1)	Yes	Yes	Yes
4	#tag (Presence or absence of #tags in tweets as 0 or 1)	Yes	Yes	Yes
5	Punctuation marks as 0 or 1	Yes	Yes	Yes
6	English words (Presence or absence of English words in the sentence as 0 or 1)	Yes	Yes	Yes
7	Any other language other than English as 0 or 1	Yes	Yes	Yes
8	Smileys (Presence or absence of smileys as 0 or 1)	Yes	Yes	Yes
9	Happy (Presence or absence as 0 or 1)	Yes	Yes	Yes
10	Sad (Presence or absence as 0 or 1)	Yes	Yes	Yes
11	Love (Presence or absence as 0 or 1)	Yes	Yes	Yes
12	Wink (Presence or absence as 0 or 1)	Yes	Yes	Yes
13	Drool (Presence or absence as 0 or 1)	Yes	Yes	Yes
14	Flirt (Presence or absence as 0 or 1)	Yes	Yes	Yes
15	Presence of text in Bengali language (0 or 1)	Yes	Yes	Yes
16	Presence of text in Hindi language (0 or 1)	Yes	Yes	Yes
17	Presence of text in Tamil language (0 or 1)	Yes	Yes	Yes
18	Presence of Date (0 or 1)	Yes	Yes	Yes
19	Total Positive polarity in a tweet	No	Yes	No
20	Total Negative polarity in a tweet	No	Yes	No
21	Total Objective polarity in a tweet	No	Yes	No
22	Total Z+ polarity score	No	Yes	No
23	Total Z- polarity score	No	Yes	No
24	Presence of word in a tweet as 0 or 1	Yes	Yes	Yes

distribution. An observation (x) can be converted into Z-score by knowing the mean (μ) and standard deviation (σ) of the set of observations [35,36].

$$z = \frac{x - \mu}{\sigma} \tag{1}$$

Random Mapping. Since most of the feature (binary, statistical, unigram presence) corresponding to each tweet is sparse in nature, a nonlinear mapping into a random Fourier feature space is done. The idea is to map the input feature space into a compact space using a randomized map $z : \Re^d \rightarrow \Re^D$ maps the features explicitly to a Euclidian inner product space [39]

$$k(x, y) = \langle \phi(x), \phi(y) \rangle \approx z(x)^T z(y) \tag{2}$$

Table 2. Feature length

Language	Feature length
Hindi	6882
Bengali	3400
Tamil	8569

Here, ϕ is an implicit mapping to higher or infinite dimensional space like in SVM. The kernel trick provides a way to work in a feature rich space without explicit mapping to higher dimension and z is an explicit mapping. However this involves high computational cost as large sized kernel matrix need to be calculated and this creates problem when the size of data increases (eg. million samples). In order to make fast kernel computation, explicit nonlinear mappings are used [39]. Thus the $z(.)$ function becomes a nonlinear map. That is, the data is nonlinearly mapped into feature space which is compact and in that space any training and classification can be performed using linear methods. One way to obtain such a $z(.)$ function is to randomly sample the Fourier transform of the chosen kernel. In this paper the kernel chose is RBF or Gaussian kernel. The Fourier transform of the Gaussian Kernel can be interpreted as expectation of $z(x)^T z(y)$ where $z(x) = \cos(\omega^T x + b)$. Here, ω is drawn from Gaussian distribution and b drawn from uniform distribution. Let the RBF kernel for a given data pair x and y.

$$k(x,y) = e^{-\gamma\|x-y\|^2} = e^{-\gamma(x-y)^T(x-y)}, \text{ where } \gamma = 1/2\sigma^2 \tag{3}$$

Since the Gaussian kernel is translation invariant, and Fourier transform of the Gaussian Kernel can be interpreted as expectation of $z(x)^T z(y)$, then $FT\,(k(x,y)) = FT\,(k(x-y)) = FT\,(k(x-y)) = E\left(e^{j(x-y)^T \Omega}\right)$ where Ω is a random variable following a Gaussian distribution. (the expected value is the mean value of the function involving random variable). From the above expression the kernel can be written as,

$$k(x,y) = (z(x)^T z(y)) \tag{4}$$

$$z(x) = \begin{bmatrix} \frac{1}{\sqrt{D}} e^{jx^T \Omega_1} \\ \frac{1}{\sqrt{D}} e^{jx^T \Omega_2} \\ . \\ \frac{1}{\sqrt{D}} e^{jx^T \Omega_D} \end{bmatrix}, z(y) = \begin{bmatrix} \frac{1}{\sqrt{D}} e^{jy^T \Omega_1} \\ \frac{1}{\sqrt{D}} e^{jy^T \Omega_2} \\ . \\ \frac{1}{\sqrt{D}} e^{jy^T \Omega_D} \end{bmatrix} \tag{5}$$

After preparing the feature matrix, using regularized least square method the weight matrix is obtained. Steps to create finite (D) dimensional $z(.)$ are

- Create D number of n-dimensional random Gaussian distributed vectors and $\Omega_i s$
- Compute $z(.)$ as per above expression

$z(.)$ is a vector with complex elements. The complex number can be avoided by creating another vector with 2 times the D elements in which the first D elements are cosines and the next D elements will be sines of $\Omega^T x$. This will not affect the value of $k(x, y)$. Once $z(.)$ is obtained for all data points, linear classifier like regularized least square method can be used for classification.

3.3 Regularized Least Square

Regularized least square based classification is a supervised method and it can be closely related with support vector machines (SVM). In this method, training of data happens by solving simple single system of linear equation. Whereas, in SVM, the training is performed by solving a convex quadratic problem [37,38]. Regularized least square method has a simpler implementation than the support vector machines. In this method, the data is mapped into another feature space, defines by the kernel, so that a simple linear classification is performed. The n-dimensional data vectors and the corresponding class labels are represented as, $\mathrm{X} = \{\mathrm{x}_1, \mathrm{x}_2, ..., \mathrm{x}_\mathrm{n}\} \in \Re^\mathrm{n}$, $Y = \{y_1, y_2, ..., y_n\} \in \Re^n$. The method of regularized least square predicts an output label value by finding a weight matrix. The sum of square differences of the predicted output label and the original label is minimized [37]. The objective function regularized least square is given as,

$$\min_{W \in ^{n \times T}} \left\{ \frac{1}{n} \|Y - WX\|_F^2 + \lambda \|W\|_F^2 \right\} \tag{6}$$

In this objective function, λ is the control parameter which decides the weightage to be given to the term. Thus the objective function is the trade-off between minimising the sum of errors and keeping an approximate weight matrix. For minimizing this objective function, it is expanded in terms of trace function. That is,

$$\frac{1}{n} Tr \|Y - WX\|_F^2 + \lambda Tr \|W\|_F^2 \tag{7}$$

$$= Tr(Y^T Y - Y^T XW - W^T X^T Y + W^T X^T XW + \lambda W^T W) \tag{8}$$

To get W, differentiate w.r.t to W and equating 0 we get,

$$\frac{\partial}{\partial W} (Tr(Y^T Y - Y^T XW - W^T X^T Y + W^T X^T XW + \lambda W^T W)) = 0 \tag{9}$$

making all the necessary substitution, the above equation can be brought to

$$- X^T Y - X^T Y + 2X^T XW + 2\lambda W = 0 \tag{10}$$

The regression equation is,

$$W = (X^T X + \lambda I)^{-1} X^T Y \tag{11}$$

For testing purpose, the test signal is projected on to the weight matrix an intermediate label vector. The test signals are taken one at a time or together as a matrix and then projected on to the weight matrix. The maximum value and the position of the obtained predicted labels are calculated and the corresponding values are subtracted with Y, the true label vector. The differences between these two label vectors are to be made minimum for efficient classification.

4 Experiments and Results

The experiment is conducted on Windows 64-bit machine with i7 core processor and 8GB RAM. Java is used as the programming langauge creating the system. The SAIL 2015 training data provided for the contest contains twitter data from three languages - Hindi, Bengali and Tamil. Since the contest was about identifying the polarity of the tweet, each training dataset was provided with three categories of polarity data namely positive, negative and neutral. Table 3 shows the details about the training data and test data. The SAIL 2015 shared task report can be found at [41]. For Bengali, 277 Positive tweets, 354 Negative tweets and 368 Neutral tweets was provided. For twitter data in Hindi language, 168 Positive tweets, 545 Negative tweets and 493 Neutral tweets was provided. For twitter data in Tamil language, number of Positive, Negative and Neutral tweets collected was 387, 316 and 400 respectively. The twitter training data provided contained duplicate tweets, poems separated by commas, multiple symbols with '?','!','*','|' etc., tweets with abuse words, retweets, for some tweet the meaning cannot understand directly. Another characteristics of the tweets were it was code-mixed, contained smileys, URL, #tags, tweet ids, numbers, different punctuations, symbols etc. The test data was not categorized into polarity categories. From Table 3 it can be observed that the number of tweets in each category of polarity is different. For training data in Hindi language, the number of twitter training data for positive polarity was less when compared to other languages. This created a challenge in identifying characteristics of positive polarity. Several other things that was used to find the polarity of the tweet apart from individual words were smileys, URL, @names, # tags, and code-mixed data. Tweets become ambiguous with the presence of both positive and negative words in the sentence. This was tried to resolve using SentiWordNet. The polarity of a single tweet was calculated using the polarity score of positive and negative words.

Table 3. Twitter training dataset for SAIL 2015 shared task

Language	Training data			Test data
	Positive	Negative	Neutral	
Bengali	277	354	368	500
Hindi	168	545	493	467
Tamil	387	316	400	560

Using the test data, the experiment was run 10 times and based on manual assessment one result was chosen and send for the evaluation at SAIL 2015. Table 4 shows the accuracy obtained for the test data provided as part of the shared task. The highest accuracy is obtained for Hindi language tweets. For tweets in Hindi and Bengali, the proposed method identified negative polarity tweets more than tweets related to positive and neutral. Since language specific

Table 4. Testing result for SAIL 2015 shared task

Language	Accuracy (in %)	Positive (in %)	Negative (in %)	Neutral (in %)
Bengali	31.4 %	27.23 %	65.56 %	0.0 %
Hindi	47.96 %	36.14 %	64.94 %	2.0 %
Tamil	32.32 %	55.77 %	40.51 %	0.51 %

features related to tweets in positive, negative and neutral were not used, the proposed method has not identified tweets with neutral polarity. Except the features generated from SentiWordNet, remaining all the features are binary and sparse in nature. This has also resulted in less accuracy for identifying polarity tweets from test data. Therefore the task was also to find a good feature space where better classification can be achieved. This lead to randomized nonlinear mapping as discussed in Sect. 3. In that feature space, the classification was performed using a simple linear projection.

Table 5 shows the training accuracy obtained for the tweets in three languages. The data used for training was given as testing data. The experiment was performed 10 times and the average accuracy was calculated.

Table 5. Training accuracy for SAIL 2015 shared task

Language	Accuracy (in %)
Bengali (Constrained)	86.08 %
Hindi (Constrained)	92.43 %
Tamil (Constrained)	92.28 %

Table 6. Overall accuracy obtained for all the selected teams under constrained/ unconstrained category

Team name	Language	Accuracy-constrained	Accuracy-unconstrained
JUTeam_KS	Bengali	41.20 %	40.40 %
	Hindi	50.75 %	48.82 %
IIT-TUDA	Bengali	43.20 %	42 %
	Hindi	49.68 %	46.25 %
ISMD_RUNS	Hindi	40.47 %	31.26 %
AMRITA-CEN	Bengali	33.60 %	-
	Hindi	55.67 %	-
	Tamil	39.28 %	-
AmritaCENNLP	Bengali	31.40 %	-
	Hindi	47.96 %	-
	Tamil	32.32 %	-

As we could not collect twitter data, we could not participate in the unconstrained category. Table 6 shows the overall accuracy obtained for all the teams participated in SAIL 2015 shared task under constrained category. Our team was the only team participated for three languages.

5 Conclusion and Future Work

The work presented in this paper is done as part of the shared task for Sentiment Analysis in Indian Languages (SAIL 2015). The shared task was to identify the twitter tweets into three polarity categories such as positive, negative and neutral. For training, twitter data from three language was provided namely Hindi, Bengali and Tamil. The features generated contains binary features, statistical features generated with the help of SentiWordNet, and unigram presence (binary feature). Since the feature was highly sparse in nature, the proposed method found random Fourier feature and used regularized least square method (linear regression) for classification. Due to the absence of language specific feature and feature related to the sentiments, the proposed method has less accuracy in correctly classifying the twitter test tweets into positive, negative and neutral polarities. Hence this issue makes the research as work in progress. Improving accuracy using good polarity specific features along with language specific features will be one of the direction of our research.

References

1. http://aci.info/2014/07/12/the-data-explosion-in-2014-minute-by-minute-info graphic. Accessed 25 August 2015
2. http://wikibon.org/blog/big-data-statistics/. Accessed 25 August 2015
3. http://www.scidev.net/global/data/feature/big-data-for-development-facts-and-figures.html. Accessed 25 August 2015
4. Seungyeon, K., Fuxin, L., Guy, L., Irfan, E.: Beyond Sentiment: The Manifold of Human Emotions (2012)
5. Amitava D., Sivaji B.: SentiWordNet for Indian languages. In: Proceedings of the 8th Workshop on Asian Language Resources (ALR), pp. 56–63, August 2010
6. Maite, T., Julian, B., Milan, T., Kimberly, V., Manfred, S.: Lexiconbased methods for sentiment analysis. Comput. Linguist. **37**(2), 267–307 (2011)
7. Abilhoa, W.D., de Castro, L.N.: A keyword extraction method from twitter messages represented as graphs. Appl. Math. Comput. **240**, 308–325 (2014)
8. Aman, S., Szpakowicz, S.: Identifying expressions of emotion in text. In: Matoušek, V., Mautner, P. (eds.) TSD 2007. LNCS (LNAI), vol. 4629, pp. 196–205. Springer, Heidelberg (2007)
9. Changhua Y., Hsin-Yih L.K., Hsin-Hsi, C.: Building emotion lexicon from weblog corpora. In: Proceedings of the 45th Annual Meeting of the ACL on Interactive Poster and Demonstration Sessions Association for Computational Linguistics, pp. 133–136 (2007)
10. Hiroya, T., Takashi, I., Manabu, O.: Extracting semantic orientations of words using spin model. In: Proceedings of the 43rd AnnualMeeting of the Association for Computational Linguistics (ACL 2005), pp. 133–140 (2005)

11. Stefano, B., Andrea, E., Fabrizio, S.: Sentiwordnet 3.0: an enhanced lexical resource for sentiment analysis and opinion mining. In: Proceedings of the 7th conference on International Language Resources and Evaluation (LREC 2010), Valletta, Malta (2010)
12. Theresa, W., Janyce, W., Paul, H.: Recognizing contextual polarity in phrase-level sentiment analysis. In: Proceedings of the HLT/EMNLP, Vancouver, Canada (2005)
13. Maite, T., Anthony, C., Voll, K.: Creating semantic orientation dictionaries. In: Proceedings of the 5th International Conference on Language Resources and Evaluation (LREC), Genoa, pp. 427–432 (2006)
14. Yoshimitsu, T., Dipankar, D., Sivaji, B., Manabu, O.: Proceedings of 2nd workshop on Computational Approaches to Subjectivity and Sentimental Analaysis, ACL-HLT, pp. 80–86 (2011)
15. Bo, P., Lee, L., Vaithyanathan S.: Thumbs up? sentiment classification using machine learning techniques. In: The Proceedings of EMNLP, pp. 79–86 (2002)
16. Kreutzer, J., Witte, N.: Opinion Mining using SentiWordNet. Uppsala University (2013)
17. Wiebe, J., Mihalcea, R.: Word sense and subjectivity. In: The Proceedings of COLING/ACL-2006, pp. 1065–1072 (2006)
18. Pooja, P., Sharvari, G.: A framework for sentiment analysis in Hindi using HSWN. IJCA **119**, 975–8887 (2015)
19. Namita, M., Basant, A., Garvit, C., Nitin, B., Prateek, P.: Sentiment analysis of Hindi review based on negation and discourse relation. In: International Joint Conference on Natural Language Processing, Nagoya, Japan, October 2013
20. Joshi, A., Balamurali, A.R., Bhattacharyya, P.: A fall-back strategy for sentiment analysis in Hindi: a case study. In: Proceedings of the 8th ICON (2010)
21. Amitava D., Sivaji, B.: SentiWordNet for Indian Languages. Asian Federation for Natural Language Processing (COLING), China, pp. 56–63, (2010)
22. Sumit K.G., Gunjan, A.: Sentiment analysis in Hindi language: a survey. In: IJMTER (2014)
23. Richa, S., Shweta, N., Rekha, J.: Opinion mining in Hindi language: a survey. IJFCST **4**(2), 41 (2014)
24. Pooja, P., Sharvari, G.: A survey of sentiment classification techniques used for Indian regional languages. IJCSA **5**(2), 13–26 (2015)
25. Anu, S.: Sentiment analyzer using Punjabi language. In: IJIRCCE, vol. 2 (2014)
26. Thangarasu, M., Manavalan, R.: Tree-based mining with sentiment analysis for discovering patterns of human interaction in meetings Tamil document. IJCII **3**(3), 151–159 (2013)
27. Arun, S., Kumar, M.A., Soman, K.P.: Sentiment analysis of Tamil movie reviews via Feature Frequency Count, in Innovations in Information, Embedded and Communication Systems (2015)
28. Neethu, M., Nair, J.P.S., Govindaru, V.: Domain specific sentence level mood extraction from malayalam text. In: Advances in Computing and Communications (2012)
29. Deepu, S. N., Jisha, P.J., Rajeev, R.R., Elizabeth, S.: SentiMa-sentiment extraction for Malayalam. In: ICACCI, pp. 1719–1723 (2014)
30. Sandeep, C., Bhadran, V.K., Santhosh, G., Manoj, K.P.: Document level Sentiment Extraction for Malayalam (Feature based Domain Independent Approach). In: IJARTET (2015)
31. Kishorjith, N., Dilipkumar, K., Wangkheimayum, H., Shinghajith, K., Sivaji, B.: Verb based Manipuri sentiment analysis. IJNLC **3**(3), 1307–2278 (2014)

32. Alekh, A., Pushpak, B.: Sentiment analysis: a new approach for effective use of linguistic knowledge and exploiting similarities in a set of documents to be classified. In: Proceedings of the International Conference on Natural Language Processing (2005)
33. Balamurali, A. R., Aditya, J., Pushpak, B.: Cross-lingual sentiment analysis for Indian languages using linked wordnets. In: COLLINS (Poster), pp. 73–82 (2012)
34. http://amitavadas.com/SAIL/index.html. Accessed on 25 August 2015
35. Olena, K., Jacques, S.: Feature selection in sentiment analysis. In: CORIA (2012)
36. Hussam, H., Patrice, B., Fredric, B.: The imapact of Zscore on twitter sentiment analysis. In: International Workshop on Semantic Evaluation, pp. 636–641 (2014)
37. Tacchetti, A., Pavan, S.M., Santoro, M., Rosasco, L.: A toolbox for regularised least squares learning, GURLS (2012)
38. Rifkin, R., Gene, Y., Tomaso, P.: Regularized least-squares classification. Nato Sci. Ser. Sub Ser. III Comput. Syst. Sci. **190**, 131–154 (2003)
39. Rahimi, A., Benjamin, R.: Random features for large-scale kernel machines. In: Advances in Neural Information Processing Systems (2007)
40. Somla, A.J., Scholkopf, B., Muller, K.R.: The connection between regularization operators and support vector kernels. Neural Netw. **11**, 637–649 (1997)
41. Patra, B.G., Das, D., Das, A., Prasath, R.: Shared Task on Sentiment Analysis in Indian Languages (SAIL) Tweets - An Overview (2015)

IIT-TUDA: System for Sentiment Analysis in Indian Languages Using Lexical Acquisition

Ayush Kumar[1(✉)], Sarah Kohail[2], Asif Ekbal[1], and Chris Biemann[2]

[1] Department of Computer Science and Engineering, IIT Patna, Patna, India
{ayush.cs12,asif}@iitp.ac.in
[2] Language Technology, Computer Science Department, Technische Universität Darmstadt, Darmstadt, Germany
{kohail,biem}@lt.informatik.tu-darmstadt.de

Abstract. Social networking platforms such as Facebook and Twitter have become a very popular communication tools among online users to share and express opinions and sentiment about the surrounding world. The availability of such opinionated text content has drawn much attention in the field of Natural Language Processing. Compared to other languages, such as English, little work has been done for Indian languages in this domain. In this paper, we present our contribution in classifying sentiment polarity for Indian tweets as a part of the shared task on Sentiment Analysis in Indian Languages (SAIL 2015). With the support of a distributional thesaurus (DTs) and sentence level co-occurrences, we expand existing Indian sentiment lexicons to reach a higher coverage on sentiment words. Our system achieves an accuracy of 43.20 % and 49.68 % for the constrained submission, and an accuracy of 42.0 % and 46.25 % for the unconstrained setup for Bengali and Hindi, respectively. This puts our system in the first position for Bengali and in the third position for Hindi, amongst six participating teams.

Keywords: Sentiment analysis · Distributional thesaurus · Co-occurrence · Indian languages

1 Introduction

Sentiment Analysis is a Natural Language Processing (NLP) task, which deals with finding the orientation of thoughts and opinions expressed in a piece of text [16]. Recently, a large body of work has been devoted to automating the process of analyzing and extracting sentiments from social media platforms and review forums [19,20]. The rapid evolution in sentiment analysis has opened up the opportunities for governments and business organization to track the public opinion about their products and services.

Most of the existing work in sentiment analysis are dedicated to processing languages such as English, German and French. Sentiment analyzers developed for such languages are not directly applicable for Indian languages, which have

R. Prasath et al. (Eds.): MIKE 2015, LNAI 9468, pp. 684–693, 2015.
DOI: 10.1007/978-3-319-26832-3_65

their own challenges with respect to language constructs, morphological variation and grammatical differences.

Sentiment Analysis in Indian Language (SAIL) [17] tweets is the first attempt to bring together the researchers for resource creation and knowledge discovery in Hindi, Bengali and Tamil. Given a set of annotated tweets in these Indian languages, the task is to classify whether the tweet is of positive, negative, or neutral sentiment, which is also called polarity classification [2]. Teams are allowed to run their systems in two modes: constrained mode and unconstrained mode. In constrained mode, the participating team is only allowed to use the resources provided by the task organizers (i.e. tagger, parser, corpus). In contrast to this, participants were allowed to use any external resource in unconstrained mode.

Probably the most important resource for polarity classification is the sentiment lexicon. The sentiment lexicon is a list of words and phrases that convey sentiment polarities. It plays an essential role in most sentiment analysis applications [9]. Considering the lack and scarcity of available sentiment lexicons for Indian languages, we introduce an unsupervised approach for expanding a (small) Indian sentiment lexicon, leveraging distributional thesauri and sentence level co-occurrence statistics. Using the new expanded lexicon, we propose a sentiment classifier based on Support Vector Machines (SVM). We have participated in the SAIL task in two languages: Hindi and Bengali.

The remainder of this paper is organized as follows: Sect. 2 presents related works. Section 3 describes our method including dataset preprocessing, feature extraction and the lexicon expansion technique. Section 4 presents and discusses our experimentation results and evaluation, followed by conclusions and future work in the last section.

2 Related Work

Trends in the last few years show the inclination of research community towards social media like Twitter to sense public opinions, in commerce to anticipate stock market trends, to predict the outcome of elections [5,14,21] and even in disaster management [13] using a variety of approaches and experimental setups. However, most of the existing work in sentiment detection involve non-Indian languages except some prior work for Bengali [7]. The authors used SentiWordNet as well as a subjectivity lexicon to generate a lexical resource containing over 35,000 Bengali entries. Using the lexicon and features like positional aspect, the supervised sentiment classifier based on Conditional Random Field (CRF) achieved a precision of 74.6 % and recall of 80.4 % in the blog domain. A fall-back strategy for sentiment analysis in Hindi is reported in [11]. The results show that in-language sentiment analysis outperforms MT-based and resource-based sentiment analysis, where e.g. Hindi texts are translated automatically to English and are subsequently classified by an English sentiment analysis system.

3 Methodology

In this section, we discuss the process of building and training our sentiment classifier for the constrained and unconstrained runs. For the machine learning setup, we choose Support Victor Machine (SVM) as the classification model [6], as it can cope well with a large number of nominal features.

3.1 Preprocessing

We replace the URL links in all tweets with 'someurl', all @username with 'someuser' and multiple white spaces with single whitespace and tokenize the tweets in order to identify word tokens.

3.2 Features

We use the following features to train the SVM classifier:

- **Character and Word Features:** Writing style features like word and character n-grams features, often incorporated in stylometry research, have also shown to be effective in sentiment analysis [1]. They are also commonly applied to non-formal texts and user-generated content. For unstructured short texts like tweets, small values for n have shown to be most effective [10]. In our experiments, word unigrams and bigrams are extracted from the dataset. We also compute the n-gram overlap at the character level on the basis of character trigrams and quad-grams for word prefixes and suffixes.
- **SentiWordNet Features:** For this task, the organizer-provided Indian sentiment lexicons [8] include a list of positive, negative, neutral and ambiguous words with the corresponding part of speech (PoS) tags. We denote the words in SentiWordNet with a score of 1 if it is found in the positive list, -1 if it occurs in negative list and 0 if the word appears in the neutral list. Based on our annotation, we count the following features:
  1. Number of tokens in the tweet with $score(w) > 0$.
  2. Number of tokens in the tweet with $score(w) < 0$.
  3. Number of tokens in the tweet with $score(w) = 0$.

3.3 Lexical Acquisition

Lexical expansion [12] is an unsupervised technique that is based on the computation of distributional thesaurus [3]. While Miller et al. [12] used a DT for lexical expansions for knowledge-based word sense disambiguation, the expansion technique can also be used in other text processing applications. For rare words and unseen instances, lexical expansion can provide a useful back-off technique [4,15].

For the unconstrained submission, we use an external dataset to generate separate lexicons for both Hindi and Bengali and run the same SVM model with additional features derived from the lexicon. We now describe this expansion technique.

We exploit the concept of distributional thesaurus and sentence level co-occurrences from large background corpora[1] to build a lexicon, denoted later as DT_COOC, assigning each entry two scores between -1 to 1: one score computed over distributional similarity and the other obtained using the co-occurrences. We also assign a third score equal to -1 (absolute negative) and 1 (absolute positive) for each word in the lexicon. For background corpora, we use a Hindi corpus containing a total of 2,358,708 sentences (45,580,789 tokens) and a Bengali corpus of 109,855 sentences (1,511,208 tokens). Both corpora are constructed from online newspapers from 2011.

3.4 Distributional Thesaurus

A Distributional Thesaurus (DT) is an automatically computed resource that relates words according to their similarity. For every sufficiently frequent word in the corpus, we find out the most similar words as computed over the similarity of contexts these words appear in. We employ an open source implementation of the DT computation as described in [3], where complete details of the computation are described.

To illustrate this, a few examples of words and their distributionally most similar words are given in Fig. 1. Our core assumption, which is backed up by data analysis, is that sentiment words are similar to other sentiment words. Moreover, while there are usually high similarities between words of positive and negative sentiment (such as 'good' and 'bad'), words tend to be similar to more words of the same sentiment.

Words	Similar Words from Distributional Thesaurus		
अतुलनीय (atulnIya)	अद्भुत (adabhuta)	महान (mahAna)	शानदार (shAnadAra)
तर्कसंगत (tarkasangata)	उचित (uchita)	सही (sahI)	गलत (galata)
धार्मिक (dhArmika)	सामाजिक (sAmAjika)	राजनीतिक (rAjanItika)	हिंदू (hindU)
ऊँची (UNchI)	ऊंची (UnchI)	लंबी (lambI)	छोटी (ChotI)
परम्परा (paramparA)	परंपरा (paramparA)	संस्कृति (sanskriti)	धर्म (dharma)

Fig. 1. Illustrative examples of words appearing in the DT expansion for Hindi.

3.5 Co-Occurrences

We obtain a list of words that co-occur significantly with the other words in a sentence [18]. Some examples of words and their most significant co-occurrences are displayed in Fig. 2. Our core assumption here is that sentence contexts are mostly either positive, negative or neutral. While this does not hold in all cases, we have observed from data analysis that words of the same polarity tend to co-occur more than words of different polarity.

[1] from http://corpora.informatik.uni-leipzig.de.

Words	Words from Co-Occurrence Lists		
अतुलनीय (atulnIya)	भारतीय (bhAratIya)	अन्य (anya)	वर्ष (warSha)
तर्कसंगत (tarkasangata)	कहना (kahanA)	ज्यादा (jyadA)	काफी (kAphI)
धार्मिक (dhArmika)	परंपराओं (paramparAon)	अपितु (apitu)	संतों (santon)
ऊँची (UNchI)	इमारत (imArata)	जाति (jAti)	जगहों (jagahon)
परम्परा (paramparA)	अनुसार (anusAra)	नई (nayI)	जीवन (jIvana)

Fig. 2. Illustrative examples of words appearing in the co-occurrence list for Hindi.

3.6 Construction of DT_COOC Lexicon

We use the given SentiWordNet for both the languages as the seed data for lexical expansion. We first filter out the candidate sentiment terms using DT expansion and then create a final lexicon using the agreement between the DT polarity list and COOC polarity list. In the subsequent sections we describe the steps in more details.

Finding the Candidate Sentiment Terms: At first, after constructing the seed corpus, we obtain the top (i.e. most similar) 125 DT expansions for each word in the seed corpus. In context of further use, we define two terms: positive expansion list and negative expansion list. The DT expansion of positive and negative words in the seed corpus results in positive and negative expansion lists, respectively. To filter out the candidate terms from the noisy tokens, we rank each word in the complete expansion list with a score (*candidateScore*).

$$candidateScore = \frac{Number\ of\ expansion\ lists\ the\ word\ appears\ in}{Frequency\ of\ the\ word\ in\ the\ DT\ corpus} \tag{1}$$

Dividing through the frequency ensures that highly frequent words, which are similar to almost every word just because they occur in so many contexts, are down-ranked.

Calculating the DT Score: Based upon *candidateScore*, we remove the 500 lowest-ranked terms for lexicon generation. Of the remaining words in the expansion, we compute another score (*score_DT*):

$$score\_DT = \frac{No.\ of\ positive\ expansions - No.\ of\ negative\ expansions}{No.\ of\ expansion\ lists\ the\ word\ appears\ in} \tag{2}$$

The DT score is a graded score between -1 and 1 that projects sentiment to new words, based on the known sentiment of distributionally similar words.

Calculating the COOC Score: From the pruned list, we calculate a score *(score_COOC)* for each word using the sentence-based co-occurrences. We define the number of *pos co-occurences* as the total number of positive seed words with which word co-occurs. Accordingly, *neg co-occurrence* is defined analogously.

$$score\_COOC = \frac{No.\ of\ pos\ co\ occurrences - No.\ of\ neg\ co\ occurrences}{No.of\ seed\ words\ with\ which\ given\ word\ co-occurs} \quad (3)$$

Generating the Final Lexicon: To construct a final expanded lexicon, we consider the agreement between the two scored lists at the absolute polarity level: for the final lexicon, only those words are added where both methods agree on polarity. The statistics of the generated lexical corpus is given in Table 1:

Table 1. Statistics of induced lexicon for both languages.

Dataset	Positive	Negative	Neutral	Total
	First Expansion			
Hindi	3980	3331	357	7668
Bengali	1205	10005	600	11810
	Final Expansion			
Hindi	5521	3926	48	9495
Bengali	7213	1461	30	8704

In principle, the expansion procedure can be iterated to bootstrap sentiment lexicons: the output of one step can serve as the input of the next expansion step. Here, we explore two levels of expansion for Hindi, using described lexicon as the new seed. However, for Bengali DT_COOC Lexicon, we note that the expansion list is too skewed: he number of negative words in the lexical corpus is much higher than the positive ones. One possibility for the skewness might be the difference in the number of positive and negative words in the seed corpus. To overcome the skewness, we balance the Bengali seed by random sampling, removing negative instances randomly until we arrive at the same number of negative positive words. Finally, we perform all the steps sequentially to obtain the expanded lexical corpus. In preliminary experiments, however, we have not found this technique to be effective for Bengali, which might be related to the corpus size.

The statistics of the final expansion lexicons for both languages, as used in our experiments, are shown in Table 1.

4 Datasets and Experimental Results

To tune and to evaluate our approach, we perform five-fold cross validation on the training set. The datasets are annotated with three classes, namely positive, negative and neutral. The overall distribution of both train and test set per class label is given in Table 2. We used classification accuracy as a measure of sentiment polarity classification performance. Based on the cross validation results, the feature combination that we use for the various runs for both languages is given in Table 3. We make use of the LibLinear[2] SVM implementation.

Table 2. Distribution of training and test set for Hindi and Bengali language.

Dataset	Positive Tweets	Negative Tweets	Neutral Tweets	Total
	Hindi			
Training Set	168 (13.75 %)	559 (45.74 %)	494 (40.46 %)	1221
Test Set	166 (35.54 %)	251 (53.74 %)	50 (10.70 %)	467
	Bengali			
Training Set	277 (27.73 %)	354 (35.43 %)	368 (36.83 %)	999
Test Set	213(42.60 %)	151 (30.20 %)	135 (27.00 %)	499

Table 3. Feature combination for different modes of submission for both languages. Description of the features: 1. Word N-Gram, 2. SentiWordNet for respective language, 3. Character N-Gram of prefizes and suffixes of size 3 and 4, 4. DT_COOC Lexicon for respective language.

Mode	Hindi	Bengali
Constrained	1 + 2	1 + 2
Unconstrained	1 + 4	1 + 2 + 3 + 4

For Bengali, our system achieves an accuracy of 43.2 % and 42.0 % for the constrained and unconstrained runs, while we score an accuracy of 49.68 % and 46.25 % for Hindi in the constrained and unconstrained setups, respectively. The confusion matrix in Table 4 shows that the classifier performs very poorly on positive instances in comparison to other two classes in Hindi. The less percentage of positive tweets (13.75 %) in the training set might be a cause for inaccurate classification. We also analyze the labeled data to determine the statistics of tokens that match in the training and test sets. The percentage of unique overlapping tokens between training and test set is 49.71 % and 41.36 % for Hindi and Bengali respectively. However, the values drop to 29.91 % and 27.07 % for the positive tweets in the two languages respectively. On further investigation, we find the

[2] http://liblinear.bwaldvogel.de/.

Table 4. Confusion Matrix for Hindi and Bengali

Class	Positive	Negative	Neutral
	Hindi		
Positive	7	65	94
Negative	2	175	74
Neutral	0	c16	34
	Bengali		
Positive	53	52	109
Negative	17	82	52
Neutral	20	40	75

Table 5. Experimental results for feature ablation for Hindi and Bengali. The values in the parenthesis denotes the deviation from the score when all the features were taken into consideration.

Features	Accuracy: Hindi	Accuracy: Bengali
All	47.96	42.00
All-SentiWordNet	47.32 (-0.64)	41.20 (-0.80)
All-Word ngram	43.25 (-4.71)	38.40 (-2.80)
All-Character ngram	47.75 (-0.21)	42.20 (+0.20)
All-DT_COOC Lexicon	49.03 (+1.07)	42.20 (+0.20)

token to have overlap between neutral tweets in the training set and positive tweets in test set, and this to be 45.21 % for Hindi which is a possibility for a majority of positive instances classified as neutral. This shows that the training data is not rich enough to capture the new positive instances effectively. We also analyze the coverage of SentiWordNet and the induced DT_COOC Lexicon on the dataset. Assuming the adjectives to be the most dominating sentiment term in the tweet, we extract the sentiment terms in Hindi tweets using a POS Tagger[3]. Only 17.57 % and 25.98 % of the adjectives in the training and test set appear in the HindiSentiWordNet list. The coverage improves to 36.56 % and 42.29 % adjectives in the training and test set, respectively, while using DT_COOC Lexicon.

To get an insight to the contribution of each feature in the development of the system, we perform feature ablation experiment. Results of the detailed feature ablation study are shown in Table 5. We find that word ngram is the most important feature in both languages which improves the accuracy by 2 %-5 %. The second most important feature is the SentiWordNet feature which helps in improving the results upto 0.8 %. However, we observe a drop in performance with the induced lexicon, probably because the external dataset is from

[3] http://sivareddy.in/downloads#hindi_tools.

a different domain or the expansion is too aggressive. Since all participating systems achieved lower scores in their unconstrained runs, this could point either at overfitting on a small training set, or a selection of the data that was biased on the provided lexical resources.

5 Conclusions and Future Work

In this paper, we developed an SVM-based classifier for Indian tweet polarity classification. Our contribution is part of a recently held evaluation challenge (SAIL 2015), which aims to instigate researchers and experts to discuss and advance sentiment analysis research for Indian languages. Our system is based on supervised classification, SVM, which is enriched by using a lexicon expansion technique based on distributional thesauri and co-occurrences. Our system achieved the highest accuracy in Bengali in the competition, and we score third for Hindi amongst the participating teams.

However, there is a lot of scope of improvement and a large headroom – classification accuracies throughout do not even come close to a quality that would be useable in industrial applications. We first and foremost attribute this to the small amount of training and test data. We find that training set in both languages contain significant percentage (3.6 % and 6.5 % in Hindi and Bengali respectively) of quasi-duplicates (tweets that differ in just URL mention or spacing between punctuation marks or @mentions or simply identical duplicates) which would have resulted in overfitting. While it is very commendable that the effort for sentiment data creation for Indian language has started, it has to be expanded significantly in order to yield reliable results in the future.

In the future, we would like to create in-domain lexicon to test the effectiveness of our method since we still believe that expanding the lexicon with statistical methods is a simple yet effective method for increasing model coverage. We also plan to investigate and implement more features specific to the languages.

References

1. Abbasi, A., Chen, H., Salem, A.: Sentiment analysis in multiple languages: Feature selection for opinion classification in web forums. ACM Trans. Inf. Syst. **26**(3), 12:1–12:34 (2008)
2. Almatrafi, O., Parack, S., Chavan, B.: Application of location-based sentiment analysis using twitter for identifying trends towards indian general elections 2014. In: Proceedings of the 9th International Conference on Ubiquitous Information Management and Communication, pp. 41:1–41:5. IMCOM 2015 (2015)
3. Biemann, C., Riedl, M.: Text: Now in 2d! a framework for lexical expansion with contextual similarity. J. Lang. Model. **1**(1), 55–95 (2013)
4. Biemann, C.: Unsupervised part-of-speech tagging in the large. Res. Lang. Comput. **7**(2–4), 101–135 (2009)
5. Bollen, J., Mao, H., Zeng, X.: Twitter mood predicts the stock market. J. Comput. Sci. **2**(1), 1–8 (2011)

6. Cortes, C., Vapnik, V.: Support-vector networks. Mach. Learn. **20**(3), 273–297 (1995)
7. Das, A., Bandyopadhyay, S.: Subjectivity detection in english and bengali: A crf-based approach. Proceeding of ICON, Hyderabad, India (2009)
8. Das, A., Bandyopadhyay, S.: Sentiwordnet for indian languages. Asian Federation for Natural Language Processing, China, pp. 56–63 (2010)
9. Feldman, R.: Techniques and applications for sentiment analysis. Commun. ACM **56**(4), 82–89 (2013)
10. Ghiassi, M., Skinner, J., Zimbra, D.: Twitter brand sentiment analysis: A hybrid system using n-gram analysis and dynamic artificial neural. Expert Syst. Appl. **40**(16), 6266–6282 (2013)
11. Joshi, A., Balamurali, A., Bhattacharyya, P.: A fall-back strategy for sentiment analysis in hindi: a case study. Proceedings of the 8th ICON, Kharagpur, India (2010)
12. Miller, T., Biemann, C., Zesch, T., Gurevych, I.: Using distributional similarity for lexical expansion in knowledge-based word sense disambiguation. In: COLING, pp. 1781–1796 (2012)
13. Nagy, A., Stamberger, J.: Crowd sentiment detection during disasters and crises. In: Proceedings of the 9th International ISCRAM Conference, Vancouver, Canada, pp. 1–9 (2012)
14. O'Connor, B., Balasubramanyan, R., Routledge, B.R., Smith, N.A.: From tweets topolls: Linking text sentiment to public opinion time series. In: Proceedings of the Fourth International AAAI Conference on Weblogs and Social Media, Washington, DC 11(122-129), pp. 1–2 (2010)
15. Panchenko, A., Beaufort, R., Naets, H., Fairon, C.: Towards detection of child sexual abuse media: categorization of the associated filenames. In: Serdyukov, P., Braslavski, P., Kuznetsov, S.O., Kamps, J., Rüger, S., Agichtein, E., Segalovich, I., Yilmaz, E. (eds.) ECIR 2013. LNCS, vol. 7814, pp. 776–779. Springer, Heidelberg (2013)
16. Pang, B., Lee, L.: Opinion mining and sentiment analysis. Found. Trends Inf. Retrieval **2**(1–2), 1–135 (2008)
17. Patra, B.G., Das, D., Das, A., Prasath, R.: Shared task on sentiment analysis in indian languages (sail) tweets - an overview. In: Mining Intelligence and Knowledge Exploration - Third International Conference, MIKE-2015. Springer, Hyderabad, India (2015)
18. Quasthoff, U., Richter, M., Biemann, C.: Corpus portal for search in monolingual corpora. In: Proceedings of the Fifth International Conference on Language Resources and Evaluation, Genoa, Italy (2006)
19. Rosenthal, S., Nakov, P., Kiritchenko, S., Mohammad, S.M., Ritter, A., Stoyanov, V.: Semeval-2015 task 10: Sentiment analysis in twitter. In: Proceedings of the 9th International Workshop on Semantic Evaluation, SemEval, Denver, Colorado (2015)
20. Rosenthal, S., Nakov, P., Ritter, A., Stoyanov, V.: Semeval-2014 task 9: Sentiment analysis in twitter. In: Proceedings of the 9th International Workshop on Semantic Evaluation, SemEval, Dublin, Ireland (2014)
21. Tumasjan, A., Sprenger, T.O., Sandner, P.G., Welpe, I.M.: Predicting elections with twitter: What 140 characters reveal about political sentiment. In: ICWSM, Washington, DC 10, pp. 178–185 (2010)

A Sentiment Analysis System for Indian Language Tweets

Kamal Sarkar[1(✉)] and Saikat Chakraborty[2]

[1] Computer Science and Engineering Department,
Jadavpur University, Kolkata 700032, India
jukamal2001@yahoo.com
[2] Computer Science and Engineering Department,
Hooghly Engineering and Technology College, Hooghly 712103, India
saikat.sc@gmail.com

Abstract. This paper reports about our work in the MIKE 2015, Shared Task on Sentiment Analysis in Indian Languages (SAIL) Tweets. We submitted runs for Hindi and Bengali. A multinomial Naïve Bayes based model has been used to implement our system. The system has been trained and tested on the dataset released for SAIL TWEET CONTEST 2015. Our system obtains accuracy of 50.75 %, 48.82 %, 41.20 %, and 40.20 % for Hindi constrained, Hindi unconstrained, Bengali constrained and Bengali unconstrained run respectively.

Keywords: Multinomial Naïve Bayes · Sentiment · Hindi · Bengali · Tweets · Indian languages

1 Introduction

Sentiment analysis is an area of study that analyzes people's sentiments, attitudes and emotions from written language. Sentiment analysis is crucial for any kind of business and social domain because opinion is one of the central forces which influence peoples' mind for taking any kind of decision or making choices or performing his own evaluation. Sentiment analysis (SA) has become a main stream research during the last two decades. The evolution of social media texts – such as blogs, micro-blogs (e.g., Twitter), and chats (e.g., Facebook messages) – has created many new opportunities for information access and new challenges. It has become one of the prime present-day research areas. This paper presents a sentiment analysis (SA) system for Indian languages tweets which is designed for the SAIL 2015 contest, the goal of which is to perform sentiment recognition and analysis on two different Indian languages tweets i.e. Hindi and Bengali.

The SAIL 2015 contest was initially defined to build the SA systems for four Indian languages – Hindi, Bengali, Tami and Telugu, but subsequently Telugu was withdrawn from the contest and for other three languages training data and test data were provided. We have participated for Hindi and Bengali languages.

There are three major approaches for sentiment classification (SC) problem: Lexicon-based approaches, machine learning (ML) based approaches and hybrid approaches [1].

R. Prasath et al. (Eds.): MIKE 2015, LNAI 9468, pp. 694–702, 2015.
DOI: 10.1007/978-3-319-26832-3_66

Lexicon-based approaches depend on a sentiment lexicon, a collection of known and precompiled sentiment terms. It is mainly divided into two approaches: dictionary-based approaches and corpus-based approaches. References [2, 3] represent the main strategy of dictionary-based approaches. One of the methods of corpus-based approaches presented in [4] which used a list of seed opinion adjectives and imposed a set of linguistic constraints to specify additional adjective opinion words and their orientations. Machine learning (ML) based approaches use syntactic and/or linguistic features with some famous ML algorithms to solve the SA as a regular text classification problem [1]. Several ML algorithm have been used for solving SC problem such as Naive Bayes classifier (NB) model [5], Support Vector Machines Classifiers (SVM) [6] etc.

This paper is organized as follows; a discussion on the training data is given in Sect. 2. The methodology is described in Sect. 3. Section 4 represents the experimental result and analysis of the result. Finally Sect. 5 concludes the paper.

2 Training Data

For each language the training data which was provided by the organizer of SAIL 2015 was in three different files containing negative, neutral and positive tweets. But the data was in unicode (UTF-8) format. We have converted the training and test data into transliterated form using ITRANS format before training and testing the system.

3 Methodology

For developing a sentiment analysis system we have used multinomial Naive Bayes as a classifier which is trained on a labelled training data to classify the tweets into one of the categories: positive, negative and neutral. The description of multinomial Naive Bayes is given in the subsequent subsection.

3.1 Multinomial Naive Bayes Classifier

Naïve Bayes multinomial classifier [7] computes class probabilities for a given text for purpose of classification. If *C* is the set of classes and N is the size of a vocabulary, Naïve Bayes multinomial classifier classifies a test document t_i to the class which has the highest class membership probability $\Pr\left(c|t_i\right)$, which can be expanded by Bayes' rule,

$$Pr\left(c|t_i\right) = \frac{Pr(c)\,Pr(t_i|c)}{Pr(t_i)},\ c \in C \tag{1}$$

where the prior probability, pr(c) is calculated as:

$$Pr(c) = \frac{\text{number of documents of class c}}{\text{total number of documents}} \tag{2}$$

The probability of a document t_i given a class c is considered as a multinomial distribution:

$$Pr(t_i|c) = \left(\sum_n f_{ni}\right)! \prod_n \frac{Pr\left(w_n|c\right)^{f_{ni}}}{f_{ni}!}, \tag{3}$$

where f_{ni} = the count of word n in our test document t_i and $Pr\left(w_n|c\right)$ = probability of word n given class c, which is estimated from the training documents as:

$$\widehat{Pr}\left(w_n|c\right) = \frac{1 + Fr_{nc}}{N + \sum_{x=1}^{N} Fr_{xc}}, \tag{4}$$

where Fr_{xc} = count of word x in all the training documents in class c and the normalization factor $Pr(t_i)$ for Eq. 1 can be computed as:

$$Pr(t_i) = \sum_{k=1}^{|C|} \Pr(k) \Pr\left(t_i|k\right). \tag{5}$$

It is obvious that the terms $\left(\sum_n f_{ni}\right)!$ and $\prod_n f_{ni}!$ in Eq. 3 are computationally expensive and neither of these depends on the class c. So, these terms can be omitted to rewrite the Eq. 3 as:

$$Pr\left(t_i|c\right) = \propto \prod_n Pr(w_n|c)^{f_{ni}}, \tag{6}$$

where α is a constant. For implementing our sentiment analysis system, we have chosen the classifier "NaiveBayesMultinomialText", included in Weka. Weka is a machine learning toolkit [8]. To apply multinomial naïve Bayes classifier we have considered each tweet as a document.

3.2 Feature Extraction

The feature extraction is a crucial component in sentiment analysis task. We have used unigrams, bigrams and trigrams extracted from the tweets as the features. The features extraction process slightly varies depending on the types of resources used for developing the system. In the contest, there were two modes of run submission for each language: constrained and unconstrained. In constrained mode, the participant team is only allowed to use the training corpus (at most SentiWordNet (ILs)). No external resource is allowed. In unconstrained mode, the participant team is allowed to use any external resource (POS tagger, NER, Parser, and additional data) to train their system.

In the constrained mode, before feature extraction, we augment the words in tweets with the sentiment tags for the corresponding word found in the SentiWordNet (ILs). If the word is not found in WordNet, then we use the tag "UNK". After this augmentation process, each word in a tweet follows a sentiment tag retrieved from SentiWordNet or UNK.

3.3 System Development

We have developed our system by choosing the multinomial Naive Bayes classifier named "NaiveBayesMultinomialText" from WEKA. Weka is a machine learning toolkit consisting of a bunch of machine learning tools.

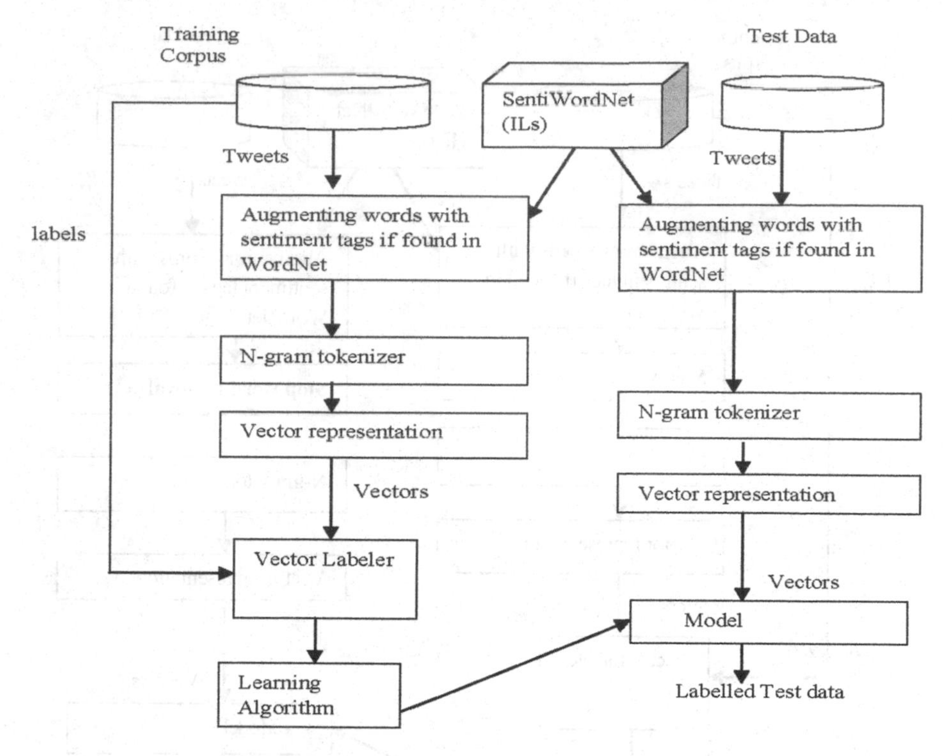

Fig. 1. System architecture for constrained mode sentiment analysis task

In unconstrained mode, we have externally supplied a stop word list to the WEKA to remove stop words from the feature set. The training data and test data are submitted to the classifier in ARFF format. *n*-gram tokenizer attached with the classifier has an option which allows us to set the value of *n*, which specifies the maximum size of the *n*-gram. We set the value for *n* to 3, that is, unigram, bigram and trigram are considered as the features. Term frequency based bag-of-terms representation is used for creating vectors for the tweets. The system architectures that we used for constrained mode sentiment analysis task and unconstrained mode sentiment analysis task are shown in Figs. 1 and 2 respectively.

4 Evaluation and Results

The test data for each language was also provided in Unicode (UTF-8) format and we had to transliterate into ITRANS format. We have submitted two runs for each language; one constrained and one unconstrained. In case of constrained task we only used the given training corpus and SentiWordNet (ILs) [9] (retrieved from http://amitavadas.com/sentiwordnet.php during the period of the contest) provided by the organizers. No external resource was used in this case. In case of unconstrained task, along with training corpus and SentiWordNet (ILs), we used only a list of stop words for each language.

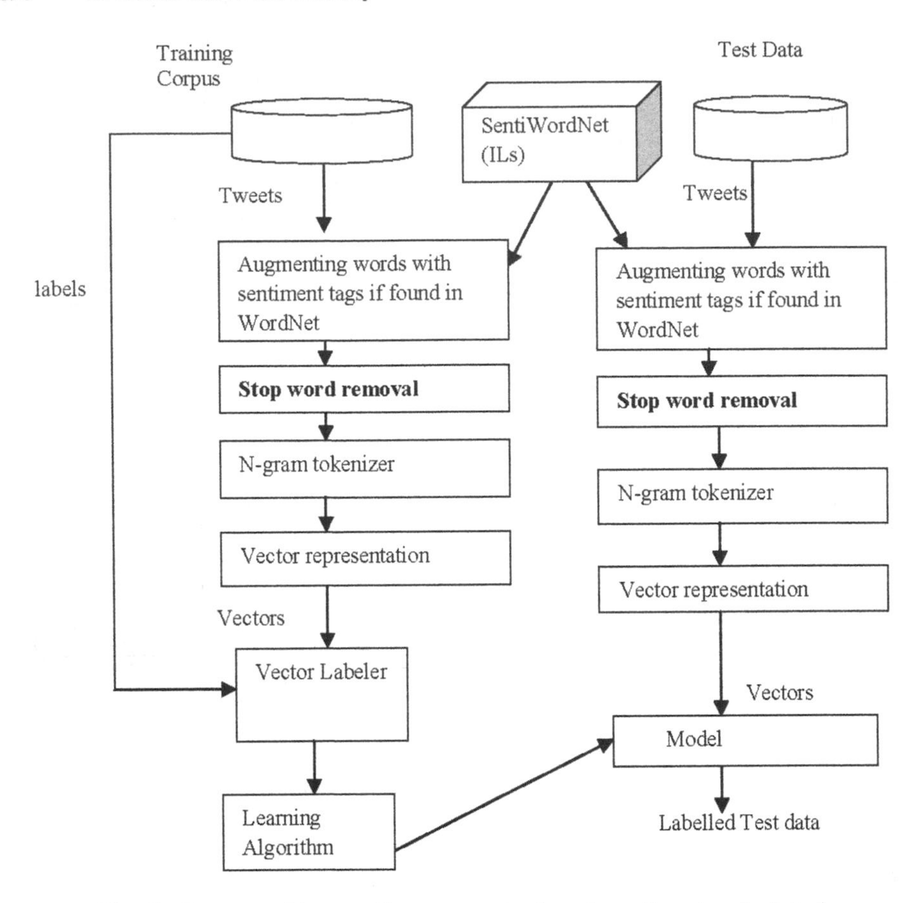

Fig. 2. System architecture for unconstrained mode sentiment analysis task

The classification outputs produced on the test data by our developed system was sent to the organizers for evaluation. The results of our submitted runs were announced and reported along with the results of the runs submitted by other participating teams. The results were published in two phases: in the first phase, only the results of the 4 systems such as "JUTeam_KS" (our system), "IIT-TUDA", "ISMD_RUNS", "AmritaCENNLP" were announced. The announcement of results of the other two systems: "AmritaCEN", "Amrita" were delayed by the organizers due to the reasons unknown to us. However, we received the results of the last 2 participating systems ("AmritaCEN", "Amrita") just before the preparation of this paper.

The outputs of the participating systems were evaluated using traditional accuracy measure i.e., percentage of correctly classified tweets for a class. The overall accuracy is also calculated based on percentage of total test tweets correctly classified. Accuracy was measured separately for constrained and unconstrained run for each language. Also the class wise accuracy for negative, neutral and positive sentiment classification was also reported.

The description of the training and test data of two Indian languages is shown in the Table 1. Table 2 shows the performances of our developed Sentiment Analysis system on the test data for the two Indian languages namely Hindi and Bengali tweets. From Table 2 it can be seen that the overall accuracy of our system for both Hindi and Bengali is around 50 %. But since the size of the training data was tiny, the outcome of this experiment cannot be justified completely.

Table 1. The description of the data for each of two Indian languages.

Language	Training data (number of tweets)			Test data (number of tweets)
	Negative	Neutral	Positive	
Hindi	615	494	168	467
Bengali	356	368	276	500

Table 2. Performance of our sentiment analysis system on the test data for Hindi and Bengali.

Language	Constrained				Unconstrained			
	Positive	Negative	Neutral	Overall accuracy	Positive	Negative	Neutral	Overall accuracy
Hindi	2.41 %	88.45 %	22 %	**50.75 %**	3.61 %	82.87 %	28 %	**48.82 %**
Bengali	23.94 %	60.26 %	47.41 %	**41.20 %**	21.13 %	63.58 %	45.18 %	**40.40 %**

4.1 Performance Comparison

In this section, we describe the results for the SAIL contest 2015. Although there were 20 registered teams, but only 6 teams submitted the runs.

Table 3. System comparisons based on the results obtained for Hindi constrained sentiment analysis task

Team name	Positive	Negative	Neutral	Overall accuracy
JUTeam_KS	2.41 %	88.45 %	22.0 %	50.75 %
IIT-TUDA	9.03 %	73.70 %	64.0 %	49.68 %
ISMD_RUNS	4.22 %	58.17 %	72.0 %	40.47 %
AmritaCENNLP	36.14 %	64.94 %	2.0 %	47.96 %
AmritaCEN	**45.79 %**	**57.37 %**	**80.0 %**	**55.67 %**
Amrita	17.47 %	54.59 %	68.0 %	42.83 %

The last two teams had some problems, so the results for these teams were not announced in the first phase. The results for these teams were announced just before preparation of this paper. However, out of 6 teams who had participated in the contest and submitted their runs, no team has participated in all the tasks defined for the contest. For example, we participated in four different tasks: two tasks for Hindi language and other two tasks for Bengali language, but we did not participate in the contest on sentiment analysis in Tamil language. Only two teams participated in the contest on sentiment analysis in Tamil language. From Tables 3, 4, 5 and 6, we compare our results obtained on the various tasks with the results obtained by the other systems participated in the same tasks.

Table 4. System comparisons based on the results obtained for Hindi unconstrained sentiment analysis task

Team name	Positive	Negative	Neutral	Overall accuracy
JUTeam_KS	**3.61 %**	**82.87 %**	**28.00 %**	**48.82 %**
IIT-TUDA	4.21 %	69.72 %	68.0 %	46.25 %
ISMD_RUNS	1.80 %	42.63 %	72.0 %	31.26 %

Table 5. System comparisons based on the results obtained for Bengali constrained sentiment analysis task

Team name	Positive	Negative	Neutral	Overall accuracy
JUTeam_KS	23.94 %	60.26 %	47.41 %	41.20 %
IIT-TUDA	**23.47 %**	**59.60 %**	**56.30 %**	**43.2 %**
AmritaCENNLP	27.23 %	65.56 %	0.0 %	31.4 %
AmritaCEN	29.58 %	34.44 %	39.26 %	33.6 %

Table 6. System comparisons based on the results obtained for Bengali unconstrained sentiment analysis task

Team name	Positive	Negative	Neutral	Overall accuracy
JUTeam_KS	21.13 %	63.58 %	45.18 %	40.40 %
IIT-TUDA	**24.88 %**	**54.30 %**	**55.55 %**	**42.0 %**

Each value shown in the table indicates accuracy obtained by the corresponding system and the "team name" column contains the code names of the teams participated in the contest. The teams with the code "AmritaCENNLP" and the code "AMRITA-CEN" participated in Tamil sentiment analysis constrained task and obtained the overall accuracy of 32.32 % and 39.28 % respectively.

As we can see from the Tables 2, 3, 4, 5 and 6, our developed system achieved the second position in Hindi language sentiment analysis task and, for Bengali language, its performance is slightly worse than the system (code "IIT-TUDA") that performs best on the bengali language sentiment analysis task.

We have only reported in this paper the results obtained by the various participating teams, but the details of the features and the classifiers which are used by the different teams for implementing their systems can be found in [10].

We have chosen multinomial Naive Bayes present in WEKA for implementing our system because our experiments with other classifiers show that they give poorer performance on SAIL data set. In Table 7 we have compared the performance of the support vector machines (SMO included under WEKA) with multinomial Naive Bayes and the table shows that multinomial Naive Bayes performs better than SVM.

Table 7. Multinomial Naive Bayes vs. SVM (SMO) for the sentiment analysis task

Classifiers	Tasks			
	Hindi constrained mode	Hindi unconstrained mode	Bengali constrained mode	Bengali unconstrained mode
Multinomial Naive Bayes	**50.75 %**	**48.82 %**	**41.20 %**	**40.40 %**
SVM (SMO)	44.11 %	42.61 %	40.89 %	38.75 %

5 Conclusion

This paper describes a sentiment analysis system for Indian languages (Hindi and Bengali) tweets. Four different runs have been performed: two runs for constrained cases (Hindi and Bengali) and two runs for unconstrained cases (Hindi and Bengali). The evaluation has been done in terms of accuracy which shows that performance of our system is comparable to other systems regarded as the best systems in the contest.

References

1. Medhat, W., Hassan, A., Korashy, H.: Sentiment analysis algorithms and applications: a survey. Ain Shams Eng. J. **5**(4), 1093–1113 (2014)
2. Minging, H., Bing, L.: Mining and summarizing customer reviews. In: Proceedings of ACM SIGKDD International Conference on Knowledge Discovery and Data Mining (KDD 2004) (2004)
3. Kim S., Hovy E.: Determining the sentiment of opinions. In: Proceedings of International Conference on Computational Linguistics (COLING 2004) (2004)
4. Hatzivassiloglou V., McKeown K.: Predicting the semantic orientation of adjectives. In: Proceedings of Annual Meeting of the Association for Computational Linguistics (ACL 1997) (1997)
5. Hanhoon, K., Joon, Y.S., Dongil, H.: Senti-lexicon and improved Naïve Bayes algorithms for sentiment analysis of restaurant reviews. Expert Syst. Appl. **39**(5), 6000–6010 (2012)
6. Chien, C.C., You-De, T.: Quality evaluation of product reviews using an information quality framework. Decis. Support Syst. **50**(4), 755–768 (2011)
7. Kibriya, A.M., Frank, E., Pfahringer, B., Holmes, G.: Multinomial naive bayes for text categorization revisited. In: Webb, G.I., Yu, X. (eds.) AI 2004. LNCS, vol. 3339, pp. 488–499. Springer, Heidelberg (2005)

8. Hall, M., Frank, E., Holmes, G., Pfahringer, B., Reutemann, P., Witten, I.H.: The WEKA data mining software: an update. SIGKDD Explor. **11**(1), 10–18 (2009)
9. Das, A., Bandyopadhyay, S.: SentiWordNet for Indian languages. In: Proceedings of 8th Workshop on Asian Language Resources (COLING 2010), Beijing, China, pp. 56–63 (2010)
10. Patra, B.G., Das, D., Das, A., Prasath, R.: Shared task on sentiment analysis in Indian languages (SAIL) tweets – an overview. In: Proceeding of the Mining Intelligence and Knowledge Exploration (MIKE-2015) (to appear)

AMRITA-CEN@SAIL2015: Sentiment Analysis in Indian Languages

Shriya Se(✉), R. Vinayakumar, M. Anand Kumar, and K.P. Soman

Centre for Excellence in Computational Engineering and Networking,
Amrita Vishwa Vidyapeetham, Ettimadai, Coimbatore, India
{shriyaseshadrik.r,vinayakumarr77}@gmail.com,
m_anandkumar@cb.amrita.edu

Abstract. The contemporary work is done as slice of the shared task in Sentiment Analysis in Indian Languages (SAIL 2015), constrained variety. Social media allows people to create and share or exchange opinions based on many perspectives such as product reviews, movie reviews and also share their thoughts through personal blogs and many more platforms. The data available in the internet is huge and is also increasing exponentially. Due to social media, the momentousness of categorizing these data has also increased and it is very difficult to categorize such huge data manually. Hence, an improvised machine learning algorithm is necessary for wrenching out the information. This paper deals with finding the sentiment of the tweets for Indian languages. These sentiments are classified using various features which are extracted using words and binary features, etc. In this paper, a supervised algorithm is used for classifying the tweets into positive, negative and neutral labels using Naive Bayes classifier.

Keywords: Sentiment analysis · Features · Social media · Machine learning · Supervised algorithm · Naive Bayes classifier

1 Introduction

Opinion plays an important role in deciding about everything in the life as millions of people express their thoughts through personal blogs, social networking sites and many more. Opinions are private states, which are not directly observable by others but expressions of opinion can be reflected through actions including written and spoken languages [1]. Sentiment analysis which is also known as opinion mining is a task in Natural Language processing which deals with the discernment and categorization of opinions in narrative [2]. Predominantly, these opinions are classified into positive, negative and neutral classes and is thus helpful in many fields including marketing, sociology, psychology etc. [3] Sentiment Analysis is popular for English languages and it is found rarely for Indian Languages [4].

Twitter, a microblogging stage, permits its clients to post short messages about any subject what's more, tail others to get their posts. Many individuals

R. Prasath et al. (Eds.): MIKE 2015, LNAI 9468, pp. 703–710, 2015.
DOI: 10.1007/978-3-319-26832-3_67

use Twitter as a platform to communicate with each other. The objective of this examination is to study client opinion communicated on Twitter and to add to a system that permits observing it in the constant. Tweet planning among others include spelling correction, equivalent word substitution, hyperlink cancellation and stop words are performed [5]. Notion is physically ordered the slant into three classes: positive, neutral and negative, so as to make the preparation set for the classifier [6]. The classified tweets are utilized to make positive, neutral and negative feeling lists. The sensational increment in the utilization of the internet as a method for correspondence has been joined by a sensational change in the way individuals express their opinion and perspective [7]. They can express their surveys online about items and administrations and also the perspectives about anything by means of social network (i.e. web journals, examination discussions). Sentiwordnet is one of the widely used lexicon resources for sentiment analysis, emotional analysis, opinion mining [8]. Sentiwordnet is an automatically created lexicon with positive and negative scores [9,10].

The contemporary work is done as slice of shared task in Sentiment Analysis in Indian Language (SAIL) 2015, constrain category. The task which contains three classes (positive, negative, neutral) of twitter data in three languages - Hindi, Bengali and Tamil is to identify the sentiment of the given tweet in a given language. The main objective of the share task is to stimulate researchers to accomplish sentiment analysis in their endemic language. Section 2 provides a view about the methodology used in the system; Sect. 3 discusses about the short analysis of the dataset provided to the work; Sect. 4 discusses about various experiments and their results.

2 Methodology

In the proposed system, feature extraction is the most crucial process as the accuracy of the classifier is based on the extracted feature. The flow of the proposed system is depicted in the Fig. 1. Generally for the text classification problem, preprocessing is to be done and is mandatory especially for twitter dataset. The preprocessing steps include normalization and tokenization. In tokenization, the tweets are further chunked into small instances called tokens. These tokens are normalized using normalization process in which superficial variations are removed from the words and are thus converted to the similar form. The common type of normalization includes case folding and stemming [11]. Predominantly, stemming is avoided for Indian languages in case of text classification as this leads to stem the useful information into its root form. Case folding is used mainly for English language as it has upper case and lower case letters [12]. This is not needed in case of Indian languages as no such case differences exist. The terms which are normalized using the system are listed in Table 1. Along with the features, the machine also learns from the training dataset which is already labelled. A small part of the data from the training dataset say about 10 % is taken and given for the validation process. This is given as an input for the Naive Bayes classifier and

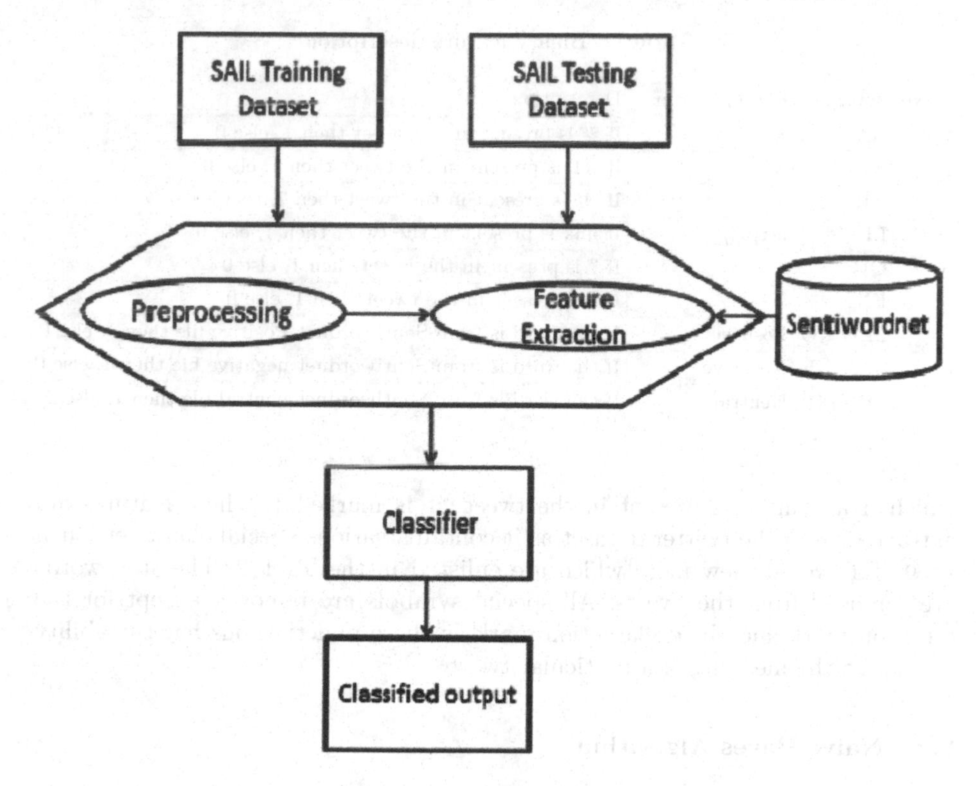

Fig. 1. Flow diagram of the proposed system

Table 1. Normalized symbol

Sl.No	Symbols	Features
1	@	User
2	#	Hash
3	1,2	Numbers
4	:-)	Emoticons
5	!,?	Punct
6	https://	URL

the classified outputs are taken into consideration. Naive Bayes classifier is chosen for the classification purpose as the size of the dataset is very small. In machine learning hunk SciPy library is used for classification.

2.1 Feature Extraction

The words of the sentiwordnet are taken as features because it contains the classified words of respective languages. The binary features are the features in

Table 2. Binary feature description

Sl.No	Symbol	Binary Features	Description
1	HA	#	If # is present in the tweet then 1, else 0
2	RT	RT	If RT is present in the tweet then 1, else 0
3	AR	@	If @ is present in the tweet then 1, else 0
4	LI	http://	If link is present in the tweet then 1, else 0
5	QU	?	If ? is present in the tweet then 1, else 0
6	EX	!	If ! is present in the tweet then 1, else 0
7	SEN_PO	Positive	If the word is from Sentiwordnet Positive file then 1, else 0
8	SEN_NE	Negative	If the word is from Sentiwordnet negative file then 1, else 0
9	SEN_NU	Neutral	If the word is from Sentiwordnet neutral file then 1, else 0

which if a symbol is present in the tweet, it is marked 1. These features are extracted from the twitter dataset as it contains various special characters such as @, RT, # and few more which are enlisted in the Table 2. The stop words are removed from the tweets. All special symbols are removed except for the question mark and the exclamation mark as these punctuations has the ability to change the meaning of a particular tweet.

2.2 Naive Bayes Algorithm

Naive Bayes has been used in information retrieval for many years and recently it has been used for many machine learning researches [13]. Multinomial Naive Bayes has been carried out for this work [14]. In recent years, the work has been focused on two basic instantiations of the classifier Bernoulli model and multinomial model [15]. Bernoulli model represents the document as a vector of binary features whereas the multinomial model uses vector of integer feature to represent documents [16]. The multinomial model works on the assumption that the probability of each word event in a document is independent of the words context and position in the document [17]. To normalize the error in the Naive Bayes, a small correction known as Laplacian Smoothing is included [18,19]. Generally the Naive Bayes is mathematically represented as,

$$P(c|d) = p(c) \prod_{1 \leq k \leq n_d \leq n} p(t_k|c) \tag{1}$$

As shown in the Fig. 2, the tweets are taken and if it includes any punctuations other than exclamations and question marks, it is removed and the tweet ids are also processed. If the tweet has any binary feature, they are marked 1.

S.No	Before Preprocessing	After Preprocessing
1	5,08771E+17 @ImJames_ நான் உள்ள இருக்கலாமான்னு கேட்டாக்க,வெளில போய்யா மொதல்லங்கறா இந்த நர்ஸூ//	AR-1 நான் உள்ள இருக்கலாமான்னு கேட்டாக்க வெளில போய்யா மொதல்லங்கறா இந்த நர்ஸூ
2	508707986391719936 चंद नोटों के लिए,-ईमान को बेआबरू होने दिया जाये,इतना सस्ता इंसान का मोल नहीं,जिसे बाजार में जाके बेच दिया जाये .... रवि कवि	चंद नोटों के लिए ईमान को बेआबरू होने दिया जाये, इतना सस्ता इंसान का मोल नहीं जिसे बाजार में जाके बेच दिया जाये रवि कवि
3	508638661265485824 ডি মারিয়া ও আলোনসোকে ছেড়ে দুর্বল হয়নি রিয়াল	ডি মারিয়া ও আলোনসোকে ছেড়ে দুর্বল হয়নি রিয়াল

Fig. 2. Tweets-before and after preprocessing

3 An Analysis of SAIL Dataset

SAIL stands for Sentiment Analysis for Indian Languages. They have released twitter dataset for three languages namely Tamil, Hindi, Bengali. The size of the training and testing dataset is shown in Table 3 in which approximately 27 % of training data of Tamil and Hindi and 54 % of the Bengali data contain URLs. Most of the Tamil tweets are regarding movie reviews and comments about some actors and actresses whereas the Hindi tweets are based on politics. The dataset has issues such as single tweet are there in both of the positive and negative training data and also there are tweets which are repeated. Many of these tweets are misspelt which affect the accuracy of the classifier. The training dataset of Tamil tweets contains more colloquial words which is not present in Sentiwordnet and hence it is not clean whereas the Hindi and Bengali tweets are conventional. In the test data, many of the ambiguous tweets are present. As these data are already ambigiuous, the accuracy drops.

Table 3. Twitter training dataset for SAIL 2015 shared task

Language	Training data				Test data
	Positive	Negative	Neutral	Total	
Tamil	387	316	400	1103	560
Hindi	168	545	493	1222	467
Bengali	277	354	368	999	500

4 Experiments and Results

The experiment is conducted on Windows 64-bit machine with i7 core processor and 8 GB RAM. The tweets from the dataset are taken. The initial step is preprocessing in which steps such as normalization and tokenization are done and the output of this step is raw tokens. These tokens are then given as an input for feature extractor. The feature extractor will take the tokens as input and extract the features from these tokens. The words, Sentiwordnet are taken as features and binary features are also included so as to improve the feature extractor. Sentiwordnet, hashtags, retweet, links, question marks, exclamatory marks are taken as binary features which means if any of these features are present then the output will be 1 else 0. The description of the binary feature is well illustrated in the given Table 2 In this paper, engrossment is given for feature extraction. The features are extracted from the training dataset and stored in a text file. Using the features that are extracted, the classification step is proceeded to. There are different algorithms that are used for the classification in Machine learning. The algorithm which is used in this paper is Naive Bayes classification algorithm. This Naive Bayes algorithm works on the principle of Bayes theorem. The training dataset and testing dataset is given as input for the classifier, in which 10 % of the training data is taken as a validation data. The data are classified using Naive Bayes and the output of the classifier will be the labelled tweets with positive, negative and neutral. The accuracy of the classifier is verified using F-score. The F-score is calculated for all the three classes. It is given below

$$F - scorepos = \frac{numberofpositivedocument}{Totalnumberofdocument} \tag{2}$$

$$F - scoreneg = \frac{numberofnegativedocument}{Totalnumberofdocument} \tag{3}$$

$$F - scoreneu = \frac{numberofneutraldocument}{Totalnumberofdocument} \tag{4}$$

The F-score and the accuracy of the system for all the three languages are given in the Table 4. Table 5 shows accuracy obtained for the proposed system by SAIL. The number of tweets that are classified into their respective classes in three languages are shown in the Fig. 3.

Table 4. Accuracy and F-score of the proposed system(Cross Validated)

Language	F-score	Accuracy
Tamil	0.4832	0.5612
Hindi	0.5219	0.5322
Bengali	0.3942	0.4171

Table 5. Testing result of the proposed system by SAIL

Language	Accuracy (in %)	Positive (in %)	Negative (in %)	Neutral (in %)
Tamil	39.28 %	29.81 %	26.58 %	59.79 %
Hindi	55.67 %	45.79 %	57.37 %	80.0 %
Bengali	33.6 %	29.58 %	34.44 %	39.26 %

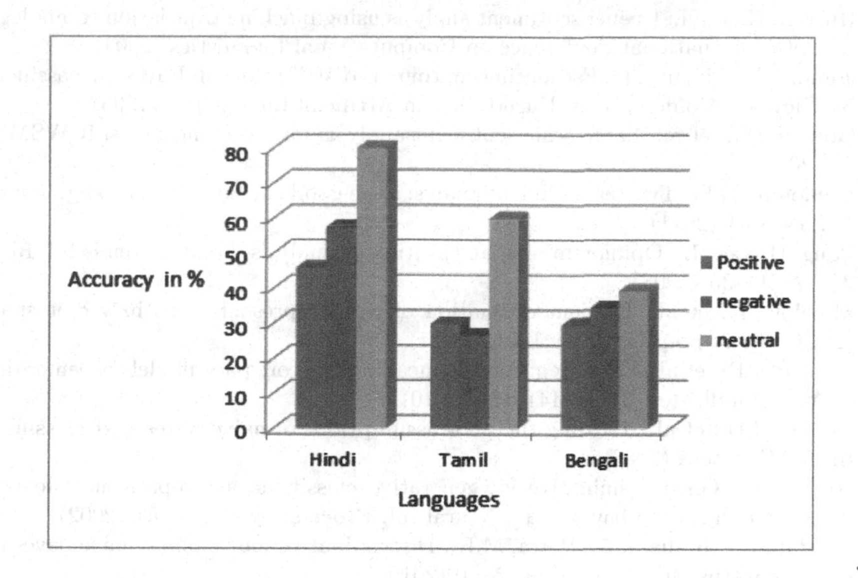

Fig. 3. Bar chart representation of the final result

5 Conclusion and Future Work

We have presented a method to classify twitter data based on the sentiment which is highly useful in the field of information retrieval (IR). In this work, we classify the tweets of the Indian languages into positive, negative and neutral classes. Generally before classifying the tweets, preprocessing is done. This is carried out in order to eliminate the unwanted symbols and also to retrieve words which are highly useful for analysing sentiment. The preprocessing step is taken extra care of and it gave better result after classification. Naive Bayes algorithm is used which gives a better classification result. This method can also be extended using SVM classifier and also the unsuupervised way of implementation can be done as future work.

References

1. Fink, C.R., et al.: Coarse- and fine-grained sentiment analysis of social media text. Johns Hopkins APL Tech. Dig. **30**(1), 22–30 (2011)
2. Balahur, A.: Sentiment analysis in social media texts. In: 4th Workshop on Computational Approaches (2013)

3. Hutto, C.J., Gilbertl, E.: Vader: a parsimonious rule-based model for sentiment analysis of social media text. In: Eighth International AAAI Conference on Weblogs and Social Media (2014)
4. Arunselvan, S.J., Anand kumar, M., et al.: Sentiment analysis of tamil moovie reviews via feature frequency count. IJAER **10**, 17934–17939 (2015)
5. Jansen, B.J., et al.: Twitter power: tweets as electronic word of mouth. J. Am. Soc. Inf. Sci. Technol. **60**(11), 2169–2188 (2009)
6. Hiroshi, K., et al.: Deeper sentiment analysis using machine translation technology. In: 20th International Conference on Computational Linguistics (2004)
7. John, G.H., Langley, P.: Estimating continuous distributions in Bayesian classifiers. In: Eleventh Conference on Uncertainty in Artificial Intelligence (1995)
8. Godbole, N., et al.: Large-scale sentiment analysis for news and blogs. ICWSM **7**, 21 (2007)
9. Kouloumpis, E.: Twitter sentiment analysis: the good the bad and the omg!. Icwsm **11**, 538–541 (2011)
10. Pang, B., Lee, L.: Opinion mining and sentiment analysis. Found. Trends Inf. Retr. **2**(1–2), 1–135 (2008)
11. Mikolov, T., et al.: Efficient estimation of word representations in vector space (2013). arXiv preprint arXiv:1301.3781
12. Turney, P.D., et al.: From frequency to meaning: vector space models of semantics. J. Artif. Intell. Res. **37**(1), 141–188 (2010)
13. Rennie, J.D., et al.: Tackling the poor assumptions of naive bayes text classifiers. In: ICML, vol. 3 (2003)
14. Jordan, A.: On discriminative vs. generative classifiers: a comparison of logistic regression and naive bayes. Adv. Neural Inf. Process. Syst. **14**, 841 (2002)
15. Panda, M., Abraham, A., Patra, M.R.: Discriminative multinomial naive bayes for network intrusion detection, pp. 5–10 (2010)
16. Juan, A., Ney, H.: Reversing and smoothing the multinomial naive bayes text classifier. In: PRIS (2002)
17. McCallum, A., Nigam, K.A.: Comparison of Event Models for Naive Bayes Text Classification. In: AAAI/ICML-98 Workshop on Learning for Text Categorization, pp. 41–48 (1998)
18. Lewis, D.D.: Naive bayes at forty: the independence assumption in information retrieval. In: Nédellec, C., Rouveirol, C. (eds.) ECML 1998. LNCS, vol. 1398. Springer, Heidelberg (1998)
19. Amor, N.B., et al.: Naive bayes vs decision trees in intrusion detection systems. In: 2004 ACM Symposium on Applied Computing (2004)

Author Index